Digital Design
Principles and Practices

DIGITAL DESIGN
Principles and Practices

Third Edition

John F. Wakerly

Cisco Systems, Inc.
Stanford University

Prentice Hall, Upper Saddle River, New Jersey 07458

Library of Congress Cataloging-in-Publication Data

Wakerly, John F.
   Digital design: principles and practices, third edition.
/John F. Wakerly.
   p. cm.
 Includes bibliographical references and index.
 ISBN 0-13-769191-2
 1. Digital integrated circuits--Design and construction.
I. Title.
TK7874.65.W34 2000
621.39'5--dc21   99-36681
                                                                 CIP

Publisher: TOM ROBBINS
Editor-in-chief: MARCIA HORTON
Production editor: IRWIN ZUCKER
Executive managing editor: VINCE O'BRIEN
Assistant managing editor: EILEEN CLARK
Manufacturing buyer: PAT BROWN
Copy editor: ROBERT LENTZ

Director of production and manufacturing: DAVID W. RICCARDI
Art director: ANN FRANCE
Associate creative director: AMY ROSEN
Cover art: ROBERT MCFADDEN
Cover designer: JOHN CHRISTIANA
Interior designer: DONNA WICKES
Editorial assistant: DAN DEPASQUALE

The author and publisher of this book have used their best efforts in preparing this book. These efforts include the development, research, and testing of the theories and programs to determine their effectiveness. The author and publisher make no warranty of any kind, expressed or implied, with regard to these programs or the documentation contained in this book. The author and publisher shall not be liable in any event for incidental or consequential damages in connection with, or arising out of, the furnishing, performance, or use of these programs.

We wish to thank Aldec, Inc. and Synopsys® for allowing the use of their software in the Xilinx Student Edition.

Trademark Information: Verilog is a trademark of Cadence Design Systems, Inc. Silos III is a trademark of Simucad Inc. Synopsys, and Foundation Express are trademarks of Synopsys, Inc. Xilinx is a trademark of Xilinx Corp. Aldec is a trademark of Aldec.

Printed in the United States of America

10 9 8 7 6 5 4 3 2 1

ISBN 0-13-769191-2

Prentice-Hall International (UK) Limited, London
Prentice-Hall of Australia Pty. Limited, Sydney
Prentice-Hall Canada Inc., Toronto
Prentice-Hall Hispanoamericana, S.A., Mexico
Prentice-Hall of India Private Limited, New Delhi
Prentice-Hall of Japan, Inc., Tokyo
Prentice-Hall (Singapore) Pte., Singapore
Editora Prentice-Hall do Brasil, Ltda., Rio de Janeiro

To my teachers

CONTENTS

FOREWORD

Moore's Law, which observes that semiconductor technology advances exponentially, has been valid for over three decades. Experts predict it will continue to hold for at least one more. When integrated circuits were introduced, logic packages had a dozen or so transistors. Today, with exponential increases in circuit density, microprocessor chips have passed the 10-million-transistor mark. In less than another decade they will reach 100 million transistors per chip.

To keep up with Moore's Law, design techniques have changed drastically. Hand-crafted logic circuits were once the norm. Now designers generate circuits from high-level descriptions. Connections on printed-circuit boards have been absorbed within chips. With programmable logic, on-chip logic functions and connections can be updated within the user environment.

How does education keep up with Moore's Law? What can we do to enable students to practice their skills today and adapt them tomorrow to new generations of devices? This is the challenge John Wakerly faced when he began this work.

His approach is multifaceted. It is grounded in basic principles of digital design that do not change with technology, such as combinational logic, sequential logic, and state machines. Wakerly weds these principles with tools and practical techniques that teach how to design for today's technology. These include how to use the ABEL and VHDL design languages, how to structure designs with large building blocks, and how to implement designs with programmable logic devices. These techniques are essential for successful design.

The most difficult goal is to help the student adapt to the inevitable changes to come. Wakerly does this by revealing what is happening underneath the logic. For example, he gives transistor models of gates and uses them to expose issues related to timing and noise. Gates may become faster and denser and may use different control voltages, but how to assure correct and reliable operation will be a continuing concern. We learn the characteristics, constraints, and failure modes and how to design to them. We learn through examples of alternate designs how to judge design quality and evaluate tradeoffs. As new technology emerges, we will be able to design to it.

Wakerly enhances the approach with presentation skills that are rare in college texts. The reader will quickly appreciate the effective graphics, entertaining writing style, and instructive exercises.

Moore's Law condemns textbooks in this field to short lives. Nevertheless, Wakerly's text is a classic.

Harold S. Stone
Princeton, New Jersey

PREFACE

This book is for everyone who wants to design and build real digital circuits. It is based on the idea that, in order to do this, you have to grasp the fundamentals, but at the same time you need to understand how things work in the real world. Hence, the "principles and practices" theme.

The material in this book is appropriate for introductory courses on digital logic design in electrical or computer engineering or computer science curricula. Computer science students who are unfamiliar with basic electronics concepts or who just aren't interested in the electrical behavior of digital devices may wish to skip Chapter 3; the rest of the book is written to be independent of this material as much as possible. On the other hand, *anyone* with a basic electronics background who wants to get up to speed on digital electronics can do so by reading Chapter 3. In addition, students with *no* electronics background can get the basics by reading Bruce M. Fleischer's "Electrical Circuits Review," a freely reproducible 20-page electronics tutorial available on this book's Web site, `www.ddpp.com`.

introductory courses

electronics concepts

Although this book's level is introductory, it contains much more material than can be taught in a typical introductory course. Once I started writing, I found that I had many important things to say that wouldn't fit into a one-quarter course at Stanford or a 400-page book. Therefore, I have followed my usual practice of including *everything* that I think is at least moderately important, and leaving it up to the instructor or reader to decide what is most important in a particular environment. To help these decisions along, though, I've marked the headings of *optional sections* with an asterisk. In general, these sections can be skipped without any loss of continuity in the non-optional sections that follow.

optional sections

Undoubtedly, some people will use this book in advanced courses and in laboratory courses. Advanced students will want to skip the basics and get right into the fun stuff. Once you know the basics, the most important and fun stuff in

advanced courses
laboratory courses
fun stuff

this book is in the sections on hardware description languages ABEL and VHDL, where you'll discover that your programming courses actually helped prepare you to design hardware.

Another use of this book is as a self-study reference for a working digital designer, who may be either of two kinds:

Novice If you're just getting started as a working digital designer, and you took a very "theoretical" logic design course in school, you should concentrate on Chapters 3, 5, 6, and 8–11 to get prepared for the real world.

Old pro If you're experienced, you may not need all of the "practices" material in this book, but the principles in Chapters 2, 4, and 7 can help you organize your thinking, and the discussions there of what's important and what's not might relieve the guilt you feel for not having used a Karnaugh map in 10 years. The examples in Chapters 6, 8, and 9 should give you additional insights into and appreciation for a variety of design methods. Finally, the ABEL and VHDL language descriptions and examples sprinkled throughout Chapters 4 through 9 may serve as your first organized introduction to HDL-based design.

All readers should make good use of the comprehensive index and of the *marginal notes* throughout the text that call attention to definitions and important topics. Maybe the highlighted topics in *this* section were more marginal than important, but I just wanted to show off my text formatting system.

Chapter Descriptions

What follows is a list of short descriptions of this book's eleven chapters and appendix. This may remind you of the section in software guides, "For People Who Hate Reading Manuals." If you read this list, then maybe you don't have to read the rest of the book.

- Chapter 1 gives a few basic definitions and lays down the ground rules for what we think is and is not important in this book.

- Chapter 2 is an introduction to binary number systems and codes. Readers who are already familiar with binary number systems from a software course should still read Sections 2.10–2.13 to get an idea of how binary codes are used by hardware. Advanced students can get a nice introduction to error-detecting codes by reading Sections 2.14 and 2.15. The material in Section 2.16.1 should be read by everyone; it is used in some design examples in Chapter 8.

- Chapter 3 describes "everything you ever wanted to know about" digital circuit operation, placing primary emphasis on the external electrical characteristics of logic devices. The starting point is a basic electronics background including voltage, current, and Ohm's law; readers unfamiliar with these concepts may wish to consult the "Electrical Circuits Review" mentioned earlier. This chapter may be omitted by readers who aren't interested in how to make real circuits work, or who have the luxury of having someone else to do the dirty work.

- Chapter 4 teaches combinational logic design principles, including switching algebra and combinational-circuit analysis, synthesis, and minimization. Introductions to ABEL and VHDL appear at the end of this chapter.

- Chapter 5 begins with a discussion of digital-system documentation standards, probably the most important practice for aspiring designers to start practicing. Next, this chapter introduces programmable logic devices (PLDs), focusing on their capability to realize combinational logic functions. The rest of the chapter describes commonly used combinational logic functions and applications. For each function, it describes standard MSI building blocks, ABEL programs for PLD realizations, and VHDL models.

- Chapter 6 is a collection of larger combinational-circuit design examples. For each example, it shows how the design can be carried out with MSI building blocks (if appropriate), ABEL and PLDs, or VHDL that can be targeted to a CPLD or FPGA.

- Chapter 7 teaches sequential logic design principles, starting with latches and flip-flops. The main emphasis in this chapter is on the analysis and design of clocked synchronous state machines. However, for the brave and daring, the chapter includes an introduction to fundamental-mode circuits and the analysis and design of feedback sequential circuits. The chapter ends with sections on ABEL and VHDL features that support sequential-circuit design.

- Chapter 8 is all about the practical design of sequential circuits. Like Chapter 5 before it, this chapter focuses on commonly used functions and gives examples using MSI building blocks, ABEL and PLDs, and VHDL. Sections 8.8 and 8.9 discuss the inevitable impediments to the ideal of fully synchronous design and address the important problem of how to live synchronously in an asynchronous world.

- Chapter 9 is a collection of state-machine and larger sequential-circuit design examples. Each example is carried out both using ABEL for a PLD and using VHDL that can be targeted to a CPLD or FPGA.

- Chapter 10 is an introduction to memory devices, CPLDs, and FPGAs. Memory coverage includes read-only memory and static and dynamic read-write memories from the points of view of both internal circuitry and functional behavior. The last two sections introduce CPLD and FPGA architecture.

- Chapter 11 discusses several miscellaneous real-world topics that are of interest to digital designers. When I started writing what I thought would be a 300-page book, I included this chapter in the outline to pad out the "core" material to a more impressive length. Well, the book is obviously long enough without it, but this material is useful just the same.

Most of the chapters contain references, drill problems, and exercises. Drill problems are typically short-answer or turn-the-crank questions that can be answered directly based on the text material, while exercises may require a little more thinking. The drill problems in Chapter 3 are particularly extensive and are designed to allow non-EE types to ease into this material.

Xilinx Foundation Tools

Xilinx, Inc. (San Jose, CA 95124) has kindly allowed us to package their Foundation Express digital-design tools on two CD-ROMs at the end of this book (in most domestic and international printings). These tools are quite comprehensive, including an ABEL compiler, VHDL and Verilog language processors, a schematic drawing package, and a simulator. Much of the software in the package is based on the popular Active-CAD™ and Active-HDL™ tools from Aldec, Inc. The package also includes FPGA Express™ software from Synopsys, which allows ABEL, VHDL, and Verilog designs to be targeted to CPLDs and FPGAs; popular Xilinx parts are supported by the included version.

The Foundation tools were very useful to me as an author. Using them, I was able to write and test all of the example programs in the text. I trust that the tools will be even more useful to the students who use the text. They will allow you to write and test your own hardware designs and download them into Xilinx CPLDs and FPGAs in a lab environment. There are no artificial design-size limits, as long as your design fits into a single device. An excellent lab-oriented manual for getting started with Xilinx devices and tools is David Van den Bout's *The Practical Xilinx Designer: Version 1.5* (Prentice Hall, 1999).

Even if you're not ready to do your own original designs, you can use the Foundation tools to try out and modify any of the examples in the text, since the source code for all of them is available at the book's Web site, discussed next.

WWW.DDPP.COM

Support materials for this book are available at the book's own dedicated Web site, www.ddpp.com. A key resource for students is the set of source listings for all of the example C, ABEL, and VHDL programs in the book. Also available are ZIP'ed Foundation project directories, including not only ABEL and VHDL source files but also schematics that are set up to use and simulate some of the example designs.

During the preparation of this edition, I was surprised and delighted to see how much digital design reference material is available on the Web, especially from device manufacturers. The DDPP Web site contains a "living" references section with links to many useful sites that you can use as a jumping-off point for your own independent study.

A couple of appendices from previous editions are available on the Web site—"Electrical Circuits Review" by Bruce M. Fleischer and "IEEE Standard Symbols." Students taking lab courses may also appreciate the four pages of handy IC pinout guides, which appeared on the inside-cover pages of previous editions.

One thing that students may or may not like is a collection of new exercises that I expect to build up as I continue to teach digital design at Stanford and as I receive contributions from others.

For Instructors

The DDPP Web site has additional materials for instructors only. This part of the site is password protected; if you plan to use it, please allow up to a week to obtain a login name and password via the procedure published there.

The instructors' area contains files with all of the figures and tables in the book. You can use these files to make presentation slides directly, or you can insert selected materials into your own customized presentations.

The site also contains answers to selected exercises—more than half of the exercises in the book, equivalent to over 200 printed pages. There are also several sample exams and solutions.

Another important resource for instructors is Xilinx' University Program (`www.xilinx.com/programs/univ`). The site offers a variety of product materials, course materials, and discounts on chips and boards that you can use in digital-design lab courses.

How This Book Was Prepared

The text of the third edition of this book was converted from the original second-edition TEX version into Adobe FrameMaker®. Figures from previous editions were converted from Cricket Draw into Adobe Illustrator® EPS files.

All of the writing, editing, drawing, and circuit designing was done on a PC running Windows 95 or 98 with 384 Mbytes of memory, which, regrettably, would still crash if too many programs or files were open at once. The good news is that this edition's use of standard programs and tools has allowed me to provide readers and instructors with a large collection of useful materials on the book's Web site, as described earlier.

Errors

Warning: This book may contain errors. The author assumes no liability for any damage—incidental, brain, or otherwise—caused by errors.

There, that should make the lawyers happy. Now, to make *you* happy, let me assure you that a great deal of care has gone into the preparation of this manuscript to make it as error free as possible. I am anxious to learn of the remaining errors so that they may be fixed in future printings, editions, and spin-offs. Therefore I will pay $5 to the first finder of each undiscovered error, be it technical, typographical, or otherwise. Please email your comments by using the link on the Web site, or by writing to me at john@wakerly.com.

Any reader can obtain an up-to-date list of discovered errors using the link at the book's Web site. It will be a very short file transfer, I hope.

Acknowledgements

Many people helped make this book possible. Most of them helped with the first and second editions and are acknowledged there. Preparation of the third edition has been a lonelier task, but it was made easier by my colleagues Mario Mazzola and Prem Jain at Cisco Systems. They and the company made it possible for me to cut back my commitment at Cisco to less than half time for the eight months that it took to prepare this edition.

For the ideas on the "principles" side of this book, I still owe great thanks to my teacher, research advisor, and friend Ed McCluskey. On the "practices" side, my personal "Digital Designers Hall of Fame" includes (in chronological order): Ed Davidson, Jim McClure, Courtenay Heater, Sam Wood, Curt Widdoes, Prem Jain, Ted Tracy, Dave Raaum, Akhil Duggal, Des Young, and Tom Edsall.

The seed that got me started writing this book and many others was planted in the early 70s by Harold Stone at Stanford. He put me to work reviewing and indexing his books, and his computer organization books inspired me to write my first software book. Now, I'd like to offer Harold my belated thanks for getting me started, and I give him special thanks for helping to pad out this edition by another two pages!

In the summer of 1997, during the early stages of this edition's planning, friend and colleague Jean-Pierre Steger took a sabbatical from the Burgdorf School of Engineering near Bern, Switzerland to help me get jump-started with VHDL, the Xilinx Foundation tools, and other topics. A number of other people contributed review comments or other materials to this edition, including John Birkner, Rebecca Farley, Don Gaubatz, John Gill, Linley Gwennap, Jesse Jenkins, and Jeff Purnell.

Xilinx, Inc. naturally deserves credit for providing the Foundation tools that are an important adjunct to this edition. On the people side, their original

University Program leader Jason Feinsmith was very helpful, and their recently appointed leader Patrick Kane has supported our efforts enthusiastically.

Since the second edition was published, I have received many helpful comments from readers. In addition to suggesting or otherwise motivating many improvements, readers have spotted dozens of typographical and technical errors whose fixes are incorporated in this third edition.

My sponsoring editor at Prentice Hall, Tom Robbins, deserves great thanks for his patience. He's the second editor to have been lured to Prentice Hall in part by the (falsely attractive) prospect of working on a project with me, only to find out upon his arrival that the project was very late. In Tom's case, though, I had known him since the early 80s when he first tried to sign me up for a project with another publisher, and we'd been trying to find a way to work together ever since. We're finally there, starting the third decade of our friendship. Tom contributed more than patience—among other things, you have him to thank for getting you the free Xilinx software at the back of this book (most printings).

Production editor Irwin Zucker also deserves credit for providing a very smooth interface with the production side of the house and for putting in long hours to help me during the final "crunch" stage of the project. If not for him, I would not be able to leave on a long-planned three-week family vacation later this morning. (I'm told that if I hadn't finished in time, our 90-pound dog would have gone to Europe in my seat instead!) Robert Lentz also did a great job as a copy editor which I could live with :-) .

Special thanks go to artist Robert McFadden, whose trippy cover painting is hanging in my home along with several other of his far-out works. His painting, which I commissioned and he completed well over a year ago, provided me with the motivation to actually get the *inside* of the book done.

It seems like some disaster always strikes just as I am completing one of these projects. For the first edition, it was the World-Series earthquake of 1989. For the second edition, it was surgery four days before the book's completion for a ruptured and very yucky-looking appendix. This time, I seemed to have dodged the bullet so far. Actually, I won't be completely done until I've finished the index, which I'll be preparing on my laptop as we ride the rails through Europe over the next few weeks. Let's hope I don't forget it on some train!

As always, I must thank my wife Kate for putting up with the late hours, frustration, crabbiness, preoccupation, and phone calls from weird people that occur when I'm engaged in a writing project like this. We hope you enjoy starting this book as much as we enjoy finishing it!

John F. Wakerly
Mountain View, California

Digital Design
Principles and Practices

Hi, I'm John . . . .

chapter 1

Introduction

W elcome to the world of digital design. Perhaps you're a computer science student who knows all about computer software and programming, but you're still trying to figure out how all that fancy hardware could possibly work. Or perhaps you're an electrical engineering student who already knows something about analog electronics and circuit design, but you wouldn't know a bit if it bit you. No matter. Starting from a fairly basic level, this book will show you how to design digital circuits and subsystems.

We'll give you the basic principles that you need to figure things out, and we'll give you lots of examples. Along with principles, we'll try to convey the flavor of real-world digital design by discussing current, practical considerations whenever possible. And I, the author, will often refer to myself as "we" in the hope that you'll be drawn in and feel that we're walking through the learning process together.

1.1 About Digital Design

Some people call it "logic design." That's OK, but ultimately the goal of design is to build systems. To that end, we'll cover a whole lot more in this text than just logic equations and theorems.

This book claims to be about principles and practices. Most of the principles that we present will continue to be important years from now; some may be applied in ways that have not even been discovered yet. As for practices, they may be a little different from what's presented here by the time you start working in the field, and they will certainly continue to change throughout your career. So you should treat the "practices" material in this book as a way to reinforce principles, and as a way to learn design methods by example.

One of the book's goals is to present enough about basic principles for you to know what's happening when you use software tools to turn the crank for you. The same basic principles can help you get to the root of problems when the tools happen to get in your way.

Listed in the box on this page are several key points that you should learn through your studies with this text. Most of these items probably make no sense to you right now, but you should come back and review them later.

Digital design is engineering, and engineering means "problem solving." My experience is that only 5%–10% of digital design is "the fun stuff"—the creative part of design, the flash of insight, the invention of a new approach. Much of the rest is just "turning the crank." To be sure, turning the crank is much easier now than it was 20 or even 10 years ago, but you still can't spend 100% or even 50% of your time on the fun stuff.

IMPORTANT THEMES IN DIGITAL DESIGN

- Good tools do not guarantee good design, but they help a lot by taking the pain out of doing things right.

- Digital circuits have analog characteristics.

- Know when to worry and when not to worry about the analog aspects of digital design.

- Always document your designs to make them understandable by yourself and others.

- Associate active levels with signal names and practice bubble-to-bubble logic design.

- Understand and use standard functional building blocks.

- Design for minimum cost at the system level, including your own engineering effort as part of the cost.

- State-machine design is like programming; approach it that way.

- Use programmable logic to simplify designs, reduce cost, and accommodate last-minute modifications.

- Avoid asynchronous design. Practice synchronous design until a better methodology comes along.

- Pinpoint the unavoidable asynchronous interfaces between different subsystems and the outside world, and provide reliable synchronizers.

- Catching a glitch in time saves nine.

Besides the fun stuff and turning the crank, there are many other areas in which a successful digital designer must be competent, including the following:

- *Debugging.* It's next to impossible to be a good designer without being a good troubleshooter. Successful debugging takes planning, a systematic approach, patience, and logic: if you can't discover where a problem *is*, find out where it *is not*!

- *Business requirements and practices.* A digital designer's work is affected by a lot of non-engineering factors, including documentation standards, component availability, feature definitions, target specifications, task scheduling, office politics, and going to lunch with vendors.

- *Risk-taking.* When you begin a design project you must carefully balance risks against potential rewards and consequences, in areas ranging from new-component selection (will it be available when I'm ready to build the first prototype?) to schedule commitments (will I still have a job if I'm late?).

- *Communication.* Eventually, you'll hand off your successful designs to other engineers, other departments, and customers. Without good communication skills, you'll never complete this step successfully. Keep in mind that communication includes not just transmitting but also receiving; learn to be a good listener!

In the rest of this chapter, and throughout the text, I'll continue to state some opinions about what's important and what is not. I think I'm entitled to do so as a moderately successful practitioner of digital design. Of course, you are always welcome to share your own opinions and experience (send email to john@wakerly.com).

1.2 Analog versus Digital

analog
digital

Analog devices and systems process time-varying signals that can take on any value across a continuous range of voltage, current, or other metric. So do *digital* circuits and systems; the difference is that we can pretend that they don't! A digital signal is modeled as taking on, at any time, only one of two discrete values, which we call *0* and *1* (or LOW and HIGH, FALSE and TRUE, negated and asserted, Sam and Fred, or whatever).

0
1

Digital computers have been around since the 1940s and have been in widespread commercial use since the 1960s. Yet only in the past 10 to 20 years has the "digital revolution" spread to many other aspects of life. Examples of once-analog systems that have now "gone digital" include the following:

- Still pictures. The majority of cameras still use silver-halide film to record images. However, the increasing density of digital memory chips has allowed the development of digital cameras which record a picture as a

640×480 or larger array of pixels, where each pixel stores the intensities of its red, green, and blue color components as 8 bits each. This large amount of data, over seven million bits in this example, may be processed and compressed in a format called JPEG down to as little as 5% of the original storage size, depending on the amount of picture detail. So, digital cameras rely on both digital storage and digital processing.

- *Video recordings.* A digital versatile disc (DVD) stores video in a highly compressed digital format called MPEG-2. This standard encodes a small fraction of the individual video frames in a compressed format similar to JPEG, and encodes each other frame as the difference between it and the previous one. The capacity of a single-layer, single-sided DVD is about 35 billion bits, sufficient for about 2 hours of high-quality video, and a two-layer, double-sided disc has four times that capacity.

- *Audio recordings.* Once made exclusively by impressing analog wave-forms onto vinyl or magnetic tape, audio recordings now commonly use digital compact discs (CDs). A CD stores music as a sequence of 16-bit numbers corresponding to samples of the original analog waveform, one sample per stereo channel every 22.7 microseconds. A full-length CD recording (73 minutes) contains over six billion bits of information.

- *Automobile carburetors.* Once controlled strictly by mechanical linkages (including clever "analog" mechanical devices that sensed temperature, pressure, etc.), automobile engines are now controlled by embedded microprocessors. Various electronic and electromechanical sensors convert engine conditions into numbers that the microprocessor can examine to determine how to control the flow of fuel and oxygen to the engine. The microprocessor's output is a time-varying sequence of numbers that operate electromechanical actuators which, in turn, control the engine.

- *The telephone system.* It started out a hundred years ago with analog microphones and receivers connected to the ends of a pair of copper wires (or was it string?). Even today, most homes still use analog telephones, which transmit analog signals to the phone company's central office (CO). However, in the majority of COs, these analog signals are converted into a digital format before they are routed to their destinations, be they in the same CO or across the world. For many years the private branch exchanges (PBXs) used by businesses have carried the digital format all the way to the desktop. Now many businesses, COs, and traditional telephony service providers are converting to integrated systems that combine digital voice with data traffic over a single IP (Internet Protocol) network.

- *Traffic lights.* Stop lights used to be controlled by electromechanical timers that would give the green light to each direction for a predetermined amount of time. Later, relays were used in controllers that could activate the lights according to the pattern of traffic detected by sensors embedded in the pavement. Today's controllers use microprocessors and can control the lights in ways that maximize vehicle throughput or, in some California cities, frustrate drivers in all kinds of creative ways.

- *Movie effects.* Special effects used to be created exclusively with miniature clay models, stop action, trick photography, and numerous overlays of film on a frame-by-frame basis. Today, spaceships, bugs, other-worldly scenes, and even babies from hell (in Pixar's animated feature *Tin Toy*) are synthesized entirely using digital computers. Might the stunt man or woman someday no longer be needed, either?

The electronics revolution has been going on for quite some time now, and the "solid-state" revolution began with analog devices and applications like transistors and transistor radios. So why has there now been a *digital* revolution? There are in fact many reasons to favor digital circuits over analog ones:

- *Reproducibility of results.* Given the same set of inputs (in both value and time sequence), a properly designed digital circuit always produces exactly the same results. The outputs of an analog circuit vary with temperature, power-supply voltage, component aging, and other factors.

- *Ease of design.* Digital design, often called "logic design," is logical. No special math skills are needed, and the behavior of small logic circuits can be visualized mentally without any special insights about the operation of capacitors, transistors, or other devices that require calculus to model.

- *Flexibility and functionality.* Once a problem has been reduced to digital form, it can be solved using a set of logical steps in space and time. For example, you can design a digital circuit that scrambles your recorded voice so that it is absolutely indecipherable by anyone who does not have your "key" (password), but it can be heard virtually undistorted by anyone who does. Try doing that with an analog circuit.

- *Programmability.* You're probably already quite familiar with digital computers and the ease with which you can design, write, and debug programs for them. Well, guess what? Much of digital design is carried out today by writing programs, too, in *hardware description languages (HDLs)*. These languages allow both structure and function of a digital circuit to be specified or *modeled*. Besides a compiler, a typical HDL also comes with simulation and synthesis programs. These software tools are used to test the hardware model's behavior before any real hardware is built, and then to synthesize the model into a circuit in a particular component technology.

  hardware description language (HDL)

  hardware model

- *Speed.* Today's digital devices are very fast. Individual transistors in the fastest integrated circuits can switch in less than 10 picoseconds, and a complete, complex device built from these transistors can examine its

SHORT TIMES   A *microsecond (μsec)* is 10^{-6} second. A *nanosecond (ns)* is just 10^{-9} second, and a *picosecond (ps)* is 10^{-12} second. In a vacuum, light travels about a foot in a nanosecond, and an inch in 85 picoseconds. With individual transistors in the fastest integrated circuits now switching in less than 10 picoseconds, the speed-of-light delay between these transistors across a half-inch-square silicon chip has become a limiting factor in circuit design.

inputs and produce an output in less than 2 nanoseconds. This means that such a device can produce 500 million or more results per second.

- *Economy.* Digital circuits can provide a lot of functionality in a small space. Circuits that are used repetitively can be "integrated" into a single "chip" and mass-produced at very low cost, making possible throw-away items like calculators, digital watches, and singing birthday cards. (You may ask, "Is this such a good thing?" Never mind!)

- *Steadily advancing technology.* When you design a digital system, you almost always know that there will be a faster, cheaper, or otherwise better technology for it in a few years. Clever designers can accommodate these expected advances during the initial design of a system, to forestall system obsolescence and to add value for customers. For example, desktop computers often have "expansion sockets" to accommodate faster processors or larger memories than are available at the time of the computer's introduction.

So, that's enough of a sales pitch on digital design. The rest of this chapter will give you a bit more technical background to prepare you for the rest of the book.

1.3 Digital Devices

gate

The most basic digital devices are called *gates*, and no, they were not named after the founder of a large software company. Gates originally got their name from their function of allowing or retarding ("gating") the flow of digital information. In general, a gate has one or more inputs and produces an output that is a function of the current input value(s). While the inputs and outputs may be analog conditions such as voltage, current, even hydraulic pressure, they are modeled as taking on just two discrete values, 0 and 1.

AND gate

Figure 1-1 shows symbols for the three most important kinds of gates. A 2-input *AND gate*, shown in (a), produces a 1 output if both of its inputs are 1; otherwise it produces a 0 output. The figure shows the same gate four times, with the four possible combinations of inputs that may be applied to it and the result-

Figure 1-1 Digital devices: (a) AND gate; (b) OR gate; (c) NOT gate or inverter.

ing outputs. A gate is called a *combinational circuit* because its output depends only on the current input combination.

A 2-input *OR gate*, shown in (b), produces a 1 output if one or both of its inputs are 1; it produces a 0 output only if both inputs are 0. Once again, there are four possible input combinations, resulting in the outputs shown in the figure.

A *NOT gate*, more commonly called an *inverter*, produces an output value that is the opposite of the input value, as shown in (c).

We called these three gates the most important for good reason. Any digital function can be realized using just these three kinds of gates. In Chapter 3 we'll show how gates are realized using transistor circuits. You should know, however, that gates have been built or proposed using other technologies, such as relays, vacuum tubes, hydraulics, and molecular structures.

A *flip-flop* is a device that stores either a 0 or 1. The *state* of a flip-flop is the value that it currently stores. The stored value can be changed only at certain times determined by a "clock" input, and the new value may further depend on the flip-flop's current state and its "control" inputs. A flip-flop can be built from a collection of gates hooked up in a clever way, as we'll show in Section 7.2.

A digital circuit that contains flip-flops is called a *sequential circuit*, because its output at any time depends not only on its current input but also on the past sequence of inputs that have been applied to it. In other words, a sequential circuit has *memory* of past events.

combinational circuit

OR gate

NOT gate
inverter

flip-flop
state

sequential circuit

memory

1.4 Electronic Aspects of Digital Design

Digital circuits are not exactly a binary version of alphabet soup—with all due respect to Figure 1-1, they don't have little 0s and 1s floating around in them. As we'll see in Chapter 3, digital circuits deal with analog voltages and currents and are built with analog components. The "digital abstraction" allows analog behavior to be ignored in most cases, so circuits can be modeled as if they really did process 0s and 1s.

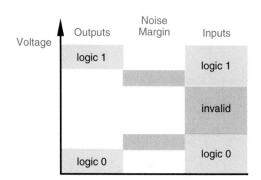

Figure 1-2
Logic values and
noise margins.

One important aspect of the digital abstraction is to associate a *range* of analog values with each logic value (0 or 1). As shown in Figure 1-2, a typical gate is not guaranteed to have a precise voltage level for a logic 0 output. Rather, it may produce a voltage somewhere in a range that is a *subset* of the range guaranteed to be recognized as a 0 by other gate inputs. The difference between *noise margin* the range boundaries is called *noise margin*—in a real circuit, a gate's output can be corrupted by this much noise and still be correctly interpreted at the inputs of other gates.

Behavior for logic 1 outputs is similar. Note in the figure that there is an "invalid" region between the input ranges for logic 0 and logic 1. Although any given digital device operating at a particular voltage and temperature will have a fairly well defined boundary (or threshold) between the two ranges, different devices may have different boundaries. Still, all properly operating devices have their boundary *somewhere* in the "invalid" range. Therefore, any signal that is within the defined ranges for 0 and 1 will be interpreted identically by different devices. This characteristic is essential for reproducibility of results.

It is the job of an *electronic* circuit designer to ensure that logic gates produce and recognize logic signals that are within the appropriate ranges. This is an analog circuit-design problem; we touch upon some of its aspects in Chapter 3. It is not possible to design a circuit that has the desired behavior under every possible condition of power-supply voltage, temperature, loading, and other factors. Instead, the electronic circuit designer or device manufacturer *specifications* provides *specifications* that define the conditions under which correct behavior is guaranteed.

As a *digital* designer, then, you need not delve into the detailed analog behavior of a digital device to ensure its correct operation. Rather, you need only examine enough about the device's operating environment to determine that it is operating within its published specifications. Granted, some analog knowledge is needed to perform this examination, but not nearly what you'd need to design a digital device starting from scratch. In Chapter 3 we'll give you just what you need.

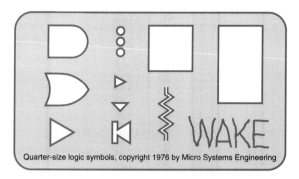

Quarter-size logic symbols, copyright 1976 by Micro Systems Engineering

Figure 1-3
A logic-design
template.

1.5 Software Aspects of Digital Design

Digital design need not involve any software tools. For example, Figure 1-3
shows the primary tool of the "old school" of digital design—a plastic template
for drawing logic symbols in schematic diagrams by hand (the designer's name
was engraved into the plastic with a soldering iron).

Today, however, software tools are an essential part of digital design.
Indeed, the availability and practicality of hardware description languages
(HDLs) and accompanying circuit simulation and synthesis tools have changed
the entire landscape of digital design over the past several years. We'll make
extensive use of HDLs throughout this book.

In *computer-aided design (CAD)* various software tools improve the
designer's productivity and help to improve the correctness and quality of
designs. In a competitive world, the use of software tools is mandatory to obtain
high-quality results on aggressive schedules. Important examples of software
tools for digital design are listed below:

computer-aided design (CAD)

- *Schematic entry.* This is the digital designer's equivalent of a word
  processor. It allows schematic diagrams to be drawn "on-line," instead
  of with paper and pencil. The more advanced schematic-entry programs
  also check for common, easy-to-spot errors, such as shorted outputs,
  signals that don't go anywhere, and so on. Such programs are discussed in
  greater detail in Section 11.1.2.

- *HDLs.* Hardware description languages, originally developed for circuit
  modeling, are now being used more and more for hardware *design.* They
  can be used to design anything from individual function modules to large,
  multichip digital systems. We'll introduce two HDLs, ABEL and VHDL,
  at the end of Chapter 4, and we'll provide examples in both languages in
  the chapters that follow.

- *HDL compilers, simulators, and synthesis tools.* A typical HDL software
  package contains several components. In a typical environment, the
  designer writes a text-based "program," and the HDL compiler analyzes

the program for syntax errors. If it compiles correctly, the designer has the option of handing it over to a synthesis tool that creates a corresponding circuit design targeted to a particular hardware technology. Most often, before synthesis the designer will use the compiler's results as input to a "simulator" to verify the behavior of the design.

- *Simulators.* The design cycle for a customized, single-chip digital integrated circuit is long and expensive. Once the first chip is built, it's very difficult, often impossible, to debug it by probing internal connections (they are really tiny), or to change the gates and interconnections. Usually, changes must be made in the original design database and a new chip must be manufactured to incorporate the required changes. Since this process can take months to complete, chip designers are highly motivated to "get it right" (or almost right) on the first try. Simulators help designers predict the electrical and functional behavior of a chip without actually building it, allowing most if not all bugs to be found before the chip is fabricated.

- Simulators are also used in the design of "programmable logic devices," introduced later, and in the overall design of systems that incorporate many individual components. They are somewhat less critical in this case because it's easier for the designer to make changes in components and interconnections on a printed-circuit board. However, even a little bit of simulation can save time by catching simple but stupid mistakes.

- *Test benches.* Digital designers have learned how to formalize circuit simulation and testing into software environments called "test benches." The idea is to build a set of programs around a design to automatically exercise its functions and check both its functional and its timing behavior. This is especially useful when small design changes are made—the test bench can be run to ensure that bug fixes or "improvements" in one area do not break something else. Test-bench programs may be written in the same HDL as the design itself, in C or C++, or in combination of languages including scripting languages like PERL.

- *Timing analyzers and verifiers.* The time dimension is very important in digital design. All digital circuits take time to produce a new output value in response to an input change, and much of a designer's effort is spent ensuring that such output changes occur quickly enough (or, in some cases, not too quickly). Specialized programs can automate the tedious task of drawing timing diagrams and specifying and verifying the timing relationships between different signals in a complex system.

- *Word processors.* Let's not forget the lowly text editor and word processor. These tools are obviously useful for creating the source code for HDL-based designs, but they have an important use in every design—to create documentation!

PROGRAMMABLE LOGIC DEVICES VERSUS SIMULATION

Later in this book you'll learn how programmable logic devices (PLDs) and field-programmable gate arrays (FPGAs) allow you to design a circuit or subsystem by writing a sort of program. PLDs and FPGAs are now available with up to millions of gates, and the capabilities of these technologies are ever increasing. If a PLD- or FPGA-based design doesn't work the first time, you can often fix it by changing the program and physically reprogramming the device, without changing any components or interconnections at the system level. The ease of prototyping and modifying PLD- and FPGA-based systems can eliminate the need for simulation in board-level design; simulation is required only for chip-level designs.

The most widely held view in industry trends says that as chip technology advances, more and more design will be done at the chip level, rather than the board level. Therefore, the ability to perform complete and accurate simulation will become increasingly important to the typical digital designer.

However, another view is possible. If we extrapolate trends in PLD and FPGA capabilities, in the next decade we will witness the emergence of devices that include not only gates and flip-flops as building blocks, but also higher-level functions such as processors, memories, and input/output controllers. At this point, most digital designers will use complex on-chip components and interconnections whose basic functions have already been tested by the device manufacturer.

In this future view, it is still possible to misapply high-level programmable functions, but it is also possible to fix mistakes simply by changing a program; detailed simulation of a design before simply "trying it out" could be a waste of time. Another, compatible view is that the PLD or FPGA is merely a full-speed simulator for the program, and this full-speed simulator is what gets shipped in the product!

Does this extreme view have any validity? To guess the answer, ask yourself the following question. How many software programmers do you know who debug a new program by "simulating" its operation rather than just trying it out?

In any case, modern digital systems are much too complex for a designer to have any chance of testing every possible input condition, with or without simulation. As in software, correct operation of digital systems is best accomplished through practices that ensure that the systems are "correct by design." It is a goal of this text to encourage such practices.

In addition to using the tools above, designers may sometimes write specialized programs in high-level languages like C or C++, or scripts in languages like PERL, to solve particular design problems. For example, Section 10.1.6 gives a couple of examples of C programs that generate the "truth tables" for complex combinational logic functions.

Although CAD tools are important, they don't make or break a digital designer. To take an analogy from another field, you couldn't consider yourself to be a great writer just because you're a fast typist or very handy with a word processor. During your study of digital design, be sure to learn and use all the tools that are available to you, such as schematic-entry programs, simulators, and HDL compilers. But remember that learning to use tools is no guarantee that you'll be able to produce good results. Please pay attention to what you're producing with them!

1.6 Integrated Circuits

integrated circuit (IC)

A collection of one or more gates fabricated on a single silicon chip is called an *integrated circuit (IC)*. Large ICs with tens of millions of transistors may be half an inch or more on a side, while small ICs may be less than one-tenth of an inch on a side.

wafer

Regardless of its size, an IC is initially part of a much larger, circular *wafer*, up to ten inches in diameter, containing dozens to hundreds of replicas of the same IC. All of the IC chips on the wafer are fabricated at the same time, like pizzas that are eventually sold by the slice, except in this case, each piece (IC

die

chip) is called a *die*. After the wafer is fabricated, the dice are tested in place on the wafer and defective ones are marked. Then the wafer is sliced up to produce the individual dice, and the marked ones are discarded. (Compare with the pizza-maker who sells all the pieces, even the ones without enough pepperoni!) Each unmarked die is mounted in a package, its pads are connected to the package pins, and the packaged IC is subjected to a final test and is shipped to a customer.

Some people use the term "IC" to refer to a silicon die. Some use "chip" to refer to the same thing. Still others use "IC" or "chip" to refer to the combination of a silicon die and its package. Digital designers tend to use the two terms interchangeably, and they really don't care what they're talking about. They don't require a precise definition, since they're only looking at the functional and elec-

IC

trical behavior of these things. In the balance of this text, we'll use the term *IC* to refer to a packaged die.

A DICEY DECISION

A reader of the second edition wrote to me to collect a $5 reward for pointing out my "glaring" misuse of "dice" as the plural of "die." According to the dictionary, she said, the plural form of "die" is "dice" *only* when describing those little cubes with dots on each side; otherwise it's "dies," and she produced the references to prove it.

Being stubborn, I asked my friends at the *Microprocessor Report* about this issue. According to the editor,

> There is, indeed, much dispute over this term. We actually stopped using the term "dice" in *Microprocessor Report* more than four years ago. I actually prefer the plural "die," … but perhaps it is best to avoid using the plural whenever possible.

So there you have it, even the experts don't agree with the dictionary! Rather than cop out, I rolled the dice and decided to use "dice" anyway.

Figure 1-4
Dual in-line pin (DIP)
packages: (a) 14-pin;
(b) 20-pin; (c) 28-pin.

In the early days of integrated circuits, ICs were classified by size—small, medium, or large—according to how many gates they contained. The simplest type of commercially available ICs are still called *small-scale integration (SSI)* and contain the equivalent of 1 to 20 gates. SSI ICs typically contain a handful of gates or flip-flops, the basic building blocks of digital design.

small-scale integration (SSI)

The SSI ICs that you're likely to encounter in an educational lab come in a 14-pin *dual in-line-pin (DIP)* package. As shown in Figure 1-4(a), the spacing between pins in a column is 0.1 inch and the spacing between columns is 0.3 inch. Larger DIP packages accommodate functions with more pins, as shown in (b) and (c). A *pin diagram* shows the assignment of device signals to package pins, or *pinout*. Figure 1-5 shows the pin diagrams for a few common SSI ICs. Such diagrams are used only for mechanical reference, when a designer needs to determine the pin numbers for a particular IC. In the schematic diagram for a

dual in-line-pin (DIP) package

pin diagram
pinout

Figure 1-5 Pin diagrams for a few 7400-series SSI ICs.

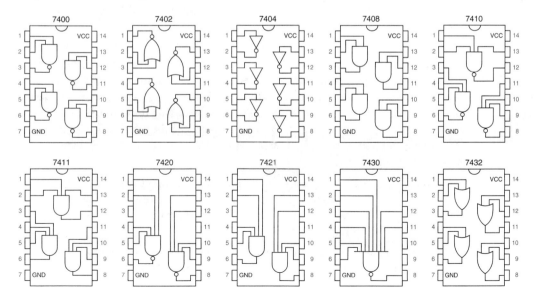

TINY-SCALE INTEGRATION

In the coming years, perhaps the most popular remaining use of SSI and MSI, especially in DIP packages, will be in educational labs. These devices will afford students the opportunity to "get their hands" dirty by "breadboarding" and wiring up simple circuits in the same way that their professors did years ago.

However, much to my surprise and delight, a segment of the IC industry has actually gone *down*scale from SSI in the past few years. The idea has been to sell individual logic gates in very small packages. These devices handle simple functions that are sometimes needed to match larger-scale components to a particular design, or in some cases they are used to work around bugs in the larger-scale components or their interfaces.

An example of such an IC is Motorola's 74VHC1G00. This chip is a single 2-input NAND gate housed in a 5-pin package (power, ground, two inputs, and one output). The entire package, including pins, measures only 0.08 inches on a side and is only 0.04 inches high! Now that's what I would call "tiny-scale integration"!

---

digital circuit, pin diagrams are not used. Instead, the various gates are grouped functionally, as we'll show in Section 5.1.

Although SSI ICs are still sometimes used as "glue" to tie together larger-scale elements in complex systems, they have been largely supplanted by programmable logic devices (PLDs), which we'll study in Sections 5.3 and 8.3.

medium-scale integration (MSI)

The next larger commercially available ICs are called *medium-scale integration (MSI)* and contain the equivalent of about 20 to 200 gates. An MSI IC typically contains a functional building block, such as a decoder, register, or counter. In Chapters 5 and 8 we'll place a strong emphasis on these building blocks. Even though the use of discrete MSI ICs is declining, the equivalent building blocks are used extensively in the design of larger ICs.

large-scale integration (LSI)

Large-scale integration (LSI) ICs are bigger still, containing the equivalent of 200 to 200,000 gates or more. LSI parts include small memories, microprocessors, programmable logic devices, and customized devices.

STANDARD LOGIC FUNCTIONS

Many standard "high-level" functions appear over and over as building blocks in digital design. Historically, these functions were first integrated in MSI circuits. Subsequently, they have appeared as components in the "macro" libraries for ASIC design, as "standard cells" in VLSI design, as "canned" functions in PLD programming languages, and as library functions in hardware-description languages such as VHDL.

Standard logic functions are introduced in Chapters 5 and 8 as 74-series MSI parts, as well as in HDL form. The discussion and examples in these chapters provide a basis for understanding and using these functions in any form.

The dividing line between LSI and *very large-scale integration (VLSI)* is fuzzy and tends to be stated in terms of transistor count rather than gate count. Any IC with over 1,000,000 transistors is definitely VLSI, and that includes most microprocessors and memories nowadays, as well as larger programmable logic devices and customized devices. In 1999, VLSI ICs with as many as 50 million transistors were being designed.

very large-scale integration (VLSI)

1.7 Programmable Logic Devices

There are a wide variety of ICs that can have their logic function "programmed" into them after they are manufactured. Most of these devices use technology that also allows the function to be *re*programmed, which means that if you find a bug in your design, you may be able to fix it without physically replacing or rewiring the device. In this book, we'll frequently refer to the design opportunities and methods for such devices.

Historically, *programmable logic arrays (PLAs)* were the first programmable logic devices. PLAs contained a two-level structure of AND and OR gates with user-programmable connections. Using this structure, a designer could accommodate any logic function up to a certain level of complexity using the well-known theory of logic synthesis and minimization that we'll present in Chapter 4.

programmable logic array (PLA)

PLA structure was enhanced and PLA costs were reduced with the introduction of *programmable array logic (PAL) devices*. Today, such devices are generically called programmable logic devices (PLDs) and are the "MSI" of the programmable logic industry. We'll have a lot to say about PLD architecture and technology in Sections 5.3 and 8.3.

programmable array logic (PAL) device
programmable logic device (PLD)

The ever-increasing capacity of integrated circuits created an opportunity for IC manufacturers to design larger PLDs for larger digital-design applications. However, for technical reasons that we'll discuss in Section 10.5, the basic two-level AND-OR structure of PLDs could not be scaled to larger sizes. Instead, IC manufacturers devised *complex PLD (CPLD)* architectures to achieve the required scale. A typical CPLD is merely a collection of multiple PLDs and an interconnection structure, all on the same chip. In addition to the individual PLDs, the on-chip interconnection structure is also programmable, providing a rich variety of design possibilities. CPLDs can be scaled to larger sizes by increasing the number of individual PLDs and the richness of the interconnection structure on the CPLD chip.

complex PLD (CPLD)

At about the same time that CPLDs were being invented, other IC manufacturers took a different approach to scaling the size of programmable logic chips. Compared to a CPLD, a field-programmable gate arrays (FPGA) contains a much larger number of smaller individual logic blocks and provides a large, distributed interconnection structure that dominates the entire chip. Figure 1-6 illustrates the difference between the two chip-design approaches.

field-programmable gate array (FPGA)

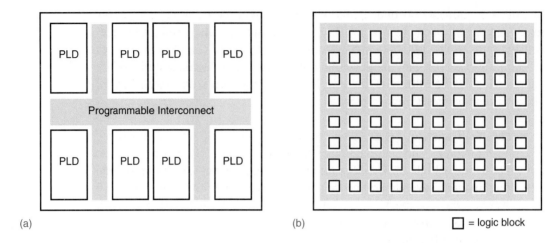

Figure 1-6 Large programmable-logic-device scaling approaches: (a) CPLD; (b) FPGA.

Proponents of one approach or the other used to get into "religious" arguments over which way was better, but the largest manufacturer of large programmable logic devices, Xilinx, Inc., acknowledges that there is a place for both approaches and manufactures both types of devices. What's more important than chip architecture is that both approaches support a style of design in which products can be moved from design concept to prototype and production in a very short time.

Also important in achieving short "time-to-market" for all kinds of PLD-based products is the use of HDLs in their design. Languages like ABEL and VHDL, and their accompanying software tools, allow a design to be compiled, synthesized, and downloaded into a PLD, CPLD, or FPGA literally in minutes. The power of highly structured, hierarchical languages like VHDL is especially important in helping designers utilize the hundreds of thousands or millions of gates that are provided in the largest CPLDs and FPGAs.

1.8 Application-Specific ICs

Perhaps the most interesting developments in IC technology for the average digital designer are not the ever-increasing chip sizes, but the ever-increasing opportunities to "design your own chip." Chips designed for a particular, limited product or application are called *semicustom ICs* or *application-specific ICs (ASICs)*. ASICs generally reduce the total component and manufacturing cost of a product by reducing chip count, physical size, and power consumption, and they often provide higher performance.

semicustom IC
application-specific IC (ASIC)

The *nonrecurring engineering (NRE) cost* for designing an ASIC can exceed the cost of a discrete design by $5,000 to $250,000 or more. NRE charges are paid to the IC manufacturer and others who are responsible for designing the

nonrecurring engineering (NRE) cost

internal structure of the chip, creating tooling such as the metal masks for manufacturing the chips, developing tests for the manufactured chips, and actually making the first few sample chips.

The NRE cost for a typical, medium-complexity ASIC with about 100,000 gates is $30,000–$50,000. An ASIC design normally makes sense only when the NRE cost can be offset by the per-unit savings over the expected sales volume of the product.

The NRE cost to design a *custom LSI* chip—a chip whose functions, internal architecture, and detailed transistor-level design is tailored for a specific customer—is very high, $250,000 or more. Thus, full custom LSI design is done only for chips that have general commercial application or that will enjoy very high sales volume in a specific application (e.g., a digital watch chip, a network interface, or a bus-interface circuit for a PC).

custom LSI

To reduce NRE charges, IC manufacturers have developed libraries of *standard cells* including commonly used MSI functions such as decoders, registers, and counters and commonly used LSI functions such as memories, programmable logic arrays, and microprocessors. In a *standard-cell design*, the logic designer interconnects functions in much the same way as in a multichip MSI/LSI design. Custom cells are created (at added cost, of course) only if absolutely necessary. All of the cells are then laid out on the chip, optimizing the layout to reduce propagation delays and minimize the size of the chip. Minimizing the chip size reduces the per-unit cost of the chip, since it increases the number of chips that can be fabricated on a single wafer. The NRE cost for a standard-cell design is typically on the order of $150,000.

standard cells

standard-cell design

Well, $150,000 is still a lot of money for most folks, so IC manufacturers have gone one step further to bring ASIC design capability to the masses. A *gate array* is an IC whose internal structure is an array of gates whose interconnections are initially unspecified. The logic designer specifies the gate types and interconnections. Even though the chip design is ultimately specified at this very low level, the designer typically works with "macrocells," the same high-level functions used in multichip MSI/LSI and standard-cell designs; software expands the high-level design into a low-level one.

gate array

The main difference between standard-cell and gate-array design is that the macrocells and the chip layout of a gate array are not as highly optimized as those in a standard-cell design, so the chip may be 25% or more larger and therefore may cost more. Also, there is no opportunity to create custom cells in the gate-array approach. On the other hand, a gate-array design can be finished faster and at lower NRE cost, ranging from about $5000 (what you're told initially) to $75,000 (what you find you've spent when you're all done).

The basic digital design methods that you'll study throughout this book apply very well to the functional design of ASICs. However, there are additional opportunities, constraints, and steps in ASIC design, which usually depend on the particular ASIC vendor and design environment.

1.9 Printed-Circuit Boards

*printed-circuit board
  (PCB)*

*printed-wiring board
  (PWB)*

PCB traces

mil

fine-line

An IC is normally mounted on a *printed-circuit board (PCB)* [or *printed-wiring board (PWB)*] that connects it to other ICs in a system. The multilayer PCBs used in typical digital systems have copper wiring etched on multiple, thin layers of fiberglass that are laminated into a single board about 1/16 inch thick.

Individual wire connections, or *PCB traces*, are usually quite narrow, 10 to 25 mils in typical PCBs. (A *mil* is one-thousandth of an inch.) In *fine-line* PCB technology, the traces are extremely narrow, as little as 4 mils wide with 4-mil spacing between adjacent traces. Thus, up to 125 connections may be routed in a one-inch-wide band on a single layer of the PCB. If higher connection density is needed, then more layers are used.

*surface-mount
  technology (SMT)*

Most of the components in modern PCBs use *surface-mount technology (SMT)*. Instead of having the long pins of DIP packages that poke through the board and are soldered to the underside, the leads of SMT IC packages are bent to make flat contact with the top surface of the PCB. Before such components are mounted on the PCB, a special "solder paste" is applied to contact pads on the PCB using a stencil whose hole pattern matches the contact pads to be soldered. Then the SMT components are placed (by hand or by machine) on the pads, where they are held in place by the solder paste (or in some cases, by glue). Finally, the entire assembly is passed through an oven to melt the solder paste, which then solidifies when cooled.

Surface-mount component technology, coupled with fine-line PCB technology, allows extremely dense packing of integrated circuits and other components on a PCB. This dense packing does more than save space. For very high-speed circuits, dense packing goes a long way toward minimizing adverse analog phenomena, including transmission-line effects and speed-of-light limitations.

*multichip module
  (MCM)*

To satisfy the most stringent requirements for speed and density, *multichip modules (MCMs)* have been developed. In this technology, IC dice are not mounted in individual plastic or ceramic packages. Instead, the IC dice for a high-speed subsystem (say, a processor and its cache memory) are bonded directly to a substrate that contains the required interconnections on multiple layers. The MCM is hermetically sealed and has its own external pins for power, ground, and just those signals that are required by the system that contains it.

1.10 Digital-Design Levels

Digital design can be carried out at several different levels of representation and abstraction. Although you may learn and practice design at a particular level, from time to time you'll need to go up or down a level or two to get the job done. Also, the industry itself and most designers have been steadily moving to higher levels of abstraction as circuit density and functionality have increased.

The lowest level of digital design is device physics and IC manufacturing processes. This is the level that is primarily responsible for the breathtaking advances in IC speed and density that have occurred over the past decades. The effects of these advances are summarized in *Moore's Law*, first stated by Intel founder Gordon Moore in 1965: that the number of transistors per square inch in an IC doubles every year. In recent years, the rate of advance has slowed down to doubling about every 18 months, but it is important to note that with each doubling of density has also come a doubling of speed.

Moore's Law

This book does not reach down to the level of device physics and IC processes, but you need to recognize the importance of that level. Being aware of likely technology advances and other changes is important in system and product planning. For example, decreases in chip geometries have recently forced a move to lower logic-power-supply voltages, causing major changes in the way designers plan and specify modular systems and upgrades.

In this book, we jump into digital design at the transistor level and go all the way up to the level of logic design using HDLs. We stop short of the next level, which includes computer design and overall system design. The "center" of our discussion is at the level of functional building blocks.

To get a preview of the levels of design that we'll cover, consider a simple design example. Suppose you are to build a "multiplexer" with two data input bits, A and B, a control input bit S, and an output bit Z. Depending on the value of S, 0 or 1, the circuit is to transfer the value of either A or B to the output Z. This idea is illustrated in the "switch model" of Figure 1-7. Let us consider the design of this function at several different levels.

Figure 1-7
Switch model for multiplexer function.

Although logic design is usually carried out at higher level, for some functions it is advantageous to optimize them by designing at the transistor level. The multiplexer is such a function. Figure 1-8 shows how the multiplexer can be designed in "CMOS" technology using specialized transistor circuit structures

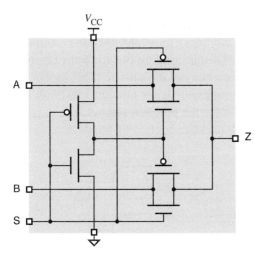

Figure 1-8
Multiplexer design using CMOS transmission gates.

S	A	B	Z
0	0	0	0
0	0	1	0
0	1	0	1
0	1	1	1
1	0	0	0
1	0	1	1
1	1	0	0
1	1	1	1

Table 1-1
Truth table for the multiplexer function.

called "transmission gates," discussed in Section 3.7.1. Using this approach, the multiplexer can be built with just six transistors. Any of the other approaches that we describe requires at least 14 transistors.

In the traditional study of logic design, we would use a "truth table" to describe the multiplexer's logic function. A truth table lists all possible combinations of input values and the corresponding output values for the function. Since the multiplexer has three inputs, it has 2^3 or 8 possible input combinations, as shown in the truth table in Table 1-1.

Once we have a truth table, traditional logic design methods, described in Section 4.3, use Boolean algebra and well-understood minimization algorithms to derive an "optimal" two-level AND-OR equation from the truth table. For the multiplexer truth table, we would derive the following equation:

$$Z = S' \cdot A + S \cdot B$$

This equation is read "Z equals not S and A or S and B." Going one step further, we can convert the equation into a corresponding set of logic gates that perform the specified logic function, as shown in Figure 1-9. This circuit requires 14 transistors if we use standard CMOS technology for the four gates shown.

A multiplexer is a very commonly used function, and most digital logic technologies provide predefined multiplexer building blocks. For example, the 74x157 is an MSI chip that performs multiplexing on two 4-bit inputs simultaneously. Figure 1-10 is a logic diagram that shows how we can hook up just one bit of this 4-bit building block to solve the problem at hand. The numbers in color are pin numbers of a 16-pin DIP package containing the device.

Figure 1-9
Gate-level logic diagram for multiplexer function.

Figure 1-10
Logic diagram for a multiplexer using an MSI building block.

We can also realize the multiplexer function as part of a programmable logic device. Languages like ABEL allow us to specify outputs using Boolean equations similar to the one on the previous page, but it's usually more convenient to use "higher-level" language elements. For example, Table 1-2 is an ABEL program for the multiplexer function. The first three lines define the name of the program module and specify the type of PLD in which the function will be realized. The next two lines specify the device pin numbers for inputs and output. The "WHEN" statement specifies the actual logic function in a way that's very easy to understand, even though we haven't covered ABEL yet.

An even higher-level language, VHDL, can be used to specify the multiplexer function in a way that is very flexible and hierarchical. Table 1-3 is an example VHDL program for the multiplexer. The first two lines specify a standard library and set of definitions to use in the design. The next four lines specify only the inputs and outputs of the function, purposely hiding any details about the way the function is realized internally. The "architecture" section of the program specifies the function's behavior. VHDL syntax takes a little getting used to, but the single "when" statement says basically the same thing that the ABEL version did. At VHDL "synthesis tool" can start with this behavioral

```
module chap1mux
title 'Two-input multiplexer example'
CHAP1MUX device 'P16V8'

A, B, S        pin 1, 2, 3;
Z              pin 13 istype 'com';

equations

WHEN S == 0 THEN Z = A;   ELSE Z = B;

end chap1mux
```

Table 1-2
ABEL program for the multiplexer.

Table 1-3
VHDL program for
the multiplexer.

```
library IEEE;
use IEEE.std_logic_1164.all;

entity Vchap1mux is
    port ( A, B, S: in  STD_LOGIC;
           Z:          out STD_LOGIC );
end Vchap1mux;

architecture Vchap1mux_arch of Vchap1mux is
begin
  Z <= A when S = '0' else B;
end Vchap1mux_arch;
```

description and produce a circuit that has this behavior in a specified target digital-logic technology.

By explicitly enforcing a separation of input/output definitions ("entity") and internal realization ("architecture"), VHDL makes it easy for designers to define alternate realizations of functions without having to make changes elsewhere in the design hierarchy. For example, a designer could specify an alternate, structural architecture for the multiplexer as shown in Table 1-4. This architecture is basically a text equivalent of the logic diagram in Figure 1-9.

Going one step further, VHDL is powerful enough that we could actually define operations that model functional behavioral at the transistor level (though we won't explore such capabilities in this book). Thus, we could come full circle by writing a VHDL program that specifies a transistor-level realization of the multiplexer equivalent to Figure 1-8.

Table 1-4
"Structural" VHDL
program for the
multiplexer.

```
architecture Vchap1mux_gate_arch of Vchap1mux is
signal SN, ASN, SB: STD_LOGIC;
begin
  U1: INV (S, SN);
  U2: AND2 (A, SN, ASN);
  U3: AND2 (S, B, SB);
  U4: OR2 (ASN, SB, Z);
end Vchap1mux_gate_arch;
```

1.11 The Name of the Game

Given the functional and performance requirements for a digital system, the
board-level design   name of the game in practical digital design is to minimize cost. For *board-level designs*—systems that are packaged on a single PCB—this usually means minimizing the number of IC packages. If too many ICs are required, they won't all fit on the PCB. "Well, just use a bigger PCB," you say. Unfortunately, PCB sizes are usually constrained by factors such as preexisting standards (e.g.,

add-in boards for PCs), packaging constraints (e.g., it has to fit in a toaster), or edicts from above (e.g., in order to get the project approved three months ago, you foolishly told your manager that it would all fit on a 3×5 inch PCB, and now you've got to deliver!). In each of these cases, the cost of using a larger PCB or multiple PCBs may be unacceptable.

Minimizing the number of ICs is usually the rule even though individual IC costs vary. For example, a typical SSI or MSI IC may cost 25 cents, while an small PLD may cost a dollar. It may be possible to perform a particular function with three SSI and MSI ICs (75 cents) or one PLD (a dollar). In most situations, the more expensive PLD solution is used, not because the designer owns stock in the IC company, but because the PLD solution uses less PCB area and is also a lot easier to change if it's not right the first time.

In *ASIC design*, the name of the game is a little different, but the importance of structured, functional design techniques is the same. Although it's easy to burn hours and weeks creating custom macrocells and minimizing the total gate count of an ASIC, only rarely is this advisable. The per-unit cost reduction achieved by having a 10% smaller chip is negligible except in high-volume applications. In applications with low to medium volume (the majority), two other factors are more important: design time and NRE cost.

ASIC design

A shorter design time allows a product to reach the market sooner, increasing revenues over the lifetime of the product. A lower NRE cost also flows right to the "bottom line" and in small companies may be the only way the project can be completed before the company runs out of money (believe me, I've been there!). If the product is successful, it's always possible and profitable to "tweak" the design later to reduce per-unit costs. The need to minimize design time and NRE cost argues in favor of a structured, as opposed to highly optimized, approach to ASIC design, using standard building blocks provided in the ASIC manufacturer's library.

The considerations in PLD, CPLD, and FPGA design are a combination of the above. The choice of a particular PLD technology and device size is usually made fairly early in the design cycle. Later, as long as the design "fits" in the selected device, there's no point in trying to optimize gate count or board area— the device has already been committed. However, if new functions or bug fixes push the design beyond the capacity of the selected device, that's when you must work very hard to modify the design to make it fit.

1.12 Going Forward

This concludes the introductory chapter. As you continue reading this book, keep in mind two things. First, the ultimate goal of digital design is to build systems that solve problems for people. While this book will give you the basic tools for design, it's still your job to keep "the big picture" in the back of your mind. Second, cost is an important factor in every design decision; and you must consider not only the cost of digital components, but also the cost of the design activity itself.

Finally, as you get deeper into the text, if you encounter something that you think you've seen before but don't remember where, please consult the index. I've tried to make it as helpful and complete as possible.

Drill Problems

1.1   Suggest some better-looking chapter-opening artwork to put on page 1 of the next edition of this book.

1.2   Give three different definitions for the word "bit" as used in this chapter.

1.3   Define the following acronyms: ASIC, CAD, CD, CO, CPLD, DIP, DVD, FPGA, HDL, IC, IP, LSI, MCM, MSI, NRE, OK, PBX, PCB, PLD, PWB, SMT, SSI, VHDL, VLSI.

1.4   Research the definitions of the following acronyms: ABEL, CMOS, JPEG, MPEG, OK, PERL, VHDL. (Is OK really an acronym?)

1.5   Excluding the topics in Section 1.2, list three once-analog systems that have "gone digital" since you were born.

1.6   Draw a digital circuit consisting of a 2-input AND gate and three inverters, where an inverter is connected to each of the AND gate's inputs and its output. For each of the four possible combinations of inputs applied to the two primary inputs of this circuit, determine the value produced at the primary output. Is there a simpler circuit that gives the same input/output behavior?

1.7   When should you use the pin diagrams of Figure 1-5 in the schematic diagram of a circuit?

1.8   What is the relationship between "die" and "dice"?

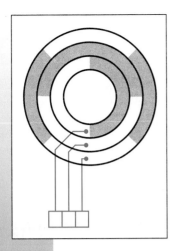

Number Systems and Codes

D igital systems are built from circuits that process binary digits—
0s and 1s—yet very few real-life problems are based on binary
numbers or any numbers at all. As a result, a digital system
designer must establish some correspondence between the
binary digits processed by digital circuits and real-life numbers,
events, and conditions. The purpose of this chapter is to show you how
familiar numeric quantities can be represented and manipulated in a digital
system, and how nonnumeric data, events, and conditions also can be
represented.

The first nine sections describe binary number systems and show how
addition, subtraction, multiplication, and division are performed in these
systems. Sections 2.10–2.13 show how other things, such as decimal
numbers, text characters, mechanical positions, and arbitrary conditions,
can be encoded using strings of binary digits.

Section 2.14 introduces "n-cubes," which provide a way to visualize
the relationship between different bit strings. The n-cubes are especially
useful in the study of error-detecting codes in Section 2.15. We conclude the
chapter with an introduction to codes for transmitting and storing data one
bit at a time.

2.1 Positional Number Systems

positional number system

weight

The traditional number system that we learned in school and use every day in business is called a *positional number system*. In such a system, a number is represented by a string of digits, where each digit position has an associated *weight*. The value of a number is a weighted sum of the digits, for example:

$$1734 = 1 \cdot 1000 + 7 \cdot 100 + 3 \cdot 10 + 4 \cdot 1$$

Each weight is a power of 10 corresponding to the digit's position. A decimal point allows negative as well as positive powers of 10 to be used:

$$5185.68 = 5 \cdot 1000 + 1 \cdot 100 + 8 \cdot 10 + 5 \cdot 1 + 6 \cdot 0.1 + 8 \cdot 0.01$$

In general, a number D of the form $d_1 d_0 . d_{-1} d_{-2}$ has the value

$$D = d_1 \cdot 10^1 + d_0 \cdot 10^0 + d_{-1} \cdot 10^{-1} + d_{-2} \cdot 10^{-2}$$

base

radix

Here, 10 is called the *base* or *radix* of the number system. In a general positional number system, the radix may be any integer $r \geq 2$, and a digit in position i has weight r^i. The general form of a number in such a system is

$$d_{p-1} d_{p-2} \cdots d_1 d_0 . d_{-1} d_{-2} \cdots d_{-n}$$

radix point

where there are p digits to the left of the point and n digits to the right of the point, called the *radix point*. If the radix point is missing, it is assumed to be to the right of the rightmost digit. The value of the number is the sum of each digit multiplied by the corresponding power of the radix:

$$D = \sum_{i=-n}^{p-1} d_i \cdot r^i$$

high-order digit

most significant digit

low-order digit

least significant digit

Except for possible leading and trailing zeroes, the representation of a number in a positional number system is unique. (Obviously, 0185.6300 equals 185.63, and so on.) The leftmost digit in such a number is called the *high-order* or *most significant digit*; the rightmost is the *low-order* or *least significant digit*.

binary digit

bit

binary radix

As we'll learn in Chapter 3, digital circuits have signals that are normally in one of only two conditions—low or high, charged or discharged, off or on. The signals in these circuits are interpreted to represent *binary digits* (or *bits*) that have one of two values, 0 and 1. Thus, the *binary radix* is normally used to represent numbers in a digital system. The general form of a binary number is

$$b_{p-1} b_{p-2} \cdots b_1 b_0 . b_{-1} b_{-2} \cdots b_{-n}$$

and its value is

$$B = \sum_{i=-n}^{p-1} b_i \cdot 2^i$$

In a binary number, the radix point is called the *binary point*. When dealing with binary and other nondecimal numbers, we use a subscript to indicate the radix of each number, unless the radix is clear from the context. Examples of binary numbers and their decimal equivalents are given below.

binary point

$$10011_2 = 1 \cdot 16 + 0 \cdot 8 + 0 \cdot 4 + 1 \cdot 2 + 1 \cdot 1 = 19_{10}$$
$$100010_2 = 1 \cdot 32 + 0 \cdot 16 + 0 \cdot 8 + 0 \cdot 4 + 1 \cdot 2 + 0 \cdot 1 = 34_{10}$$
$$101.001_2 = 1 \cdot 4 + 0 \cdot 2 + 1 \cdot 1 + 0 \cdot 0.5 + 0 \cdot 0.25 + 1 \cdot 0.125 = 5.125_{10}$$

The leftmost bit of a binary number is called the *high-order* or *most significant bit (MSB)*; the rightmost is the *low-order* or *least significant bit (LSB)*.

MSB
LSB

2.2 Octal and Hexadecimal Numbers

Radix 10 is important because we use it in everyday business, and radix 2 is important because binary numbers can be processed directly by digital circuits. Numbers in other radices are not often processed directly but may be important for documentation or other purposes. In particular, the radices 8 and 16 provide convenient shorthand representations for multibit numbers in a digital system.

The *octal number system* uses radix 8, while the *hexadecimal number system* uses radix 16. Table 2-1 shows the binary integers from 0 to 1111 and their octal, decimal, and hexadecimal equivalents. The octal system needs 8 digits, so it uses digits 0–7 of the decimal system. The hexadecimal system needs 16 digits, so it supplements decimal digits 0–9 with the letters *A–F*.

octal number system
hexadecimal number system

hexadecimal digits A–F

The octal and hexadecimal number systems are useful for representing multibit numbers because their radices are powers of 2. Since a string of three bits can take on eight different combinations, it follows that each 3-bit string can be uniquely represented by one octal digit, according to the third and fourth columns of Table 2-1. Likewise, a 4-bit string can be represented by one hexadecimal digit according to the fifth and sixth columns of the table.

Thus, it is very easy to convert a binary number to octal. Starting at the binary point and working left, we simply separate the bits into groups of three and replace each group with the corresponding octal digit:

binary-to-octal conversion

$$100011001110_2 = 100\ 011\ 001\ 110_2 = 4316_8$$
$$11101101110101001_2 = 011\ 101\ 101\ 110\ 101\ 001_2 = 355651_8$$

The procedure for binary-to-hexadecimal conversion is similar, except we use groups of four bits:

binary-to-hexadecimal conversion

$$100011001110_2 = 1000\ 1100\ 1110_2 = 8CE_{16}$$
$$11101101110101001_2 = 00011101\ 1011\ 1010\ 1001_2 = 1DBA9_{16}$$

In these examples we have freely added zeroes on the left to make the total number of bits a multiple of 3 or 4 as required.

Table 2-1
Binary, decimal, octal, and hexadecimal numbers.

Binary	Decimal	Octal	3-Bit String	Hexadecimal	4-Bit String
0	0	0	000	0	0000
1	1	1	001	1	0001
10	2	2	010	2	0010
11	3	3	011	3	0011
100	4	4	100	4	0100
101	5	5	101	5	0101
110	6	6	110	6	0110
111	7	7	111	7	0111
1000	8	10	—	8	1000
1001	9	11	—	9	1001
1010	10	12	—	A	1010
1011	11	13	—	B	1011
1100	12	14	—	C	1100
1101	13	15	—	D	1101
1110	14	16	—	E	1110
1111	15	17	—	F	1111

If a binary number contains digits to the right of the binary point, we can convert them to octal or hexadecimal by starting at the binary point and working right. Both the lefthand and righthand sides can be padded with zeroes to get multiples of three or four bits, as shown in the example below:

$$10.1011001011_2 = 010 . 101\ 100\ 101\ 100_2 = 2.5454_8$$
$$= 0010 . 1011\ 0010\ 1100_2 = 2.B2C_{16}$$

octal- or hexadecimal-to-binary conversion

Converting in the reverse direction, from octal or hexadecimal to binary, is very easy. We simply replace each octal or hexadecimal digit with the corresponding 3- or 4-bit string, as shown below:

$$1357_8 = 001\ 011\ 101\ 111_2$$
$$2046.17_8 = 010\ 000\ 100\ 110 . 001\ 111_2$$
$$BEAD_{16} = 1011\ 1110\ 1010\ 1101_2$$
$$9F.46C_{16} = 1001\ 111 . 0100\ 0110\ 1100_2$$

The octal number system was quite popular 25 years ago because of certain minicomputers that had their front-panel lights and switches arranged in groups of three. However, the octal number system is not used much today, because of the preponderance of machines that process 8-bit *bytes*. It is difficult to extract individual byte values in multibyte quantities in the octal representation; for

byte

WHEN I'M 64   As you grow older, you'll find that the hexadecimal number system is useful for more than just computers. When I turned 40, I told friends that I had just turned 28_{16}. The "$_{16}$" was whispered under my breath, of course. At age 50, I'll be only 32_{16}.

People get all excited about decennial birthdays like 20, 30, 40, 50, ..., but you should be able to convince your friends that the decimal system is of no fundamental significance. More significant life changes occur around birthdays 2, 4, 8, 16, 32, and 64, when you add a most significant bit to your age. Why do you think the Beatles sang "When I'm sixty-*four*"?

---

example, what are the octal values of the four 8-bit bytes in the 32-bit number with octal representation 12345670123_8?

In the hexadecimal system, two digits represent an 8-bit byte, and $2n$ digits represent an n-byte word; each pair of digits constitutes exactly one byte. For example, the 32-bit hexadecimal number $5678ABCD_{16}$ consists of four bytes with values 56_{16}, 78_{16}, AB_{16}, and CD_{16}. In this context, a 4-bit hexadecimal digit is sometimes called a *nibble*; a 32-bit (4-byte) number has eight nibbles. Hexa- *nibble* decimal numbers are often used to describe a computer's memory address space. For example, a computer with 16-bit addresses might be described as having read/write memory installed at addresses 0–$EFFF_{16}$, and read-only memory at addresses $F000$–$FFFF_{16}$. Many computer programming languages use the prefix "0x" to denote a hexadecimal number, for example, 0xBFC0000. *0x prefix*

2.3 General Positional-Number-System Conversions

In general, conversion between two radices cannot be done by simple substitutions; arithmetic operations are required. In this section, we show how to convert a number in any radix to radix 10 and vice versa, using radix-10 arithmetic.

In Section 2.1, we indicated that the value of a number in any radix is given   *radix-r-to-decimal* by the formula   *conversion*

$$D = \sum_{i=-n}^{p-1} d_i \cdot r^i$$

where r is the radix of the number and there are p digits to the left of the radix point and n to the right. Thus, the value of the number can be found by converting each digit of the number to its radix-10 equivalent and expanding the formula using radix-10 arithmetic. Some examples are given below:

$$
\begin{aligned}
1CE8_{16} &= 1 \cdot 16^3 + 12 \cdot 16^2 + 14 \cdot 16^1 + 8 \cdot 16^0 = 7400_{10} \\
F1A3_{16} &= 15 \cdot 16^3 + 1 \cdot 16^2 + 10 \cdot 16^1 + 3 \cdot 16^0 = 61859_{10} \\
436.5_8 &= 4 \cdot 8^2 + 3 \cdot 8^1 + 6 \cdot 8^0 + 5 \cdot 8^{-1} = 286.625_{10} \\
132.3_4 &= 1 \cdot 4^2 + 3 \cdot 4^1 + 2 \cdot 4^0 + 3 \cdot 4^{-1} = 30.75_{10}
\end{aligned}
$$

nested expansion formula

A shortcut for converting whole numbers to radix 10 can be obtained by rewriting the expansion formula in a nested fashion:

$$D = ((\cdots((d_{p-1}) \cdot r + d_{p-2}) \cdot r + \cdots) \cdots r + d_1) \cdot r + d_0$$

That is, we start with a sum of 0; beginning with the leftmost digit, we multiply the sum by r and add the next digit to the sum, repeating until all digits have been processed. For example, we can write

$$F1AC_{16} = (((15) \cdot 16 + 1) \cdot 16 + 10) \cdot 16 + 12$$

decimal-to-radix-r conversion

This formula is used in iterative, programmed conversion algorithms (such as Table 4-38 on page 279). It is also the basis of a very convenient method of converting a decimal number D to a radix r. Consider what happens if we divide the formula by r. Since the parenthesized part of the formula is evenly divisible by r, the quotient will be

$$Q = (\cdots((d_{p-1}) \cdot r + d_{p-2}) \cdot r + \cdots) \cdot r + d_1$$

and the remainder will be d_0. Thus, d_0 can be computed as the remainder of the long division of D by r. Furthermore, the quotient Q has the same form as the original formula. Therefore, successive divisions by r yield successive digits of D from right to left, until all the digits of D have been derived. Examples are given below:

$$
\begin{aligned}
179 \div 2 &= 89 \text{ remainder } 1 \quad \text{(LSB)} \\
\div 2 &= 44 \text{ remainder } 1 \\
\div 2 &= 22 \text{ remainder } 0 \\
\div 2 &= 11 \text{ remainder } 0 \\
\div 2 &= 5 \text{ remainder } 1 \\
\div 2 &= 2 \text{ remainder } 1 \\
\div 2 &= 1 \text{ remainder } 0 \\
\div 2 &= 0 \text{ remainder } 1 \quad \text{(MSB)}
\end{aligned}
$$

$$179_{10} = 10110011_2$$

$$
\begin{aligned}
467 \div 8 &= 58 \text{ remainder } 3 \quad \text{(least significant digit)} \\
\div 8 &= 7 \text{ remainder } 2 \\
\div 8 &= 0 \text{ remainder } 7 \quad \text{(most significant digit)}
\end{aligned}
$$

$$467_{10} = 723_8$$

$$
\begin{aligned}
3417 \div 16 &= 213 \text{ remainder } 9 \quad \text{(least significant digit)} \\
\div 16 &= 13 \text{ remainder } 5 \\
\div 16 &= 0 \text{ remainder } 13 \quad \text{(most significant digit)}
\end{aligned}
$$

$$3417_{10} = D59_{16}$$

Table 2-2 summarizes methods for converting among the most common radices.

■ **Table 2-2** Conversion methods for common radices.

Conversion	Method	Example
Binary to		
Octal	Substitution	$10111011001_2 = 10\ 111\ 011\ 001_2 = 2731_8$
Hexadecimal	Substitution	$10111011001_2 = 101\ 1101\ 1001_2 = 5D9_{16}$
Decimal	Summation	$10111011001_2 = 1 \cdot 1024 + 0 \cdot 512 + 1 \cdot 256 + 1 \cdot 128 + 1 \cdot 64$ $+\ 0 \cdot 32 + 1 \cdot 16 + 1 \cdot 8 + 0 \cdot 4 + 0 \cdot 2 + 1 \cdot 1 = 1497_{10}$
Octal to		
Binary	Substitution	$1234_8 = 001\ 010\ 011\ 100_2$
Hexadecimal	Substitution	$1234_8 = 001\ 010\ 011\ 100_2 = 0010\ 1001\ 1100_2 = 29C_{16}$
Decimal	Summation	$1234_8 = 1 \cdot 512 + 2 \cdot 64 + 3 \cdot 8 + 4 \cdot 1 = 668_{10}$
Hexadecimal to		
Binary	Substitution	$C0DE_{16} = 1100\ 0000\ 1101\ 1110_2$
Octal	Substitution	$C0DE_{16} = 1100\ 0000\ 1101\ 1110_2 = 1\ 100\ 000\ 011\ 011\ 110_2 = 140336_8$
Decimal	Summation	$C0DE_{16} = 12 \cdot 4096 + 0 \cdot 256 + 13 \cdot 16 + 14 \cdot 1 = 49374_{10}$
Decimal to		
Binary	Division	$108_{10} \div 2 = 54$ remainder 0 (LSB) $\div 2 = 27$ remainder 0 $\div 2 = 13$ remainder 1 $\div 2 = 6$ remainder 1 $\div 2 = 3$ remainder 0 $\div 2 = 1$ remainder 1 $\div 2 = 0$ remainder 1 (MSB) $108_{10} = 1101100_2$
Octal	Division	$108_{10} \div 8 = 13$ remainder 4 (least significant digit) $\div 8 = 1$ remainder 5 $\div 8 = 0$ remainder 1 (most significant digit) $108_{10} = 154_8$
Hexadecimal	Division	$108_{10} \div 16 = 6$ remainder 12 (least significant digit) $\div 16 = 0$ remainder 6 (most significant digit) $108_{10} = 6C_{16}$

Table 2-3
Binary addition and
subtraction table.

c_{in} **or** b_{in}	x	y	c_{out}	s	b_{out}	d
0	0	0	0	0	0	0
0	0	1	0	1	1	1
0	1	0	0	1	0	1
0	1	1	1	0	0	0
1	0	0	0	1	1	1
1	0	1	1	0	1	0
1	1	0	1	0	0	0
1	1	1	1	1	1	1

2.4 Addition and Subtraction of Nondecimal Numbers

Addition and subtraction of nondecimal numbers by hand uses the same technique that we learned in grammar school for decimal numbers; the only catch is
that the addition and subtraction tables are different.

binary addition   Table 2-3 is the addition and subtraction table for binary digits. To add two
binary numbers X and Y, we add together the least significant bits with an initial
carry (c_{in}) of 0, producing carry (c_{out}) and sum (s) bits according to the table. We
continue processing bits from right to left, adding the carry out of each column
into the next column's sum.

Two examples of decimal additions and the corresponding binary additions
are shown in Figure 2-1, using a colored arrow to indicate a carry of 1. The same
examples are repeated below along with two more, with the carries shown as a
bit string C:

C		101111000		C		001011000
X	190	10111110		X	173	10101101
Y	+141	+ 10001101		Y	+ 44	+ 00101100
$X + Y$	331	101001011		$X + Y$	217	11011001

C		011111110		C		000000000
X	127	01111111		X	170	10101010
Y	+ 63	+ 00111111		Y	+ 85	+ 01010101
$X + Y$	190	10111110		$X + Y$	255	11111111

binary subtraction   Binary subtraction is performed similarly, using borrows (b_{in} and b_{out})
instead of carries between steps, and producing a difference bit d. Two examples
minuend   of decimal subtractions and the corresponding binary subtractions are shown in
subtrahend   Figure 2-2. As in decimal subtraction, the binary minuend values in the columns
are modified when borrows occur, as shown by the colored arrows and bits. The

$$
\begin{array}{lrl}
 & & \quad 1\;1\;1\;1 \\
X & 190 & 1\;0\;1\;1\;1\;1\;1\;0 \\
Y & +141 & +\;1\;0\;0\;0\;1\;1\;0\;1 \\
\hline
X+Y & 331 & 1\;0\;1\;0\;0\;1\;0\;1\;1
\end{array}
\qquad
\begin{array}{lrl}
 & & 1\quad 1\;1 \\
X & 173 & 1\;0\;1\;0\;1\;1\;0\;1 \\
Y & +44 & +\;0\;0\;1\;0\;1\;1\;0\;0 \\
\hline
X+Y & 217 & 1\;1\;0\;1\;1\;0\;0\;1
\end{array}
$$

Figure 2-1   Examples of decimal and corresponding binary additions.

examples from the figure are repeated below along with two more, this time showing the borrows as a bit string B:

$$
\begin{array}{lrl}
B & & 001111100 \\
X & 229 & 11100101 \\
Y & -46 & -\;00101110 \\
\hline
X-Y & 183 & 10110111
\end{array}
\qquad
\begin{array}{lrl}
B & & 011011010 \\
X & 210 & 11010010 \\
Y & -109 & -\;01101101 \\
\hline
X-Y & 101 & 01100101
\end{array}
$$

$$
\begin{array}{lrl}
B & & 010101010 \\
X & 170 & 10101010 \\
Y & -85 & -\;01010101 \\
\hline
X-Y & 85 & 01010101
\end{array}
\qquad
\begin{array}{lrl}
B & & 000000000 \\
X & 221 & 11011101 \\
Y & -76 & -\;01001100 \\
\hline
X-Y & 145 & 10010001
\end{array}
$$

A very common use of subtraction in computers is to compare two numbers. For example, if the operation $X - Y$ produces a borrow out of the most significant bit position, then X is less than Y; otherwise, X is greater than or equal to Y. The relationship between carries and borrows in adders and subtractors will be explored in Section 5.10.

comparing numbers

Addition and subtraction tables can be developed for octal and hexadecimal digits, or any other desired radix. However, few computer engineers bother to memorize these tables. If you rarely need to manipulate nondecimal numbers,

Figure 2-2
Examples of decimal and corresponding binary subtractions.

then it's easy enough on those occasions to convert them to decimal, calculate results, and convert back. On the other hand, if you must perform calculations in binary, octal, or hexadecimal frequently, then you should ask Santa for a programmer's "hex calculator" from Texas Instruments or Casio.

If the calculator's battery wears out, some mental shortcuts can be used to facilitate nondecimal arithmetic. In general, each column addition (or subtraction) can be done by converting the column digits to decimal, adding in decimal, and converting the result to corresponding sum and carry digits in the nondecimal radix. (A carry is produced whenever the column sum equals or exceeds the radix.) Since the addition is done in decimal, we rely on our knowledge of the decimal addition table; the only new thing that we need to learn is the conversion from decimal to nondecimal digits and vice versa. The sequence of steps for mentally adding two hexadecimal numbers is shown below:

hexadecimal addition

C	1 1 0 0	1	1	0	0
X	$1\ 9\ B\ 9_{16}$	1	9	11	9
Y	$+\ C\ 7\ E\ 6_{16}$	$+12$	7	14	6
$X+Y$	$E\ 1\ 9\ F_{16}$	14	17	25	15
		14	$16+1$	$16+9$	15
		E	1	9	F

2.5 Representation of Negative Numbers

So far, we have dealt only with positive numbers, but there are many ways to represent negative numbers. In everyday business we use the signed-magnitude system, discussed next. However, most computers use one of the complement number systems that we introduce later.

2.5.1 Signed-Magnitude Representation

signed-magnitude system

In the *signed-magnitude system*, a number consists of a magnitude and a symbol indicating whether the magnitude is positive or negative. Thus, we interpret decimal numbers +98, −57, +123.5, and −13 in the usual way, and we also assume that the sign is "+" if no sign symbol is written. There are two possible representations of zero, "+0" and "−0", but both have the same value.

sign bit

The signed-magnitude system is applied to binary numbers by using an extra bit position to represent the sign (the *sign bit*). Traditionally, the most significant bit (MSB) of a bit string is used as the sign bit (0 = plus, 1 = minus), and the lower-order bits contain the magnitude. Thus, we can write several 8-bit signed-magnitude integers and their decimal equivalents:

$$01010101_2 = +85_{10} \qquad 11010101_2 = -85_{10}$$
$$01111111_2 = +127_{10} \qquad 11111111_2 = -127_{10}$$
$$00000000_2 = +0_{10} \qquad 10000000_2 = -0_{10}$$

The signed-magnitude system has an equal number of positive and negative integers. An n-bit signed-magnitude integer lies within the range $-(2^{n-1} - 1)$ through $+(2^{n-1} - 1)$, and there are two possible representations of zero.

Now suppose that we wanted to build a digital logic circuit that adds signed-magnitude numbers. The circuit must examine the signs of the addends to determine what to do with the magnitudes. If the signs are the same, it must add the magnitudes and give the result the same sign. If the signs are different, it must compare the magnitudes, subtract the smaller from the larger, and give the result the sign of the larger. All of these "ifs," "adds," "subtracts," and "compares" translate into a lot of logic-circuit complexity. Adders for complement number systems are much simpler, as we'll show next. Perhaps the one redeeming feature of a signed-magnitude system is that, once we know how to build a signed-magnitude adder, a signed-magnitude subtractor is almost trivial to build—it need only change the sign of the subtrahend and pass it along with the minuend to an adder.

signed-magnitude adder

signed-magnitude subtractor

2.5.2 Complement Number Systems

While the signed-magnitude system negates a number by changing its sign, a *complement number system* negates a number by taking its complement as defined by the system. Taking the complement is more difficult than changing the sign, but two numbers in a complement number system can be added or subtracted directly without the sign and magnitude checks required by the signed-magnitude system. We shall describe two complement number systems, called the "radix complement" and the "diminished radix-complement."

complement number system

In any complement number system, we normally deal with a fixed number of digits, say n. (However, we can increase the number of digits by "sign extension" as shown in Exercise 2.23, and decrease the number by truncating high-order digits as shown in Exercise 2.24.) We further assume that the radix is r, and that numbers have the form

$$D = d_{n-1}d_{n-2}\cdots d_1 d_0.$$

The radix point is on the right and so the number is an integer. If an operation produces a result that requires more than n digits, we throw away the extra high-order digit(s). If a number D is complemented twice, the result is D.

2.5.3 Radix-Complement Representation

In a *radix-complement system*, the complement of an n-digit number is obtained by subtracting it from r^n. In the decimal number system, the radix complement is called the *10's complement*. Some examples using 4-digit decimal numbers (and subtraction from 10,000) are shown in Table 2-4.

radix-complement system

10's complement

By definition, the radix complement of an n-digit number D is obtained by subtracting it from r^n. If D is between 1 and $r^n - 1$, this subtraction produces

Table 2-4
Examples of 10's and
9s' complements.

Number	10's Complement	9s' Complement
1849	8151	8150
2067	7933	7932
100	9900	9899
7	9993	9992
8151	1849	1848
0	10000 (= 0)	9999

another number between 1 and $r^n - 1$. If D is 0, the result of the subtraction is r^n, which has the form $100 \cdots 00$, where there are a total of $n + 1$ digits. We throw away the extra high-order digit and get the result 0. Thus, there is only one representation of zero in a radix-complement system.

It seems from the definition that a subtraction operation is needed to compute the radix complement of D. However, this subtraction can be avoided by rewriting r^n as $(r^n - 1) + 1$ and $r^n - D$ as $((r^n - 1) - D) + 1$. The number $r^n - 1$ has the form $mm \cdots mm$, where $m = r - 1$ and there are n m's. For example, 10,000 equals $9,999 + 1$. If we define the complement of a digit d to be $r - 1 - d$, then $(r^n - 1) - D$ is obtained by complementing the digits of D. Therefore, the radix complement of a number D is obtained by complementing the individual

computing the radix complement

Table 2-5
Digit complements.

Digit	Complement			
	Binary	Octal	Decimal	Hexadecimal
0	1	7	9	F
1	0	6	8	E
2	–	5	7	D
3	–	4	6	C
4	–	3	5	B
5	–	2	4	A
6	–	1	3	9
7	–	0	2	8
8	–	–	1	7
9	–	–	0	6
A	–	–	–	5
B	–	–	–	4
C	–	–	–	3
D	–	–	–	2
E	–	–	–	1
F	–	–	–	0

digits of D and adding 1. For example, the 10's complement of 1849 is $8150 + 1$, or 8151. You should confirm that this trick also works for the other 10's-complement examples above. Table 2-5 lists the digit complements for binary, octal, decimal, and hexadecimal numbers.

2.5.4 Two's-Complement Representation

For binary numbers, the radix complement is called the *two's complement*. The MSB of a number in this system serves as the sign bit; a number is negative if and only if its MSB is 1. The decimal equivalent for a two's-complement binary number is computed the same way as for an unsigned number, except that the weight of the MSB is -2^{n-1} instead of $+2^{n-1}$. The range of representable numbers is $-(2^{n-1})$ through $+(2^{n-1} - 1)$. Some 8-bit examples are shown below:

two's complement

weight of MSB

$$17_{10} = 00010001_2$$
$$\Downarrow \quad \text{complement bits}$$
$$11101110$$
$$+1$$
$$\overline{11101111_2} = -17_{10}$$

$$-99_{10} = 10011101_2$$
$$\Downarrow \quad \text{complement bits}$$
$$01100010$$
$$+1$$
$$\overline{01100011_2} = 99_{10}$$

$$119_{10} = 01110111$$
$$\Downarrow \quad \text{complement bits}$$
$$10001000$$
$$+1$$
$$\overline{10001001_2} = -119_{10}$$

$$-127_{10} = 10000001$$
$$\Downarrow \quad \text{complement bits}$$
$$01111110$$
$$+1$$
$$\overline{01111111_2} = 127_{10}$$

$$0_{10} = 00000000_2$$
$$\Downarrow \quad \text{complement bits}$$
$$11111111$$
$$+1$$
$$\overline{1\ 00000000_2} = 0_{10}$$

$$-128_{10} = 10000000_2$$
$$\Downarrow \quad \text{complement bits}$$
$$01111111$$
$$+1$$
$$\overline{10000000_2} = -128_{10}$$

A carry out of the MSB position occurs in one case, as shown in color above. As in all two's-complement operations, this bit is ignored and only the low-order n bits of the result are used.

In the two's-complement number system, zero is considered positive because its sign bit is 0. Since two's complement has only one representation of zero, we end up with one extra negative number, $-(2^{n-1})$, that doesn't have a positive counterpart.

extra negative number

We can convert an n-bit two's-complement number X into an m-bit one, but some care is needed. If $m > n$, we must append $m - n$ copies of X's sign bit to the left of X (see Exercise 2.23). That is, we pad a positive number with 0s and a negative one with 1s; this is called *sign extension*. If $m < n$, we discard X's $n - m$

sign extension

leftmost bits; however, the result is valid only if all of the discarded bits are the same as the sign bit of the result (see Exercise 2.24).

Most computers and other digital systems use the two's-complement system to represent negative numbers. However, for completeness, we'll also describe the diminished radix-complement and ones'-complement systems.

*2.5.5 Diminished Radix-Complement Representation

diminished radix-complement system

In a *diminished radix-complement system*, the complement of an n-digit number D is obtained by subtracting it from $r^n - 1$. This can be accomplished by complementing the individual digits of D, *without* adding 1 as in the radix-complement system. In decimal, this is called the *9s' complement*; some examples are given in the last column of Table 2-4 on page 36.

9s' complement

*2.5.6 Ones'-Complement Representation

ones' complement

The diminished radix-complement system for binary numbers is called the *ones' complement*. As in two's complement, the most significant bit is the sign, 0 if positive and 1 if negative. Thus there are two representations of zero, positive zero $(00 \cdots 00)$ and negative zero $(11 \cdots 11)$. Positive-number representations are the same for both ones' and two's complements. However, negative-number representations differ by 1. A weight of $-(2^{n-1} - 1)$, rather than -2^{n-1}, is given to the most significant bit when computing the decimal equivalent of a ones'-complement number. The range of representable numbers is $-(2^{n-1} - 1)$ through $+(2^{n-1} - 1)$. Some 8-bit numbers and their ones' complements are shown below:

$$17_{10} = 00010001_2 \qquad\qquad -99_{10} = 10011100_2$$
$$\Downarrow \qquad\qquad\qquad \Downarrow$$
$$11101110_2 = -17_{10} \qquad\qquad 01100011_2 = 99_{10}$$

$$119_{10} = 01110111_2 \qquad\qquad -127_{10} = 10000000_2$$
$$\Downarrow \qquad\qquad\qquad \Downarrow$$
$$10001000_2 = -119_{10} \qquad\qquad 01111111_2 = 127_{10}$$

$$0_{10} = 00000000_2 \text{ (positive zero)}$$
$$\Downarrow$$
$$11111111_2 = 0_{10} \text{ (negative zero)}$$

The main advantages of the ones'-complement system are its symmetry and the ease of complementation. However, the adder design for ones'-complement numbers is somewhat trickier than a two's-complement adder (see Exercise 7.72). Also, zero-detecting circuits in a ones'-complement system

---

* Throughout this book, *optional sections* are marked with an asterisk.

either must check for both representations of zero, or must always convert $11 \cdots 11$ to $00 \cdots 00$.

*2.5.7 Excess Representations

Yes, the number of different systems for representing negative numbers *is* excessive, but there's just one more for us to cover. In *excess-B representation*, an *m*-bit string whose unsigned integer value is M $(0 \le M < 2^m)$ represents the signed integer $M - B$, where B is called the *bias* of the number system.

excess-B representation

bias

For example, an *excess-2^{m-1} system* represents any number X in the range -2^{m-1} through $+2^{m-1} - 1$ by the *m*-bit binary representation of $X + 2^{m-1}$ (which is always nonnegative and less than 2^m). The range of this representation is exactly the same as that of *m*-bit two's-complement numbers. In fact, the representations of any number in the two systems are identical except for the sign bits, which are always opposite. (Note that this is true only when the bias is 2^{m-1}.)

excess-2^{m-1} system

The most common use of excess representations is in floating-point number systems (see References).

2.6 Two's-Complement Addition and Subtraction

2.6.1 Addition Rules

A table of decimal numbers and their equivalents in different number systems, Table 2-6, reveals why the two's complement is preferred for arithmetic operations. If we start with 1000_2 (-8_{10}) and count up, we see that each successive two's-complement number all the way to 0111_2 $(+7_{10})$ can be obtained by adding 1 to the previous one, ignoring any carries beyond the fourth bit position. The same cannot be said of signed-magnitude and ones'-complement numbers. Because ordinary addition is just an extension of counting, two's-complement numbers can thus be added by ordinary binary addition, ignoring any carries beyond the MSB. The result will always be the correct sum as long as the range of the number system is not exceeded. Some examples of decimal addition and the corresponding 4-bit two's-complement additions confirm this:

two's-complement addition

$$
\begin{array}{rc@{\qquad}rc}
+3 & 0011 & -2 & 1110 \\
+\ +4 & +\ 0100 & +\ -6 & +\ 1010 \\
\hline
+7 & 0111 & -8 & 1\,1000 \\[1em]
+6 & 0110 & +4 & 0100 \\
+\ -3 & +\ 1101 & +\ -7 & +\ 1001 \\
\hline
+3 & 1\,0011 & -3 & 1101 \\
\end{array}
$$

Table 2-6 Decimal and 4-bit numbers.

Decimal	Two's Complement	Ones' Complement	Signed Magnitude	Excess 2^{m-1}
−8	1000	—	—	0000
−7	1001	1000	1111	0001
−6	1010	1001	1110	0010
−5	1011	1010	1101	0011
−4	1100	1011	1100	0100
−3	1101	1100	1011	0101
−2	1110	1101	1010	0110
−1	1111	1110	1001	0111
0	0000	1111 or 0000	1000 or 0000	1000
1	0001	0001	0001	1001
2	0010	0010	0010	1010
3	0011	0011	0011	1011
4	0100	0100	0100	1100
5	0101	0101	0101	1101
6	0110	0110	0110	1110
7	0111	0111	0111	1111

2.6.2 A Graphical View

Another way to view the two's-complement system uses the 4-bit "counter" shown in Figure 2-3. Here we have shown the numbers in a circular or "modular" representation. The operation of this counter very closely mimics that of a real up/down counter circuit, which we'll study in Section 8.4. Starting

Figure 2-3
A modular counting representation of 4-bit two's-complement numbers.

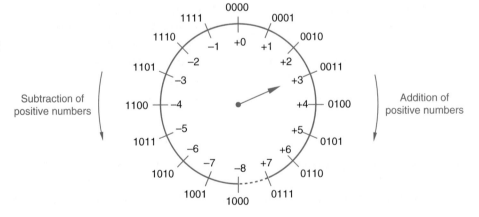

with the arrow pointing to any number, we can add $+n$ to that number by counting up n times, that is, by moving the arrow n positions clockwise. It is also evident that we can subtract n from a number by counting down n times, that is, by moving the arrow n positions counterclockwise. Of course, these operations give correct results only if n is small enough that we don't cross the discontinuity between -8 and $+7$.

What is most interesting is that we can also subtract n (or add $-n$) by moving the arrow $16 - n$ positions clockwise. Notice that the quantity $16 - n$ is what we defined to be the 4-bit two's complement of n, that is, the two's-complement representation of $-n$. This graphically supports our earlier claim that a negative number in two's-complement representation may be added to another number simply by adding the 4-bit representations using ordinary binary addition. Adding a number in Figure 2-3 is equivalent to moving the arrow a corresponding number of positions clockwise.

2.6.3 Overflow

If an addition operation produces a result that exceeds the range of the number system, *overflow* is said to occur. In the modular counting representation of Figure 2-3, overflow occurs during addition of positive numbers when we count past $+7$. Addition of two numbers with different signs can never produce overflow, but addition of two numbers of like sign can, as shown by the following examples:

overflow

$$
\begin{array}{rl}
-3 & 1101 \\
+\ -6 & +\ 1010 \\
\hline
-9 & 1\,0111 = +7
\end{array}
\qquad
\begin{array}{rl}
+5 & 0101 \\
+\ +6 & +\ 0110 \\
\hline
+11 & 1011 = -5
\end{array}
$$

$$
\begin{array}{rl}
-8 & 1000 \\
+\ -8 & +\ 1000 \\
\hline
-16 & 1\,0000 = +0
\end{array}
\qquad
\begin{array}{rl}
+7 & 0111 \\
+\ +7 & +\ 0111 \\
\hline
+14 & 1110 = -2
\end{array}
$$

Fortunately, there is a simple rule for detecting overflow in addition: An addition overflows if the signs of the addends are the same and the sign of the sum is different from the addends' sign. The overflow rule is sometimes stated in terms of carries generated during the addition operation: An addition overflows if the carry bits c_{in} into and c_{out} out of the sign position are different. Close examination of Table 2-3 on page 32 shows that the two rules are equivalent—there are only two cases where $c_{in} \neq c_{out}$, and these are the only two cases where $x = y$ and the sum bit is different.

overflow rules

2.6.4 Subtraction Rules

Two's-complement numbers may be subtracted as if they were ordinary unsigned binary numbers, and appropriate rules for detecting overflow may be formulated. However, most subtraction circuits for two's-complement numbers

two's-complement subtraction

do not perform subtraction directly. Rather, they negate the subtrahend by taking its two's complement, and then add it to the minuend using the normal rules for addition.

Negating the subtrahend and adding the minuend can be accomplished with only one addition operation as follows: Perform a bit-by-bit complement of the subtrahend and add the complemented subtrahend to the minuend with an initial carry (c_{in}) of 1 instead of 0. Examples are given below:

$$
\begin{array}{rrr}
 & & 1 - c_{in} \\
+4 & 0100 & 0100 \\
-\ +3 & -\ 0011 & +\ 1100 \\
\hline
+3 & & 1\,0001 \\
\end{array}
\qquad
\begin{array}{rrr}
 & & 1 - c_{in} \\
+3 & 0011 & 0011 \\
-\ +4 & -\ 0100 & +\ 1011 \\
\hline
-1 & & 1111 \\
\end{array}
$$

$$
\begin{array}{rrr}
 & & 1 - c_{in} \\
+3 & 0011 & 0011 \\
-\ -4 & -\ 1100 & +\ 0011 \\
\hline
+7 & & 0111 \\
\end{array}
\qquad
\begin{array}{rrr}
 & & 1 - c_{in} \\
-3 & 1101 & 1101 \\
-\ -4 & -\ 1100 & +\ 0011 \\
\hline
+1 & & 1\,0001 \\
\end{array}
$$

Overflow in subtraction can be detected by examining the signs of the minuend and the *complemented* subtrahend, using the same rule as in addition. Or, using the technique in the preceding examples, the carries into and out of the sign position can be observed and overflow detected irrespective of the signs of inputs and output, again using the same rule as in addition.

An attempt to negate the "extra" negative number results in overflow according to the rules above, when we add 1 in the complementation process:

$$
\begin{array}{rl}
-(-8) = -1000 = & 0111 \\
& +\ 0001 \\
\hline
& 1000 \ = \ -8 \\
\end{array}
$$

However, this number can still be used in additions and subtractions as long as the final result does not exceed the number range:

$$
\begin{array}{rr}
+4 & 0100 \\
+\ -8 & +\ 1000 \\
\hline
-4 & 1100 \\
\end{array}
\qquad
\begin{array}{rrr}
 & & 1 - c_{in} \\
-3 & 1101 & 1101 \\
-\ -8 & -\ 1000 & +\ 0111 \\
\hline
+5 & & 1\,0101 \\
\end{array}
$$

2.6.5 Two's-Complement and Unsigned Binary Numbers

Since two's-complement numbers are added and subtracted by the same basic binary addition and subtraction algorithms as unsigned numbers of the same length, a computer or other digital system can use the same adder circuit to handle numbers of both types. However, the results must be interpreted differently, depending on whether the system is dealing with signed numbers (e.g., −8 through +7) or unsigned numbers (e.g., 0 through 15).

signed vs. unsigned numbers

We introduced a graphical representation of the 4-bit two's-complement system in Figure 2-3. We can relabel this figure as shown in Figure 2-4 to obtain a representation of the 4-bit unsigned numbers. The binary combinations occupy the same positions on the wheel, and a number is still added by moving the arrow a corresponding number of positions clockwise, and subtracted by moving the arrow counterclockwise.

An addition operation can be seen to exceed the range of the 4-bit unsigned-number system in Figure 2-4 if the arrow moves clockwise through the discontinuity between 0 and 15. In this case a *carry* out of the most significant bit position is said to occur.

carry

Likewise a subtraction operation exceeds the range of the number system if the arrow moves counterclockwise through the discontinuity. In this case a *borrow* out of the most significant bit position is said to occur.

borrow

From Figure 2-4 it is also evident that we may subtract an unsigned number n by counting *clockwise* $16 - n$ positions. This is equivalent to *adding* the 4-bit two's-complement of n. The subtraction produces a borrow if the corresponding addition of the two's complement *does not* produce a carry.

In summary, in unsigned addition the carry or borrow in the most significant bit position indicates an out-of-range result. In signed, two's-complement addition the overflow condition defined earlier indicates an out-of-range result. The carry from the most significant bit position is irrelevant in signed addition in the sense that overflow may or may not occur independently of whether or not a carry occurs.

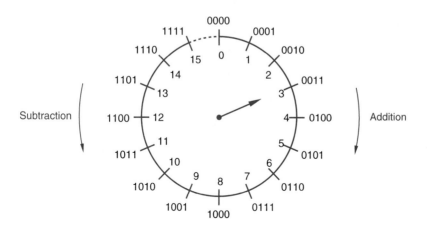

Figure 2-4
A modular counting representation of 4-bit unsigned numbers.

*2.7 Ones'-Complement Addition and Subtraction

Another look at Table 2-6 helps to explain the rule for adding ones'-complement numbers. If we start at 1000_2 (-7_{10}) and count up, we obtain each successive ones'-complement number by adding 1 to the previous one, *except* at the transition from 1111_2 (negative 0) to 0001_2 ($+1_{10}$). To maintain the proper count, we must add 2 instead of 1 whenever we count past 1111_2. This suggests a technique for adding ones'-complement numbers: Perform a standard binary addition, but add an extra 1 whenever we count past 1111_2.

Counting past 1111_2 during an addition can be detected by observing the carry out of the sign position. Thus, the rule for adding ones'-complement numbers can be stated quite simply:

ones'-complement addition

- Perform a standard binary addition; if there is a carry out of the sign position, add 1 to the result.

end-around carry

This rule is often called *end-around carry*. Examples of ones'-complement addition are given below; the last three include an end-around carry:

$$
\begin{array}{rr}
+3 & 0011 \\
+ \ +4 & + \ 0100 \\
\hline
+7 & 0111
\end{array}
\qquad
\begin{array}{rr}
+4 & 0100 \\
+ \ -7 & + \ 1000 \\
\hline
3 & 1100
\end{array}
\qquad
\begin{array}{rr}
+5 & 0101 \\
+ \ -5 & + \ 1010 \\
\hline
-0 & 1111
\end{array}
$$

$$
\begin{array}{rr}
-2 & 1101 \\
+ \ -5 & + \ 1010 \\
\hline
-7 & 1\,0111 \\
& + \quad 1 \\
\hline
& 1000
\end{array}
\qquad
\begin{array}{rr}
+6 & 0110 \\
+ \ -3 & + \ 1100 \\
\hline
+3 & 1\,0010 \\
& + \quad 1 \\
\hline
& 0011
\end{array}
\qquad
\begin{array}{rr}
-0 & 1111 \\
+ \ -0 & + \ 1111 \\
\hline
-0 & 1\,1110 \\
& + \quad 1 \\
\hline
& 1111
\end{array}
$$

Following the two-step addition rule above, the addition of a number and its ones' complement produces negative 0. In fact, an addition operation using this rule can never produce positive 0 unless both addends are positive 0.

ones'-complement subtraction

As with two's complement, the easiest way to do ones'-complement subtraction is to complement the subtrahend and add. Overflow rules for ones'-complement addition and subtraction are the same as for two's complement.

Table 2-7 summarizes the rules that we presented in this and previous sections for negation, addition, and subtraction in binary number systems.

Table 2-7 Summary of addition and subtraction rules for binary numbers.

Number System	Addition Rules	Negation Rules	Subtraction Rules
Unsigned	Add the numbers. Result is out of range if a carry out of the MSB occurs.	Not applicable	Subtract the subtrahend from the minuend. Result is out of range if a borrow out of the MSB occurs.
Signed magnitude	(same sign) Add the magnitudes; overflow occurs if a carry out of MSB occurs; result has the same sign. (opposite sign) Subtract the smaller magnitude from the larger; overflow is impossible; result has the sign of the larger.	Change the number's sign bit.	Change the sign bit of the subtrahend and proceed as in addition.
Two's complement	Add, ignoring any carry out of the MSB. Overflow occurs if the carries into and out of MSB are different.	Complement all bits of the number; add 1 to the result.	Complement all bits of the subtrahend and add to the minuend with an initial carry of 1.
Ones' complement	Add; if there is a carry out of the MSB, add 1 to the result. Overflow occurs if carries into and out of MSB are different.	Complement all bits of the number.	Complement all bits of the subtrahend and proceed as in addition.

*2.8 Binary Multiplication

In grammar school we learned to multiply by adding a list of shifted multiplicands computed according to the digits of the multiplier. The same method can be used to obtain the product of two unsigned binary numbers. Forming the shifted multiplicands is trivial in binary multiplication, since the only possible values of the multiplier digits are 0 and 1. An example is shown below:

shift-and-add multiplication

unsigned binary multiplication

```
     11              1011    multiplicand
   × 13           ×  1101    multiplier
   ────           ───────
     33              1011
     11              0000
   ────                     shifted multiplicands
    143              1011
                     1011
                  ───────
                  10001111   product
```

partial product

Instead of listing all the shifted multiplicands and then adding, in a digital system it is more convenient to add each shifted multiplicand as it is created to a *partial product*. Applying this technique to the previous example, four additions and partial products are used to multiply 4-bit numbers:

	11		1011	multiplicand
$\times$	13	$\times$	1101	multiplier
			0000	partial product
			1011	shifted multiplicand
			01011	partial product
			0000↓	shifted multiplicand
			001011	partial product
			1011↓↓	shifted multiplicand
			0110111	partial product
			1011↓↓↓	shifted multiplicand
			10001111	product

In general, when we multiply an n-bit number by an m-bit number, the resulting product requires at most $n + m$ bits to express. The shift-and-add algorithm requires m partial products and additions to obtain the result, but the first addition is trivial, since the first partial product is zero. Although the first partial product has only n significant bits, after each addition step the partial product gains one more significant bit, since each addition may produce a carry. At the same time, each step yields one more partial product bit, starting with the rightmost and working toward the left, that does not change. The shift-and-add algorithm can be performed by a digital circuit that includes a shift register, an adder, and control logic, as shown in Section 8.7.2.

signed multiplication

Multiplication of signed numbers can be accomplished using unsigned multiplication and the usual grammar-school rules: Perform an unsigned multiplication of the magnitudes and make the product positive if the operands had the same sign, negative if they had different signs. This is very convenient in signed-magnitude systems, since the sign and magnitude are separate.

In the two's-complement system, obtaining the magnitude of a negative number and negating the unsigned product are nontrivial operations. This leads us to seek a more efficient way of performing two's-complement multiplication, described next.

two's-complement multiplication

Conceptually, unsigned multiplication is accomplished by a sequence of unsigned additions of the shifted multiplicands; at each step, the shift of the multiplicand corresponds to the weight of the multiplier bit. The bits in a two's-complement number have the same weights as in an unsigned number, except for the MSB, which has a negative weight (see Section 2.5.4). Thus, we can perform two's-complement multiplication by a sequence of two's-complement additions of shifted multiplicands, except for the last step, in which the shifted multiplicand corresponding to the MSB of the multiplier must be negated before it is added to the partial product. Our previous example is repeated below, this time interpreting the multiplier and multiplicand as two's-complement numbers:

$$
\begin{array}{rrl}
-5 & 1011 & \text{multiplicand} \\
\times\ -3 & \times\quad 1101 & \text{multiplier} \\ \hline
& 00000 & \text{partial product} \\
& 11011 & \text{shifted multiplicand} \\ \hline
& 111011 & \text{partial product} \\
& 00000\downarrow & \text{shifted multiplicand} \\ \hline
& 1111011 & \text{partial product} \\
& 11011\downarrow\downarrow & \text{shifted multiplicand} \\ \hline
& 11100111 & \text{partial product} \\
& 00101\downarrow\downarrow\downarrow & \text{shifted and negated multiplicand} \\ \hline
& 00001111 & \text{product}
\end{array}
$$

Handling the MSBs is a little tricky because we gain one significant bit at each step and we are working with signed numbers. Therefore, before adding each shifted multiplicand and k-bit partial product, we change them to $k + 1$ significant bits by sign extension, as shown in color above. Each resulting sum has $k + 1$ bits; any carry out of the MSB of the $k + 1$-bit sum is ignored.

*2.9 Binary Division

The simplest binary division algorithm is based on the shift-and-subtract method that we learned in grammar school. Table 2-8 gives examples of this method for unsigned decimal and binary numbers. In both cases, we mentally compare the reduced dividend with multiples of the divisor to determine which multiple of

shift-and-subtract division

unsigned division

$$
\begin{array}{rrrl}
& 19 & 10011 & \text{quotient} \\
11\ \overline{)217} & & 1011\ \overline{)11011001} & \text{dividend} \\
& 11 & 1011 & \text{shifted divisor} \\ \hline
& 107 & 0101 & \text{reduced dividend} \\
& 99 & 0000 & \text{shifted divisor} \\ \hline
& 8 & 1010 & \text{reduced dividend} \\
& & 0000 & \text{shifted divisor} \\ \hline
& & 10100 & \text{reduced dividend} \\
& & 1011 & \text{shifted divisor} \\ \hline
& & 10011 & \text{reduced dividend} \\
& & 1011 & \text{shifted divisor} \\ \hline
& & 1000 & \text{remainder}
\end{array}
$$

Table 2-8
Example of long division.

the shifted divisor to subtract. In the decimal case, we first pick 11 as the greatest multiple of 11 less than 21, and then pick 99 as the greatest multiple less than 107. In the binary case, the choice is somewhat simpler, since the only two choices are zero and the divisor itself.

Division methods for binary numbers are somewhat complementary to binary multiplication methods. A typical division algorithm takes an $(n+m)$-bit dividend and an n-bit divisor, and produces an m-bit quotient and an n-bit *division overflow* remainder. A division *overflows* if the divisor is zero or the quotient would take more than m bits to express. In most computer division circuits, $n = m$.

signed division   Division of signed numbers can be accomplished using unsigned division and the usual grammar school rules: Perform an unsigned division of the magnitudes and make the quotient positive if the operands had the same sign, negative if they had different signs. The remainder should be given the same sign as the dividend. As in multiplication, there are special techniques for performing division directly on two's-complement numbers; these techniques are often implemented in computer division circuits (see References).

2.10 Binary Codes for Decimal Numbers

Even though binary numbers are the most appropriate for the internal computations of a digital system, most people still prefer to deal with decimal numbers. As a result, the external interfaces of a digital system may read or display decimal numbers, and some digital devices actually process decimal numbers directly.

The human need to represent decimal numbers doesn't change the basic nature of digital electronic circuits—they still process signals that take on one of only two states that we call 0 and 1. Therefore, a decimal number is represented in a digital system by a string of bits, where different combinations of bit values in the string represent different decimal numbers. For example, if we use a 4-bit string to represent a decimal number, we might assign bit combination 0000 to decimal digit 0, 0001 to 1, 0010 to 2, and so on.

code   A set of n-bit strings in which different bit strings represent different num-
code word   bers or other things is called a *code*. A particular combination of n bit-values is called a *code word*. As we'll see in the examples of decimal codes in this section, there may or may not be an arithmetic relationship between the bit values in a code word and the thing that it represents. Furthermore, a code that uses n-bit strings need not contain 2^n valid code words.

At least four bits are needed to represent the ten decimal digits. There are billions and billions of different ways to choose ten 4-bit code words, but some of the more common decimal codes are listed in Table 2-9.

binary-coded decimal   Perhaps the most "natural" decimal code is *binary-coded decimal (BCD)*,
(BCD)   which encodes the digits 0 through 9 by their 4-bit unsigned binary representations, 0000 through 1001. The code words 1010 through 1111 are not used.

Table 2-9 Decimal codes.

Decimal digit	BCD (8421)	2421	Excess-3	Biquinary	1-out-of-10
0	0000	0000	0011	0100001	1000000000
1	0001	0001	0100	0100010	0100000000
2	0010	0010	0101	0100100	0010000000
3	0011	0011	0110	0101000	0001000000
4	0100	0100	0111	0110000	0000100000
5	0101	1011	1000	1000001	0000010000
6	0110	1100	1001	1000010	0000001000
7	0111	1101	1010	1000100	0000000100
8	1000	1110	1011	1001000	0000000010
9	1001	1111	1100	1010000	0000000001
Unused code words					
	1010	0101	0000	0000000	0000000000
	1011	0110	0001	0000001	0000000011
	1100	0111	0010	0000010	0000000101
	1101	1000	1101	0000011	0000000110
	1110	1001	1110	0000101	0000000111
	1111	1010	1111	. . .	. . .

Conversions between BCD and decimal representations are trivial, a direct substitution of four bits for each decimal digit. Some computer programs place two BCD digits in one 8-bit byte in *packed-BCD representation*; thus, one byte may represent the values from 0 to 99 as opposed to 0 to 255 for a normal unsigned 8-bit binary number. BCD numbers with any desired number of digits may be obtained by using one byte for each two digits.

packed-BCD representation

As with binary numbers, there are many possible representations of negative BCD numbers. Signed BCD numbers have one extra digit position for

BINOMIAL COEFFICIENTS   The number of different ways to choose m items from a set of n items is given by a *binomial coefficient*, denoted $\binom{n}{m}$, whose value is $\dfrac{n!}{m! \cdot (n-m!)}$. For a 4-bit decimal code, there are $\binom{16}{10}$ different ways to choose 10 out of 16 4-bit code words, and 10! ways to assign each different choice to the 10 digits. So there are $\dfrac{16!}{10! \cdot 6!} \cdot 10!$ or 29,059,430,400 different 4-bit decimal codes.

BCD addition

the sign. Both the signed-magnitude and 10's-complement representations are popular. In signed-magnitude BCD, the encoding of the sign bit string is arbitrary; in 10's-complement, 0000 indicates plus and 1001 indicates minus.

Addition of BCD digits is similar to adding 4-bit unsigned binary numbers, except that a correction must be made if a result exceeds 1001. The result is corrected by adding 6; examples are shown below:

$$
\begin{array}{rl}
5 & 0101 \\
+ \ 9 & + \ 1001 \\
\hline
14 & 1110 \\
& + \ 0110 \quad \text{— correction} \\
\hline
10+4 & 1\ 0100
\end{array}
\qquad
\begin{array}{rl}
4 & 0100 \\
+ \ 5 & + \ 0101 \\
\hline
9 & 1001
\end{array}
$$

$$
\begin{array}{rl}
8 & 1000 \\
+ \ 8 & + \ 1000 \\
\hline
-16 & 1\ 0000 \\
& + \ 0110 \quad \text{— correction} \\
\hline
10+6 & 1\ 0110
\end{array}
\qquad
\begin{array}{rl}
9 & 1001 \\
+ \ 9 & + \ 1001 \\
\hline
18 & 1\ 0010 \\
& + \ 0110 \quad \text{— correction} \\
\hline
10+8 & 1\ 1000
\end{array}
$$

Notice that the addition of two BCD digits produces a carry into the next digit position if either the initial binary addition or the correction-factor addition produces a carry. Many computers perform packed-BCD arithmetic using special instructions that handle the carry correction automatically.

weighted code

Binary-coded decimal is a *weighted code* because each decimal digit can be obtained from its code word by assigning a fixed weight to each code-word bit. The weights for the BCD bits are 8, 4, 2, and 1, and for this reason the code is sometimes called the *8421 code*. Another set of weights results in the *2421 code* shown in Table 2-9. This code has the advantage that it is *self-complementing*, that is, the code word for the 9s' complement of any digit may be obtained by complementing the individual bits of the digit's code word.

8421 code
2421 code
self-complementing code

excess-3 code

Another self-complementing code shown in Table 2-9 is the *excess-3 code*. Although this code is not weighted, it has an arithmetic relationship with the BCD code—the code word for each decimal digit is the corresponding BCD code word plus 0011_2. Because the code words follow a standard binary counting sequence, standard binary counters can easily be made to count in excess-3 code, as we'll show in Figure 8-37 on page 700.

biquinary code

Decimal codes can have more than four bits; for example, the *biquinary code* in Table 2-9 uses seven. The first two bits in a code word indicate whether the number is in the range 0–4 or 5–9, and the last five bits indicate which of the five numbers in the selected range is represented.

One potential advantage of using more than the minimum number of bits in a code is an error-detecting property. In the biquinary code, if any one bit in a code word is accidentally changed to the opposite value, the resulting code word does not represent a decimal digit and can therefore be flagged as an error. Out of 128 possible 7-bit code words, only 10 are valid and recognized as decimal digits; the rest can be flagged as errors if they appear.

A *1-out-of-10 code*, such as the one shown in the last column of Table 2-9, is the sparsest encoding for decimal digits, using 10 out of 1024 possible 10-bit code words.

1-out-of-10 code

2.11 Gray Code

In electromechanical applications of digital systems—such as machine tools, automotive braking systems, and copiers—it is sometimes necessary for an input sensor to produce a digital value that indicates a mechanical position. For example, Figure 2-5 is a conceptual sketch of an encoding disk and a set of contacts that produce one of eight 3-bit binary-coded values depending on the rotational position of the disk. The dark areas of the disk are connected to a signal source corresponding to logic 1, and the light areas are unconnected, which the contacts interpret as logic 0.

The encoder in Figure 2-5 has a problem when the disk is positioned at certain boundaries between the regions. For example, consider the boundary between the 001 and 010 regions of the disk; two of the encoded bits change here. What value will the encoder produce if the disk is positioned right on the theoretical boundary? Since we're on the border, both 001 and 010 are acceptable. However, because the mechanical assembly is not perfect, the two righthand contacts may both touch a "1" region, giving an incorrect reading of 011. Likewise, a reading of 000 is possible. In general, this sort of problem can occur at any boundary where more than one bit changes. The worst problems occur when all three bits are changing, as at the 000–111 and 011–100 boundaries.

The encoding-disk problem can be solved by devising a digital code in which only one bit changes between each pair of successive code words. Such a code is called a *Gray code*; a 3-bit Gray code is listed in Table 2-10. We've

Gray code

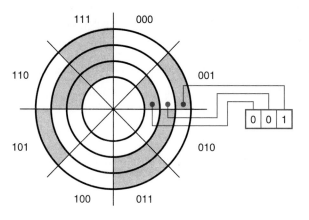

Figure 2-5
A mechanical encoding disk using a 3-bit binary code.

Table 2-10
A comparison of 3-bit binary code and Gray code.

Decimal Number	Binary Code	Gray Code
0	000	000
1	001	001
2	010	011
3	011	010
4	100	110
5	101	111
6	110	101
7	111	100

redesigned the encoding disk using this code as shown in Figure 2-6. Only one bit of the new disk changes at each border, so borderline readings give us a value on one side or the other of the border.

There are two convenient ways to construct a Gray code with any desired number of bits. The first method is based on the fact that Gray code is a *reflected code*; it can be defined (and constructed) recursively using the following rules:

reflected code

1. A 1-bit Gray code has two code words, 0 and 1.
2. The first 2^n code words of an $(n+1)$-bit Gray code equal the code words of an n-bit Gray code, written in order with a leading 0 appended.
3. The last 2^n code words of an $(n+1)$-bit Gray code equal the code words of an n-bit Gray code, but written in reverse order with a leading 1 appended.

If we draw a line between rows 3 and 4 of Table 2-10, we can see that rules 2 and 3 are true for the 3-bit Gray code. Of course, to construct an n-bit Gray code for an arbitrary value of n with this method, we must also construct a Gray code of each length smaller than n.

Figure 2-6
A mechanical encoding disk using a 3-bit Gray code.

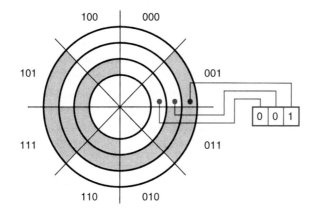

The second method allows us to derive an n-bit Gray-code code word directly from the corresponding n-bit binary code word:

1. The bits of an n-bit binary or Gray-code code word are numbered from right to left, from 0 to $n - 1$.

2. Bit i of a Gray-code code word is 0 if bits i and $i + 1$ of the corresponding binary code word are the same, else bit i is 1. (When $i + 1 = n$, bit n of the binary code word is considered to be 0.)

Again, inspection of Table 2-10 shows that this is true for the 3-bit Gray code.

*2.12 Character Codes

As we showed in the preceding section, a string of bits need not represent a number, and in fact most of the information processed by computers is nonnumeric. The most common type of nonnumeric data is *text*, strings of characters from some character set. Each character is represented in the computer by a bit string according to an established convention.

text

The most commonly used character code is *ASCII* (pronounced *ASS key*), the American Standard Code for Information Interchange. ASCII represents each character with a 7-bit string, yielding a total of 128 different characters shown in Table 2-11. The code contains the uppercase and lowercase alphabet, numerals, punctuation, and various nonprinting control characters. Thus, the text string "Yeccch!" is represented by a rather innocuous-looking list of seven 7-bit numbers:

ASCII

1011001   1100101   1100011   1100011   1100011   1101000   0100001

2.13 Codes for Actions, Conditions, and States

The codes that we've described so far are generally used to represent things that we would probably consider to be "data"—things like numbers, positions, and characters. Programmers know that dozens of different data types can be used in a single computer program.

In digital system design, we often encounter nondata applications where a string of bits must be used to control an action, to flag a condition, or to represent the current state of the hardware. Probably the most commonly used type of code for such an application is a simple binary code. If there are n different actions, conditions, or states, we can represent them with a b-bit binary code with $b = \lceil \log_2 n \rceil$ bits. (The brackets $\lceil \ \rceil$ denote the *ceiling function*—the smallest integer greater than or equal to the bracketed quantity. Thus, b is the smallest integer such that $2^b \geq n$.)

$\lceil \ \rceil$

ceiling function

For example, consider a simple traffic-light controller. The signals at the intersection of a north-south (N-S) and an east-west (E-W) street might be in any

Table 2-11 American Standard Code for Information Interchange (ASCII), Standard No. X3.4-1968 of the American National Standards Institute.

$b_3b_2b_1b_0$	Row (hex)	000 0	001 1	010 2	011 3	100 4	101 5	110 6	111 7
					$b_6b_5b_4$ (column)				
0000	0	NUL	DLE	SP	0	@	P	`	p
0001	1	SOH	DC1	!	1	A	Q	a	q
0010	2	STX	DC2	"	2	B	R	b	r
0011	3	ETX	DC3	#	3	C	S	c	s
0100	4	EOT	DC4	$	4	D	T	d	t
0101	5	ENQ	NAK	%	5	E	U	e	u
0110	6	ACK	SYN	&	6	F	V	f	v
0111	7	BEL	ETB	'	7	G	W	g	w
1000	8	BS	CAN	(	8	H	X	h	x
1001	9	HT	EM	)	9	I	Y	i	y
1010	A	LF	SUB	*	:	J	Z	j	z
1011	B	VT	ESC	+	;	K	[	k	{
1100	C	FF	FS	,	<	L	\	l	|
1101	D	CR	GS	-	=	M	]	m	}
1110	E	SO	RS	.	>	N	^	n	~
1111	F	SI	US	/	?	O	_	o	DEL

Control codes

NUL	Null	DLE	Data link escape
SOH	Start of heading	DC1	Device control 1
STX	Start of text	DC2	Device control 2
ETX	End of text	DC3	Device control 3
EOT	End of transmission	DC4	Device control 4
ENQ	Enquiry	NAK	Negative acknowledge
ACK	Acknowledge	SYN	Synchronize
BEL	Bell	ETB	End transmitted block
BS	Backspace	CAN	Cancel
HT	Horizontal tab	EM	End of medium
LF	Line feed	SUB	Substitute
VT	Vertical tab	ESC	Escape
FF	Form feed	FS	File separator
CR	Carriage return	GS	Group separator
SO	Shift out	RS	Record separator
SI	Shift in	US	Unit separator
SP	Space	DEL	Delete or rubout

Table 2-12 States in a traffic-light controller.

	Lights						
State	N-S Green	N-S Yellow	N-S Red	E-W Green	E-W Yellow	E-W Red	Code Word
N-S go	ON	off	off	off	off	ON	000
N-S wait	off	ON	off	off	off	ON	001
N-S delay	off	off	ON	off	off	ON	010
E-W go	off	off	ON	ON	off	off	100
E-W wait	off	off	ON	off	ON	off	101
E-W delay	off	off	ON	off	off	ON	110

of the six states listed in Table 2-12. These states can be encoded in three bits, as shown in the last column of the table. Only six of the eight possible 3-bit code words are used, and the assignment of the six chosen code words to states is arbitrary, so many other encodings are possible. An experienced digital designer chooses a particular encoding to minimize circuit cost or to optimize some other parameter (like design time—there's no need to try billions and billions of possible encodings).

Another application of a binary code is illustrated in Figure 2-7(a). Here, we have a system with n devices, each of which can perform a certain action. The characteristics of the devices are such that they may be enabled to operate only one at a time. The control unit produces a binary-coded "device-select" word with $\lceil \log_2 n \rceil$ bits to indicate which device is enabled at any time. The "device-select code word is applied to each device, which compares it with its own "device ID" to determine whether it is enabled. Although its code words have the minimum number of bits, a binary code isn't always the best choice for encoding actions, conditions, or states. Figure 2-7(b) shows how to control n devices with a *1-out-of-n code*, an n-bit code in which valid code words have one bit equal to 1 and the rest of the bits equal to 0. Each bit of the 1-out-of-n code word is connected directly to the enable input of a corresponding device. This simplifies the design of the devices, since they no longer have device IDs; they need only a single "enable" input bit.

1-out-of-n code

The code words of a 1-out-of-10 code were listed in Table 2-9. Sometimes an all-0s word may also be included in a 1-out-of-n code, to indicate that no device is selected. Another common code is an *inverted 1-out-of-n code*, in which valid code words have one 0 bit and the rest of the bits equal to 1.

inverted 1-out-of-n code

In complex systems, a combination of coding techniques may be used. For example, consider a system similar to Figure 2-7(b), in which each of the n devices contains up to s subdevices. The control unit could produce a device-

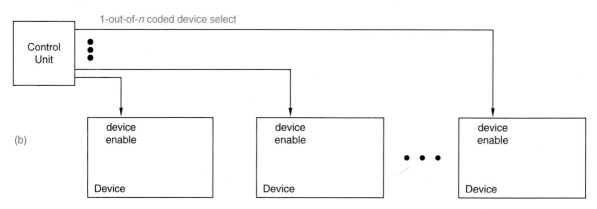

Figure 2-7 Control structure for a digital system with *n* devices:
(a) using a binary code; (b) using a 1-out-of-*n* code.

select code word with a 1-out-of-*n* coded field to select a device, and a $\lceil \log_2 s \rceil$-bit binary-coded field to select one of the *s* subdevices of the selected device.

m-out-of-n code
        An *m-out-of-n code* is a generalization of the 1-out-of-*n* code in which valid code words have *m* bits equal to 1 and the rest of the bits equal to 0. An *m*-out-of-*n* code word can be detected with an *m*-input AND gate, which produces a 1 output if all of its inputs are 1. This is fairly simple and inexpensive to do, yet for most values of *m*, an *m*-out-of-*n* code typically has far more valid code words than a 1-out-of-*n* code. The total number of code words is given by the binomial coefficient $\binom{n}{m}$, which has the value $\dfrac{n!}{m! \cdot (n-m)!}$. Thus, a 2-out-of-4 code has 6 valid code words, and a 3-out-of-10 code has 120.

8B10B code
        An important variation of an *m-out-of-n* code is the *8B10B code* used in the 802.3z Gigabit Ethernet standard. This code uses 10 bits to represent 256 valid code words, or 8 bits worth of data. Most code words use a 5-out-of-10 coding. However, since $\binom{5}{10}$ is only 252, some 4- and 6-out-of-10 words are also used to complete the code in a very interesting way; more on this in Section 2.16.2.

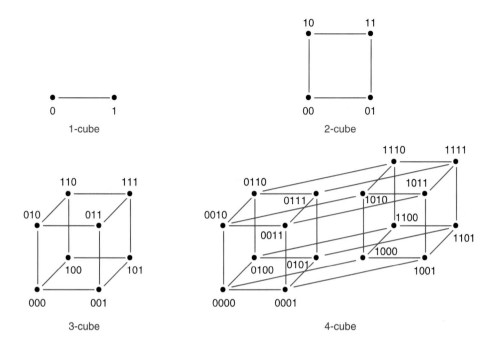

Figure 2-8
n-cubes for $n = 1, 2,$
3, and 4.

*2.14 *n*-Cubes and Distance

An *n*-bit string can be visualized geometrically, as a vertex of an object called an   *n-cube*
n-cube. Figure 2-8 shows *n*-cubes for $n = 1, 2, 3, 4$. An *n*-cube has 2^n vertices,
each of which is labeled with an *n*-bit string. Edges are drawn so that each vertex
is adjacent to *n* other vertices whose labels differ from the given vertex in only
one bit. Beyond $n = 4$, *n*-cubes are really tough to draw.

For reasonable values of *n*, *n*-cubes make it easy to visualize certain coding
and logic-minimization problems. For example, the problem of designing an
n-bit Gray code is equivalent to finding a path along the edges of an *n*-cube, a
path that visits each vertex exactly once. The paths for 3- and 4-bit Gray codes
are shown in Figure 2-9.

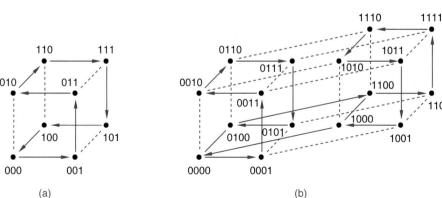

Figure 2-9
Traversing *n*-cubes
in Gray-code order:
(a) 3-cube;
(b) 4-cube.

distance
Hamming distance

Cubes also provide a geometrical interpretation for the concept of *distance*, also called *Hamming distance*. The distance between two *n*-bit strings is the number of bit positions in which they differ. In terms of an *n*-cube, the distance is the minimum length of a path between the two corresponding vertices. Two adjacent vertices have distance 1; vertices 001 and 100 in the 3-cube have distance 2. The concept of distance is crucial in the design and understanding of error-detecting codes, discussed in the next section.

m-subcube

An *m-subcube* of an *n*-cube is a set of 2^m vertices in which $n - m$ of the bits have the same value at each vertex, and the remaining *m* bits take on all 2^m combinations. For example, the vertices (000, 010, 100, 110) form a 2-subcube of the 3-cube. This subcube can also be denoted by a single string, xx0, where "x" denotes that a particular bit is a *don't-care*; any vertex whose bits match in the non-x positions belongs to this subcube. The concept of subcubes is particularly useful in visualizing algorithms that minimize the cost of combinational logic functions, as we'll show in Section 4.4.

don't-care

*2.15 Codes for Detecting and Correcting Errors

error
failure
temporary failure
permanent failure

An *error* in a digital system is the corruption of data from its correct value to some other value. An error is caused by a physical *failure*. Failures can be either temporary or permanent. For example, a cosmic ray or alpha particle can cause a temporary failure of a memory circuit, changing the value of a bit stored in it. Letting a circuit get too hot or zapping it with static electricity can cause a permanent failure, so that it never works correctly again.

error model
independent error model
single error
multiple error

The effects of failures on data are predicted by *error models*. The simplest error model, which we consider here, is called the *independent error model*. In this model, a single physical failure is assumed to affect only a single bit of data; the corrupted data is said to contain a *single error*. Multiple failures may cause *multiple errors*—two or more bits in error—but multiple errors are normally assumed to be less likely than single errors.

2.15.1 Error-Detecting Codes

Recall from our definitions in Section 2.10 that a code that uses *n*-bit strings need not contain 2^n valid code words; this is certainly the case for the codes that we now consider. An *error-detecting code* has the property that corrupting or garbling a code word will likely produce a bit string that is not a code word (a *noncode word*).

error-detecting code

noncode word

A system that uses an error-detecting code generates, transmits, and stores only code words. Thus, errors in a bit string can be detected by a simple rule—if the bit string is a code word, it is assumed to be correct; if it is a noncode word, it contains an error.

An *n*-bit code and its error-detecting properties under the independent error model are easily explained in terms of an *n*-cube. A code is simply a subset

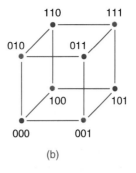

● = code word
● = noncode word

Figure 2-10
Code words in two
different 3-bit codes:
(a) minimum distance
= 1, does not detect
all single errors;
(b) minimum distance
= 2, detects all single
errors.

of the vertices of the n-cube. In order for the code to detect all single errors, no
code-word vertex can be immediately adjacent to another code-word vertex.

For example, Figure 2-10(a) shows a 3-bit code with five code words.
Code word 111 is immediately adjacent to code words 110, 011 and 101. Since
a single failure could change 111 to 110, 011 or 101 this code does not detect all
single errors. If we make 111 a noncode word, we obtain a code that does have
the single-error-detecting property, as shown in (b). No single error can change
one code word into another.

The ability of a code to detect single errors can be stated in terms of the
concept of distance introduced in the preceding section:

- A code detects all single errors if the *minimum distance* between all possible pairs of code words is 2.

  minimum distance

In general, we need $n + 1$ bits to construct a single-error-detecting code
with $2n$ code words. The first n bits of a code word, called *information bits*, may
be any of the $2n$ n-bit strings. To obtain a minimum-distance-2 code, we add one
more bit, called a *parity bit*, that is set to 0 if there are an even number of 1s
among the information bits, and to 1 otherwise. This is illustrated in the first two
columns of Table 2-13 for a code with three information bits. A valid $(n+1)$-bit
code word has an even number of 1s, and this code is called an *even-parity code*.

information bit

parity bit

even-parity code

Information Bits	Even-parity Code	Odd-parity Code
000	000 0	000 1
001	001 1	001 0
010	010 1	010 0
011	011 0	011 1
100	100 1	100 0
101	101 0	101 1
110	110 0	110 1
111	111 1	111 0

Table 2-13
Distance-2 codes with
three information bits.

odd-parity code
1-bit parity code

We can also construct a code in which the total number of 1s in a valid $(n + 1)$-bit code word is odd; this is called an *odd-parity code* and is shown in the third column of the table. These codes are also sometimes called *1-bit parity codes*, since they each use a single parity bit.

The 1-bit parity codes do not detect 2-bit errors, since changing two bits does not affect the parity. However, the codes can detect errors in any *odd* number of bits. For example, if three bits in a code word are changed, then the resulting word has the wrong parity and is a noncode word. This doesn't help us much, though. Under the independent error model, 3-bit errors are much less likely than 2-bit errors, which are not detectable. Thus, practically speaking, the 1-bit parity codes' error-detection capability stops after 1-bit errors. Other codes, with minimum distance greater than 2, can be used to detect multiple errors.

2.15.2 Error-Correcting and Multiple-Error-Detecting Codes

check bits

By using more than one parity bit, or *check bits*, according to some well-chosen rules, we can create a code whose minimum distance is greater than 2. Before showing how this can be done, let's look at how such a code can be used to correct single errors or detect multiple errors.

Suppose that a code has a minimum distance of 3. Figure 2-11 shows a fragment of the n-cube for such a code. As shown, there are at least two noncode words between each pair of code words. Now suppose we transmit code words

Figure 2-11
Some code words
and noncode words in
a 7-bit, distance-3
code.

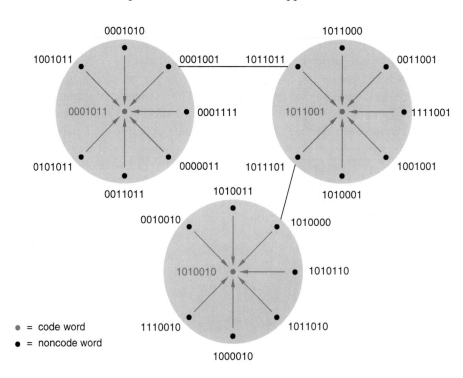

and assume that failures affect at most one bit of each received code word. Then a received noncode word with a 1-bit error will be closer to the originally transmitted code word than to any other code word. Therefore, when we receive a noncode word, we can *correct* the error by changing the received noncode word to the nearest code word, as indicated by the arrows in the figure. Deciding which code word was originally transmitted to produce a received word is called *decoding*, and the hardware that does this is an error-correcting *decoder*.

error correction

decoding
decoder

A code that is used to correct errors is called an *error-correcting code*. In general, if a code has minimum distance $2c + 1$, it can be used to correct errors that affect up to c bits ($c = 1$ in the preceding example). If a code's minimum distance is $2c + d + 1$, it can be used to correct errors in up to c bits and to detect errors in up to d additional bits.

error-correcting code

For example, Figure 2-12(a) shows a fragment of the n-cube for a code with minimum distance 4 ($c = 1$, $d = 1$). Single-bit errors that produce noncode words 00101010 and 11010011 can be corrected. However, an error that produces 10100011 cannot be corrected, because no single-bit error can produce this noncode word, and either of two 2-bit errors could have produced it. So the code can detect a 2-bit error, but it cannot correct it.

When a noncode word is received, we don't know which code word was originally transmitted; we only know which code word is closest to what we've received. Thus, as shown in Figure 2-12(b), a 3-bit error may be "corrected" to the wrong value. The possibility of making this kind of mistake may be acceptable if 3-bit errors are very unlikely to occur. On the other hand, if we are concerned about 3-bit errors, we can change the decoding policy for the code. Instead of trying to correct errors, we just flag all noncode words as uncorrectable errors. Thus, as shown in (c), we can use the same distance-4 code to detect up to 3-bit errors but correct no errors ($c = 0$, $d = 3$).

2.15.3 Hamming Codes

In 1950, R. W. Hamming described a general method for constructing codes with a minimum distance of 3, now called *Hamming codes*. For any value of i, his method yields a $(2^i - 1)$-bit code with i check bits and $2^i - 1 - i$ information bits. Distance-3 codes with a smaller number of information bits are obtained by deleting information bits from a Hamming code with a larger number of bits.

Hamming code

The bit positions in a Hamming code word can be numbered from 1 through $2^i - 1$. In this case, any position whose number is a power of 2 is a check bit, and the remaining positions are information bits. Each check bit is grouped with a subset of the information bits, as specified by a *parity-check matrix*. As

parity-check matrix

DECISIONS,
DECISIONS   The names *decoding* and *decoder* make sense, since they are just distance-1 perturbations of *deciding* and *decider*.

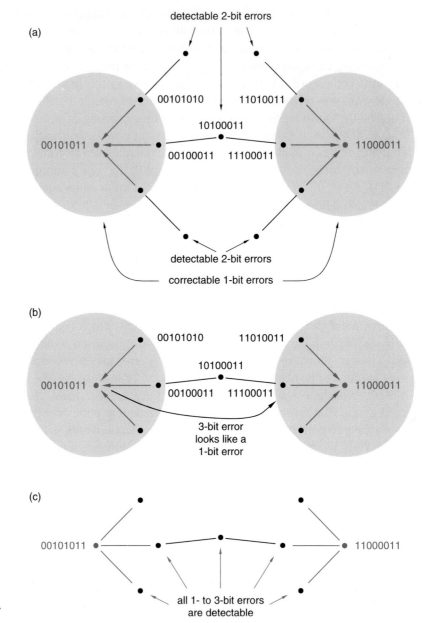

Figure 2-12
Some code words and
noncode words in an 8-bit,
distance-4 code:
(a) correcting 1-bit and
detecting 2-bit errors;
(b) incorrectly "correcting"
a 3-bit error;
(c) correcting no errors but
detecting up to 3-bit errors.

shown in Figure 2-13(a), each check bit is grouped with the information positions whose numbers have a 1 in the same bit when expressed in binary. For
example, check bit 2 (010) is grouped with information bits 3 (011), 6 (110), and
7 (111). For a given combination of information-bit values, each check bit is
chosen to produce even parity, that is, so the total number of 1s in its group is
even.

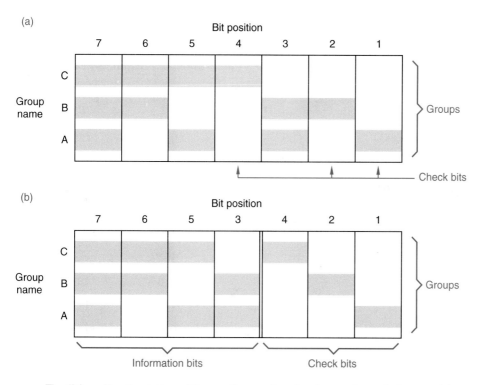

Figure 2-13
Parity-check
matrices for 7-bit
Hamming codes:
(a) with bit positions
in numerical order;
(b) with check bits
and information bits
separated.

Traditionally, the bit positions of a parity-check matrix and the resulting code words are rearranged so that all of the check bits are on the right, as in Figure 2-13(b). The first two columns of Table 2-14 list the resulting code words.

We can prove that the minimum distance of a Hamming code is 3 by proving that at least a 3-bit change must be made to a code word to obtain another code word. That is, we'll prove that a 1-bit or 2-bit change in a code word yields a noncode word.

If we change one bit of a code word, in position j, then we change the parity of every group that contains position j. Since every information bit is contained in at least one group, at least one group has incorrect parity, and the result is a noncode word.

What happens if we change two bits, in positions j and k? Parity groups that contain both positions j and k will still have correct parity, since parity is not affected when an even number of bits are changed. However, since j and k are different, their binary representations differ in at least one bit, corresponding to one of the parity groups. This group has only one bit changed, resulting in incorrect parity and a noncode word.

If you understand this proof, you should also understand how the position-numbering rules for constructing a Hamming code are a simple consequence of the proof. For the first part of the proof (1-bit errors), we required that the position numbers be nonzero. And for the second part (2-bit errors), we required

Table 2-14 Code words in distance-3 and distance-4 Hamming codes with four information bits.

Minimum-Distance-3 Code		Minimum-Distance-4 Code	
Information Bits	Parity Bits	Information Bits	Parity Bits
0000	000	0000	0000
0001	011	0001	0111
0010	101	0010	1011
0011	110	0011	1100
0100	110	0100	1101
0101	101	0101	1010
0110	011	0110	0110
0111	000	0111	0001
1000	111	1000	1110
1001	100	1001	1001
1010	010	1010	0101
1011	001	1011	0010
1100	001	1100	0011
1101	010	1101	0100
1110	100	1110	1000
1111	111	1111	1111

that no two positions have the same number. Thus, with an i-bit position number, you can construct a Hamming code with up to $2^i - 1$ bit positions.

error-correcting decoder

The proof also suggests how we can design an *error-correcting decoder* for a received Hamming code word. First, we check all of the parity groups; if all have even parity, then the received word is assumed to be correct. If one or more groups have odd parity, then a single error is assumed to have occurred. The

syndrome

pattern of groups that have odd parity (called the *syndrome*) must match one of the columns in the parity-check matrix; the corresponding bit position is assumed to contain the wrong value and is complemented. For example, using the code defined by Figure 2-13(b), suppose we receive the word 0101011. Groups B and C have odd parity, corresponding to position 6 of the parity-check matrix (the syndrome is 110, or 6). By complementing the bit in position 6 of the received word, we determine that the correct word is 0001011.

A distance-3 Hamming code can easily be modified to increase its minimum distance to 4. We simply add one more check bit, chosen so that the parity of all the bits, including the new one, is even. As in the 1-bit even-parity code, this bit ensures that all errors affecting an odd number of bits are detectable. In particular, any 3-bit error is detectable. We already showed that 1- and 2-bit errors are detected by the other parity bits, so the minimum distance of the modified code must be 4.

Distance-3 and distance-4 Hamming codes are commonly used to detect and correct errors in computer memory systems, especially in large mainframe computers where memory circuits account for the bulk of the system's failures. These codes are especially attractive for very wide memory words, since the required number of parity bits grows slowly with the width of the memory word, as shown in Table 2-15.

Table 2-15 Word sizes of distance-3 and distance-4 Hamming codes.

	Minimum-Distance-3 Codes		Minimum-Distance-4 Codes	
Information Bits	Parity Bits	Total Bits	Parity Bits	Total Bits
1	2	3	3	4
≤ 4	3	≤ 7	4	≤ 8
≤ 11	4	≤ 15	5	≤ 16
≤ 26	5	≤ 31	6	≤ 32
≤ 57	6	≤ 63	7	≤ 64
≤ 120	7	≤ 127	8	≤ 128

2.15.4 CRC Codes

Beyond Hamming codes, many other error-detecting and -correcting codes have been developed. The most important codes, which happen to include Hamming codes, are the *cyclic-redundancy-check (CRC) codes*. A rich theory has been developed for these codes, focused both on their error-detecting and correcting properties and on the design of inexpensive encoders and decoders for them (see References).

cyclic-redundancy-check (CRC) code

Two important applications of CRC codes are in disk drives and in data networks. In a disk drive, each block of data (typically 512 bytes) is protected by a CRC code, so that errors within a block can be detected and, in some drives, corrected. In a data network, each packet of data ends with check bits in a CRC code. The CRC codes for both applications were selected because of their burst-error detecting properties. In addition to single-bit errors, they can detect multibit errors that are clustered together within the disk block or packet. Such errors are more likely than errors of randomly distributed bits, because of the likely physical causes of errors in the two applications—surface defects in disk drives and noise bursts in communication links.

2.15.5 Two-Dimensional Codes

two-dimensional code

Another way to obtain a code with large minimum distance is to construct a *two-dimensional code*, as illustrated in Figure 2-14(a). The information bits are conceptually arranged in a two-dimensional array, and parity bits are provided to check both the rows and the columns. A code C_{row} with minimum distance d_{row} is used for the rows, and a possibly different code C_{col} with minimum distance d_{col} is used for the columns. That is, the row-parity bits are selected so that each row is a code word in C_{row} and the column-parity bits are selected so that each column is a code word in C_{col}. (The "corner" parity bits can be chosen according to either code.) The minimum distance of the two-dimensional code is the product of d_{row}

product code

and d_{col}; in fact, two-dimensional codes are sometimes called *product codes*.

As shown in Figure 2-14(b), the simplest two-dimensional code uses 1-bit even-parity codes for the rows and columns and has a minimum distance of $2 \cdot 2$,

Figure 2-14
Two-dimensional codes:
(a) general structure;
(b) using even parity for
both the row and column
codes to obtain
minimum distance 4;
(c) typical pattern of an
undetectable error.

(a)

(b)

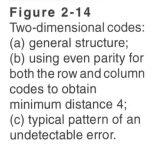

Columns are code words
in 1-bit even-parity code

(c)

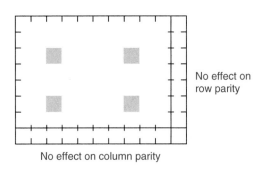

or 4. You can easily prove that the minimum distance is 4 by convincing yourself that any pattern of one, two, or three bits in error causes incorrect parity of a row or a column or both. In order to obtain an undetectable error, at least four bits must be changed in a rectangular pattern as in (c).

The error-detecting and correcting procedures for this code are straightforward. Assume we are reading information one row at a time. As we read each row, we check its row code. If an error is detected in a row, we can't tell which bit is wrong from the row check alone. However, assuming only one row is bad, we can reconstruct it by forming the bit-by-bit Exclusive OR of the columns, omitting the bad row, but including the column-check row.

To obtain an even larger minimum distance, a distance-3 or -4 Hamming code can be used for the row or column code or both. It is also possible to construct a code in three or more dimensions, with minimum distance equal to the product of the minimum distances in each dimension.

An important application of two-dimensional codes is in RAID storage systems. *RAID* stands for "redundant array of inexpensive disks." In this scheme, $n + 1$ identical disk drives are used to store n disks worth of data. For example, eight 8-gigabyte drives could be use to store 64 gigabytes of non-redundant data, and a ninth 8-gigabyte drive would be used to store checking information.

RAID

Figure 2-15 shows the general scheme of a two-dimensional code for a RAID system; each disk drive is considered to be a row in the code. Each drive stores m blocks of data, where a block typically contains 512 bytes. For example, an 8-gigabyte drive would store about 16 million blocks. As shown in the figure, each block includes its own check bits in a CRC code, to detect errors within that block. The first n drives store the nonredundant data. Each block in drive $n + 1$ stores parity bits for the corresponding blocks in the first n drives. That is, each

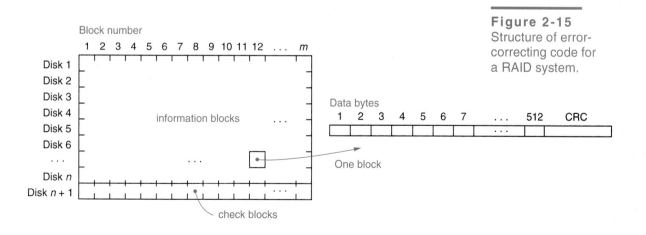

Figure 2-15
Structure of error-correcting code for a RAID system.

bit i in drive $n+1$, block b, is chosen so that there are an even number of 1s in block b, bit position i, across all the drives.

In operation, errors in the information blocks are detected by the CRC code. Whenever an error is detected in a block on one of the drives, the correct contents of that block can be constructed simply by computing the parity of the corresponding blocks in all the other drives, including drive $n+1$. Although this requires n extra disk-read operations, it's better than losing your data! Write operations require extra disk accesses as well, to update the corresponding check block when an information block is written (see Exercise 2.46). Since disk writes are much less frequent than reads in typical applications, this overhead usually is not a problem.

2.15.6 Checksum Codes

The parity-checking operation that we've used in the previous subsections is essentially modulo-2 addition of bits—the sum modulo 2 of a group of bits is 0 if the number of 1s in the group is even, and 1 if it is odd. The technique of modular addition can be extended to other bases besides 2 to form check digits.

checksum
checksum code

For example, a computer stores information as a set of 8-bit bytes. Each byte may be considered to have a decimal value from 0 to 255. Therefore, we can use modulo-256 addition to check the bytes. We form a single check byte, called a *checksum*, that is the sum modulo 256 of all the information bytes. The resulting *checksum code* can detect any single *byte* error, since such an error will cause a recomputed sum of bytes to disagree with the checksum.

ones'-complement
checksum code

Checksum codes can also use a different modulus of addition. In particular, checksum codes using modulo-255, ones'-complement addition are important because of their special computational and error-detecting properties, and because they are used to check packet headers in the ubiquitous Internet Protocol (IP) (see References).

2.15.7 *m*-out-of-*n* Codes

The 1-out-of-n and m-out-of-n codes that we introduced in Section 2.13 have a minimum distance of 2, since changing only one bit changes the total number of 1s in a code word and therefore produces a noncode word.

unidirectional error

These codes have another useful error-detecting property—they detect unidirectional multiple errors. In a *unidirectional error*, all of the erroneous bits change in the same direction (0s change to 1s, or vice versa). This property is very useful in systems where the predominant error mechanism tends to change all bits in the same direction.

2.16 Codes for Serial Data Transmission and Storage

2.16.1 Parallel and Serial Data

Most computers and other digital systems transmit and store data in a *parallel* format. In parallel data transmission, a separate signal line is provided for each bit of a data word. In parallel data storage, all of the bits of a data word can be written or read simultaneously.

parallel data

Parallel formats are not cost effective for some applications. For example, parallel transmission of data bytes over the telephone network would require eight phone lines, and parallel storage of data bytes on a magnetic disk would require a disk drive with eight separate read/write heads. *Serial* formats allow data to be transmitted or stored one bit at a time, reducing system cost in many applications.

serial data

Figure 2-16 illustrates some of the basic ideas in serial data transmission. A repetitive clock signal, named CLOCK in the figure, defines the rate at which bits are transmitted, one bit per clock cycle. Thus, the *bit rate* in bits per second (bps) numerically equals the clock frequency in cycles per second (hertz, or Hz).

bit rate, bps

The reciprocal of the bit rate is called the *bit time* and numerically equals the clock period in seconds (s). This amount of time is reserved on the serial data line (named SERDATA in the figure) for each bit that is transmitted. The time occupied by each bit is sometimes called a *bit cell*. The format of the actual signal that appears on the line during each bit cell depends on the *line code*. In the simplest line code, called *Non-Return-to-Zero (NRZ)*, a 1 is transmitted by placing a 1 on the line for the entire bit cell, and a 0 is transmitted as a 0. More complex line codes have other rules, as discussed in the next subsection.

bit time

bit cell
line code
Non-Return-to-Zero (NRZ)

Figure 2-16 Basic concepts for serial data transmission.

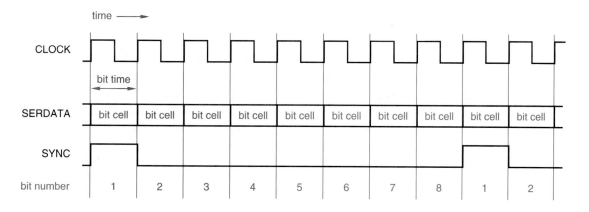

Regardless of the line code, a serial data-transmission or storage system needs some way of identifying the significance of each bit in the serial stream. For example, suppose that 8-bit bytes are transmitted serially. How can we tell which *synchronization signal* is the first bit of each byte? A *synchronization signal*, named SYNC in Figure 2-16, provides the necessary information; it is 1 for the first bit of each byte.

Evidently, we need a minimum of three signals to recover a serial data stream: a clock to define the bit cells, a synchronization signal to define the word boundaries, and the serial data itself. In some applications, like the interconnection of modules in a computer or telecommunications system, a separate wire is used for each of these signals, since reducing the number of wires per connection from *n* to three is savings enough. We'll give an example of a 3-wire serial data system in Section 8.5.4.

In many applications, the cost of having three separate signals is still too high (e.g., three phone lines, three read/write heads). Such systems typically combine all three signals into a single serial data stream and use sophisticated analog and digital circuits to recover the clock and synchronization information from the data stream.

*2.16.2 Serial Line Codes

The most commonly used line codes for serial data are illustrated in Figure 2-17. In the NRZ code, each bit value is sent on the line for the entire bit cell. This is the simplest and most reliable coding scheme for short-distance transmission. However, it generally requires a clock signal to be sent along with the data to define the bit cells. Otherwise, it is not possible for the receiver to determine how many 0s or 1s are represented by a continuous 0 or 1 level. For example, without a clock to define the bit cells, the NRZ waveform in Figure 2-17 might be erroneously interpreted as 01010.

Figure 2-17
Commonly used line codes for serial data.

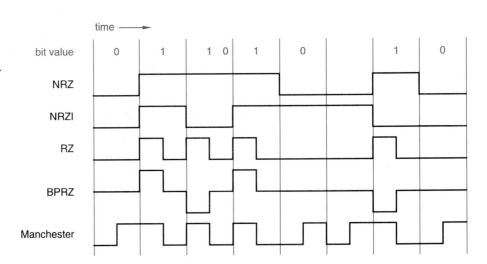

A *digital phase-locked loop (DPLL)* is an analog/digital circuit that can be used to recover a clock signal from a serial data stream. The DPLL works only if the serial data stream contains enough 0-to-1 and 1-to-0 transitions to give the DPLL "hints" about when the original clock transitions took place. With NRZ-coded data, the DPLL works only if the data does not contain any long, continuous streams of 1s or 0s.

digital phase-locked loop (DPLL)

Some serial transmission and storage media are *transition sensitive*; they cannot transmit or store absolute 0 or 1 levels, only transitions between two discrete levels. For example, a magnetic disk or tape stores information by changing the polarity of the medium's magnetization in regions corresponding to the stored bits. When the information is recovered, it is not feasible to determine the absolute magnetization polarity of a region, only that the polarity changes between one region and the next.

transition-sensitive media

Data stored in NRZ format on transition-sensitive media cannot be recovered unambiguously; the data in Figure 2-17 might be interpreted as 01110010 or 10001101. The *Non-Return-to-Zero Invert-on-1s (NRZI)* code overcomes this limitation by sending a 1 as the opposite of the level that was sent during the previous bit cell, and a 0 as the same level. A DPLL can recover the clock from NRZI-coded data as long as the data does not contain any long, continuous streams of 0s.

Non-Return-to-Zero Invert-on-1s (NRZI)

The Return-to-Zero (RZ) code is similar to NRZ except that, for a 1 bit, the 1 level is transmitted only for a fraction of the bit time, usually 1/2. With this code, data patterns that contain a lot of 1s create lots of transitions for a DPLL to use to recover the clock. However, as in the other line codes, a string of 0s has no transitions, and a long string of 0s makes clock recovery impossible.

Return-to-Zero (RZ)

Another requirement of some transmission media, such as high-speed fiber-optic links, is that the serial data stream be *DC balanced*. That is, it must have an equal number of 1s and 0s; any long-term DC component in the stream (created by having a lot more 1s than 0s or vice versa) creates a bias at the receiver that reduces its ability to distinguish reliably between 1s and 0s.

DC balance

Ordinarily, NRZ, NRZI or RZ data has no guarantee of DC balance; there's nothing to prevent a user data stream from having a long string of words with more than 1s than 0s or vice versa. However, DC balance can still be achieved by using a few extra bits to code the user data in a *balanced code*, in which each code word has an equal number of 1s and 0s, and then sending these code words in NRZ format.

balanced code

For example, in Section 2.13 we introduced the 8B10B code, which codes 8 bits of user data into 10 bits in a mostly 5-out-of-10 code. Recall that there are only 252 5-out-of-10 code words, but there are another $\binom{4}{10} = 210$ 4-out-of-10 code words and an equal number of 6-out-of-10 code words. Of course, these code words aren't quite DC balanced. The 8B10B code solves this problem by associating with each 8-bit value to be encoded a *pair* of unbalanced code words,

---

KILO-, MEGA-, GIGA-, TERA-

The prefixes K (kilo-), M (mega-), G (giga-), and T (tera-) mean 10^3, 10^6, 10^9, and 10^{12}, respectively, when referring to bps, hertz, ohms, watts, and most other engineering quantities. However, when referring to memory sizes, the prefixes mean 2^{10}, 2^{20}, 2^{30}, and 2^{40}. Historically, the prefixes were co-opted for this purpose because memory sizes are normally powers of 2, and 2^{10} (1024) is very close to 1000,

Now, when somebody offers you 50 kilobucks a year for your first engineering job, it's up to you to negotiate what the prefix means!

---

running disparity

one 4-out-of-10 ("light") and the other 6-out-of-10 ("heavy"). The coder also keeps track of the *running disparity*, a single bit of information indicating whether the last unbalanced code word that it transmitted was heavy or light. When it comes time to transmit another unbalanced code word, the coder selects the one of the pair with the opposite weight. This simple trick makes available 252 + 210 = 462 code words for the 8B10B to encode 8 bits of user data. Some of the "extra" code words are used to conveniently encode nondata conditions on the serial line, such as IDLE, SYNC, and ERROR. Not all the unbalanced code words are used. Also, some of the balanced code words, such as 0000011111, are not used either, in favor of unbalanced pairs that contain more transitions.

Bipolar Return-to-Zero (BPRZ)

Alternate Mark Inversion (AMI)

All of the preceding codes transmit or store only two signal levels. The *Bipolar Return-to-Zero (BPRZ)* code transmits three signal levels: +1, 0, and −1. The code is like RZ except that 1s are alternately transmitted as +1 and −1; for this reason, the code is also known as *Alternate Mark Inversion (AMI)*.

The big advantage of BPRZ over RZ is that it's DC balanced. This makes it possible to send BPRZ streams over transmission media that cannot tolerate a DC component, such as transformer-coupled phone lines. In fact, the BPRZ code has been used in T1 digital telephone links for decades, where analog speech signals are carried as streams of 8000 8-bit digital samples per second that are transmitted in BPRZ format on 64-Kbps serial channels.

As with RZ, it is possible to recover a clock signal from a BPRZ stream as long as there aren't too many 0s in a row. Although TPC (The Phone Company) has no control over what you say (at least, not yet), they still have a simple way of limiting runs of 0s. If one of the 8-bit bytes that results from sampling your analog speech pattern is all 0s, they simply change second-least significant bit to 1! This is called *zero-code suppression* and I'll bet you never noticed it. And this is also why, in many data applications of T1 links, you get only 56 Kbps of usable data per 64-Kbps channel; the LSB of each byte is always set to 1 to prevent zero-code suppression from changing the other bits.

zero-code suppression

Manchester

diphase

The last code in Figure 2-17 is called *Manchester* or *diphase* code. The major strength of this code is that, regardless of the transmitted data pattern, it provides at least one transition per bit cell, making it very easy to recover the clock. As shown in the figure, a 0 is encoded as a 0-to-1 transition in the middle

ABOUT TPC Watch the 1967 James Coburn movie, *The President's Analyst,* for an amusing view of TPC. With the growing pervasiveness of digital technology and cheap wireless communications, the concept of universal, *personal* connectivity to the phone network presented in the movie's conclusion has become much less far-fetched.

of the bit cell, and a 1 is encoded as a 1-to-0 transition. The Manchester code's major strength is also its major weakness. Since it has more transitions per bit cell than other codes, it also requires more media bandwidth to transmit a given bit rate. Bandwidth is not a problem in coaxial cable, however, which was used in the original Ethernet local area networks to carry Manchester-coded serial data at the rate of 10 Mbps (megabits per second).

References

The presentation in the first nine sections of this chapter is based on Chapter 4 of *Microcomputer Architecture and Programming*, by John F. Wakerly (Wiley, 1981). Precise, thorough, and entertaining discussions of these topics can also be found in Donald E. Knuth's *Seminumerical Algorithms*, third edition (Addison-Wesley, 1997). Mathematically inclined readers will find Knuth's analysis of the properties of number systems and arithmetic to be excellent, and all readers should enjoy the insights and history sprinkled throughout the text.

Descriptions of digital logic circuits for arithmetic operations, as well as an introduction to properties of various number systems, appear in *Computer Arithmetic* by Kai Hwang (Wiley, 1979). *Decimal Computation* by Hermann Schmid (Wiley, 1974) contains a thorough description of techniques for BCD arithmetic.

An introduction to algorithms for binary multiplication and division and to floating-point arithmetic appears in *Microcomputer Architecture and Programming: The 68000 Family* by John F. Wakerly (Wiley, 1989). A more thorough discussion of arithmetic techniques and floating-point number systems can be found in *Introduction to Arithmetic for Digital Systems Designers* by Shlomo Waser and Michael J. Flynn (Holt, Rinehart and Winston, 1982).

CRC codes are based on the theory of *finite fields,* which was developed by *finite fields* French mathematician Évariste Galois (1811–1832) shortly before he was killed in a duel with a political opponent. The classic book on error-detecting and error-correcting codes is *Error-Correcting Codes* by W. W. Peterson and E. J. Weldon, Jr. (MIT Press, 1972, second edition); however, this book is recommended only for mathematically sophisticated readers. A more accessible introduction to coding can be found in *Error Control Coding: Fundamentals and Applications* by S. Lin and D. J. Costello, Jr. (Prentice Hall, 1983). Another communication-oriented introduction to coding theory can be found in *Error-*

Control Techniques for Digital Communication by A. M. Michelson and A. H. Levesque (Wiley-Interscience, 1985). Hardware applications of codes in computer systems are discussed in *Error-Detecting Codes, Self-Checking Circuits, and Applications* by John F. Wakerly (Elsevier/North-Holland, 1978).

As shown in the above reference by Wakerly, ones'-complement checksum codes have the ability to detect long bursts of unidirectional errors; this is useful in communication channels where errors all tend to be in the same direction. The special computational properties of these codes also make them quite amenable to efficient checksum calculation by software programs, important for their use in the Internet Protocol; see RFC-1071 and RFC-1141. RFCs (Requests for Comments) are archived in many places on the Web; just search for "RFC".

An introduction to coding techniques for serial data transmission, including mathematical analysis of the performance and bandwidth requirements of several codes, appears in *Introduction to Communications Engineering* by R. M. Gagliardi (Wiley-Interscience, 1988, second edition). A nice introduction to the serial codes used in magnetic disks and tapes is given in *Computer Storage Systems and Technology* by Richard Matick (Wiley-Interscience, 1977).

The structure of the 8B10B code and the rationale behind it is explained nicely in the original IBM patent by Peter Franaszek and Albert Widmer, U.S. patent number 4,486,739 (1984). This and almost all U.S. patents issued after 1971 can be found on the Web at `www.patents.ibm.com`.

Drill Problems

2.1   Perform the following number system conversions:

(a) $1101011_2 = ?_{16}$ (b) $174003_8 = ?_2$

(c) $10110111_2 = ?_{16}$ (d) $67.24_8 = ?_2$

(e) $10100.1101_2 = ?_{16}$ (f) $F3A5_{16} = ?_2$

(g) $11011001_2 = ?_8$ (h) $AB3D_{16} = ?_2$

(i) $101111.0111_2 = ?_8$ (j) $15C.38_{16} = ?_2$

2.2   Convert the following octal numbers into binary and hexadecimal:

(a) $1023_8 = ?_2 = ?_{16}$ (b) $761302_8 = ?_2 = ?_{16}$

(c) $163417_8 = ?_2 = ?_{16}$ (d) $552273_8 = ?_2 = ?_{16}$

(e) $5436.15_8 = ?_2 = ?_{16}$ (f) $13705.207_8 = ?_2 = ?_{16}$

2.3   Convert the following hexadecimal numbers into binary and octal:

(a) $1023_{16} = ?_2 = ?_8$ (b) $7E6A_{16} = ?_2 = ?_8$

(c) $ABCD_{16} = ?_2 = ?_8$ (d) $C350_{16} = ?_2 = ?_8$

(e) $9E36.7A_{16} = ?_2 = ?_8$ (f) $DEAD.BEEF_{16} = ?_2 = ?_8$

2.4   What are the octal values of the four 8-bit bytes in the 32-bit number with octal representation 123456701238?

2.5   Convert the following numbers into decimal:

(a) $1101011_2 = ?_{10}$ (b) $174003_8 = ?_{10}$

(c) $10110111_2 = ?_{10}$ (d) $67.24_8 = ?_{10}$

(e) $10100.1101_2 = ?_{10}$ (f) $F3A5_{16} = ?_{10}$

(g) $12010_3 = ?_{10}$ (h) $AB3D_{16} = ?_{10}$

(i) $7156_8 = ?_{10}$ (j) $15C.38_{16} = ?_{10}$

2.6   Perform the following number-system conversions:

(a) $125_{10} = ?_2$ (b) $3489_{10} = ?_8$

(c) $209_{10} = ?_2$ (d) $9714_{10} = ?_8$

(e) $132_{10} = ?_2$ (f) $23851_{10} = ?_{16}$

(g) $727_{10} = ?_5$ (h) $57190_{10} = ?_{16}$

(i) $1435_{10} = ?_8$ (j) $65113_{10} = ?_{16}$

2.7   Add the following pairs of binary numbers, showing all carries:

(a)   110101   (b)   101110   (c)   11011101   (d)   1110010
       + 11001   + 100101   + 1100011   + 1101101

2.8   Repeat Drill 2.7 using subtraction instead of addition, and showing borrows instead of carries.

2.9   Add the following pairs of octal numbers:

(a)   1372   (b)   47135   (c)   175214   (d)   110321
       + 4631   + 5125   + 152405   + 56573

2.10   Add the following pairs of hexadecimal numbers:

(a)   1372   (b)   4F1A5   (c)   F35B   (d)   1B90F
       + 4631   + B8D5   + 27E6   + C44E

2.11   Write the 8-bit signed-magnitude, two's-complement, and ones'-complement representations for each of these decimal numbers: +18, +115, +79, −49, −3, −100.

2.12   Indicate whether or not overflow occurs when adding the following 8-bit two's-complement numbers:

(a)   11010100   (b)   10111001   (c)   01011101   (d)   00100110
       + 10101011   + 11010110   + 00100001   + 01011010

2.13   How many errors can be detected by a code with minimum distance d?

2.14   What is the minimum number of parity bits required to obtain a distance-4, two-dimensional code with n information bits?

Exercises

2.15   Here's a problem to whet your appetite. What is the hexadecimal equivalent of 61453_{10}?

2.16   Each of the following arithmetic operations is correct in at least one number system. Determine possible radices of the numbers in each operation.

(a) $1234 + 5432 = 6666$ (b) $41 / 3 = 13$

(c) $33/3 = 11$ (d) $23+44+14+32 = 223$

(e) $302/20 = 12.1$ (f) $14 = 5$

2.17   The first expedition to Mars found only the ruins of a civilization. From the artifacts and pictures, the explorers deduced that the creatures who produced this civilization were four-legged beings with a tentacle that branched out at the end with a number of grasping "fingers." After much study, the explorers were able to translate Martian mathematics. They found the following equation:

$$5x^2 - 50x + 125 = 0$$

with the indicated solutions $x = 5$ and $x = 8$. The value $x = 5$ seemed legitimate enough, but $x = 8$ required some explanation. Then the explorers reflected on the way in which Earth's number system developed, and found evidence that the Martian system had a similar history. How many fingers would you say the Martians had? (From *The Bent of Tau Beta Pi*, February 1956.)

2.18   Suppose a $4n$-bit number B is represented by an n-digit hexadecimal number H. Prove that the two's complement of B is represented by the 16's complement of H. Make and prove true a similar statement for octal representation.

2.19   Repeat Exercise 2.18 using the ones' complement of B and the 15s' complement of H.

2.20   Given an integer x in the range $-2^{n-1} \le x \le 2^{n-1} - 1$, we define $[x]$ to be the two's-complement representation of x, expressed as a positive number: $[x] = x$ if $x \ge 0$ and $[x] = 2n - |x|$ if $x < 0$, where $|x|$ is the absolute value of x. Let y be another integer in the same range as x. Prove that the two's-complement addition rules given in Section 2.6 are correct by proving that the following equation is always true:

$$[x + y] = ([x] + [y]) \text{ modulo } 2^n$$

(*Hints:* Consider four cases based on the signs of x and y. Without loss of generality, you may assume that $|x| \ge |y|$.)

2.21   Repeat Exercise 2.20 using appropriate expressions and rules for ones'-complement addition.

2.22   State an overflow rule for addition of two's-complement numbers in terms of counting operations in the modular representation of Figure 2-3.

2.23   Show that a two's-complement number can be converted to a representation with more bits by *sign extension*. That is, given an n-bit two's-complement number X, show that the m-bit two's-complement representation of X, where $m > n$, can be obtained by appending $m - n$ copies of X's sign bit to the left of the n-bit representation of X.

2.24   Show that a two's-complement number can be converted to a representation with fewer bits by removing higher-order bits. That is, given an n-bit two's-complement number X, show that the m-bit two's-complement number Y obtained by discarding the d leftmost bits of X represents the same number as X if and only if the discarded bits all equal the sign bit of Y.

2.25   Why is the punctuation of "two's complement" and "ones' complement" inconsistent? (See the first two citations in the References.)

2.26   A n-bit binary adder can be used to perform an n-bit unsigned subtraction operation $X - Y$, by performing the operation $X + \overline{Y} + 1$, where X and Y are n-bit unsigned numbers and $\overline{Y}$ represents the bit-by-bit complement of Y. Demonstrate this fact as follows. First, prove that $(X - Y) = (X + \overline{Y} + 1) - 2^n$. Second, prove that the carry out of the n-bit adder is the opposite of the borrow from the n-bit subtraction. That is, show that the operation $X - Y$ produces a borrow out of the MSB position if and only if the operation $X + \overline{Y} + 1$ *does not* produce a carry out of the MSB position.

2.27   In most cases, the product of two n-bit two's-complement numbers requires fewer than $2n$ bits to represent it. In fact, there is only one case in which $2n$ bits are needed—find it.

2.28   Prove that a two's-complement number can be multiplied by 2 by shifting it one bit position to the left, with a carry of 0 into the least significant bit position and disregarding any carry out of the most significant bit position, assuming no overflow. State the rule for detecting overflow.

2.29   State and prove correct a technique similar to the one described in Exercise 2.28, for multiplying a ones'-complement number by 2.

2.30   Show how to subtract BCD numbers, by stating the rules for generating borrows and applying a correction factor. Show how your rules apply to each of the following subtractions: $9 - 3, 5 - 7, 4 - 9, 1 - 8$.

2.31   How many different 3-bit binary state encodings are possible for the traffic-light controller of Table 2-12?

2.32   List all of the "bad" boundaries in the mechanical encoding disk of Figure 2-5, where an incorrect position may be sensed.

2.33   As a function of n, how many "bad" boundaries are there in a mechanical encoding disk that uses an n-bit binary code?

2.34   On-board altitude transponders on commercial and private aircraft use Gray code to encode the altitude readings that are transmitted to air traffic controllers. Why?

2.35   An incandescent light bulb is stressed every time it is turned on, so in some applications the lifetime of the bulb is limited by the number of on/off cycles rather than the total time it is illuminated. Use your knowledge of codes to suggest a way to double the lifetime of 3-way bulbs in such applications.

2.36   As a function of n, how many different distinct subcubes of an n-cube are there?

2.37   Find a way to draw a 3-cube on a sheet of paper (or other two-dimensional object) so that none of the lines cross, or prove that it's impossible.

2.38   Repeat Exercise 2.37 for a 4-cube.

2.39   Write a formula that gives the number of m-subcubes of an n-cube for a specific value of m. (Your answer should be a function of n and m.)

2.40   Define parity groups for a distance-3 Hamming code with 11 information bits.

2.41   Write the code words of a Hamming code with one information bit.

2.42   Exhibit the pattern for a 3-bit error that is not detected if the "corner" parity bits are not included in the two-dimensional codes of Figure 2-14.

2.43   The *rate of a code* is the ratio of the number of information bits to the total number of bits in a code word. High rates, approaching 1, are desirable for efficient transmission of information. Construct a graph comparing the rates of distance-2 parity codes and distance-3 and -4 Hamming codes for up to 100 information bits.

2.44   Which type of distance-4 code has a higher rate—a two-dimensional code or a Hamming code? Support your answer with a table in the style of Table 2-15, including the rate as well as the number of parity and information bits of each code for up to 100 information bits.

2.45   Show how to construct a distance-6 code with four information bits. Write a list of its code words.

2.46   Describe the operations that must be performed in a RAID system to write new data into information block b in drive d, so the data can be recovered in the event of an error in block b in any drive. Minimize the number of disk accesses required.

2.47   In the style of Figure 2-17, draw the waveforms for the bit pattern 10101110 when sent serially using the NRZ, NRZI, RZ, BPRZ, and Manchester codes, assuming that the bits are transmitted in order from left to right.

chapter 3

Digital Circuits

Marketing hype notwithstanding, we live in an analog world, not a digital one. Voltages, currents, and other physical quantities in real circuits take on values that are infinitely variable, depending on properties of the real devices that comprise the circuits. Because real values are continuously variable, we could use a physical quantity such as a signal voltage in a circuit to represent a real number (e.g., 3.14159265358979 volts represents the mathematical constant *pi* to 14 decimal digits of precision).

Unfortunately, stability and accuracy in physical quantities are difficult to obtain in real circuits. They are affected by manufacturing tolerances, temperature, power-supply voltage, cosmic rays, and noise created by other circuits, among other things. If we used an analog voltage to represent *pi*, we might find that instead of being an absolute mathematical constant, *pi* varied over a range of 10% or more.

Also, many mathematical and logical operations can be difficult or impossible to perform with analog quantities. While it is possible with some cleverness to build an analog circuit whose output voltage is the square root of its input voltage, no one has ever built a 100-input, 100-output analog circuit whose outputs are a set of voltages identical to the set of input voltages, but sorted arithmetically.

The purpose of this chapter is to give you a solid working knowledge of the electrical aspects of digital circuits, enough for you to understand and build real circuits and systems. We'll see in later chapters that with modern software tools, it's possible to "build" circuits in the abstract, using hardware design languages to specify their design and simulators to test their operation. Still, to build real, production-quality circuits, either at the board level or the chip level, you need to understand most of the material in this chapter. However, if you're anxious to start designing and simulating abstract circuits, you can just read the first section of this chapter and come back to the rest of it later.

3.1 Logic Signals and Gates

digital logic

logic values

Digital logic hides the pitfalls of the analog world by mapping the infinite set of real values for a physical quantity into two subsets corresponding to just two possible numbers or *logic values*—0 and 1. As a result, digital logic circuits can be analyzed and designed functionally, using switching algebra, tables, and other abstract means to describe the operation of well-behaved 0s and 1s in a circuit.

binary digit

bit

A logic value, 0 or 1, is often called a *binary digit*, or *bit*. If an application requires more than two discrete values, additional bits may be used, with a set of n bits representing 2^n different values.

Examples of the physical phenomena used to represent bits in some modern (and not-so-modern) digital technologies are given in Table 3-1. With most phenomena, there is an undefined region between the 0 and 1 states (e.g., voltage = 1.8 V, dim light, capacitor slightly charged, etc.). This undefined region is needed so that the 0 and 1 states can be unambiguously defined and reliably detected. Noise can more easily corrupt results if the boundaries separating the 0 and 1 states are too close.

LOW

HIGH

When discussing electronic logic circuits such as CMOS and TTL, digital designers often use the words "LOW" and "HIGH" in place of "0" and "1" to remind them that they are dealing with real circuits, not abstract quantities:

LOW   A signal in the range of algebraically lower voltages, which is interpreted as a logic 0.

HIGH   A signal in the range of algebraically higher voltages, which is interpreted as a logic 1.

positive logic

negative logic

Note that the assignments of 0 and 1 to LOW and HIGH are somewhat arbitrary. Assigning 0 to LOW and 1 to HIGH seems most natural and is called *positive logic*. The opposite assignment, 1 to LOW and 0 to HIGH, is not often used and is called *negative logic*.

buffer amplifier

Because a wide range of physical values represent the same binary value, digital logic is highly immune to component and power-supply variations and noise. Furthermore, *buffer-amplifier* circuits can be used to regenerate "weak" values into "strong" ones, so that digital signals can be transmitted over arbitrary distances without loss of information. For example, a buffer amplifier for CMOS

Table 3-1 Physical states representing bits in different computer logic and memory technologies.

Technology	State Representing Bit	
	0	1
Pneumatic logic	Fluid at low pressure	Fluid at high pressure
Relay logic	Circuit open	Circuit closed
Complementary metal-oxide semiconductor (CMOS) logic	0–1.5 V	3.5–5.0 V
Transistor-transistor logic (TTL)	0–0.8 V	2.0–5.0 V
Fiber optics	Light off	Light on
Dynamic memory	Capacitor discharged	Capacitor charged
Nonvolatile, erasable memory	Electrons trapped	Electrons released
Bipolar read-only memory	Fuse blown	Fuse intact
Bubble memory	No magnetic bubble	Bubble present
Magnetic tape or disk	Flux direction "north"	Flux direction "south"
Polymer memory	Molecule in state A	Molecule in state B
Read-only compact disc	No pit	Pit
Rewriteable compact disc	Dye in crystalline state	Dye in noncrystalline state

logic converts any HIGH input voltage into an output very close to 5.0 V, and any LOW input voltage into an output very close to 0.0 V.

A logic circuit can be represented with a minimum amount of detail simply as a "black box" with a certain number of inputs and outputs. For example, Figure 3-1 shows a logic circuit with three inputs and one output. However, this representation does not describe how the circuit responds to input signals.

From the point of view of electronic circuit design, it takes a lot of information to describe the precise electrical behavior of a circuit. However, since the inputs of a digital logic circuit can be viewed as taking on only discrete 0 and 1 values, the circuit's "logical" operation can be described with a table that ignores electrical behavior and lists only discrete 0 and 1 values.

Figure 3-1
"Black-box" representation
of a three-input, one-output
logic circuit.

combinational circuit

truth table

A logic circuit whose outputs depend only on its current inputs is called a *combinational circuit*. Its operation is fully described by a *truth table* that lists all combinations of input values and the output value(s) produced by each one. Table 3-2 is the truth table for a logic circuit with three inputs X, Y, and Z and a single output F.

Table 3-2
Truth table for a combinational logic circuit.

X	Y	Z	F
0	0	0	0
0	0	1	1
0	1	0	0
0	1	1	0
1	0	0	0
1	0	1	0
1	1	0	1
1	1	1	1

sequential circuit

state table

A circuit with memory, whose outputs depend on the current input *and* the sequence of past inputs, is called a *sequential circuit*. The behavior of such a circuit may be described by a *state table* that specifies its output and next state as functions of its current state and input. Sequential circuits will be introduced in Chapter 7.

As we'll show in Section 4.1, just three basic logic functions, AND, OR, and NOT, can be used to build any combinational digital logic circuit. Figure 3-2 shows the truth tables and symbols for logic "gates" that perform these functions. The symbols and truth tables for AND and OR may be extended to gates with any number of inputs. The gates' functions are easily defined in words:

AND gate

- An *AND gate* produces a 1 output if and only if all of its inputs are 1.

OR gate

- An *OR gate* produces a 1 if and only if one or more of its inputs are 1.

NOT gate

inverter

- A *NOT gate*, usually called an *inverter*, produces an output value that is the opposite of its input value.

Figure 3-2
Basic logic elements:
(a) AND; (b) OR;
(c) NOT (inverter).

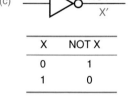

(a)
X
Y
X AND Y
X · Y

X	Y	X AND Y
0	0	0
0	1	0
1	0	0
1	1	1

(b)
X
Y
X OR Y
X + Y

X	Y	X OR Y
0	0	0
0	1	1
1	0	1
1	1	1

(c)
X
NOT X
X'

X	NOT X
0	1
1	0

Figure 3-3
Inverting gates:
(a) NAND; (b) NOR.

The circle on the inverter symbol's output is called an *inversion bubble* and is   *inversion bubble*
used in this and other gate symbols to denote "inverting" behavior.

Notice that in the definitions of AND and OR functions, we only had to
state the input conditions for which the output is 1, because there is only one
possibility when the output is not 1—it must be 0.

Two more logic functions are obtained by combining NOT with an AND or
OR function in a single gate. Figure 3-3 shows the truth tables and symbols for
these gates; Their functions are also easily described in words:

- A *NAND gate* produces the opposite of an AND gate's output, a 0 if and   *NAND gate*
  only if all of its inputs are 1.

- A *NOR gate* produces the opposite of an OR gate's output, a 0 if and only   *NOR gate*
  if one or more of its inputs are 1.

As with AND and OR gates, the symbols and truth tables for NAND and NOR
may be extended to gates with any number of inputs.

Figure 3-4 is a logic circuit using AND, OR, and NOT gates that functions
according to the truth table of Table 3-2. In Chapter 4 you'll learn how to go
from a truth table to a logic circuit, and vice versa, and you'll also learn about the
switching-algebra notation used in Figures 3-2 through 3-4.

Real logic circuits also function in another analog dimension—time. For
example, Figure 3-5 is a *timing diagram* that shows how the circuit of Figure 3-4   *timing diagram*
might respond to a time-varying pattern of input signals. The timing diagram
shows that the logic signals do not change between 0 and 1 instantaneously, and
also that there is a lag between an input change and the corresponding output

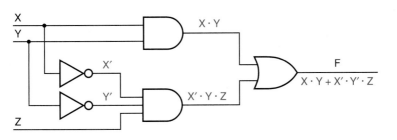

Figure 3-4
Logic circuit with the
truth table of
Table 3-2.

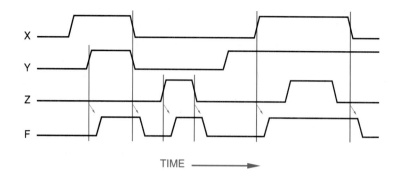

Figure 3-5
Timing diagram for a
logic circuit.

change. Later in this chapter you'll learn some of the reasons for this timing behavior, and how it is specified and handled in real circuits. And once again, you'll learn in a later chapter how this analog timing behavior can be generally ignored in most sequential circuits, and instead the circuit can be viewed as moving between discrete states at precise intervals defined by a clock signal.

Thus, even if you know nothing about analog electronics, you should be able to understand the logical behavior of digital circuits. However, there comes a time in design and debugging when every digital logic designer must throw out "the digital abstraction" temporarily and consider the analog phenomena that limit or disrupt digital performance. The rest of this chapter prepares you for that day by discussing the electrical characteristics of digital logic circuits.

THERE'S HOPE FOR NON-EE'S   If all of this electrical "stuff" bothers you, don't worry, at least for now. The rest of this book is written to be as independent of this stuff as possible. But you'll need it later, if you ever have to design and build digital systems in the real world.

3.2 Logic Families

There are many, many ways to design an electronic logic circuit. The first electrically controlled logic circuits, developed at Bell Laboratories in 1930s, were based on relays. In the mid-1940s, the first electronic digital computer, the Eniac, used logic circuits based on vacuum tubes. The Eniac had about 18,000 tubes and a similar number of logic gates, not a lot by today's standards of microprocessor chips with tens of millions of transistors. However, the Eniac could hurt you a lot more than a chip could if it fell on you—it was 100 feet long, 10 feet high, 3 feet deep, and consumed 140,000 watts of power!

semiconductor diode
bipolar junction
transistor

The inventions of the *semiconductor diode* and the *bipolar junction transistor* allowed the development of smaller, faster, and more capable computers in the late 1950s. In the 1960s, the invention of the *integrated circuit (IC)* allowed multiple diodes, transistors, and other components to be fabricated on a single chip, and computers got still better.

integrated circuit (IC)

The 1960s also saw the introduction of the first integrated-circuit logic families. A *logic family* is a collection of different integrated-circuit chips that have similar input, output, and internal circuit characteristics, but that perform different logic functions. Chips from the same family can be interconnected to perform any desired logic function. On the other hand, chips from differing families may not be compatible; they may use different power-supply voltages or may use different input and output conditions to represent logic values.

logic family

The most successful *bipolar logic family* (one based on bipolar junction transistors) is *transistor-transistor logic (TTL)*. First introduced in the 1960s, TTL now is actually a family of logic families that are compatible with each other but differ in speed, power consumption, and cost. Digital systems can mix components from several different TTL families, according to design goals and constraints in different parts of the system. Although TTL was largely replaced by CMOS in the 1990s, you're still likely to encounter TTL components in academic labs; therefore, we introduce TTL families in Section 3.10.

bipolar logic family
transistor-transistor logic (TTL)

Ten years *before* the bipolar junction transistor was invented, the principles of operation were patented for another type of transistor, called the *metal-oxide semiconductor field-effect transistor (MOSFET)*, or simply *MOS transistor.* However, MOS transistors were difficult to fabricate in the early days, and it wasn't until the 1960s that a wave of developments made MOS-based logic and memory circuits practical. Even then, MOS circuits lagged bipolar circuits considerably in speed and were attractive only in selected applications because of their lower power consumption and higher levels of integration.

metal-oxide semiconductor field-effect transistor (MOSFET)

MOS transistor

Beginning in the mid-1980s, advances in the design of MOS circuits, in particular *complementary MOS (CMOS)* circuits, vastly increased their performance and popularity. By far the majority of new large-scale integrated circuits, such as microprocessors and memories, use CMOS. Likewise, small- to medium-scale applications, for which TTL was once the logic family of choice, are now likely to use CMOS devices with equivalent functionality but higher speed and lower power consumption. CMOS circuits now account for the vast majority of the worldwide IC market.

complementary MOS (CMOS)

CMOS logic is both the most capable and the easiest to understand commercial digital logic technology. Beginning in the next section, we describe the basic structure of CMOS logic circuits and introduce the most commonly used commercial CMOS logic families.

GREEN STUFF   Nowadays, the acronym "MOS" is usually spoken as "moss," rather than spelled out. And "CMOS" has always been spoken as "sea moss."

As a consequence of the industry's transition from TTL to CMOS over a long period of time, many CMOS families were designed to be somewhat compatible with TTL. In Section 3.12 we show how TTL and CMOS families can be mixed within a single system.

3.3 CMOS Logic

The functional behavior of a CMOS logic circuit is fairly easy to understand, even if your knowledge of analog electronics is not particularly deep. The basic (and typically only) building blocks in CMOS logic circuits are MOS transistors, described shortly. Before introducing MOS transistors and CMOS logic circuits, we must talk about logic levels.

3.3.1 CMOS Logic Levels

Abstract logic elements process binary digits, 0 and 1. However, real logic circuits process electrical signals such as voltage levels. In any logic circuit, there is a range of voltages (or other circuit conditions) that is interpreted as a logic 0, and another, nonoverlapping range that is interpreted as a logic 1.

A typical CMOS logic circuit operates from a 5-volt power supply. Such a circuit may interpret any voltage in the range 0–1.5 V as a logic 0, and in the range 3.5–5.0 V as a logic 1. Thus, the definitions of LOW and HIGH for 5-volt CMOS logic are as shown in Figure 3-6. Voltages in the intermediate range (1.5–3.5 V) are not expected to occur except during signal transitions, and yield undefined logic values (i.e., a circuit may interpret them as either 0 or 1). CMOS circuits using other power-supply voltages, such as 3.3 or 2.7 volts, partition the voltage range similarly.

3.3.2 MOS Transistors

A MOS transistor can be modeled as a 3-terminal device that acts like a voltage-controlled resistance. As suggested by Figure 3-7, an input voltage applied to one terminal controls the resistance between the remaining two terminals. In digital logic applications, a MOS transistor is operated so its resistance is always either very high (and the transistor is "off") or very low (and the transistor is "on").

Figure 3-6
Logic levels for typical
CMOS logic circuits.

Figure 3-7
The MOS transistor as
a voltage-controlled
resistance.

There are two types of MOS transistors, *n*-channel and *p*-channel; the names refer to the type of semiconductor material used for the resistance-controlled terminals. The circuit symbol for an *n-channel MOS (NMOS) transistor* is shown in Figure 3-8. The terminals are called *gate, source,* and *drain.* (Note that the "gate" of a MOS transistor has nothing to do with a "logic gate.") As you might guess from the orientation of the circuit symbol, the drain is normally at a higher voltage than the source.

*n-channel MOS
(NMOS) transistor*
gate
source
drain

Figure 3-8
Circuit symbol for an
n-channel MOS (NMOS)
transistor.

The voltage from gate to source (V_{gs}) in an NMOS transistor is normally zero or positive. If $V_{gs} = 0$, then the resistance from drain to source (R_{ds}) is very high, at least a megohm (10^6 ohms) or more. As we increase V_{gs} (i.e., increase the voltage on the gate), R_{ds} decreases to a very low value, 10 ohms or less in some devices.

The circuit symbol for a *p-channel MOS (PMOS) transistor* is shown in Figure 3-9. Operation is analogous to that of an NMOS transistor, except that the source is normally at a higher voltage than the drain, and V_{gs} is normally zero or negative. If V_{gs} is zero, then the resistance from source to drain (R_{ds}) is very high. As we algebraically decrease V_{gs} (i.e., *decrease* the voltage on the gate), R_{ds} decreases to a very low value.

*p-channel MOS
(PMOS) transistor*

The gate of a MOS transistor has a very high impedance. That is, the gate is separated from the source and the drain by an insulating material with a very high resistance. However, the gate voltage creates an electric field that enhances or retards the flow of current between source and drain. This is the "field effect" in the "MOSFET" name.

Regardless of gate voltage, almost no current flows from the gate to source, or from the gate to drain for that matter. The resistance between the gate and the

Figure 3-9
Circuit symbol for a
p-channel MOS (PMOS)
transistor.

leakage current

other terminals of the device is extremely high, well over a megohm. The small amount of current that flows across this resistance is very small, typically less than one microampere (μA, 10^{-6} A), and is called a *leakage current*.

The MOS transistor symbol itself reminds us that there is no connection between the gate and the other two terminals of the device. However, the gate of a MOS transistor is capacitively coupled to the source and drain, as the symbol might suggest. In high-speed circuits, the power needed to charge and discharge this capacitance on each input-signal transition accounts for a nontrivial portion of a circuit's power consumption.

3.3.3 Basic CMOS Inverter Circuit

CMOS logic

NMOS and PMOS transistors are used together in a complementary way to form *CMOS logic*. The simplest CMOS circuit, a logic inverter, requires only one of each type of transistor, connected as shown in Figure 3-10(a). The power-supply voltage, V_{DD}, typically may be in the range 2–6 V and is most often set at 5.0 V for compatibility with TTL circuits.

Ideally, the functional behavior of the CMOS inverter circuit can be characterized by just two cases tabulated in Figure 3-10(b):

1. V_{IN} is 0.0 V. In this case, the bottom, *n*-channel transistor *Q1* is off, since its V_{gs} is 0, but the top, *p*-channel transistor *Q2* is on, since its V_{gs} is a large negative value (−5.0 V). Therefore, *Q2* presents only a small resistance between the power-supply terminal (V_{DD}, +5.0 V) and the output terminal (V_{OUT}), and the output voltage is 5.0 V.

Figure 3-10
CMOS inverter:
(a) circuit diagram;
(b) functional behavior;
(c) logic symbol.

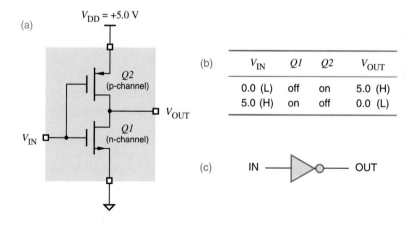

V_{IN}	$Q1$	$Q2$	V_{OUT}
0.0 (L)	off	on	5.0 (H)
5.0 (H)	on	off	0.0 (L)

2. V_{IN} is 5.0 V. Here, $Q1$ is on, since its V_{gs} is a large positive value (+5.0 V), but $Q2$ is off, since its V_{gs} is 0. Thus, $Q1$ presents a small resistance between the output terminal and ground, and the output voltage is 0 V.

With the foregoing functional behavior, the circuit clearly behaves as a logical inverter, since a 0-volt input produces a 5-volt output, and vice versa

Another way to visualize CMOS operation uses switches. As shown in Figure 3-11(a), the n-channel (bottom) transistor is modeled by a normally-open switch, and the p-channel (top) transistor by a normally-closed switch. Applying a HIGH voltage changes each switch to the opposite of its normal state, as shown in (b).

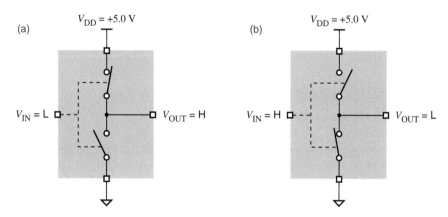

Figure 3-11
Switch model for CMOS inverter: (a) LOW input; (b) HIGH input.

The switch model gives rise to a way of drawing CMOS circuits that makes their logical behavior more readily apparent. As shown in Figure 3-12, different symbols are used for the p- and n-channel transistors to reflect their logical behavior. The n-channel transistor ($Q1$) is switched "on," and current flows between source and drain, when a HIGH voltage is applied to its gate; this seems natural enough. The p-channel transistor ($Q2$) has the opposite behavior. It is

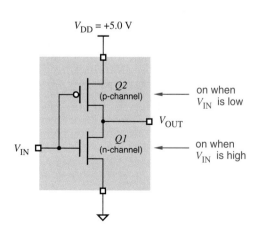

Figure 3-12
CMOS inverter logical operation.

"on" when a LOW voltage is applied; the inversion bubble on its gate indicates this inverting behavior.

3.3.4 CMOS NAND and NOR Gates

Both NAND and NOR gates can be constructed using CMOS. A k-input gate uses k p-channel and k n-channel transistors. Figure 3-13 shows a 2-input CMOS NAND gate. If either input is LOW, the output Z has a low-impedance connection to V_{DD} through the corresponding "on" p-channel transistor, and the path to ground is blocked by the corresponding "off" n-channel transistor. If both inputs are HIGH, the path to V_{DD} is blocked, and Z has a low-impedance connection to ground. Figure 3-14 shows the switch model for the NAND gate's operation.

Figure 3-15 shows a CMOS NOR gate. If both inputs are LOW, then the output Z has a low-impedance connection to V_{DD} through the "on" p-channel transistors, and the path to ground is blocked by the "off" n-channel transistors. If either input is HIGH, the path to V_{DD} is blocked, and Z has a low-impedance connection to ground.

Figure 3-13
CMOS 2-input NAND gate:
(a) circuit diagram;
(b) function table;
(c) logic symbol.

A	B	Q1	Q2	Q3	Q4	Z
L	L	off	on	off	on	H
L	H	off	on	on	off	H
H	L	on	off	off	on	H
H	H	on	off	on	off	L

NAND VS. NOR   CMOS NAND and NOR gates do not have identical performance. For a given silicon area, an *n*-channel transistor has lower "on" resistance than a *p*-channel transistor. Therefore, when transistors are put in series, *k* *n*-channel transistors have lower "on" resistance than do *k* *p*-channel ones. As a result, a *k*-input NAND gate is generally faster than and preferred over a *k*-input NOR gate.

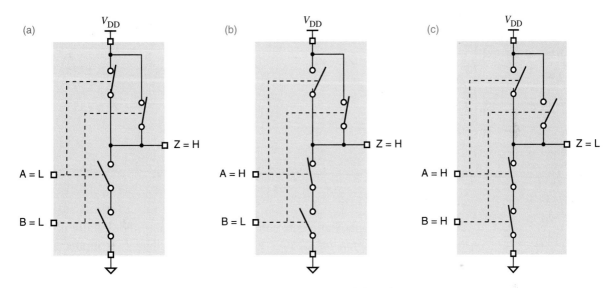

Figure 3-14 Switch model for CMOS 2-input NAND gate: (a) both inputs LOW; (b) one input HIGH; (c) both inputs HIGH.

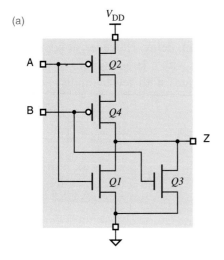

A	B	*Q1*	*Q2*	*Q3*	*Q4*	Z
L	L	off	on	off	on	H
L	H	off	on	on	off	L
H	L	on	off	off	on	L
H	H	on	off	on	off	L

Figure 3-15
CMOS 2-input
NOR gate:
(a) circuit diagram;
(b) function table;
(c) logic symbol.

(a)

(b)

A	B	C	Q1	Q2	Q3	Q4	Q5	Q6	Z
L	L	L	off	on	off	on	off	on	H
L	L	H	off	on	off	on	on	off	H
L	H	L	off	on	on	off	off	on	H
L	H	H	off	on	on	off	on	off	H
H	L	L	on	off	off	on	off	on	H
H	L	H	on	off	off	on	on	off	H
H	H	L	on	off	on	off	off	on	H
H	H	H	on	off	on	off	on	off	L

(c)

Figure 3-16 CMOS 3-input NAND gate: (a) circuit diagram; (b) function table; (c) logic symbol.

3.3.5 Fan-In

fan-in

The number of inputs that a gate can have in a particular logic family is called the logic family's *fan-in*. CMOS gates with more than two inputs can be obtained by extending series-parallel designs on Figures 3-13 and 3-15 in the obvious manner. For example, Figure 3-16 shows a 3-input CMOS NAND gate.

In principle, you could design a CMOS NAND or NOR gate with a very large number of inputs. In practice, however, the additive "on" resistance of series transistors limits the fan-in of CMOS gates, typically to 4 for NOR gates and 6 for NAND gates.

As the number of inputs is increased, designers of CMOS gate circuits may compensate by increasing the size of the series transistors to reduce their resistance and the corresponding switching delay. However, at some point this becomes inefficient or impractical. Gates with a large number of inputs can be made faster and smaller by cascading gates with fewer inputs. For example,

Figure 3-17
Logic diagram equivalent to the internal structure of an 8-input CMOS NAND gate.

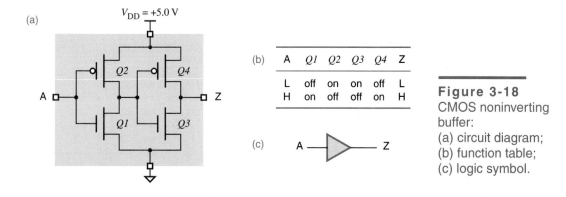

Figure 3-18
CMOS noninverting
buffer:
(a) circuit diagram;
(b) function table;
(c) logic symbol.

Figure 3-17 shows the logical structure of an 8-input CMOS NAND gate. The total delay through a 4-input NAND, a 2-input NOR, and an inverter is typically less than the delay of a one-level 8-input NAND circuit.

3.3.6 Noninverting Gates

In CMOS, and in most other logic families, the simplest gates are inverters, and the next simplest are NAND gates and NOR gates. A logical inversion comes "for free," and it typically is not possible to design a noninverting gate with a smaller number of transistors than an inverting one.

CMOS noninverting buffers and AND and OR gates are obtained by connecting an inverter to the output of the corresponding inverting gate. Thus, Figure 3-18 shows a noninverting buffer and Figure 3-19 shows an AND gate. Combining Figure 3-15(a) with an inverter yields an OR gate.

Figure 3-19 CMOS 2-input AND gate: (a) circuit diagram; (b) function table; (c) logic symbol.

(a)

(b)

A	B	C	D	Q1	Q2	Q3	Q4	Q5	Q6	Q7	Q8	Z
L	L	L	L	off	on	off	on	off	on	off	on	H
L	L	L	H	off	on	off	on	off	on	on	off	H
L	L	H	L	off	on	off	on	on	off	off	on	H
L	L	H	H	off	on	off	on	on	off	on	off	L
L	H	L	L	off	on	on	off	off	on	off	on	H
L	H	L	H	off	on	on	off	off	on	on	off	H
L	H	H	L	off	on	on	off	on	off	off	on	H
L	H	H	H	off	on	on	off	on	off	on	off	L
H	L	L	L	on	off	off	on	off	on	off	on	H
H	L	L	H	on	off	off	on	off	on	on	off	H
H	L	H	L	on	off	off	on	on	off	off	on	H
H	L	H	H	on	off	off	on	on	off	on	off	L
H	H	L	L	on	off	on	off	off	on	off	on	L
H	H	L	H	on	off	on	off	off	on	on	off	L
H	H	H	L	on	off	on	off	on	off	off	on	L
H	H	H	H	on	off	on	off	on	off	on	off	L

Figure 3-20 CMOS AND-OR-INVERT gate: (a) circuit diagram; (b) function table.

3.3.7 CMOS AND-OR-INVERT and OR-AND-INVERT Gates

AND-OR-INVERT (AOI) gate

CMOS circuits can perform two levels of logic with just a single "level" of transistors. For example, the circuit in Figure 3-20(a) is a two-wide, two-input CMOS *AND-OR-INVERT (AOI) gate*. The function table for this circuit is shown in (b) and a logic diagram for this function using AND and NOR gates is shown in Figure 3-21. Transistors can be added to or removed from this circuit to obtain an AOI function with a different number of ANDs or a different number of inputs per AND.

The contents of each of the *Q1–Q8* columns in Figure 3-20(b) depends only on the input signal connected to the corresponding transistor's gate. The last column is constructed by examining each input combination and determining whether Z is connected to V_{DD} or ground by "on" transistors for that input combination. Note that Z is never connected to *both* V_{DD} and ground for any input combination; in such a case the output would be a nonlogic value

Figure 3-21
Logic diagram for CMOS
AND-OR-INVERT gate.

(a)   (b)

A	B	C	D	Q1	Q2	Q3	Q4	Q5	Q6	Q7	Q8	Z
L	L	L	L	off	on	off	on	off	on	off	on	H
L	L	L	H	off	on	off	on	off	on	on	off	H
L	L	H	L	off	on	off	on	on	off	off	on	H
L	L	H	H	off	on	off	on	on	off	on	off	H
L	H	L	L	off	on	on	off	off	on	off	on	H
L	H	L	H	off	on	on	off	off	on	on	off	L
L	H	H	L	off	on	on	off	off	off	off	on	L
L	H	H	H	off	on	on	off	off	off	on	off	L
H	L	L	L	on	off	off	on	off	on	off	on	H
H	L	L	H	on	off	off	on	off	on	on	off	L
H	L	H	L	on	off	off	on	on	off	off	on	L
H	L	H	H	on	off	off	on	on	off	on	off	L
H	H	L	L	on	off	on	off	off	on	off	on	H
H	H	L	H	on	off	on	off	off	on	on	off	L
H	H	H	L	on	off	on	off	on	off	off	on	L
H	H	H	H	on	off	on	off	on	off	on	off	L

Figure 3-22 CMOS OR-AND-INVERT gate: (a) circuit diagram; (b) function table.

somewhere between LOW and HIGH, and the output structure would consume excessive power due to the low-impedance connection between V_{DD} and ground.

A circuit can also be designed to perform an OR-AND-INVERT function. For example, Figure 3-22(a) is a two-wide, two-input CMOS *OR-AND-INVERT (OAI) gate*. The function table for this circuit is shown in (b); the values in each column are determined just as we did for the CMOS AOI gate. A logic diagram for the OAI function using OR and NAND gates is shown in Figure 3-23.

OR-AND-INVERT (AOI) gate

The speed and other electrical characteristics of a CMOS AOI or OAI gate are quite comparable to those of a single CMOS NAND or NOR gate. As a result, these gates are very appealing because they can perform two levels of logic (AND-OR or OR-AND) with just one level of delay. Most digital designers don't bother to use AOI gates in their discrete designs. However, CMOS VLSI devices often use these gates internally, since many HDL synthesis tools can automatically convert AND/OR logic into AOI gates when appropriate.

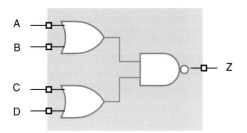

Figure 3-23
Logic diagram for CMOS OR-AND-INVERT gate.

3.4 Electrical Behavior of CMOS Circuits

engineering design margins

The next three sections discuss electrical, not logical, aspects of CMOS circuit operation. It's important to understand this material when you design real circuits using CMOS or other logic families. Most of this material is aimed at providing a framework for ensuring that the "digital abstraction" is really valid for a given circuit. In particular, a circuit or system designer must provide in a number of areas adequate engineering design margins—insurance that the circuit will work properly even under the worst of conditions.

3.4.1 Overview

The topics that we discuss in Sections 3.5–3.7 include the following:

- *Logic voltage levels.* CMOS devices operating under normal conditions are guaranteed to produce output voltage levels within well-defined LOW and HIGH ranges. And they recognize LOW and HIGH input voltage levels over somewhat wider ranges. CMOS manufacturers specify these ranges and operating conditions very carefully to ensure compatibility among different devices in the same family and to provide a degree of interoperability (if you're careful) among devices in different families.

- *DC noise margins.* Nonnegative DC noise margins ensure that the highest LOW voltage produced by an output is always lower than the highest voltage that an input can reliably interpret as LOW, and that the lowest HIGH voltage produced by an output is always higher than the lowest voltage that an input can reliably interpret as HIGH. A good understanding of noise margins is especially important in circuits that use devices from a number of different families.

- *Fanout.* This refers to the number and type of inputs that are connected to a given output. If too many inputs are connected to an output, the DC noise margins of the circuit may be inadequate. Fanout may also affect the speed at which the output changes from one state to another.

- *Speed.* The time that it takes a CMOS output to change from the LOW state to the HIGH state, or vice versa, depends on both the internal structure of the device and the characteristics of the other devices that it drives, even to the extent of being affected by the wire or printed-circuit-board traces connected to the output. We'll look at two separate components of "speed"—transition time and propagation delay.

- *Power consumption.* The power consumed by a CMOS device depends on a number of factors, including not only its internal structure, but also the input signals that it receives, the other devices that it drives, and how often its output changes between LOW and HIGH.

- *Noise.* The main reason for providing engineering design margins is to ensure proper circuit operation in the presence of noise. Noise can be generated by a number of sources; several of them are listed below, from the least likely to the (perhaps surprisingly) most likely:
  - Cosmic rays.
  - Magnetic fields from nearby machinery.
  - Power-supply disturbances.
  - The switching action of the logic circuits themselves.

- *Electrostatic discharge.* Would you believe that you can destroy a CMOS device just by touching it?

- *Open-drain outputs.* Some CMOS outputs omit the usual *p*-channel pull-up transistors. In the HIGH state, such an output behaves essentially like a "no-connection," which is useful in some applications.

- *Three-state outputs.* Some CMOS devices have an extra "output enable" control input that can be used to disable both the *p*-channel pull-up transistors and the *n*-channel pull-down transistors. Many such device outputs can be tied together to create a multisource bus, as long as the control logic is arranged so that at most one output is enabled at a time.

3.4.2 Data Sheets and Specifications

The manufacturers of real-world devices provide *data sheets* that specify the devices' logical and electrical characteristics. The electrical specifications portion of a minimal data sheet for a simple CMOS device, the 54/74HC00 quadruple NAND gate, is shown in Table 3-3. Different manufacturers typically specify additional parameters, and they may vary in how they specify even the "standard" parameters shown in the table. Thus, they usually also show the test circuits and waveforms that they use to define various parameters, for example as shown in Figure 3-24. Note that this figure contains information for some parameters in addition to those used with the 54/74HC00.

data sheet

Most of the terms in the data sheet and the waveforms in the figure are probably meaningless to you at this point. However, after reading the next three sections you should know enough about the electrical characteristics of CMOS circuits that you'll be able to understand the salient points of this or any other data sheet. As a logic designer, you'll need this knowledge to create reliable and robust real-world circuits and systems.

DON'T BE AFRAID   Computer science students and other non-EE readers should not have undue fear of the material in the next three sections. Only a basic understanding of electronics, at about the level of Ohm's law, is required.

Table 3-3 Manufacturer's data sheet for a typical CMOS device, a 54/74HC00 quad NAND gate.

DC ELECTRICAL CHARACTERISTICS OVER OPERATING RANGE

The following conditions apply unless otherwise specified:

Commercial: $T_A = -40°C$ to $+85°C$, $V_{CC} = 5.0V \pm 5\%$;   Military: $T_A = -55°C$ to $+125°C$, $V_{CC} = 5.0$ V $\pm 10\%$

Sym.	Parameter	Test Conditions[1]		Min.	Typ.[2]	Max.	Unit
V_{IH}	Input HIGH level	Guaranteed logic HIGH level		3.15	—	—	V
V_{IL}	Input LOW level	Guaranteed logic LOW level		—	—	1.35	V
I_{IH}	Input HIGH current	$V_{CC} =$ Max., $V_I = V_{CC}$		—	—	1	µA
I_{IL}	Input LOW current	$V_{CC} =$ Max., $V_I = 0$ V		—	—	−1	µA
V_{IK}	Clamp diode voltage	$V_{CC} =$ Min., $I_N = -18$ mA		—	−0.7	−1.2	V
I_{IOS}	Short-circuit current	$V_{CC} =$ Max.,[3] $V_O =$ GND		—	—	−35	mA
V_{OH}	Output HIGH voltage	$V_{CC} =$ Min., $V_{IN} = V_{IL}$	$I_{OH} = -20$ µA	4.4	4.499	—	V
			$I_{OH} = -4$ mA	3.84	4.3	—	V
V_{OL}	Output LOW voltage	$V_{CC} =$ Min. $V_{IN} = V_{IH}$	$I_{OL} = 20$ µA	—	.001	0.1	V
			$I_{OL} = 4$ mA		0.17	0.33	
I_{CC}	Quiescent power supply current	$V_{CC} =$ Max. $V_{IN} =$ GND or V_{CC}, $I_O = 0$		—	2	10	µA

SWITCHING CHARACTERISTICS OVER OPERATING RANGE, $C_L = 50$ pF

Sym.	Parameter[4]	Test Conditions		Min.	Typ.	Max.	Unit
t_{PD}	Propagation delay	A or B to Y		—	9	19	ns
C_I	Input capacitance	$V_{IN} = 0$ V		—	3	10	pF
C_{pd}	Power dissipation capacitance per gate	No load		—	22	—	pF

NOTES:

1. For conditions shown as Max. or Min., use appropriate value specified under Electrical Characteristics.

2. Typical values are at $V_{CC} = 5.0$ V, $+25°C$ ambient.

3. Not more than one output should be shorted at a time. Duration of short-circuit test should not exceed one second.

4. This parameter is guaranteed but not tested.

WHAT'S IN A NUMBER?   Two different prefixes, "74" and "54," are used in the part numbers of CMOS and TTL devices. These prefixes simply distinguish between commercial and military versions. A 74HC00 is the commercial part and the 54HC00 is the military version.

TEST CIRCUIT FOR ALL OUTPUTS

LOADING

Parameter		R_L	C_L	S1	S2
t_{en}	t_{pZH}	1 KΩ	50 pF or 150 pF	Open	Closed
	t_{pZL}			Closed	Open
t_{dis}	t_{pHZ}	1 KΩ		Open	Closed
	t_{pLZ}			Closed	Open
t_{pd}		—	50 pF or 150 pF	Open	Open

DEFINITIONS:

C_L = Load capacitance, includes jig and probe capacitance.
R_T = Termination resistance, should equal Z_{OUT} of the Pulse Generator.

SETUP, HOLD, AND RELEASE TIMES

PULSE WIDTH

PROPAGATION DELAY

THREE-STATE ENABLE AND DISABLE TIMES

Figure 3-24 Test circuits and waveforms for HC-series logic.

3.5 CMOS Steady-State Electrical Behavior

This section discusses the steady-state behavior of CMOS circuits, that is, the circuits' behavior when inputs and outputs are not changing. The next section discusses dynamic behavior, including speed and power dissipation.

3.5.1 Logic Levels and Noise Margins

The table in Figure 3-10(b) on page 88 defined the CMOS inverter's behavior only at two discrete input voltages; other input voltages may yield different output voltages. The complete input-output transfer characteristic can be

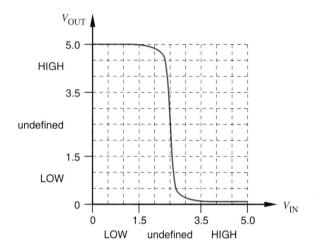

Figure 3-25
Typical input-output
transfer characteristic
of a CMOS inverter.

described by a graph such as Figure 3-25. In this graph, the input voltage is varied from 0 to 5 V, as shown on the X axis; the Y axis plots the output voltage.

If we believed the curve in Figure 3-25, we could define a CMOS LOW input level as any voltage under 2.4 V, and a HIGH input level as anything over 2.6 V. Only when the input is between 2.4 and 2.6 V does the inverter produce a nonlogic output voltage under this definition.

Unfortunately, the typical transfer characteristic shown in Figure 3-25 is just that—typical, but not guaranteed. It varies greatly under different conditions of power-supply voltage, temperature, and output loading. The transfer characteristic may even vary depending on when the device was fabricated. For example, after months of trying to figure out why gates made on some days were good and on other days were bad, legend has it that one manufacturer discovered that the bad gates were victims of airborne contamination by a particularly noxious perfume worn by one of its production-line workers!

Sound engineering practice dictates that we use more conservative specifications for LOW and HIGH. The conservative specs for a typical CMOS logic family (HC-series) are depicted in Figure 3-26. These parameters are specified by CMOS device manufacturers in data sheets like Table 3-3, and are defined as follows:

V_{OHmin} The minimum output voltage in the HIGH state.

V_{IHmin} The minimum input voltage guaranteed to be recognized as a HIGH.

V_{ILmax} The maximum input voltage guaranteed to be recognized as a LOW.

V_{OLmax} The maximum output voltage in the LOW state.

The input voltages are determined mainly by switching thresholds of the two transistors, while the output voltages are determined mainly by the "on" resistance of the transistors.

Figure 3-26
Logic levels and
noise margins
for the HC-series
CMOS logic family.

All of the parameters in Figure 3-26 are guaranteed by CMOS manufacturers over a range of temperature and output loading. Parameters are also guaranteed over a range of power-supply voltage V_{CC}, typically 5.0 V±10%.

The data sheet in Table 3-3 on page 98 specifies values for each of these parameters for HC-series CMOS. Notice that there are two values specified for V_{OHmin} and V_{OLmax}, depending on whether the output current (I_{OH} or I_{OL}) is large or small. When the device outputs are connected only to other CMOS inputs, the output current is low (e.g., $I_{OL} \leq 20\ \mu A$), so there's very little voltage drop across the output transistors. In the next few subsections we'll focus on these "pure" CMOS applications.

The power-supply voltage V_{CC} and ground are often called the *power-supply rails*. CMOS levels are typically a function of the power-supply rails: *power-supply rails*

V_{OHmin} $V_{CC} - 0.1$ V

V_{IHmin} 70% of V_{CC}

V_{ILmax} 30% of V_{CC}

V_{OLmax} ground + 0.1 V

Notice in Table 3-3 that V_{OHmin} is specified as 4.4 V. This is only a 0.1-V drop from V_{CC}, since the worst-case number is specified with V_{CC} at its minimum value of 5.0 – 10% = 4.5 V.

DC noise margin is a measure of how much noise it takes to corrupt a   *DC noise margin* worst-case output voltage into a value that may not be recognized properly by an input. For HC-series CMOS in the LOW state, V_{ILmax} (1.35 V) exceeds V_{OLmax} (0.1 V) by 1.25 V so the LOW-state DC noise margin is 1.25 V. Likewise, there is DC noise margin of 1.25 V in the HIGH state. In general, CMOS outputs have excellent DC noise margins when driving other CMOS inputs.

Regardless of the voltage applied to the input of a CMOS inverter, the input consumes very little current, only the leakage current of the two transistors' gates. The maximum amount of current that can flow is also specified by the device manufacturer:

I_{IH} The maximum current that flows into the input in the LOW state.

I_{IL} The maximum current that flows into the input in the HIGH state.

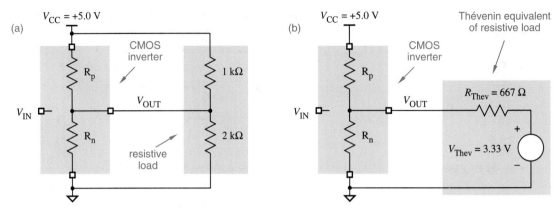

Figure 3-27 Resistive model of a CMOS inverter with a resistive load: (a) showing actual load circuit; (b) using Thévenin equivalent of load.

The input current shown in Table 3-3 for the 'HC00 is only ±1 μA. Thus, it takes very little power to maintain a CMOS input in one state or the other. This is in sharp contrast to bipolar logic circuits like TTL and ECL, whose inputs may consume significant current (and power) in one or both states.

3.5.2 Circuit Behavior with Resistive Loads

As mentioned previously, CMOS gate inputs have very high impedance and consume very little current from the circuits that drive them. There are other devices, however, which require nontrivial amounts of current to operate. When such a device is connected to a CMOS output, we call it a *resistive load* or a *DC load.* Here are some examples of resistive loads:

resistive load
DC load

- Discrete resistors may be included to provide transmission-line termination, discussed in Section 11.4.

- Discrete resistors may not really be present in the circuit, but the load presented by one or more TTL or other non-CMOS inputs may be modeled by a simple resistor network.

- The resistors may be part of or may model a current-consuming device such as a light-emitting diode (LED) or a relay coil.

When the output of a CMOS circuit is connected to a resistive load, the output behavior is not nearly as ideal as we described previously. In either logic state, the CMOS output transistor that is "on" has a nonzero resistance, and a load connected to the output terminal will cause a voltage drop across this resistance. Thus, in the LOW state, the output voltage may be somewhat higher than 0.1 V, and in the HIGH state it may be lower than 4.4 V. The easiest way to see how this happens is look at a resistive model of the CMOS circuit and load.

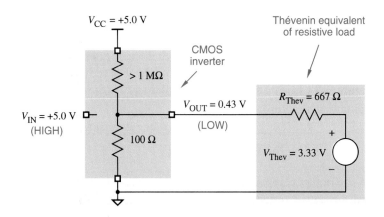

Figure 3-28
Resistive model for
CMOS LOW output
with resistive load.

Figure 3-27(a) shows the resistive model. The *p*-channel and *n*-channel transistors have resistances R_p and R_n, respectively. In normal operation, one resistance is high (> 1 MΩ) and the other is low (perhaps 100 Ω), depending on whether the input voltage is HIGH or LOW. The load in this circuit consists of two resistors attached to the supply rails; a real circuit may have any resistor values, or an even more complex resistive network. In any case, a resistive load, consisting only of resistors and voltage sources, can always be modeled by a Thévenin equivalent network, such as the one shown in Figure 3-27(b).

When the CMOS inverter has a HIGH input, the output should be LOW; the actual output voltage can be predicted using the resistive model shown in Figure 3-28. The *p*-channel transistor is "off" and has a very high resistance, high enough to be negligible in the calculations that follow. The *n*-channel

**REMEMBERING
THÉVENIN**

Any two-terminal circuit consisting of only voltage sources and resistors can be modeled by a *Thévenin equivalent* consisting of a single voltage source in series with a single resistor. The *Thévenin voltage* is the open-circuit voltage of the original circuit, and the *Thévenin resistance* is the Thévenin voltage divided by the short-circuit current of the original circuit.

In the example of Figure 3-27, the Thévenin voltage of the resistive load, including its connection to V_{CC}, is established by the 1-kΩ and 2-kΩ resistors, which form a voltage divider:

$$V_{Thev} = \frac{2\ \text{k}\Omega}{2\ \text{k}\Omega + 1\ \text{k}\Omega} \cdot 5.0\ \text{V} = 3.33\ \text{V}$$

The short-circuit current is (5.0 V)/(1 kΩ) = 5 mA, so the Thévenin resistance is (3.33 V)/(5 mA) = 667 Ω. Experienced readers may recognize this as the parallel resistance of the 1-kΩ and 2-Ω resistors.

transistor is "on" and has a low resistance, which we assume to be 100 Ω (The actual "on" resistance depends on the CMOS family and other characteristics such as operating temperature and whether or not the device was manufactured on a good day.) The "on" transistor and the Thévenin-equivalent resistor R_{Thev} in Figure 3-28 form a simple voltage divider. The resulting output voltage can be calculated as follows:

$$V_{OUT} = 3.33 \text{ V} \cdot [100/(100 + 667)]$$
$$= 0.43 \text{ V}$$

Similarly, when the inverter has a LOW input, the output should be HIGH, and the actual output voltage can be predicted with the model in Figure 3-29. We'll assume that the p-channel transistor's "on" resistance is 200 Ω. Once again, the "on" transistor and the Thévenin-equivalent resistor R_{Thev} in the figure form a simple voltage divider, and the resulting output voltage can be calculated as follows:

$$V_{OUT} = 3.33 \text{ V} + (5 \text{ V} - 3.33 \text{ V}) \cdot [667/(200 + 667)]$$
$$= 4.61 \text{ V}$$

In practice, it's seldom necessary to calculate output voltages as in the preceding examples. In fact, IC manufacturers usually don't specify the equivalent resistances of the "on" transistors, so you wouldn't have the necessary information to make the calculation anyway. Instead, IC manufacturers specify a maximum load for the output in each state (HIGH or LOW), and guarantee a worst-case output voltage for that load. The load is specified in terms of current:

I_{OLmax} The maximum current that the output can sink in the LOW state while still maintaining an output voltage no greater than V_{OLmax}.

I_{OHmax} The maximum current that the output can source in the HIGH state while still maintaining an output voltage no less than V_{OHmin}.

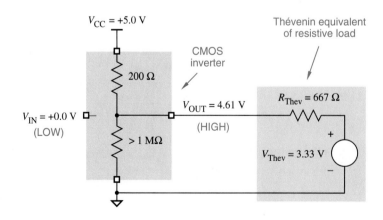

Figure 3-29
Resistive model for
CMOS HIGH output
with resistive load.

Figure 3-30 Circuit definitions of (a) I_{OLmax}; (b) I_{OHmax}.

These definitions are illustrated in Figure 3-30. A device output is said to *sink* *sinking current* current when current flows from the power supply, through the load, and through the device output to ground as in (a). The output is said to *source current* *sourcing current* when current flows from the power supply, out of the device output, and through the load to ground as in (b).

Most CMOS devices have two sets of loading specifications. One set is for "CMOS loads," where the device output is connected to other CMOS inputs, which consume very little current. The other set is for "TTL loads," where the output is connected to resistive loads such as TTL inputs or other devices that consume significant current. For example, the specifications for HC-series CMOS outputs were shown in Table 3-3 and are repeated in Table 3-4.

Notice in the table that the output current in the HIGH state is shown as a negative number. By convention, the *current flow* measured at a device terminal *current flow* is positive if positive current flows *into* the device; in the HIGH state, current flows *out* of the output terminal.

Table 3-4 Output loading specifications for HC-series CMOS with a 5-volt supply.

	CMOS Load		TTL Load	
Parameter	*Name*	*Value*	*Name*	*Value*
Maximum LOW-state output current (mA)	I_{OLmaxC}	0.02	I_{OLmaxT}	4.0
Maximum LOW-state output voltage (V)	V_{OLmaxC}	0.1	V_{OLmaxT}	0.33
Maximum HIGH-state output current (mA)	I_{OHmaxC}	−0.02	I_{OHmaxT}	−4.0
Minimum HIGH-state output voltage (V)	V_{OHminC}	4.4	V_{OHminT}	3.84

As the table shows, with CMOS loads, the CMOS gate's output voltage is maintained within 0.1 V of the power-supply rail. With TTL loads, the output voltage may degrade quite a bit. Also notice that for the same output current (±4 mA) the maximum voltage drop with respect to the power-supply rail is twice as much in the HIGH state (0.66 V) as in the LOW state (0.33 V). This suggests that the p-channel transistors in HC-series CMOS have a higher "on" resistance than the n-channel transistors do. This is natural, since in any CMOS circuit, a p-channel transistor has over twice the "on" resistance of an n-channel transistor with the same area. Equal voltage drops in both states could be obtained by making the p-channel transistors much larger than the n-channel transistors, but for various reasons this was not done.

Ohm's law can be used to determine how much current an output sources or sinks in a given situation. In Figure 3-28 on page 103, the "on" n-channel transistor modeled by a 100-Ω resistor has a 0.43-V drop across it; therefore it sinks (0.43 V)/(100 Ω) = 4.3 mA of current. Similarly, the "on" p-channel transistor in Figure 3-29 sources (0.39 V)/(200 Ω) = 1.95 mA.

The actual "on" resistances of CMOS output transistors usually aren't published, so it's not always possible to use the exact models of the previous paragraphs. However, you can estimate "on" resistances using the following equations, which rely on specifications that are always published:

$$R_{p(on)} = \frac{V_{DD} - V_{OHminT}}{|I_{OHmaxT}|}$$

$$R_{n(on)} = \frac{V_{OLmaxT}}{I_{OLmaxT}}$$

These equations use Ohm's law to compute the "on" resistance as the voltage drop across the "on" transistor divided by the current through it with a worst-case resistive load. Using the numbers given for HC-series CMOS in Table 3-4, we can calculate $R_{p(on)} = 175\ \Omega$ and $R_{n(on)} = 82.5\ \Omega$.

Very good *worst-case* estimates of output current can be made by assuming that there is *no* voltage drop across the "on" transistor. This assumption simplifies the analysis, and yields a conservative result that is almost always good enough for practical purposes. For example, Figure 3-31 shows a CMOS inverter driving the same Thévenin-equivalent load that we've used in previous examples. The resistive model of the output structure is not shown, because it is no longer needed; we assume that there is no voltage drop across the "on" CMOS transistor. In (a), with the output LOW, the entire 3.33-V Thévenin-equivalent voltage source appears across R_{Thev}, and the estimated sink current is (3.33 V)/(667 Ω) = 5.0 mA. In (b), with the output HIGH and assuming a 5.0-V supply, the voltage drop across R_{Thev} is 1.67 V, and the estimated source current is (1.67 V)/(667 Ω) = 2.5 mA.

Figure 3-31 Estimating sink and source current: (a) output LOW; (b) output HIGH.

An important feature of the CMOS inverter (or any CMOS circuit) is that the output structure by itself consumes very little current in either state, HIGH or LOW. In either state, one of the transistors is in the high-impedance "off" state. All of the current flow that we've been talking about occurs when a resistive load is connected to the CMOS output. If there's no load, then there's no current flow, and the power consumption is zero. With a load, however, current flows through both the load and the "on" transistor, and power is consumed in both.

THE TRUTH ABOUT POWER CONSUMPTION	As we've stated elsewhere, an "off" transistor's resistance is over one megohm, but it's not infinite. Therefore, a very tiny leakage current actually does flow in "off" transistors, and the CMOS output structure does have a correspondingly tiny but nonzero power consumption. In most applications, this power consumption is tiny enough to ignore. It is usually significant only in "standby mode" in battery-powered devices, such as the laptop computer on which this chapter was first prepared.

3.5.3 Circuit Behavior with Nonideal Inputs

So far, we have assumed that the HIGH and LOW inputs to a CMOS circuit are ideal voltages, very close to the power-supply rails. However, the behavior of a real CMOS inverter circuit depends on the input voltage as well as on the characteristics of the load. If the input voltage is not close to the power-supply rail, then the "on" transistor may not be fully "on" and its resistance may increase. Likewise, the "off" transistor may not be fully "off" and its resistance may be quite a bit less than one megohm. These two effects combine to move the output voltage away from the power-supply rail.

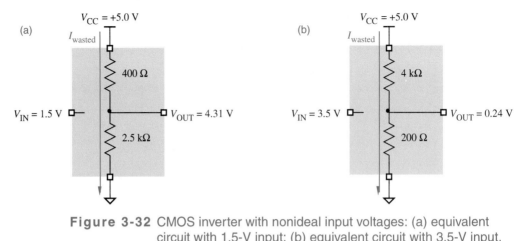

Figure 3-32 CMOS inverter with nonideal input voltages: (a) equivalent circuit with 1.5-V input; (b) equivalent circuit with 3.5-V input.

For example, Figure 3-32(a) shows a CMOS inverter's possible behavior with a 1.5-V input. The *p*-channel transistor's resistance has doubled at this point, and that the *n*-channel transistor is beginning to turn on. (These values are simply assumed for the purposes of illustration; the actual values depend on the detailed characteristics of the transistors.)

In the figure, the output at 4.31 V is still well within the valid range for a HIGH signal, but not quite the ideal of 5.0 V. Similarly, with a 3.5-V input in (b), the LOW output is 0.24 V, not 0 V. The slight degradation of output voltage is generally tolerable; what's worse is that the output structure is now consuming a nontrivial amount of power. The current flow with the 1.5-V input is

$$I_{\text{wasted}} = 5.0 \text{ V}/400 \text{ } \Omega + 2.5 \text{ k}\Omega = 1.72 \quad A \qquad m$$

and the power consumption is

$$P_{\text{wasted}} = 5.0 \text{ V} \cdot I_{\text{wasted}} = 8.62 \text{ mW}$$

The output voltage of a CMOS inverter deteriorates further with a resistive load. Such a load may exist for any of a variety of reasons discussed previously. Figure 3-33 shows a CMOS inverter's possible behavior with a resistive load. With a 1.5-V input, the output at 3.98 V is still within the valid range for a HIGH signal, but it is far from the ideal of 5.0 V. Similarly, with a 3.5-V input as shown in Figure 3-34, the LOW output is 0.93 V, not 0 V.

In "pure" CMOS systems, all of the logic devices in a circuit are CMOS. Since CMOS inputs have a very high impedance, they present very little resistive load to the CMOS outputs that drive them. Therefore, the CMOS output levels all remain very close to the power-supply rails (0 V and 5 V), and none of the devices waste power in their output structures. On the other hand, if TTL outputs or other nonideal logic signals are connected to CMOS inputs, then the CMOS

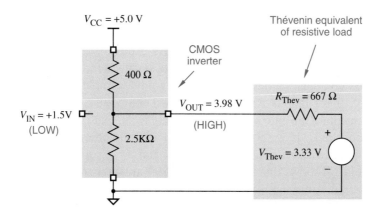

Figure 3-33
CMOS inverter with load and nonideal 1.5-V input.

outputs use power in the way depicted in this subsection; this is formalized in the box at the top of page 139. In addition, if TTL inputs or other resistive loads are connected to CMOS outputs, then the CMOS outputs use power in the way depicted in the preceding subsection.

3.5.4 Fanout

The *fanout* of a logic gate is the number of inputs that the gate can drive without *fanout* exceeding its worst-case loading specifications. The fanout depends not only on the characteristics of the output, but also on the inputs that it is driving. Fanout must be examined for both possible output states, **HIGH** and **LOW**.

For example, we showed in Table 3-4 on page 105 that the maximum **LOW**-state output current I_{OLmaxC} for an HC-series CMOS gate driving CMOS inputs is 0.02 mA (20 μA). We also stated previously that the maximum input current I_{Imax} for an HC-series CMOS input in any state is ± 1 μA. Therefore, the *LOW*-*state fanout* for an HC-series output driving HC-series inputs is 20. *LOW*-state fanout Table 3-4 also showed that the maximum **HIGH**-state output current I_{OHmaxC} is

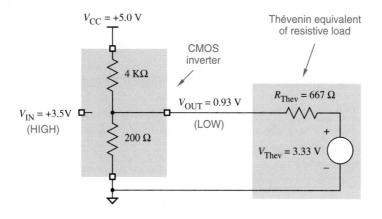

Figure 3-34
CMOS inverter with load and nonideal 3.5-V input.

HIGH-state fanout

-0.02 mA $(-20 \mu A)$ Therefore, the *HIGH-state fanout* for an HC-series output driving HC-series inputs is also 20.

overall fanout

Note that the HIGH-state and LOW-state fanouts of a gate aren't necessarily equal. In general, the *overall fanout* of a gate is the minimum of its HIGH-state and LOW-state fanouts, 20 in the foregoing example.

In the fanout example that we just completed, we assumed that we needed to maintain the gate's output at CMOS levels, that is, within 0.1 V of the power-supply rails. If we were willing to live with somewhat degraded, TTL output levels, then we could use I_{OLmaxT} and I_{OHmaxT} in the fanout calculation. Table 3-4 shows that these specifications are 4.0 mA and -4.0 mA, respectively. Therefore, the fanout of an HC-series output driving HC-series inputs at TTL levels is 4000—virtually unlimited, apparently.

DC fanout

Well, not quite. The calculations that we've just carried out give the *DC fanout,* defined as the number of inputs that an output can drive *with the output in a constant state* (HIGH or LOW). Even if the DC fanout specification is met, a CMOS output driving a large number of inputs may not behave satisfactorily on transitions, LOW-to-HIGH or vice versa.

During transitions, the CMOS output must charge or discharge the stray capacitance associated with the inputs that it drives. If this capacitance is too large, the transition from LOW to HIGH (or vice versa) may be too slow, causing improper system operation.

AC fanout

The ability of an output to charge and discharge stray capacitance is sometimes called *AC fanout,* though it is seldom calculated as precisely as DC fanout. As you'll see in Section 3.6.1, it's more a matter of deciding how much speed degradation you're willing to live with.

3.5.5 Effects of Loading

Loading an output beyond its rated fanout has several effects:

- In the LOW state, the output voltage (V_{OL}) may increase beyond V_{OLmax}.

- In the HIGH state, the output voltage (V_{OH}) may fall below V_{OHmin}.

- Propagation delay to the output may increase beyond specifications.

- Output rise and fall times may increase beyond their specifications.

- The operating temperature of the device may increase, thereby reducing the reliability of the device and eventually causing device failure.

The first four effects reduce the DC noise margins and timing margins of the circuit. Thus, a slightly overloaded circuit may work properly in ideal conditions, but experience says that it will fail once it's out of the friendly environment of the engineering lab.

Figure 3-35 Unused inputs: (a) tied to another input; (b) NAND pulled up; (c) NOR pulled down.

3.5.6 Unused Inputs

Sometimes not all of the inputs of a logic gate are used. In a real design problem, you may need an n-input gate but have only an $(n + 1)$-input gate available. Tying together two inputs of the $(n + 1)$-input gate gives it the functionality of an n-input gate. You can convince yourself of this fact intuitively now, or use switching algebra to prove it after you've studied Section 4.1. Figure 3-35(a) shows a NAND gate with its inputs tied together.

You can also tie unused inputs to a constant logic value. An unused AND or NAND input should be tied to logic 1, as in (b), and an unused OR or NOR input should be tied to logic 0, as in (c). In high-speed circuit design, it's usually better to use method (b) or (c) rather than (a), which increases the capacitive load on the driving signal and may slow things down. In (b) and (c), a resistor value in the range 1–10 kΩ is typically used, and a single pull-up or pull-down resistor can serve multiple unused inputs. It is also possible to tie unused inputs directly to the appropriate power-supply rail.

Unused CMOS inputs should never be left unconnected (or *floating*). On one hand, such an input will behave as if it had a LOW signal applied to it and will normally show a value of 0 V when probed with an oscilloscope or voltmeter. So you might think that an unused OR or NOR input can be left floating, because it will act as if a logic 0 is applied and not affect the gate's output.

floating input

SUBTLE BUGS Floating CMOS inputs are often the cause of mysterious circuit behavior, as an unused input erratically changes its effective state based on noise and conditions elsewhere in the circuit. When you're trying to debug such a problem, the extra capacitance of an oscilloscope probe touched to the floating input is often enough to damp out the noise and make the problem go away. This can be especially baffling if you don't realize that the input is floating!

However, since CMOS inputs have such high impedance, it takes only a small amount of circuit noise to temporarily make a floating input look HIGH, creating some very nasty intermittent circuit failures.

3.5.7 Current Spikes and Decoupling Capacitors

When a CMOS output switches between LOW and HIGH, current flows from V_{CC} to ground through the partially-on p- and n-channel transistors. These currents, often called *current spikes* because of their brief duration, may show up as noise on the power-supply and ground connections in a CMOS circuit, especially when multiple outputs are switched simultaneously.

current spikes

decoupling capacitors

For this reason, systems that use CMOS circuits require *decoupling capacitors* between V_{CC} and ground. These capacitors must be distributed throughout the circuit, at least one within an inch or so of each chip, to supply current during transitions. The large *filtering capacitors* typically found in the power supply itself don't satisfy this requirement, because stray wiring inductance prevents them from supplying the current fast enough, hence the need for a *physically distributed* system of decoupling capacitors.

filtering capacitors

3.5.8 How to Destroy a CMOS Device

Hit it with a sledge hammer. Or simply walk across a carpet and then touch an input pin with your finger. Because CMOS device inputs have such high impedance, they are subject to damage from *electrostatic discharge (ESD)*.

electrostatic discharge (ESD)

ESD occurs when a buildup of charge on one surface arcs through a dielectric to another surface with the opposite charge. In the case of a CMOS input, the dielectric is the insulation between an input transistor's gate and its source and drain. ESD may damage this insulation, causing a short-circuit between the device's input and output.

The input structures of modern CMOS devices use various measures to reduce their susceptibility to ESD damage, but no device is completely immune. Therefore, to protect CMOS devices from ESD damage during shipment and handling, manufacturers normally package their devices in conductive bags, tubes, or foam. To prevent ESD damage when handling loose CMOS devices, circuit assemblers and technicians usually wear conductive wrist straps that are connected by a coil cord to earth ground; this prevents a static charge from building up on their bodies as they move around the factory or lab.

latch-up

Once a CMOS device is installed in a system, another possible source of damage is *latch-up*. The physical input structure of just about any CMOS device contains parasitic bipolar transistors between V_{CC} and ground configured as a "silicon-controlled rectifier (SCR)." In normal operation, this "parasitic SCR" has no effect on device operation. However, an input voltage that is less than ground or more than V_{CC} can "trigger" the SCR, creating a virtual short-circuit between V_{CC} and ground. Once the SCR is triggered, the only way to turn it off

ELIMINATE RUDE, SHOCKING BEHAVIOR!	Some design engineers consider themselves above such inconveniences, but to be safe you should follow several ESD precautions in the lab:

- Before handling a CMOS device, touch the grounded metal case of a plugged-in instrument or another source of earth ground.

- Before transporting a CMOS device, insert it in conductive foam.

- When carrying a circuit board containing CMOS devices, handle the board by the edges, and touch a ground terminal on the board to earth ground before poking around with it.

- When handing over a CMOS device to a partner, especially on a dry winter day, touch the partner first. He or she will thank you for it.

is to turn off the power supply. Before you have a chance to do this, enough power may be dissipated to destroy the device (i.e., you may see smoke).

One possible trigger for latch-up is "undershoot" on high-speed HIGH-to-LOW signal transitions, discussed in Section 11.4. In this situation, the input signal may go several volts below ground for several nanoseconds before settling into the normal LOW range. However, modern CMOS logic circuits are fabricated with special structures that prevent latch-up in this transient case.

Latch-up can also occur when CMOS inputs are driven by the outputs of another system or subsystem with a separate power supply. If a HIGH input is applied to a CMOS gate before power is present, the gate may come up in the "latched-up" state when power is applied. Again, modern CMOS logic circuits are fabricated with special structures that prevent this in most cases. However, if the driving output is capable of sourcing lots of current (e.g., tens of mA), latch-up is still possible. One solution to this problem is to apply power before hooking up input cables.

3.6 CMOS Dynamic Electrical Behavior

Both the speed and the power consumption of a CMOS device depend to a large extent on AC or dynamic characteristics of the device and its load, that is, what happens when the output changes between states. As part of the internal design of CMOS ASICs, logic designers must carefully examine the effects of output loading and redesign where the load is too high. Even in board-level design, the effects of loading must be considered for clocks, buses, and other signals that have high fanout or long interconnections.

Speed depends on two characteristics, transition time and propagation delay, discussed in the next two subsections. Power dissipation is discussed in the third subsection.

Figure 3-36
Transition times:
(a) ideal case of
zero-time switching;
(b) a more realistic
approximation;
(c) actual timing,
showing rise and fall
times.

3.6.1 Transition Time

transition time

The amount of time that the output of a logic circuit takes to change from one state to another is called the *transition time*. Figure 3-36(a) shows how we might like outputs to change state—in zero time. However, real outputs cannot change instantaneously, because they need time to charge the stray capacitance of the wires and other components that they drive. A more realistic view of a circuit's output is shown in (b). An output takes a certain time, called the *rise time* (t_r), to change from LOW to HIGH, and a possibly different time, called the *fall time* (t_f), to change from HIGH to LOW.

rise time (t_r)
fall time (t_f)

Even Figure 3-36(b) is not quite accurate, because the rate of change of the output voltage does not change instantaneously, either. Instead, the beginning and the end of a transition are smooth, as shown in (c). To avoid difficulties in defining the endpoints, rise and fall times are normally measured at the boundaries of the valid logic levels as indicated in the figure.

With the convention in (c), the rise and fall times indicate how long an output voltage takes to pass through the "undefined" region between LOW and HIGH. The initial part of a transition is not included in the rise- or fall-time number. Instead, the initial part of a transition contributes to the "propagation delay" number discussed in the next subsection.

The rise and fall times of a CMOS output depend mainly on two factors, the "on" transistor resistance and the load capacitance. A large capacitance increases transition times; since this is undesirable, it is very rare for a logic designer to purposely connect a capacitor to a logic circuit's output. However, *stray capacitance* is present in every circuit; it comes from at least three sources:

stray capacitance

1. Output circuits, including a gate's output transistors, internal wiring, and packaging, have some capacitance associated with them, in the range of 2–10 picofarads (pF) in typical logic families, including CMOS.

2. The wiring that connects an output to other inputs has capacitance, about 1 pF per inch or more, depending on the wiring technology.

3. Input circuits, including transistors, internal wiring, and packaging, have capacitance, from 2 to 15 pF per input in typical logic families.

Stray capacitance is sometimes called a *capacitive load* or an *AC load*. *capacitive load*
  AC load

A CMOS output's rise and fall times can be analyzed using the equivalent circuit shown in Figure 3-37. As in the preceding section, the *p*-channel and *n*-channel transistors are modeled by resistances R_p and R_n, respectively. In normal operation, one resistance is high and the other is low, depending on the output's state. The output's load is modeled by an *equivalent load circuit* with three components: *equivalent load circuit*

R_L, V_L These two components represent the DC load. They determine the voltages and currents that are present when the output has settled into a stable HIGH or LOW state. The DC load doesn't have too much effect on transition times when the output changes states.

C_L This capacitance represents the AC load. It determines the voltages and currents that are present while the output is changing, and how long it takes to change from one state to the other

When a CMOS output drives only CMOS inputs, the DC load is negligible. To simplify matters, we'll analyze only this case, with $R_L = \infty$ and $V_L = 0$, in the remainder of this subsection. The presence of a nonnegligible DC load would affect the results, but not dramatically (see Exercise 3.68).

We can now analyze the transition times of a CMOS output. For the purpose of this analysis, we'll assume $C_L = 100$ pF, a moderate capacitive load. Also, we'll assume that the "on" resistances of the *p*-channel and *n*-channel transistors are 200 Ω and 100 Ω, respectively, as in the preceding subsection. The rise and fall times depend on how long it takes to charge or discharge the capacitive load C_L.

Figure 3-37
Equivalent circuit for analyzing transition times of a CMOS output.

Figure 3-38 Model of a CMOS HIGH-to-LOW transition: (a) in the HIGH state; (b) after *p*-channel transistor turns off and *n*-channel transistor turns on.

First, we'll look at fall time. Figure 3-38(a) shows the electrical conditions in the circuit when the output is in a steady HIGH state. (R_L and V_L are not drawn; they have no effect, since we assume $R_L = \infty$.) For the purposes of our analysis, we'll assume that when CMOS transistors change between "on" and "off," they do so instantaneously. We'll assume that at time $t = 0$ the CMOS output changes to the LOW state, resulting in the situation depicted in (b).

At time $t = 0$, V_{OUT} is still 5.0 V. (A useful electrical engineering maxim is that the voltage across a capacitor cannot change instantaneously.) At time $t = \infty$, the capacitor must be fully discharged and V_{OUT} will be 0 V. In between, the value of V_{OUT} is governed by an exponential law:

$$V_{OUT} = V_{DD} \cdot e^{-t/R_n C_L}$$

$$= 5.0 \cdot e^{-t(100 \cdot 100 \cdot 10^{-12})}$$

$$= 5.0 \cdot e^{-t/(10 \cdot 10^{-9})} \text{ V}$$

RC time constant

The factor $R_n C_L$ has units of seconds and is called an *RC time constant*. The preceding calculation shows that the *RC* time constant for HIGH-to-LOW transitions is 10 nanoseconds (ns).

Figure 3-39 plots V_{OUT} as a function of time. To calculate fall time, recall that 1.5 V and 3.5 V are the defined boundaries for LOW and HIGH levels for CMOS inputs being driven by the CMOS output. To obtain the fall time, we must solve the preceding equation for $V_{OUT} = 3.5$ and $V_{OUT} = 1.5$, yielding:

$$t = -R_n C_L \cdot \ln \frac{V_{OUT}}{V_{DD}} = -10 \cdot 10^{-9} \cdot \ln \frac{V_{OUT}}{5.0}$$

$$t_{3.5} = 3.57 \text{ ns}$$

$$t_{1.5} = 12.04 \text{ ns}$$

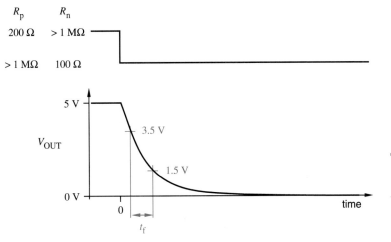

Figure 3-39
Fall time for a HIGH-to-LOW transition of a CMOS output.

The fall time t_f is the difference between these two numbers, or about 8.5 ns.

Rise time can be calculated in a similar manner. Figure 3-40(a) shows the conditions in the circuit when the output is in a steady **LOW** state. If at time $t = 0$ the CMOS output changes to the **HIGH** state, the situation depicted in (b) results. Once again, V_{OUT} cannot change instantly, but at time $t = \infty$, the capacitor will be fully charged and V_{OUT} will be 5.0 V. Once again, the value of V_{OUT} in between is governed by an exponential law:

$$
\begin{aligned}
V_{OUT} &= V_{DD} \cdot (1 - e^{-t/R_p C_L}) \\
&= 5.0 \cdot (1 - e^{-t(200 \cdot 100 \cdot 10^{-12})}) \\
&= 5.0 \cdot (1 - e^{-t/(20 \cdot 10^{-9})}) \text{ V}
\end{aligned}
$$

Figure 3-40 Model of a CMOS LOW-to-HIGH transition: (a) in the LOW state; (b) after *n*-channel transistor turns off and *p*-channel transistor turns on.

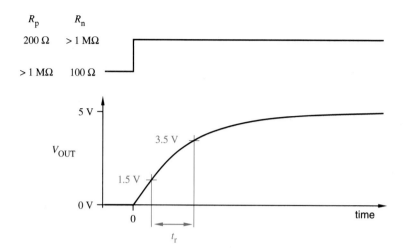

Figure 3-41
Rise time for a LOW-
to-HIGH transition of
a CMOS output.

The RC time constant in this case is 20 ns. Figure 3-41 plots V_{OUT} as a function of time. To obtain the rise time, we must solve the preceding equation for $V_{OUT} = 1.5$ and $V_{OUT} = 3.5$, yielding

$$t = -RC \cdot \ln \frac{V_{DD} - V_{OUT}}{V_{DD}}$$

$$= -20 \cdot 10^{-9} \cdot \ln \frac{5.0 - V_{OUT}}{5.0}$$

$$t_{1.5} = 7.13 \text{ ns}$$

$$t_{3.5} = 24.08 \text{ ns}$$

The rise time t_r is the difference between these two numbers, or about 17 ns.

The foregoing example assumes that the p-channel transistor has twice the resistance of the n-channel one, and as a result the rise time is twice as long as the fall time. It takes longer for the "weak" p-channel transistor to pull the output up than it does for the "strong" n-channel transistor to pull it down; the output's drive capability is "asymmetric." High-speed CMOS devices are sometimes fabricated with larger p-channel transistors to make the transition times more nearly equal and output drive more symmetric.

Regardless of the transistors' characteristics, an increase in load capacitance causes an increase in the RC time constant and a corresponding increase in the transition times of the output. Thus, it is a goal of high-speed circuit designers to minimize load capacitance, especially on the most timing-critical signals. This can be done by minimizing the number of inputs driven by the signal, by creating multiple copies of the signal, and by careful physical layout of the circuit.

When working with real digital circuits, it's often useful to estimate transition times, without going through a detailed analysis. A useful rule of thumb is that the transition time approximately equals the *RC* time constant of the charging or discharging circuit. For example, estimates of 10 and 20 ns for fall and rise time in the preceding example would have been pretty much on target, especially considering that most assumptions about load capacitance and transistor "on" resistances are approximate to begin with.

Manufacturers of commercial CMOS circuits typically do not specify transistor "on" resistances on their data sheets. If you search carefully, you might find this information published in the manufacturers' application notes. In any case, you can estimate an "on" resistance as the voltage drop across the "on" transistor divided by the current through it with a worst-case resistive load, as we showed in Section 3.5.2:

$$R_{p(on)} = \frac{V_{DD} - V_{OHminT}}{\left| I_{OHmaxT} \right|}$$

$$R_{n(on)} = \frac{V_{OLmaxT}}{\left| I_{OLmaxT} \right|}$$

THERE'S A
CATCH!   Calculated transition times are actually quite sensitive to the choice of logic levels. In the examples in this subsection, if we used 2.0 V and 3.0 V instead of 1.5 V and 3.5 V as the thresholds for **LOW** and **HIGH**, we would calculate shorter transition times. On the other hand, if we used 0.0 and 5.0 V, the calculated transition times would be infinity! You should also be aware that in some logic families (most notably TTL), the thresholds are not symmetric around the voltage midpoint. Still, it is the author's experience that the "time-constant-equals-transition-time" rule of thumb usually works for practical circuits.

3.6.2 Propagation Delay

Rise and fall times only partially describe the dynamic behavior of a logic element; we need additional parameters to relate output timing to input timing. A *signal path* is the electrical path from a particular input signal to a particular output signal of a logic element. The *propagation delay* t_p of a signal path is the amount of time that it takes for a change in the input signal to produce a change in the output signal.

signal path

propagation delay t_p

A complex logic element with multiple inputs and outputs may specify a different value of t_p for each different signal path. Also, different values may be specified for a particular signal path, depending on the direction of the output change. Ignoring rise and fall times, Figure 3-42(a) shows two different

(a)

(b)

Figure 3-42
Propagation delays
for a CMOS inverter:
(a) ignoring rise and
fall times; (b) measured at
midpoints of transitions.

propagation delays for the input-to-output signal path of a CMOS inverter, depending on the direction of the output change:

t_{pHL}

t_{pHL} The time between an input change and the corresponding output change when the output is changing from HIGH to LOW.

t_{pLH}

t_{pLH} The time between an input change and the corresponding output change when the output is changing from LOW to HIGH.

Several factors lead to nonzero propagation delays. In a CMOS device, the rate at which transistors change state is influenced both by the semiconductor physics of the device and by the circuit environment, including input-signal transition rate, input capacitance, and output loading. Multistage devices such as noninverting gates or more complex logic functions may require several internal transistors to change state before the output can change state. And even when the output begins to change state, with nonzero rise and fall times it takes quite some time to cross the region between states, as we showed in the preceding subsection. All of these factors are included in propagation delay.

To factor out the effect of rise and fall times, manufacturers usually specify propagation delays at the midpoints of input and output transitions, as shown in Figure 3-42(b). However, sometimes the delays are specified at the logic-level boundary points, especially if the device's operation may be adversely affected by slow rise and fall times. For example, Figure 3-43 shows how the minimum input pulse width for an SR latch (discussed in Section 7.2.1) might be specified.

Figure 3-43
Worst-case timing
specified using logic-
level boundary points.

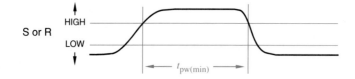

In addition, a manufacturer may specify absolute maximum input rise and fall times that must be satisfied to guarantee proper operation. High-speed CMOS circuits may consume excessive current or oscillate if their input transitions are too slow.

3.6.3 Power Consumption

The power consumption of a CMOS circuit whose output is not changing is called *static power dissipation* or *quiescent power dissipation*. (The words *consumption* and *dissipation* are used pretty much interchangeably when discussing how much power a device uses.) Most CMOS circuits have very low static power dissipation. This is what makes them so attractive for laptop computers and other low-power applications—when computation pauses, very little power is consumed. A CMOS circuit consumes significant power only during transitions; this is called *dynamic power dissipation.*

static power dissipation

quiescent power dissipation

dynamic power dissipation

One source of dynamic power dissipation is the partial short-circuiting of the CMOS output structure. When the input voltage is not close to one of the power supply rails (0 V or V_{CC}), both the *p*-channel and *n*-channel output transistors may be partially "on," creating a series resistance of 600 Ω or less. In this case, current flows through the transistors from V_{CC} to ground. The amount of power consumed in this way depends on both the value of V_{CC} and the rate at which output transitions occur, according to the formula

$$P_T = C_{PD} \cdot V_{CC}^2 \cdot f$$

The following variables are used in the formula:

P_T The circuit's internal power dissipation due to output transitions.

V_{CC} The power-supply voltage. As all electrical engineers know, power dissipation across a resistive load (the partially-on transistors) is proportional to the *square* of the voltage.

f The *transition frequency* of the output signal. This specifies the number of power-consuming output transitions per second. (But note that frequency is defined as the number of transitions divided by 2.)

transition frequency

C_{PD} The *power-dissipation capacitance*. This constant, normally specified by the device manufacturer, completes the formula. C_{PD} turns out to have units of capacitance, but does not represent an actual output capacitance. Rather, it embodies the dynamics of current flow through the changing output-transistor resistances during a single pair of output transitions, HIGH-to-LOW and LOW-to-HIGH. For example, C_{PD} for HC-series CMOS gates is typically 20–24 pF, even though the actual output capacitance is much less.

power-dissipation capacitance

The P_T formula is valid only if input transitions are fast enough, leading to fast output transitions. If the input transitions are too slow, then the output transistors stay partially on for a longer time, and power consumption increases. Device manufacturers usually recommend a maximum input rise and fall time, below which the value specified for C_{PD} is valid.

C_L

A second, and often more significant, source of CMOS power consumption is the capacitive load (C_L) on the output. During a LOW-to-HIGH transition, current flows through a p-channel transistor to charge C_L. Likewise, during a HIGH-to-LOW transition, current flows through an n-channel transistor to discharge C_L. In each case, power is dissipated in the "on" resistance of the transistor. We'll use P_L to denote the total amount of power dissipated by charging and discharging C_L.

P_L

The units of P_L are power, or energy usage per unit time. The energy for one transition could be determined by calculating the current through the charging transistor as a function of time (using the RC time constant as in Section 3.6.1), squaring this function, multiplying by the "on" resistance of the charging transistor, and integrating over time. An easier way is described below.

During a transition, the voltage across the load capacitance C_L changes by $\pm V_{CC}$. According to the definition of capacitance, the total amount of charge that must flow to make a voltage change of V_{CC} across C_L is $C_L \cdot V_{CC}$. The total amount of energy used in one transition is charge times the average voltage change. The first little bit of charge makes a voltage change of V_{CC}, while the last bit of charge makes a vanishingly small voltage change, hence the average change is $V_{CC}/2$. The total energy per transition is therefore $C_L \cdot V_{CC}^2 /2$. If there are $2f$ transitions per second, the total power dissipated due to the capacitive load is

$$P_L = C_L \cdot (V_{CC}^2 /2) \cdot 2f$$
$$= C_L \cdot V_{CC}^2 \cdot f$$

The total dynamic power dissipation of a CMOS circuit is the sum of P_T and P_L:

$$P_D = P_T + P_L$$
$$= C_{PD} \cdot V_{CC}^2 \cdot f + C_L \cdot V_{CC}^2 \cdot f$$
$$= (C_{PD} + C_L) \cdot V_{CC}^2 \cdot f$$

CV²f power

Based on this formula, dynamic power dissipation is often called *CV²f power*. In most applications of CMOS circuits, *CV²f* power is by far the major contributor to total power dissipation. Note that *CV²f* power is also consumed by bipolar logic circuits like TTL and ECL, but at low to moderate frequencies it is insignificant compared to the static (DC or quiescent) power dissipation of bipolar circuits.

Figure 3-44
CMOS transmission gate.

3.7 Other CMOS Input and Output Structures

Circuit designers have modified the basic CMOS circuit in many ways to produce gates that are tailored for specific applications. This section describes some of the more common variations in CMOS input and output structures.

3.7.1 Transmission Gates

A *p*-channel and *n*-channel transistor pair can be connected together to form a logic-controlled switch. Shown in Figure 3-44(a), this circuit is called a CMOS *transmission gate*.

transmission gate

A transmission gate is operated so that its input signals EN and EN_L are always at opposite levels. When EN is HIGH and EN_L is LOW, there is a low-impedance connection (as low as 2–5 Ω) between points A and B. When EN is LOW and /EN is HIGH, points A and B are disconnected.

Once a transmission gate is enabled, the propagation delay from A to B (or vice versa) is very short. Because of their short delays and conceptual simplicity, transmission gates are often used internally in larger-scale CMOS devices such as multiplexers and flip-flops. For example, Figure 3-45 shows how transmission gates can be used to create a "2-input multiplexer." When S is LOW, the X "input" is connected to the Z "output"; when S is HIGH, Y is connected to Z.

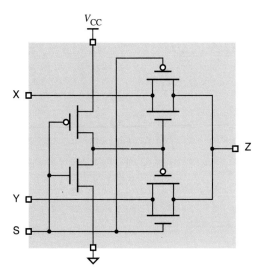

Figure 3-45
Two-input multiplexer using CMOS transmission gates.

At least one commercial manufacturer (Integrated Device Technology) makes a variety of logic functions based on transmission gates. In their multiplexer devices, it takes several nanoseconds for a change in the "select" inputs (such as in Figure 3-45) to affect the input-output path (X or Y to Z). Once a path is set up, however, the propagation delay from input to output is specified to be at most 0.25 ns; this is the fastest discrete CMOS multiplexer you can buy.

3.7.2 Schmitt-Trigger Inputs

Schmitt-trigger input

The input-output transfer characteristic for a typical CMOS gate was shown in Figure 3-25 on page 100. The corresponding transfer characteristic for a gate with *Schmitt-trigger inputs* is shown in Figure 3-46(a). A Schmitt trigger is a special circuit that uses feedback internally to shift the switching threshold depending on whether the input is changing from LOW to HIGH or from HIGH to LOW.

For example, suppose the input of a Schmitt-trigger inverter is initially at 0 V, a solid LOW. Then the output is HIGH, close to 5.0 V. If the input voltage is increased, the output will not go LOW until the input voltage reaches about 2.9 V. However, once the output is LOW, it will not go HIGH again until the input is decreased to about 2.1 V. Thus, the switching threshold for positive-going input changes, denoted V_{T+}, is about 2.9 V, and for negative-going input changes, denoted V_{T-}, is about 2.1 V. The difference between the two thresholds is called

hysteresis

hysteresis. The Schmitt-trigger inverter provides about 0.8 V of hysteresis.

To demonstrate the usefulness of hysteresis, Figure 3-47(a) shows an input signal with long rise and fall times and about 0.5 V of noise on it. An ordinary inverter, without hysteresis, has the same switching threshold for both positive-going and negative-going transitions, $V_T \approx 2.5$ V. Thus, the ordinary inverter responds to the noise as shown in (b), producing multiple output changes each time the noisy input voltage crosses the switching threshold. However, as shown in (c), a Schmitt-trigger inverter does not respond to the noise, because its hysteresis is greater than the noise amplitude.

Figure 3-46
A Schmitt-trigger inverter: (a) input-output transfer characteristic; (b) logic symbol.

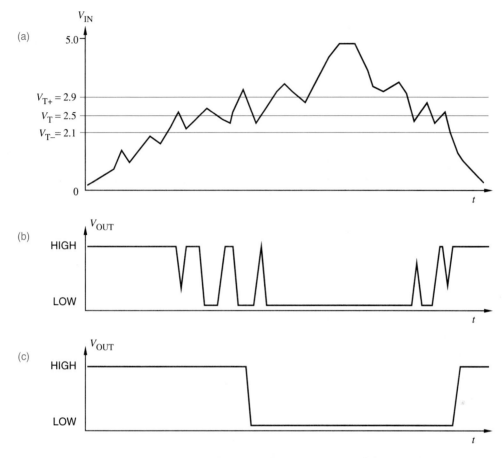

Figure 3-47 Device operation with slowly changing inputs: (a) a noisy, slowly
changing input; (b) output produced by an ordinary inverter;
(c) output produced by an inverter with 0.8 V of hysteresis.

FIXING YOUR
TRANSMISSION

Schmitt-trigger inputs have better noise immunity than ordinary gate inputs for
signals that contain transmission-line reflections, discussed in Section 11.4, or that
have long rise and fall times. Such signals typically occur in physically long connections, such as input-output buses and computer interface cables. Noise immunity is
important in these applications, since long signal lines are more likely to have
reflections or to pick up noise from adjacent signal lines, circuits, and appliances.

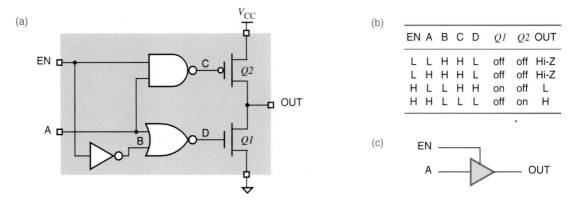

Figure 3-48 CMOS three-state buffer: (a) circuit diagram; (b) function table; (c) logic symbol.

3.7.3 Three-State Outputs

high-impedance state

Hi-Z state

floating state

Logic outputs have two normal states, LOW and HIGH, corresponding to logic values 0 and 1. However, some outputs have a third electrical state that is not a logic state at all, called the *high-impedance, Hi-Z,* or *floating state.* In this state, the output behaves as if it isn't even connected to the circuit, except for a small leakage current that may flow into or out of the output pin. Thus, an output can have one of three states—logic 0, logic 1, and Hi-Z.

three-state output

tri-state output

An output with three possible states is called (surprise!) a *three-state output* or, sometimes, a *tri-state output.* Three-state devices have an extra input, usually called "output enable" or "output disable," for placing the device's output(s) in the high-impedance state.

three-state bus

A *three-state bus* is created by wiring several three-state outputs together. Control circuitry for the "output enables" must ensure that at most one output is enabled (not in its Hi-Z state) at any time. The single enabled device can transmit logic levels (HIGH and LOW) on the bus. Examples of three-state bus design are given in Section 5.6.

three-state buffer

A circuit diagram for a CMOS *three-state buffer* is shown in Figure 3-48(a). To simplify the diagram, the internal NAND, NOR, and inverter functions are shown in functional rather than transistor form; they actually use a total of 10 transistors (see Exercise 3.80). As shown in the function table (b), when the enable (EN) input is LOW, both output transistors are off, and the output is in the Hi-Z state. Otherwise, the output is HIGH or LOW as controlled by

LEGAL NOTICE   "TRI-STATE" is a trademark of National Semiconductor Corporation. Their lawyer thought you'd like to know.

the "data" input A. Logic symbols for three-state buffers and gates are normally drawn with the enable input coming into the top, as shown in (c).

In practice, the three-state control circuit may be different from what we have shown, in order to provide proper dynamic behavior of the output transistors during transitions to and from the Hi-Z state. In particular, devices with three-state outputs are normally designed so that the output-enable delay (Hi-Z to LOW or HIGH) is somewhat longer than the output-disable delay (LOW or HIGH to Hi-Z). Thus, if a control circuit activates one device's output-enable input at the same time that it deactivates a second's, the second device is guaranteed to enter the Hi-Z state before the first places a HIGH or LOW level on the bus.

If two three-state outputs on the same bus are enabled at the same time and try to maintain opposite states, the situation is similar to tying standard active-pull-up outputs together as in Figure 3-56 on page 133—a nonlogic voltage is produced on the bus. If fighting is only momentary, the devices probably will not be damaged, but the large current drain through the tied outputs can produce noise pulses that affect circuit behavior elsewhere in the system.

There is a leakage current of up to 10 μA associated with a CMOS three-state output in its Hi-Z state. This current, as well as the input currents of receiving gates, must be taken into account when calculating the maximum number of devices that can be placed on a three-state bus. That is, in the LOW or HIGH state, an enabled three-state output must be capable of sinking or sourcing up to 10 μA of leakage current for every other three-state output on the bus, as well as handling the current required by every input on the bus. As with standard CMOS logic, separate LOW-state and HIGH-state calculations must be made to ensure that the fanout requirements of a particular circuit configuration are met.

*3.7.4 Open-Drain Outputs

The *p*-channel transistors in CMOS output structures are said to provide *active pull-up*, since they actively pull up the output voltage on a LOW-to-HIGH transition. These transistors are omitted in gates with *open-drain outputs*, such as the NAND gate in Figure 3-49(a). The drain of the topmost *n*-channel transistor is left unconnected internally, so if the output is not LOW it is "open," as indicated in (b). The underscored diamond in the symbol in (c) is sometimes used to indicate an open-drain output. A similar structure, called an "open-collector output," is provided in TTL logic families as described in Section 3.10.5.

An open-drain output requires an external *pull-up resistor* to provide *passive pull-up* to the HIGH level. For example, Figure 3-50 shows an open-drain CMOS NAND gate, with its pull-up resistor, driving a load.

For the highest possible speed, an open-drain output's pull-up resistor should be as small as possible; this minimizes the *RC* time constant for LOW-to-

active pull-up
open-drain output

pull-up resistor
passive pull-up

---

*Throughout this book, optional sections are marked with an asterisk.

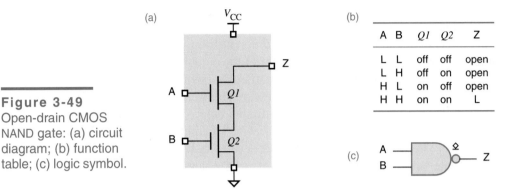

(b)

A	B	Q1	Q2	Z
L	L	off	off	open
L	H	off	on	open
H	L	on	off	open
H	H	on	on	L

Figure 3-49
Open-drain CMOS
NAND gate: (a) circuit
diagram; (b) function
table; (c) logic symbol.

HIGH transitions (rise time). However, the pull-up resistance cannot be arbitrarily small; the minimum resistance is determined by the open-drain output's maximum sink current, I_{OLmax}. For example, in HC- and HCT-series CMOS, I_{OLmax} is 4 mA, and the pull-up resistor can be no less than 5.0 V/4 mA, or 1.25 kΩ. Since this is an order of magnitude greater than the "on" resistance of the p-channel transistors in a standard CMOS gate, the LOW-to-HIGH output transitions are much slower for an open-drain gate than for standard gate with active pull-up.

As an example, let us assume that the open-drain gate in Figure 3-50 is HC-series CMOS, the pull-up resistance is 1.5 kΩ, and the load capacitance is 100 pF. We showed in Section 3.5.2 that the "on" resistance of an HC-series CMOS output in the LOW state is about 80 Ω. Thus, the RC time constant for a HIGH-to-LOW transition is about 80 Ω · 100 pF = 8 ns, and the output's fall time is about 8 ns. However, the RC time constant for a LOW-to-HIGH transition is about 1.5 kΩ · 100 pF = 150 ns, and the rise time is about 150 ns. This relatively slow rise time is contrasted with the much faster fall time in Figure 3-51. A friend of the author calls such slow rising transitions *ooze*.

ooze

Figure 3-50
Open-drain CMOS
NAND gate driving
a load.

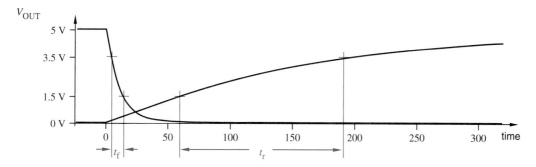

Figure 3-51 Rising and falling transitions of an open-drain CMOS output.

So why use open-drain outputs? Despite slow rise times, they can be useful in at least three applications: driving light-emitting diodes (LEDs) and other devices; performing wired logic; and driving multisource buses.

*3.7.5 Driving LEDs

An open-drain output can drive an LED as shown in Figure 3-52. If either input A or B is LOW, the corresponding n-channel transistor is off and the LED is off. When A and B are both HIGH, both transistors are on, the output Z is LOW, and the LED is on. The value of the pull-up resistor R is chosen so that the proper amount of current flows through the LED in the "on" state.

Typical LEDs require 10 mA for normal brightness. HC- and HCT-series CMOS outputs are only specified to sink or source 4 mA and are not normally used to drive LEDs. However, the outputs in advanced CMOS families such as 74AC and 74ACT can sink 24 mA or more and can be used quite effectively to drive LEDs.

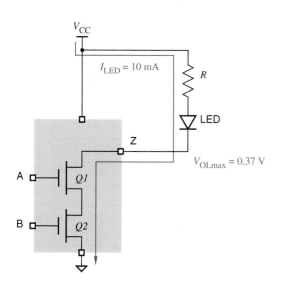

Figure 3-52
Driving an LED with an
open-drain output.

Three pieces of information are needed to calculate the proper value of the pull-up resistor R:

1. The LED current I_{LED} needed for the desired brightness, 10 mA for typical LEDs.

2. The voltage drop V_{LED} across the LED in the "on" condition, about 1.6 V for typical LEDs.

3. The output voltage V_{OL} of the open-drain output that sinks the LED current. In the 74AC and 74ACT CMOS families, V_{OLmax} is 0.37 V. If an output can sink I_{LED} and maintain a lower voltage, say 0.2 V, then the calculation below yields a resistor value that is a little too low, but normally with no harm done. A little more current than I_{LED} will flow and the LED will be just a little brighter than expected.

Using the above information, we can write the following equation:

$$V_{OL} + V_{LED} + (I_{LED} \cdot R) = V_{CC}$$

Assuming $V_{CC} = 5.0$ V and the other typical values above, we can solve for the required value of R:

$$R = \frac{V_{CC} - V_{OL} - V_{LED}}{I_{LED}}$$

$$= (5.0 - 0.37 - 1.6) \text{ V}/10 \text{ mA} = 303 \text{ }\Omega$$

Note that you don't have to use an open-drain output to drive an LED. Figure 3-53(a) shows an LED driven by ordinary an CMOS NAND-gate output with active pull-up. If both inputs are HIGH, the bottom (n-channel) transistors pull the output LOW as in the open-drain version. If either input is LOW, the output is HIGH; although one or both of the top (p-channel) transistors is on, no current flows through the LED.

With some CMOS families, you can turn an LED "on" when the output is in the HIGH state, as shown in Figure 3-53(b). This is possible if the output can source enough current to satisfy the LED's requirements. However, method (b) isn't used as often as method (a), because most CMOS and TTL outputs cannot source as much current in the HIGH state as they can sink in the LOW state.

RESISTOR VALUES   In most applications, the precise value of LED series resistors is unimportant, as long as groups of nearby LEDs have similar drivers and resistors to give equal apparent brightness. In the example in this subsection, one might use an off-the-shelf resistor value of 270, 300, or 330 ohms, whatever is readily available.

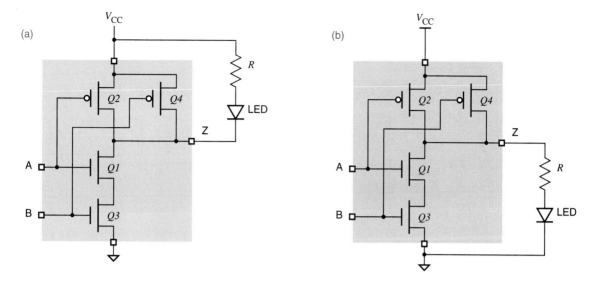

Figure 3-53 Driving an LED with an ordinary CMOS output: (a) sinking current, "on" in the LOW state; (b) sourcing current, "on" in the HIGH state.

*3.7.6 Multisource Buses

Open-drain outputs can be tied together to allow several devices, one at a time, *open-drain bus*
to put information on a common bus. At any time all but one of the outputs on
the bus are in their HIGH (open) state. The remaining output either stays in the
HIGH state or pulls the bus LOW, depending on whether it wants to transmit a
logical 1 or a logical 0 on the bus. Control circuitry selects the particular device
that is allowed to drive the bus at any time.

For example, in Figure 3-54, eight 2-input open-drain NAND-gate outputs
drive a common bus. The top input of each NAND gate is a data bit, and the

Figure 3-54 Eight open-drain outputs driving a bus.

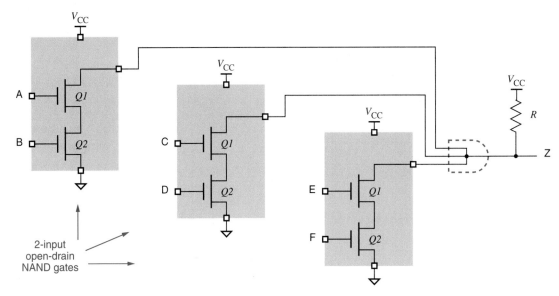

Figure 3-55 Wired-AND function on three open-drain NAND-gate outputs.

bottom input of each is a control bit. At most one control bit is HIGH at any time, enabling the corresponding data bit to be passed through to the bus. (Actually, the complement of the data bit is placed on the bus.) The other gate outputs are HIGH, that is, "open," so the data input of the enabled gate determines the value on the bus.

*3.7.7 Wired Logic

wired logic

wired AND

fighting

If the outputs of several open-drain gates are tied together with a single pull-up resistor, then *wired logic* is performed. (That's *wired*, not *weird*!) An AND function is obtained, since the wired output is HIGH if and only if all of the individual gate outputs are HIGH (actually, open); any output going LOW is sufficient to pull the wired output LOW. For example, a three-input *wired AND* function is shown in Figure 3-55. If any of the individual 2-input NAND gates has both inputs HIGH, it pulls the wired output LOW; otherwise, the pull-up resistor R pulls the wired output HIGH.

Note that wired logic cannot be performed using gates with active pull-up. Two such outputs wired together and trying to maintain opposite logic values result in a very high current flow and an abnormal output voltage. Figure 3-56 shows this situation, which is sometimes called *fighting*. The exact output voltage depends on the relative "strengths" of the fighting transistors, but with 5-V CMOS devices it is typically about 1–2 V, almost always a nonlogic voltage. Worse, if outputs fight continuously for more than a few seconds, the chips can get hot enough to sustain internal damage *and* to burn your fingers!

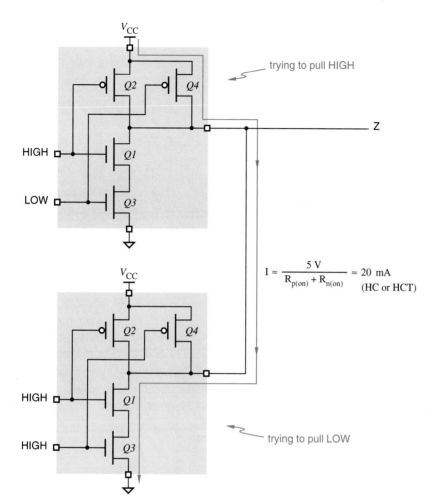

trying to pull HIGH

$$I \approx \frac{5\text{ V}}{R_{p(on)} + R_{n(on)}} \approx 20\text{ mA}$$
(HC or HCT)

trying to pull LOW

Figure 3-56
Two CMOS outputs
trying to maintain
opposite logic values
on the same line.

*3.7.8 Pull-Up Resistors

A proper choice of value for the pull-up resistor R must be made in open-drain   *pull-up- resistor*
applications. Two calculations are made to bracket the allowable values of R: *calculation*

Minimum   The sum of the current through R in the LOW state and the LOW-
            state input currents of the gates driven by the wired outputs must not
            exceed the LOW-state driving capability of the active output, 4 mA
            for HC and HCT, 24 mA for AC and ACT.

Maximum   The voltage drop across R in the HIGH state must not reduce the
            output voltage below 2.4 V, which is V_{IHmin} for typical driven gates
            plus a 400-mV noise margin. This drop is produced by the HIGH-
            state output leakage current of the wired outputs and the HIGH-state
            input currents of the driven gates.

Figure 3-57
Four open-drain
outputs driving two
inputs in the LOW
state.

For example, suppose that four HCT open-drain outputs are wired together
and drive two LS-TTL inputs (Section 3.11) as shown in Figure 3-57. A **LOW**
output must sink 0.4 mA from each LS-TTL input as well as sink the current
through the pull-up resistor R. For the total current to stay within the HCT I_{OLmax}
spec of 4 mA, the current through R may be no more than

$$I_{R(max)} = 4 - (2 \cdot 0.4) = 3.2 \text{ mA}$$

Figure 3-58
Four open-drain
outputs driving two
inputs in the HIGH
state.

Assuming that V_{OL} of the open-drain output is 0.0 V, the minimum value of R is

$$R_{min} = (5.0 - 0.0)/I_{R(max)} = 1562.5 \ \Omega$$

In the HIGH state, typical open-drain outputs have a maximum leakage current of 5 μA, and typical LS-TTL inputs require 20 μA of source current. Hence, the HIGH-state current requirement as shown in Figure 3-58 is

$$I_{R(leak)} = (4 \cdot 5) + (2 \cdot 20) = 60 \ \mu A$$

This current produces a voltage drop across R, and must not lower the output voltage below $V_{OHmin} = 2.4$ V; thus the maximum value of R is

$$R_{max} = (5.0 - 2.4)/I_{R(leak)} = 43.3 \ \Omega$$

Hence, any value of R between 1562.5 Ω and 43.3 kΩ may be used. Higher values reduce power consumption and improve the LOW-state noise margin, while lower values increase power consumption but improve both the HIGH-state noise margin and the speed of LOW-to-HIGH output transitions.

OPEN-DRAIN ASSUMPTION   In our open-drain resistor calculations, we assume that the output voltage can be as low as 0.0 V rather than 0.4 V (V_{OLmax}) in order to obtain a worst-case result. That is, even if the open-drain output is so strong that it can pull the output voltage all the way down to 0.0 V (it's only required to pull down to 0.4 V), we'll never allow it to sink more than 4 mA, so it doesn't get overstressed. Some designers prefer to use 0.4 V in this calculation, figuring that if the output is so good that it can pull lower than 0.4 V, a little bit of excess sink current beyond 4 mA won't hurt it.

3.8 CMOS Logic Families

The first commercially successful CMOS family was *4000-series CMOS*.   *4000-series CMOS* Although 4000-series circuits offered the benefit of low power dissipation, they were fairly slow and were not easy to interface with the most popular logic family of the time, bipolar TTL. Thus, the 4000 series was supplanted in most applications by the more capable CMOS families discussed in this section.

All of the CMOS devices that we discuss have part numbers of the form "74FAM*nn*," where "FAM" is an alphabetic family mnemonic and *nn* is a numeric function designator. Devices in different families with the same value of *nn* perform the same function. For example, the 74HC30, 74HCT30, 74AC30, 74ACT30, and 74AHC30 are all 8-input NAND gates.

The prefix "74" is simply a number that was used by an early, popular supplier of TTL devices, Texas Instruments. The prefix "54" is used for identical parts that are specified for operation over a wider range of temperature and power-supply voltage, for use in military applications. Such parts are usually

Figure 3-59 Input and output levels for CMOS devices using a 5-V supply: (a) HC; (b) HCT.

fabricated in the same way as their 74-series counterparts, except that they are tested, screened, and marked differently, a lot of extra paperwork is generated, and a higher price is charged, of course.

3.8.1 HC and HCT

HC (High-speed CMOS)

HCT (High-speed CMOS, TTL compatible)

The first two 74-series CMOS families are *HC (High-speed CMOS)* and *HCT (High-speed CMOS, TTL compatible)*. Compared with the original 4000 family, HC and HCT both have higher speed and better current sinking and sourcing capability. The HCT family uses a power-supply voltage V_{CC} of 5 V and can be intermixed with TTL devices, which also use a 5-V supply.

The HC family is optimized for use in systems that use CMOS logic exclusively, and can use any power-supply voltage between 2 and 6 V. A higher voltage is used for higher speed, and a lower voltage for lower power dissipation. Lowering the supply voltage is especially effective, since most CMOS power dissipation is proportional to the square of the voltage (CV^2f power).

Even when used with a 5-V supply, HC devices are not quite compatible with TTL. In particular, HC circuits are designed to recognize CMOS input levels. Assuming a supply voltage of 5.0 V, Figure 3-59(a) shows the input and output levels of HC devices. The output levels produced by TTL devices do not quite match this range, so HCT devices use the different input levels shown in (b). These levels are established in the fabrication process by making transistors with different switching thresholds, producing the different transfer characteristics shown in Figure 3-60.

Figure 3-60
Transfer characteristics of HC and HCT circuits under typical conditions.

We'll have more to say about CMOS/TTL interfacing in Section 3.12. For now, it is useful simply to note that HC and HCT are essentially identical in their output specifications; only their input levels differ.

3.8.2 VHC and VHCT

Several new CMOS families were introduced in the 1980s and the 1990s. Two of the most recent and probably the most versatile are *VHC (Very High-Speed CMOS)* and *VHCT (Very High-Speed CMOS, TTL compatible)*. These families are about twice as fast as HC/HCT while maintaining backward compatibility with their predecessors. Like HC and HCT, the VHC and VHCT families differ from each other only in the input levels that they recognize; their output characteristics are the same.

VHC (Very High-speed CMOS)

VHCT (Very High-speed CMOS, TTL compatible)

Also like HC/HCT, VHC/VHCT outputs have *symmetric output drive*. That is, an output can sink or source equal amounts of current; the output is just as "strong" in both states. Other logic families, including the FCT and TTL families introduced later, have *asymmetric output drive*; they can sink much more current in the LOW state than they can source in the HIGH state.

symmetric output drive

asymmetric output drive

3.8.3 HC, HCT, VHC, and VHCT Electrical Characteristics

Electrical characteristics of the HC, HCT, VHC, and VHCT families are summarized in this subsection. The specifications assume that the devices are used with a nominal 5-V power supply, although (derated) operation is possible with any supply voltage in the range 2–5.5 V (up to 6 V for HC/HCT). We'll take a closer look at low-voltage and mixed-voltage operation in Section 3.13.

Commercial (74-series) parts are intended to be operated at temperatures between 0°C and 70°C, while military (54-series) parts are characterized for operation between −55°C and 125°C. The specs in Table 3-5 assume an operating temperature of 25°C. A full manufacturer's data sheet provides additional specifications for device operation over the entire temperature range.

Most devices within a given logic family have the same electrical specifications for inputs and outputs, typically differing only in power consumption and propagation delay. Table 3-5 includes specifications for a 74x00 two-input NAND gate and a 74x138 3-to-8 decoder in the HC, HCT, VHC, and VHCT families. The '00 NAND gate is included as the smallest logic-design building block in each family, while the '138 is a "medium-scale" part containing the equivalent of about 15 NAND gates. (The '138 spec is included to allow

---

VERY = ADVANCED, SORT OF The VHC and VHCT logic families are manufactured by several companies, including Motorola, Fairchild, and Toshiba. Compatible families with similar but not identical specifications are manufactured by Texas Instruments and Philips; they are called AHC and AHCT, where the "A" stands for "Advanced."

Table 3-5 Speed and power characteristics of CMOS families operating at 5 V.

Description	Part	Symbol	Condition	Family			
				HC	HCT	VHC	VHCT
Typical propagation delay (ns)	'00	t_{PD}		9	10	5.2	5.5
	'138			18	20	7.2	8.1
Quiescent power-supply current (µA)	'00	I_{CC}	$V_{in} = 0$ or V_{CC}	2.5	2.5	5.0	5.0
	'138		$V_{in} = 0$ or V_{CC}	40	40	40	402
Quiescent power dissipation (mW)	'00		$V_{in} = 0$ or V_{CC}	0.0125	0.0125	0.025	0.025
	'138		$V_{in} = 0$ or V_{CC}	0.2	0.2	0.2	0.2
Power-dissipation capacitance (pF)	'00	C_{PD}		22	15	19	17
	'138	C_{PD}		55	51	34	49
Dynamic power dissipation (mW/MHz)	'00			0.55	0.38	0.48	0.43
	'138			1.38	1.28	0.85	1.23
Total power dissipation (mW)	'00		$f = 100$ kHz	0.068	0.050	0.073	0.068
	'00		$f = 1$ MHz	0.56	0.39	0.50	0.45
	'00		$f = 10$ MHz	5.5	3.8	4.8	4.3
	'138		$f = 100$ kHz	0.338	0.328	0.285	0.323
	'138		$f = 1$ MHz	1.58	1.48	1.05	1.43
	'138		$f = 10$ MHz	14.0	13.0	8.7	12.5
Speed-power product (pJ)	'00		$f = 100$ kHz	0.61	0.50	0.38	0.37
	'00		$f = 1$ MHz	5.1	3.9	2.6	2.5
	'00		$f = 10$ MHz	50	38	25	24
	'138		$f = 100$ kHz	6.08	6.55	2.05	2.61
	'138		$f = 1$ MHz	28.4	29.5	7.56	11.5
	'138		$f = 10$ MHz	251	259	63	101

comparison with the faster FCT family in Section 3.8.4; '00 gates are not manufactured in the FCT family.)

The first row of Table 3-5 specifies propagation delay. As discussed in Section 3.6.2, two numbers, t_{pHL} and t_{pLH}, may be used to specify delay; the number in the table is the worst case of the two. Skipping ahead to Table 3-11 on page 167, you can see that HC and HCT are about the same speed as LS TTL, and that VHC and VHCT are almost as fast as ALS TTL. The propagation delay

NOTE ON NOTATION   The "x" in the notation "74x00" takes the place of a family designator such as HC, HCT, VHC, VHCT, FCT, LS, ALS, AS, or F. We may also refer to such a generic part simply as a " '00" and leave off the "74x."

QUIETLY GETTING
MORE DISS'ED HCT and VHCT circuits can also be driven by TTL devices, which may produce HIGH output levels as low as 2.4 V. As we explained in Section 3.5.3, a CMOS output may draw additional current from the power supply if any of the inputs are nonideal. In the case of an HCT or VHCT inverter with a HIGH input of 2.4 V, the bottom, *n*-channel output transistor is fully "on." However, the top, *p*-channel transistor is also partially "on." This allows the additional quiescent current flow, specified as ΔI_{CC} or I_{CCT} in the data sheet, which can be as much as 2–3 mA per nonideal input in HCT and VHCT devices.

for the '138 is somewhat longer than for the '00, since signals must travel through three or four levels of gates internally.

The second and third rows of the table show that the quiescent power dissipation of these CMOS devices is practically nil, well under a milliwatt (mW) if the inputs have CMOS levels—0 V for LOW and V_{CC} for HIGH. (Note that in the table, the quiescent power dissipation numbers given for the '00 are per gate, while for the '138 they apply to the entire MSI device.)

As we discussed in Section 3.6.3, the dynamic power dissipation of a CMOS gate depends on the voltage swing of the output (usually V_{CC}), the output transition frequency (f), and the capacitance that is being charged and discharged on transitions, according to the formula

$$P_D = (C_L + C_{PD}) \cdot V_{DD}^2 \cdot f$$

Here, C_{PD} is the power-dissipation capacitance of the device and C_L is the capacitance of the load attached to the CMOS output in a given application. The table lists both C_{PD} and an equivalent dynamic power-dissipation factor in units of milliwatts per megahertz, assuming that $C_L = 0$. Using this factor, the total power dissipation is computed at various frequencies as the sum of the dynamic power dissipation at that frequency and the quiescent power dissipation.

Shown next in the table, the *speed-power product* is simply the product of the propagation delay and power consumption of a typical gate; the result is measured in picojoules (pJ). Recall from physics that the joule is a unit of energy, so the speed-power product measures a sort of efficiency—how much energy a logic gate uses to switch its output. In this day and age, it's obvious that the lower the energy usage, the better.

speed-power product

SAVING ENERGY There are practical as well as geopolitical reasons for saving energy in digital systems. Lower energy consumption means lower cost of power supplies and cooling systems. Also, a digital system's reliability is improved more by running it cooler than by any other single reliability improvement strategy.

Table 3-6 Input specifications for CMOS families with V_{CC} between 4.5 and 5.5 V.

Description	Symbol	Condition	Family HC	HCT	VHC	VHCT
Input leakage current (μA)	I_{Imax}	V_{in} = any	±1	±1	±1	±1
Maximum input capacitance (pF)	C_{INmax}		10	10	10	10
LOW-level input voltage (V)	V_{ILmax}		1.35	0.8	1.35	0.8
HIGH-level input voltage (V)	V_{IHmin}		3.85	2.0	3.85	2.0

Table 3-6 gives the input specs of typical CMOS devices in each of the families. Some of the specs assume that the 5-V supply has a ±10% margin; that is, V_{CC} can be anywhere between 4.5 and 5.5 V. These parameters were discussed in previous sections, but for reference purposes their meanings are summarized here:

I_{Imax} The maximum input current for any value of input voltage. This spec states that the current flowing into or out of a CMOS input is 1 μA or less for any value of input voltage. In other words, CMOS inputs create almost no DC load on the circuits that drive them.

C_{INmax} The maximum capacitance of an input. This number can be used when figuring the AC load on an output that drives this and other inputs. Most manufacturers also specify a lower, typical input capacitance of about 5 pF, which gives a good estimate of AC load if you're not unlucky.

V_{ILmax} The maximum voltage that an input is guaranteed to recognize as LOW. Note that the values are different for HC/VHC versus HCT/VHCT. The "CMOS" value, 1.35 V, is 30% of the minimum power-supply voltage, while the "TTL" value is 0.8 V for compatibility with TTL families.

CMOS VS. TTL POWER DISSIPATION

At high transition frequencies (f), CMOS families actually use more power than TTL. For example, compare HCT CMOS in Table 3-5 at f = 10 MHz with LS TTL in Table 3-11; a CMOS gate uses three times as much power as a TTL gate at this frequency. Both HCT and LS may be used in systems with maximum "clock" frequencies of up to about 20 MHz, so you might think that CMOS is not so good for high-speed systems. However, the transition frequencies of most outputs in typical systems are much less than the maximum frequency present in the system (e.g., see Exercise 3.77). Thus, typical CMOS systems have a lower total power dissipation than they would have if they were built with TTL.

Table 3-7 Output specifications for CMOS families operating with V_{CC} between 4.5 and 5.5 V.

Description	Symbol	Condition	Family			
			HC	HCT	VHC	VHCT
LOW-level output current (mA)	I_{OLmaxC}	CMOS load	0.02	0.02	0.05	0.05
	I_{OLmaxT}	TTL load	4.0	4.0	8.0	8.0
LOW-level output voltage (V)	V_{OLmaxC}	$I_{out} \le I_{OLmaxC}$	0.1	0.1	0.1	0.1
	V_{OLmaxT}	$I_{out} \le I_{OLmaxT}$	0.33	0.33	0.44	0.44
HIGH-level output current (mA)	I_{OHmaxC}	CMOS load	−0.02	−0.02	−0.05	−0.05
	I_{OHmaxT}	TTL load	−4.0	−4.0	−8.0	−8.0
HIGH-level output voltage (V)	V_{OHminC}	$|I_{out}| \le |I_{OHmaxC}|$	4.4	4.4	4.4	4.4
	V_{OHminT}	$|I_{out}| \le |I_{OHmaxT}|$	3.84	3.84	3.80	3.80

V_{IHmin} The minimum voltage that an input is guaranteed to recognize as HIGH. The "CMOS" value, 3.85 V, is 70% of the maximum power-supply voltage, while the "TTL" value is 2.0 V for compatibility with TTL families. (Unlike CMOS levels, TTL input levels are not symmetric with respect to the power-supply rails.)

The specifications for TTL-compatible CMOS outputs usually have two sets of output parameters; one set or the other is used depending on how an output is loaded. A *CMOS load* is one that requires the output to sink and source very little DC current, 20 μA for HC/HCT and 50 μA for VHC/VHCT. This is, of course, the case when the CMOS outputs drive only CMOS inputs. With CMOS loads, CMOS outputs maintain an output voltage within 0.1 V of the supply rails, 0 and V_{CC}. (A worst-case $V_{CC} = 4.5$ V is used for the table entries; hence, $V_{OHminC} = 4.4$ V.)

CMOS load

A *TTL load* can consume much more sink and source current, up to 4 mA from an HC/HCT output and 8 mA from a VHC/VHCT output. In this case, a higher voltage drop occurs across the "on" transistors in the output circuit, but the output voltage is still guaranteed to be within the normal range of TTL output levels.

TTL load

Table 3-7 lists CMOS output specifications for both CMOS and TTL loads. These parameters have the following meanings:

I_{OLmaxC} The maximum current that an output can supply in the LOW state while driving a CMOS load. Since this is a positive value, current flows *into* the output pin.

I_{OLmaxT} The maximum current that an output can supply in the LOW state while driving a TTL load.

V_{OLmaxC} The maximum voltage that a LOW output is guaranteed to produce while driving a CMOS load, that is, as long as I_{OLmaxC} is not exceeded.

V_{OLmaxT} The maximum voltage that a LOW output is guaranteed to produce while driving a TTL load, that is, as long as I_{OLmaxT} is not exceeded.

I_{OHmaxC} The maximum current that an output can supply in the HIGH state while driving a CMOS load. Since this is a negative value, positive current flows out of the output pin.

I_{OHmaxT} The maximum current that an output can supply in the HIGH state while driving a TTL load.

V_{OHminC} The minimum voltage that a HIGH output is guaranteed to produce while driving a CMOS load, that is, as long as I_{OHmaxC} is not exceeded.

V_{OHminT} The minimum voltage that a HIGH output is guaranteed to produce while driving a TTL load, that is, as long as I_{OHmaxT} is not exceeded.

The voltage parameters above determine DC noise margins. The LOW-state DC noise margin is the difference between V_{OLmax} and V_{ILmax}. This depends on the characteristics of both the driving output and the driven inputs. For example, the LOW-state DC noise margin of HCT driving a few HCT inputs (a CMOS load) is $0.8 - 0.1 = 0.7$ V. With a TTL load, the noise margin for the HCT inputs drops to $0.8 - 0.33 = 0.47$ V. Similarly, the HIGH-state DC noise margin is the difference between V_{OHmin} and V_{IHmin}. In general, when different families are interconnected, you have to compare the appropriate V_{OLmax} and V_{OHmin} of the driving gate with V_{ILmax} and V_{IHmin} of all the driven gates to determine the worst-case noise margins.

The I_{OLmax} and I_{OHmax} parameters in the table determine fanout capability and are especially important when an output drives inputs in one or more different families. Two calculations must be performed to determine whether an output is operating within its rated fanout capability:

HIGH-state fanout   The I_{IHmax} values for all of the driven inputs are added. The sum must be less than I_{OHmax} of the driving output.

LOW-state fanout   The I_{ILmax} values for all of the driven inputs are added. The sum must be less than I_{OLmax} of the driving output

Note that the input and output characteristics of specific components may vary from the representative values given in Table 3-7, so you must always consult the manufacturers' data sheets when analyzing a real design.

*3.8.4 FCT and FCT-T

FCT (Fast CMOS, TTL compatible)

In the early 1990s, yet another CMOS family was launched. The key benefit of the *FCT (Fast CMOS, TTL compatible)* family was its ability to meet or exceed the speed and the output drive capability of the best TTL families while reducing power consumption and maintaining full compatibility with TTL.

The original FCT family had the drawback of producing a full 5-V CMOS V_{OH}, creating enormous CV^2f power dissipation and circuit noise as its outputs swung from 0 V to almost 5 V in high-speed (25 MHz+) applications. A variation of the family, *FCT-T (Fast CMOS, TTL compatible with TTL V_{OH})*, was quickly introduced with circuit innovations to reduce the HIGH-level output voltage, thereby reducing both power consumption and switching noise while maintaining the same high operating speed as the original FCT. A suffix of "T" is used on part numbers to denote the FCT-T output structure, for example, 74FCT138T versus 74FCT138.

FCT-T (Fast CMOS, TTL compatible with TTL V_{OH}

The FCT-T family remains very popular today. A key application of FCT-T is driving buses and other heavy loads. Compared with other CMOS families, it can source or sink gobs of current, up to 64 mA in the LOW state.

*3.8.5 FCT-T Electrical Characteristics

Electrical characteristics of the 5-V FCT-T family are summarized in Table 3-8. The family is specifically designed to be intermixed with TTL devices, so its operation is only specified with a nominal 5-V supply and TTL logic levels. Some manufacturers are beginning to sell parts with similar capabilities using a 3.3-V supply, and using the FCT designation. However, they are different devices with different part numbers.

Individual logic gates are not manufactured in the FCT family. Perhaps the simplest FCT logic element is a 74FCT138T decoder, which has six inputs, eight outputs, and contains the equivalent of about a dozen 4-input gates internally. (This function is described later, in Section 5.4.4.) Comparing its propagation delay and power consumption in Table 3-8 with the corresponding HCT and VHCT numbers in Table 3-5 on page 138, you can see that the FCT-T family is superior in both speed and power dissipation. When comparing, note that FCT-T manufacturers specify only maximum, not typical propagation delays.

Unlike other CMOS families, FCT-T does not have a C_{PD} specification. Instead, it has an I_{CCD} specification:

I_{CCD} Dynamic power-supply current, in units of mA/MHz. This is the amount of additional power-supply current that flows when one input is changing at the rate of 1 MHz.

EXTREME SWITCHING   Device outputs in the FCT and FCT-T families have very low impedance and as a consequence extremely fast rise and fall times. In fact, they are so fast that they are often a major source of "analog" problems, including switching noise and "ground bounce," so extra care must be taken in the analog and physical design of printed-circuit boards using these and other extremely high-speed parts. To reduce the effects of transmission-line reflections (Section 11.4), another high-speed design worry, some FCT-T outputs have built-in 25-Ω series resistors.

Table 3-8 Specifications for a 74FCT138T decoder in the FCT-T logic family.

Description	Symbol	Condition	Value
Maximum propagation delay (ns)	t_{PD}		5.8
Quiescent power-supply current (μA)	I_{CC}	$V_{in} = 0$ or V_{CC}	200
Quiescent power dissipation (mW)		$V_{in} = 0$ or V_{CC}	1.0
Dynamic power-supply current (mA/MHz)	I_{CCD}	Outputs open, one input changing	0.12
Quiescent power-supply current per TTL input (mA)	ΔI_{CC}	$V_{in} = 3.4$ V	2.0
Total power dissipation (mW)		$f = 100$ kHz $f = 1$ MHz $f = 10$ MHz	0.60 1.06 1.6
Speed-power product (pJ)		$f = 100$ kHz $f = 1$ MHz $f = 10$ MHz	6.15 9.3 41
Input leakage current (μA)	I_{Imax}	$V_{in} = $ any	±5
Typical input capacitance (pF)	C_{INtyp}		5
LOW-level input voltage (V)	V_{ILmax}		0.8
HIGH-level input voltage (V)	V_{IHmin}		2.0
LOW-level output current (mA)	I_{OLmax}		64
LOW-level output voltage (V)	V_{OLmax}	$I_{out} \leq I_{OLmax}$	0.55
HIGH-level output current (mA)	I_{OHmax}		−15
HIGH-level output voltage (V)	V_{OHmin} V_{OHtyp}	$|I_{out}| \leq |I_{OHmax}|$ $|I_{out}| \leq |I_{OHmax}|$	2.4 3.3

The I_{CCD} specification gives the same information as C_{PD}, but in a different way. The circuit's internal power dissipation due to transitions at a given frequency f can be calculated by the formula

$$P_T = V_{CC} \cdot I_{CCD} \cdot f$$

Thus, I_{CCD}/V_{CC} is algebraically equivalent to the C_{PD} specification of other CMOS families (see Exercise 3.83). FCT-T also has a ΔI_{CC} specification for the extra quiescent current that is consumed with nonideal HIGH inputs (see box at the top of page 139).

3.9 Bipolar Logic

Bipolar logic families use semiconductor diodes and bipolar junction transistors as the basic building blocks of logic circuits. The simplest bipolar logic elements use diodes and resistors to perform logic operations; this is called diode logic. *diode logic* Most TTL logic gates use diode logic internally and boost their output drive capability using transistor circuits. Some TTL gates use parallel configurations of transistors to perform logic functions. ECL gates, described in Section 3.14, use transistors as current switches to achieve very high speed.

This section covers the basic operation of bipolar logic circuits made from diodes and transistors, and the next section covers TTL circuits in detail. Although TTL is the most commonly used bipolar logic family, it has been largely supplanted by the CMOS families that we studied in previous sections.

Still, it is useful to study basic TTL operation for the occasional application that requires TTL/CMOS interfacing, discussed in Section 3.12. Also, an understanding of TTL may give you insight into the fortuitous similarity of logic levels that allowed the industry to migrate smoothly from TTL to 5-V CMOS logic, and now to lower-voltage, higher-performance 3.3-V CMOS logic, as described in Section 3.13. If you're not interested in all the gory details of TTL, you can skip to Section 3.11 for an overview of TTL families.

3.9.1 Diodes

A *semiconductor diode* is fabricated from two types of semiconductor material, *semiconductor diode* called *p*-type and *n*-type, that are brought into contact with each other as shown *p-type material* in Figure 3-61(a). This is basically the same material that is used in *p*-channel *n-type material* and *n*-channel MOS transistors. The point of contact between the *p* and *n* materials is called a *pn junction.* (Actually, a diode is normally fabricated from a *pn junction* single monolithic crystal of semiconductor material in which the two halves are "doped" with different impurities to give them *p*-type and *n*-type properties.)

The physical properties of a *pn* junction are such that positive current can easily flow from the *p*-type material to the *n*-type. Thus, if we build the circuit shown in Figure 3-61(b), the *pn* junction acts almost like a short circuit. However, the physical properties also make it very difficult for positive current to

Figure 3-61 Semiconductor diodes: (a) the *pn* junction; (b) forward-biased junction allowing current flow; (c) reverse-biased junction blocking current flow.

Figure 3-62 Diodes: (a) symbol; (b) transfer characteristic of an ideal diode; (c) transfer characteristic of a real diode.

diode action

flow in the opposite direction, from n to p. Thus, in the circuit of Figure 3-61(c), the pn junction behaves almost like an open circuit. This is called *diode action*.

diode

Although it's possible to build vacuum tubes and other devices that exhibit diode action, modern systems use pn junctions—semiconductor diodes—which we'll henceforth call simply *diodes*. Figure 3-62(a) shows the schematic symbol for a diode. As we've shown, in normal operation significant amounts of current

anode
cathode

can flow only in the direction indicated by the two arrows, from *anode* to *cathode*. In effect, the diode acts like a short circuit as long as the voltage across the anode-to-cathode junction is nonnegative. If the anode-to-cathode voltage is negative, the diode acts like an open circuit and no current flows.

The transfer characteristic of an ideal diode shown in Figure 3 -62(b) further illustrates this principle. If the anode-to-cathode voltage, V, is negative,

reverse-biased diode
forward-biased diode

the diode is said to be *reverse biased* and the current I through the diode is zero. If V is nonnegative, the diode is said to be *forward biased* and I can be an arbitrarily large positive value. In fact, V can never get larger than zero, because an ideal diode acts like a zero-resistance short circuit when forward biased.

A nonideal, real diode has a resistance that is less than infinity when reverse biased, and greater than zero when forward biased, so the transfer characteristic looks like Figure 3-62(c). When forward biased, the diode acts

YES, THERE ARE TWO ARROWS   . . . in Figure 3-62(a). The second arrow is built into the diode symbol to help you remember the direction of current flow. Once you know this, there are many ways to remember which end is called the anode and which is the cathode. Aficionados of vacuum-tube hi-fi amplifiers may remember that electrons travel from the hot cathode to the anode, and therefore positive current flow is from anode to cathode. Those of us who were still watching "Sesame Street" when most vacuum tubes went out of style might like to think in terms of the alphabet—current flows alphabetically from A to C. (Thanks to Big-Bird-fan Rick Casey for suggesting this memory crutch.)

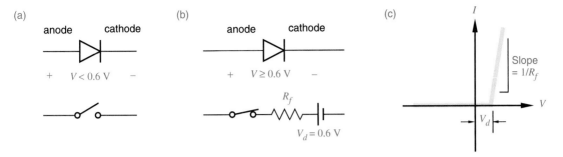

Figure 3-63 Model of a real diode: (a) reverse biased; (b) forward biased;
(c) transfer characteristic of forward-biased diode.

like a small nonlinear resistance; its voltage drop increases as current increases, but not strictly proportionally. When the diode is reverse biased, a small amount of negative *leakage current* flows. If the voltage is made too negative, the diode *breaks down,* and large amounts of negative current can flow; in most applications, this type of operation is avoided.

leakage current

diode breakdown

A real diode can be modeled very simply as shown in Figure 3-63(a) and (b). When the diode is reverse biased, it acts like an open circuit; we ignore leakage current. When the diode is forward biased, it acts like a small resistance, R_f, in series with V_d, a small voltage source. R_f is called the *forward resistance* of the diode, and V_d is called a *diode-drop.*

forward resistance

diode-drop

Careful choice of values for R_f and V_d yields a reasonable piecewise-linear approximation to the real diode transfer characteristic, as in Figure 3-63(c). In a typical small-signal diode such as a 1N914, the forward resistance R_f is about 25 Ω and the diode-drop V_d is about 0.6 V.

In order to get a feel for diodes, you should remember that a real diode does not actually contain the 0.6-V source that appears in the model. It's just that, due to the nonlinearity of the real diode's transfer characteristic, significant amounts of current do not begin to flow until the diode's forward voltage V has reached about 0.6 V. Also note that in typical applications, the 25-Ω forward resistance of the diode is small compared to other resistances in the circuit, so that very little additional voltage drop occurs across the forward-biased diode once V has reached 0.6 V. Thus, for practical purposes, a forward-biased diode may be considered to have a fixed drop of 0.6 V or so .

ZENER DIODES *Zener diodes* take advantage of diode breakdown, in particular the steepness of the $V–I$ slope in the breakdown region. A Zener diode can function as a voltage regulator when used with a resistor to limit the breakdown current. A wide variety of Zeners with different breakdown voltages are produced for voltage-regulator applications.

Table 3-9
Logic levels in a
simple diode logic
system.

Signal Level	Designation	Binary Logic Value
0–2 volts	LOW	0
2–3 volts	noise margin	undefined
3–5 volts	HIGH	1

3.9.2 Diode Logic

LOW

HIGH

diode AND gate

Diode action can be exploited to perform logical operations. Consider a logic system with a 5-V power supply and the characteristics shown in Table 3-9. Within the 5-volt range, signal voltages are partitioned into two ranges, LOW and HIGH, with a 1-volt noise margin between. A voltage in the LOW range is considered to be a logic 0, and a voltage in the HIGH range is a logic 1.

With these definitions, a *diode AND gate* can be constructed as shown in Figure 3-64(a). In this circuit, suppose that both inputs X and Y are connected to HIGH voltage sources, say 4 V, so that VX and VY are both 4 V as in (b). Then both diodes are forward biased, and the output voltage VZ is one diode-drop above 4 V, or about 4.6 V. A small amount of current, determined by the value of R, flows from the 5-V supply through the two diodes and into the 4-V sources. The colored arrows in the figure show the path of this current flow.

Figure 3-64
Diode AND gate:
(a) electrical circuit;
(b) both inputs HIGH;
(c) one input HIGH,
one LOW; (d) function
table; (e) truth table.

Figure 3-65
Two AND gates:
(a) logic diagram;
(b) electrical circuit.

Now suppose that *VX* drops to 1 V as in Figure 3-64(c). In the diode AND gate, the output voltage equals the lower of the two input voltages plus a diode-drop. Thus, *VZ* drops to 1.6 V, and diode *D2* is reverse biased (the anode is at 1.6 V and the cathode is still at 4 V). The single LOW input "pulls down" the output of the diode AND gate to a LOW value. Obviously, two LOW inputs create a LOW output as well. This functional operation is summarized in (d) and is repeated in terms of binary logic values in (e); clearly, this is an AND gate.

Figure 3-65(a) shows a logic circuit with two AND gates connected together; Figure 3-65(b) shows the equivalent electrical circuit with a particular set of input values. This example shows the necessity of diodes in the AND circuit: *D3* allows the output Z of the first AND gate to remain HIGH while the output C of the second AND gate is being pulled LOW by input B through *D4*.

When diode logic gates are cascaded as in Figure 3-65, the voltage levels of the logic signals move away from the power-supply rails and toward the undefined region. Thus, in practice, a diode AND gate normally must be followed by a transistor amplifier to restore the logic levels; this is the scheme used in TTL NAND gates, described in Section 3.10.1. However, logic designers are occasionally tempted to use discrete diodes to perform logic under special circumstances; for example, see Exercise 3.95.

3.9.3 Bipolar Junction Transistors

A *bipolar junction transistor* is a three-terminal device that, in most logic circuits, acts like a current-controlled switch. If we put a small current into one of the terminals, called the *base*, then the switch is "on"—current may flow between the other two terminals, called the *emitter* and the *collector*. If no current is put into the base, then the switch is "off"—no current flows between the emitter and the collector.

bipolar junction transistor

base

emitter

collector

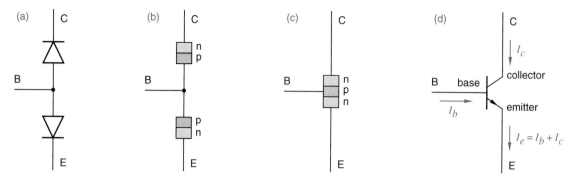

Figure 3-66 Development of an *npn* transistor: (a) back-to-back diodes; (b) equivalent *pn* junctions; (c) structure of an *npn* transistor; (d) *npn* transistor symbol.

To study the operation of a transistor, we first consider the operation of a pair of diodes connected as shown in Figure 3-66(a). In this circuit, current can flow from node B to node C or node E, when the appropriate diode is forward biased. However, no current can flow from C to E, or vice versa, since for any choice of voltages on nodes B, C, and E, one or both diodes will be reverse biased. The *pn* junctions of the two diodes in this circuit are shown in (b).

Now suppose that we fabricate the back-to-back diodes so that they share a common *p*-type region, as shown in Figure 3-66(c). The resulting structure is called an *npn transistor* and has an amazing property. (At least, the physicists working on transistors back in the 1950s thought it was amazing!) If we put current across the base-to-emitter *pn* junction, then current is also enabled to flow across the collector-to-base *np* junction (which is normally impossible) and from there to the emitter.

npn transistor

The circuit symbol for the *npn* transistor is shown in Figure 3-66(d). Notice that the symbol contains a subtle arrow in the direction of positive current flow. This also reminds us that the base-to-emitter junction is a *pn* junction, the same as a diode whose symbol has an arrow pointing in the same direction.

pnp transistor

It is also possible to fabricate a *pnp transistor,* as shown in Figure 3-67. However, *pnp* transistors are seldom used in digital circuits, so we won't discuss them any further.

The current I_e flowing out of the emitter of an *npn* transistor is the sum of the currents I_b and I_c flowing into the base and the collector. A transistor is often used as a signal *amplifier*, because over a certain operating range (the *active region*) the collector current is equal to a fixed constant times the base current $(I_c = \beta \cdot I_b)$. However, in digital circuits, we normally use a transistor as a simple switch that's always fully "on" or fully "off," as explained next.

amplifier
active region

common-emitter
configuration

Figure 3-68 shows the *common-emitter configuration* of an *npn* transistor, which is most often used in digital switching applications. This configuration

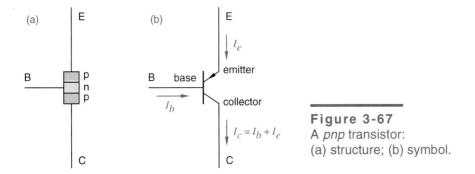

Figure 3-67
A *pnp* transistor:
(a) structure; (b) symbol.

uses two discrete resistors, *R1* and *R2*, in addition to a single *npn* transistor. In this circuit, if V_{IN} is 0 or negative, then the base-to-emitter diode junction is reverse biased, and no base current (I_b) can flow. If no base current flows, then no collector current (I_c) can flow, and the transistor is said to be *cut off (OFF)*.   *cut off (OFF)*

Since the base-to-emitter junction is a *real* diode, as opposed to an ideal one, V_{IN} must reach at least +0.6 V (one diode-drop) before any base current can flow. Once this happens, Ohm's law tells us that

$$I_b = (V_{IN} - 0.6) / R1$$

(We ignore the forward resistance R_f of the forward-biased base-to-emitter junction, which is usually small compared to the base resistor *R1*.) When base current flows, then collector current can flow in an amount proportional to I_b, that is,

$$I_c = \beta \cdot I_b$$

The constant of proportionality, β, is called the *gain* of the transistor, and is in β
the range of 10 to 100 for typical transistors.   *gain*

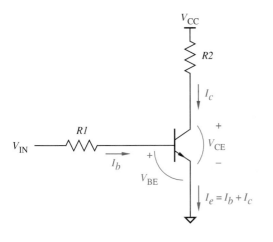

Figure 3-68
Common-emitter
configuration of an
npn transistor.

Although the base current I_b controls the collector current flow I_c, it also indirectly controls the voltage V_{CE} across the collector-to-emitter junction, since V_{CE} is just the supply voltage V_{CC} minus the voltage drop across resistor $R2$:

$$V_{CE} = V_{CC} - I_c \cdot R2$$
$$= V_{CC} - \beta \cdot I_b \cdot R2$$
$$= V_{CC} - (V_{IN} - 0.6) \cdot R2 / R1$$

However, in an ideal transistor V_{CE} can never be less than zero (the transistor cannot just create a negative potential), and in a real transistor V_{CE} can never be less than $V_{CE(sat)}$, a transistor parameter that is typically about 0.2 V.

saturation region
saturated (ON)

If the values of V_{IN}, β, $R1$, and $R2$ are such that the above equation predicts a value of V_{CE} that is less than $V_{CE(sat)}$, then the transistor cannot be operating in the active region and the equation does not apply. Instead, the transistor is operating in the *saturation region,* and is said to be *saturated (ON).* No matter how much current I_b we put into the base, V_{CE} cannot drop below $V_{CE(sat)}$, and the collector current I_c is determined mainly by the load resistor $R2$:

$$I_c = (V_{CC} - V_{CE(sat)}) / (R2 + R_{CE(sat)})$$

saturation resistance

Here, $R_{CE(sat)}$ is the *saturation resistance* of the transistor. Typically, $R_{CE(sat)}$ is 50 Ω or less and is insignificant compared with $R2$.

transistor simulation

Computer scientists might like to imagine an *npn* transistor as a device that continuously looks at its environment and executes the program in Table 3-10.

3.9.4 Transistor Logic Inverter

Figure 3-69 shows that we can make a logic inverter from an *npn* transistor in the common-emitter configuration. When the input voltage is LOW, the output voltage is HIGH, and vice versa.

In digital switching applications, bipolar transistors are often operated so they are always either cut off or saturated. That is, digital circuits such as the

Figure 3-69
Transistor inverter:
(a) logic symbol;
(b) circuit diagram;
(c) transfer characteristic.

Table 3-10 A C program that simulates the function of an *npn* transistor in the common-emitter configuration.

```
/* Transistor parameters */
#define DIODEDROP 0.6 /* volts */
#define BETA 10;
#define VCE_SAT 0.2    /* volts */
#define RCE_SAT 50     /* ohms  */

main()
{
    float Vcc, Vin, R1, R2;  /* circuit parameters */
    float Ib, Ic, Vce;       /* circuit conditions */

    if (Vin < DIODEDROP) {   /* cut off */
       Ib  = 0.0;
       Ic  = 0.0;
       Vce = Vcc;
    }
    else {                    /* active or saturated */
       Ib = (Vin - DIODEDROP) / R1;
       if ((Vcc - ((BETA * Ib) * R2)) >= VCE_SAT) {   /* active */
          Ic  = BETA * Ib;
          Vce = Vcc - (Ic * R2);
       }
       else {                 /* saturated */
          Vce = VCE_SAT;
          Ic  = (Vcc - Vce) / (R2 + RCE_SAT);
       }
    }
}
```

Figure 3-70 Normal states of an *npn* transistor in a digital switching circuit: (a) transistor symbol and currents; (b) equivalent circuit for a cut-off (OFF) transistor; (c) equivalent circuit for a saturated (ON) transistor.

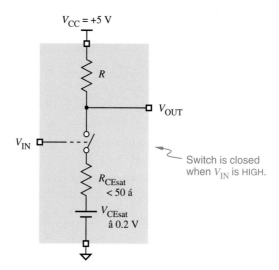

$V_{CC} = +5$ V

R

V_{OUT}

V_{IN}

Switch is closed
when V_{IN} is HIGH.

R_{CEsat}
< 50 á

V_{CEsat}
â 0.2 V

Figure 3-71
Switch model for a
transistor inverter.

inverter in Figure 3-69 are designed so that their transistors are always (well, almost always) in one of the states depicted in Figure 3-70. When the input voltage V_{IN} is LOW, it is low enough that I_b is zero and the transistor is cut off; the collector-emitter junction looks like an open circuit. When V_{IN} is HIGH, it is high enough (and $R1$ is low enough and β is high enough) that the transistor will be saturated for any reasonable value of $R2$; the collector-emitter junction looks almost like a short circuit. Input voltages in the undefined region between LOW and HIGH are not allowed, except during transitions. This undefined region corresponds to the noise margin that we discussed in conjunction with Table 3-1.

Another way to visualize the operation of a transistor inverter is shown in Figure 3-71. When V_{IN} is HIGH, the transistor switch is closed, and the output terminal is connected to ground, definitely a LOW voltage. When V_{IN} is LOW, the transistor switch is open and the output terminal is pulled to +5 V through a resistor; the output voltage is HIGH unless the output terminal is too heavily loaded (i.e., improperly connected through a low impedance to ground).

3.9.5 Schottky Transistors

storage time

When the input of a saturated transistor is changed, the output does not change immediately; it takes extra time, called *storage time*, to come out of saturation. In fact, storage time accounts for a significant portion of the propagation delay in the original TTL logic family.

Schottky diode
Schottky-clamped transistor
Schottky transistor

Storage time can be eliminated and propagation delay can be reduced by ensuring that transistors do not saturate in normal operation. Contemporary TTL logic families do this by placing a *Schottky diode* between the base and collector of each transistor that might saturate, as shown in Figure 3-72. The resulting transistors, which do not saturate, are called *Schottky-clamped transistors* or *Schottky transistors* for short.

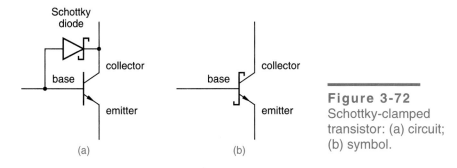

Figure 3-72
Schottky-clamped
transistor: (a) circuit;
(b) symbol.

When forward biased, a Schottky diode's voltage drop is much less than a standard diode's, 0.25 V vs. 0.6 V. In a standard saturated transistor, the base-to-collector voltage is 0.4 V, as shown in Figure 3-73(a). In a Schottky transistor, the Schottky diode shunts current from the base into the collector before the transistor goes into saturation, as shown in (b). Figure 3-74 is the circuit diagram of a simple inverter using a Schottky transistor.

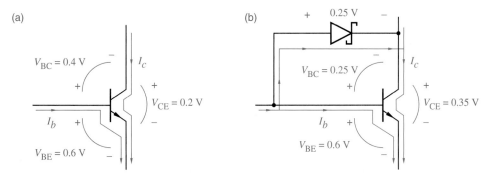

Figure 3-73 Operation of a transistor with large base current: (a) standard saturated transistor; (b) transistor with Schottky diode to prevent saturation.

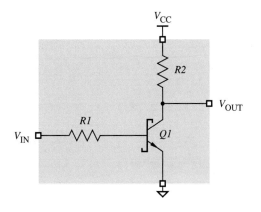

Figure 3-74
Inverter using Schottky
transistor.

3.10 Transistor-Transistor Logic

The most commonly used bipolar logic family is transistor-transistor logic. Actually, there are many different TTL families, with a range of speed, power consumption, and other characteristics. The circuit examples in this section are based on a representative TTL family, Low-power Schottky (LS or LS-TTL).

TTL families use basically the same logic levels as the TTL-compatible CMOS families in previous sections. We'll use the following definitions of LOW and HIGH in our discussions of TTL circuit behavior:

LOW   0–0.8 volts.

HIGH   2.0–5.0 volts.

3.10.1 Basic TTL NAND Gate

The circuit diagram for a two-input LS-TTL NAND gate, part number 74LS00, is shown in Figure 3-75. The NAND function is obtained by combining a diode AND gate with an inverting buffer amplifier. The circuit's operation is best understood by dividing it into the three parts that are shown in the figure and discussed in the next three paragraphs:

- Diode AND gate and input protection.
- Phase splitter.
- Output stage.

diode AND gate

clamp diode

Diodes *D1X* and *D1Y* and resistor *R1* in Figure 3-75 form a *diode AND gate,* as in Section 3.9.2. *Clamp diodes D2X* and *D2Y* do nothing in normal operation, but limit undesirable negative excursions on the inputs to a single diode-drop. Such negative excursions may occur on HIGH-to-LOW input transitions as a result of transmission-line effects, discussed in Section 11.4.

hase splitter

Transistor *Q2* and the surrounding resistors form a *phase splitter* that controls the output stage. Depending on whether the diode AND gate produces a "low" or a "high" voltage at V_A, *Q2* is either cut off or turned on.

WHERE IN THE WORLD IS *Q1*?   Notice that there is no transistor *Q1* in Figure 3-75, but the other transistors are named in a way that's traditional; some TTL devices do in fact have a transistor named *Q1*. Instead of diodes like *D1X* and *D1Y*, these devices use a multiple-emitter transistor *Q1* to perform logic. This transistor has one emitter per logic input, as shown in the figure to the right. Pulling any one of the emitters LOW is sufficient to turn the transistor ON and thus pull V_A LOW.

Figure 3-75
Circuit diagram of
two-input LS-TTL
NAND gate.

The *output stage* has two transistors, *Q4* and *Q5*, only one of which is on at
any time. The TTL output stage is sometimes called a *totem-pole* or *push-pull*
output. Similar to the *p*-channel and *n*-channel transistors in CMOS, *Q4* and *Q5*
provide active pull-up and pull-down to the HIGH and LOW states, respectively.

The functional operation of the TTL NAND gate is summarized in
Figure 3-76(a). The gate does indeed perform the NAND function, with the truth
table and logic symbol shown in (b) and (c). TTL NAND gates can be designed
with any desired number of inputs simply by changing the number of diodes in

output stage
totem-pole output
push-pull output

Figure 3-76
Functional operation
of a TTL two-input
NAND gate:
(a) function table;
(b) truth table;
(c) logic symbol.

(a)

X	Y	V_A	Q2	Q3	Q4	Q5	Q6	V_Z	Z
L	L	≤1.05	off	on	on	off	off	2.7	H
L	H	≤1.05	off	on	on	off	off	2.7	H
H	L	≤1.05	off	on	on	off	off	2.7	H
H	H	1.2	on	off	off	on	on	≤0.35	L

(b)

X	Y	Z
0	0	1
0	1	1
1	0	1
1	1	0

(c)

the diode AND gate in the figure. Commercially available TTL NAND gates have as many as 13 inputs. A TTL inverter is designed as a 1-input NAND gate, omitting diodes *D1Y* and *D2Y* in Figure 3-75.

Since the output transistors *Q4* and *Q5* are normally complementary—one ON and the other OFF—you might question the purpose of the 120-Ω resistor *R5* in the output stage. A value of 0 Ω would give even better driving capability in the HIGH state. This is certainly true from a DC point of view. However, when the TTL output is changing from HIGH to LOW or vice versa, there is a short time when both transistors may be on. The purpose of *R5* is to limit the amount of current that flows from VCC to ground during this time. Even with a 120 Ω resistor in the TTL output stage, higher-than-normal currents called current spikes flow when TTL outputs are switched. These are similar to the current spikes that occur when high-speed CMOS outputs switch.

So far we have shown the input signals to a TTL gate as ideal voltage sources. Figure 3-77 shows the situation when a TTL input is driven LOW by the output of another TTL gate. Transistor *Q5A* in the driving gate is ON, and thereby provides a path to ground for the current flowing out of the diode *D1XB* in the driven gate. When current flows *into* a TTL output in the LOW state, as in this case, the output is said to be *sinking current*.

sinking current

Figure 3-78 shows the same circuit with a HIGH output. In this case, *Q4A* in the driving gate is turned on enough to supply the small amount of leakage current flowing through reverse-biased diodes *D1XB* and *D2XB* in the driven gate. When current flows out of a TTL output in the HIGH state, the output is said to be *sourcing current*.

sourcing current

3.10.2 Logic Levels and Noise Margins

At the beginning of this section, we indicated that we would consider TTL signals between 0 and 0.8 V to be LOW, and signals between 2.0 and 5.0 V to be HIGH. Actually, we can be more precise by defining TTL input and output levels in the same way as we did for CMOS:

V_{OHmin} The minimum output voltage in the HIGH state, 2.7 V for most TTL families.

V_{IHmin} The minimum input voltage guaranteed to be recognized as a HIGH, 2.0 V for all TTL families.

CURRENT SPIKES AGAIN   Current spikes can show up as noise on the power-supply and ground connections in TTL and CMOS circuits, especially when multiple outputs are switched simultaneously. For this reason, reliable circuits require decoupling capacitors between V_{CC} and ground, distributed throughout the circuit so that there is a capacitor within an inch or so of each chip. Decoupling capacitors supply the instantaneous current needed during transitions.

Figure 3-77 A TTL output driving a TTL input LOW.

Figure 3-78 A TTL output driving a TTL input HIGH.

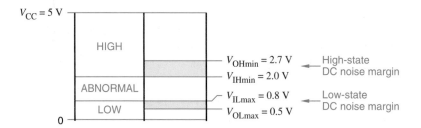

Figure 3-79
Noise margins for
popular TTL logic
families (74LS, 74S,
74ALS, 74AS, 74F).

V_{ILmax} The maximum input voltage guaranteed to be recognized as a **LOW**, 0.8 V for most TTL families.

V_{OLmax} The maximum output voltage in the **LOW** state, 0.5 V for most families.

These noise margins are illustrated in Figure 3-79.

DC noise margin

In the **HIGH** state, the V_{OHmin} specification of most TTL families exceeds V_{IHmin} by 0.7 V, so TTL has a *DC noise margin* of 0.7 V in the **HIGH** state. That is, it takes at least 0.7 V of noise to corrupt a worst-case **HIGH** output into a voltage that is not guaranteed to be recognizable as a **HIGH** input. In the **LOW** state, however, V_{ILmax} exceeds V_{OLmax} by only 0.3 V, so the DC noise margin in the **LOW** state is only 0.3 V. In general, TTL and TTL-compatible circuits tend to be more sensitive to noise in the **LOW** state than in the **HIGH** state.

3.10.3 Fanout

fanout

As we defined it previously in Section 3.5.4, *fanout* is a measure of the number of gate inputs that are connected to (and driven by) a single gate output. As we showed in that section, the DC fanout of CMOS outputs driving CMOS inputs is virtually unlimited, because CMOS inputs require almost no current in either state, **HIGH** or **LOW**. This is not the case with TTL inputs. As a result, there are very definite limits on the fanout of TTL or CMOS outputs driving TTL inputs, as you'll learn in the paragraphs that follow.

current flow

As in CMOS, the *current flow* in a TTL input or output lead is defined to be positive if the current actually flows *into* the lead, and negative if current flows *out* of the lead. As a result, when an output is connected to one or more inputs, the algebraic sum of all the input and output currents is 0.

The amount of current required by a TTL input depends on whether the input is **HIGH** or **LOW**, and is specified by two parameters:

I_{ILmax} The maximum current that an input requires to pull it **LOW**. Recall from the discussion of Figure 3-77 that positive current is actually flowing from V_{CC}, through $R1B$, through diode $D1XB$, out of the input lead, through the driving output transistor $Q5A$, and into ground.

LOW-state unit load

Since current flows out of a TTL input in the **LOW** state, I_{ILmax} has a negative value. Most LS-TTL inputs have $I_{ILmax} = -0.4$ mA, which is sometimes called a *LOW-state unit load* for LS-TTL.

I_{IHmax} The maximum current that an input requires to pull it HIGH. As shown in Figure 3-78 on page 159, positive current flows from V_{CC}, through *R5A* and *Q4A* of the driving gate, and *into* the driven input, where it leaks to ground through reverse-biased diodes *D1XB* and *D2XB*.

Since current flows *into* a TTL input in the HIGH state, I_{IHmax} has a positive value. Most LS-TTL inputs have $I_{\text{IHmax}} = 20\ \mu\text{A}$, which is sometimes called a *HIGH-state unit load* for LS-TTL.   *HIGH-state unit load*

Like CMOS outputs, TTL outputs can source or sink a certain amount of current depending on the state, HIGH or LOW:

I_{OLmax} The maximum current an output can sink in the LOW state while maintaining an output voltage no more than V_{OLmax}. Since current flows into the output, I_{OLmax} has a positive value, 8 mA for most LS-TTL outputs.

I_{OHmax} The maximum current an output can source in the HIGH state while maintaining an output voltage no less than V_{OHmin}. Since current flows out of the output, I_{OHmax} has a negative value, $-400\ \mu\text{A}$ for most LS-TTL outputs.

Notice that the value of I_{OLmax} for typical LS-TTL outputs is exactly 20 times the absolute value of I_{ILmax}. As a result, LS-TTL is said to have a *LOW-state fanout* of 20, because an output can drive up to 20 inputs in the LOW state.   *LOW-state fanout*
Similarly, the absolute value of I_{OHmax} is exactly 20 times I_{IHmax}, so LS-TTL is said to have a *HIGH-state fanout* of 20 also. The *overall fanout* is the lesser of the LOW- and HIGH-state fanouts.   *HIGH-state fanout*   *overall fanout*

Loading a TTL output with more than its rated fanout has the same deleterious effects that were described for CMOS devices in Section 3.5.5 on page 110. That is, DC noise margins may be reduced or eliminated, transition times and delays may increase, and the device may overheat.

TTL OUTPUT ASYMMETRY	Although LS-TTL's numerical fanouts for HIGH and LOW states are equal, LS-TTL and other TTL families have a definite asymmetry in current driving capability—an LS-TTL output can sink 8 mA in the LOW state, but can source only 400 μA in the HIGH state.

        This asymmetry is no problem when TTL outputs drive other TTL inputs, because it is matched by a corresponding asymmetry in TTL input current requirements (I_{ILmax} is large, while I_{IHmax} is small). However, it is a limitation when TTL is used to drive LEDs, relays, solenoids, or other devices requiring large amounts of current, often tens of milliamperes. Circuits using these devices must be designed so that current flows (and the driven device is "on") when the TTL output is in the LOW state, and so little or no current flows in the HIGH state. Special TTL buffer/driver gates are made that can sink up to 60 mA in the LOW state, but that still have a rather puny current-sourcing capability in the HIGH state (2.4 mA).

BURNED FINGERS   If a TTL or CMOS output is forced to sink a lot more than I_{OLmax}, the device may be damaged, especially if high current is allowed to flow for more than a second or so. For example, suppose that a TTL output in the **LOW** state is short-circuited directly to the 5 V supply. The **ON** resistance, $R_{CE(sat)}$, of the saturated $Q5$ transistor in a typical TTL output stage is less than 10 Ω. Thus, $Q5$ must dissipate about $5^2/10$ or 2.5 watts. Don't try this yourself unless you're prepared to deal with the consequences! That's enough heat to destroy the device (and burn your finger) in a very short time.

In general, two calculations must be carried out to confirm that an output is not being overloaded:

HIGH state   The I_{IHmax} values for all of the driven inputs are added. This sum must be less than or equal to the absolute value of I_{OHmax} for the driving output.

LOW state   The I_{ILmax} values for all of the driven inputs are added. The absolute value of this sum must be less than or equal to I_{OLmax} for the driving output.

For example, suppose you designed a system in which a certain LS-TTL output drives ten LS-TTL and three S-TTL gate inputs. In the HIGH state, a total of $10 \cdot 20 + 3 \cdot 50$ μA = 350 μA is required. This is within an LS-TTL output's HIGH-state current-sourcing capability of 400 μA. But in the LOW state, a total of $10 \cdot 0.4 + 3 \cdot 2.0$ mA = 10.0 mA is required. This is more than an LS-TTL output's LOW-state current-sinking capability of 8 mA, so the output is overloaded.

3.10.4 Unused Inputs

Unused inputs of TTL gates can be handled in the same way as we described for CMOS gates in Section 3.5.6 on page 111. That is, unused inputs may be tied to used ones, or unused inputs may be pulled HIGH or LOW as is appropriate for the logic function.

The resistance value of a pull-up or pull-down resistor is more critical with TTL gates than CMOS gates, because TTL inputs draw significantly more current, especially in the LOW state. If the resistance is too large, the voltage drop across the resistor may result in a gate input voltage beyond the normal LOW or HIGH range.

For example, consider the pull-down resistor shown in Figure 3-80. The pull-down resistor must sink 0.4 mA of current from each of the unused LS-TTL inputs that it drives. Yet the voltage drop across the resistor must be no more than 0.5 V in order to have a LOW input voltage no worse than that produced by a normal gate output. If the resistor drives n LS-TTL inputs, then we must have

$$n \cdot 0.4 \text{ mA} \cdot R_{pd} < 0.5 \text{ V}$$

FLOATING TTL INPUTS   Analysis of the TTL input structure shows that unused inputs left unconnected (or *floating*) behave as if they have a HIGH voltage applied—they are pulled HIGH by base resistor *R1* in Figure 3-75 on page 157. However, *R1*'s pull-up is much weaker than that of a TTL output driving the input. As a result, a small amount of circuit noise, such as that produced by other gates when they switch, can make a floating input spuriously behave like it's LOW. Therefore, for the sake of reliability, unused TTL inputs should be tied to a stable HIGH or LOW voltage source.

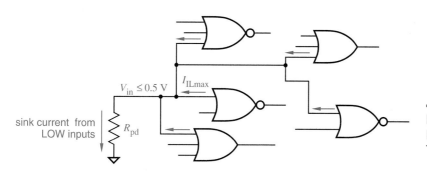

Figure 3-80
Pull-down resistor for TTL inputs.

Thus, if the resistor must pull 10 LS-TTL inputs LOW, then we must have $R_{pd} < 0.5 / (10 \cdot 4 \cdot 10^{-3})$, or $R_{pd} < 125\ \Omega$.

Similarly, consider the pull-up resistor shown in Figure 3-81. It must source 20 μA of current to each unused input while producing a HIGH voltage no worse than that produced by a normal gate output, 2.7 V. Therefore, the voltage drop across the resistor must be no more than 2.3 V; if n LS-TTL input are driven, we must have

$$n \cdot 20\ \mu A \cdot R_{pu} < 2.3\ V$$

Thus, if 10 LS-TTL inputs are pulled up, then $R_{pu} < 2.3 / (10 \cdot 20 \cdot 10^{-6})$, or $R_{pu} < 11.5\ K\Omega$.

Figure 3-81
Pull-up resistor for TTL inputs.

WHY USE A RESISTOR?

You might be asking yourself, "Why use a pull-up or pull-down resistor, when a direct connection to ground or the 5-V power supply should be a perfectly good source of LOW or HIGH?"

Well, for a HIGH source, a direct connection to the 5-V power supply is not recommended, since an input transient of over 5.5 V can damage some TTL devices, ones that use a multiple-emitter transistor in the input stage. The pull-up resistor limits current and prevents damage in this case.

For a LOW source, a direct connection to ground without the pull-down resistor is actually OK in most cases. You'll see many examples of this sort of connection throughout this book. However, as explained in Section 11.2.2 on page 904, the pull-down resistor is still desirable in some cases, so that the "constant" LOW signal it produces can be overridden and driven HIGH for system-testing purposes.

3.10.5 Additional TTL Gate Types

Although the NAND gate is the "workhorse" of the TTL family, other types of gates can be built with the same general circuit structure.

The circuit diagram for an LS-TTL NOR gate is shown in Figure 3-82. If either input X or Y is HIGH, the corresponding phase-splitter transistor $Q2X$ or $Q2Y$ is turned on, which turns off $Q3$ and $Q4$ while turning on $Q5$ and $Q6$, and the output is LOW. If both inputs are LOW, then both phase-splitter transistors are off, and the output is forced HIGH. This functional operation is summarized in Figure 3-83.

The LS-TTL NOR gate's input circuits, phase splitter, and output stage are almost identical to those of an LS-TTL NAND gate. The difference is that an LS-TTL NAND gate uses diodes to perform the AND function, while an LS-TTL NOR gate uses parallel transistors in the phase splitter to perform the OR function.

The speed, input, and output characteristics of a TTL NOR gate are comparable to those of a TTL NAND. However, an n-input NOR gate uses more transistors and resistors and is thus more expensive in silicon area than an n-input NAND. Also, internal leakage current limits the number of $Q2$ transistors that can be placed in parallel, so NOR gates have poor fan-in. (The largest discrete TTL NOR gate has only 5 inputs, compared with a 13-input NAND.) As a result, NOR gates are less commonly used than NAND gates in TTL designs.

The most "natural" TTL gates are inverting gates like NAND and NOR. Noninverting TTL gates include an extra inverting stage, typically between the input stage and the phase splitter. As a result, noninverting TTL gates are typically larger and slower than the inverting gates on which they are based.

Like CMOS, TTL gates can be designed with three-state outputs. Such gates have an "output-enable" or "output-disable" input that allows the output to be placed in a high-impedance state where neither output transistor is turned on.

Figure 3-82 Circuit diagram of a two-input LS-TTL NOR gate.

(a)

X	Y	V_{AX}	Q2X	V_{AY}	Q2Y	Q3	Q4	Q5	Q6	V_Z	Z
L	L	≤ 1.05	off	≤ 1.05	off	on	on	off	off	≥ 2.7	H
L	H	≤ 1.05	off	1.2	on	off	off	on	on	≤ 0.35	L
H	L	1.2	on	≤ 1.05	off	off	off	on	on	≤ 0.35	L
H	H	1.2	on	1.2	on	off	off	on	on	≤ 0.35	L

Figure 3-83
Two-input LS-TTL
NOR gate:
(a) function table;
(b) truth table;
(c) logic symbol.

(b)

X	Y	Z
0	0	1
0	1	0
1	0	0
1	1	0

(c)

Some TTL gates are also available with *open-collector outputs*. Such gates omit the entire upper half of the output stage in Figure 3-75, so that only passive pull-up to the HIGH state is provided by an external resistor. The applications and required calculations for TTL open-collector gates are similar to those for CMOS gates with open-drain outputs.

open-collector output

3.11 TTL Families

TTL families have evolved over the years in response to the demands of digital designers for better performance. As a result, three TTL families have come and gone, and today's designers have five surviving families from which to choose. All of the TTL families are compatible in that they use the same power-supply voltage and logic levels, but each family has its own advantages in terms of speed, power consumption, and cost.

3.11.1 Early TTL Families

The original TTL family of logic gates was introduced by Sylvania in 1963. It was popularized by Texas Instruments, whose "7400-series" part numbers for gates and other TTL components quickly became an industry standard.

As in 7400-series CMOS, devices in a given TTL family have part numbers of the form 74FAMnn, where "FAM" is an alphabetic family mnemonic and nn is a numeric function designator. Devices in different families with the same value of nn perform the same function. In the original TTL family, "FAM" is null and the family is called *74-series TTL.*

74-series TTL

74H (High-speed TTL)

74L (Low-power TTL)

Resistor values in the original TTL circuit were changed to obtain two more TTL families with different performance characteristics. The *74H (High-speed TTL)* family used lower resistor values to reduce propagation delay at the expense of increased power consumption. The *74L (Low-power TTL)* family used higher resistor values to reduce power consumption at the expense of propagation delay.

The availability of three TTL families allowed digital designers in the 1970s to make a choice between high speed and low power consumption for their circuits. However, like many people in the 1970s, they wanted to "have it all, now." The development of Schottky transistors provided this opportunity and made 74, 74H, and 74L TTL obsolete. The characteristics of better-performing, contemporary TTL families are discussed in the rest of this section.

3.11.2 Schottky TTL Families

74S (Schottky TTL)

Historically, the first family to make use of Schottky transistors was *74S (Schottky TTL)*. With Schottky transistors and low resistor values, this family has much higher speed, but higher power consumption, than the original 74-series TTL.

74LS (Low-power Schottky TTL)

Perhaps the most widely used and certainly the least expensive TTL family is *74LS (Low-power Schottky TTL)*, introduced shortly after 74S. By combining Schottky transistors with higher resistor values, 74LS TTL matches the speed of 74-series TTL but has about one-fifth of its power consumption. Thus, 74LS is a preferred logic family for new TTL designs.

74AS (Advanced Schottky TTL)

Subsequent IC processing and circuit innovations gave rise to two more Schottky logic families. The *74AS (Advanced Schottky TTL)* family offers speeds approximately twice as fast as 74S with approximately the same power

Table 3-11 Characteristics of gates in TTL families.

Description	Symbol	Family 74S	74LS	74AS	74ALS	74F
Maximum propagation delay (ns)		3	9	1.7	4	3
Power consumption per gate (mW)		19	2	8	1.2	4
Speed-power product (pJ)		57	18	13.6	4.8	12
LOW-level input voltage (V)	V_{ILmax}	0.8	0.8	0.8	0.8	0.8
LOW-level output voltage (V)	V_{OLmax}	0.5	0.5	0.5	0.5	0.5
HIGH-level input voltage (V)	V_{IHmin}	2.0	2.0	2.0	2.0	2.0
HIGH-level output voltage (V)	V_{OHmin}	2.7	2.7	2.7	2.7	2.7
LOW-level input current (mA)	I_{ILmax}	−2.0	−0.4	−0.5	−0.2	−0.6
LOW-level output current (mA)	I_{OLmax}	20	8	20	8	20
HIGH-level input current (μA)	I_{IHmax}	50	20	20	20	20
HIGH-level output current (μA)	I_{OHmax}	−1000	−400	−2000	−400	−1000

consumption. The *74ALS (Advanced Low-power Schottky TTL)* family offers both lower power and higher speeds than 74LS, and rivals 74LS in popularity for general-purpose requirements in new TTL designs. The *74F (Fast TTL)* family is positioned between 74AS and 74ALS in the speed/power tradeoff, and is probably the most popular choice for high-speed requirements in new TTL designs.

74ALS (Advanced Low-power Schottky TTL)

74F (Fast TTL)

3.11.3 Characteristics of TTL Families

The important characteristics of contemporary TTL families are summarized in Table 3-11. The first two rows of the table list the propagation delay (in nano-seconds) and the power consumption (in milliwatts) of a typical 2-input NAND gate in each family.

One figure of merit of a logic family is its *speed-power product* listed in the third row of the table. As discussed previously, this is simply the product of the propagation delay and power consumption of a typical gate. The speed-power product measures a sort of efficiency—how much energy a logic gate uses to switch its output.

The remaining rows in Table 3-11 describe the input and output parameters of typical TTL gates in each of the families. Using this information, you can analyze the external behavior of TTL gates without knowing the details of the internal TTL circuit design. These parameters were defined and discussed in Sections 3.10.2 and 3.10.3. As always, the input and output characteristics of specific components may vary from the representative values given in Table 3-11, so you must always consult the manufacturer's data book when analyzing a real design.

3.11.4 A TTL Data Sheet

Table 3-12 shows the part of a typical manufacturer's data sheet for the 74LS00. The 54LS00 listed in the data sheet is identical to the 74LS00, except that it is specified to operate over the full "military" temperature and voltage range, and it costs more. Most TTL parts have corresponding 54-series (military) versions. Three sections of the data sheet are shown in the table:

recommended operating conditions

- *Recommended operating conditions* specify power-supply voltage, input-voltage ranges, DC output loading, and temperature values under which the device is normally operated.

electrical characteristics

- *Electrical characteristics* specify additional DC voltages and currents that are observed at the device inputs and output when it is operated under the recommended conditions:

    I_I Maximum input current for a very high HIGH input voltage.

    I_OS Output current with HIGH output shorted to ground.

    I_CCH Power-supply current when all outputs (on four NAND gates) are HIGH. (The number given is for the entire package, which contains four NAND gates, so the current per gate is one-fourth of the specified amount.)

    I_CCL Power-supply current when all outputs (on four NAND gates) are LOW.

switching characteristics

- *Switching characteristics* give maximum and typical propagation delays under "typical" operating conditions of $V_\text{CC} = 5$ V and $T_\text{A} = 25°\text{C}$. A conservative designer must increase these delays by 5%–10% to account for different power-supply voltages and temperatures, and even more under heavy loading conditions.

A fourth section is also included in the manufacturer's data book:

absolute maximum ratings

- *Absolute maximum ratings* indicate the worst-case conditions for operating or storing the device without damage.

A complete data book also shows test circuits that are used to measure the parameters when the device is manufactured, and graphs that show how the typical parameters vary with operating conditions such as power-supply voltage (V_CC), ambient temperature (T_A), and load (R_L, C_L).

Table 3-12 Typical manufacturer's data sheet for the 74LS00.

RECOMMENDED OPERATING CONDITIONS

Parameter	Description	SN54LS00			SN74LS00			Unit
		Min.	Nom.	Max.	Min.	Nom.	Max.	
V_{CC}	Supply voltage	4.5	5.0	5.5	4.75	5.0	5.25	V
V_{IH}	High-level input voltage	2.0			2.0			V
V_{IL}	Low-level input voltage			0.7			0.8	V
I_{OH}	High-level output current			−0.4			−0.4	mA
I_{OL}	Low-level output current			4			8	mA
T_A	Operating free-air temperature	−55		125	0		70	°C

ELECTRICAL CHARACTERISTICS OVER RECOMMENDED FREE-AIR TEMPERATURE RANGE

Parameter	Test Conditions[1]	SN54LS00			SN74LS00			Unit
		Min.	Typ.[2]	Max.	Min.	Typ.[2]	Max.	
V_{IK}	V_{CC} = Min., I_N = −18 mA			−1.5			−1.5	V
V_{OH}	V_{CC} = Min., V_{IL} = Max., I_{OH} = −0.4 mA	2.5	3.4		2.7	3.4		V
V_{OL}	V_{CC} = Min., V_{IH} = 2.0 V, I_{OL} = 4 mA		0.25	0.4		0.25	0.4	V
	V_{CC} = Min., V_{IH} = 2.0 V, I_{OL} = 8 mA					0.35		
I_I	V_{CC} = Max., V_I = 7.0 V			0.1			0.1	mA
I_{IH}	V_{CC} = Max. V_I = 2.7 V			20			20	μA
I_{IL}	V_{CC} = Max. V_I = 0.4 V			−0.4			−0.4	mA
I_{IOS}[3]	V_{CC} = Max.	−20		−100	−20		−100	mA
I_{CCH}	V_{CC} = Max., V_I = 0 V		0.8	1.6		0.8	1.6	mA
I_{CCL}	V_{CC} = Max., V_I = 4.5 V		2.4	4.4		2.4	4.4	mA

SWITCHING CHARACTERISTICS, V_{CC} = 5.0 V, T_A = 25°C

Parameter	From (Input)	To (Output)	Test Conditions	Min.	Typ.	Max.	Unit
t_{PLH}	A or B	Y	R_L = 2 kΩ, C_L = 15 pF		9	15	ns
t_{PHL}					10	15	

NOTES:

1. For conditions shown as Max. or Min., use appropriate value specified under Recommended Operating Conditions.
2. All typical values are at V_{CC} = 5.0 V, T_A = 25°C.
3. Not more than one output should be shorted at a time; duration of short-circuit should not exceed one second.

*3.12 CMOS/TTL Interfacing

A digital designer selects a "default" logic family to use in a system, based on general requirements of speed, power, cost, and so on. However, the designer may select devices from other families in some cases because of availability or other special requirements. (For example, not all 74LS part numbers are available in 74HCT, and vice versa.) Thus, it's important for a designer to understand the implications of connecting TTL outputs to CMOS inputs, and vice versa.

There are several factors to consider in TTL/CMOS interfacing, and the first is noise margin. The LOW-state DC noise margin depends on V_{OLmax} of the driving output and V_{ILmax} of the driven input, and equals $V_{ILmax} - V_{OLmax}$. Similarly, the HIGH-state DC noise margin equals $V_{OHmin} - V_{IHmin}$. Figure 3-84 shows the relevant numbers for TTL and CMOS families.

For example, the LOW-state DC noise margin of HC or HCT driving TTL is $0.8 - 0.33 = 0.47$ V, and the HIGH-state is $3.84 - 2.0 = 1.84$ V. On the other hand, the HIGH-state margin of TTL driving HC or VHC is $2.7 - 3.85 = -1.15$ V. In other words, TTL driving HC or AC doesn't work, unless the TTL HIGH output happens to be higher and the CMOS HIGH input threshold happens to be lower by a total of 1.15 V compared to their worst-case specs. To drive CMOS inputs properly from TTL outputs, the CMOS devices should be HCT, VHCT. or FCT rather than HC or VHC.

The next factor to consider is fanout. As with pure TTL (Section 3.10.3), a designer must sum the input current requirements of devices driven by an output and compare with the output's capabilities in both states. Fanout is not a problem when TTL drives CMOS, since CMOS inputs require almost no current in either state. On the other hand, TTL inputs, especially in the LOW state, require

Figure 3-84
Output and input levels for interfacing TTL and CMOS families. (Note that HC and VHC inputs are not TTL compatible.)

substantial current, especially compared to HC and HCT output capabilities. For example, an HC or HCT output can drive 10 LS or only two S-TTL inputs.

The last factor is capacitive loading. We've seen that load capacitance increases both the delay and the power dissipation of logic circuits. Increases in delay are especially noticeable with HC and HCT outputs, whose transition times increase about 1 ns for each 5 pF of load capacitance. The transistors in FCT outputs have very low "on" resistances, so their transition times increase only about 0.1 ns for each 5 pF of load capacitance.

For a given load capacitance, power-supply voltage, and application, all of the CMOS families have similar dynamic power dissipation, since each variable in the CV^2f equation is the same. On the other hand, TTL outputs have somewhat lower dynamic power dissipation, since the voltage swing between TTL HIGH and LOW levels is smaller.

*3.13 Low-Voltage CMOS Logic and Interfacing

Two important factors have led the IC industry to move toward lower power-supply voltages in CMOS devices:

- In most applications, CMOS output voltages swing from rail to rail, so the V in the CV^2f equation is the power-supply voltage. Cutting power-supply voltage reduces dynamic power dissipation more than proportionally.

- As the industry moves toward ever-smaller transistor geometries, the oxide insulation between a CMOS transistor's gate and its source and drain is getting ever thinner, and thus incapable of insulating voltage potentials as "high" as 5 V.

As a result, JEDEC, an IC industry standards group, selected $3.3\,V \pm 0.3V$, $2.5\,V \pm 0.2V$, and $1.8\,V \pm 0.15V$ as the next "standard" logic power-supply voltages. JEDEC standards specify the input and output logic voltage levels for devices operating with these power-supply voltages.

The migration to lower voltages has occurred in stages and will continue to do so. For discrete logic families, the trend has been to produce parts that operate and produce outputs at the lower voltage but can also tolerate inputs at the higher voltage. This approach has allowed 3.3-V CMOS families to operate with 5-V CMOS and TTL families, as we'll see in the next section.

Many ASICs and microprocessors have followed a similar approach, but another approach is often used as well. These devices are large enough that it can make sense to provide them with two power-supply voltages. A low voltage, such as 2.5 V, is supplied to operate the chip's internal gates, or *core logic*. A higher voltage, such as 3.3 V, is supplied to operate the external input and output circuits, or *pad ring*, for compatibility with older-generation devices in the system. Special buffer circuits are used internally to translate safely and quickly between the core-logic and the pad-ring logic voltages.

core logic

pad ring

*3.13.1 3.3-V LVTTL and LVCMOS Logic

The relationships among signal levels for standard TTL and low-voltage CMOS devices operating at their nominal power-supply voltages are illustrated nicely in Figure 3-85, adapted from a Texas Instruments application note. The original, symmetric signal levels for pure 5-V CMOS families such as HC and VHC are shown in (a). TTL-compatible CMOS families such as HCT, VHCT, and FCT shift the voltage levels downward for compatibility with TTL as shown in (b).

LVCMOS (low-voltage CMOS)

LVTTL (low-voltage TTL)

The first step in the progression of lower CMOS power-supply voltages was 3.3 V. The JEDEC standard for 3.3-V logic actually defines two sets of levels. *LVCMOS (low-voltage CMOS)* levels are used in pure CMOS applications where outputs have light DC loads (less than 100 µA), so V_{OL} and V_{OH} are maintained within 0.2 V of the power-supply rails. *LVTTL (low-voltage TTL)* levels, shown in (c), are used in applications where outputs have significant DC loads, so V_{OL} can be as high as 0.4 V and V_{OH} can be as low as 2.4 V.

The positioning of TTL's logic levels at the low end of the 5-V range was really quite fortuitous. As shown in Figure 3-85(b) and (c), it was possible to define the LVTTL levels to match up with TTL levels exactly. Thus, an LVTTL output can drive a TTL input with no problem, as long as its output current specifications (I_{OLmax}, I_{OHmax}) are respected. Similarly, a TTL output can drive an LVTTL input, except for the problem of driving it beyond LVTTL's 3.3-V V_{CC}, as discussed next.

Figure 3-85 Comparison of logic levels: (a) 5-V CMOS; (b) 5-V TTL, including 5-V TTL-compatible CMOS; (c) 3.3-V LVTTL; (d) 2.5-V CMOS; (e) 1.8-V CMOS.

*3.13.2 5-V Tolerant Inputs

The inputs of a gate won't necessarily tolerate voltages greater than V_{CC}. This is a problem when both 5-V and 3.3-V logic families are used in a system. For example, 5-V CMOS devices easily produce 4.9-V outputs when lightly loaded, and both CMOS and TTL devices routinely produce 4.0-V outputs even when moderately loaded. The inputs of 3.3-V devices may not like these high voltages.

The maximum voltage V_{Imax} that can be tolerated by an input is listed in the "absolute maximum ratings" section of the manufacturer's data sheet. For HC devices, V_{Imax} equals V_{CC}. Thus, if an HC device is powered by a 3.3-V supply, its cannot be driven by any 5-V CMOS or TTL outputs. For VHC devices, on the other hand, V_{Imax} is 7 V; thus, VHC devices with a 3.3-V power supply may be used to convert 5-V outputs to 3.3-V levels for use with 3.3-V microprocessors, memories, and other devices in a pure 3.3-V subsystem.

Figure 3-86 explains why some inputs are 5-V tolerant and others are not. As shown in (a), the HC and HCT input structure actually contains two reverse-biased *clamp diodes*, which we haven't shown before, between each input signal and V_{CC} and ground. The purpose of these diodes is specifically to shunt any transient input signal value less than 0 through *D1* or greater than V_{CC} through *D2* to the corresponding power-supply rail. Such transients can occur as a result of transmission-line reflections, as described in Section 11.4. Shunting the so-called "undershoot" or "overshoot" to ground or V_{CC} reduces the magnitude and duration of reflections.

clamp diode

Of course, diode *D2* can't distinguish between transient overshoot and a persistent input voltage greater than V_{CC}. Hence, if a 5-V output is connected to one of these inputs, it will not see the very high impedance normally associated with a CMOS input. Instead, it will see a relatively low impedance path to V_{CC} through the now forward-biased diode *D2*, and excessive current will flow.

Figure 3-86(b) shows a 5-V tolerant CMOS input. This input structure simply omits *D2*; diode *D1* is still provided to clamp undershoot. The VHC and AHC families use this input structure.

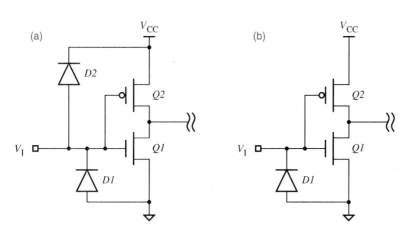

Figure 3-86
CMOS input structures:
(a) non-5-V tolerant HC;
(b) 5-V tolerant VHC.

The kind of input structure shown in Figure 3-86(b) is necessary but not sufficient to create 5-V tolerant inputs. The transistors in a device's particular fabrication process must also be able to withstand voltage potentials higher than V_{CC}. On this basis, V_{Imax} in the VHC family is limited to 7.0 V. In many 3.3-V ASIC processes, it's not possible to get 5-V tolerant inputs, even if you're willing to give up the transmission-line benefits of diode $D2$.

*3.13.3 5-V Tolerant Outputs

Five-volt tolerance must also be considered for outputs, in particular, when both 3.3-V and 5-V three-state outputs are connected to a bus. When the 3.3-V output is in the disabled, Hi-Z state, a 5-V device may be driving the bus, and a 5-V signal may appear on the 3.3-V device's *output*.

In this situation, Figure 3-87 explains why some outputs are 5-V tolerant and others are not. As shown in (a), the standard CMOS three-state output has an *n*-channel transistor $Q1$ to ground and a *p*-channel transistor $Q2$ to V_{CC}. When the output is disabled, circuitry (not shown) holds the gate of $Q1$ near 0 V, and the gate of $Q2$ near V_{CC}, so both transistors are off and Y is Hi-Z.

Now consider what happens if V_{CC} is 3.3 V and a different device applies a 5-V signal to the output pin Y in (a). Then the drain of $Q2$ (Y) is at 5 V while the gate (V_2) is still at only 3.3 V. With the gate at a lower potential than the drain, $Q2$ will begin to conduct and provide a relatively low-impedance path from Y to V_{CC}, and excessive current will flow. Both HC and VHC three-state outputs have this structure and therefore are not 5-V tolerant.

Figure 3-87(b) shows a 5-V tolerant output structure. An extra *p*-channel transistor $Q3$ is used to prevent $Q2$ from turning on when it shouldn't. When V_{OUT} is greater than V_{CC}, $Q3$ turns on. This forms a relatively low impedance path from Y to the gate of $Q2$, which now stays off because its gate voltage V_2 can no longer be below the drain voltage. This output structure is used in Texas Instruments' LVC (Low-Voltage CMOS) family.

Figure 3-87
CMOS three-state output structures:
(a) non-5-V tolerant HC and VHC;
(b) 5-V tolerant LVC.

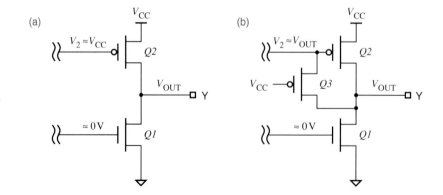

*3.13.4 TTL/LVTTL Interfacing Summary

Based on the information in the preceding subsections, TTL (5-V) and LVTTL (3.3-V) devices can be mixed in the same system subject to just three rules:

1. LVTTL outputs can drive TTL inputs directly, subject to the usual constraints on output current (I_{OLmax}, I_{OHmax}) of the driving devices.

2. TTL outputs can drive LVTTL inputs if the inputs are 5-V tolerant.

3. TTL and LVTTL three-state outputs can drive the same bus if the LVTTL outputs are 5-V tolerant.

*3.13.5 2.5-V and 1.8-V Logic

The transition from 3.3-V to 2.5-V logic will not be so easy. It is true that 3.3-V outputs can drive 2.5-V inputs as long as the inputs are 3.3-V tolerant. However, a quick look at Figure 3-85(c) and (d) on page 172 shows that V_{OH} of a 2.5-V output equals V_{IH} of a 3.3-V input. In other words, there is zero HIGH-state DC noise margin when a 2.5-V output drives a 3.3-V input—not a good situation.

The solution to this problem is to use a *level translator* or *level shifter*, a device which is powered by both supply voltages and which internally boosts the lower logic levels (2.5 V) to the higher ones (3.3 V). Many of today's ASICs and microprocessors contain level translators internally, allowing them to operate with a 2.5-V or 2.7-V core and a 3.3-V pad ring, as we discussed at the beginning of this section. If and when 2.5-V discrete devices become popular, we can expect the major semiconductor vendors to produce level translators as stand-alone components as well.

level translator
level shifter

The next step will be a transition from 2.5-V to 1.8-V logic. Referring to Figure 3-85(d) and (e), you can see that the HIGH-state DC noise margin is actually negative when a 1.8-V output drives a 2.5-V input, so level translators will be needed in this case also.

*3.14 Emitter-Coupled Logic

The key to reducing propagation delay in a bipolar logic family is to prevent a gate's transistors from saturating. In Section 3.9.5, we learned how Schottky diodes prevent saturation in TTL gates. However, it is also possible to prevent saturation by using a radically different circuit structure, called *current-mode logic (CML)* or *emitter-coupled logic (ECL)*.

current-mode logic (CML)

emitter-coupled logic (ECL)

Unlike the other logic families in this chapter, CML does not produce a large voltage swing between the LOW and HIGH levels. Instead, it has a small voltage swing, less than a volt, and it internally switches current between two possible paths, depending on the output state.

The first CML logic family was introduced by General Electric in 1961. The concept was soon refined by Motorola and others to produce the still popular 10K and 100K *emitter-coupled logic (ECL)* families. These families are

emitter-coupled logic (ECL)

extremely fast, offering propagation delays as short as 1 ns. The newest ECL family, ECLinPS (literally, ECL in picoseconds), offers maximum delays under 0.5 ns (500 ps), including the signal delay getting on and off of the IC package. Throughout the evolution of digital circuit technology, some type of ECL has always been the fastest technology for discrete, packaged logic components.

Still, commercial ECL families aren't nearly as popular as CMOS and TTL, mainly because they consume much more power. In fact, high power consumption made the design of ECL supercomputers, such as the Cray-1 and Cray-2, as much of a challenge in cooling technology as in digital design. Also, ECL has a poor speed-power product, does not provide a high level of integration, has fast edge rates requiring design for transmission-line effects in most applications, and is not directly compatible with TTL and CMOS. Nevertheless, ECL still finds its place as a logic and interface technology in very high-speed communications gear, including fiber-optic transceiver interfaces for gigabit Ethernet and Asynchronous Transfer Mode (ATM) networks.

*3.14.1 Basic CML Circuit

differential amplifier

The basic idea of current-mode logic is illustrated by the inverter/buffer circuit in Figure 3-88. This circuit has both an inverting output (OUT1) and a noninverting output (OUT2). Two transistors are connected as a *differential amplifier* with a common emitter resistor. The supply voltages for this example are $V_{CC} = 5.0$, $V_{BB} = 4.0$, and $V_{EE} = 0$ V, and the input LOW and HIGH levels are defined to be 3.6 and 4.4 V. This circuit actually produces output LOW and HIGH levels that are 0.6 V higher (4.2 and 5.0 V), but this is corrected in real ECL circuits.

Figure 3-88
Basic CML inverter/buffer circuit with input HIGH.

When V_{IN} is HIGH, as shown in the figure, transistor $Q1$ is on, but not saturated, and transistor $Q2$ is OFF. This is true because of a careful choice of resistor values and voltage levels. Thus, V_{OUT2} is pulled to 5.0 V (HIGH) through $R2$, and it can be shown that the voltage drop across $R1$ is about 0.8 V, so that V_{OUT1} is about 4.2 V (LOW).

When V_{IN} is LOW, as shown in Figure 3-89, transistor $Q2$ is on, but not saturated, and transistor $Q1$ is OFF. Thus, V_{OUT1} is pulled to 5.0 V through $R1$, and it can be shown that V_{OUT2} is about 4.2 V.

The outputs of this inverter are called *differential outputs* because they are always complementary, and it is possible to determine the output state by looking at the difference between the output voltages ($V_{OUT1} - V_{OUT2}$) rather than their absolute values. That is, the output is 1 if ($V_{OUT1} - V_{OUT2}$) > 0, and it is 0 if ($V_{OUT1} - V_{OUT2}$) > 0. It is possible to build input circuits with two wires per logical input that define the logical signal value in this way; these are called *differential inputs*.

Differential signals are used in most ECL "interfacing" and "clock distribution" applications because of their low skew and high noise immunity. They are "low skew" because the timing of a 0-to-1 or 1-to-0 transition does not depend critically on voltage thresholds, which may change with temperature or between devices. Instead, the timing depends only on when the voltages cross over relative to each other. Similarly, the "relative" definition of 0 and 1 provides outstanding noise immunity, since noise created by variations in the power supply or coupled from external sources tends to be a *common-mode signal* that affect both differential signals similarly, leaving the difference value unchanged.

differential outputs

differential inputs

common-mode signal

Figure 3-89
Basic CML inverter/buffer circuit with input LOW.

single-ended input

It is also possible, of course, to determine the logic value by sensing the absolute voltage level of one input signal, called a *single-ended input*. Single-ended signals are used in most ECL "logic" applications to avoid the obvious expense of doubling the number of signal lines. The basic CML inverter in Figure 3-89 has a single-ended input. It always has both "outputs" available internally; the circuit is actually either an inverter or a noninverting buffer, depending on whether we use OUT1 or OUT2.

To perform logic with the basic circuit of Figure 3-89, we simply place additional transistors in parallel with $Q1$, similar to the approach in a TTL NOR gate. For example, Figure 3-90 shows a 2-input CML OR/NOR gate. If any input is HIGH, the corresponding input transistor is active, and V_{OUT1} is LOW (NOR output). At the same time, $Q3$ is OFF, and V_{OUT2} is HIGH (OR output).

Recall that the input levels for the inverter/buffer are defined to be 3.6 and 4.4 V, while the output levels that it produces are 4.2 and 5.0 V. This is obviously

Figure 3-90 CML 2-input OR/NOR gate: (a) circuit diagram; (b) function table; (c) logic symbol; (d) truth table.

(b)

X	Y	V_X	V_Y	$Q1$	$Q2$	$Q3$	V_E	V_{OUT1}	V_{OUT2}	OUT1	OUT2
L	L	3.6	3.6	OFF	OFF	on	3.4	5.0	4.2	H	L
L	H	3.6	4.4	OFF	on	OFF	3.8	4.2	5.0	L	H
H	L	4.4	3.6	on	OFF	OFF	3.8	4.2	5.0	L	H
H	H	4.4	4.4	on	on	OFF	3.8	4.2	5.0	L	H

(d)

X	Y	OUT1	OUT2
0	0	1	0
0	1	0	1
1	0	0	1
1	1	0	1

a problem. We could put a diode in series with each output to lower it by 0.6 V to match the input levels, but that still leaves another problem—the outputs have poor fanout. A HIGH output must supply base current to the inputs that it drives, and this current creates an additional voltage drop across $R1$ or $R2$, reducing the output voltage (and we don't have much margin to work with). These problems are solved in commercial ECL families, such as the 10K family described next.

*3.14.2 ECL 10K/10H Families

The packaged components in today's most popular ECL family have 5-digit part numbers of the form "10xxx" (e.g., 10102, 10181, 10209), so the family is generically called *ECL 10K*. This family has several improvements over the basic CML circuit described previously:

ECL 10K family

- An emitter-follower output stage shifts the output levels to match the input levels and provides very high current-driving capability, up to 50 mA per output. It is also responsible for the family's name, "emitter-coupled" logic.

- An internal bias network provides V_{BB} without the need for a separate, external power supply.

- The family is designed to operate with $V_{CC} = 0$ (ground) and $V_{EE} = -5.2$ V. In most applications, ground signals are more noise-free than the power-supply signals. In ECL, the logic signals are referenced to the algebraically higher power-supply voltage rail, so the family's designers decided to make that 0 V (the "clean" ground) and use a negative voltage for V_{EE}. The power-supply noise that does appear on V_{EE} is a "common-mode" signal that is rejected by the input structure's differential amplifier.

- Parts with a 10H prefix (the *ECL 10H family*) are fully voltage compensated, so they will work properly with power-supply voltages other than $V_{EE} = -5.2$ V, as we'll discuss in Section 3.14.4.

ECL 10H family

Logic LOW and HIGH levels are defined in the ECL 10K family as shown in Figure 3-91. Note that even though the power supply is negative, ECL assigns the names LOW and HIGH to the *algebraically* lower and higher voltages, respectively.

DC noise margins in ECL 10K are much less than in CMOS and TTL, only 0.155 V in the LOW state and 0.125 V in the HIGH state. However, ECL gates do not need as much noise margin as these families. Unlike CMOS and TTL, an ECL gate generates very little power-supply and ground noise when it changes state; its current requirement remains constant as it merely steers current from one path to another. Also, ECL's emitter-follower outputs have very low impedance in either state, and it is difficult to couple noise from an external source into a signal line driven by such a low-impedance output.

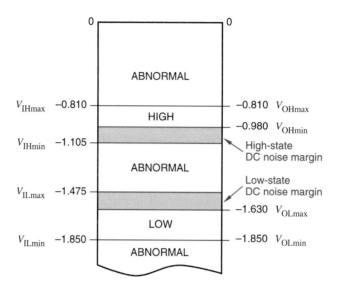

Figure 3-91
ECL 10K logic levels.

Figure 3-92(a) is the circuit for an ECL OR/NOR gate, one section of a quad OR/NOR gate with part number 10102. A pull-down resistor on each input ensures that if the input is left unconnected, it is treated as LOW. The bias network has component values selected to generate $V_{BB} = -1.29$ V for proper operation of the differential amplifier. Each output transistor, using the emitter-follower configuration, maintains its emitter voltage at one diode-drop below its base voltage, thereby achieving the required output-level shift. Figure 3-92(b) summarizes the electrical operation of the gate.

The emitter-follower outputs used in ECL 10K require external pull-down resistors, as shown in the figure. The 10K family is designed to use external rather than internal pull-down resistors for good reason. The rise and fall times of ECL output transitions are so fast (typically 2 ns) that any connection longer than a few inches must be treated as a transmission line and must be terminated as discussed in Section 11.4. Rather than waste power with an internal pull-down resistor, ECL 10K allows the designer to select an external resistor that satisfies both pull-down and transmission-line termination requirements. The simplest termination, sufficient for short connections, is to connect a resistor in the range of 270 Ω to 2 kΩ from each output to V_{EE}.

A typical ECL 10K gate has a propagation delay of 2 ns, comparable to 74AS TTL. With its outputs left unconnected, a 10K gate consumes about 26 mW of power, also comparable to a 74AS TTL gate, which consumes about 20 mW. However, the termination required by ECL 10K also consumes power, from 10 to 150 mW per output depending on the type of termination circuit. A 74AS TTL output may or may not require a power-consuming termination circuit, depending on the physical characteristics of the application.

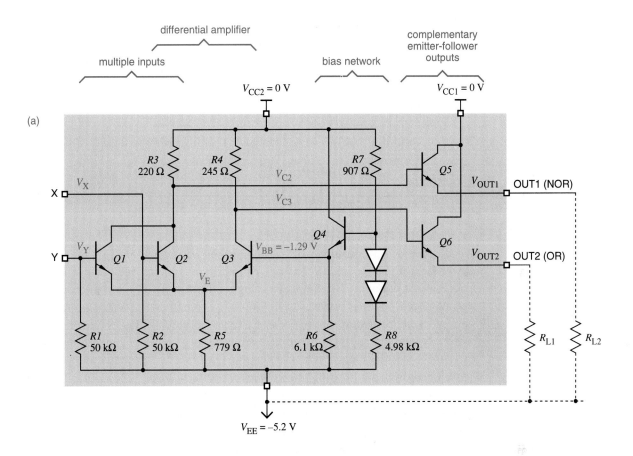

X	Y	V_X	V_Y	Q1	Q2	Q3	V_E	V_{C2}	V_{C3}	V_{OUT1}	V_{OUT2}	OUT1	OUT2
L	L	−1.8	−1.8	OFF	OFF	on	−1.9	−0.2	−1.2	−0.9	−1.8	H	L
L	H	−1.8	−0.9	OFF	on	OFF	−1.5	−1.2	−0.2	−1.8	−0.9	L	H
H	L	−0.9	−1.8	on	OFF	OFF	−1.5	−1.2	−0.2	−1.8	−0.9	L	H
H	H	−0.9	−0.9	on	on	OFF	−1.5	−1.2	−0.2	−1.8	−0.9	L	H

(c)

X	Y	OUT1	OUT2
0	0	1	0
0	1	0	1
1	0	0	1
1	1	0	1

(d)

Figure 3-92 Two-input 10K ECL OR/NOR gate: (a) circuit diagram; (b) function table; (c) truth table; (d) logic symbol.

*3.14.3 ECL 100K Family

ECL 100K family

Members of the *ECL 100K family* have 6-digit part numbers of the form "100xxx" (e.g., 100101, 100117, 100170), but in general their functions are different from those of 10K parts with similar numbers. The 100K family has the following major differences from the 10K family:

- Reduced power-supply voltage, $V_{EE} = -4.5$ V.
- Different logic levels, as a consequence of the different supply voltage.
- Shorter propagation delays, typically 0.75 ns.
- Shorter transition times, typically 0.70 ns.
- Higher power consumption, typically 40 mW per gate.

*3.14.4 Positive ECL (PECL)

We described the advantage of noise immunity provided by ECL's negative power supply ($V_{EE} = -5.2$ V or -4.5 V), but there's also a big disadvantage—today's most popular CMOS and TTL logic families, ASICs, and microprocessors all use a positive power-supply voltage, typically +5.0 V but trending to +3.3 V. Systems incorporating both ECL and CMOS/TTL devices therefore require two power supplies. In addition, interfacing between standard, negative ECL 10K or 100K logic levels and positive CMOS/TTL levels requires special level-translation components that connect to both supplies.

positive ECL (PECL)

Positive ECL (*PECL*, pronounced "peckle") uses a standard +5.0-V power supply. Note that there's nothing in the ECL 10K circuit design of Figure 3-92 that requires V_{CC} to be grounded and V_{EE} to be connected to a -5.2-V supply. The circuit will function exactly the same with V_{EE} connected to ground, and V_{CC} to a +5.2-V supply.

Thus, PECL components are nothing more than standard ECL components with V_{EE} connected to ground and V_{CC} to a +5.0-V supply. The voltage between V_{EE} and V_{CC} is a little less than with standard 10K ECL and more than with standard 100K ECL, but the 10H-series and 100K parts are voltage compensated, designed to still work well with the supply voltage being a little high or low.

Like ECL logic levels, PECL levels are referenced to V_{CC}, so the PECL HIGH level is about $V_{CC} - 0.9$ V, and LOW is about $V_{CC} - 1.7$ V, or about 4.1 V and 3.3 V with a nominal 5-V V_{CC}. Since these levels are referenced to V_{CC}, they move up and down with any variations in V_{CC}. Thus, PECL designs require particularly close attention to power-distribution issues, to prevent noise on V_{CC} from corrupting the logic levels transmitted and received by PECL devices.

Recall that CML/ECL devices produce differential outputs and can have differential inputs. A differential input is relatively insensitive to the absolute voltage levels of an input-signal pair, and sensitive only to their difference. Therefore, differential signals can be used quite effectively in PECL applications to ease the noise concerns raised in the preceding paragraph.

It is also quite common to provide differential PECL-compatible inputs and outputs on CMOS devices, allowing a direct interface between the CMOS device and a device such as a fiber-optic transceiver that expects ECL or PECL levels. In fact, as CMOS circuits have migrated to 3.3-V power supplies, it has even been possible to build PECL-like differential inputs and outputs that are simply referenced to the 3.3-V supply instead of a 5-V supply.

References

Students who need to study the basics may wish to consult "Electrical Circuits Review" by Bruce M. Fleischer. This 20-page tutorial covers all of the basic circuit concepts that are used in this chapter. It appears both as an appendix in this book's first edition and as a .pdf file on this book's Web page, www.ddpp.com.

After seeing the results of last few decades' amazing pace of development in digital electronics, it's easy to forget that logic circuits had an important place in technologies that came before the transistor. In Chapter 5 of *Introduction to the Methodology of Switching Circuits* (Van Nostrand, 1972), George J. Klir shows how logic can be (and has been) performed by a variety of physical devices, including relays, vacuum tubes, and pneumatic systems.

If you're interested in more history, a nice introduction to all of the early bipolar logic families can be found in *Logic Design with Integrated Circuits* by William E. Wickes (Wiley-Interscience, 1968). The classic introduction to TTL electrical characteristics appeared in *The TTL Applications Handbook,* edited by Peter Alfke and Ib Larsen (Fairchild Semiconductor, 1973). Early logic designers also enjoyed *The TTL Cookbook* by Don Lancaster.

For another perspective on the electronics material in this chapter, you can consult almost any modern electronics text. Many contain a much more analytical discussion of digital circuit operation; for example, see *Microelectronics* by J. Millman and A. Grabel (McGraw-Hill, 1987, second edition). Another good introduction to ICs and important logic families can be found in *VLSI System Design* by Saburo Muroga (Wiley, 1982). For NMOS and CMOS circuits in particular, two good books are *Introduction to VLSI Systems* by Carver Mead and Lynn Conway (Addison-Wesley, 1980) and *Principles of CMOS VLSI Design* by Neil H. E. Weste and Kamran Eshraghian (Addison-Wesley, 1993).

A light-hearted and very readable introduction to digital circuits can be found in Clive Maxfield's *Bebop to the Boolean Boogie* (LLH Technology Publishing, www.llh-publishing.com). Some people think that the seafood gumbo recipe in Appendix H is alone worth the price! Even without the recipe, the book is a well-illustrated classic that guides you through the basics of digital electronics fundamentals, components, and processes.

A sound understanding of the electrical aspects of digital circuit operation is mandatory for successful high-speed circuit design. Unquestionably the best book on this subject is *High-Speed Digital Design: A Handbook of Black Magic,*

Table 3-13 Manufacturers' logic data books.

Manufacturer	Order Number	Topics	Title	Year
Texas Instruments	SDLD001	74, 74S, 74LS TTL	*TTL Logic Data Book*	1988
Texas Instruments	SDAD001C	74AS, 74ALS TTL	*ALS/AS Logic Data Book*	1995
Texas Instruments	SDFD001B	74F TTL	*F Logic Data Book*	1994
Texas Instruments	SCLD001D	74HC, 74HCT CMOS	*HC/HCT Logic Data Book*	1997
Texas Instruments	SCAD001D	74AC, 74ACT CMOS	*AC/ACT Logic Data Book*	1997
Texas Instruments	SCLD003A	74AHC, 74AHCT CMOS	*AHC/AHCT Logic Data Book*	1997
Motorola	DL121/D	74F, 74LSTTL	*Fast and LSTTL Data*	1989
Motorola	DL129/D	74HC, 74HCT	*High-Speed CMOS Data*	1988
Motorola	DL138/D	74AC, 74ACT	*FACT Data*	1988
Motorola	DL122/D	10K ECL	*MECL Device Data*	1989

by Howard Johnson and Martin Graham (Prentice Hall, 1993). It combines solid electronics principles with tremendous insight and experience in the design of practical digital systems.

Characteristics of today's logic families can be found in the data books published by the device manufacturers. Both Texas Instruments and Motorola publish comprehensive data books for TTL and CMOS devices, as listed in Table 3-13. Both manufacturers keep up-to-date versions of their data books on the Web, at www.ti.com and www.mot.com. Motorola also provides a nice introduction to ECL design in the *MECL System Design Handbook* (publ. HB205, rev. 1, 1988).

The JEDEC (Joint Electron Device Engineering Council) standards for digital logic levels can be found on JEDEC's Web site, www.jedec.org. The JEDEC standards for 3.3-V, 2.5-V, and 1.8-V logic were published in 1994, 1995, and 1997, respectively.

Drill Problems

3.1   A particular logic family defines a LOW signal to be in the range 0.0–0.8 V and a HIGH signal to be in the range 2.0–3.3 V. Under a positive-logic convention, indicate the logic value associated with each of the following signal levels:

(a)   0.0 V   (b)   3.0 V   (c)   0.8 V   (d)   1.9 V

(e)   2.0 V   (f)   5.0 V   (g)   −0.7 V   (h)   −3.0 V

3.2   Repeat Drill 3.1 using a negative-logic convention.

3.3   Discuss how a logic buffer amplifier is different from an audio amplifier.

3.4   Is a buffer amplifier equivalent to a 1-input AND gate or a 1-input OR gate?

3.5   True or false: For a given set of input values, a NAND gate produces the opposite output as a NOR gate.

3.6   Write two completely different definitions of "gate" used in this chapter.

3.7   What kind of transistors are used in CMOS gates?

3.8   (Hobbyists only.) Draw an equivalent circuit for a CMOS inverter using a single-pole, double-throw relay.

3.9   For a given silicon area, which is likely to be faster, a CMOS NAND gate or a CMOS NOR?

3.10   Define "fan-in" and "fanout." Which one are you likely to have to calculate?

3.11   Draw the circuit diagram, function table, and logic symbol for a 3-input CMOS NOR gate in the style of Figure 3-16.

3.12   Draw switch models in the style of Figure 3-14 for a 2-input CMOS NOR gate for all four input combinations.

3.13   Draw a circuit diagram, function table, and logic symbol for a CMOS OR gate in the style of Figure 3-19.

3.14   Which has fewer transistors, a CMOS inverting gate or a noninverting gate?

3.15   Name and draw the logic symbols of four different 4-input CMOS gates that each use 8 transistors.

3.16   The circuit in Figure X3.16(a) is a type of CMOS AND-OR-INVERT gate. Write a function table for this circuit in the style of Figure 3-15(b), and a corresponding logic diagram using AND and OR gates and inverters.

3.17   The circuit in Figure X3.16(b) is a type of CMOS OR-AND-INVERT gate. Write function table for this circuit in the style of Figure 3-15(b), and a corresponding logic diagram using AND and OR gates and inverters.

3.18   How is it that perfume can be bad for digital designers?

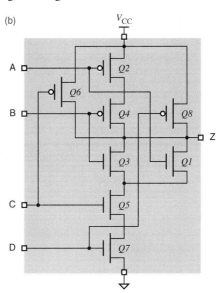

Figure X3.16

3.19   How much high-state DC noise margin is available in a CMOS inverter whose transfer characteristic under worst-case conditions looks like Figure 3-25? How much low-state DC noise margin is available? (Assume standard 1.5-V and 3.5-V thresholds for LOW and HIGH.)

3.20   Using the data sheet in Table 3-3, determine the worst-case LOW-state and HIGH-state DC noise margins of the 74HC00. State any assumptions required by your answer.

3.21   Section 3.5 defines seven different electrical parameters for CMOS circuits. Using the data sheet in Table 3-3, determine the worst-case value of each of these for the 74HC00. State any assumptions required by your answer.

3.22   Based on the conventions and definitions in Section 3.4, if the current at a device output is specified as a negative number, is the output sourcing current or sinking current?

3.23   For each of the following resistive loads, determine whether the output drive specifications of the 74HC00 over the commercial operating range are exceeded. (Refer to Table 3-3, and use $V_{OHmin} = 2.4$ V and $V_{CC} = 5.0$ V.)

(a)   120 Ω to V_{CC} (b)   270 Ω to V_{CC} and 330 Ω to GND

(c)   1 KΩ to GND   (d)   150 Ω to V_{CC} and 150 Ω to GND

(e)   100 Ω to V_{CC} (f)   75 Ω to V_{CC} and 150 Ω to GND

(g)   75 Ω to V_{CC} (h)   270 Ω to V_{CC} and 150 Ω to GND

3.24   Across the range of valid HIGH input levels, 2.0–5.0 V, at what input level would you expect the 74FCT257T (Table 3-3) to consume the most power?

3.25   Determine the LOW-state and HIGH-state DC fanout of the 74FCT257T when it drives 74LS00-like inputs. (Refer to Tables 3-3 and 3-12.)

3.26   Estimate the "on" resistances of the p-channel and n-channel output transistors of the 74FCT257T using information in Table 3-3.

3.27   Under what circumstances is it safe to allow an unused CMOS input to float?

3.28   Explain "latch up" and the circumstances under which it occurs.

3.29   Explain why putting all the decoupling capacitors in one corner of a printed-circuit board is not a good idea.

3.30   When is it important to hold hands with a friend?

3.31   Name the two components of CMOS logic gate's delay. Which one is most affected by load capacitance?

3.32   Determine the RC time constant for each of the following resistor-capacitor combinations:

(a) $R = 100\ \Omega, C = 50$ pF   (b) $R = 330\ \Omega, C = 150$ pF

(c) $R = 1$ K$\Omega, C = 30$ pF   (d) $R = 4.7$ K$\Omega, C = 100$ pF

3.33   Besides delay, what other characteristic(s) of a CMOS circuit are affected by load capacitance?

3.34   Explain the I_C formula in footnote 5 in Table 3-3 in terms of concepts presented in Sections 3.5 and 3.6.

3.35 It is possible to operate 74AC CMOS devices with a 3.3-volt power supply. How much power does this typically save, compared to 5-volt operation?

3.36 A particular Schmitt-trigger inverter has $V_{ILmax} = 0.8$ V, $V_{IHmin} = 2.0$ V, $V_{T+} = 1.6$ V, and $V_{T-} = 1.3$ V. How much hysteresis does it have?

3.37 Why are three-state outputs usually designed to turn off faster than they turn on?

3.38 Discuss the pros and cons of larger vs. smaller pull-up resistors for open-drain CMOS outputs or open-collector TTL outputs.

3.39 A particular LED has a voltage drop of about 2.0 V in the "on" state and requires about 5 mA of current for normal brightness. Determine an appropriate value for the pull-up resistor when the LED is connected as shown in Figure 3-52.

3.40 How does the answer for Drill 3.39 change if the LED is connected as shown in Figure 3-53(a)?

3.41 A wired-AND function is obtained simply by tying two open-drain or open-collector outputs together, without going through another level of transistor circuitry. How is it, then, that a wired-AND function can actually be slower than a discrete AND gate? (*Hint:* Recall the title of a Teenage Mutant Ninja Turtles movie.)

3.42 Which CMOS or TTL logic family in this chapter has the strongest output driving capability?

3.43 Concisely summarize the difference between HC and HCT logic families. The same concise statement should apply to AC versus ACT.

3.44 Why don't the specifications for FCT devices include parameters like V_{OLmaxC} that apply to CMOS loads, as HCT and ACT specifications do?

3.45 How do FCT-T devices reduce power consumption compared to FCT devices?

3.46 How many diodes are required for an *n*-input diode AND gate?

3.47 True or false: A TTL NOR gate uses diode logic.

3.48 Are TTL outputs more capable of sinking current or sourcing current?

3.49 Compute the maximum fanout for each of the following cases of a TTL output driving multiple TTL inputs. Also indicate how much "excess" driving capability is available in the LOW or HIGH state for each case.

(a)   74LS driving 74LS   (b)   74LS driving 74S
(c)   74S driving 74AS   (d)   74F driving 74S
(e)   74AS driving 74AS   (f)   74AS driving 74F
(g)   74ALS driving 74F   (h)   74AS driving 74ALS

3.50 Which resistor dissipates more power, the pull-down for an unused LS-TTL NOR-gate input, or the pull-up for an unused LS-TTL NAND-gate input? Use the minimum allowable resistor value in each case.

3.51 Which would you expect to be faster, a TTL AND gate or a TTL AND-OR-INVERT gate? Why?

3.52 Describe the key benefit and the key drawback of Schottky transistors in TTL.

3.53 Using the data sheet in Table 3-12, determine the worst-case LOW-state and HIGH-state DC noise margins of the 74LS00.

3.54 Sections 3.10.2 and 3.10.3 define eight different electrical parameters for TTL circuits. Using the data sheet in Table 3-12, determine the worst-case value of each of these for the 74LS00.

3.55 For each of the following resistive loads, determine whether the output drive specifications of the 74LS00 over the commercial operating range are exceeded. (Refer to Table 3-12, and use $V_{OLmax} = 0.5$ V and $V_{CC} = 5.0$ V.)

(a)   470 Ω to V_{CC} (b)   330 Ω to V_{CC} and 470 Ω to GND

(c)   10 KΩ to GND   (d)   390 Ω to V_{CC} and 390 Ω to GND

(e)   600 Ω to V_{CC} (f)   510 Ω to V_{CC} and 510 Ω to GND

(g)   4.7 KΩ to GND   (h)   220 Ω to V_{CC} and 330 Ω to GND

3.56 Compute the LOW-state and HIGH-state DC noise margins for each of the following cases of a TTL output driving a TTL-compatible CMOS input, or vice versa.

(a)   74HCT driving 74LS   (b)   74VHCT driving 74AS

(c)   74LS driving 74HCT   (d)   74S driving 74VHCT

3.57 Compute the maximum fanout for each of the following cases of a TTL-compatible CMOS output driving multiple inputs in a TTL logic family. Also indicate how much "excess" driving capability is available in the LOW or HIGH state for each case.

(a)   74HCT driving 74LS   (b)   74HCT driving 74S

(c)   74VHCT driving 74AS   (d)   74VHCT driving 74LS

3.58 For a given load capacitance and transition rate, which logic family in this chapter has the lowest dynamic power dissipation?

Exercises

3.59 Design a CMOS circuit that has the functional behavior shown in Figure X3.59. (*Hint:* Only six transistors are required.)

3.60 Design a CMOS circuit that has the functional behavior shown in Figure X3.60. (*Hint:* Only six transistors are required.)

3.61 Draw a circuit diagram, function table, and logic symbol in the style of Figure 3-19 for a CMOS gate with two inputs A and B and an output Z, where Z=1 if A=0 and B=1, and Z=0 otherwise. (*Hint:* Only six transistors are needed.)

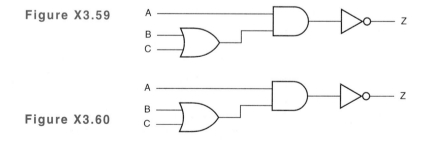

Figure X3.59

Figure X3.60

3.62 Draw a circuit diagram, function table, and logic symbol in the style of Figure 3-19 for a CMOS gate with two inputs A and B and an output Z, where Z=0 if A=1 and B=0, and Z=1 otherwise. (*Hint:* Only six transistors are needed.)

3.63 Draw a figure showing the logical structure of an 8-input CMOS NOR gate, assuming that at most 4-input gate circuits are practical. Using your general knowledge of CMOS characteristics, select a circuit structure that minimizes the NOR gate's propagation delay for a given silicon area, and explain why this is so.

3.64 The circuit designers of TTL-compatible CMOS families presumably could have made the voltage drop across the "on" transistor under load in the HIGH state as little as it is in the LOW state, simply by making the *p*-channel transistors bigger. Why do you suppose they didn't bother to do this?

3.65 How much current and power are "wasted" in Figure 3-32(b)?

3.66 Perform a detailed calculation of V_{OUT} in Figures 3-34 and 3-33. (*Hint:* Create a Thévenin equivalent for the CMOS inverter in each figure.)

3.67 Consider the dynamic behavior of a CMOS output driving a given capacitive load. If the resistance of the charging path is double the resistance of the discharging path, is the rise time exactly twice the fall time? If not, what other factors affect the transition times?

3.68 Analyze the fall time of the CMOS inverter output of Figure 3-37, with $R_L=1$ kΩ and $V_L=2.5$ V. Compare your result with the result in Section 3.6.1 and explain.

3.69 Repeat Exercise 3.68 for rise time.

3.70 Assuming that the transistors in an FCT CMOS three-state buffer are perfect zero-delay on-off devices that switch at an input threshold of 1.5 V, determine the value of t_{PLZ} for the test circuit and waveforms in Figure 3-24. (*Hint:* You have to determine the time using an *RC* time constant.) Explain the difference between your result and the specifications in Table 3-3.

3.71 Repeat Exercise 3.70 for t_{PHZ}.

3.72 Using the specifications in Table 3-6, estimate the "on" resistances of the *p*-channel and *n*-channel transistors in 74AC-series CMOS logic.

3.73 Create a $4 \times 4 \times 2 \times 2$ matrix of worst-case DC noise margins for the following CMOS interfacing situations: an (HC, HCT, VHC, or VHCT) output driving an (HC, HCT, VHC, or VHCT) input with a (CMOS, TTL) load in the (LOW, HIGH) state; Figure X3.73 illustrates. (*Hints:* There are 64 different combinations, but many give identical results. Some combinations yield negative margins.)

Figure X3.73

		Input							
		HC		HCT		VHC		VHCT	
Output	HC	CL	TL	CL	TL	CL	TL	CL	TL
		CH	TH	CH	TH	CH	TH	CH	TH
	HCT	CL	TL	CL	TL	CL	TL	CL	TL
		CH	TH	CH	TH	CH	TH	CH	TH
	VHC	CL	TL	CL	TL	CL	TL	CL	TL
		CH	TH	CH	TH	CH	TH	CH	TH
	VHCT	CL	TL	CL	TL	CL	TL	CL	TL
		CH	TH	CH	TH	CH	TH	CH	TH

Key:
CL = CMOS load, LOW
CH = CMOS load, HIGH
TL = TTL load, LOW
TH = TTL load, HIGH

3.74 Using Figure 3-85, determine the DC noise margins for 5-V-tolerant, 3.3-V CMOS driving 5-V CMOS logic with TTL input levels, and vice versa.

3.75 Using Figure 3-85, determine the DC noise margins for 3.3-V-tolerant, 2.5-V CMOS driving 3.3-V CMOS, and vice versa.

3.76 In the LED example in Section 3.7.5, a designer chose a resistor value of 300 Ω and found that the open-drain gate was able to maintain its output at 0.1 V while driving the LED. How much current flows through the LED, and how much power is dissipated by the pull-up resistor in this case?

3.77 Consider a CMOS 8-bit binary counter (Section 8.4) clocked at 16 MHz. For the purpose of computing the counter's dynamic power dissipation, what is the transition frequency of the least significant bit? Of the most significant bit? For the purpose of determining the dynamic power dissipation of the eight output bits, what frequency should be used?

3.78 Using only AND and NOR gates, draw a logic diagram for the logic function performed by the circuit in Figure 3-55.

3.79 Calculate the approximate output voltage at Z in Figure 3-56, assuming that the gates are HCT-series CMOS.

3.80 Redraw the circuit diagram of a CMOS 3-state buffer in Figure 3-48 using actual transistors instead of NAND, NOR, and inverter symbols. Can you find a circuit for the same function that requires a smaller total number of transistors? If so, draw it.

3.81 Modify the CMOS 3-state buffer circuit in Figure 3-48 so that the output is in the High-Z state when the enable input is HIGH. The modified circuit should require no more transistors than the original.

3.82 Using information in Table 3-3, estimate how much current can flow through each output pin if the outputs of two different 74HC00s are fighting.

3.83 Show that at a given power-supply voltage, an FCT-type I_{CCD} specification can be derived from an HCT/ACT-type C_{PD} specification, and vice versa.

3.84 If both VZ and V_B in Figure 3-65(b) are 4.6 V, can we get $V_C = 5.2$ V? Explain.

3.85 Modify the program in Table 3-10 to account for leakage current in the OFF state.

3.86 Assuming "ideal" conditions, what is the minimum voltage that will be recognized as a HIGH in the TTL NAND gate in Figure 3-75 with one input LOW and the other HIGH?

3.87 Assuming "ideal" conditions, what is the maximum voltage that will be recognized as a LOW in the TTL NAND gate in Figure 3-75 with both inputs HIGH?

3.88 Find a commercial TTL part that can source 40 mA in the HIGH state. What is its application?

3.89 What happens if you try to drive an LED with its cathode grounded and its anode connected to a TTL totem-pole output, analogous to Figure 3-53 for CMOS?

3.90 What happens if you try to drive a 12-volt relay with a TTL totem-pole output?

3.91 Suppose that a single pull-up resistor to +5 V is used to provide a constant-1 logic source to 15 different 74LS00 inputs. What is the maximum value of this resistor? How much HIGH-state DC noise margin are you providing in this case?

Figure X3.92

Figure X3.92

3.92 The circuit in Figure X3.92 uses open-collector NAND gates to perform "wired logic." Write a truth table for output signal F and, if you've read Section 4.2, a logic expression for F as a function of the circuit inputs.

3.93 What is the maximum allowable value for *R1* in Figure X3.92? Assume that a 0.7-V HIGH-state noise margin is required. The 74LS01 has the specs shown in the 74LS column of Table 3-11, except that I_{OHmax} is 100 μA, a leakage current that flows *into* the output in the HIGH state.

3.94 Suppose that the output signal F in Figure X3.92 drives the inputs of two 74S04 inverters. Compute the minimum and maximum allowable values of *R2*, assuming that a 0.7 V HIGH-state noise margin is required.

3.95 A logic designer found a problem in a certain circuit's function after the circuit had been released to production and 1000 copies of it built. A portion of the circuit is shown in Figure X3.95 in black; all of the gates are 74LS00 NAND gates. The logic designer fixed the problem by adding the two diodes shown in color. What do the diodes do? Describe both the logical effects of this change on the circuit's function and the electrical effects on the circuit's noise margins.

3.96 A 74LS125 is a buffer with a three-state output. When enabled, the output can sink 24 mA in the LOW state and source 2.6 mA in the HIGH state. When disabled, the output has a leakage current of ±20 μA (the sign depends on the output voltage—plus if the output is pulled HIGH by other devices, minus if it's LOW). Suppose a system is designed with multiple modules connected to a bus, where each module has a single 74LS125 to drive the bus and one 74LS04 to receive information on the bus. What is the maximum number of modules that can be connected to the bus without exceeding the 74LS125's specs?

3.97 Repeat Exercise 3.97, this time assuming that a single pull-up resistor is connected from the bus to +5 V to guarantee that the bus is HIGH when no device is driving it. Calculate the maximum possible value of the pull-up resistor, and the number of modules that can be connected to the bus.

3.98 Find the circuit design in a TTL data book for an actual three-state gate, and explain how it works.

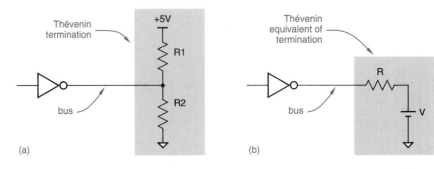

Figure X3.99   (a)   (b)

3.99   A *Thévenin termination* for an open-collector or three-state bus has the structure shown in Figure X3.99(a). The idea is that, by selecting appropriate values of $R1$ and $R2$, a designer can obtain a circuit equivalent to the termination in (b) for any desired values of V and R. The value of V determines the voltage on the bus when no device is driving it, and the value of R is selected to match the characteristic impedance of the bus for transmission-line purposes (Section 11.4). For each of the following pairs of V and R, determine the required values of $R1$ and $R2$.

(a) $V = 2.75, R = 148.5$ (b) $V = 2.7, R = 180$

(c) $V = 3.0, R = 130$ (d) $V = 2.5, R = 75$

3.100   For each of the $R1$ and $R2$ pairs in Exercise 3.99, determine whether the termination can be properly driven by a three-state output in each of the following logic families: 74LS, 74S, 74ACT. For proper operation, the family's I_{OL} and I_{OH} specs must not be exceeded when $V_{OL} = V_{OLmax}$ and $V_{OH} = V_{OHmin}$, respectively.

3.101   Using the graphs in a TTL data book, develop some rules of thumb for derating the maximum-propagation-delay specification of LS-TTL under nonoptimal conditions of power-supply voltage, temperature, and loading.

3.102   Determine the total power dissipation of the circuit in Figure X3.102 as function of transition frequency f for two realizations: (a) using 74LS gates; (b) using 74HC gates. Assume that input capacitance is 3 pF for a TTL gate and 7 pF for a CMOS gate, that a 74LS gate has an internal power-dissipation capacitance of 20 pF, and that there is an additional 20 pF of stray wiring capacitance in the circuit. Also assume that the X, Y, and Z inputs are always HIGH, and that input C is driven with a CMOS-level square wave with frequency f. Other information that you need for this problem can be found in Tables 3-5 and 3-11. State any other assumptions that you make. At what frequency does the TTL circuit dissipate less power than the CMOS circuit?

Figure X3.102

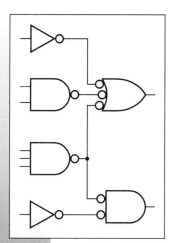

chapter 4

Combinational Logic Design Principles

Logic circuits are classified into two types, "combinational" and "sequential." A *combinational logic circuit* is one whose outputs depend only on its current inputs. The rotary channel selector knob on an old-fashioned television is like a combinational circuit—its "output" selects a channel based only on the current position of the knob ("input").

The outputs of a *sequential logic circuit* depend not only on the current inputs but also on the past sequence of inputs, possibly arbitrarily far back in time. The channel selector controlled by the up and down pushbuttons on a TV or VCR is a sequential circuit—the channel selection depends on the past sequence of up/down pushes, at least since when you started viewing 10 hours before, and perhaps as far back as when you first powered up the device. Sequential circuits are discussed in Chapters 7 and 8.

A combinational circuit may contain an arbitrary number of logic gates and inverters but no feedback loops. A *feedback loop* is a signal path of a circuit that allows the output of a gate to propagate back to the input of that same gate; such a loop generally creates sequential circuit behavior.

In combinational circuit *analysis* we start with a logic diagram and proceed to a formal description of the function performed by that circuit, such as a truth table or a logic expression. In *synthesis* we do the reverse, starting with a formal description and proceeding to a logic diagram.

193

SYNTHESIS VS.
DESIGN

Logic circuit design is a superset of synthesis, since in a real design problem we usually start out with an informal (word or thought) description of the circuit. Often the most challenging and creative part of design is to formalize the circuit description, defining the circuit's input and output signals and specifying its functional behavior by means of truth tables and equations. Once we've created the formal circuit description, we can usually follow a "turn-the-crank" synthesis procedure to obtain a logic diagram for a circuit with the required functional behavior. The material in the first four sections of this chapter is the basis for "turn-the-crank" procedures, whether the crank is turned by hand or by a computer. The last two sections describe actual design languages, ABEL and VHDL. When we create a design using one of these languages, a computer program can perform the synthesis steps for us. In later chapters we'll encounter many examples of the real design process.

Combinational circuits may have one or more outputs. Most analysis and synthesis techniques can be extended in an obvious way from single-output to multiple-output circuits (e.g., "Repeat these steps for each output"). We'll also point out how some techniques can be extended in a not-so-obvious way for improved effectiveness in the multiple-output case.

The purpose of this chapter is to give you a solid theoretical foundation for the analysis and synthesis of combinational logic circuits, a foundation that will be doubly important later when we move on to sequential circuits. Although most of the analysis and synthesis procedures in this chapter are automated nowadays by computer-aided design tools, you need a basic understanding of the fundamentals to use the tools and to figure out what's wrong when you get unexpected or undesirable results.

With the fundamentals well in hand, it is appropriate next to understand how combinational functions can be expressed and analyzed using hardware description languages (HDLs). So, the last two sections of this chapter introduce basic features of ABEL and VHDL, which we'll use to design all kinds of logic circuits throughout the balance of the text.

Before launching into a discussion of combinational logic circuits, we must introduce switching algebra, the fundamental mathematical tool for analyzing and synthesizing logic circuits of all types.

4.1 Switching Algebra

Formal analysis techniques for digital circuits have their roots in the work of an English mathematician, George Boole. In 1854, he invented a two-valued algebraic system, now called *Boolean algebra*, to "give expression ... to the fundamental laws of reasoning in the symbolic language of a Calculus." Using this system, a philosopher, logician, or inhabitant of the planet Vulcan can

Boolean algebra

formulate propositions that are true or false, combine them to make new propositions, and determine the truth or falsehood of the new propositions. For example, if we agree that "People who haven't studied this material are either failures or not nerds," and "No computer designer is a failure," then we can answer questions like "If you're a nerdy computer designer, then have you already studied this?"

Long after Boole, in 1938, Bell Labs researcher Claude E. Shannon showed how to adapt Boolean algebra to analyze and describe the behavior of circuits built from relays, the most commonly used digital logic elements of that time. In Shannon's *switching algebra*, the condition of a relay contact, open or closed, is represented by a variable X that can have one of two possible values, 0 or 1. In today's logic technologies, these values correspond to a wide variety of physical conditions—voltage HIGH or LOW, light off or on, capacitor discharged or charged, fuse blown or intact, and so on—as we detailed in Table 3-1 on page 81.

switching algebra

In the remainder of this section we develop the switching algebra directly, using "first principles" and what we already know about the behavior of logic elements (gates and inverters). For more historical and/or mathematical treatments of this material, consult the References.

4.1.1 Axioms

In switching algebra we use a symbolic variable, such as X, to represent the condition of a logic signal. A logic signal is in one of two possible conditions—low or high, off or on, and so on, depending on the technology. We say that X has the value "0" for one of these conditions and "1" for the other.

For example, with the CMOS and TTL logic circuits in Chapter 3, the *positive-logic convention* dictates that we associate the value "0" with a LOW voltage and "1" with a HIGH voltage. The *negative-logic convention* makes the opposite association: 0 = HIGH and 1 = LOW. However, the choice of positive or negative logic has no effect on our ability to develop a consistent algebraic description of circuit behavior; it only affects details of the physical-to-algebraic abstraction, as we'll explain later in our discussion of "duality." For the moment, we may ignore the physical realities of logic circuits and pretend that they operate directly on the logic symbols 0 and 1.

positive-logic convention

negative-logic convention

The *axioms* (or *postulates*) of a mathematical system are a minimal set of basic definitions that we assume to be true, from which all other information about the system can be derived. The first two axioms of switching algebra embody the "digital abstraction" by formally stating that a variable X can take on only one of two values:

axiom

postulate

$$(A1) \quad X = 0 \quad \text{if } X \neq 1 \qquad (A1') \quad X = 1 \quad \text{if } X \neq 0$$

Notice that we stated these axioms as a pair, the only difference between A1 and A1′ being the interchange of the symbols 0 and 1. This is a characteristic of all

complement
prime (´)

the axioms of switching algebra and is the basis of the "duality" principle that we'll study later.

In Section 3.3.3 we showed the design of an inverter, a logic circuit whose output signal level is the opposite (or *complement*) of its input signal level. We use a *prime (´)* to denote an inverter's function. That is, if a variable X denotes an inverter's input signal, then X´ denotes the value of a signal on the inverter's output. This notation is formally specified in the second pair of axioms:

$$(A2) \quad \text{If } X = 0, \text{ then } X' = 1 \qquad (A2') \quad \text{If } X = 1, \text{ then } X' = 0$$

algebraic operator
expression
NOT operation

As shown in Figure 4-1, the output of an inverter with input signal X may have an arbitrary signal name, say Y. Algebraically, however, we write Y = X´ to say "signal Y always has the opposite value as signal X." The prime (´) is an *algebraic operator*, and X´ is an *expression*, which you can read as "X prime" or "NOT X." This usage is analogous to what you've learned in programming languages, where if J is an integer variable, then −J is an expression whose value is 0 − J. Although this may seem like a small point, you'll learn that the distinction between signal names (X, Y), expressions (X´), and equations (Y = X´) is very important when we study documentation standards and software tools for logic design. In the logic diagrams in this book we maintain this distinction by writing signal names in black and expressions in color.

Figure 4-1
Signal naming and algebraic
notation for an inverter.

$$X \longrightarrow Y = X'$$

logical multiplication
multiplication dot (·)

In Section 3.3.6 we showed how to build a 2-input CMOS AND gate, a circuit whose output is 1 if both of its inputs are 1. The function of a 2-input AND gate is sometimes called *logical multiplication* and is symbolized algebraically by a *multiplication dot (·)*. That is, an AND gate with inputs X and Y has an output signal whose value is X · Y, as shown in Figure 4-2(a). Some authors, especially mathematicians and logicians, denote logical multiplication with a wedge (X ∨ Y). We follow standard engineering practice by using the dot (X ·Y). When

NOTE ON NOTATION

The notations X, ~X, and ¬X are also used by some authors to denote the complement of X. The overbar notation ($\overline{X}$) is probably the most widely used and the best looking typographically. However, we use the prime notation to get you used to writing logic expressions on a single text line without the more graphical overbar, and to force you to parenthesize complex complemented subexpressions—because this is what you'll have to do when you use HDLs and other tools.

Figure 4-2
Signal naming and
algebraic notation:
(a) AND gate;
(b) OR gate.

we study hardware description languages (HDLs), we'll encounter several other symbols that are used to denote the same thing.

We also described in Section 3.3.6 how to build a 2-input CMOS OR gate, a circuit whose output is 1 if either of its inputs is 1. The function of a 2-input OR gate is sometimes called *logical addition* and is symbolized algebraically by a *logical addition* plus sign (+). An OR gate with inputs X and Y has an output signal whose value is $X + Y$, as shown in Figure 4-2(b). Some authors denote logical addition with a vee ($X \wedge Y$), but we follow the standard engineering practice of using the plus sign ($X + Y$). Once again, other symbols may be used in HDLs. By convention, in a logic expression involving both multiplication and addition, multiplication has *precedence*, just as in integer expressions in conventional programming *precedence* languages. That is, the expression $W \cdot X + Y \cdot Z$ is equivalent to $(W \cdot X) + (Y \cdot Z)$.

The last three pairs of axioms state the formal definitions of the AND and *AND operation* OR operations by listing the output produced by each gate for each possible *OR operation* input combination:

(A3)	$0 \cdot 0 = 0$	(A3′)	$1 + 1 = 1$
(A4)	$1 \cdot 1 = 1$	(A4′)	$0 + 0 = 0$
(A5)	$0 \cdot 1 = 1 \cdot 0 = 0$	(A5′)	$1 + 0 = 0 + 1 = 1$

The five pairs of axioms, A1–A5 and A1′–A5′, completely define switching algebra. All other facts about the system can be proved using these axioms as a starting point.

JUXT A MINUTE...   Older texts use simple *juxtaposition* (XY) to denote logical multiplication, but we don't. In general, juxtaposition is a clear notation only when signal names are limited to a single character. Otherwise, is XY a logical product or a two-character signal name? One-character variable names are common in algebra, but in real digital design problems we prefer to use multicharacter signal names that mean something. Thus, we need a separator between names, and the separator might just as well be a multiplication dot rather than a space. The HDL equivalent of the multiplication dot (often * or &) is absolutely required when logic formulas are written in a hardware-description language.

Table 4-1	(T1)	$X + 0 = X$	(T1′)	$X \cdot 1 = X$	(Identities)
Switching-algebra	(T2)	$X + 1 = 1$	(T2′)	$X \cdot 0 = 0$	(Null elements)
theorems with one	(T3)	$X + X = X$	(T3′)	$X \cdot X = X$	(Idempotency)
variable.	(T4)	$(X')' = X$			(Involution)
	(T5)	$X + X' = 1$	(T5′)	$X \cdot X' = 0$	(Complements)

4.1.2 Single-Variable Theorems

theorem

During the analysis or synthesis of logic circuits, we often write algebraic expressions that characterize a circuit's actual or desired behavior. Switching-algebra *theorems* are statements, known to be always true, that allow us to manipulate algebraic expressions to allow simpler analysis or more efficient synthesis of the corresponding circuits. For example, the theorem $X + 0 = X$ allows us to substitute every occurrence of $X + 0$ in an expression with X.

Table 4-1 lists switching-algebra theorems involving a single variable X. How do we know that these theorems are true? We can either prove them ourselves or take the word of someone who has. OK, we're in college now, let's learn how to prove them.

perfect induction

Most theorems in switching algebra are exceedingly simple to prove using a technique called *perfect induction*. Axiom A1 is the key to this technique— since a switching variable can take on only two different values, 0 and 1, we can prove a theorem involving a single variable X by proving that it is true for both $X = 0$ and $X = 1$. For example, to prove theorem T1, we make two substitutions:

$$[X = 0] \quad 0 + 0 = 0 \quad \text{true, according to axiom A4}'$$
$$[X = 1] \quad 1 + 0 = 1 \quad \text{true, according to axiom A5}'$$

All of the theorems in Table 4-1 can be proved using perfect induction, as you're asked to do in the Drills 4.2 and 4.3.

4.1.3 Two- and Three-Variable Theorems

Switching-algebra theorems with two or three variables are listed in Table 4-2. Each of these theorems is easily proved by perfect induction, by evaluating the theorem statement for the four possible combinations of X and Y, or the eight possible combinations of X, Y, and Z.

The first two theorem pairs concern commutativity and associativity of logical addition and multiplication and are identical to the commutative and associative laws for addition and multiplication of integers and reals. Taken together, they indicate that the parenthesization or order of terms in a logical sum or logical product is irrelevant. For example, from a strictly algebraic point of view, an expression such as $W \cdot X \cdot Y \cdot Z$ is ambiguous; it should be written as $(W \cdot (X \cdot (Y \cdot Z)))$ or $(((W \cdot X) \cdot Y) \cdot Z)$ or $(W \cdot X) \cdot (Y \cdot Z)$ (see Exercise 4.34).

Table 4-2 Switching-algebra theorems with two or three variables.

(T6)	$X + Y = Y + X$	(T6′) $X \cdot Y = Y \cdot X$	(Commutativity)
(T7)	$(X + Y) + Z = X + (Y + Z)$	(T7′) $(X \cdot Y) \cdot Z = X \cdot (Y \cdot Z)$	(Associativity)
(T8)	$X \cdot Y + X \cdot Z = X \cdot (Y + Z)$	(T8′) $(X + Y) \cdot (X + Z) = X + Y \cdot Z$	(Distributivity)
(T9)	$X + X \cdot Y = X$	(T9′) $X \cdot (X + Y) = X$	(Covering)
(T10)	$X \cdot Y + X \cdot Y' = X$	(T10′) $(X + Y) \cdot (X + Y') = X$	(Combining)
(T11)	$X \cdot Y + X' \cdot Z + Y \cdot Z = X \cdot Y + X' \cdot Z$		(Consensus)
(T11′)	$(X + Y) \cdot (X' + Z) \cdot (Y + Z) = (X + Y) \cdot (X' + Z)$		

But the theorems tell us that the ambiguous form of the expression is OK because we get the same results in any case. We even could have rearranged the order of the variables (e.g., $X \cdot Z \cdot Y \cdot W$) and gotten the same results.

As trivial as this discussion may seem, it is very important, because it forms the theoretical basis for using logic gates with more than two inputs. We defined $\cdot$ and $+$ as *binary operators*—operators that combine *two* variables. Yet *binary operator* we use 3-input, 4-input, and larger AND and OR gates in practice. The theorems tell us we can connect gate inputs in any order; in fact, many printed-circuit-board and ASIC layout programs take advantage of this. We can use either one n-input gate or $(n-1)$ 2-input gates interchangeably, though propagation delay and cost are likely to be higher with multiple 2-input gates.

Theorem T8 is identical to the distributive law for integers and reals—that is, logical multiplication distributes over logical addition. Hence, we can "multiply out" an expression to obtain a sum-of-products form, as in the example below:

$$V \cdot (W + X) \cdot (Y + Z) = V \cdot W \cdot Y + V \cdot W \cdot Z + V \cdot X \cdot Y + V \cdot X \cdot Z$$

However, switching algebra also has the unfamiliar property that the reverse is true—logical addition distributes over logical multiplication—as demonstrated by theorem T8′. Thus, we can also "add out" an expression to obtain a product-of-sums form:

$$(V \cdot W \cdot X) + (Y \cdot Z) = (V + Y) \cdot (V + Z) \cdot (W + Y) \cdot (W + Z) \cdot (X + Y) \cdot (X + Z)$$

Theorems T9 and T10 are used extensively in the minimization of logic functions. For example, if the subexpression $X + X \cdot Y$ appears in a logic expression, the *covering theorem* T9 says that we need only include X in the *covering theorem* expression; X is said to *cover* $X \cdot Y$. The *combining theorem* T10 says that if the *cover* subexpression $X \cdot Y + X \cdot Y'$ appears in an expression, we can replace it with X. *combining theorem* Since Y must be 0 or 1, either way the original subexpression is 1 if and only if X is 1.

Although we could easily prove T9 by perfect induction, the truth of T9 is more obvious if we prove it using the other theorems that we've proved so far:

$$
\begin{aligned}
X + X \cdot Y &= X \cdot 1 + X \cdot Y &&\text{(according to T1$'$)} \\
&= X \cdot (1 + Y) &&\text{(according to T8)} \\
&= X \cdot 1 &&\text{(according to T2)} \\
&= X &&\text{(according to T1$'$)}
\end{aligned}
$$

Likewise, the other theorems can be used to prove T10, where the key step is to use T8 to rewrite the lefthand side as $X \cdot (Y + Y')$.

consensus theorem

consensus

Theorem T11 is known as the *consensus theorem*. The $Y \cdot Z$ term is called the *consensus* of $X \cdot Y$ and $X' \cdot Z$. The idea is that if $Y \cdot Z$ is 1, then either $X \cdot Y$ or $X' \cdot Z$ must also be 1, since Y and Z are both 1 and either X or X' must be 1. Thus, the $Y \cdot Z$ term is redundant and may be dropped from the righthand side of T11. The consensus theorem has two important applications. It can be used to eliminate certain timing hazards in combinational logic circuits, as we'll see in Section 4.5. And it also forms the basis of the iterative-consensus method of finding prime implicants (see References).

In all of the theorems, it is possible to replace each variable with an arbitrary logic expression. A simple replacement is to complement one or more variables:

$$
(X + Y') + Z' = X + (Y' + Z') \quad \text{(based on T7)}
$$

But more complex expressions may be substituted as well:

$$
(V' + X) \cdot (W \cdot (Y' + Z)) + (V' + X) \cdot (W \cdot (Y' + Z))' = V' + X \quad \text{(based on T10)}
$$

4.1.4 *n*-Variable Theorems

Several important theorems, listed in Table 4-3, are true for an arbitrary number of variables, n. Most of these theorems can be proved using a two-step method called *finite induction*—first proving that the theorem is true for $n = 2$ (the *basis step*) and then proving that if the theorem is true for $n = i$, then it is also true for $n = i + 1$ (the *induction step*). For example, consider the generalized idempotency theorem T12. For $n = 2$, T12 is equivalent to T3 and is therefore true. If it is true for a logical sum of i X's, then it is also true for a sum of $i + 1$ X's, according to the following reasoning:

finite induction

basis step

induction step

$$
\begin{aligned}
X + X + X + \cdots + X &= X + (X + X + \cdots + X) &&\text{($i + 1$ X's on either side)} \\
&= X + (X) &&\text{(if T12 is true for $n = i$)} \\
&= X &&\text{(according to T3)}
\end{aligned}
$$

Thus, the theorem is true for all finite values of n.

DeMorgan's theorems

DeMorgan's theorems (T13 and T13$'$) are probably the most commonly used of all the theorems of switching algebra. Theorem T13 says that an *n*-input

Table 4-3 Switching-algebra theorems with n variables.

(T12)	$X + X + \cdots + X = X$	(Generalized idempotency)
(T12')	$X \cdot X \cdot \; \cdots \; \cdot X = X$	
(T13)	$(X_1 \cdot X_2 \cdot \; \cdots \; \cdot X_n)' = X_1' + X_2' + \cdots + X_n'$	(DeMorgan's theorems)
(T13')	$(X_1 + X_2 + \cdots + X_n)' = X_1' \cdot X_2' \cdot \; \cdots \; \cdot X_n'$	
(T14)	$[F(X_1, X_2, ..., X_n, +, \cdot)]' = F(X_1', X_2', ..., X_n', \cdot, +)$	(Generalized DeMorgan's theorem)
(T15)	$F(X_1, X_2, ..., X_n) = X_1 \cdot F(1, X_2, ..., X_n) + X_1' \cdot F(0, X_2, ..., X_n)$	(Shannon's expansion theorems)
(T15')	$F(X_1, X_2, ..., X_n) = [X_1 + F(0, X_2, ..., X_n)] \cdot [X_1' + F(1, X_2, ..., X_n)]$	

AND gate whose output is complemented is equivalent to an n-input OR gate whose inputs are complemented. That is, the circuits of Figure 4-3(a) and (b) are equivalent.

In Section 3.3.4 we showed how to build a CMOS NAND gate. The output of a NAND gate for any set of inputs is the complement of an AND gate's output for the same inputs, so a NAND gate can have the logic symbol in Figure 4-3(c). However, the CMOS NAND circuit is not designed as an AND gate followed by a transistor inverter (NOT gate); it's just a collection of transistors that happens to perform the AND-NOT function. In fact, theorem T13 tells us that the logic symbol in (d) denotes the same logic function (bubbles on the OR-gate inputs indicate logical inversion). That is, a NAND gate may be viewed as performing a NOT-OR function.

By observing the inputs and output of a NAND gate, it is impossible to determine whether it has been built internally as an AND gate followed by an inverter, as inverters followed by an OR gate, or as a direct CMOS realization, *because all NAND circuits perform precisely the same logic function.* Although the choice of symbol has no bearing on the functionality of a circuit, we'll show in Section 5.1 that the proper choice can make the circuit's function much easier to understand.

Figure 4-3 Equivalent circuits according to DeMorgan's theorem T13:
(a) AND-NOT; (b) NOT-OR; (c) logic symbol for a NAND gate;
(d) equivalent symbol for a NAND gate.

(a) X, Y into AND gate → X + Y into inverter → $Z = (X + Y)'$

(c) X, Y into NOR gate → $Z = (X + Y)'$

(b) X into inverter → X', Y into inverter → Y', both into OR gate → $Z = X' \cdot Y'$

(d) X, Y into gate with bubbled inputs → $Z = X' \cdot Y'$

Figure 4-4 Equivalent circuits according to DeMorgan's theorem T13′:
(a) OR-NOT; (b) NOT-AND; (c) logic symbol for a NOR gate;
(d) equivalent symbol for a NOR gate.

A similar symbolic equivalence can be inferred from theorem T13′. As shown in Figure 4-4, a NOR gate may be realized as an OR gate followed by an inverter, or as inverters followed by an AND gate.

generalized DeMorgan's theorem

complement of a logic expression

Theorems T13 and T13′ are just special cases of a *generalized DeMorgan's theorem*, T14, that applies to an arbitrary logic expression F. By definition, the *complement of a logic expression*, denoted (F)′, is an expression whose value is the opposite of F's for every possible input combination. Theorem T14 is very important because it gives us a way to manipulate and simplify the complement of an expression.

Theorem T14 states that, given any *n*-variable logic expression, its complement can be obtained by swapping + and · and complementing all variables. For example, suppose that we have

$$F(W, X, Y, Z) = (W' \cdot X) + (X \cdot Y) + (W \cdot (X' + Z'))$$
$$= ((W)' \cdot X) + (X \cdot Y) + (W \cdot ((X)' + (Z)'))$$

In the second line we have enclosed complemented variables in parentheses to remind you that the ′ is an operator, not part of the variable name. By applying theorem T14, we obtain

$$[F(W, X, Y, Z)]' = ((W')' + X') \cdot (X' + Y') \cdot (W' + ((X)' \cdot (Z)'))$$

Using theorem T4, this can be simplified to

$$[F(W, X, Y, Z)]' = (W) + X') \cdot (X' + Y') \cdot (W' + (X \cdot (Z))$$

In general, we can use theorem T14 to complement a parenthesized expression by swapping + and ·, complementing all uncomplemented variables, and uncomplementing all complemented ones.

The generalized DeMorgan's theorem T14 can be proved by showing that all logic functions can be written as either a sum or a product of subfunctions, and then applying T13 and T13′ recursively. However, a much more enlightening and satisfying proof can be based on the principle of duality, explained next.

4.1.5 Duality

We stated all of the axioms of switching algebra in pairs. The primed version of each axiom (e.g., A5′) is obtained from the unprimed version (e.g., A5) by simply swapping 0 and 1 and, if present, · and +. As a result, we can state the following *metatheorem*, a theorem about theorems:

metatheorem

Principle of Duality   Any theorem or identity in switching algebra remains true if 0 and 1 are swapped and · and + are swapped throughout.

The metatheorem is true because the duals of all the axioms are true, so duals of all switching-algebra theorems can be proved using duals of the axioms.

After all, what's in a name, or in a symbol for that matter? If the software that was used to typeset this book had a bug, one that swapped $0 \leftrightarrow 1$ and $\cdot \leftrightarrow +$ throughout this chapter, you still would have learned exactly the same switching algebra; only the nomenclature would have been a little weird, using words like "product" to describe an operation that uses the symbol "+".

Duality is important because it doubles the usefulness of everything that you learn about switching algebra and manipulation of switching functions. Stated more practically, from a student's point of view, it halves the amount that you have to learn! For example, once you learn how to synthesize two-stage AND-OR logic circuits from sum-of-products expressions, you automatically know a dual technique to synthesize OR-AND circuits from product-of-sums expressions.

There is just one convention in switching algebra where we did not treat · and + identically, so duality does not necessarily hold true—can you figure out what it is before reading the answer below? Consider the following statement of theorem T9 and its clearly absurd "dual":

$$X + X \cdot Y = X \qquad \text{(theorem T9)}$$
$$X \cdot X + Y = X \qquad \text{(after applying the principle of duality)}$$
$$X + Y = X \qquad \text{(after applying theorem T3′)}$$

Obviously the last line above is false—where did we go wrong? The problem is in operator precedence. We were able to write the lefthand side of the first line without parentheses because of our convention that · has precedence. However, once we applied the principle of duality, we should have given precedence to + instead, or written the second line as $X \cdot (X + Y) = X$. The best way to avoid problems like this is to parenthesize an expression fully before taking its dual.

Let us formally define the *dual of a logic expression*. If $F(X_1, X_2, ..., X_n, +, \cdot, ')$ is a fully parenthesized logic expression involving the

dual of a logic expression

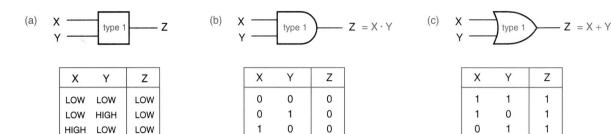

Figure 4-5 A "type-1"logic gate: (a) electrical function table; (b) logic function table and symbol with positive logic; (c) logic function table and symbol with negative logic.

variables $X_1, X_2, ..., X_n$ and the operators $+$, $\cdot$, and $'$, then the dual of F, written F^D, is the same expression with $+$ and $\cdot$ swapped:

$$F^D(X_1, X_2, ..., X_n, +, \cdot, ') = F(X_1, X_2, ..., X_n, \cdot, +, ')$$

You already knew this, of course, but we wrote the definition in this way just to highlight the similarity between duality and the generalized DeMorgan's theorem T14, which may now be restated as follows:

$$[F(X_1, X_2, ..., X_n)]' = F^D(X_1', X_2', ..., X_n')$$

Let's examine this statement in terms of a physical network.

Figure 4-5(a) shows the electrical function table for a logic element that we'll simply call a "type-1" gate. Under the positive-logic convention ($LOW = 0$ and $HIGH = 1$), this is an AND gate, but under the negative-logic convention ($LOW = 1$ and $HIGH = 0$), it is an OR gate, as shown in (b) and (c). We can also imagine a "type-2" gate, shown in Figure 4-6, that is a positive-logic OR or a negative-logic AND. Similar tables can be developed for gates with more than two inputs.

Figure 4-6 A "type-2" logic gate: (a) electrical function table; (b) logic function table and symbol with positive logic; (c) logic function table and symbol with negative logic.

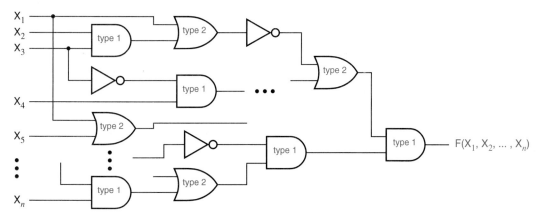

Figure 4-7 Circuit for a logic function using inverters and type-1 and type-2 gates under a positive-logic convention.

Suppose that we are given an arbitrary logic expression, $F(X_1, X_2, ..., X_n)$. Following the positive-logic convention, we can build a circuit corresponding to this expression using inverters for **NOT** operations, type-1 gates for **AND**, and type-2 gates for **OR**, as shown in Figure 4-7. Now suppose that, without changing this circuit, we simply change the logic convention from positive to negative. Then we should redraw the circuit as shown in Figure 4-8. Clearly, for every possible combination of input voltages (**HIGH** and **LOW**), the circuit still produces the same output voltage. However, from the point of view of switching algebra, the output value—0 or 1—is the opposite of what it was under the positive-logic convention. Likewise, each input value is the opposite of what it was. Therefore, for each possible input combination to the circuit in Figure 4-7, the output is the opposite of that produced by the opposite input combination applied to the circuit in Figure 4-8:

$$F(X_1, X_2, ..., X_n) = [F^D(X_1', X_2', ..., X_n')]'$$

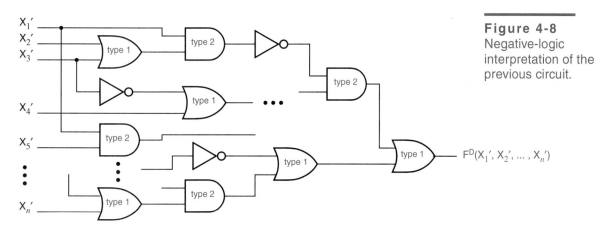

Figure 4-8
Negative-logic interpretation of the previous circuit.

By complementing both sides, we get the generalized DeMorgan's theorem:

$$[F(X_1, X_2, ..., X_n)]' = F^D(X_1', X_2', ..., X_n')$$

Amazing!

DUALITY IS GOOD FOR STUDENTS AND AUTHORS	You've seen that duality is the basis for the generalized DeMorgan's theorem. Going forward, it will halve the number of methods you must learn to manipulate and simplify logic functions. It also halved the amount of original material I had to write in these sections!

4.1.6 Standard Representations of Logic Functions

Before moving on to analysis and synthesis of combinational logic functions we'll introduce some necessary nomenclature and notation.

truth table

The most basic representation of a logic function is the *truth table*. Similar in philosophy to the perfect-induction proof method, this brute-force representation simply lists the output of the circuit for every possible input combination. Traditionally, the input combinations are arranged in rows in ascending binary counting order, and the corresponding output values are written in a column next to the rows. The general structure of a 3-variable truth table is shown below in Table 4-4.

The rows are numbered 0–7, corresponding to the binary input combinations, but this numbering is not an essential part of the truth table. The truth table for a particular 3-variable logic function is shown in Table 4-5. Each distinct pattern of 0s and 1s in the output column yields a different logic function; there are 2^8 such patterns. Thus, the logic function in Table 4-5 is one of 2^8 different logic functions of three variables.

The truth table for an *n*-variable logic function has 2^n rows. Obviously, truth tables are practical to write only for logic functions with a small number of variables, say, 10 for students and about 4–5 for everyone else.

Table 4-4
General truth table structure for a 3-variable logic function, F(X, Y, Z).

Row	X	Y	Z	F
0	0	0	0	F(0,0,0)
1	0	0	1	F(0,0,1)
2	0	1	0	F(0,1,0)
3	0	1	1	F(0,1,1)
4	1	0	0	F(1,0,0)
5	1	0	1	F(1,0,1)
6	1	1	0	F(1,1,0)
7	1	1	1	F(1,1,1)

Row	X	Y	Z	F
0	0	0	0	1
1	0	0	1	0
2	0	1	0	0
3	0	1	1	1
4	1	0	0	1
5	1	0	1	0
6	1	1	0	1
7	1	1	1	1

Table 4-5
Truth table for a
particular 3-variable
logic function, $F(X, Y, Z)$.

The information contained in a truth table can also be conveyed algebraically. To do so, we first need some definitions:

- A *literal* is a variable or the complement of a variable. Examples: X, Y, X', Y'.   *literal*

- A *product term* is a single literal or a logical product of two or more   *product term*
literals. Examples: Z', $W \cdot X \cdot Y$, $X \cdot Y' \cdot Z$, $W' \cdot Y' \cdot Z$.

- A *sum-of-products expression* is a logical sum of product terms. Example: *sum-of-products*
$Z' + W \cdot X \cdot Y + X \cdot Y' \cdot Z + W' \cdot Y' \cdot Z$.   *expression*

- A *sum term* is a single literal or a logical sum of two or more literals.   *sum term*
Examples: Z', $W + X + Y$, $X + Y' + Z$, $W' + Y' + Z$.

- A *product-of-sums expression* is a logical product of sum terms. Example: *product-of-sums*
$Z' \cdot (W + X + Y) \cdot (X + Y' + Z) \cdot (W' + Y' + Z)$.   *expression*

- A *normal term* is a product or sum term in which no variable appears more   *normal term*
than once. A nonnormal term can always be simplified to a constant or a
normal term using one of theorems T3, T3', T5, or T5'. Examples of non-
normal terms: $W \cdot X \cdot X \cdot Y'$, $W + W + X' + Y$, $X \cdot X' \cdot Y$. Examples of
normal terms: $W \cdot X \cdot Y'$, $W + X' + Y$.

- An n-variable *minterm* is a normal product term with n literals. There are   *minterm*
2^n such product terms. Examples of 4-variable minterms:
$W' \cdot X' \cdot Y' \cdot Z'$, $W \cdot X \cdot Y' \cdot Z$, $W' \cdot X' \cdot Y \cdot Z'$.

- An n-variable *maxterm* is a normal sum term with n literals. There are 2^n *maxterm*
such sum terms. Examples of 4-variable maxterms: $W' + X' + Y' + Z'$,
$W + X' + Y' + Z$, $W' + X' + Y + Z'$.

There is a close correspondence between the truth table and minterms and
maxterms. A minterm can be defined as a product term that is 1 in exactly one
row of the truth table. Similarly, a maxterm can be defined as a sum term that is
0 in exactly one row of the truth table. Table 4-6 shows this correspondence for a
3-variable truth table.

Table 4-6
Minterms and maxterms for a 3-variable logic function, F(X, Y, Z).

Row	X	Y	Z	F	*Minterm*	*Maxterm*
0	0	0	0	F(0,0,0)	$X' \cdot Y' \cdot Z'$	$X + Y + Z$
1	0	0	1	F(0,0,1)	$X' \cdot Y' \cdot Z$	$X + Y + Z'$
2	0	1	0	F(0,1,0)	$X' \cdot Y \cdot Z'$	$X + Y' + Z$
3	0	1	1	F(0,1,1)	$X' \cdot Y \cdot Z$	$X + Y' + Z'$
4	1	0	0	F(1,0,0)	$X \cdot Y' \cdot Z'$	$X' + Y + Z$
5	1	0	1	F(1,0,1)	$X \cdot Y' \cdot Z$	$X' + Y + Z'$
6	1	1	0	F(1,1,0)	$X \cdot Y \cdot Z'$	$X' + Y' + Z$
7	1	1	1	F(1,1,1)	$X \cdot Y \cdot Z$	$X' + Y' + Z'$

minterm number
minterm i

An *n*-variable minterm can be represented by an *n*-bit integer, the *minterm number*. We'll use the name *minterm i* to denote the minterm corresponding to row *i* of the truth table. In minterm *i*, a particular variable appears complemented if the corresponding bit in the binary representation of *i* is 0; otherwise, it is uncomplemented. For example, row 5 has binary representation 101 and the corresponding minterm is $X \cdot Y' \cdot Z$. As you might expect, the correspondence for maxterms is just the opposite: in *maxterm i*, a variable is complemented if the corresponding bit in the binary representation of *i* is 1. Thus, maxterm 5 (101) is $X' + Y + Z'$. Note that all of this makes sense only if we know the number of variables in the truth table, three in the examples.

maxterm i

Based on the correspondence between the truth table and minterms, we can easily create an algebraic representation of a logic function from its truth table. The *canonical sum* of a logic function is a sum of the minterms corresponding to truth-table rows (input combinations) for which the function produces a 1 output. For example, the canonical sum for the logic function in Table 4-5 on page 207 is

canonical sum

$$F = \Sigma_{X,Y,Z}(0, 3, 4, 6, 7)$$
$$= X' \cdot Y' \cdot Z' + X' \cdot Y \cdot Z + X \cdot Y' \cdot Z' + X \cdot Y \cdot Z' + X \cdot Y \cdot Z$$

minterm list

on-set

Here, the notation $\Sigma_{X,Y,Z}(0, 3, 4, 6, 7)$ is a *minterm list* and means "the sum of minterms 0, 3, 4, 6, and 7 with variables X, Y, and Z." The minterm list is also known as the *on-set* for the logic function. You can visualize that each minterm "turns on" the output for exactly one input combination. Any logic function can be written as a canonical sum.

canonical product

The *canonical product* of a logic function is a product of the maxterms corresponding to input combinations for which the function produces a 0 output. For example, the canonical product for the logic function in Table 4-5 is

$$F = \Pi_{X,Y,Z}(1, 2, 5)$$
$$= (X + Y + Z') \cdot (X + Y' + Z) \cdot (X' + Y + Z')$$

Here, the notation $\prod_{X,Y,Z}(1,2,5)$ is a *maxterm list* and means "the product of *maxterm list* maxterms 1, 2, and 5 with variables X, Y, and Z." The maxterm list is also known as the *off-set* for the logic function. You can visualize that each maxterm "turns *off-set* off" the output for exactly one input combination. Any logic function can be written as a canonical product.

It's easy to convert between a minterm list and a maxterm list. For a function of n variables, the possible minterm and maxterm numbers are in the set $\{0, 1, \ldots, 2^n - 1\}$; a minterm or maxterm list contains a subset of these numbers. To switch between list types, take the set complement, for example,

$$\Sigma_{A,B,C}(0,1,2,3) = \prod_{A,B,C}(4,5,6,7)$$
$$\Sigma_{X,Y}(1) = \prod_{X,Y}(0,2,3)$$
$$\Sigma_{W,X,Y,Z}(0,1,2,3,5,7,11,13) = \prod_{W,X,Y,Z}(4,6,8,9,10,12,14,15)$$

We have now learned five possible representations for a combinational logic function:

1. A truth table.
2. An algebraic sum of minterms, the canonical sum.
3. A minterm list using the Σ notation.
4. An algebraic product of maxterms, the canonical product.
5. A maxterm list using the $\prod$ notation.

Each one of these representations specifies exactly the same information; given any one of them, we can derive the other four using a simple mechanical process.

4.2 Combinational-Circuit Analysis

We analyze a combinational logic circuit by obtaining a formal description of its logic function. Once we have a description of the logic function, a number of other operations are possible:

- We can determine the behavior of the circuit for various input combinations.
- We can manipulate an algebraic description to suggest different circuit structures for the logic function.
- We can transform an algebraic description into a standard form corresponding to an available circuit structure. For example, a sum-of-products expression corresponds directly to the circuit structure used in PLDs (programmable logic devices).
- We can use an algebraic description of the circuit's functional behavior in the analysis of a larger system that includes the circuit.

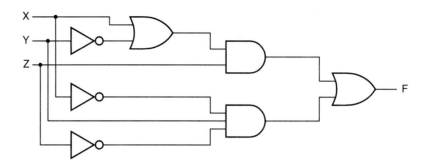

Figure 4-9
A three-input, one-output logic circuit.

Given a logic diagram for a combinational circuit, such as Figure 4-9, there are a number of ways to obtain a formal description of the circuit's function. The most primitive functional description is the truth table.

Using only the basic axioms of switching algebra, we can obtain the truth table of an n-input circuit by working our way through all 2^n input combinations. For each input combination, we determine all of the gate outputs produced by that input, propagating information from the circuit inputs to the circuit outputs. Figure 4-10 applies this "exhaustive" technique to our example circuit. Written on each signal line in the circuit is a sequence of eight logic values, the values present on that line when the circuit inputs XYZ are 000, 001, ..., 111. The truth table can be written by transcribing the output sequence of the final OR gate, as

Figure 4-10 Gate outputs created by all input combinations.

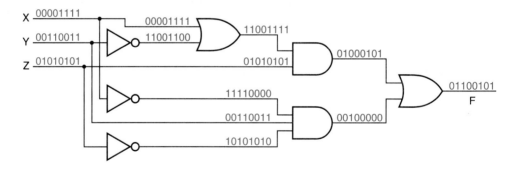

A LESS EXHAUSTING WAY TO GO   You can easily obtain the results in Figure 4-10 with typical logic design tools that include a logic simulator. First, you draw the schematic. Then, you apply the outputs of a 3-bit binary counter to the X, Y, and Z inputs. (Most simulators have such counter outputs built in for just this sort of exercise.) The counter repeatedly cycles through the eight possible input combinations, in the same order that we've shown in the figure. The simulator allows you to graph the resulting signal values at any point in the schematic, including the intermediate points as well as the output.

Row	X	Y	Z	F
0	0	0	0	0
1	0	0	1	1
2	0	1	0	1
3	0	1	1	0
4	1	0	0	0
5	1	0	1	1
6	1	1	0	0
7	1	1	1	1

Table 4-7
Truth table for the
logic circuit of
Figure 4-9.

shown in Table 4-7. Once we have the truth table for the circuit, we can also directly write a logic expression—the canonical sum or product—if we wish.

The number of input combinations of a logic circuit grows exponentially with the number of inputs, so the exhaustive approach can quickly become exhausting. Instead, we normally use an algebraic approach whose complexity is more linearly proportional to the size of the circuit. The method is simple—we build up a parenthesized logic expression corresponding to the logic operators and structure of the circuit. We start at the circuit inputs and propagate expressions through gates toward the output. Using the theorems of switching algebra, we may simplify the expressions as we go, or we may defer all algebraic manipulations until an output expression is obtained.

Figure 4-11 applies the algebraic technique to our example circuit. The output function is given on the output of the final OR gate:

$$F = ((X+Y') \cdot Z) + (X' \cdot Y \cdot Z')$$

No switching-algebra theorems were used to obtain this expression. However, we can use theorems to transform this expression into another form. For example, a sum of products can be obtained by "multiplying out":

$$F = X \cdot Z + Y' \cdot Z + X' \cdot Y \cdot Z'$$

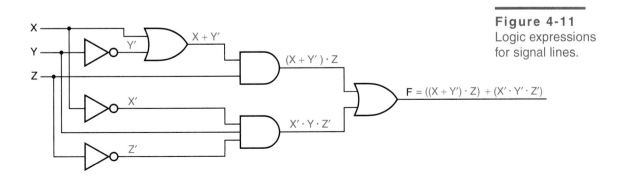

Figure 4-11
Logic expressions
for signal lines.

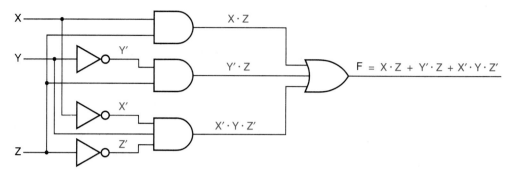

Figure 4-12 Two-level AND-OR circuit.

The new expression corresponds to a different circuit for the same logic function, as shown in Figure 4-12.

Similarly, we can "add out" the original expression to obtain a product of sums:

$$
\begin{aligned}
F &= ((X + Y') \cdot Z) + (X' \cdot Y \cdot Z') \\
&= (X + Y' + X') \cdot (X + Y' + Y) \cdot (X + Y' + Z') \cdot (Z + X') \cdot (Z + Y) \cdot (Z + Z') \\
&= 1 \cdot 1 \cdot (X + Y' + Z') \cdot (X' + Z) \cdot (Y + Z) \cdot 1 \\
&= (X + Y' + Z') \cdot (X' + Z) \cdot (Y + Z)
\end{aligned}
$$

The corresponding logic circuit is shown in Figure 4-13.

Our next example of algebraic analysis uses a circuit with NAND and NOR gates, shown in Figure 4-14. This analysis is a little messier than the previous example, because each gate produces a complemented subexpression, not just a simple sum or product. However, the output expression can be simplified by repeated application of the generalized DeMorgan's theorem:

Figure 4-13 Two-level OR-AND circuit.

Figure 4-14
Algebraic analysis of a logic circuit with NAND and NOR gates.

$$F = [((W \cdot X')' \cdot Y)' + (W' + X + Y')' + (W + Z)']'$$
$$= ((W' + X)' + Y')' \cdot (W \cdot X' \cdot Y)' \cdot (W' \cdot Z')'$$
$$= ((W \cdot X')' \cdot Y) \cdot (W' + X + Y') \cdot (W + Z)$$
$$= ((W' + X) \cdot Y) \cdot (W' + X + Y') \cdot (W + Z)$$

Quite often, DeMorgan's theorem can be applied *graphically* to simplify algebraic analysis. Recall from Figures 4-3 and 4-4 that NAND and NOR gates each have two equivalent symbols. By judiciously redrawing Figure 4-14, we make it possible to cancel out some of the inversions during the analysis by using theorem T4 $[(X')' = X]$, as shown in Figure 4-15. This manipulation leads us to a simplified output expression directly:

$$F = ((W' + X) \cdot Y) \cdot (W' + X + Y') \cdot (W + Z)$$

Figures 4-14 and 4-15 were just two different ways of drawing the same physical logic circuit. However, when we simplify a logic expression using the theorems of switching algebra, we get an expression corresponding to a different physical circuit. For example, the simplified expression above corresponds to the circuit of Figure 4-16, which is physically different from the one in the previous two figures. Furthermore, we could multiply out and add out the

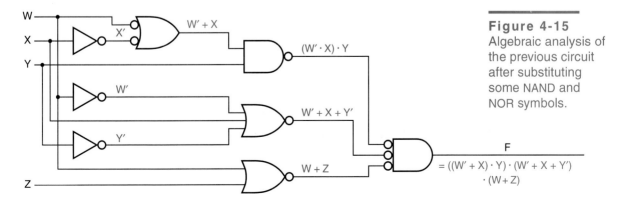

Figure 4-15
Algebraic analysis of the previous circuit after substituting some NAND and NOR symbols.

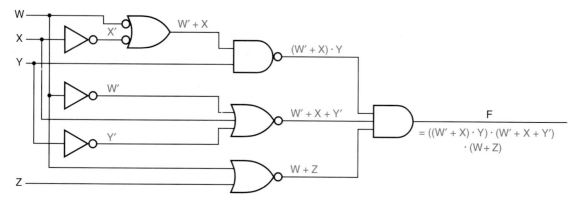

Figure 4-16 A different circuit for same logic function.

expression to obtain sum-of-products and product-of-sums expressions corresponding to two more physically different circuits for the same logic function.

Although we used logic expressions above to convey information about the physical structure of a circuit, we don't always do this. For example, we might use the expression $G(W, X, Y, Z) = W \cdot X \cdot Y + Y \cdot Z$ to describe any one of the circuits in Figure 4-17. Normally, the only sure way to determine a circuit's structure is to look at its schematic drawing. However, for certain restricted classes of circuits, structural information can be inferred from logic expressions. For example, the circuit in (a) could be described without reference to the drawing as "a two-level AND-OR circuit for $W \cdot X \cdot Y + Y \cdot Z$," while the circuit in (b) could be described as "a two-level NAND-NAND circuit for $W \cdot X \cdot Y + Y \cdot Z$."

Figure 4-17 Three circuits for $G(W, X, Y, Z) = W \cdot X \cdot Y + Y \cdot Z$: (a) two-level AND-OR; (b) two-level NAND-NAND; (c) ad hoc.

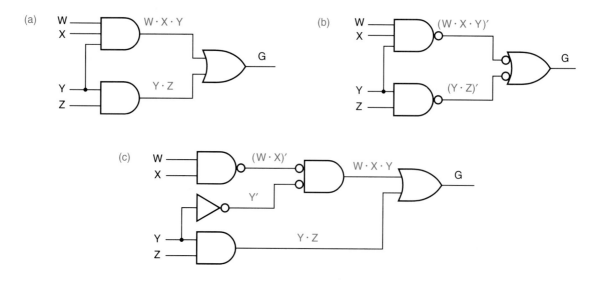

4.3 Combinational-Circuit Synthesis

4.3.1 Circuit Descriptions and Designs

What is the starting point for designing combinational logic circuits? Usually, we are given a word description of a problem or we develop one ourselves. Occasionally, the description is a list of input combinations for which a signal should be on or off, the verbal equivalent of a truth table or the Σ or $\prod$ notation introduced previously. For example, the description of a 4-bit prime-number detector might be, "Given a 4-bit input combination $N = N_3N_2N_1N_0$, this function produces a 1 output for $N = 1, 2, 3, 5, 7, 11, 13$, and 0 otherwise." A logic function described in this way can be designed directly from the canonical sum or product expression. For the prime-number detector, we have

$$
\begin{aligned}
F &= \Sigma_{N_3,N_2,N_1,N_0}(1, 2, 3, 5, 7, 11, 13) \\
&= N_3' \cdot N_2' \cdot N_1' \cdot N_0 + N_3' \cdot N_2' \cdot N_1 \cdot N_0' + N_3' \cdot N_2' \cdot N_1 \cdot N_0 + N_3' \cdot N_2 \cdot N_1' \cdot N_0 \\
&\quad + N_3' \cdot N_2 \cdot N_1 \cdot N_0 + N_3 \cdot N_2' \cdot N_1 \cdot N_0 + N_3 \cdot N_2 \cdot N_1' \cdot N_0
\end{aligned}
$$

The corresponding circuit is shown in Figure 4-18.

More often, we describe a logic function using the English-language connectives "and," "or," and "not." For example, we might describe an alarm circuit by saying, "The ALARM output is 1 if the PANIC input is 1, or if the ENABLE input is 1, the EXITING input is 0, and the house is not secure; the

Figure 4-18 Canonical-sum design for 4-bit prime-number detector.

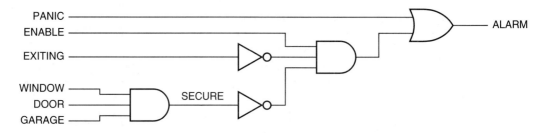

Figure 4-19 Alarm circuit derived from logic expression.

house is secure if the WINDOW, DOOR, and GARAGE inputs are all 1." Such a description can be translated directly into algebraic expressions:

ALARM = PANIC + ENABLE · EXITING′ · SECURE′

SECURE = WINDOW · DOOR · GARAGE

ALARM = PANIC + ENABLE · EXITING′ · (WINDOW · DOOR · GARAGE)′

Notice that we used the same method in switching algebra as in ordinary algebra to formulate a complicated expression—we defined an auxiliary variable **SECURE** to simplify the first equation, developed an expression for **SECURE**, and used substitution to get the final expression. We can easily draw a circuit using AND, OR, and NOT gates that realizes the final expression, as shown in Figure 4-19. A circuit *realizes* ["makes real"] an expression if its output function equals that expression, and the circuit is called a *realization* of the function.

realize
realization

Once we have an expression, any expression, for a logic function, we can do other things besides building a circuit directly from the expression. We can manipulate the expression to get different circuits. For example, the ALARM expression above can be multiplied out to get the sum-of-products circuit in Figure 4-20. Or, if the number of variables is not too large, we can construct the truth table for the expression and use any of the synthesis methods that apply to truth tables, including the canonical sum or product method described earlier and the minimization methods described later.

Figure 4-20 Sum-of-products version of alarm circuit.

(b)

(a)

(c)

Figure 4-21
Alternative sum-of-products realizations:
(a) AND-OR;
(b) AND-OR with extra inverter pairs;
(c) NAND-NAND.

In general, it's easier to describe a circuit in words using logical connectives and to write the corresponding logic expressions than it is to write a complete truth table, especially if the number of variables is large. However, sometimes we have to work with imprecise word descriptions of logic functions, for example, "The ERROR output should be 1 if the GEARUP, GEARDOWN, and GEARCHECK inputs are inconsistent." In this situation, the truth-table approach is best because it allows us to determine the output required for every input combination, based on our knowledge and understanding of the problem environment (e.g., the brakes cannot be applied unless the gear is down).

4.3.2 Circuit Manipulations

The design methods that we've described so far use AND, OR, and NOT gates. We might like to use NAND and NOR gates, too—they're faster than ANDs and ORs in most technologies. However, most people don't develop logical propositions in terms of NAND and NOR connectives. That is, you probably wouldn't say, "I won't date you if you're not clean or not wealthy and also you're not smart or not friendly." It would be more natural for you to say, "I'll date you if you're clean and wealthy, or if you're smart and friendly." So, given a "natural" logic expression, we need ways to translate it into other forms.

We can translate any logic expression into an equivalent sum-of-products expression, simply by multiplying it out. As shown in Figure 4-21(a), such an expression may be realized directly with AND and OR gates. The inverters required for complemented inputs are not shown.

As shown in Figure 4-21(b), we may insert a pair of inverters between each AND-gate output and the corresponding OR-gate input in a two-level AND-OR

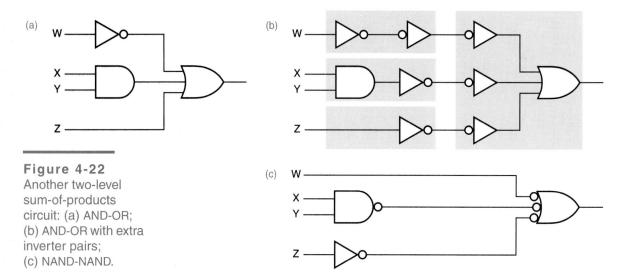

Figure 4-22
Another two-level
sum-of-products
circuit: (a) AND-OR;
(b) AND-OR with extra
inverter pairs;
(c) NAND-NAND.

circuit. According to theorem T4, these inverters have no effect on the output function of the circuit. In fact, we've drawn the second inverter of each pair with its inversion bubble on its input to provide a graphical reminder that the inverters cancel. However, if these inverters are absorbed into the AND and OR gates, we wind up with AND-NOT gates at the first level and a NOT-OR gate at the second level. These are just two different symbols for the same type of gate—a NAND gate. Thus, a two-level AND-OR *circuit* may be converted to a two-level NAND-NAND *circuit* simply by substituting gates.

AND-OR circuit

NAND-NAND circuit

Figure 4-23
Realizations of a
product-of-sums
expression:
(a) OR-AND;
(b) OR-AND with extra
inverter pairs;
(c) NOR-NOR.

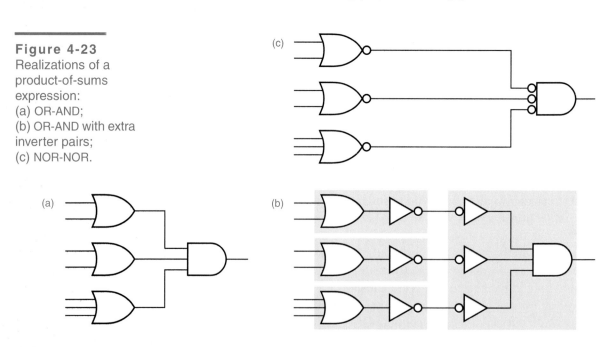

If any product terms in the sum-of-products expression contain just a single literal, then we may gain or lose inverters in the transformation from AND-OR to NAND-NAND. For example, Figure 4-22 is an example where an inverter on the W input is no longer needed, but an inverter must be added to the Z input.

We have shown that any sum-of-products expression can be realized in either of two ways—as an AND-OR circuit or as a NAND-NAND circuit. The dual of this statement is also true: any product-of-sums expression can be realized as an OR-AND *circuit* or as a NOR-NOR *circuit*. Figure 4-23 shows an example. Any logic expression can be translated into an equivalent product-of-sums expression by adding it out, and hence has both OR-AND and NOR-NOR circuit realizations.

OR-AND circuit
NOR-NOR circuit

The same kind of manipulations can be applied to arbitrary logic circuits. For example, Figure 4-24(a) shows a circuit built from AND and OR gates. After adding pairs of inverters, we obtain the circuit in (b). However, one of the gates, a 2-input AND gate with a single inverted input, is not a standard type. We can use a discrete inverter as shown in (c) to obtain a circuit that uses only standard gate types—NAND, AND, and inverters. Actually, a better way to use the inverter is shown in (d); one level of gate delay is eliminated, and the bottom gate becomes a NOR instead of AND. In most logic technologies, inverting gates like NAND and NOR are faster than noninverting gates like AND and OR.

Figure 4-24 Logic-symbol manipulations: (a) original circuit; (b) transformation with a nonstandard gate; (c) inverter used to eliminate nonstandard gate; (d) preferred inverter placement.

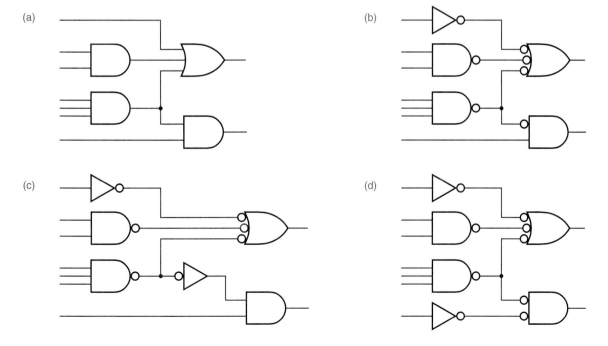

WHY MINIMIZE?   Minimization is an important step in both ASIC design and in PLD-based design. Extra gates and gate inputs require more area in an ASIC chip and thereby increase cost. The number of gates in a PLD is fixed, so you might think that extra gates are free—and they are, until you run out of them and have to upgrade to a bigger, slower, more expensive PLD. Fortunately, most software tools for both ASIC and PLD design have a minimization program built in. The purpose of Sections 4.3.3 through 4.3.8 is to give you a feel for how minimization works.

4.3.3 Combinational-Circuit Minimization

It's often uneconomical to realize a logic circuit directly from the first logic expression that pops into your head. Canonical sum and product expressions are especially expensive because the number of possible minterms or maxterms (and hence gates) grows exponentially with the number of variables. We *minimize* *minimize* a combinational circuit by reducing the number and size of gates that are needed to build it.

The traditional combinational-circuit-minimization methods that we'll study have as their starting point a truth table or, equivalently, a minterm list or maxterm list. If we are given a logic function that is not expressed in this form, then we must convert it to an appropriate form before using these methods. For example, if we are given an arbitrary logic expression, then we can evaluate it for every input combination to construct the truth table. The minimization methods reduce the cost of a two-level AND-OR, OR-AND, NAND-NAND, or NOR-NOR circuit in three ways:

1. By minimizing the number of first-level gates.
2. By minimizing the number of inputs on each first-level gate.
3. By minimizing the number of inputs on the second-level gate. This is actually a side effect of the first reduction.

However, the minimization methods do not consider the cost of input inverters; they assume that both true and complemented versions of all input variables are available. While this is not always the case in gate-level or ASIC design, it's very appropriate for PLD-based design; PLDs have both true and complemented versions of all input variables available "for free."

Most minimization methods are based on a generalization of the combining theorems, T10 and T10′:

$$\text{given product term} \cdot Y + \text{given product term} \cdot Y' = \text{given product term}$$
$$\text{given sum term} + Y) \cdot \text{given sum term} + Y') = \text{given sum term}$$

That is, if two product or sum terms differ only in the complementing or not of one variable, we can combine them into a single term with one less variable. So we save one gate and the remaining gate has one fewer input.

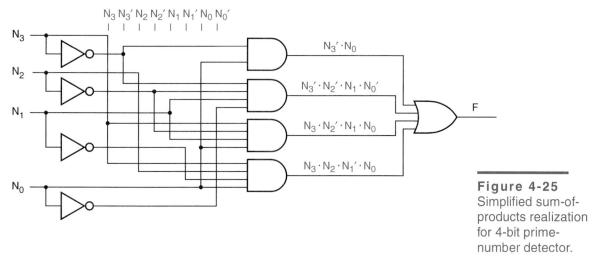

Figure 4-25
Simplified sum-of-products realization for 4-bit prime-number detector.

We can apply this algebraic method repeatedly to combine minterms 1, 3, 5, and 7 of the prime-number detector shown in Figure 4-18 on page 215:

$$
\begin{aligned}
F &= \Sigma_{N_3,N_2,N_1,N_0}(1, 3, 5, 7, 2, 11, 13) \\
&= N_3' \cdot N_2' N_1' N_0 + N_3' \cdot N_2' \cdot N_1 \cdot N_0 + N_3' \cdot N_2 \cdot N_1' \cdot N_0 + N_3' \cdot N_2 \cdot N_1 \cdot N_0 + \ldots \\
&= (N_3' \cdot N_2' \cdot N_1' \cdot N_0 + N_3' \cdot N_2' \cdot N_1 \cdot N_0) + (\cdot N_3' \cdot N_2 \cdot N_1' \cdot N_0 + N_3' \cdot N_2 \cdot N_1 \cdot N_0) + \ldots \\
&= N_3' N_2' \cdot N_0 + N_3' \cdot N_2 \cdot N_0 + \ldots \\
&= N_3' \cdot N_0 + \ldots
\end{aligned}
$$

The resulting circuit is shown in Figure 4-25; it has three fewer gates, and one of the remaining gates has two fewer inputs.

If we had worked a little harder on the preceding expression, we could have saved a couple more first-level gate inputs, though not any gates. It's difficult to find terms that can be combined in a jumble of algebraic symbols. In the next subsection we'll begin to explore a minimization method that is more fit for human consumption. Our starting point will be the graphical equivalent of a truth table.

4.3.4 Karnaugh Maps

A *Karnaugh map* is a graphical representation of a logic function's truth table. *Karnaugh map*
Figure 4-26 shows Karnaugh maps for logic functions of 2, 3, and 4 variables. The map for an *n*-input logic function is an array with 2^n cells, one for each possible input combination or minterm.

The rows and columns of a Karnaugh map are labeled so that the input combination for any cell is easily determined from the row and column headings for that cell. The small number inside each cell is the corresponding minterm number in the truth table, assuming that the truth-table inputs are labeled alphabetically from left to right (e.g., X, Y, Z) and the rows are numbered in binary

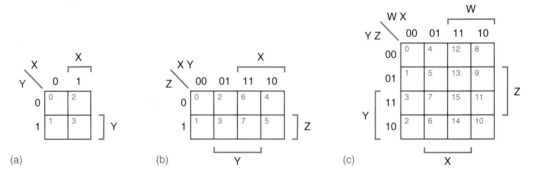

Figure 4-26 Karnaugh maps: (a) 2-variable; (b) 3-variable; (c) 4-variable.

counting order, like all the examples in this text. For example, cell 13 in the 4-variable map corresponds to the truth-table row in which W X Y Z = 1101.

When we draw the Karnaugh map for a given function, each cell of the map contains the information from the like-numbered row of the function's truth table—a 0 if the function is 0 for that input combination, a 1 otherwise.

In this text we use two redundant labelings for map rows and columns. For example, consider the 4-variable map in Figure 4-26(c). The columns are labeled with the four possible combinations of W and X, W X = 00, 01, 11, and 10. Similarly, the rows are labeled with the Y Z combinations. These labels give us all the information we need. However, we also use brackets to associate four regions of the map with the four variables. Each bracketed region is the part of the map in which the indicated variable is 1. Obviously, the brackets convey the same information that is given by the row and column labels.

When we draw a map by hand, it is much easier to draw the brackets than to write out all of the labels. However, we retain the labels in the text's Karnaugh maps as an additional aid to understanding. In any case, you must be sure to label the rows and columns in the proper order to preserve the correspondence between map cells and truth-table row numbers shown in Figure 4-26.

To represent a logic function on a Karnaugh map, we simply copy 1s and 0s from the truth table or equivalent to the corresponding cells of the map. Figures 4-27(a) and (b) show the truth table and Karnaugh map for a logic function that we analyzed (beat to death?) in Section 4.2. From now on, we'll reduce the clutter in maps by copying only the 1s or the 0s, not both.

4.3.5 Minimizing Sums of Products

By now you must be wondering about the "strange" ordering of the row and column numbers in a Karnaugh map. There is a very important reason for this ordering—each cell corresponds to an input combination that differs from each of its immediately adjacent neighbors in only one variable. For example, cells 5 and 13 in the 4-variable map differ only in the value of W. In the 3- and

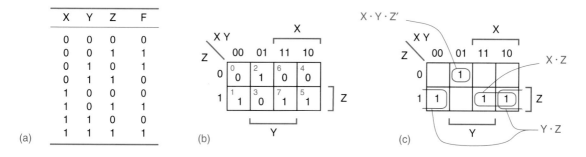

Figure 4-27 $F = \Sigma_{X,Y,Z}(1,2,5,7)$: (a) truth table; (b) Karnaugh map; (c) combining adjacent 1-cells.

4-variable maps, corresponding cells on the left/right or top/bottom borders are less obvious neighbors; for example, cells 12 and 14 in the 4-variable map are adjacent because they differ only in the value of Y.

Each input combination with a "1" in the truth table corresponds to a minterm in the logic function's canonical sum. Since pairs of adjacent "1" cells in the Karnaugh map have minterms that differ in only one variable, the minterm pairs can be combined into a single product term using the generalization of theorem T10, term $\cdot$ Y + term $\cdot$ Y' = term. Thus, we can use a Karnaugh map to simplify the canonical sum of a logic function.

For example, consider cells 5 and 7 in Figure 4-27(b) and their contribution to the canonical sum for this function:

$$
\begin{aligned}
F &= \cdots + X \cdot Y' \cdot Z + X \cdot Y \cdot Z \\
&= \cdots + (X \cdot Z) \cdot Y' + (X \cdot Z) \cdot Y \\
&= \cdots + X \cdot Z
\end{aligned}
$$

Remembering wraparound, we see that cells 1 and 5 in Figure 4-27(b) are also adjacent and can be combined:

$$
\begin{aligned}
F &= X' \cdot Y' \cdot Z + X \cdot Y' \cdot Z + \cdots \\
&= X' \cdot (Y' \cdot Z) + X \cdot (Y' \cdot Z) + \cdots \\
&= Y' \cdot Z + \ldots
\end{aligned}
$$

In general, we can simplify a logic function by combining pairs of adjacent 1-cells (minterms) whenever possible and writing a sum of product terms that covers all of the 1-cells. Figure 4-27(c) shows the result for our example logic function. We circle a pair of 1s to indicate that the corresponding minterms are combined into a single product term. The corresponding AND-OR circuit is shown in Figure 4-28.

In many logic functions the cell-combining procedure can be extended to combine more than two 1-cells into a single product term. For example, consider

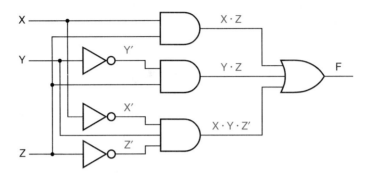

Figure 4-28
Minimized AND-OR circuit.

the canonical sum for the logic function $F = \Sigma_{X,Y,Z}(0, 1, 4, 5, 6)$. We can use the algebraic manipulations of the previous examples iteratively to combine four of the five minterms:

$$
\begin{aligned}
F &= X' \cdot Y' \cdot Z' + X' \cdot Y' \cdot Z + X \cdot Y' \cdot Z' + X \cdot Y' \cdot Z + X \cdot Y \cdot Z' \\
&= [(X' \cdot Y') \cdot Z' + (X' \cdot Y') \cdot Z] + [(X \cdot Y') \cdot Z' + (X \cdot Y') \cdot Z] + X \cdot Y \cdot Z' \\
&= X' \cdot Y' + X \cdot Y' + X \cdot Y \cdot Z' \\
&= [X' \cdot (Y') + X \cdot (Y')] + X \cdot Y \cdot Z' \\
&= Y' + X \cdot Y \cdot Z'
\end{aligned}
$$

In general, 2^i 1-cells may be combined to form a product term containing $n - i$ literals, where n is the number of variables in the function.

A precise mathematical rule determines how 1-cells may be combined and the form of the corresponding product term:

- A set of 2^i 1-cells may be combined if there are i variables of the logic function that take on all 2^i possible combinations within that set, while the remaining $n - i$ variables have the same value throughout that set. The corresponding product term has $n - i$ literals, where a variable is complemented if it appears as 0 in all of the 1-cells, and uncomplemented if it appears as 1.

rectangular sets of 1s   Graphically, this rule means that we can circle *rectangular* sets of 2^n 1s, literally as well as figuratively stretching the definition of rectangular to account for wraparound at the edges of the map. We can determine the literals of the corresponding product terms directly from the map; for each variable we make the following determination:

- If a circle covers only areas of the map where the variable is 0, then the variable is complemented in the product term.
- If a circle covers only areas of the map where the variable is 1, then the variable is uncomplemented in the product term.
- If a circle covers both areas of the map where the variable is 0 and areas where it is 1, then the variable does not appear in the product term.

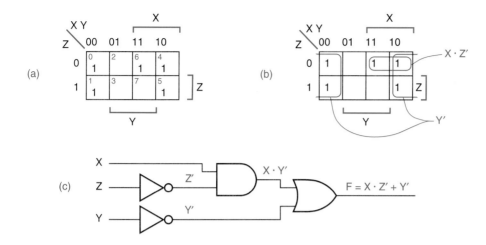

Figure 4-29
$F = \Sigma_{X,Y,Z}(0,1,4,5,6)$:
(a) initial Karnaugh
map; (b) Karnaugh
map with circled
product terms;
(c) AND/OR circuit.

A sum-of-products expression for a function must contain product terms (circled sets of 1-cells) that cover all of the 1s and none of the 0s on the map.

The Karnaugh map for our most recent example, $F = \Sigma_{X,Y,Z}(0, 1, 4, 5, 6)$, is shown in Figure 4-29(a) and (b). We have circled one set of four 1s, corresponding to the product term Y', and a set of two 1s corresponding to the product term $X \cdot Z'$. Notice that the second product term has one less literal than the corresponding product term in our algebraic solution ($X \cdot Y \cdot Z'$). By circling the largest possible set of 1s containing cell 6, we have found a less expensive realization of the logic function, since a 2-input AND gate should cost less than a 3-input one. The fact that two different product terms now cover the same 1-cell (4) does not affect the logic function, since for logical addition $1 + 1 = 1$, not 2! The corresponding two-level AND/OR circuit is shown in (c).

As another example, the prime-number detector circuit that we introduced in Figure 4-18 on page 215 can be minimized as shown in Figure 4-30.

At this point, we need some more definitions to clarify what we're doing:

- A *minimal sum* of a logic function $F(X_1,...,X_n)$ is a sum-of-products   *minimal sum*
  expression for F such that no sum-of-products expression for F has fewer
  product terms, and any sum-of-products expression with the same number of
  product terms has at least as many literals.

That is, the minimal sum has the fewest possible product terms (first-level gates and second-level gate inputs) and, within that constraint, the fewest possible literals (first-level gate inputs). Thus, among our three prime-number detector circuits, only the one in Figure 4-30 on the next page realizes a minimal sum.

The next definition says precisely what the word "imply" means when we talk about logic functions:

- A logic function $P(X_1,...,X_n)$ *implies* a logic function $F(X_1,...,X_n)$ if for every   *imply*
  input combination such that $P = 1$, then $F = 1$ also.

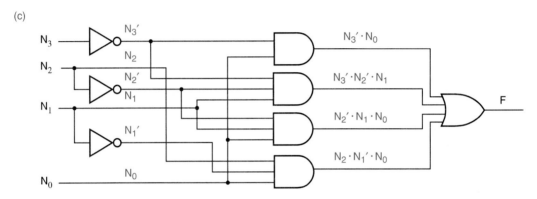

Figure 4-30 Prime-number detector: (a) initial Karnaugh map; (b) circled product terms; (c) minimized circuit.

That is, if P implies F, then F is 1 for every input combination that P is 1, and maybe some more. We may write the shorthand $P \Rightarrow F$. We may also say that "F *includes* P," or that "F *covers* P."

includes

covers

prime implicant

• A *prime implicant* of a logic function $F(X_1,...,X_n)$ is a normal product term $P(X_1,...,X_n)$ that implies F, such that if any variable is removed from P, then the resulting product term does not imply F.

In terms of a Karnaugh map, a prime implicant of F is a circled set of 1-cells satisfying our combining rule, such that if we try to make it larger (covering twice as many cells), it covers one or more 0s.

Now comes the most important part, a theorem that limits how much work we must do to find a minimal sum for a logic function:

Prime-Implicant Theorem   A minimal sum is a sum of prime implicants.

That is, to find a minimal sum, we need not consider any product terms that are not prime implicants. This theorem is easily proved by contradiction. Suppose

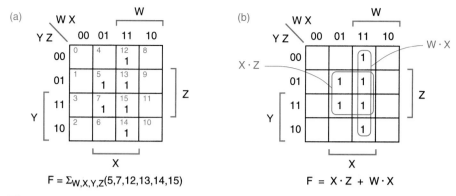

Figure 4-31 $F = \Sigma_{W,X,Y,Z}(5,7,12,13,14,15)$: (a) Karnaugh map; (b) prime implicants.

that a product term P in a "minimal" sum is *not* a prime implicant. Then according to the definition of prime implicant, if P is not one, it is possible to remove some literal from P to obtain a new product term P* that still implies F. If we replace P with P* in the presumed "minimal" sum, the resulting sum still equals F but has one fewer literal. Therefore, the presumed "minimal" sum was not minimal after all.

Another minimization example, this time a 4-variable function, is shown in Figure 4-31. There are just two prime implicants, and it's quite obvious that both of them must be included in the minimal sum in order to cover all of the 1-cells on the map. We didn't draw the logic diagram for this example because you should know how to do that yourself by now.

The sum of all the prime implicants of a logic function is called the *complete sum*. Although the complete sum is always a legitimate way to realize a logic function, it's not always minimal. For example, consider the logic function shown in Figure 4-32. It has five prime implicants, but the minimal sum includes

complete sum

Figure 4-32 $F = \Sigma_{W,X,Y,Z}(1,3,4,5,9,11,12,13,14,15)$: (a) Karnaugh map; (b) prime implicants and distinguished 1-cells.

Figure 4-33 $F = \Sigma_{W,X,Y,Z}(2,3,4,5,6,7,11,13,15)$: (a) Karnaugh map; (b) prime implicants and distinguished 1-cells.

only three of them. So, how can we systematically determine which prime implicants to include and which to leave out? Two more definitions are needed:

distinguished 1-cell

- A *distinguished 1-cell* of a logic function is an input combination that is covered by only one prime implicant.

essential prime implicant

- An *essential prime implicant* of a logic function is a prime implicant that covers one or more distinguished 1-cells.

Since an essential prime implicant is the *only* prime implicant that covers some 1-cell, it *must* be included in every minimal sum for the logic function. So, the first step in the prime-implicant selection process is simple—we identify distinguished 1-cells and the corresponding prime implicants and include the essential prime implicants in the minimal sum. Then we need only determine how to cover the 1-cells, if any, that are not covered by the essential prime implicants. In the example of Figure 4-32, the three distinguished 1-cells are shaded, and the corresponding essential prime implicants are circled with heavier lines. All of the 1-cells in this example are covered by essential prime implicants, so we need go no further. Likewise, Figure 4-33 shows an example where all of the prime implicants are essential, and so all are included in the minimal sum.

A logic function in which not all the 1-cells are covered by essential prime implicants is shown in Figure 4-34. By removing the essential prime implicants and the 1-cells they cover, we obtain a reduced map with only a single 1-cell and two prime implicants that cover it. The choice in this case is simple—we use the $W' \cdot Z$ product term because it has fewer inputs and therefore lower cost.

For more complex cases, we need yet another definition:

eclipse

- Given two prime implicants P and Q in a reduced map, P is said to eclipse Q (written P … Q) if P covers at least all the 1-cells covered by Q.

If P costs no more than Q and eclipses Q, then removing Q from consideration cannot prevent us from finding a minimal sum; that is, P is at least as good as Q.

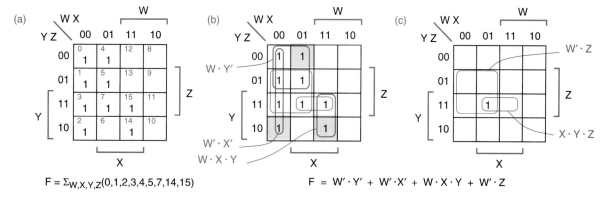

F = $\Sigma_{W,X,Y,Z}(0,1,2,3,4,5,7,14,15)$

F = W′·Y′ + W′·X′ + W·X·Y + W′·Z

Figure 4-34 F = $\Sigma_{W,X,Y,Z}(0,1,2,3,4,5,7,14,15)$: (a) Karnaugh map; (b) prime implicants and distinguished 1-cells; (c) reduced map after removal of essential prime implicants and covered 1-cells.

An example of eclipsing is shown in Figure 4-35. After removing essential prime implicants, we are left with two 1-cells, each of which is covered by two prime implicants. However, X · Y · Z eclipses the other two prime implicants, which therefore may be removed from consideration. The two 1-cells are then covered only by X · Y · Z, which is a *secondary essential prime implicant* that must be included in the minimal sum.

secondary essential prime implicant

Figure 4-36 shows a more difficult case—a logic function with no essential prime implicants. By trial and error we can find two different minimal sums for this function.

We can also approach the problem systematically using the *branching method*. Starting with any 1-cell, we arbitrarily select one of the prime implicants that covers it, and we include it as if it were essential. This simplifies the

branching method

Figure 4-35 F = $\Sigma_{W,X,Y,Z}(2,6,7,9,13,15)$: (a) Karnaugh map; (b) prime implicants and distinguished 1-cells; (c) reduced map after removal of essential prime implicants and covered 1-cells.

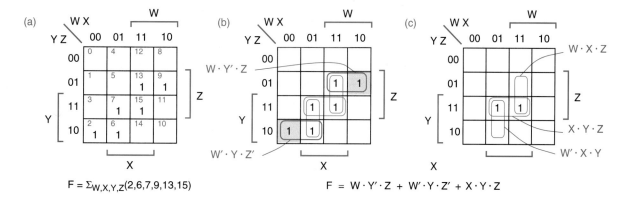

F = $\Sigma_{W,X,Y,Z}(2,6,7,9,13,15)$

F = W·Y′·Z + W′·Y·Z′ + X·Y·Z

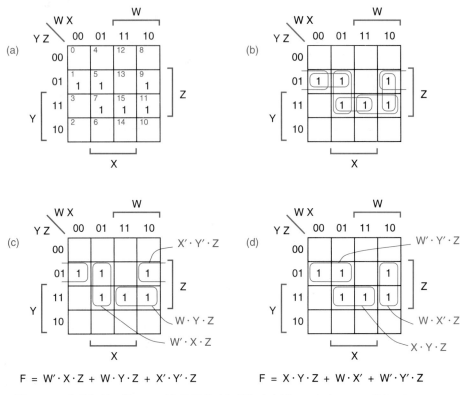

Figure 4-36 $F = \Sigma_{W,X,Y,Z}(1,5,7,9,11,15)$: (a) Karnaugh map; (b) prime implicants; (c) a minimal sum; (d) another minimal sum.

remaining problem, which we can complete in the usual way to find a tentative minimal sum. We repeat this process starting with all other prime implicants that cover the starting 1-cell, generating a different tentative minimal sum from each starting point. We may get stuck along the way and have to apply the branching method recursively. Finally, we examine all of the tentative minimal sums that we generated in this way and select one that is truly minimal.

4.3.6 Simplifying Products of Sums

Using the principle of duality, we can minimize product-of-sums expressions by looking at the 0s on a Karnaugh map. Each 0 on the map corresponds to a maxterm in the canonical product of the logic function. The entire process in the preceding subsection can be reformulated in a dual way, including the rules for writing sum terms corresponding to circled sets of 0s, in order to find a *minimal product*.

minimal product

Fortunately, once we know how to find minimal sums, there's an easier way to find the minimal product for a given logic function F. The first step is to complement F to obtain F'. Assuming that F is expressed as a minterm list or a truth table, complementing is very easy; the 1s of F' are just the 0s of F. Next, we find a minimal sum for F' using the method of the preceding subsection. Finally, we complement the result using the generalized DeMorgan's theorem, which yields a minimal product for $(F')' = F$. (Note that if you simply "add out" the minimal-sum expression for the original function, the resulting product-of-sums expression is not necessarily minimal; for example, see Exercise 4.61.)

In general, to find the lowest-cost two-level realization of a logic function, we have to find both a minimal sum and a minimal product and compare them. If a minimal sum for a logic function has many terms, then a minimal product for the same function may have few terms. As a trivial example of this tradeoff, consider a 4-input OR function:

$$F = (W) + (X) + (Y) + (Z) \text{ (a sum of four trivial product terms)}$$
$$= (W + X + Y + Z) \text{ (a product of one sum term)}$$

For a nontrivial example, you're invited to find the minimal product for the function that we minimized in Figure 4-33 on page 228; it has just two sum terms.

The opposite situation is also sometimes true, as trivially illustrated by a 4-input AND:

$$F = (W) \cdot (X) \cdot (Y) \cdot (Z) \text{ (a product of four trivial sum terms)}$$
$$= (W \cdot X \cdot Y \cdot Z) \text{ (a sum of one product term)}$$

A nontrivial example with a higher-cost product-of-sums is the function in Figure 4-29 on page 225.

For some logic functions, both minimal forms are equally costly. For example, consider a 3-input "Exclusive OR" function; both minimal expressions have four terms, and each term has three literals:

$$F = \Sigma_{X,Y,Z}(1, 2, 4, 7)$$
$$= (X' \cdot Y' \cdot Z) + (X' \cdot Y \cdot Z') + (X \cdot Y' \cdot Z') + (X \cdot Y \cdot Z)$$
$$= (X + Y + Z) \cdot (X + Y' + Z') \cdot (X' + Y + Z') \cdot (X' + Y' + Z)$$

Still, in most cases, one form or the other will give better results. Looking at both forms is especially useful in PLD-based designs.

PLD MINIMIZATION   Typical PLDs have an AND-OR array corresponding to a sum-of-products form, so you might think that only the minimal sum-of-products is relevant to a PLD-based design. However, most PLDs also have a programmable inverter/buffer at the output of the AND-OR array, which can either invert or not. Thus, the PLD can utilize the equivalent of the minimal sum by using the AND-OR array to realize the complement of the desired function and then programming the inverter/buffer to invert. Most logic-minimization programs for PLDs automatically find both the minimal sum and the minimal product and select the one that requires fewer terms.

$$F = \Sigma_{N3,N2,N1,N0}(1,2,3,5,7) + d(10,11,12,13,14,15) \qquad\qquad F = N_3' \cdot N_0 + N_2' \cdot N_1$$

Figure 4-37 Prime BCD-digit detector: (a) initial Karnaugh map;
(b) Karnaugh map with prime implicants and distinguished 1-cells.

*4.3.7 "Don't-Care" Input Combinations

don't-care

Sometimes the specification of a combinational circuit is such that its output doesn't matter for certain input combinations, called *don't-cares*. This may be true because the outputs really don't matter when these input combinations occur, or because these input combinations never occur in normal operation. For example, suppose we wanted to build a prime-number detector whose 4-bit input $N = N_3N_2N_1N_0$ is always a BCD digit; then minterms 10–15 should never occur. A prime BCD-digit detector function may therefore be written as follows:

$$F \;=\; \Sigma_{N_3,N_2,N_1,N_0}(1,2,3,5,7) + d(10,11,12,13,14,15)$$

d-set

The d(...) list specifies the don't-care input combinations for the function, also known as the *d-set*. Here F must be 1 for input combinations in the on-set $(1,2,3,5,7)$, F can have any values for inputs in the d-set $(10,11,12,13,14,15)$, and F must be 0 for all other input combinations (in the 0-set).

Figure 4-37 shows how to find a minimal sum-of-products realization for the prime BCD-digit detector, including don't-cares. The d's in the map denote the don't-care input combinations. We modify the procedure for circling sets of 1s (prime implicants) as follows:

- Allow d's to be included when circling sets of 1s, to make the sets as large as possible. This reduces the number of variables in the corresponding prime implicants. Two such prime implicants ($N_2 \cdot N_0$ and $N_2' \cdot N_1$) appear in the example.

- Do not circle any sets that contain only d's. Including the corresponding product term in the function would unnecessarily increase its cost. Two such product terms ($N_3 \cdot N_2$ and $N_3 \cdot N_1$) are circled in the example.

- Just a reminder: As usual, do not circle any 0s.

*Throughout this book, optional sections are marked with an asterisk.

The remainder of the procedure is the same. In particular, we look for distinguished 1-cells and *not* distinguished d-cells, and we include only the corresponding essential prime implicants and any others that are needed to cover all the 1s on the map. In Figure 4-37, the two essential prime implicants are sufficient to cover all of the 1s on the map. Two of the d's also happen to be covered, so F will be 1 for don't-care input combinations 10 and 11, and 0 for the other don't-cares.

Some HDLs, including ABEL, provide a means for the designer to specify don't-care inputs, and the logic-minimization program takes these into account when computing a minimal sum.

*4.3.8 Multiple-Output Minimization

Most practical combinational logic circuits require more than one output. We can always handle a circuit with n outputs as n independent single-output design problems. However, in doing so, we may miss some opportunities for optimization. For example, consider the following two logic functions:

$$F = \Sigma_{X,Y,Z}(3,6,7)$$
$$G = \Sigma_{X,Y,Z}(0,1,3)$$

Figure 4-38 shows the design of F and G as two independent single-output functions. However, as shown in Figure 4-39, we can also find a pair of sum-of-products expressions that share a product term, such that the resulting circuit has one fewer gate than our original design.

Figure 4-38 Treating a 2-output design as two independent single-output designs: (a) Karnaugh maps; (b) "minimal" circuit.

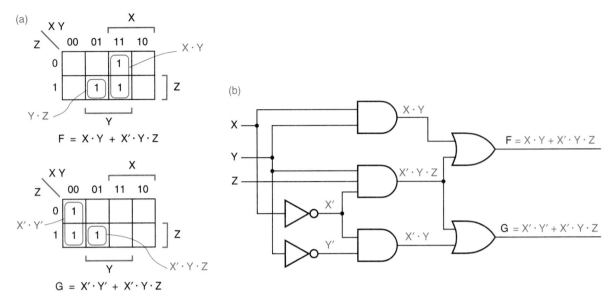

Figure 4-39 Multiple-output minimization for a 2-output circuit: (a) minimized maps including a shared term; (b) minimal multiple-output circuit

When we design multiple-output combinational circuits using discrete gates, as in an ASIC, product-term sharing obviously reduces circuit size and cost. In addition, PLDs contain multiple copies of the sum-of-products structure that we've been learning how to minimize, one per output, and some PLDs allow product terms to be shared among multiple outputs. Thus, the ideas introduced in this subsection are used in many logic-minimization programs.

You probably could have "eyeballed" the Karnaugh maps for F and G in Figure 4-39 and discovered the minimal solution. However, larger circuits can be minimized only with a formal multiple-output minimization algorithm. We'll outline the ideas in such an algorithm here; details can be found in the References.

The key to successful multiple-output minimization of a set of n functions is to consider not only the n original single-output functions, but also "product functions." An *m-product function* of a set of n functions is the product of m of the functions, where $2 \le m \le n$. There are $2^n - n - 1$ such functions. Fortunately, $n = 2$ in our example and there is only one product function, F · G, to consider. The Karnaugh maps for F, G, and F · G are shown in Figure 4-40; in general, the map for an m-product function is obtained by ANDing the maps of its m components.

A *multiple-output prime implicant* of a set of n functions is a prime implicant of one of the n functions or of one of the product functions. The

m-product function

multiple-output prime implicant

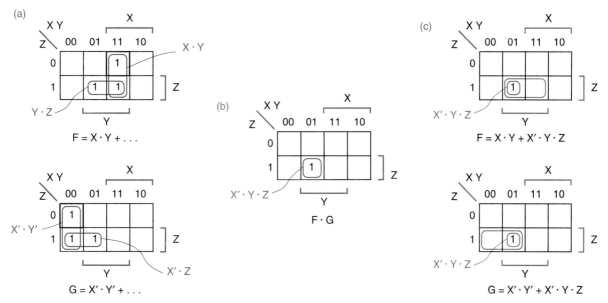

Figure 4-40 Karnaugh maps for a set of two functions: (a) maps for F and G; (b) 2-product map for F · G; (c) reduced maps for F and G after removal of essential prime implicants and covered 1-cells.

first step in multiple-output minimization is to find all of the multiple-output prime implicants. Each prime implicant of an m-product function is a possible term to include in the corresponding m outputs of the circuit. If we were trying to minimize a set of 8 functions, we would have to find the prime implicants for $2^8 - 8 - 1 = 247$ product functions as well as for the 8 given functions. Obviously, multiple-output minimization is not for the faint-hearted!

Once we have found the multiple-output prime implicants, we try to simplify the problem by identifying the essential ones. A *distinguished 1-cell* of a particular single-output function F is a 1-cell that is covered by exactly one prime implicant of F or of the product functions involving F. The distinguished 1-cells in Figure 4-39 are shaded. An *essential prime implicant* of a particular single-output function is one that contains a distinguished 1-cell. As in single-output minimization, the essential prime implicants must be included in a minimum-cost solution. Only the 1-cells that are not covered by essential prime implicants are considered in the remainder of the algorithm.

The final step is to select a minimal set of prime implicants to cover the remaining 1-cells. In this step we must consider all n functions simultaneously, including the possibility of sharing; details of this procedure are discussed in the References. In the example of Figure 4-40(c), we see that there exists a single, shared product term that covers the remaining 1-cell in both F and G.

distinguished 1-cell
essential prime
implicant

*4.4 Programmed Minimization Methods

Obviously, logic minimization can be a very involved process. In real logic-design applications you are likely to encounter only two kinds of minimization problems: functions of a few variables that you can "eyeball" using the methods of the previous section, and more complex, multiple-output functions that are hopeless without the use of a minimization program.

Quine-McCluskey
algorithm

We know that minimization can be performed visually for functions of a few variables using the Karnaugh-map method. We'll show in this section that the same operations can be performed for functions of an arbitrarily large number of variables (at least in principle) using a tabular method called the *Quine-McCluskey algorithm*. Like all algorithms, the Quine-McCluskey algorithm can be translated into a computer program. And like the map method, the algorithm has two steps: (a) finding all prime implicants of the function, and (b) selecting a minimal set of prime implicants that covers the function.

The Quine-McCluskey algorithm is often described in terms of handwritten tables and manual check-off procedures. However, since no one ever uses these procedures manually, it's more appropriate for us to discuss the algorithm in terms of data structures and functions in a high-level programming language. The goal of this section is to give you an appreciation for computational complexity involved in a large minimization problem. We consider only fully specified, single-output functions; don't-cares and multiple-output functions can be handled by fairly straightforward modifications to the single-output algorithms, as discussed in the References.

*4.4.1 Representation of Product Terms

The starting point for the Quine-McCluskey minimization algorithm is the truth table or, equivalently, the minterm list of a function. If the function is specified differently, it must first be converted into this form. For example, an arbitrary n-variable logic expression can be multiplied out (perhaps using DeMorgan's theorem along the way) to obtain a sum-of-products expression. Once we have a sum-of-products expression, each p-variable product term produces 2^{n-p} minterms in the minterm list.

We showed in Section 4.1.6 that a minterm of an n-variable logic function can be represented by an n-bit integer (the minterm number), where each bit indicates whether the corresponding variable is complemented or uncomplemented. However, a minimization algorithm must also deal with product terms that are not minterms, where some variables do not appear at all. Thus, we must represent three possibilities for each variable in a general product term:

1   Uncomplemented.

0   Complemented.

x   Doesn't appear.

These possibilities are represented by a string of *n* of the above digits in the *cube representation* of a product term. For example, if we are working with product terms of up to eight variables, X7, X6, ..., X1, X0, we can write the following product terms and their cube representations:

$$X7' \cdot X6 \cdot X5 \cdot X4' \cdot X3 \cdot X2 \cdot X1 \cdot X0' \equiv 01101110$$
$$X3 \cdot X2 \cdot X1 \cdot X0' \equiv \text{xxxx}1110$$
$$X7 \cdot X5' \cdot X4 \cdot X3 \cdot X2' \cdot X1 \equiv 1\text{x}01101\text{x}$$
$$X6 \equiv \text{x}1\text{xxxxxx}$$

Notice that for convenience, we named the variables just like the bit positions in *n*-bit binary integers.

In terms of the *n*-cube and *m*-subcube nomenclature of Section 2.14, the string 1x01101x represents a 2-subcube of an 8-cube, and the string 01101110 represents a 0-subcube of an 8-cube. However, in the minimization literature, the maximum dimension *n* of a cube or subcube is usually implicit, and an *m*-subcube is simply called an "*m*-cube" or a "cube" for short; we'll follow this practice in this section.

To represent a product term in a computer program, we can use a data structure with *n* elements, each of which has three possible values. In C, we might make the following declarations:

```
typedef enum {complemented, uncomplemented, doesntappear} TRIT;
typedef TRIT[16] CUBE;  /* Represents a single product
                           term with up to 16 variables */
```

However, these declarations do not lead to a particularly efficient internal representation of cubes. As we'll see, cubes are easier to manipulate using conventional computer instructions if an *n*-variable product term is represented by two *n*-bit computer words, as suggested by the following declarations:

```
#define MAX_VARS 16      /* Max # of variables in a product term */
typedef unsigned short WORD;   /* Use 16-bit words */
struct cube {
  WORD t;  /* Bits 1 for uncomplemented variables. */
  WORD f;  /* Bits 1 for complemented variables.   */
};
typedef struct cube CUBE;
CUBE P1, P2, P3;    /* Allocate three cubes for use by program. */
```

Here, a WORD is a 16-bit integer, and a 16-variable product term is represented by a record with two WORDs, as shown in Figure 4-41(a). The first word in a CUBE has a 1 for each variable in the product term that appears uncomplemented (or "true," t), and the second has a 1 for each variable that appears complemented (or "false," f). If a particular bit position has 0s in both WORDs, then the corresponding variable does not appear, while the case of a particular bit

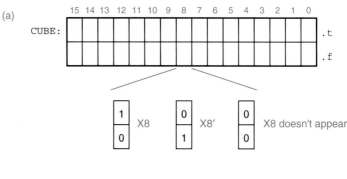

Figure 4-41
Internal representation
of 16-variable product terms
in a Pascal program:
(a) general format; (b) P1 =
X15·X12′·X10′·X9·X4′·X1·X0

position having 1s in both WORDs is not used. Thus, the program variable P1 in (b) represents the product term P1 = X15 · X12′ · X10′ · X9 · X4′ · X1 · X0. If we wished to represent a logic function F of up to 16 variables, containing up to 100 product terms, we could declare an array of 100 CUBEs:

```
CUBE F[100];        /* Storage for a logic function
                       with up to 100 product terms. */
```

Using the foregoing cube representation, it is possible to write short, efficient C functions that manipulate product terms in useful ways. Table 4-8 shows several such functions. Corresponding to two of the functions, Figure 4-42 depicts how two cubes can be compared and combined if possible

Figure 4-42 Cube manipulations: (a) determining whether two cubes are combinable using theorem T10, term · X + term · X′ = term; (b) combining cubes using theorem T10.

Table 4-8 Cube comparing and combining functions used in minimization program.

```
int EqualCubes(CUBE C1, CUBE C2)          /* Returns true if C1 and C2 are identical.  */
{
  return ( (C1.t == C2.t) && (C1.f == C2.f) );
}


int Oneone(WORD w)           /* Returns true if w has exactly one 1 bit.     */
{                            /* Optimizing the speed of this routine is critical   */
  int ones, b;              /*   and is left as an exercise for the hacker.    */
  ones = 0;
  for (b=0; b<MAX_VARS; b++) {
    if (w & 1) ones++;
    w = w>>1;
  }
  return((ones==1));
}


int Combinable(CUBE C1, CUBE C2)
{                            /* Returns true if C1 and C2 differ in only one variable, */
  WORD twordt, twordf;      /* which appears true in one and false in the other.    */

  twordt = C1.t ^ C2.t;
  twordf = C1.f ^ C2.f;
  return( (twordt==twordf) && Oneone(twordt) );
}


void Combine(CUBE C1, CUBE C2, CUBE *C3)
{                            /* Combines C1 and C2 using theorem T10, and stores the  */
{                            /*   result in C3.  Assumes Combinable(C1,C2) is true.   */
  C3->t = C1.t & C2.t;
  C3->f = C1.f & C2.f;
}
```

using theorem T10, term $\cdot$ X + term $\cdot$ X′ = term. This theorem says that two product terms can be combined if they differ in only one variable that appears complemented in one term and uncomplemented in the other. Combining two m-cubes yields an $(m + 1)$-cube. Using cube representation, we can apply the combining theorem to a few examples:

$$010 + 000 = 0\text{x}0$$
$$00111001 + 00111000 = 0011100\text{x}$$
$$101\text{xx}0\text{x}0 + 101\text{xx}1\text{x}0 = 101\text{xxxx}0$$
$$\text{x}111\text{xx}00110\text{x}000\text{x} + \text{x}111\text{xx}00010\text{x}000\text{x} = \text{x}111\text{xx}00\text{x}10\text{x}000\text{x}$$

*4.4.2 Finding Prime Implicants by Combining Product Terms

The first step in the Quine-McCluskey algorithm is to determine all of the prime implicants of the logic function. With a Karnaugh map, we do this visually by identifying "largest possible rectangular sets of 1s." In the algorithm, this is done by systematic, repeated application of theorem T10 to combine minterms, then 1-cubes, 2-cubes, and so on, creating the largest possible cubes (smallest possible product terms) that cover only 1s of the function.

The C program in Table 4-9 applies the algorithm to functions with up to 16 variables. It uses 2-dimensional arrays, cubes[m][j] and covered[m][j], to keep track of MAX_VARS m-cubes. The 0-cubes (minterms) are supplied by the user. Starting with the 0-cubes, the program examines every pair of cubes at each level and combines them when possible into cubes at the next level. Cubes that are combined into a next-level cube are marked as "covered"; cubes that are not covered are prime implicants.

Even though the program in Table 4-9 is short, an experienced programmer could become very pessimistic just looking at its structure. The inner for loop is nested four levels deep, and the number of times it might be executed is of the order of MAX_VARS · MAX_CUBES^3. That's right, that's an exponent, not a footnote! We picked the value maxCubes = 1000 somewhat arbitrarily (in fact, too optimistically for many functions), but if you believe this number, then the inner loop can be executed *billions and billions* of times.

The maximum number of minterms of an n-variable function is 2^n, of course, and so by all rights the program in Table 4-9 should declare maxCubes to be 2^{16}, at least to handle the maximum possible number of 0-cubes. Such a declaration would not be overly pessimistic. If an n-variable function has a product term equal to a single variable, then 2^{n-1} minterms are in fact needed to cover that product term.

For larger cubes, the situation is actually worse. The number of possible m-subcubes of an n-cube is $\binom{n}{m} \times 2^{n-m}$, where the binomial coefficient $\binom{n}{m}$ is the number of ways to choose m variables to be x's, and 2^{n-m} is the number of ways to assign 0s and 1s to the remaining variables. For 16-variable functions, the worst case occurs with $m = 5$; there are 8,945,664 possible 5-subcubes of a 16-cube. The total number of distinct m-subcubes of an n-cube, over all values of m, is 3^n. So a general minimization program might require a *lot* more memory than we've allocated in Table 4-9.

There are a few things that we can do to optimize the storage space and execution time required in Table 4-9 (see Exercises 4.77–4.80), but they are piddling compared to the overwhelming combinatorial complexity of the problem. Thus, even with today's fast computers and huge memories, direct application of the Quine-McCluskey algorithm for generating prime implicants is generally limited to functions with only a few variables (fewer than 15–20).

Table 4-9 A C program that finds prime implicants using the Quine-McCluskey algorithm.

```
#define TRUE    1
#define FALSE   0
#define MAX_CUBES 50

void main()
{
  CUBE cubes[MAX_VARS+1][MAX_CUBES];
  int covered[MAX_VARS+1][MAX_CUBES];
  int numCubes[MAX_VARS+1];
  int m;          /* Value of m in an m-cube, i.e., ''level m.'' */
  int j, k, p;    /* Indices into the cubes or covered array.    */
  CUBE tempCube;
  int found;

  /* Initialize number of m-cubes at each level m. */
  for (m=0; m<MAX_VARS+1; m++) numCubes[m] = 0;

  /* Read a list of minterms (0-cubes) supplied by the user, storing them  */
  /* in the cubes[0,j] subarray, setting covered[0,j] to false for each     */
  /* minterm, and setting numCubes[0] to the total number of minterms read. */
  ReadMinterms;

  for (m=0; m<MAX_VARS; m++)           /* Do for all levels except the last */
    for (j=0; j<numCubes[m]; j++)      /* Do for all cubes at this level     */
      for (k=j+1; k<numCubes[m]; k++)  /* Do for other cubes at this level   */
        if (Combinable(cubes[m][j], cubes[m][k])) {
          /* Mark the cubes as covered. */
          covered[m][j] = TRUE;  covered[m][k] = TRUE;
          /* Combine into an (m+1)-cube, store in tempCube. */
          Combine(cubes[m][j], cubes[m][k], &tempCube);
          found = FALSE;  /* See if we've generated this one before. */
          for (p=0; p<numCubes[m+1]; p++)
            if (EqualCubes(cubes[m+1][p],tempCube)) found = TRUE;
          if (!found) {  /* Add the new cube to the next level. */
            numCubes[m+1] = numCubes[m+1] + 1;
            cubes[m+1][numCubes[m+1]-1] = tempCube;
            covered[m+1][numCubes[m+1]-1] = FALSE;
          }
        }
  for (m=0; m<MAX_VARS; m++)           /* Do for all levels                */
    for (j=0; j<numCubes[m]; j++)  /* Do for all cubes at this level */
      /* Print uncovered cubes -- these are the prime implicants. */
      if (!covered[m][j]) PrintCube(cubes[m][j]);
}
```

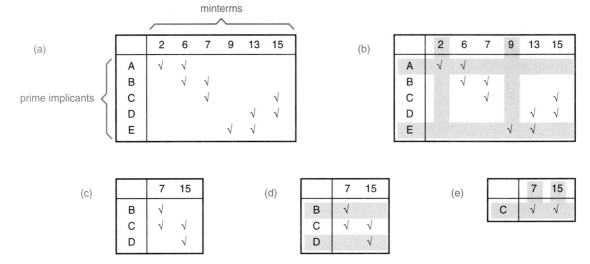

Figure 4-43 Prime-implicant tables: (a) original table; (b) showing distinguished 1-cells and essential prime implicants; (c) after removal of essential prime implicants; (d) showing eclipsed rows; (e) after removal of eclipsed rows, showing secondary essential prime implicant.

*4.4.3 Finding a Minimal Cover Using a Prime-Implicant Table

The second step in minimizing a combinational logic function, once we have a list of all its prime implicants, is to select a minimal subset of them to cover all the 1s of the function. The Quine-McCluskey algorithm uses a two-dimensional

prime-implicant table   array called a *prime-implicant table* to do this. Figure 4-43(a) shows a small but representative prime-implicant table, corresponding to the Karnaugh-map minimization problem of Figure 4-35. There is one column for each minterm of the function, and one row for each prime implicant. Each entry is a bit that is 1 if and only if the prime implicant for that row covers the minterm for that column (shown in the figure as a check).

The steps for selecting prime implicants with the table are analogous to the steps that we used in Section 4.3.5 with Karnaugh maps:

1. Identify distinguished 1-cells. These are easily identified in the table as columns with a single 1, as shown in Figure 4-43(b).

2. Include all essential prime implicants in the minimal sum. A row that contains a check in one or more distinguished-1-cell columns corresponds to an essential prime implicant.

3. Remove from consideration the essential prime implicants and the 1-cells (minterms) that they cover. In the table, this is done by deleting the corresponding rows and columns, marked in color in Figure 4-43(b). If any

rows have no checks remaining, they are also deleted; the corresponding prime implicants are *redundant*, that is, completely covered by essential prime implicants. This step leaves the reduced table shown in (c).

redundant prime implicant

4. Remove from consideration any prime implicants that are "eclipsed" by others with equal or lesser cost. In the table, this is done by deleting any rows whose checked columns are a proper subset of another row's, and deleting all but one of a set of rows with identical checked columns. This is shown in color in (d) and leads to the further reduced table in (e).

   When a function is realized in a PLD, all of its prime implicants may be considered to have equal cost, because all of the AND gates in a PLD have all of the inputs available. Otherwise, the prime implicants must be sorted and selected according to the number of AND-gate inputs.

5. Identify distinguished 1-cells and include all secondary essential prime implicants in the minimal sum. As before, any row that contains a check in one or more distinguished-1-cell columns corresponds to a secondary essential prime implicant.

6. If all remaining columns are covered by the secondary essential prime implicants, as in (e), we're done. Otherwise, if any secondary essential prime implicants were found in the previous step, we go back to step 3 and iterate. Otherwise, the branching method must be used, as described in Section 4.3.5. This involves picking rows one at a time, treating them as if they were essential, and recursing (and cursing) on steps 3–6.

Although a prime-implicant table allows a fairly straightforward prime-implicant selection algorithm, the data structure required in a corresponding computer program is huge, since it requires on the order of $p \cdot 2^n$ bits, where p is the number of prime implicants and n is the number of input bits (assuming that the given function produces a 1 output for most input combinations). Worse, executing the steps that we so blithely described in a few sentences above requires a huge amount of computation.

*4.4.4 Other Minimization Methods

Although the previous subsections form an introduction to logic-minimization algorithms, the methods they describe are by no means the latest and greatest. Spurred on by the ever-increasing density of VLSI chips, many researchers have discovered more effective ways to minimize combinational logic functions. Their results fall roughly into three categories:

1. *Computational improvements.* Improved algorithms typically use clever data structures or rearrange the order of the steps to reduce the memory requirements and execution time of the classical algorithms.

2. *Heuristic methods.* Some minimization problems are just too big to be solved using an "exact" algorithm. These problems can be attacked using

shortcuts and well-educated guesses to reduce memory size and execution time to a fraction of what an "exact" algorithm would require. However, rather than finding a provably minimal expression for a logic function, heuristic methods attempt to find an "almost minimal" one.

Even for problems that can be solved by an "exact" method, a heuristic method typically finds a good solution ten times faster. The most successful heuristic program, Espresso-II, does in fact produce minimal or near-minimal results for the majority of problems (within one or two product terms), including problems with dozens of inputs and hundreds of product terms.

3. *Looking at things differently.* As we mentioned earlier, multiple-output minimization can be handled by straightforward, fairly mechanical modifications to single-output minimization methods. However, by looking at multiple-output minimization as a problem in multivalued (nonbinary) logic, the designers of the Espresso-MV algorithm were able to make substantial performance improvements over Espresso-II.

More information on these methods can be found in the References.

*4.5 Timing Hazards

steady-state behavior

The analysis methods that we developed in Section 4.2 ignore circuit delay and predict only the *steady-state behavior* of combinational logic circuits. That is, they predict a circuit's output as a function of its inputs under the assumption that the inputs have been stable for a long time, relative to the delays in the circuit's electronics. However, we showed in Section 3.6 that the actual delay from an input change to the corresponding output change in a real logic circuit is nonzero and depends on many factors.

transient behavior

glitch
hazard

Because of circuit delays, the *transient behavior* of a logic circuit may differ from what is predicted by a steady-state analysis. In particular, a circuit's output may produce a short pulse, often called a *glitch*, at a time when steady-state analysis predicts that the output should not change. A *hazard* is said to exist when a circuit has the possibility of producing such a glitch. Whether or not the glitch actually occurs depends on the exact delays and other electrical characteristics of the circuit. Since such parameters are difficult to control in production circuits, a logic designer must be prepared to eliminate hazards (the *possibility* of a glitch) even though a glitch may occur only under a worst-case combination of logical and electrical conditions.

*4.5.1 Static Hazards

static-1 hazard

A *static-1 hazard* is the possibility of a circuit's output producing a 0 glitch when we would expect the output to remain at a nice steady 1 based on a static analysis of the circuit function. A formal definition is given as follows.

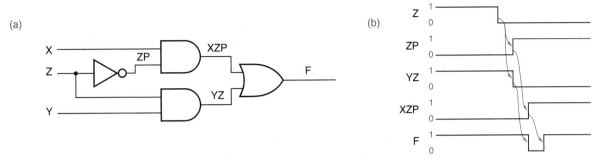

Figure 4-44 Circuit with a static-1 hazard: (a) logic diagram; (b) timing diagram.

Definition: A static-1 hazard is a pair of input combinations that: (a) differ in only one input variable and (b) both give a 1 output; such that it is possible for a momentary 0 output to occur during a transition in the differing input variable.

For example, consider the logic circuit in Figure 4-44(a). Suppose that X and Y are both 1 and that Z is changing from 1 to 0. Then (b) shows the timing diagram, assuming that the propagation delay through each gate or inverter is one unit time. Even though "static" analysis predicts that the output is 1 for both input combinations X,Y,Z = 111 and X,Y,Z = 110, the timing diagram shows that F goes to 0 for one unit time during a 1-0 transition on Z, because of the delay in the inverter that generates Z'.

A *static-0 hazard* is the possibility of a 1 glitch when we expect the circuit to have a steady 0 output: *static-0 hazard*

Definition: A static-0 hazard is a pair of input combinations that: (a) differ in only one input variable and (b) both give a 0 output; such that it is possible for a momentary 1 output to occur during a transition in the differing input variable.

Since a static-0 hazard is just the dual of a static-1 hazard, an **OR-AND** circuit that is the dual of Figure 4-44(a) would have a static-0 hazard.

An **OR-AND** circuit with four static-0 hazards is shown in Figure 4-45(a). One of the hazards occurs when W,X,Y = 000 and Z is changed, as shown in (b). You should be able to find the other three hazards and eliminate all of them after studying the next subsection.

*4.5.2 Finding Static Hazards Using Maps

A Karnaugh map can be used to detect static hazards in a two-level sum-of-products or product-of-sums circuit. The existence or nonexistence of static hazards depends on the circuit design for a logic function.

A properly designed two-level sum-of-products (**AND-OR**) circuit has no static-0 hazards. A static-0 hazard would exist in such a circuit only if both a

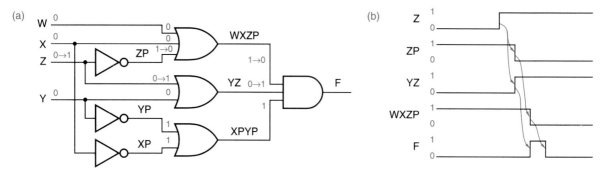

Figure 4-45 Circuit with static-0 hazards: (a) logic diagram; (b) timing diagram.

variable and its complement were connected to the same AND gate, which would be silly. However, the circuit *may* have static-1 hazards. Their existence can be predicted from a Karnaugh map where the product terms corresponding to the AND gates in the circuit are circled.

Figure 4-46(a) shows the Karnaugh map for the circuit of Figure 4-44. It is clear from the map that there is no single product term that covers both input combinations $X,Y,Z = 111$ and $X,Y,Z = 110$. Thus, intuitively, it is possible for the output to "glitch" momentarily to 0 if the AND gate output that covers one of the combinations goes to 0 before the AND gate output covering the other input combination goes to 1. The way to eliminate the hazard is also quite apparent: Simply include an extra product term (AND gate) to cover the hazardous input pair, as shown in Figure 4-46(b). The extra product term, it turns out, is the *consensus* *consensus* of the two original terms; in general, we must add consensus terms to eliminate hazards. The corresponding hazard-free circuit is shown in Figure 4-47.

Another example is shown in Figure 4-48. In this example, three product terms must be added to eliminate the static-1 hazards.

A properly designed two-level product-of-sums (OR-AND) circuit has no static-1 hazards. It *may* have static-0 hazards, however. These hazards can be detected and eliminated by studying the adjacent 0s in the Karnaugh map, in a manner dual to the foregoing.

Figure 4-46
Karnaugh map for the circuit of Figure 4-44:
(a) as originally designed;
(b) with static-1 hazard eliminated.

$$F = X \cdot Z' + Y \cdot Z$$

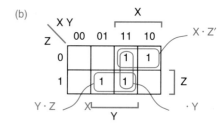

$$F = X \cdot Z' + Y \cdot Z + X \cdot Y$$

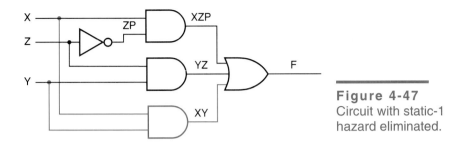

Figure 4-47
Circuit with static-1 hazard eliminated.

*4.5.3 Dynamic Hazards

A *dynamic hazard* is the possibility of an output changing more than once as the result of a single input transition. Multiple output transitions can occur if there are multiple paths with different delays from the changing input to the changing output.

dynamic hazard

For example, consider the circuit in Figure 4-49; it has three different paths from input X to the output F. One of the paths goes through a slow OR gate, and another goes through an OR gate that is even slower. If the input to the circuit is W, X, Y, Z = 0, 0, 0, 1, then the output will be 1, as shown. Now suppose we change the X input to 1. Assuming that all of the gates except the two marked "slow" and "slower" are very fast, the transitions shown in black occur next, and the output goes to 0. Eventually, the output of the "slow" OR gate changes, creating the transitions shown in nonitalic color, and the output goes to 1. Finally, the output of the "slower" OR gate changes, creating the transitions shown in italic color, and the output goes to its final state of 0.

Dynamic hazards do not occur in a properly designed two-level AND-OR or OR-AND circuit, that is, one in which no variable and its complement are con-

Figure 4-48 Karnaugh map for another sum-of-products circuit: (a) as originally designed; (b) with extra product terms to cover static-1 hazards.

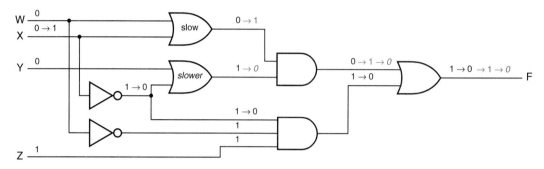

Figure 4-49 Circuit with a dynamic hazard.

nected to the same first-level gate. In multilevel circuits, dynamic hazards can be discovered using a method described in the References.

*4.5.4 Designing Hazard-Free Circuits

Oonly a few situations, such as the design of feedback sequential circuits, require hazard-free combinational circuits. Techniques for finding hazards in arbitrary circuits, described in the References, are rather difficult to use. So, when you require a hazard-free design, it's best to use a circuit structure that is easy to analyze.

In particular, we have indicated that a properly designed two-level AND-OR circuit has no static-0 or dynamic hazards. Static-1 hazards may exist in such a circuit, but they can be found and eliminated using the map method described earlier. If cost is not a problem, then a brute-force method of obtaining a hazard-free realization is to use the complete sum—the sum of all of the prime implicants of the logic function (see Exercise 4.84). In a dual manner, a hazard-free two-level OR-AND circuit can be designed for any logic function. Finally, note that everything we've said about AND-OR circuits naturally applies to the corresponding NAND-NAND designs, and about OR-AND applies to NOR-NOR.

| **MOST HAZARDS ARE NOT HAZARDOUS!** | Any combinational circuit can be analyzed for the presence of hazards. However, a well-designed, *synchronous* digital system is structured so that hazard analysis is not needed for most of its circuits. In a synchronous system, all of the inputs to a combinational circuit are changed at a particular time, and the outputs are not "looked at" until they have had time to settle to a steady-state value. Hazard analysis and elimination are typically needed only in the design of asynchronous sequential circuits, such as the feedback sequential circuits discussed in Section 7.9. You'll rarely have reason to design such a circuit, but if you do, an understanding of hazards will be absolutely essential for a reliable result. |

4.6 The ABEL Hardware Description Language

ABEL is a hardware descriptipn language (HDL) that was invented to allow designers to specify logic functions for realization in PLDs. An ABEL program is a text file containing several elements:

- Documentation, including program name and comments.
- Declarations that identify the inputs and outputs of the logic functions to be performed.
- Statements that specify the logic functions to be performed.
- Usually, a declaration of the type of PLD or other targeted device in which the specified logic functions are to be performed.
- Usually, "test vectors" that specify the logic functions' expected outputs for certain inputs.

ABEL is supported by an *ABEL language processor*, which we'll simply call an *ABEL compiler*. The compiler's job is to translate the ABEL text file into a "fuse pattern" that can be downloaded into a physical PLD. Even though most PLDs can be physically programmed only with patterns corresponding to sum-of-products expressions, ABEL allows PLD functions to be expressed with truth tables or nested "IF" statements as well as by any algebraic expression format. The compiler manipulates these formats and minimizes the resulting equations to fit, if possible, into the available PLD structure.

ABEL language processor

ABEL compiler

We'll talk about PLD structures, fuse patterns, and related topics later, in Section 5.3, and we'll show how to target ABEL programs to specific PLDs. In the meantime, we'll show how ABEL can be used to specify combinational logic functions without necessarily having to declare the targeted device type. Later, in Section 7.11, we'll do the same for sequential logic functions.

4.6.1 ABEL Program Structure

Table 4-10 shows the typical structure of an ABEL program, and Table 4-11 shows an actual program exhibiting the following language features:

- *Identifiers* must begin with a letter or underscore, may contain up to 31 letters, digits, and underscores, and are case sensitive.

identifier

- A program file begins with a *module* statement, which associates an identifier (`Alarm_Circuit`) with the program module. Large programs can have multiple modules, each with its own local title, declarations, and equations. Note that keywords such as "`module`" are not case sensitive.

module

LEGAL NOTICE   ABEL (Advanced Boolean Equation Language) is a trademark of Data I/O Corporation (Redmond, WA 98073).

Table 4-10	module *module name*
Typical structure of an ABEL program.	title *string*
	deviceID device *deviceType* ;
	pin declarations
	other declarations
	equations
	equations
	test_vectors
	test vectors
	end *module name*

title • The `title` statement specifies a title string that will be inserted into the documentation files that are created by the compiler.

string • A *string* is a series of characters enclosed by single quotes.

device • The optional `device` declaration includes a device identifier (ALARMCKT) and a string that denotes the device type (`'P16V8C'` for a GAL16V8). The compiler uses the device identifier in the names of documentation files that it generates, and it uses the device type to determine whether the device can really perform the logic functions specified in the program.

comment • *Comments* begin with a double quote and end with another double quote or the end of the line, whichever comes first.

pin declarations • *Pin declarations* tell the compiler about symbolic names associated with the device's external pins. If the signal name is preceded with the **NOT** prefix (!), then the complement of the named signal will appear on the pin. Pin declarations may or may not include pin numbers; if none are given, the compiler assigns them based on the capabilities of the targeted device.

istype • The *istype* keyword precedes a list of one or properties, separated by
com commas. This tells the compiler the type of output signal. The "com" keyword indicates a combinational output. If no `istype` keyword is given, the compiler generally assumes that the signal is an input unless it appears on the lefthand side of an equation, in which case it tries to figure out the output's properties from the context. For your own protection, it's best just to use the `istype` keyword for all outputs!

other declarations • *Other declarations* allow the designer to define constants and expressions to improve program readability and to simplify logic design.

equations • The *equations* statement indicates that logic equations defining output signals as functions of input signals will follow.

equations • *Equations* are written like assignment statements in a conventional programming language. Each equation is terminated by a semicolon. ABEL uses the following symbols for logical operations:

Table 4-11 An ABEL program for the alarm circuit of Figure 4-11.

```
module Alarm_Circuit
title 'Alarm Circuit Example
J. Wakerly, Micro Systems Engineering'
ALARMCKT device 'P16V8C';

" Input pins
PANIC, ENABLEA, EXITING      pin 1, 2, 3;
WINDOW, DOOR, GARAGE         pin 4, 5, 6;
" Output pins
ALARM                        pin 11 istype 'com';

" Constant definition
X = .X.;

" Intermediate equation
SECURE = WINDOW & DOOR & GARAGE;

equations
ALARM = PANIC # ENABLEA & !EXITING & !(WINDOW & DOOR & GARAGE);

test_vectors
([PANIC,ENABLEA,EXITING,WINDOW,DOOR,GARAGE] -> [ALARM])
[    1,    .X.,    .X.,    .X., .X.,    .X.] -> [    1];
[    0,    0,      .X.,    .X., .X.,    .X.] -> [    0];
[    0,    1,      1,      .X., .X.,    .X.] -> [    0];
[    0,    1,      0,      0, .X.,      .X.] -> [    1];
[    0,    1,      0,      .X., 0,      .X.] -> [    1];
[    0,    1,      0,      .X., .X.,      0] -> [    1];
[    0,    1,      0,      1, 1,         1] -> [    0];

end Alarm_Circuit
```

&	AND.	*& (AND)*
#	OR.	*# (OR)*
!	NOT (used as a prefix).	*! (NOT)*
$	XOR.	*$ (XOR)*
!$	XNOR.	*!$ (XNOR)*

As in conventional programming languages, AND (&) has precedence over OR (#) in expressions. The *@ALTERNATE* directive makes the compiler recognize an alternate set of symbols for these operations: +, *, /, :+:, and :*:, respectively. This book uses the default symbols throughout.   *@ALTERNATE*

- The optional *test_vectors* statement indicates that test vectors follow.   *test_vectors*
- *Test vectors* associate input combinations with expected output values; they are used for simulation and testing as explained in Section 4.6.7.   *test vectors*

.X.

end

unclocked assignment
operator, =

intermediate equation

- The compiler recognizes several special constants, including *.X.*, a single bit whose value is "don't-care."

- The *end* statement marks the end of the module.

Equations for combinational outputs use the *unclocked assignment operator,* =. The lefthand side of an equation normally contains a signal name. The righthand side is a logic expression, not necessarily in sum-of-products form. The signal name on the lefthand side of an equation may be optionally preceded by the **NOT** operator !; this is equivalent to complementing the righthand side. The compiler's job is to generate a fuse pattern such that the signal named on the lefthand side realizes the logic expression on the righthand side.

4.6.2   ABEL Compiler Operation

The program in Table 4-11 realizes the alarm function that we described on page 215. The signal named ENABLE has been coded as ENABLEA because ENABLE is a reserved word in ABEL.

Notice that not all of the equations appear under the equations statement. An intermediate equation for an identifier SECURE appears earlier. This equation is merely a definition that associates an expression with the identifier SECURE. The ABEL compiler substitutes this expression for the identifier SECURE in every place that SECURE appears after its definition.

In Figure 4-19 on page 216 we realized the alarm circuit directly from the SECURE and ALARM expressions, using multiple levels of logic. The ABEL compiler doesn't use expressions to interconnect gates in this way. Rather, it "crunches" the expressions to obtain a minimal two-level sum-of-products result appropriate for realization in a PLD. Thus, when compiled, Table 4-11 should yield a result equivalent to the AND-OR circuit that we showed in Figure 4-20 on page 216, which happens to be minimal.

In fact, it does. Table 4-12 shows the synthesized equations file created by the ABEL compiler. Notice that the compiler creates equations only for the ALARM signal, the only output. The SECURE signal does not appear anywhere.

The compiler finds a minimal sum-of-products expression for both ALARM and its complement, !ALARM. As mentioned previously, many PLDs have the ability selectively to invert or not to invert their AND-OR output. The "reverse-polarity equation" in Table 4-12 is a sum-of-products realization of !ALARM and would be used if output inversion were selected.

In this example, the reverse-polarity equation has one less product term than the normal-polarity equation for ALARM, so the compiler will select this equation if the targeted device has selectable output inversion. A user can also force the compiler to use either normal or reverse polarity for a signal by including the keyword "buffer" or "invert," respectively, in the signal's istype property list. (With some ABEL compilers, keywords "pos" and "neg" can be used for this purpose, but see Section 4.6.6.)

Table 4-12 Synthesized equations file produced by ABEL for program in Table 4-11.

```
ABEL 6.30

Design alarmckt created Tue Nov 24 1998

Title: Alarm Circuit Example
Title: J. Wakerly, Micro Systems Engineering

P-Terms   Fan-in  Fan-out  Type  Name (attributes)
---------  ------  -------  ----  -----------------
   4/3       6        1     Pin   ALARM
=========
   4/3              Best P-Term Total: 3
                         Total Pins: 7
                        Total Nodes: 0
              Average P-Term/Output: 3


Equations:

ALARM = (ENABLEA & !EXITING & !DOOR
     # ENABLEA & !EXITING & !WINDOW
     # ENABLEA & !EXITING & !GARAGE
     # PANIC);


Reverse-Polarity Equations:

!ALARM = (!PANIC & WINDOW & DOOR & GARAGE
     # !PANIC & EXITING
     # !PANIC & !ENABLEA);
```

4.6.3 WHEN Statements and Equation Blocks

In addition to equations, ABEL provides the *WHEN statement* as another means *WHEN statement* to specify combinational logic functions in the *equations* section of an ABEL program. Table 4-13 shows the general structure of a WHEN statement, similar to an IF statement in a conventional programming language. The ELSE clause is optional. Here *LogicExpression* is an expression which results in a value of true (1) or false (0). Either *TrueEquation* or *FalseEquation* is "executed" depending

WHEN *LogicExpression* THEN
   TrueEquation;
ELSE
   FalseEquation;

Table 4-13
Structure of an ABEL
WHEN statement.

on the value of *LogicExpression*. But we need to be a little more precise about what we mean by "executed," as discussed below.

In the simplest case, *TrueEquation* and the optional *FalseEquation* are assignment statements, as in the first two WHEN statements in Table 4-14 (for X1 and X2). In this case, *LogicExpression* is logically ANDed with the righthand side of *TrueEquation*, and the complement of *LogicExpression* is ANDed with the righthand side of *FalseEquation*. Thus, the equations for X1A and X2A produce the same results as the corresponding WHEN statements but do not use WHEN.

Notice in the first example that X1 appears in the *TrueEquation*, but there is no *FalseEquation*. So, what happens to X1 when *LogicExpression* (!A#B) is false? You might think that X1's value should be don't-care for these input combinations, but it's not, as explained below.

Formally, the unclocked assignment operator, =, specifies input combinations that should be added to the on-set for the output signal appearing on the lefthand side of the equation. An output's on-set starts out empty and is augmented each time that the output appears on the lefthand side of an equation. That is, the righthand sides of all equations for the same (uncomplemented) output are ORed together. (If the output appears complemented on the lefthand side, the righthand side is complemented before being ORed.) Thus, the value of X1 is 1 only for the input combinations for which *LogicExpression* (!A#B) is true and the righthand side of *TrueEquation* (C&!D) is also true.

In the second example, X2 appears on the lefthand side of two equations, so the equivalent equation shown for X2A is obtained by ORing two righthand sides after ANDing each with the appropriate condition.

The *TrueEquation* and the optional *FalseEquation* in a WHEN statement can be any equation. In addition, WHEN statements can be "nested" by using another WHEN statement as the *FalseEquation*. When statements are nested, all of the conditions leading to an "executed" statement are ANDed. The equation for X3 and its WHEN-less counterpart for X3A in Table 4-14 illustrate the concept.

The *TrueEquation* can be another WHEN statement if it's enclosed in braces, as shown in the X4 example in the table. This is just one instance of the general use of braces described shortly.

Although all of our WHEN examples have assigned values to the same output within each part of a given WHEN statement, this does not have to be the case. The second-to-last WHEN statement in Table 4-14 is such an example.

equation block

It's often useful to make more than one assignment in *TrueEquation* or *FalseEquation* or both. For this purpose, ABEL supports equation blocks anywhere that it supports a single equation. An *equation block* is just a sequence of statements enclosed in braces, as shown in the last WHEN statement in the table. The individual statements in the sequence may be simple assignment statements, or they may be WHEN statements or nested equation blocks. A semicolon is not used after a block's closing brace. Just for fun, Table 4-15 shows the equations that the ABEL compiler produces for the entire example program.

Table 4-14 Examples of WHEN statements.

```
module WhenEx
title 'WHEN Statement Examples'

" Input pins
A, B, C, D, E, F                    pin;

" Output pins
X1, X1A, X2, X2A, X3, X3A, X4     pin istype 'com';
X5, X6, X7, X8, X9, X10           pin istype 'com';

equations

WHEN (!A # B) THEN X1 = C & !D;

X1A = (!A # B) & (C & !D);

WHEN (A & B) THEN X2 = C # D;
ELSE X2 = E # F;

X2A = (A & B) & (C # D)
    # !(A & B) & (E # F);

WHEN (A) THEN X3 = D;
ELSE WHEN (B) THEN X3 = E;
ELSE WHEN (C) THEN X3 = F;

X3A = (A) & (D)
    # !(A) & (B) & (E)
    # !(A) & !(B) & (C) & (F);

WHEN (A) THEN
  {WHEN (B) THEN X4 = D;}
ELSE X4 = E;

WHEN (A & B) THEN X5 = D;
ELSE WHEN (A # !C) THEN X6 = E;
ELSE WHEN (B # C) THEN X7 = F;

WHEN (A) THEN {
    X8 = D & E & F;
    WHEN (B) THEN X8 = 1; ELSE {X9 = D; X10 = E;}
} ELSE {
    X8 = !D # !E;
    WHEN (D) THEN X9 = 1;
    {X10 = C & D;}
}

end WhenEx
```

Table 4-15 Synthesized equations file produced by ABEL for program in Table 4-14.

```
ABEL 6.30

Design whenex created Wed Dec 2 1998

Title: WHEN Statement Examples

 P-Terms  Fan-in Fan-out Type  Name
---------  ------ ------- ----  -----
   2/3       4       1    Pin   X1
   2/3       4       1    Pin   X1A
   6/3       6       1    Pin   X2
   6/3       6       1    Pin   X2A
   3/4       6       1    Pin   X3
   3/4       6       1    Pin   X3A
   2/3       4       1    Pin   X4
   1/3       3       1    Pin   X5
   2/3       4       1    Pin   X6
   1/3       3       1    Pin   X7
   4/4       5       1    Pin   X8
   2/2       3       1    Pin   X9
   2/4       5       1    Pin   X10
=========
  36/42       Best P-Term Total: 30
                    Total Pins: 19
                   Total Nodes: 0
       Average P-Term/Output: 2
```

```
Equations:

X1 = (C & !D & !A
    # C & !D & B);

X1A = (C & !D & !A
     # C & !D & B);

X2 = (D & A & B
    # C & A & B
    # !B & E
    # !A & E
    # !B & F
    # !A & F);

X2A = (D & A & B
     # C & A & B
     # !B & E
     # !A & E
     # !B & F
     # !A & F);

X3 = (C & !A & !B & F
    # !A & B & E
    # D & A);

X3A = (C & !A & !B & F
     # !A & B & E
     # D & A);

X4 = (D & A & B
    # !A & E);

X5 = (D & A & B);

X6 = (A & !B & E
    # !C & !A & E);

X7 = (C & !A & F);

X8 = (D & A & E & F
    # A & B
    # !A & !E
    # !D & !A);

X9 = (D & !A
    # D & !B);

X10 = (C & D & !A
     # A & !B & E);
```

```
Reverse-Polarity Eqns:

!X1 = (A & !B
     # D
     # !C);

!X1A = (A & !B
      # D
      # !C);

!X2 = (!C & !D & A & B
     # !B & !E & !F
     # !A & !E & !F);

!X2A = (!C & !D & A & B
      # !B & !E & !F
      # !A & !E & !F);

!X3 = (!C & !A & !B
     # !A & B & !E
     # !D & A
     # !A & !B & !F);

!X3A = (!C & !A & !B
      # !A & B & !E
      # !D & A
      # !A & !B & !F);

!X4 = (A & !B
     # !D & A
     # !A & !E);

!X5 = (!A
     # !D
     # !B);

!X6 = (A & B
     # C & !A
     # !E);

!X7 = (A
     # !C
     # !F);

!X8 = (A & !B & !F
     # D & !A & E
     # A & !B & !E
     # !D & A & !B);

!X9 = (!D
     # A & B);

!X10 = (A & B
      # !D & !A
      # !C & !A
      # A & !E);
```

```
truth_table  (input-list -> output-list)
              input-value -> output-value;
              . . .
              input-value -> output-value;
```

Table 4-16
Structure of an ABEL
truth table.

4.6.4 Truth Tables

ABEL provides one more way to specify combinational logic functions—
the *truth table*, with the general format shown in Table 4-16. The keyword
`truth_table` introduces a truth table. The *input-list* and *output-list* give the
names of the input signals and the outputs that they affect. Each of these lists is
either a single signal name or a *set*; sets are described fully in Section 4.6.5.
Following the truth-table introduction are a series of statements, each of which
specifies an input value and a required output value using the "->" operator. For
example, the truth table for an inverter is shown below:

truth table
truth_table
input-list
output-list

*unclocked truth-table
operator, ->*

```
truth_table (X -> NOTX)
            0 -> 1;
            1 -> 0;
```

The list of input values does not need to be complete; only the on-set of the
function needs to be specified unless don't-care processing is enabled (see
Section 4.6.6). Table 4-17 shows how the prime-number-detector function
described on page 215 can be specified using an ABEL program. For convenience, the identifier NUM is defined as a synonym for the set of four input bits
[N3,N2,N1,N0], allowing a 4-bit input value to be written as a decimal integer.

Table 4-17 An ABEL program for the prime-number detector.

```
module PrimeDet
title '4-Bit Prime Number Detector'

" Input and output pins
N0, N1, N2, N3                  pin;
F                               pin istype 'com';

" Definition
NUM = [N3,N2,N1,N0];

truth_table (NUM -> F)
              1 -> 1;
              2 -> 1;
              3 -> 1;
              5 -> 1;
              7 -> 1;
             11 -> 1;
             13 -> 1;
end PrimeDet
```

Both truth tables and equations can be used within the same ABEL program. The `equations` keyword introduces a sequence of equations, while the `truth_table` keyword introduces a single truth table.

4.6.5 Ranges, Sets, and Relations

Most digital systems include buses, registers, and other circuits that handle a group of two or more signals in an identical fashion. ABEL provides several shortcuts for conveniently defining and using such signals.

range   The first shortcut is for naming similar, numbered signals. As shown in the pin definitions in Table 4-18, a *range* of signal names can be defined by stating the first and last names in the range, separated by "`..`". For example, writing "`N3..N0`" is the same as writing "`N3,N2,N1,N0`." Notice in the table that the range can be ascending or descending.

Next, we need a facility for writing equations more compactly when a group of signals are all handled identically, in order to reduce the chance of *set*   errors and inconsistencies. An ABEL *set* is simply a defined collection of signals that is handled as a unit. When a logical operation such as AND, OR, or assignment is applied to a set, it is applied to each element of the set.

Each set is defined at the beginning of the program by associating a set name with a bracketed list of the set elements (e.g., `N=[N3,N2,N1,N0]` in Table 4-18). The set element list may use shortcut notation (`YOUT=[Y1..Y4]`), but the element names need not be similar or have any correspondence with the set name (`COMP=[EQ,GE]`). Set elements can also be constants (`GT=[0,1]`). In any case, the number and order of elements in a set are significant, as we'll see.

Most of ABEL's operators, can be applied to sets. When an operation is applied to two or more sets, all of the sets must have the same number of elements, and the operation is applied individually to set elements in like positions, regardless of their names or numbers. Thus, the equation "`YOUT = N & M`" is equivalent to four equations:

```
Y1 = N3 & M3;
Y2 = N2 & M2;
Y3 = N1 & M1;
Y4 = N0 & M0;
```

When an operation includes both set and nonset variables, the nonset variables are combined individually with set elements in each position. Thus, the equation "`ZOUT = (SEL & N) # (!SEL & M)`" is equivalent to four equations of the form "`Zi = (SEL & Ni) # (!SEL & Mi)`" for i equal 0 to 3.

relation   Another important feature is ABEL's ability to convert "relations" into *relational operator*   logic expressions. A *relation* is a pair of operands combined with one of the *relational operators* listed in Table 4-19. The compiler converts a relation into a logic expression that is 1 if and only if the relation is true.

The operands in a relation are treated as unsigned integers, and either operand may be an integer or a set. If the operand is a set, it is treated as an unsigned

Table 4-18 Examples of ABEL ranges, sets, and relations.

```
module SetOps
title 'Set Operation Examples'

" Input and output pins
N3..N0, M3..M0, SEL                      pin;
Y1..Y4, Z0..Z3, EQ, GE, GTR, LTH, UNLUCKY    pin istype 'com';

" Definitions
N    = [N3,N2,N1,N0];
M    = [M3,M2,M1,M0];
YOUT = [Y1..Y4];
ZOUT = [Z3..Z0];

COMP = [EQ,GE];
GT   = [ 0, 1];
LT   = [ 0, 0];

equations

YOUT = N & M;
ZOUT = (SEL & N) # (!SEL & M);
EQ = (N == M);
GE = (N >= M);
GTR = (COMP == GT);
LTH = (COMP == LT);
UNLUCKY = (N == 13) # (M == ^hD) # ((N + M) == ^b1101);

end SetOps
```

binary integer with the leftmost variable representing the most significant bit. By default, numbers in ABEL programs are assumed to be base-10. Hexadecimal and binary numbers are denoted by a prefix of "^h" or "^b," respectively, as shown in the last equation in Table 4-18.

^h hexadecimal prefix
^b binary prefix

    ABEL sets and relations allow a lot of functionality to be expressed in very few lines of code. For example, the equations in Table 4-18 generate minimized equations with 69 product terms, as shown in the summary in Table 4-20.

Symbol	Relation
==	equal
!=	not equal
<	less than
<=	less than or equal
>	greater than
>=	greater than or equal

Table 4-19
Relational operators in ABEL.

Table 4-20 Synthesized equations summary produced by ABEL for program in Table 4-18.

P-Terms	Fan-in	Fan-out	Type	Name (attributes)
1/2	2	1	Pin	Y1
1/2	2	1	Pin	Y2
1/2	2	1	Pin	Y3
1/2	2	1	Pin	Y4
2/2	3	1	Pin	Z0
2/2	3	1	Pin	Z1
2/2	3	1	Pin	Z2
2/2	3	1	Pin	Z3
16/8	8	1	Pin	EQ
23/15	8	1	Pin	GE
1/2	2	1	Pin	GTR
1/2	2	1	Pin	LTH
16/19	8	1	Pin	UNLUCKY

```
=========
   69/62          Best P-Term Total: 53
                       Total Pins: 22
                      Total Nodes: 0
               Average P-Term/Output: 4
```

*4.6.6 Don't-Care Inputs

Some versions of the ABEL compiler have a limited ability to handle don't-care inputs. As mentioned previously, ABEL equations specify input combinations that belong to the on-set of a logic function; the remaining combinations are assumed to belong to the off-set. If some input combinations can instead be assigned to the d-set, then the program may be able to use these don't-care inputs to do a better job of minimization.

The ABEL language defines two mechanisms for assigning input combinations to the d-set. In order to use either mechanism, you must include the compiler directive @DCSET in your program or include "dc" in the istype property list of the outputs for which you want don't-cares to be considered.

@DCSET

dc

?= don't-care unclocked assignment operator

The first mechanism is the *don't-care unclocked assignment operator*, ?=. This operator is used instead of = in equations to indicate that input combinations matching the righthand side should be put into the d-set instead of the on-set. Although this operator is documented in the ABEL compiler that I use, unfortunately it seems to be broken, so I'm not going to talk about it any more.

The second mechanism is the truth table. When don't-care processing is enabled, any input combinations that are not explicitly listed in the truth table are put into the d-set. Thus, the prime BCD-digit detector described on page 232 can be specified in ABEL as shown in Table 4-21. A don't-care value is implied for input combinations 10–15 because these combinations do not appear in the truth table and the @DCSET directive is in effect.

```
module DontCare
title 'Dont Care Examples'
@DCSET

" Input and output pins
N3..N0, A, B                    pin;
F, Y                           pin istype 'com';

NUM = [N3..N0];
X = .X.;

truth_table (NUM->F)
            0->0;
            1->1;
            2->1;
            3->1;
            4->0;
            5->1;
            6->0;
            7->1;
            8->0;
            9->0;

truth_table ([A,B]->Y)
            [0,0]->0;
            [0,1]->X;
            [1,0]->X;
            [1,1]->1;


end DontCare
```

Table 4-21
ABEL program using
don't-cares.

It's also possible to specify don't-care combinations explicitly, as shown in the second truth table. As introduced at the very beginning of this section, ABEL recognizes .X. as a special one-bit constant whose value is "don't-care." In Table 4-21, the identifier "X" has been equated to this constant just to make it easier to type don't-cares in the truth table. The minimized equations resulting from Table 4-21 are shown in Table 4-22. Notice that the two equations for F are not equal; the compiler has selected different values for the don't-cares.

```
Equations:
F = (!N2 & N1
     # !N3 & N0);
Y = (B);

Reverse-Polarity Equations:
!F = (N2 & !N0
      # N3
      # !N1 & !N0);
!Y = (!B);
```

Table 4-22
Minimized equations
derived from
Table 4-21.

Table 4-23
Structure of ABEL
test vectors.

```
test_vectors  (input-list -> output-list)
                input-value -> output-value;
                . . .
                input-value -> output-value;
```

4.6.7 Test Vectors

test_vectors

input-list
output-list

ABEL programs may contain optional test vectors, as we showed in Table 4-11 on page 251. The general format of test vectors is very similar to a truth table and is shown in Table 4-23. The keyword *test_vectors* introduces a truth table. The *input-list* and *output-list* give the names of the input signals and the outputs that they affect. Each of these lists is either a single signal name or a set. Following the test-vector introduction are a series of statements, each of which specifies an input value and an expected output value using the "->" operator.

ABEL test vectors have two main uses and purposes:

1. After the ABEL compiler translates the program into "fuse pattern" for a particular device, it simulates the operation of the final programmed device by applying the test-vector inputs to a software model of the device and comparing its outputs with the corresponding test-vector outputs. The designer may specify a series of test vectors in order to double-check that device will behave as expected for some or all input combinations.

2. After a PLD is physically programmed, the programming unit applies the test-vector inputs to the physical device and compares the device outputs with the corresponding test-vector outputs. This is done to check for correct device programming and operation.

Unfortunately, ABEL test vectors seldom do a very good job at either one of these tasks, as we'll explain.

The test vectors from Table 4-11 are repeated in Table 4-24, except that for readability we've assumed that the identifier X has been equated to the don't-care constant .X., and we've added comments to number the test vectors.

Table 4-24 actually appears to be a pretty good set of test vectors. From the designer's point of view, these vectors fully cover the expected operation of the alarm circuit, as itemized vector-by-vector below:

1. If PANIC is 1, then the alarm output (F) should be on regardless of the other input values. All of the remaining vectors cover cases where PANIC is 0.

2. If the alarm is not enabled, then the output should be off.

3. If the alarm is enabled but we're exiting, then the output should be off.

4–6. If the alarm is enabled and we're not exiting, then the output should be on if any of the sensor signals WINDOW, DOOR, or GARAGE is 0.

7. If the alarm is enabled, we're not exiting, and all of the sensor signals are 1, then the output should be off.

```
test_vectors
([PANIC,ENABLEA,EXITING,WINDOW,DOOR,GARAGE] -> [ALARM])
[    1,       X,       X,       X,    X,       X] -> [     1];    1
[    0,       0,       X,       X,    X,       X] -> [     0];    2
[    0,       1,       1,       X,    X,       X] -> [     0];    3
[    0,       1,       0,       0,    X,       X] -> [     1];    4
[    0,       1,       0,       X,    0,       X] -> [     1];    5
[    0,       1,       0,       X,    X,       0] -> [     1];    6
[    0,       1,       0,       1,    1,       1] -> [     0];    7
```

Table 4-24
Test vectors for the
alarm circuit program
in Table 4-11.

The problem is that ABEL doesn't handle don't-cares in test-vector inputs the way that it should. For example, by all rights, test vector 1 should test 32 distinct input combinations corresponding to all 32 possible combinations of don't-care inputs ENABLEA, EXITING, WINDOW, DOOR, and GARAGE. But it doesn't. In this situation, the ABEL compiler interprets "don't care" as "the user doesn't care what input value I use," and it just assigns 0 to all don't-care inputs in a test vector. In this example, you could have incorrectly written the output equation as "F = PANIC & !ENABLEA # ENABLEA & ..."; the test vectors would still pass, even though the panic button would work only when the system is disabled.

The second use of test vectors is in physical device testing. Most physical defects in logic devices can be detected using the *single stuck-at fault model,* which assumes that any physical defect is equivalent to having a single gate input or output stuck at a logic 0 or 1 value. Just putting together a set of test vectors that seems to exercise a circuit's functional specifications, as we did in Table 4-24, doesn't guarantee that all single stuck-at faults can be detected. The test vectors have to be chosen so that every possible stuck-at fault causes an incorrect value at the circuit output for some test-vector input combination.

single stuck-at fault model

Table 4-25 shows a complete set of test vectors for the alarm circuit when it is realized as a two-level sum-of-products circuit. The first four vectors check for stuck-at-1 faults on the OR gate, and the last three check for stuck-at-0 faults on the AND gates; it turns out that this is sufficient to detect all single stuck-at faults. If you know something about fault testing you can generate test vectors for small circuits by hand (as I did in this example), but most designers use automated third-party tools to create high-quality test vectors for their PLD designs.

```
test_vectors
([PANIC,ENABLEA,EXITING,WINDOW,DOOR,GARAGE] -> [ALARM])
[    1,       0,       1,       1,    1,       1] -> [     1];    1
[    0,       1,       0,       0,    1,       1] -> [     1];    2
[    0,       1,       0,       1,    0,       1] -> [     1];    3
[    0,       1,       0,       1,    1,       0] -> [     1];    4
[    0,       0,       0,       0,    0,       0] -> [     0];    5
[    0,       1,       1,       0,    0,       0] -> [     0];    6
[    0,       1,       0,       1,    1,       1] -> [     0];    7
```

Table 4-25
Single stuck-at fault
test vectors for the
minimal sum-of-
products realization
of the alarm circuit.

4.7 The VHDL Hardware Description Language

VHDL

In the mid-1980s, the U.S. Department of Defense (DoD) and the IEEE sponsored the development of a highly capable hardware-description language called *VHDL*. The language started out with and still has the following features:

- Designs may be decomposed hierarchically.
- Each design element has both a well-defined interface (for connecting it to other elements) and a precise behavioral specification (for simulating it).
- Behavioral specifications can use either an algorithm or an actual hardware structure to define an element's operation. For example, an element can be defined initially by an algorithm, to allow design verification of higher-level elements that use it; later, the algorithmic definition can be replaced by a hardware structure.
- Concurrency, timing, and clocking can all be modeled. VHDL handles asynchronous as well as synchronous sequential-circuit structures.
- The logical operation and timing behavior of a design can be simulated.

Thus, VHDL started out as a documentation and modeling language, allowing the behavior of digital-system designs to be precisely specified and simulated.

VHDL synthesis tools

While the VHDL language and simulation environment were important innovations by themselves, VHDL's utility and popularity took a quantum leap with the commercial development of *VHDL synthesis tools*. These programs can create logic-circuit structures directly from VHDL behavioral descriptions. Using VHDL, you can design, simulate, and synthesize anything from a simple combinational circuit to a complete microprocessor system on a chip.

VHDL-87
VHDL-93

VHDL was standardized by the IEEE in 1987 (*VHDL-87*) and extended in 1993 (*VHDL-93*). In this section we'll a subset of language features that are legal under either standard. We'll describe additional features for sequential logic design in Section 7.12.

4.7.1 Design Flow

design flow

It's useful to understand the overall VHDL design environment before jumping into the language itself. There are several steps in a VHDL-based design process, often called the *design flow*. These steps are applicable to any HDL-based design process and are outlined in Figure 4-50 on page 266.

THE MEANING OF VHDL   "VHDL" stands for "VHSIC Hardware Description Language." VHSIC, in turn, stands for "Very High Speed Integrated Circuit," which was a U.S. Department of Defense program to encourage research on high-performance IC technology (using Very Healthy Sums of Instant Cash!).

VERILOG AND VHDL

At about the same time that VHDL was developing, a different hardware design language appeared on the scene. *Verilog HDL*, or simply *Verilog*, was introduced by Gateway Design Automation in 1984 as a proprietary hardware description and simulation language. The subsequent introduction of Verilog-based synthesis tools in 1988 by then-fledgling Synopsys and the 1989 acquisition of Gateway by Cadence Design Systems was a winning combination that led to widespread use of the language.

Today, VHDL and Verilog both enjoy widespread use and share the logic synthesis market roughly 50/50. Verilog has its syntactic roots in C and is in some respects in easier language to learn and use, while VHDL is more like Ada (a DoD-sponsored software programming language) and has more features that support large project development.

Comparing the pros and cons of starting out with one language versus the other, David Pellerin and Douglas Taylor probably put it best in their book, *VHDL Made Easy!* (Prentice Hall, 1997):

> Both languages are easy to learn and hard to master. And once you have learned one of these languages, you will have no trouble transitioning to the other.

The so-called "front end" begins with figuring out the basic approach and building blocks at the block-diagram level. Large logic designs, like software programs, are usually hierarchical, and VHDL gives you a good framework for defining modules and their interfaces and filling in the details later.

The next step is the actual writing of VHDL code for modules, their interfaces, and their internal details. Since VHDL is a text-based language, in principle you can use any text editor for this part of the job. However, most design environments include a specialized *VHDL text editor* that makes the job a little easier. Such editors include features like automatic highlighting of VHDL keywords, automatic indenting, built-in templates for frequently used program structures, and built-in syntax checking and one-click access to the compiler.

VHDL text editor

Once you've written some code, you will want to compile it, of course. A *VHDL compiler* analyzes your code for syntax errors and also checks it for compatibility with other modules on which it relies. It also creates the internal information that is needed for a simulator to process your design later. As in other programming endeavors, you probably shouldn't wait until the very end of coding to compile all of your code. Doing a piece at a time can prevent you from

THE MEANING OF VERILOG

"Verilog" isn't an acronym, but it has some interesting anagrams, including "I, Glover" (Danny?), "G.I. lover," "Go, liver!" and "I grovel." Oh, I suppose it could also be a contraction of "VERIfy LOGic."

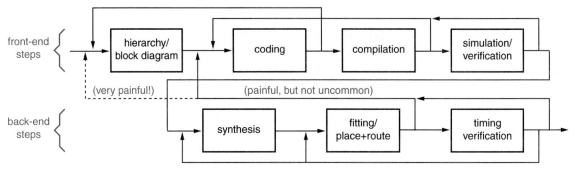

Figure 4-50 Steps in a VHDL or other HDL-based design flow.

proliferating syntax errors, inconsistent names, and so on, and can certainly give you a much-needed sense of progress when the project end is far from sight!

VHDL simulator

Perhaps the most satisfying step comes next—simulation. A *VHDL simulator* allows you to define and apply inputs to your design, and to observe its outputs, without ever having to build the physical circuit. In small projects, the kind you might do as homework in a digital-design class, you would probably generate inputs and observe outputs manually. But for larger projects, VHDL gives you the ability to create "test benches" that automatically apply inputs and compare them with expected outputs.

verification

Actually, simulation is just one piece of a larger step called *verification*. Sure, it is satisfying to watch your simulated circuit produce simulated outputs, but the purpose of simulation is larger—it is to *verify* that the circuit works as desired. In a typical large project, a substantial amount of effort is expended both during and after the coding stage to define test cases that exercise the circuit over a wide range of logical operating conditions. Finding design bugs at this stage has a high value; if bugs are found later, all of the so-called "back-end" steps must typically be repeated.

functional verification

Note that there are at least two dimensions to verification. In *functional verification*, we study the circuit's logical operation independent of timing considerations; gate delays and other timing parameters are considered to be zero. In *timing verification*, we study the circuit's operation including estimated delays, and we verify that the setup, hold, and other timing requirements for sequential devices like flip-flops are met. It is customary to perform thorough *functional* verification before starting the back-end steps. However, our ability to do *timing* verification at this stage is often limited, since timing may be heavily dependent on the results of synthesis and fitting. We may do preliminary timing verification to gain some comfort with the overall design approach, but detailed timing verification must wait until the end.

timing verification

After verification, we are ready to move into the "back-end" stage. The nature of and tools for this stage vary somewhat, depending on the target technology for the design, but there are three basic steps. The first is *synthesis*, converting the VHDL description into a set of primitives or components that can

synthesis

be assembled in the target technology. For example, with PLDs or CPLDs, the synthesis tool may generate two-level sum-of-products equations. With ASICs, it may generate a list of gates and a *netlist* that specifies how they should be interconnected. The designer may "help" the synthesis tool by specifying certain technology-specific *constraints*, such as the maximum number of logic levels or the strength of logic buffers to use.

netlist

constraints

In the *fitting* step, a fitting tool or *fitter* maps the synthesized primitives or components onto available device resources. For a PLD or CPLD, this may mean assigning equations to available AND-OR elements. For an ASIC, it may mean laying down individual gates in a pattern and finding ways to connect them within the physical constraints of the ASIC die; this is called the *place-and-route* process. The designer can usually specify additional constraints at this stage, such as the placement of modules with a chip or the pin assignments of external input and output pins.

fitting
fitter

place and route

The "final" step is timing verification of the fitted circuit. It is only at this stage that the actual circuit delays due to wire lengths, electrical loading, and other factors can be calculated with reasonable precision. It is usual during this step to apply the same test cases that were used in functional verification, but in this step they are run against the circuit as it will actually be built.

As in any other creative process, you may occasionally take two steps forward and one step back (or worse!). As suggested in the figure, during coding you may encounter problems that force you to go back and rethink your hierarchy, and you will almost certainly have compilation and simulation errors that force you to rewrite parts of the code.

The most painful problems are the ones that you encounter in the back end of the design flow. For example, if the synthesized design doesn't fit into an available FPGA or doesn't meet timing requirements, you may have to go back as far as rethinking your whole design approach. That's worth remembering—excellent tools are still no substitute for careful thought at the outset of a design.

IT WORKS!?   As a long-time logic designer and system builder, I always thought I knew what it means when someone says about their circuit, "It works!" It means you can go into the lab, power-up a prototype without seeing smoke, and push a reset button and use an oscilloscope or logic analyzer to watch the prototype go through its paces.

But over the years, the meaning of "It works" has changed, at least for some people. When I took a new job a few years ago, I was very pleased to hear that several key ASICs for an important new product were all "working." But later (just a short time later) I figured out that the ASICs were working only in simulation, and that the design team still had to do several arduous months of synthesis, fitting, timing verification, and repeating, before they could order any prototypes. "It works!"—sure. Just like my kids' homework—"It's done!"

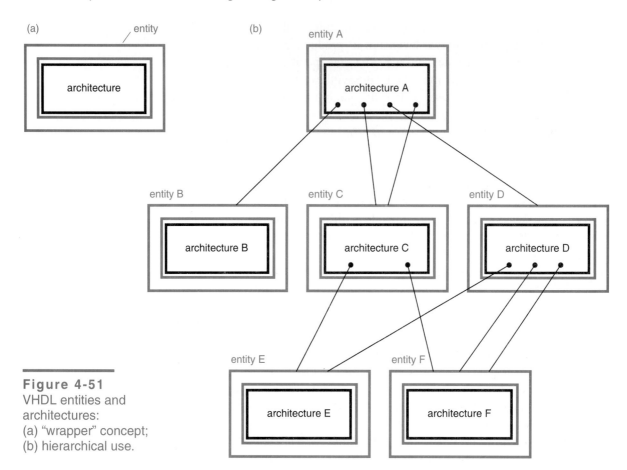

Figure 4-51
VHDL entities and
architectures:
(a) "wrapper" concept;
(b) hierarchical use.

4.7.2 Program Structure

VHDL was designed with principles of structured programming in mind,
borrowing ideas from the Pascal and Ada software programming languages. A
key idea is to define the interface of a hardware module while hiding its internal
details. Thus, a VHDL *entity* is simply a declaration of a module's inputs and
outputs, while a VHDL *architecture* is a detailed description of the module's
internal structure or behavior.

entity

architecture

Figure 4-51(a) illustrates the concept. Many designers like to think of a
VHDL entity declaration as a "wrapper" for the architecture, hiding the details
of what's inside while providing the "hooks" for other modules to use it. This
forms the basis for hierarchical system design—the architecture of a top-level
entity may use (or "instantiate") other entities, while hiding the architectural
details of lower-level entities from the higher-level ones. As shown in (b), a
higher-level architecture may use a lower-level entity multiple times, and
multiple top-level architectures may use the same lower-level one. In the figure,
architectures B, E and F stand alone; they do not use any other entities.

GO CONFIGURE! VHDL actually allows you to define multiple architectures for a single entity, and it provides a configuration management facility that allows you to specify which one to use during a particular compilation or synthesis run. This lets you try out a different architectural approach without throwing away or hiding your other efforts. However, we won't use or further discuss this facility in this text.

text file (e.g., `mydesign.vhd`)

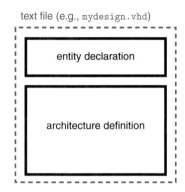

Figure 4-52
VHDL program file structure.

In the text file of a VHDL program, the *entity declaration* and *architecture definition* are separated, as shown in Figure 4-52. For example, Table 4-26 is a very simple VHDL program for a 2-input "inhibit" gate. In large projects, entities and architectures are sometimes defined in separate files, which the compiler matches up according to their declared names.

entity declaration
architecture definition

Like other high-level programming languages, VHDL generally ignores spaces and line breaks, and these may be provided as desired for readability. *Comments* begin with two hyphens (--) and end at the end of a line.

comments

VHDL defines many special character strings, called *reserved words* or *keywords*. Our example includes several—`entity`, `port`, `is`, `in`, `out`, `end`, `architecture`, `begin`, `when`, `else`, and `not`. User-defined *identifiers* begin with a letter and contain letters, digits, and underscores. (An underscore may not follow another underscore or be the last character in an identifier.) Identifiers in the example are `Inhibit`, `X`, `Y`, `BIT`, `Z`, and `Inhibit_arch`. "BIT" is a built-in identifier for a predefined type; it's not considered a reserved word because it can be redefined. Reserved words and identifiers are not case sensitive.

reserved words
keywords
identifiers

```
entity Inhibit is      -- also known as 'BUT-NOT'
  port (X,Y: in BIT;    -- as in 'X but not Y'
        Z:   out BIT);  -- (see [Klir, 1972])
end Inhibit;

architecture Inhibit_arch of Inhibit is
begin
  Z <= '1' when X='1' and Y='0' else '0';
end Inhibit_arch;
```

Table 4-26
VHDL program for an "inhibit" gate.

Table 4-27
Syntax of a VHDL
entity declaration.

```
entity entity-name is
  port (signal-names : mode signal-type;
        signal-names : mode signal-type;
        . . .
        signal-names : mode signal-type);
end entity-name;
```

A basic entity declaration has the syntax shown in Table 4-27. Besides naming the entity, the purpose of the entity declaration is to define its external interface signals or *ports* in its *port declaration* part. In addition to the keywords `entity`, `is`, `port`, and `end`, an entity declaration has the following elements:

port
port declaration

entity-name   A user-selected identifier to name the entity.

signal-names   A comma-separated list of one or more user-selected identifiers to name external-interface signals.

mode   One of four reserved words, specifying the signal direction:

in   The signal is an input to the entity.

out   The signal is an output of the entity. Note that the value of such a signal cannot be "read" inside the entity's architecture, only by other entities that use it.

buffer   The signal is an output of the entity, and its value can also be read inside the entity's architecture.

inout   The signal can be used as an input or an output of the entity. This mode is typically used for three-state input/output pins on PLDs.

signal-type   A built-in or user-defined signal type. We'll have a lot to say about types in the next subsection.

Note that there is no semicolon after the final *signal-type*; swapping the closing parenthesis with the semicolon after it is a common syntax error for beginning VHDL programmers.

An entity's ports and their modes and types are all that is seen by other modules that use it. The entity's internal operation is specified in its *architecture definition*, whose general syntax is shown in Table 4-28. The *entity-name* in this definition must be the same as the one given previously in the entity declaration. The *architecture-name* is a user-selected identifier, usually related to the entity name; it can be the same as the entity name if desired.

architecture definition

An architecture's external interface signals (ports) are inherited from the port-declaration part of its corresponding entity declaration. An architecture may also include signals and other declarations that are local to that architecture, similar to other high-level languages. Declarations common to multiple entities can be made in a separate "package" used by all entities, as discussed later.

```
architecture architecture-name of entity-name is
    type declarations
    signal declarations
    constant declarations
    function definitions
    procedure definitions
    component declarations
begin
    concurrent-statement
    ...
    concurrent-statement
end architecture-name;
```

Table 4-28
Syntax of a VHDL
architecture definition.

The declarations in Table 4-28 can appear in any order. In due course we'll discuss many different kinds of declarations and statements that can appear in the architecture definition, but the easiest to start with is the *signal declaration*. It gives the same information about a signal as in a port declaration, except that no mode is specified:

signal declaration

signal *signal-names* : *signal-type*;

Zero or more signals can be defined within an architecture, and they roughly correspond to named wires in a logic diagram. They can be read or written within the architecture definition and, like other local objects, can be referenced only within the encompassing architecture definition.

VHDL *variables* are similar to signals, except that they usually don't have physical significance in a circuit. In fact, notice that Table 4-28 has no provision for "variable declarations" in an architecture definition. Rather, variables are used in VHDL functions, procedures, and processes, each of which we'll discuss later. Within these program elements, the syntax of a *variable declaration* is just like that of a signal declaration, except that the variable keyword is used:

variable

variable declaration

variable *variable-names* : *variable-type*;

4.7.3 Types and Constants

All signals, variables, and constants in a VHDL program must have an associated "type." The *type* specifies the set or range of values that the object can take on, and there is also typically a set of operators (such as add, AND, and so on) associated with a given type.

type

VHDL has just a few *predefined types*, listed in Table 4-29. In the rest of this book, the only predefined types that we'll use are integer, character, and boolean. You would think that types with names "bit" and "bit_vector"

predefined types

bit	character	severity_level
bit_vector	integer	string
boolean	real	time

Table 4-29
VHDL predefined
types.

Table 4-30
Predefined operators
for VHDL's integer
and boolean types.

integer **Operators**		boolean **Operators**	
+	addition	and	AND
–	subtraction	or	OR
*	multiplication	nand	NAND
/	division	nor	NOR
mod	modulo division	xor	Exclusive OR
rem	modulo remainder	xnor	Exclusive NOR
abs	absolute value	not	complementation
**	exponentiation		

would be essential in digital design, but it turns out that user-defined versions of these types are more useful, as discussed shortly.

integer

Type *integer* is defined as the range of integers including at least the range $-2,147,483,647$ through $+2,147,483,647$ ($-2^{31}+1$ through $+2^{31}-1$); VHDL implementations may extend this range. Type *boolean* has two values, *true* and *false*. The *character* type contains all of the characters in the ISO 8-bit character set; the first 128 are the ASCII characters. Built-in operators for the integer and boolean types are listed in Table 4-30.

boolean
true, false
character

user-defined type
enumerated type

The most commonly used types in typical VHDL programs are *user-defined types*, and the most common of these are *enumerated types*, which are defined by listing their values. Predefined types boolean and character are enumerated types. A type declaration for an enumerated type has the format shown in the first line of Table 4-31. Here, *value-list* is a comma-separated list (enumeration) of all possible values of the type. The values may be user-defined identifiers or characters (where a "character" is an ISO character enclosed in

Table 4-31
Syntax of VHDL
type and constant
declarations.

```
type type-name is (value-list);

subtype subtype-name is type-name start to end;
subtype subtype-name is type-name start downto end;

constant constant-name: type-name := value;
```

STRONG TYPING

Unlike C, VHDL is a strongly typed language. This means that the compiler does not allow you to assign a value to a signal or variable unless the type of the value precisely matches the declared type of the signal or variable.

Strong typing is both a blessing and a curse. It makes your programs more reliable and easier to debug, because it makes it difficult for you to make "dumb errors" where you assign a value of the wrong type or size. On the other hand, it can be exasperating at times. Even simple operations, such as reinterpreting a 2-bit signal as an integer (for example, to select one of four outcomes in a "case" statement), may require you to call a type-conversion function explicitly.

```
type STD_ULOGIC is ( 'U',   -- Uninitialized
                     'X',   -- Forcing  Unknown
                     '0',   -- Forcing  0
                     '1',   -- Forcing  1
                     'Z',   -- High Impedance
                     'W',   -- Weak     Unknown
                     'L',   -- Weak     0
                     'H',   -- Weak     1
                     '-'    -- Don't care
                   );
subtype STD_LOGIC is resolved STD_ULOGIC;
```

Table 4-32
Definition of VHDL
std_logic type
(see Section 5.6.4
for discussion of
"resolved").

single quotes). The first style is used most often to define cases or states for a state machine, for example,

```
type traffic_light_state is (reset, stop, wait, go);
```

The second style is used in the very important case of a standard user-defined logic type *std_logic*, shown in Table 4-32 and part of the IEEE 1164 standard package, discussed in Section 4.7.5. This type includes not only '0' and '1', but seven other values that have been found useful in simulating a logic signal (bit) in a real logic circuit, as explained in more detail in Section 5.6.4.

std_logic

    VHDL also allows users to create *subtypes* of a type, using the syntax shown in Table 4-31. The values in the subtype must be a contiguous range of values of the base type, from *start* to *end*. For an enumerated type, "contiguous" refers to positions in the original, defining *value-list*. Some examples of subtype definitions are shown below:

subtypes

```
subtype twoval_logic is std_logic range '0' to '1';
subtype fourval_logic is std_logic range 'X' to 'Z';
subtype negint is integer range -2147483647 to 1;
subtype bitnum is integer range 31 downto 0;
```

Notice that the order of a range may be specified in ascending or descending order, depending on whether *to* or *downto* is used. There are certain attributes of subtypes for which this distinction is significant, but we don't use them in this book and we won't discuss this further.

to

downto

WHAT A CHARACTER!

You may be wondering why the values in the std_logic type are defined as characters rather than one-letter identifiers. Certainly "U", "X", and so on would be easier to type than "'U'", "'X'", and so on. Well, that would require a one-letter identifier other than "–" to be used for don't-care, but that's no big deal. The main reason for using characters is that "0" and "1" could not be used, because they're already recognized as integer constants. This goes back to VHDL's strong typing; it was not deemed advisable to let the compiler perform an automatic type conversion depending on the context.

Table 4-33
Syntax of VHDL
array declarations.

type *type-name* is array (*start* to *end*) of *element-type*;

type *type-name* is array (*start* downto *end*) of *element-type*;

type *type-name* is array (*range-type*) of *element-type*;

type *type-name* is array (*range-type* range *start* to *end*) of *element-type*;

type *type-name* is array (*range-type* range *start* downto *end*) of *element-type*;

VHDL has two predefined integer subtypes, defined below:

subtype natural is integer range 0 to *highest-integer*;
subtype positive is integer range 1 to *highest-integer*;

constants
constant declaration

Constants contribute to the readability, maintainability, and portability of programs in any language. The syntax of a *constant declaration* in VHDL is shown in the last line of Table 4-31; examples are shown below:

```
constant BUS_SIZE: integer := 32;      -- width of component
constant MSB: integer := BUS_SIZE-1;   -- bit number of MSB
constant Z: character := 'Z';          -- synonym for Hi-Z value
```

Notice that the value of a constant can be a simple expression. Constants can be used anywhere the corresponding value can be used, and they can be put to especially good use in type definitions, as we'll soon show.

array types
array
array index

Another very important category of user-defined types are *array types*. Like other languages, VHDL defines an *array* as an ordered set of elements of the same type, where each element is selected by an *array index*. Table 4-33 shows several versions of the syntax for declaring an array in VHDL. In the first two versions, *start* and *end* are integers that define the possible range of the array index and hence the total number of array elements. In the last three versions, all or a subset of the values of an existing type (*range-type*) are the range of the array index.

Examples of array declarations are given in Table 4-34. The first pair of examples are very ordinary and show both ascending and descending ranges. The next example shows how a constant, WORD_LEN, can be used with an array declaration, showing also that a range value can be a simple expression. The

---

UNNATURAL ACTS

Although VHDL defines the subtype "natural" as being nonnegative integers starting with 0, most mathematicians consider and define the natural numbers to begin with 1. After all, in early history people began counting with 1, and the concept of "0" arrived much later. Still, there is some discussion and perhaps controversy on the subject, especially as the computer age has led more of us to think with 0 as a starting number. For the latest thinking, search the Web for "natural numbers."

Table 4-34
Examples of VHDL
array declarations.

```
type monthly_count is array (1 to 12) of integer;
type byte is array (7 downto 0) of STD_LOGIC;

constant WORD_LEN: integer := 32;
type word is array (WORD_LEN-1 downto 0) of STD_LOGIC;

constant NUM_REGS: integer := 8;
type reg_file is array (1 to NUM_REGS) of word;

type statecount is array (traffic_light_state) of integer;
```

third example shows that an array element may itself be an array, thus creating a two-dimensional array. The last example shows that an enumerated type (or a subtype) may be specified as the array element range; the array in this example has four elements, based on our previous definition of traffic_light_state.

Array elements are considered to be ordered from left to right, in the same direction as index range. Thus, the leftmost elements of arrays of types monthly_count, byte, word, reg_file, and statecount have indices 1, 7, 31, 1, and reset, respectively.

Within VHDL program statements, individual array elements are accessed using the array name and the element's index in parentheses. For example, if M, B, W, R, and S are signals or variables of the five array types defined in Table 4-34, then M(11), B(5), W(WORD_LEN-5), R(0,0), R(0), and S(reset) are all valid elements.

Array literals can be specified by listing the element values in parentheses. *array literal*
For example, the byte variable B could be set to all ones by the statement

```
B := ('1','1','1','1','1','1','1','1');
```

VHDL also has a shorthand notation that allows you to specify values by index. For example, to set word variable W to all ones except for zeroes in the LSB of each byte, you can write

```
W := (0=>'0',8=>'0',16=>'0',24=>'0',others=>'1');
```
others

The methods just described work for arrays with any *element-type*, but the easiest way write a literal of type STD_LOGIC is to use a "string." A VHDL *string* *string* is a sequence of ISO characters enclosed in double quotes, such as "Hi there". A string is just an array of characters; as a result, a STD_LOGIC array of a given length can be assigned the value of a string of the same length, as long as the characters in the string are taken from the set of nine characters defined as the possible values of the STD_LOGIC elements—'0', '1', 'U', and so on. Thus, the previous two examples can be rewritten as follows:

```
B := "11111111";
W := "11111110111111101111111011111110";
```

array slice

It is also possible to refer to a contiguous subset or *slice* of an array by specifying the starting and ending indices of the subset, for example, M(6 to 9), B(3 downto 0), W(15 downto 8), R(0,7 downto 0), R(1 to 2), S(stop to go). Notice that the slice's direction must be the same as the original array's.

&, concatenation operator

Finally, you can combine arrays or array elements using the *concatenation operator &*, which joins arrays and elements in the order written, from left to right. For example, '0'&'1'&"1Z" is equivalent to "011Z", and the expression B(6 downto 0) & B(7) yields a 1-bit left circular shift of the 8-bit array B.

std_logic_vector

The most important array type in typical VHDL programs is the IEEE 1164 standard user-defined logic type std_logic_vector, which defines an ordered set of std_logic bits. The definition of this type is:

```
type STD_LOGIC_VECTOR is array ( natural range <> ) of STD_LOGIC;
```

unconstrained array type

This is an example of an *unconstrained array type*—the range of the array is unspecified, except that it must be a subrange of a defined type, in this case, natural. This VHDL feature allows us to develop architectures, functions, and other program elements in a more general way, somewhat independent of the array size or its range of index values. An actual range is specified when a signal or variable is assigned this type. We'll see examples in the next subsection.

4.7.4 Functions and Procedures

function
arguments
result
function definition
formal parameters
actual parameters

Like a function in a high-level programming language, a VHDL *function* accepts a number of *arguments* and returns a *result*. Each of the arguments and the result in a VHDL function definition or function call have a predetermined type.

The syntax of a *function definition* is shown in Table 4-35. After giving the name of the function, it lists zero or more *formal parameters* which are used within the function body. When the function is called, the *actual parameters* in the function call are substituted for the formal parameters. Following VHDL's

Table 4-35
Syntax of a VHDL function definition.

```
function function-name (
      signal-names : signal-type;
      signal-names : signal-type;
      . . .
      signal-names : signal-type
) return return-type is
   type declarations
   constant declarations
   variable declarations
   function definitions
   procedure definitions
begin
   sequential-statement
   . . .
   sequential-statement
end function-name;
```

strong-typing policy, the actual parameters must be the same type or a subtype of the formal parameters. When the function is called from within an architecture, a value of the type *return-type* is returned in place of the function call.

As shown in the table, a function may define its own local types, constants, variables, and nested functions and procedures. The keywords begin and end enclose a series of "sequential statements" that are executed when the function is called. We'll take a closer look at different kinds of sequential statements and their syntax later, but you should be able to understand the examples here based on your previous programming experience.

The VHDL "inhibit-gate" architecture of Table 4-26 on page 269 is modified in Table 4-36 to use a function. Within the function definition, the keyword return indicates when the function should return to the caller, and it is followed by an expression with the value to be returned to the caller. The type resulting from this expression must match the *return-type* in the function declaration.

return

The IEEE 1164 standard logic package defines many functions that operate on the standard types std_logic and std_logic_vector. Besides specifying a number of user-defined types, the package also defines the basic logic operations on these types such as and and or. This takes advantage of VHDL's ability to *overload* operators. This facility allows the user to specify a function that is invoked whenever a built-in operator symbol (and, or, +, etc.) is used with a matching set of operand types. There may be several definitions for a given operator symbol; the compiler automatically picks the definition that matches the operand types in each use of the operator.

operator overloading

For example, Table 4-37 contains code, taken from the IEEE package, that shows how the "and" operation is defined for std_logic operands. This code may look complicated, but we've already introduced all of the basic language elements that it uses (except for "resolved", which we describe in connection with three-state logic in Section 5.6.4).

The inputs to the function may be of type std_ulogic or its subtype std_logic. Another subtype UX01 is defined to be used as the function's return type; even if one of the "and" inputs is a nonlogic value ('Z', 'W', etc.), the

---

```
architecture Inhibit_archf of Inhibit is

function ButNot (A, B: bit) return bit is
begin
  if B = '0' then return A;
  else return '0';
  end if;
end ButNot;

begin
  Z <= ButNot(X,Y);
end Inhibit_archf;
```

Table 4-36
VHDL program for an "inhibit" function.

Table 4-37
Definitions relating to the "and" operation on STD_LOGIC values in IEEE 1164.

```
SUBTYPE UX01 IS resolved std_ulogic RANGE 'U' TO '1';
                                       -- ('U','X','0','1')
TYPE stdlogic_table IS ARRAY(std_ulogic, std_ulogic) OF std_ulogic;

-- truth table for "and" function
CONSTANT and_table : stdlogic_table := (
--      ------------------------------------------------------
--      |  U    X    0    1    Z    W    L    H    -     |   |
--      ------------------------------------------------------
       ( 'U', 'U', '0', 'U', 'U', 'U', '0', 'U', 'U' ),  -- | U |
       ( 'U', 'X', '0', 'X', 'X', 'X', '0', 'X', 'X' ),  -- | X |
       ( '0', '0', '0', '0', '0', '0', '0', '0', '0' ),  -- | 0 |
       ( 'U', 'X', '0', '1', 'X', 'X', '0', '1', 'X' ),  -- | 1 |
       ( 'U', 'X', '0', 'X', 'X', 'X', '0', 'X', 'X' ),  -- | Z |
       ( 'U', 'X', '0', 'X', 'X', 'X', '0', 'X', 'X' ),  -- | W |
       ( '0', '0', '0', '0', '0', '0', '0', '0', '0' ),  -- | L |
       ( 'U', 'X', '0', '1', 'X', 'X', '0', '1', 'X' ),  -- | H |
       ( 'U', 'X', '0', 'X', 'X', 'X', '0', 'X', 'X' )   -- | - |
);

FUNCTION "and" ( L : std_ulogic; R : std_ulogic ) RETURN UX01 IS
BEGIN
     RETURN (and_table(L, R));
END "and";
```

function will return one of only four possible values. Type stdlogic_table defines a two-dimensional, 9×9 array indexed by a pair of std_ulogic values. For the and_table, the table entries are arranged so that if either index is '0' or 'L' (a weak '0'), the entry is '0'. A '1' entry is found only if both inputs are '1' or 'H' (a weak '1'). Otherwise, a 'U' or 'X' entry appears.

In the function definition itself, double quotes around the function name indicate operator overloading. The "executable" part of the function is just a single statement that returns the table element indexed by the two inputs, L and R, of the "and" function.

Because of VHDL's strong typing requirements, it's often necessary to convert a signal from one type to another, and the IEEE 1164 package contains several conversion functions—for example, from BIT to STD_LOGIC or vice versa. A commonly needed conversion is from STD_LOGIC_VECTOR into a corresponding integer value. IEEE 1164 does not include such a conversion function, because different designs may need to use different number interpretations—for example, signed versus unsigned. However, we can write our own conversion function as shown in Table 4-38.

The CONV_INTEGER function uses a simple iterative algorithm equivalent to the nested expansion formula on page 30. We won't be describing the FOR, CASE, and WHEN statements that it uses until Section 4.7.8, but you should get the idea. The *null statement* is easy—it means "do nothing." The range of the FOR loop is specified by "X'range", where the single quote after a signal name

null statement

```
function CONV_INTEGER (X: STD_LOGIC_VECTOR) return INTEGER is
  variable RESULT: INTEGER;
begin
  RESULT := 0;
  for i in X'range loop
    RESULT := RESULT * 2;
    case X(i) is
      when '0' | 'L'  => null;
      when '1' | 'H'  => RESULT := RESULT + 1;
      when others     => null;
    end case;
  end loop;
  return RESULT;
end CONV_INTEGER;
```

Table 4-38
VHDL function
for converting
STD_LOGIC_VECTOR
to INTEGER.

means "attribute," and *range* is a built-in attribute identifier that applies only to arrays and means "range of this array's index, from left to right."

range attribute

In the other direction, we can convert an integer to a STD_LOGIC_VECTOR as shown in Table 4-39. Here we must specify not only the integer value to be converted (ARG), but also the number of bits in the desired result (SIZE). Notice that the function declares a local variable "result", a STD_LOGIC_VECTOR whose index range is dependent on SIZE. For this reason, SIZE must be a constant or other value that is known when CONV_STD_LOGIC_VECTOR is compiled. To perform the conversion, the function uses the successive-division algorithm that was described on page 30.

A VHDL *procedure* is similar to a function, except it does not return a result. Whereas a function call can be used in the place of an expression, a procedure call can be used in the place of a statement. VHDL procedures allow their arguments to be specified with type out or inout, so it is actually possible to for a procedure to "return" a result. However, we don't use VHDL procedures in the rest of this book, so we won't discuss them further.

procedure

```
function CONV_STD_LOGIC_VECTOR(ARG: INTEGER; SIZE: INTEGER)
    return STD_LOGIC_VECTOR is
  variable result: STD_LOGIC_VECTOR (SIZE-1 downto 0);
  variable temp: integer;
begin
  temp := ARG;
  for i in 0 to SIZE-1 loop
    if (temp mod 2) = 1 then result(i) := '1';
    else result(i) := '0';
    end if;
    temp := temp / 2;
  end loop;
  return result;
end;
```

Table 4-39
VHDL function
for converting
INTEGER to
STD_LOGIC_VECTOR.

4.7.5 Libraries and Packages

library

A VHDL *library* is a place where the VHDL compiler stores information about a particular design project, including intermediate files that are used in the analysis, simulation, and synthesis of the design. The location of the library within a host computer's file system is implementation dependent. For a given VHDL design, the compiler automatically creates and uses a library named "`work`".

A complete VHDL design usually has multiple files, each containing different design units including entities and architectures. When the VHDL compiler analyzes each file in the design, it places the results in the "`work`" library, and it also searches this library for needed definitions, such as other entities. Because of this feature, a large design can be broken up into multiple files, yet the compiler will find external references as needed.

Not all of the information needed in a design may be in the "`work`" library. For example, a designer may rely on common definitions or functional modules across a family of different projects. Each project has its own "`work`" library (typically a subdirectory within that project's overall directory), but it must also refer to a common library containing the shared definitions. Even small projects may use a standard library such as the one containing IEEE standard definitions.

library clause

The designer can specify the name of such a library using a *library clause* at the beginning of the design file. For example, we can specify the IEEE library:

```
library ieee;
```

The clause "`library work;`" is included implicitly at the beginning of every VHDL design file.

Specifying a library name in a design gives it access to any previously analyzed entities and architectures stored in the library, but it does not give access to type definitions and the like. This is the function of "packages" and "use clauses," described next.

package

A VHDL *package* is a file containing definitions of objects that can be used in other programs. The kind of objects that can be put into a package include signal, type, constant, function, procedure, and component declarations.

Signals that are defined in a package are "global" signals, available to any VHDL entity that uses the package. Types and constants defined in a package are known in any file that uses the package. Likewise, functions and procedures defined in a package can be called in files that use the package, and components (described in the next subsection) can be "instantiated" in architectures that use the package.

use clause

A design can "use" a package by including a *use clause* at the beginning of the design file. For example, to use all of the definitions in the IEEE standard 1164 package, we would write

```
use ieee.std_logic_1164.all;
```

Here, "ieee" is the name of a library which has been previously given in a library clause. Within this library, the file named "std_logic_1164" contains the desired definitions. The suffix "all" tells the compiler to use all of the definitions in this file. Instead of "all", you can write the name of a particular object to use just its definition, for example,

```
use ieee.std_logic_1164.std_ulogic
```

This clause would make available just the definition of the std_ulogic type in Table 4-32 on page 273, without all of the related types and functions. However, multiple "use" clauses can be written to use additional definitions.

Defining packages is not limited to standards bodies. Anyone can write a package, using the syntax shown in Table 4-40. All of the objects declared between "package" and the first "end" statement are visible in any design file that uses the package; objects following the "package body" keyword are local. In particular, notice that the first part includes "function declarations," not definitions. A *function declaration* lists only the function name, arguments, and type, up to but not including the "is" keyword in Table 4-35 on page 276. The complete function definition is given in the package body and is not visible to function users.

function declaration

IEEE VHDL STANDARDS

VHDL has excellent capabilities for extending its data types and functions. This is important, because the language's built-in BIT and BIT_VECTOR actually are quite inadequate for modeling real circuits that also handle three-state, unknown, don't-care, and varying-strength signals.

As a result, soon after the language was formalized as IEEE standard 1076, commercial vendors began to introduce their own built-in data types to deal with logic values other than 0 and 1. Of course, each vendor had different definitions for these extended types, creating a potential "Tower of Babel."

To avoid this situation, the IEEE developed the 1164 standard logic package (std_logic_1164) with a nine-valued logic system that satisfies most designers' needs. This was later followed by standard 1076-3, discussed in Section 5.9.6, which includes several packages with standard types and operations for vectors of STD_LOGIC components that are interpreted as signed or unsigned integers. The packages include std_logic_arith, std_logic_signed, and std_logic_unsigned.

By using IEEE standards, designers can ensure a high degree of portability and interoperability among their designs. This is increasingly important, as the deployment of very large ASICs necessitates cooperation not only among multiple designers but also among multiple vendors who may each contribute different pieces of a "system-on-a-chip" design.

Table 4-40
Syntax of a VHDL
package definition.

```
package package-name is
    type declarations
    signal declarations
    constant declarations
    component declarations
    function declarations
    procedure declarations
end package-name;
package body package-name is
    type declarations
    constant declarations
    function definitions
    procedure definitions
end package-name;
```

4.7.6 Structural Design Elements

We're finally ready to look at the guts of a VHDL design, the "executable" portion of an architecture. Recall from Table 4-28 on page 271 that the body of an architecture is a series of concurrent statements. In VHDL, each *concurrent statement* executes simultaneously with the other concurrent statements in the same architecture body.

concurrent statement

This behavior is markedly different from that of statements in conventional software programming languages, where statements execute sequentially. Concurrent statements are necessary to simulate the behavior of hardware, where connected elements affect each other continuously, not just at particular, ordered time steps. Thus, in a VHDL architecture body, if the last statement updates a signal that is used by the first statement, then the simulator will go back to that first statement and update its results according to the signal that just changed. In fact, the simulator will keep propagating changes and updating results until the simulated circuit stabilizes; we'll discuss this in more detail in Section 4.7.9.

VHDL has several different concurrent statements, as well as a mechanism for bundling a set of sequential statements to operate as a single concurrent statement. Used in different ways, these statements give rise to three somewhat distinct styles of circuit design and description, which we cover in this and the next two subsections.

The most basic of VHDL's concurrent statements is the `component statement`, whose basic syntax is shown in Table 4-41. Here, *component-name* is the name of a previously defined entity that is to be used, or *instantiated*, within the current architecture body. One instance of the named entity is created for

instantiate

Table 4-41 Syntax of a VHDL `component` statement.

label: *component-name* `port map`(*signal1*, *signal2*, ..., *signaln*);

label: *component-name* `port map`(*port1=>signal1*, *port2=>signal2*, ..., *portn=>signaln*);

```
component component-name
   port (signal-names : mode signal-type;
         signal-names : mode signal-type;
         ...
         signal-names : mode signal-type);
end component;
```

Table 4-42
Syntax of a
VHDL component
declaration.

each component statement that invokes its name, and each instance must be named by a unique *label*.

The *port map* keywords introduce a list that associates ports of the named entity with signals in the current architecture. The list may be written in either of two different styles. The first is a positional style; as in conventional programming languages, the signals in the list are associated with the entity's ports in the same order in which they appear in the entity's definition. The second is an explicit style; each of the entity's ports is connected to a signal using the "=>" operator, and these associations may be listed in any order.

port map

Before being instantiated in an architecture, a component must be declared in a *component declaration* in the architecture's definition (see Table 4-28 on page 271). As shown in Table 4-42, a component declaration is essentially the same as the port-declaration part of the corresponding entity declaration—it lists the name, mode, and type of each of its ports.

component declaration

The components used in an architecture may be ones that were previously defined as part of a design, or they may be part of a library. Table 4-43 is an example of a VHDL entity and its architecture that uses components, a "prime-number detector" that is structurally identical to the gate-level circuit in Figure 4-30(c) on page 226. The entity declaration names the inputs and the output of the circuit. The declarations section of the architecture defines all of the signal names and the components that are used internally. The components, INV, AND2, AND3, and OR4, are predefined in the design environment in which this example was created and compiled (Xilinx Foundation 1.5, see References).

Note that component statements in Table 4-43 execute *concurrently*. Even if the statements were listed in a different order, the same circuit would be synthesized, and the simulated circuit operation would be the same.

A VHDL architecture that uses components is often called a *structural description* or *structural design*, because it defines the precise interconnection structure of signals and entities that realize the entity. In this regard, a pure structural description is equivalent to a schematic or a net list for the circuit.

structural description
structural design

In some applications it is necessary to create multiple copies of a particular structure within an architecture. For example, we'll see in Section 5.10.2 that an n-bit "ripple adder" can be created by cascading n "full adders." VHDL includes a *generate statement* that allows you to create such repetitive structures using a kind of "for loop," without having to write out all of the component instantiations individually.

Table 4-43 Structural VHDL program for a prime-number detector.

```
library IEEE;
use IEEE.std_logic_1164.all;

entity prime is
    port ( N: in STD_LOGIC_VECTOR (3 downto 0);
           F: out STD_LOGIC );
end prime;

architecture prime1_arch of prime is
signal N3_L, N2_L, N1_L: STD_LOGIC;
signal N3L_N0, N3L_N2L_N1, N2L_N1_N0, N2_N1L_N0: STD_LOGIC;
component INV port (I: in STD_LOGIC; O: out STD_LOGIC); end component;
component AND2 port (I0,I1: in STD_LOGIC; O: out STD_LOGIC); end component;
component AND3 port (I0,I1,I2: in STD_LOGIC; O: out STD_LOGIC); end component;
component OR4 port (I0,I1,I2,I3: in STD_LOGIC; O: out STD_LOGIC); end component;
begin
  U1: INV port map (N(3), N3_L);
  U2: INV port map (N(2), N2_L);
  U3: INV port map (N(1), N1_L);
  U4: AND2 port map (N3_L, N(0), N3L_N0);
  U5: AND3 port map (N3_L, N2_L, N(1), N3L_N2L_N1);
  U6: AND3 port map (N2_L, N(1), N(0), N2L_N1_N0);
  U7: AND3 port map (N(2), N1_L, N(0), N2_N1L_N0);
  U8: OR4 port map (N3L_N0, N3L_N2L_N1, N2L_N1_N0, N2_N1L_N0, F);
end prime1_arch;
```

The syntax of a simple iterative `generate` loop is shown in Table 4-44. The *identifier* is implicitly declared as a variable with type compatible with the *range*. The *concurrent statement* is executed once for each possible value of the *identifier* within the range, and *identifier* may be used within the concurrent statement. For example, Table 4-45 shows how an 8-bit inverter can be created.

The value of a constant must be known at the time that a VHDL program is compiled. In many applications it is useful to design and compile an entity and its architecture while leaving some of its parameters, such as bus width, unspecified. VHDL's "generic" facility lets you do this.

generic constant
generic declaration

One or more *generic constants* can be defined in an entity declaration with a *generic declaration* before the port declaration, using the syntax shown in Table 4-46. Each of the named constants can be used within the architecture definition for the entity, and the value of the constant is deferred until the entity is instantiated using a component statement within another architecture. Within that component statement, values are assigned to the generic constants using a

generic map

`generic map` clause in the same style as the `port map` clause. Table 4-47 is an example that combines generic and generate statements to define a "bus inverter" with a user-specifiable width. Multiple copies of this inverter, each with a different width, are instantiated in the program in Table 4-48.

```
label: for identifier in range generate
                concurrent-statement
            end generate;
```

Table 4-44
Syntax of a VHDL
for-generate loop.

```
library IEEE;
use IEEE.std_logic_1164.all;

entity inv8 is
    port ( X: in STD_LOGIC_VECTOR (1 to 8);
           Y: out STD_LOGIC_VECTOR (1 to 8) );
end inv8;

architecture inv8_arch of inv8 is
component INV port (I: in STD_LOGIC; O: out STD_LOGIC); end component;
begin
  g1: for b in 1 to 8 generate
       U1: INV port map (X(b), Y(b));
     end generate;
end inv8_arch;
```

Table 4-45
VHDL entity and
architecture for an
8-bit inverter.

```
entity entity-name is
    generic (constant-names : constant-type;
                 constant-names : constant-type;
                     . . .
                 constant-names : constant-type);
    port (signal-names : mode signal-type;
          signal-names : mode signal-type;
              . . .
          signal-names : mode signal-type);
end entity-name;
```

Table 4-46
Syntax of a VHDL ge-
neric declaration with-
in an entity
declaration.

```
library IEEE;
use IEEE.std_logic_1164.all;

entity businv is
    generic (WIDTH: positive);
    port ( X: in STD_LOGIC_VECTOR (WIDTH-1 downto 0);
           Y: out STD_LOGIC_VECTOR (WIDTH-1 downto 0) );
end businv;

architecture businv_arch of businv is
component INV port (I: in STD_LOGIC; O: out STD_LOGIC); end component;
begin
  g1: for b in WID-1 downto 0 generate
       U1: INV port map (X(b), Y(b));
     end generate;
end businv_arch;
```

Table 4-47
VHDL entity and
architecture for an
arbitrary-width bus
inverter.

```
library IEEE;
use IEEE.std_logic_1164.all;

entity businv_example is
    port ( IN8: in STD_LOGIC_VECTOR (7 downto 0);
            OUT8: out STD_LOGIC_VECTOR (7 downto 0);
            IN16: in STD_LOGIC_VECTOR (15 downto 0);
            OUT16: out STD_LOGIC_VECTOR (15 downto 0);
            IN32: in STD_LOGIC_VECTOR (31 downto 0);
            OUT32: out STD_LOGIC_VECTOR (31 downto 0) );
end businv_example;

architecture businv_ex_arch of businv_example is
component businv
    generic (WIDTH: positive);
    port ( X: in STD_LOGIC_VECTOR (WIDTH-1 downto 0);
            Y: out STD_LOGIC_VECTOR (WIDTH-1 downto 0) );
end component;
begin
U1: businv generic map (WIDTH=>8) port map (IN8, OUT8);
U2: businv generic map (WIDTH=>16) port map (IN16, OUT16);
U3: businv generic map (WIDTH=>32) port map (IN32, OUT32);
end businv_ex_arch;
```

4.7.7 Dataflow Design Elements

If component statements were its only concurrent statements, then VHDL would be little more than a strongly typed, hierarchical net-list description language. Several additional concurrent statements allow VHDL to describe a circuit in terms of the flow of data and operations on it within the circuit. This style is called a *dataflow description* or *dataflow design*.

dataflow description
dataflow design
concurrent signal-
*   assignment statement*

Two additional concurrent statements used in dataflow designs are shown in Table 4-49. The first of these is the most often used and is called a *concurrent signal-assignment statement.* You can read this as "*signal-name* gets *expression.*" Because of VHDL's strong typing, the type of *expression* must be compatible with that of *signal-name*. In general, this means that either the types must be identical or *expression*'s type is a subtype of *signal-name*'s. In the case of arrays, both the element type and the length must match; however, the index range and direction need not match.

signal-name <= *expression*;

signal-name <= *expression* when *boolean-expression* else
           expression when *boolean-expression* else
           . . .
           expression when *boolean-expression* else
           expression;

```
architecture prime2_arch of prime is
signal N3L_N0, N3L_N2L_N1, N2L_N1_N0, N2_N1L_N0: STD_LOGIC;
begin
  N3L_N0      <= not N(3)                               and N(0);
  N3L_N2L_N1 <= not N(3) and not N(2) and     N(1)          ;
  N2L_N1_N0  <=              not N(2) and     N(1) and N(0);
  N2_N1L_N0  <=                   N(2) and not N(1) and N(0);
  F <= N3L_N0 or N3L_N2L_N1 or N2L_N1_N0 or N2_N1L_N0;
end prime2_arch;
```

Table 4-50
Dataflow VHDL architecture for the prime-number detector.

Table 4-50 shows an architecture for the prime-number detector entity (Table 4-43 on page 284) written in dataflow style. In this style we don't show the explicit gates and their connections; rather, we use VHDL's built-in and, or, and not operators. (Actually, these operators are not built in for signals of type STD_LOGIC, but they are defined and overloaded by the IEEE 1164 package.) Note that the not operator has the highest precedence, so no parentheses are required around subexpressions like "not N(3)" to get the intended result.

We can also use the second, *conditional* form of the concurrent signal-assignment statement, using the keywords when and else as shown in Table 4-49. Here, a *boolean-expression* combines individual boolean terms using VHDL's built-in boolean operators such as and, or, and not. Boolean terms are typically boolean variables or results of comparisons using *relational operators* =, /= (not equal), >, >=, <, and <=.

Table 4-51 is an example using conditional concurrent assignment statements. Each of the comparisons of a individual STD_LOGIC bit such as N(3) is made against a character literal '1' or '0' and returns a value of type boolean. These comparison results are combined in the boolean expression between the when and else keywords in each statement. The else clauses are generally required; the combined set of conditions in a single statement should cover all possible input combinations.

Another kind of concurrent assignment statement is the *selected signal assignment*, whose syntax is shown in Table 4-52. This statement evaluates the given *expression*, and when the value matches one of the *choices*, it assigns the corresponding *signal-value* to *signal-name*. The *choices* in each when clause

conditional signal-assignment statement
when
else
relational operators
=, /=, >, >=, <, <=

selected signal-assignment statement

```
architecture prime3_arch of prime is
signal N3L_N0, N3L_N2L_N1, N2L_N1_N0, N2_N1L_N0: STD_LOGIC;
begin
  N3L_N0      <= '1' when N(3)='0' and N(0)='1' else '0';
  N3L_N2L_N1 <= '1' when N(3)='0' and N(2)='0' and N(1)='1' else '0';
  N2L_N1_N0  <= '1' when N(2)='0' and N(1)='1' and N(0)='1' else '0';
  N2_N1L_N0  <= '1' when N(2)='1' and N(1)='0' and N(0)='1' else '0';
  F <= N3L_N0 or N3L_N2L_N1 or N2L_N1_N0 or N2_N1L_N0;
end prime3_arch;
```

Table 4-51
Prime-number-detector architecture using conditional assignments.

Table 4-52
Syntax of VHDL
selected signal-
assignment statement.

```
with expression select
    signal-name <= signal-value when choices,
                   signal-value when choices,
                   . . .
                   signal-value when choices;
```

others

may be a single value of *expression* or a list of values separated by vertical bars
(|). The *choices* for the entire statement must be mutually exclusive and all
inclusive. The keyword *others* can be used in the last when clause to denote all
values of *expression* that have not yet been covered.

Table 4-53 is an architecture for the prime-number detector that uses a
selected signal-assignment statement. All of the *choices* for which F is '1' could
have been written in a single when clause, but multiple clauses are shown just for
instructional purposes. In this example, the selected signal-assignment statement reads somewhat like a listing of the on-set of the function F.

We can modify the previous architecture slightly to take advantage of the
numeric interpretation of N in the function definition. Using the CONV_INTEGER
function that we defined previously, Table 4-54 writes the *choices* in terms of
integers, which we can readily see are prime as required. We can think of this
version of the architecture as a "behavioral" description, because it describes the
desired function in such a way that its behavior is quite evident.

Table 4-53
Prime-number
detector architecture
using selected signal
assignment.

```
architecture prime4_arch of prime is
begin
  with N select
    F <= '1' when "0001",
         '1' when "0010",
         '1' when "0011" | "0101" | "0111",
         '1' when "1011" | "1101",
         '0' when others;
end prime4_arch;
```

**COVERING
ALL THE
CASES**

Conditional and selected signal assignments require all possible conditions to be
covered. In a conditional signal assignment, the final "else *expression*" covers
missing conditions. In a selected signal assignment, "others" can be used in the
final when clause to pick up the remaining conditions.

In Table 4-53, you might think that instead of writing "others" in the
final when clause, we could have written the nine remaining 4-bit combinations,
"0000", "0100", and so on. But that's not true! Remember that STD_LOGIC is
a nine-valued system, so a 4-bit STD_LOGIC_VECTOR actually has 9^4 possible
values. So "others" in this example is really covering 6,554 cases!

```
architecture prime5_arch of prime is
begin
  with CONV_INTEGER(N) select
    F <= '1' when 1 | 2 | 3 | 5 | 7 | 11 | 13,
         '0' when others;
end prime5_arch;
```

Table 4-54
A more behavioral
description of the
prime-number
detector.

4.7.8 Behavioral Design Elements

As we saw in the last example, it is sometimes possible to directly describe a desired logic-circuit behavior using a concurrent statement. This is a good thing, as the ability to create a *behavioral design* or *behavioral description* is one of the key benefits of hardware-description languages in general and VHDL in particular. However, for most behavioral descriptions, we need to employ some additional language elements described in this subsection.

behavioral design
behavioral description

VHDL's key behavioral element is the "process." A *process* is a collection of "sequential" statements (described shortly) that executes in parallel with other concurrent statements and other processes. Using a process, you can specify a complex interaction of signals and events in a way that executes in essentially zero simulated time during simulation and that gives rise to a synthesized combinational or sequential circuit that performs the modeled operation directly.

process

A VHDL *process statement* can be used anywhere that a concurrent statement can be used. A process statement is introduced by the keyword *process* and has the syntax shown in Table 4-55. Since a process statement is written within the scope of an enclosing architecture, it has visibility of the types, signals, constants, functions, and procedures that are declared or are otherwise visible in the enclosing architecture. However, you can also define types, signals, constants, functions, and procedures that are local to the process.

process statement
process

Note that a process may not declare signals, only "variables." A VHDL *variable* keeps track of the state within a process and is not visible outside of the process. Depending on its use, it may or may not give rise to a corresponding signal in a physical realization of the modeled circuit. The syntax for defining a

variable

```
process (signal-name, signal-name, ..., signal-name)
    type declarations
    variable declarations
    constant declarations
    function definitions
    procedure definitions
begin
    sequential-statement
    ...
    sequential-statement
end process;
```

Table 4-55
Syntax of a VHDL
process statement.

variable

variable within a process is similar to the syntax for a signal declaration within an architecture, except that the keyword *variable* is used:

variable *variable-names* : *variable-type*;

running process
suspended process
sensitivity list

A VHDL process is always either *running* or *suspended*. The list of signals in the process definition, called the *sensitivity list*, determines when the process runs. A process initially is suspended; when any signal in its sensitivity list changes value, the process resumes execution, starting with its first sequential statement and continuing until the end. If any signal in the sensitivity list change value as a result of running the process, it runs again. This continues until the process runs without any of these signals changing value. In simulation, all of this happens in zero simulated time.

Upon resumption, a properly written process will suspend after one or a few runs. However, it is possible to write an incorrect process that never suspends. For example, consider a process with just one sequential statement, "X <= not X" and a sensitivity list of "(X)". Since X changes on every pass, the process will run forever in zero simulated time—not very useful! In practice, simulators have safeguards that normally can detect such unwanted behavior, terminating the misbehaving process after a thousand or so passes.

The sensitivity list is optional; a process without a sensitivity list starts running at time zero in simulation. One application of such a process is to generate input waveforms in a test bench, as in Table 4-65 on page 296.

sequential signal-
* assignment statement*

variable-assignment
* statement*

:=

VHDL has several kinds of sequential statements. The first is a *sequential signal-assignment statement*; this has the same syntax as the concurrent version (*signal-name* <= *expression*;), but it occurs within the body of a process rather than an architecture. An analogous statement for variables is the *variable-assignment statement*, which has the syntax "*variable-name* := *expression*;". Notice that a different assignment operator, :=, is used for variables.

For instruction purposes, the dataflow architecture of the prime-number detector in Table 4-50 is rewritten as a process in Table 4-56. Notice that we're still working off the same original entity declaration of prime that appeared in

Table 4-56
Process-based dataflow VHDL architecture for the prime-number detector.

```
architecture prime6_arch of prime6 is
begin
  process(N)
    variable N3L_N0, N3L_N2L_N1, N2L_N1_N0, N2_N1L_N0: STD_LOGIC;
  begin
    N3L_N0      := not N(3)                              and N(0);
    N3L_N2L_N1 := not N(3) and not N(2) and     N(1)           ;
    N2L_N1_N0  :=              not N(2) and     N(1) and N(0);
    N2_N1L_N0  :=                  N(2) and not N(1) and N(0);
    F <= N3L_N0 or N3L_N2L_N1 or N2L_N1_N0 or N2_N1L_N0;
  end process;
end prime6_arch;
```

WEIRD BEHAVIOR	Remember that the statements within a process are executed *sequentially*. Suppose that for some reason we wrote the last statement in Table 4-56 (the signal assignment to F) as the first. Then we would see rather weird behavior from this process. The first time the process was run, the simulator would complain that the values of the variables were being read before any value was assigned to them. On subsequent resumptions, a value would be assigned to F based on the *previous* values of the variables, which are remembered while the process is suspended. New values would then be assigned to the variables and remembered until the next resumption. So the circuit's output value would always be one input-change behind.

Table 4-43. Within the new architecture (prime6_arch), we have just one concurrent statement, which is a process. The process sensitivity list contains just N, the primary inputs of the desired combinational logic function. The AND-gate outputs must be defined as variables rather than signals, since signal definitions are not allowed within a process. Otherwise, the body of the process is very similar to that of the original architecture. In fact, a typical synthesis tool would probably create the same circuit from either description.

Other sequential statements, beyond simple assignment, can give us more creative control in expressing circuit behavior. The *if statement*, with the syntax *if statement* shown in Table 4-57, is probably the most familiar of these. In the first and simplest form of the statement, a *boolean-expression* is tested, and a *sequential-statement* is executed if the expression's value is true. In the second form,

Table 4-57
Syntax of a VHDL
if statement.

```
if boolean-expression then sequential-statement
end if;

if boolean-expression then sequential-statement
else sequential-statement
end if;

if boolean-expression then sequential-statement
elsif boolean-expression then sequential-statement
...
elsif boolean-expression then sequential-statement
end if;

if boolean-expression then sequential-statement
elsif boolean-expression then sequential-statement
...
elsif boolean-expression then sequential-statement
else sequential-statement
end if;
```

Table 4-58
Prime-number-
detector architecture
using an if statement.

```
architecture prime7_arch of prime is
begin
  process(N)
    variable NI: INTEGER;
  begin
    NI := CONV_INTEGER(N);
    if NI=1 or NI=2 then F <= '1';
    elsif NI=3 or NI=5 or NI=7 or NI=11 or NI=13 then F <= '1';
    else F <= '0';
    end if;
  end process;
end prime7_arch;
```

else

we've added an "*else*" clause with another *sequential-statement* that's executed if the expression's value is false.

elsif

To create nested if-then-else statements, VHDL uses a special keyword *elsif*, which introduces the "middle" clauses. An elsif clause's *sequential-statement* is executed if its *boolean-expression* is true and all of the preceding *boolean-expressions* were false. The optional final else-clause's *sequential-statement* is executed if all of the preceding *boolean-expressions* were false.

Table 4-58 is a version of the prime-number-detector architecture that uses an if statement. A local variable NI is used to hold a converted, integer version of the input N, so that the comparisons in the if statement can be written using integer values.

The boolean expressions in Table 4-58 are nonoverlapping; that is, only one of them is true at a time. For this application we really didn't need the full power of nested if statements. In fact, a synthesis engine might create a circuit that evaluates the boolean expressions in series, with slower operation than might otherwise be possible. When we need to select among multiple alterna-

case statement

tives based on the value of just one signal or expression, a *case statement* is usually more readable and may yield a better synthesized circuit.

Table 4-59 shows the syntax of a case statement. This statement evaluates the given *expression*, finds a matching value in one of the *choices*, and executes the corresponding *sequential-statements*. Note that one or more sequential statements can be written for each set of *choices*. The *choices* may take the form of a single value or of multiple values separated by vertical bars (|). The *choices* must be mutually exclusive and include all possible values of *expression*'s type; the keyword others can be used as the last *choices* to denote all values that have not yet been covered.

Table 4-59
Syntax of a VHDL
case statement.

```
case expression is
  when choices => sequential-statements
    ...
  when choices => sequential-statements
end case;
```

```
architecture prime8_arch of prime is
begin
  process(N)
  begin
    case CONV_INTEGER(N) is
      when 1 => F <= '1';
      when 2 => F <= '1';
      when 3 | 5 | 7 | 11 | 13 => F <= '1';
      when others => F <= '0';
    end case;
  end process;
end prime8_arch;
```

Table 4-60
Prime-number-
detector architecture
using a case statement.

Table 4-60 is yet another architecture for the prime-number detector, this time coded with a case statement. Like the concurrent version, the select statement in Table 4-54 on page 289, the case statement makes it very easy to see the desired functional behavior.

Another important class of sequential statements are the *loop statements*, The simplest of these has the syntax shown in Table 4-61 and creates an infinite loop. Although infinite loops are undesirable in conventional software programming languages, we'll show in Section 7.12 how such a loop can be very useful in hardware modeling.

loop statement

```
loop
  sequential-statement
  . . .
  sequential-statement
end loop;
```

Table 4-61
Syntax of a basic
VHDL loop
statement.

A more familiar loop, one that we've seen before, is the *for loop*, with the syntax shown in Table 4-62. Note that the loop variable, *identifier*, is declared implicitly by its appearance in the for loop and has the same type as *range*. This variable may be used within the loop's sequential statements, and it steps through all of the values in *range*, from left to right, one per iteration.

for loop

Two more useful sequential statements that can be executed within a loop are "*exit*" and "*next*". When executed, *exit* transfers control to the statement immediately following the loop end. On the other hand, *next* causes any remaining statements in the loop to be bypassed and begins the next iteration of the loop.

exit statement
next statement

```
for identifier in range loop
  sequential-statement
  . . .
  sequential-statement
end loop;
```

Table 4-62
Syntax of a VHDL
for loop.

Table 4-63
Prime-number-
detector architecture
using a for statement.

```
library IEEE;
use IEEE.std_logic_1164.all;

entity prime9 is
      port ( N: in STD_LOGIC_VECTOR (15 downto 0);
              F: out STD_LOGIC );
end prime9;

architecture prime9_arch of prime9 is
begin
  process(N)
  variable NI: INTEGER;
  variable prime: boolean;
  begin
    NI := CONV_INTEGER(N);
    prime := true;
    if NI=1 or NI=2 then null; -- take care of boundary cases
    else for i in 2 to 253 loop
            if NI mod i = 0 then
               prime := false; exit;
            end if;
         end loop;
    end if;
    if prime then F <= '1'; else F <= '0'; end if;
  end process;
end prime9_arch;
```

Our good old prime-number detector is coded one more time in Table 4-63, this time using a for loop. The striking thing about this example is that it is truly a behavioral description—we have actually used VHDL to compute whether the input N is a prime number. We've also increased the size of N to 16 bits, just to emphasize the fact that we were able to create a compact model for the circuit without having to explicitly list hundreds of primes.

BAD DESIGN

Table 4-63 has a good example of a for loop, but is a bad example of how to design a circuit. Although VHDL is a powerful programming language, design descriptions that use its full power may be inefficient or unsynthesizable.

The culprit in Table 4-63 is the mod operator. This operation requires an integer division, and most VHDL tools are unable to synthesize division circuits except for special cases, such as division by a power of two (realized as a shift).

Even if the tools could synthesize a divider, we wouldn't want to specify a prime number detector in this way. The description in Table 4-63 implies a combinational circuit, and the tools would have to create 252 combinational dividers, one for each value of i, to "unroll" the for loop and realize the circuit!

```
while boolean-expression loop
    sequential-statement
    ...
    sequential-statement
end loop;
```

Table 4-64
Syntax of a VHDL
`while` loop.

while loop

The last kind of `loop` statement is the *while loop*, with the syntax shown in Table 4-64. In this form, *boolean-expression* is tested before each iteration of the loop, and the loop is executed only if the value of the expression is `true`.

We can use processes to write behavioral descriptions of both combinational and sequential circuits. Many more examples of combinational-circuit descriptions appear in the VHDL subsections of Chapter 5. A few additional language features are needed to describe sequential circuits; these are described in Section 7.12, and sequential examples appear in the VHDL subsections of Chapter 8.

4.7.9 The Time Dimension and Simulation

None of the examples that we've dealt with so far models the time dimension of circuit operation—everything happens in zero simulated time. However, VHDL has excellent facilities for modeling the time, and it is indeed another significant dimension of the language. In this book we won't go into detail on this subject, but we'll introduce just a few ideas here.

VHDL allows you to specify a time delay using the keyword *after* in any signal-assignment statement, including sequential, concurrent, conditional, and selected assignments. For example, in the inhibit-gate architecture of Table 4-26 on page 269 you could write

after keyword

```
Z <= '1' after 4 ns when X='1' and Y='0' else '0' after 3 ns;
```

This allows you to model an inhibit gate that has 4 ns of delay on a 0-to-1 output transition and only 3 ns on a 1-to-0 transition. In typical ASIC design environments, the VHDL models for all of the low-level components in the component library include such delay parameters. Using these estimates, a VHDL simulator can predict the approximate timing behavior of a larger circuit that uses these components.

Another way to invoke the time dimension is with *wait*, a sequential statement. This statement can be used to suspend a process for a specified time period. Table 4-65 is an example program that uses `wait` to create simulated input waveforms to test the operation of the inhibit gate for four different input combinations at 10-ns time steps.

wait statement

Once you have a VHDL program that whose syntax and semantics are correct, you can use a VHDL simulator to observe its operation. Although we won't go into great detail, it's useful to have a basic understanding of how such a simulator works.

Table 4-65
Using the VHDL `wait` statement to generate input waveforms in a test-bench program.

```
entity InhibitTestBench is
end InhibitTB_arch;

architecture InhibitTB_arch of InhibitTestBench is
component Inhibit port (X,Y: in BIT; Z: out BIT); end component;
signal XT, YT, ZT: BIT;
begin
  U1: Inhibit port map (XT, YT, ZT);
  process
  begin
    XT <= '0'; YT <= '0';
    wait for 10 ns;
    XT <= '0'; YT <= '1';
    wait for 10 ns;
    XT <= '1'; YT <= '0';
    wait for 10 ns;
    XT <= '1'; YT <= '1';
    wait; -- this suspends the process indefinitely
  end process;
end InhibitTB_arch;
```

simulation time

Simulator operation begins at *simulation time* of zero. At this time, the simulator initializes all signals to a default value (which you shouldn't depend on!). It also initializes any signals or variables for which initial values have been declared explicitly (we haven't shown you how to do this). Next, the simulator begins the execution of all the processes and concurrent statements in the design.

Of course, the simulator can't really simulate all of the processes and concurrent statements simultaneously, but it can pretend that it does, using a time-based "event list" and a "signal-sensitivity matrix." Note that each concurrent statement is equivalent to one process.

At simulation time zero, all of the processes are scheduled for execution, and one of these is selected. All of its sequential statements are executed, including any looping behavior that is specified. When the execution of this process is completed, another one is selected, and so on, until all of the processes have been executed. This completes one *simulation cycle*.

simulation cycle

event list

During its execution, a process may assign new values to signals. The new values are not assigned immediately; rather, they are placed on the *event list* and scheduled to become effective at a certain time. If the assignment has an explicit simulation time associated with it (for example, after a delay specified by an `after` clause), then it is scheduled on the event list to occur at that time. Otherwise, it is supposed to occur "immediately"; however, it is actually scheduled to occur at the current simulation time plus one "delta delay." The *delta delay* is an infinitesimally short time, such that the current simulation time plus any number of delta delays still equals the current simulation time. This concept allows processes to execute multiple times if necessary in zero simulated time.

delta delay

After a simulation cycle completes, the event list is scanned for the signal or signals that change at the next earliest time on the list. This may be as little as one delta delay later, or it may be a real circuit delay later, in which case the simulation time is advanced to this time. In any case, the scheduled signal changes are made. Some processes may be sensitive to the changing signals, as indicated by the signal-sensitivity matrix. It indicates, for each signal, which processes have that signal in their sensitivity list. (The process equivalent of a concurrent statement has *all* of its control and data signals in its sensitivity list.) All of the processes that are sensitive to a signal that just changed are scheduled for execution in the next simulation cycle, which now begins.

The simulator's two-phase operation of a simulation cycle followed by scanning the event list and making the next scheduled signal assignments goes on indefinitely, until the list is empty. At this point the simulation is complete.

The event-list mechanism makes it possible to simulate the operation of concurrent processes even though the simulator runs on a single computer with a single thread of execution. And the delta-delay mechanism ensures correct operation even though a process or set of processes may require multiple executions, spanning several delta delays, before changing signals settle down to a stable value. This mechanism is also used to detect runaway processes (such as "X <= not X"); if a thousand simulation cycles occur over a thousand delta delays without advancing simulation time by any "real" amount, it's most likely that something's amiss.

4.7.10 Synthesis

As we mentioned at the beginning of this section, VHDL was originally invented as a logic circuit description and simulation language and was later adapted to synthesis. The language has many features and constructs that cannot be synthesized. However, the subset of the language and the style of programs that we've presented in this section are generally synthesizable by most tools.

Still, the code that you write can have a big effect on the quality of the synthesized circuits that you get. A few examples are listed below:

- "Serial" control structures like if-elsif-elsif-else can result in a corresponding serial chain of logic gates to test conditions. It's better to use a case or select statement if the conditions are mutually exclusive.

- Loops in processes are generally "unwound" to create multiple copies of combinational logic to execute the statements in the loop. If you want to use just one copy of the combinational logic in a sequence of steps, then you have to design a sequential circuit, as discussed in later chapters.

- When using conditional statements in a process, failing to state an outcome for some input combination will cause the compiler to create a "latch" to hold the old value of a signal that might otherwise change. Such latches are generally not intended.

In addition, some language features and constructs may just be unsynthesizable, depending on the tool. Naturally, you have to consult the documentation to find out what's disallowed, allowed, and recommended for a particular tool.

For the foreseeable future, digital designers who use synthesis tools will need to pay reasonably close attention to their coding style in order to obtain good results. And for the moment, the definition of "good coding style" depends somewhat on both the synthesis tool and the target technology. The examples in the rest of this book, while syntactically and semantically correct, hardly scratch the surface of coding methods for large HDL-based designs. The art and practice of large HDL-based hardware design is still very much evolving.

References

A historical description of Boole's development of "the science of Logic" appears in *The Computer from Pascal to von Neumann* by Herman H. Goldstine (Princeton University Press, 1972). Claude E. Shannon showed how Boole's work could be applied to logic circuits in "A Symbolic Analysis of Relay and Switching Circuits" (*Trans. AIEE,* Vol. 57, 1938, pp. 713–723).

Although the two-valued Boolean algebra is the basis for switching algebra, a Boolean algebra need not have only two values. Boolean algebras with 2^n values exist for every integer n; for example, see *Discrete Mathematical Structures and Their Applications* by Harold S. Stone (SRA, 1973). Such

Huntington postulates

algebras may be formally defined using the so-called *Huntington postulates* devised by E. V. Huntington in 1907; for example, see *Digital Design* by M. Morris Mano (Prentice Hall, 1984). A mathematician's development of Boolean algebra based on a more modern set of postulates appears in *Modern Applied Algebra* by G. Birkhoff and T. C. Bartee (McGraw-Hill, 1970). Our engineering-style, "direct" development of switching algebra follows that of Edward J. McCluskey in his *Introduction to the Theory of Switching Circuits* (McGraw-Hill, 1965) and *Logic Design Principles* (Prentice Hall, 1986).

The prime-implicant theorem was proved by W. V. Quine in "The Problem of Simplifying Truth Functions" (*Am. Math. Monthly*, Vol. 59, No. 8, 1952, pp. 521–531). In fact it is possible to prove a more general prime-implicant theorem showing that there exists at least one minimal sum that is a sum of prime implicants even if we remove the constraint on the number of literals in the definition of "minimal."

A graphical method for simplifying Boolean functions was proposed by E. W. Veitch in "A Chart Method for Simplifying Boolean Functions" (*Proc. ACM*, May 1952, pp. 127–133). His *Veitch diagram*, shown in Figure 4-53, actually reinvented a chart proposed by an English archaeologist, A. Marquand ("On Logical Diagrams for *n* Terms," *Philosophical Magazine* XII, 1881, pp. 266–270). The Veitch diagram or Marquand chart uses "natural" binary counting order for its rows and columns, with the result that some adjacent rows and columns differ in more than one value, and product terms do not always cover adjacent cells. M. Karnaugh showed how to fix the problem in "A Map Method for Synthesis of Combinational Logic Circuits" (*Trans. AIEE, Comm. and Electron.*, Vol. 72, Part I, November 1953, pp. 593–599). On the other hand, George J. Klir, in his book *Introduction to the Methodology of Switching Circuits* (Van Nostrand, 1972), claims that binary counting order is just as good as, perhaps better than, Karnaugh-map order for minimizing logic functions.

At this point, the Karnaugh vs. Veitch argument is of course irrelevant, because no one draws charts any more to minimize logic circuits. Instead, we use computer programs running logic-minimization algorithms. The first of such algorithms was described by Quine in "A Way to Simplify Truth Functions" (*Am. Math. Monthly*, Vol. 62, No. 9, 1955, pp. 627–631) and modified by E. J. McCluskey in "Minimization of Boolean Functions" (*Bell Sys. Tech. J.*, Vol. 35, No. 5, November 1956, pp. 1417–1444). The Quine-McCluskey algorithm is fully described in McCluskey's books cited earlier.

McCluskey's 1965 book also covers the iterative consensus algorithm for finding prime implicants and proves that it works. The starting point for this algorithm is a sum-of-products expression, or equivalently, a list of cubes. The product terms *need not* be minterms or prime implicants, but *may* be either or anything in between. In other words, the cubes in the list may have any and all dimensions, from 0 to *n* in an *n*-variable function. Starting with the list of cubes, the algorithm generates a list of all the prime-implicant cubes of the function, without ever having to generate a full minterm list.

The iterative consensus algorithm was first published by T. H. Mott, Jr., in "Determination of the Irredundant Normal Forms of a Truth Function by Iterated Consensus of the Prime Implicants" (*IRE Trans. Electron. Computers*, Vol. EC-9, No. 2, 1960, pp. 245–252). A generalized consensus algorithm was published by Pierre Tison in "Generalization of Consensus Theory and Application to the Minimization of Boolean Functions" (*IEEE Trans. Electron. Computers*, Vol. EC-16, No. 4, 1967, pp. 446–456). All of these algorithms are described by Thomas Downs in *Logic Design with Pascal* (Van Nostrand Reinhold, 1988).

As we explained in Section 4.4.4, the huge number of prime implicants in some logic functions makes it impractical or impossible deterministically to find them all or select a minimal cover. However, efficient heuristic methods can find

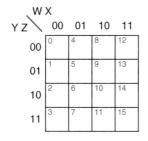

Figure 4-53
A 4-variable Veitch diagram or Marquand chart.

solutions that are close to minimal. The Espresso-II method is described in *Logic Minimization Algorithms for VLSI Synthesis* by R. K. Brayton, C. McMullen, G. D. Hachtel, and A. Sangiovanni-Vincentelli (Kluwer Academic Publishers, 1984). The more recent Espresso-MV and Espresso-EXACT algorithms are described in "Multiple-Valued Minimization for PLA Optimization" by R. L. Rudell and A. Sangiovanni-Vincentelli (*IEEE Trans. CAD*, Vol. CAD-6, No. 5, 1987, pp. 727–750).

0-set
1-set
P-set
S-set

In this chapter we described a map method for finding static hazards in two-level AND-OR and OR-AND circuits, but any combinational circuit can be analyzed for hazards. In both his 1965 and 1986 books, McCluskey defines the *0-set* and *1-sets* of a circuit and shows how they can be used to find static hazards. He also defines *P-sets* and *S-sets* and shows how they can be used to find dynamic hazards.

Many deeper and varied aspects of switching theory have been omitted from this book but have been beaten to death in other books and literature. A good starting point for an academic study of classical switching theory is Zvi Kohavi's book, *Switching and Finite Automata Theory*, second edition (McGraw-Hill, 1978), which includes material on set theory, symmetric networks, functional decomposition, threshold logic, fault detection, and path sensitization. Another area of great academic interest (but little commercial

multiple-valued logic

activity) is nonbinary *multiple-valued logic*, in which each signal line can take on more than two values. In his 1986 book, McCluskey gives a good introduction to multiple-valued logic, explaining its pros and cons and why it has seen little commercial development.

Over the years, I've struggled to find a readily accessible and definitive reference on the ABEL language, and I finally found it—Appendix A of *Digital Design Using ABEL*, by David Pellerin and Michael Holley (Prentice Hall, 1994). It makes sense that this would be the definitive work—Pellerin and Holley invented the language and wrote the original compiler code! Pellerin also wrote, with co-author Douglas Taylor, a very nice stand-alone introduction to VHDL, titled *VHDL Made Easy!* (Prentice Hall, 1997).

VHDL is a "Very Huge Design Language," and we cover only a subset of its features and capabilities in this book. There are many language options (such as statement labels) that we don't use, but these are covered in dedicated VHDL tutorial and reference books. One of the best is *The Designer's Guide to VHDL* (Morgan Kaufmann, 1996). Though it's big (700 pages), it covers the whole language quite systematically and includes a very helpful and concise appendix on the complete language syntax of both VHDL-87 and VHDL-93. In this text we use a subset of the language that's compatible with both standards.

A somewhat shorter VHDL introduction is *VHDL for Designers* by Stefan Sjoholm and Lennart Lindh (Prentice Hall, 1997). This book is recommended for its nice emphasis on practical design issues including synthesis and test benches. Another practical book is *VHDL for Programmable Logic*, by Kevin Skahill of Cypress Semiconductor (Addison-Wesley, 1996).

All of the ABEL and VHDL examples in this chapter and throughout the text were compiled and in most cases simulated using Foundation 1.5 Student Edition software from Xilinx, Inc. (San Jose, CA 95124, www.xilinx.com). The Foundation software integrates a schematic editor, HDL editor, compilers for ABEL, VHDL and Verilog, and a simulator from Aldec, Inc. (Henderson, NV 89014, www.aldec.com) along with Xilinx' own specialized tools for CPLD and FPGA design and programming. It also contains an excellent on-line help system, including reference manuals for both ABEL and VHDL. The Foundation software is packaged on CD-ROM with some editions of this book.

The IEEE standard VHDL packages are important elements of any VHDL design environment. A listing of their type and function definitions is included as an appendix in some VHDL texts, but if you're really curious about them, you can find the complete source files in the library subsystem of any VHDL design system, including Foundation software.

We briefly discussed device testing in the context of ABEL test vectors. There is a large, well-established body of literature on digital device testing, and a good starting point for study is McCluskey's 1986 book. Generating a set of test vectors that completely tests a large circuit such as a PLD is a task best left to a program. At least one company's entire business is focused on programs that automatically create test vectors for PLD testing (ACUGEN Software, Inc., Nashua, NH 03063, www.acugen.com).

Drill Problems

4.1   Using variables NERD, DESIGNER, FAILURE, and STUDIED, write a boolean expression that is 1 for successful designers who never studied and for nerds who studied all the time.

4.2   Prove theorems T2–T5 using perfect induction.

4.3   Prove theorems T1′–T3′ and T5′ using perfect induction.

4.4   Prove theorems T6–T9 using perfect induction.

4.5   According to DeMorgan's theorem, the complement of $X + Y \cdot Z$ is $X' \cdot Y' + Z'$. Yet both functions are 1 for $XYZ = 110$. How can both a function and its complement be 1 for the same input combination? What's wrong here?

4.6   Use the theorems of switching algebra to simplify each of the following logic functions:

(a) $F = W \cdot X \cdot Y \cdot Z \cdot (W \cdot X \cdot Y \cdot Z' + W \cdot X' \cdot Y \cdot Z + W' \cdot X \cdot Y \cdot Z + W \cdot X \cdot Y' \cdot Z)$

(b) $F = A \cdot B + A \cdot B \cdot C' \cdot D + A \cdot B \cdot D \cdot E' + A \cdot B \cdot C' \cdot E + C' \cdot D \cdot E$

(c) $F = M \cdot N \cdot O + Q' \cdot P' \cdot N' + P \cdot R \cdot M + Q' \cdot O \cdot M \cdot P' + M \cdot R$

4.7   Write the truth table for each of the following logic functions:

(a) $F = X' \cdot Y + X' \cdot Y' \cdot Z$ (b) $F = W' \cdot X + Y' \cdot Z' + X' \cdot Z$

(c) $F = W + X' \cdot (Y' + Z)$ (d) $F = A \cdot B + B' \cdot C + C' \cdot D + D' \cdot A$

(e) $F = V \cdot W + X' \cdot Y' \cdot Z$ (f) $F = (A' + B' + C \cdot D) \cdot (B + C' + D' \cdot E')$

(g) $F = (W \cdot X)' \cdot (Y' + Z')'$ (h) $F = (((A + B)' + C')' + D)'$

(i) $F = (A' + B + C) \cdot (A + B' + D') \cdot (B + C' + D') \cdot (A + B + C + D)$

4.8   Write the truth table for each of the following logic functions:

(a) $F = X' \cdot Y' \cdot Z' + X \cdot Y \cdot Z + X \cdot Y' \cdot Z$

(b) $F = M' \cdot N' + M \cdot P + N' \cdot P$

(c) $F = A \cdot B + A \cdot B' \cdot C' + A' \cdot B \cdot C$

(d) $F = A' \cdot B \cdot (C \cdot B \cdot A' + B \cdot C')$

(e) $F = X \cdot Y \cdot (X' \cdot Y \cdot Z + X \cdot Y' \cdot Z + X \cdot Y \cdot Z' + X' \cdot Y' \cdot Z)$

(f) $F = M \cdot N + M' \cdot N' \cdot P'$

(g) $F = (A + A') \cdot B + B \cdot A \cdot C' + C \cdot (A + B') \cdot (A' + B)$

(h) $F = X \cdot Y' + Y \cdot Z + Z' \cdot X$

4.9   Write the canonical sum and product for each of the following logic functions:

(a) $F = \Sigma_{X,Y}(1,2)$ (b) $F = \Pi_{A,B}(0,1,2)$

(c) $F = \Sigma_{A,B,C}(2,4,6,7)$ (d) $F = \Pi_{W,X,Y}(0,1,3,4,5)$

(e) $F = X + Y' \cdot Z'$ (f) $F = V' + (W' \cdot X)'$

4.10   Write the canonical sum and product for each of the following logic functions:

(a) $F = \Sigma_{X,Y,Z}(0,3)$ (b) $F = \Pi_{A,B,C}(1,2,4)$

(c) $F = \Sigma_{A,B,C,D}(1,2,5,6)$ (d) $F = \Pi_{M,N,P}(0,1,3,6,7)$

(e) $F = X' + Y \cdot Z' + Y \cdot Z$ (f) $F = A'B + B'C + A$

4.11   If the canonical sum for an n-input logic function is also a minimal sum, how many literals are in each product term of the sum? Might there be any other minimal sums in this case?

4.12   Give two reasons why the cost of inverters is not included in the definition of "minimal" for logic minimization.

4.13   Using Karnaugh maps, find a minimal sum-of-products expression for each of the following logic functions. Indicate the distinguished 1-cells in each map.

(a) $F = \Sigma_{X,Y,Z}(1,3,5,6,7)$ (b) $F = \Sigma_{W,X,Y,Z}(1,4,5,6,7,9,14,15)$

(c) $F = \Pi_{W,X,Y}(0,1,3,4,5)$ (d) $F = \Sigma_{W,X,Y,Z}(0,2,5,7,8,10,13,15)$

(e) $F = \Pi_{A,B,C,D}(1,7,9,13,15)$ (f) $F = \Sigma_{A,B,C,D}(1,4,5,7,12,14,15)$

4.14 Find a minimal product-of-sums expression for each function in Drill 4.13 using the method of Section 4.3.6.

4.15 Find a minimal product-of-sums expression for the function in each of the following figures and compare its cost with the previously found minimal sum-of-products expression: (a) Figure 4-27; (b) Figure 4-29; (c) Figure 4-33.

4.16 Using Karnaugh maps, find a minimal sum-of-products expression for each of the following logic functions. Indicate the distinguished 1-cells in each map.

(a) $F = \Sigma_{A,B,C}(0,1,2,4)$ (b) $F = \Sigma_{W,X,Y,Z}(1,4,5,6,11,12,13,14)$

(c) $F = \Pi_{A,B,C}(1,2,6,7)$ (d) $F = \Sigma_{W,X,Y,Z}(0,1,2,3,7,8,10,11,15)$

(e) $F = \Sigma_{W,X,Y,X}(1,2,4,7,8,11,13,14)$ (f) $F = \Pi_{A,B,C,D}(1,3,4,5,6,7,9,12,13,14)$

4.17 Find a minimal product-of-sums expression for each function in Drill 4.16 using the method of Section 4.3.6.

4.18 Find the complete sum for the logic functions in Drill 4.16(d) and (e).

4.19 Using Karnaugh maps, find a minimal sum-of-products expression for each of the following logic functions. Indicate the distinguished 1-cells in each map.

(a) $F = \Sigma_{W,X,Y,Z}(0,1,3,5,14) + d(8,15)$ (b) $F = \Sigma_{W,X,Y,Z}(0,1,2,8,11) + d(3,9,15)$

(c) $F = \Sigma_{A,B,C,D}(1,5,9,14,15) + d(11)$ (d) $F = \Sigma_{A,B,C,D}(1,5,6,7,9,13) + d(4,15)$

(e) $F = \Sigma_{W,X,Y,Z}(3,5,6,7,13) + d(1,2,4,12,15)$

4.20 Repeat Drill 4.19, finding a minimal product-of-sums expression for each logic function.

4.21 For each logic function in the two preceding exercises, determine whether the minimal sum-of-products expression equals the minimal product-of-sums expression. Also compare the circuit cost for realizing each of the two expressions.

4.22 For each of the following logic expressions, find all of the static hazards in the corresponding two-level AND-OR or OR-AND circuit, and design a hazard-free circuit that realizes the same logic function.

(a) $F = W \cdot X + W'Y'$ (b) $F = W \cdot X' \cdot Y' + X \cdot Y' \cdot Z + X \cdot Y$

(c) $F = W' \cdot Y + X' \cdot Y' + W \cdot X \cdot Z$ (d) $F = W' \cdot X + Y' \cdot Z + W \cdot X \cdot Y \cdot Z + W \cdot X' \cdot Y \cdot Z'$

(e) $F = (W + X + Y) \cdot (X' + Z')$ (f) $F = (W + Y' + Z') \cdot (W' + X' + Z') \cdot (X' + Y + Z)$

(g) $F = (W + Y + Z') \cdot (W + X' + Y + Z) \cdot (X' + Y') \cdot (X + Z)$

4.23 Write s set of ABEL test vectors for prime-number detector in Table 4-17.

4.24 Write a structural VHDL program (entity and architecture) for the alarm circuit in Figure 4-19.

4.25 Repeat Drill 4.24, using the dataflow style of description.

4.26 Repeat Drill 4.24, using a process and a behavioral description style.

4.27 Write a srtructural VHDL program corresponding to the NAND-gate based logic circuit in Figure 5-17.

Exercises

4.28 Design a non-trivial-looking logic circuit that contains a feedback loop but has an output that depends only on its current input.

4.29 Prove the combining theorem T10 without using perfect induction, but assuming that theorems T1–T9 and T1′–T9′ are true.

4.30 Show that the combining theorem T10 is just a special case of consensus (T11) used with covering (T9).

4.31 Prove that $(X + Y') \cdot Y = X \cdot Y$ *without* using perfect induction. You may assume that theorems T1–T11 and T1′–T11′ are true.

4.32 Prove that $(X + Y) \cdot (X' + Z) = X \cdot Z + X' \cdot Y$ *without* using perfect induction. You may assume that theorems T1–T11 and T1′–T11′ are true.

4.33 Show that an n-input AND gate can be replaced by $(n-1)$ 2-input AND gates. Can the same statement be made for NAND gates? Justify your answer.

4.34 How many physically different ways are there to realize $V \cdot W \cdot X \cdot Y \cdot Z$ using four 2-input AND gates (4/4 of a 74x08)? Justify your answer.

4.35 Use switching algebra to prove that tying together two inputs of an $(n+1)$-input AND or OR gate gives it the functionality of an n-input gate.

4.36 Prove DeMorgan's theorems (T13 and T13′) using finite induction.

4.37 Which logic symbol more closely approximates the internal realization of a TTL NOR gate, Figure 4-4(c) or (d)? Why?

4.38 Use the theorems of switching algebra to rewrite the following expression using as few inversions as possible (complemented parentheses are allowed):

$$B' \cdot C + A \cdot C \cdot D' + A' \cdot C + E \cdot B' + E \cdot (A + C) \cdot (A' + D')$$

4.39 Prove or disprove the following propositions:
   (a) Let A and B be switching-algebra *variables*. Then $A \cdot B = 0$ and $A + B = 1$ implies that $A = B'$.
   (b) Let X and Y be switching-algebra *expressions*. Then $X \cdot Y = 0$ and $X + Y = 1$ implies that $X = Y'$.

4.40 Prove Shannon's expansion theorems. (*Hint:* Don't get carried away; it's easy.)

generalized Shannon-expansion theorems

4.41 Shannon's expansion theorems can be generalized to "pull out" not just one but i variables so that a logic function can be expressed as a sum or product of 2^i terms. State the generalized Shannon expansion theorems.

4.42 Show how the generalized Shannon expansion theorems lead to the canonical sum and canonical product representations of logic functions.

Exclusive OR (XOR) gate

4.43 An *Exclusive OR (XOR) gate* is a 2-input gate whose output is 1 if and only if exactly one of its inputs is 1. Write a truth table, sum-of-products expression, and corresponding AND-OR circuit for the Exclusive OR function.

4.44 From the point of view of switching algebra, what is the function of a 2-input XOR gate whose inputs are tied together? How might the output behavior of a real XOR gate differ?

4.45 After completing the design and fabrication of an SSI-based digital system, a designer finds that one more inverter is required. However, the only spare gates in the system are a 3-input OR, a 2-input AND, and a 2-input XOR. How should the designer realize the inverter function without adding another IC?

4.46 Any set of logic-gate types that can realize any logic function is called a *complete set* of logic gates. For example, 2-input AND gates, 2-input OR gates, and inverters are a complete set, because any logic function can be expressed as a sum of products of variables and their complements, and AND and OR gates with any number of inputs can be made from 2-input gates. Do 2-input NAND gates form a complete set of logic gates? Prove your answer.

complete set

4.47 Do 2-input NOR gates form a complete set of logic gates? Prove your answer.

4.48 Do 2-input XOR gates form a complete set of logic gates? Prove your answer.

4.49 Define a two-input gate, other than NAND, NOR, or XOR, that forms a complete set of logic gates if the constant inputs 0 and 1 are allowed. Prove your answer.

4.50 Some people think that there are *four* basic logic functions, AND, OR, NOT, and *BUT*. Figure X4.50 is a possible symbol for a 4-input, 2-output *BUT* gate. Invent a useful, nontrivial function for the BUT gate to perform. The function should have something to do with the name (BUT). Keep in mind that, due to the symmetry of the symbol, the function should be symmetric with respect to the A and B inputs of each section and with respect to sections 1 and 2. Describe your BUT's function and write its truth table.

BUT
BUT gate

Figure X4.50

4.51 Write logic expressions for the Z1 and Z2 outputs of the BUT gate you designed in the preceding exercise, and draw a corresponding logic diagram using AND gates, OR gates, and inverters.

4.52 Most students have no problem using theorem T8 to "multiply out" logic expressions, but many develop a mental block if they try to use theorem T8′ to "add out" a logic expression. How can duality be used to overcome this problem?

4.53 How many different logic functions are there of n variables?

4.54 How many different 2-variable logic functions $F(X,Y)$ are there? Write a simplified algebraic expression for each of them.

4.55 A *self-dual logic function* is a function F such that $F = F^D$. Which of the following functions are self-dual? (The symbol $\oplus$ denotes the Exclusive OR (XOR) operation.)

self-dual logic function
$\oplus$

(a) $F = X$

(b) $F = \Sigma_{X,Y,Z}(0,3,5,6)$

(c) $F = X \cdot Y' + X' \cdot Y$

(d) $F = W \cdot (X \oplus Y \oplus Z) + W' \cdot (X \oplus Y \oplus Z)'$

(e) A function F of 7 variables such that $F = 1$ if and only if 4 or more of the variables are 1

(f) A function F of 10 variables such that $F = 1$ if and only if 5 or more of the variables are 1

4.56 How many self-dual logic functions of n input variables are there? (*Hint:* Consider the structure of the truth table of a self-dual function.)

4.57   Prove that any n-input logic function $F(X_1, \ldots, X_n)$ that can be written in the form $F = X_1 \cdot G(X_2, \ldots, X_n) + X_1' \cdot G^D(X_2, \ldots, X_n)$ is self-dual.

4.58   Assuming that an inverting gate has a propagation delay of 5 ns, and that a non-inverting gate has a propagation delay of 8 ns, compare the speeds of the circuits in Figure 4-24(a), (c), and (d).

4.59   Find the minimal product-of-sums expressions for the logic functions in Figures 4-27 and 4-29.

4.60   Use switching algebra to show that the logic functions obtained in Exercise 4.59 equal the AND-OR functions obtained in Figures 4-27 and 4-29.

4.61   Determine whether the product-of-sums expressions obtained by "adding out" the minimal sums in Figure 4-27 and 4-29 are minimal.

4.62   Prove that the rule for combining 2^i 1-cells in a Karnaugh map is true, using the axioms and theorems of switching algebra.

irredundant sum   4.63   An *irredundant sum* for a logic function F is a sum of prime implicants for F such that if any prime implicant is deleted, the sum no longer equals F. This sounds a lot like a minimal sum, but an irredundant sum is not necessarily minimal. For example, the minimal sum of the function in Figure 4-35 has only three product terms, but there is an irredundant sum with four product terms. Find the irredundant sum and draw a map of the function, circling only the prime implicants in the irredundant sum.

4.64   Find another logic function in Section 4.3 that has one or more nonminimal irredundant sums, and draw its map, circling only the prime implicants in the irredundant sum.

4.65   Draw a Karnaugh map and assign variables to the inputs of the AND-XOR circuit in Figure X4.65 so that its output is $F = \Sigma_{W,X,Y,Z}(6,7,12,13)$. Note that the output gate is a 2-input XOR rather than an OR.

Figure X4.65

4.66   A 3-bit "comparator" circuit receives two 3-bit numbers, $P = P_2 P_1 P_0$ and $Q = Q_2 Q_1 Q_0$. Design a minimal sum-of-products circuit that produces a 1 output if and only if $P > Q$.

4.67   Find minimal multiple-output sum-of-products expressions for $F = \Sigma_{X,Y,Z}(0,1,2)$, $G = \Sigma_{X,Y,Z}(1,4,6)$, and $H = \Sigma_{X,Y,Z}(0,1,2,4,6)$.

4.68   Prove whether or not the following expression is a minimal sum. Do it the easiest way possible (algebraically, not using maps).

$$F = T' \cdot U \cdot V \cdot W \cdot X + T' \cdot U \cdot V' \cdot X \cdot Z + T' \cdot U \cdot W \cdot X \cdot Y' \cdot Z$$

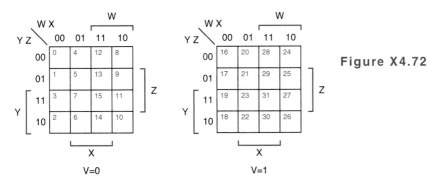

Figure X4.72

4.69 The text states that a truth table or equivalent is the starting point for traditional combinational minimization methods. A Karnaugh map itself contains the same information as a truth table. Given a sum-of-products expression, it is possible to write the 1s corresponding to each product term directly on the map without developing an explicit truth table or minterm list, and then proceed with the map-minimization procedure. In this way, find a minimal sum-of-products expression for each of the following logic functions:

(a) $F = X' \cdot Z + X \cdot Y + X \cdot Y' \cdot Z$

(b) $F = A' \cdot C' \cdot D + B' \cdot C \cdot D + A \cdot C' \cdot D + B \cdot C \cdot D$

(c) $F = W \cdot X \cdot Z' + W \cdot X' \cdot Y \cdot Z + X \cdot Z$

(d) $F = (X' + Y') \cdot (W' + X' + Y) \cdot (W' + X + Z)$

(e) $F = A \cdot B \cdot C' \cdot D' + A' \cdot B \cdot C' + A \cdot B \cdot D + A' \cdot C \cdot D + B \cdot C \cdot D'$

4.70 Repeat Exercise 4.69, finding a minimal product-of-sums expression for each logic function.

4.71 Derive the minimal product-of-sums expression for the prime BCD-digit-detector function of Figure 4-37. Determine whether or not the expression algebraically equals the minimal sum-of-products expression and explain your result.

4.72 A Karnaugh map for a 5-variable function can be drawn as shown in Figure X4.72. In such a map, cells that occupy the same relative position in the $V = 0$ and $V = 1$ submaps are considered to be adjacent. (Many worked examples of 5-variable Karnaugh maps appear in Sections 7.4.4 and 7.4.5.) Find a minimal sum-of-products expression for each of the following functions using a 5-variable map: *5-variable Karnaugh map*

(a) $F = \Sigma_{V,W,X,Y,Z}(5,7,13,15,16,20,25,27,29,31)$

(b) $F = \Sigma_{V,W,X,Y,Z}(0,7,8,9,12,13,15,16,22,23,30,31)$

(c) $F = \Sigma_{V,W,X,Y,Z}(0,1,2,3,4,5,10,11,14,20,21,24,25,26,27,28,29,30)$

(d) $F = \Sigma_{V,W,X,Y,Z}(0,2,4,6,7,8,10,11,12,13,14,16,18,19,29,30)$

(e) $F = \Pi_{V,W,X,Y,Z}(4,5,10,12,13,16,17,21,25,26,27,29)$

(f) $F = \Sigma_{V,W,X,Y,Z}(4,6,7,9,11,12,13,14,15,20,22,25,27,28,30) + d(1,5,29,31)$

Figure X4.74

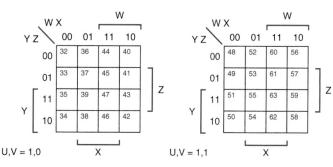

4.73   Repeat Exercise 4.72, finding a minimal product-of-sums expression for each logic function.

6-variable Karnaugh map

4.74   A Karnaugh map for a 6-variable function can be drawn as shown in Figure X4.74. In such a map, cells that occupy the same relative position in adjacent submaps are considered to be adjacent. Minimize the following functions using 6-variable maps:

(a) $F = \Sigma_{U,V,W,X,Y,Z}(1,5,9,13,21,23,29,31,37,45,53,61)$

(b) $F = \Sigma_{U,V,W,X,Y,Z}(0,4,8,16,24,32,34,36,37,39,40,48,50,56)$

(c) $F = \Sigma_{U,V,W,X,Y,Z}(2,4,5,6,12\text{–}21,28\text{–}31,34,38,50,51,60\text{–}63)$

4.75   There are $2n$ m-subcubes of an n-cube for the value $m = n - 1$. Show their text representations and the corresponding product terms. (You may use ellipses as required, e.g., 1, 2, …, n.)

4.76   There is just one m-subcube of an n-cube for the value $m = n$; its text representation is xx…xx. Write the product term corresponding to this cube.

4.77   The C program in Table 4-9 uses memory inefficiently because it allocates memory for a maximum number of cubes at each level, even if this maximum is never used. Redesign the program so that the cubes and used arrays are one-dimensional arrays, and each level uses only as many array entries as needed. (*Hint:* You can still allocate cubes sequentially, but keep track of the starting point in the array for each level.)

4.78   As a function of *m*, how many times is each distinct *m*-cube rediscovered in Table 4-9, only to be found in the inner loop and thrown away? Suggest some ways to eliminate this inefficiency.

4.79   The third for-loop in Table 4-9 tries to combine all *m*-cubes at a given level with all other *m*-cubes at that level. In fact, only *m*-cubes with x's in the same positions can be combined, so it is possible to reduce the number of loop iterations by using a more sophisticated data structure. Design a data structure that segregates the cubes at a given level according to the position of their x's, and determine the maximum size required for various elements of the data structure. Rewrite Table 4-9 accordingly.

4.80   Estimate whether the savings in inner-loop iterations achieved in Exercise 4.80 outweighs the overhead of maintaining a more complex data structure. Try to make reasonable assumptions about how cubes are distributed at each level, and indicate how your results are affected by these assumptions.

4.81   Optimize the Oneone function in Table 4-8. An obvious optimization is to drop out of the loop early, but other optimizations exist that eliminate the for loop entirely. One is based on table look-up and another uses a tricky computation involving complementing, Exclusive ORing, and addition.

4.82   Extend the C program in Table 4-9 to handle don't-care conditions. Provide another data structure, dc[MAX_VARS+1][MAX_CUBES], that indicates whether a given cube contains only don't-cares, and update it as cubes are read and generated.

4.83   (*Hamlet circuit.*) Complete the timing diagram and explain the function of the circuit in Figure X4.83. Where does the circuit get its name?

Figure X4.83

4.84   Prove that a two-level AND-OR circuit corresponding to the complete sum of a logic function is always hazard free.

4.85   Find a four-variable logic function whose minimal sum-of-products realization is not hazard free, but for which there exists a hazard-free sum-of-products realization with fewer product terms than the complete sum.

4.86   Starting with the WHEN statements in the ABEL program in Table 4-14, work out the logic equations for variables X4 through X10 in the program. Explain any discrepancies between your results and the equations in Table 4-15.

4.87   Draw a circuit diagram corresponding to the minimal two-level sum-of-products equations for the alarm circuit, as given in Table 4-12. On each inverter, AND gate, and OR gate input and output, write a pair of numbers $(t0, t1)$, where $t0$ is the test number from Table 4-25 that detects a stuck-at-0 fault on that line, and $t1$ is the test number that detects a stuck-at-1 fault.

4.88   Write a dataflow-style VHDL program (entity and architecture) corresponding to the full-adder circuit in Figure 5-86.

4.89   Using the entity that you designed in Exercise 4.88, write a structural VHDL program for a 4-bit ripple adder using the structure of Figure 5-87.

4.90   Using the entity that you defined in Exercise 4.88, write a structural VHDL program for a 16-bit ripple adder along the lines of Figure 5-87. Use a generate statement to create the 16 full adders and their signal connection.

4.91   Rewrite the prime-number-detector architecture of Table 4-63 using a while statement.

chapter 5

Combinational Logic Design Practices

T he preceding chapter described the theoretical principles used in combinational logic design. In this chapter we'll build on that foundation and describe many of the devices, structures, and methods used by engineers to solve practical digital design problems.

A practical combinational circuit may have dozens of inputs and outputs and could require hundreds, thousands, even millions of terms to describe as a sum of products, and *billions and billions* of rows to describe in a truth table. Thus, most real combinational logic design problems are too large to solve by "brute-force" application of theoretical techniques.

But wait, you say, how could any human being conceive of such a complex logic circuit in the first place? The key is structured thinking. A complex circuit or system is conceived as a collection of smaller subsystems, each of which has a much simpler description.

In combinational logic design, there are several straightforward structures—decoders, multiplexers, comparators, and the like—that turn up quite regularly as building blocks in larger systems. The most important of these structures are described in this chapter. We describe each structure generally and then we give examples and applications using 74-series components, ABEL, and VHDL.

Before launching into these combinational building blocks, we need to discuss several important topics. The first topic is documentation standards

THE IMPORTANCE OF 74-SERIES LOGIC	Later in this chapter we'll look at commonly used 74-series ICs that perform well-structured logic functions. These parts are important building blocks in a digital designer's toolbox, because their level of functionality often matches a designer's level of thinking when partitioning a large problem into smaller chunks. Even when you design for PLDs, FPGAs, or ASICs, understanding 74-series MSI functions is important. In PLD-based design, standard MSI functions can be used as a starting point for developing logic equations for more specialized functions. And in FPGA and ASIC design, the basic building blocks (or "standard cells" or "macros") provided by the FPGA or ASIC manufacturer may actually be defined as 74-series MSI functions, even to the extent of having similar descriptive numbers.

that are used by digital designers to ensure that their designs are correct, manufacturable, and maintainable. Next we discuss circuit timing, a crucial element for successful digital design. Third, we describe the internal structure of combinational PLDs, which we use later as "universal" building blocks.

5.1 Documentation Standards

Good documentation is essential for correct design and efficient maintenance of digital systems. In addition to being accurate and complete, documentation must be somewhat instructive, so that a test engineer, maintenance technician, or even the original design engineer (six months after designing the circuit) can figure out how the system works just by reading the documentation.

Although the type of documentation depends on system complexity and the engineering and manufacturing environments, a documentation package should generally contain at least the following six items:

circuit specification
1. A *specification* describes exactly what the circuit or system is supposed to do, including a description of all inputs and outputs ("interfaces") and the functions that are to be performed. Note that the "spec" doesn't have to specify *how* the system achieves its results, just *what* the results are supposed to be. However, in many companies it is common practice also to incorporate one or more of the documents below into the spec to describe how the system works at the same time.

block diagram
2. A *block diagram* is an informal pictorial description of the system's major functional modules and their basic interconnections.

schematic diagram
3. A *schematic diagram* is a formal specification of the electrical components of the system, their interconnections, and all of the details needed to construct the system, including IC types, reference designators, and pin numbers. We've been using the term *logic diagram* for an informal drawing that does not have quite this level of detail. Most schematic

logic diagram

drawing programs have the ability to generate a *bill of materials (BOM)* from the schematic; this tells the purchasing department what electrical components they have to order to build the system. *bill of materials (BOM)*

4. A *timing diagram* shows the values of various logic signals as a function of time, including the cause-and-effect delays between critical signals. *timing diagram*

5. A *structured logic device description* describes the internal function of a programmable logic device (PLD), field-programmable gate array (FPGA), or application-specific integrated circuit (ASIC). It is normally written in a hardware description language (HDL) such as ABEL or VHDL, but it may be in the form of logic equations, state tables, or state diagrams. In some cases, a conventional programming language such as C may be used to model the operation of a circuit or to specify its behavior. *structured logic device description*

6. A *circuit description* is a narrative text document that, in conjunction with the other documentation, explains how the circuit works internally. The circuit description should list any assumptions and potential pitfalls in the circuit's design and operation, and point out the use of any nonobvious design "tricks." A good circuit description also contains definitions of acronyms and other specialized terms and has references to related documents. *circuit description*

You've probably already seen block diagrams in many contexts. We present a few rules for drawing them in the next subsection, and then in the rest of this section we concentrate on schematics for combinational logic circuits. Section 5.2.1 introduces timing diagrams. Structured logic descriptions in the form of ABEL and VHDL programs were covered in Sections 4.6 and 4.7. In Section 10.1.6 we'll show how a C program can be used to generate the contents of a read-only memory.

The last area of documentation, the circuit description, is very important in practice. Just as an experienced programmer creates a program design document before beginning to write code, an experienced logic designer starts writing the circuit description before drawing the schematic. Unfortunately, the circuit description is sometimes the last document to be created, and sometimes it's never written at all. A circuit without a description is difficult to debug, manufacture, test, maintain, modify, and enhance.

**DON'T FORGET
TO WRITE!**   In order to create great products, logic designers must develop their language and writing skills, especially in the area of *logical* outlining and organization. The most successful logic designers (and later, project leaders, system architects, and entrepreneurs) are the ones who communicate their ideas, proposals, and decisions effectively to others. Even though it's a lot of fun to tinker in the digital design lab, don't use that as an excuse to shortchange your writing and communications courses and projects!

5.1.1 Block Diagrams

block diagram   A *block diagram* shows the inputs, outputs, functional modules, internal data paths, and important control signals of a system. In general, it should not be so detailed that it occupies more than one page, yet it must not be too vague. A small block diagram may have three to six blocks, while a large one may have 10 to 15, depending on system complexity. In any case, the block diagram must

Figure 5-1
Block diagram for a
digital design project.

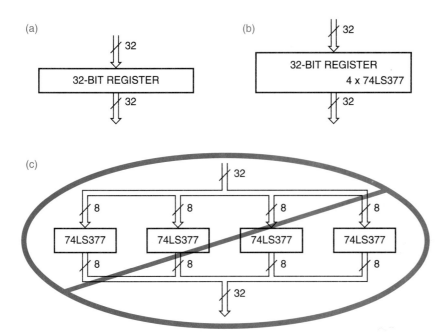

Figure 5-2
A 32-bit register block: (a) realization unspecified; (b) chips specified; (c) too much detail.

show the most important system elements and how they work together. Large systems may require additional block diagrams of individual subsystems, but there should always be a "top-level" diagram showing the entire system.

Figure 5-1 shows a sample block diagram. Each block is labeled with the function of the block, not the individual chips that comprise it. As another example, Figure 5-2(a) shows the block-diagram symbol for a 32-bit register. If the register is to be built using four 74LS377 8-bit registers, and this information is important to someone reading the diagram (e.g., for cost reasons), then it can be conveyed as shown in (b). However, splitting the block to show individual chips as in (c) is incorrect.

A *bus* is a collection of two or more related signal lines. In a block diagram, *bus* buses are drawn with a double or heavy line. A slash and a number may indicate how many individual signal lines are contained in a bus. Alternatively, size may be denoted in the bus name (e.g., INBUS[31..0] or INBUS[31:0]). Active levels (defined later) and inversion bubbles may or may not appear in block diagrams; in most cases, they are unimportant at this level of detail. However, important control signals and buses should have names, usually the same names that appear in the more detailed schematic.

The flow of control and data in a block diagram should be clearly indicated. Logic diagrams are generally drawn with signals flowing from left to right, but in block diagrams this ideal is more difficult to achieve. Inputs and outputs may be on any side of a block, and the direction of signal flow may be arbitrary. Arrowheads are used on buses and ordinary signal lines to eliminate ambiguity.

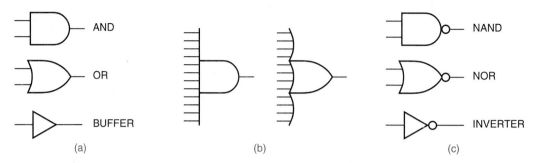

Figure 5-3 Shapes for basic logic gates: (a) AND, OR, and buffers; (b) expansion of inputs; (c) inversion bubbles.

5.1.2 Gate Symbols

The symbol shapes for AND and OR gates and buffers are shown in Figure 5-3(a). (Recall from Chapter 3 that a buffer is a circuit that simply converts "weak" logic signals into "strong" ones.) To draw logic gates with more than a few inputs, we expand the AND and OR symbols as shown in (b). A small *inversion bubble* circle, called an *inversion bubble*, denotes logical inversion or complementing and is used in the symbols for NAND and NOR gates and inverters in (c).

Using the generalized DeMorgan's theorem, we can manipulate the logic expressions for gates with complemented outputs. For example, if X and Y are the inputs of a NAND gate with output Z, then we can write

$$Z = (X \cdot Y)'$$
$$= X' + Y'$$

This gives rise to two different but equally correct symbols for a NAND gate, as we demonstrated in Figure 4-3 on page 201. In fact, this sort of manipulation may be applied to gates with uncomplemented inputs as well. For example, consider the following equations for an AND gate:

$$Z = X \cdot Y$$
$$= ((X \cdot Y)')'$$
$$= (X' + Y')'$$

IEEE STANDARD LOGIC SYMBOLS

Together with the American National Standards Institute (ANSI), the Institute of Electrical and Electronic Engineers (IEEE) has developed a standard set of logic symbols. The most recent revision of the standard is ANSI/IEEE Std 91-1984, *IEEE Standard Graphic Symbols for Logic Functions*, and it allows both rectangular- and distinctive-shape symbols for logic gates. We have been using and will continue to use the distinctive-shape symbols in this book, but the rectangular-shape symbols are described in a guide to IEEE symbols on the web at www.ddpp.com.

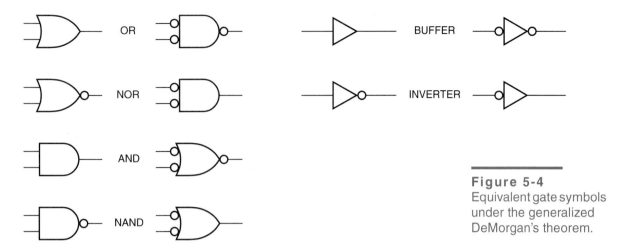

Figure 5-4
Equivalent gate symbols under the generalized DeMorgan's theorem.

Thus, an AND gate may be symbolized as an OR gate with inversion bubbles on its inputs and output.

Equivalent symbols for standard gates that can be obtained by these manipulations are summarized in Figure 5-4. Even though both symbols in a pair represent the same logic function, the choice of one symbol or the other in a logic diagram is not arbitrary, at least not if we are adhering to good documentation standards. As we'll show in the next three subsections, proper choices of signal names and gate symbols can make logic diagrams much easier to use and understand.

5.1.3 Signal Names and Active Levels

Each input and output signal in a logic circuit should have a descriptive alphanumeric label, the signal's name. Most computer-aided design systems for drawing logic circuits also allow certain special characters, such as *, _, and !, to be included in signal names. In the analysis and synthesis examples in Chapter 4, we used mostly single-character signal names (X, Y, etc.) because the circuits didn't do much. However, in a real system, well-chosen signal names convey information to someone reading the logic diagram the same way that variable names in a software program do. A signal's name indicates an action that is controlled (GO, PAUSE), a condition that it detects (READY, ERROR), or data that it carries (INBUS[31:0]).

Each signal name should have an *active level* associated with it. A signal is *active high* if it performs the named action or denotes the named condition when it is HIGH or 1. (Under the positive-logic convention, which we use throughout this book, "HIGH" and "1" are equivalent.) A signal is *active low* if it performs the named action or denotes the named condition when it is LOW or 0. A signal is said to be *asserted* when it is at its active level. A signal is said to be *negated* (or, sometimes, *deasserted*) when it is not at its active level.

active level
active high

active low
assert
negate
deassert

Table 5-1
Each line shows a different naming convention for active levels.

Active Low	Active High
READY–	READY+
ERROR.L	ERROR.H
ADDR15(L)	ADDR15(H)
RESET*	RESET
ENABLE~	ENABLE
~GO	GO
/RECEIVE	RECEIVE
TRANSMIT_L	TRANSMIT

active-level naming convention

_L suffix

signal name
logic expression

logic equation

The active level of each signal in a circuit is normally specified as part of its name, according to some convention. Examples of several different *active-level naming conventions* are shown in Table 5-1. The choice of one of these or other signal-naming conventions is sometimes just a matter of personal preference, but more often it is constrained by the engineering environment. Since the active-level designation is part of the signal name, the naming convention must be compatible with the input requirements of any computer-aided design tools that will process the signal names, such as schematic editors, HDL compilers, and simulators. In this text, we'll use the last convention in the table: An active-low signal name has a suffix of _L, and an active-high signal has no suffix. The _L suffix may be read as if it were a prefix "not."

It's extremely important for you to understand the difference between signal names, expressions, and equations. A *signal name* is just a name—an alphanumeric label. A *logic expression* combines signal names using the operators of switching algebra—AND, OR, and NOT—as we explained and used throughout Chapter 4. A *logic equation* is an assignment of a logic expression to a signal name—it describes one signal's function in terms of other signals.

The distinction between signal names and logic expressions can be related to a concept used in computer programming languages: The lefthand side of an assignment statement contains a variable *name*, and the righthand side contains an *expression* whose value will be given to the named variable (e.g., $Z = -(X+Y)$ in C). In a programming language, you can't put an expression on the lefthand side of an assignment statement. In logic design, you can't use a logic expression as a signal name.

Logic signals may have names like X, READY, and GO_L. The "_L" in GO_L is just part of the signal's name, like an underscore in a variable name in a C program. There is *no* signal whose name is READY′—this is an expression, since ′ is an operator. However, there may be two signals named READY and READY_L such that READY_L = READY′ during normal operation of the circuit. We are very careful in this book to distinguish between signal names, which are always printed in black, and logic expressions, which are always printed in color when they are written near the corresponding signal lines.

Figure 5-5 Logic symbols: (a) AND, OR, and a larger-scale logic element;
(b) the same elements with active-low inputs and outputs.

5.1.4 Active Levels for Pins

When we draw the outline of an **AND** or **OR** symbol, or a rectangle representing a larger-scale logic element, we think of the given logic function as occurring *inside* that symbolic outline. In Figure 5-5(a), we show the logic symbols for an **AND** and **OR** gate and for a larger-scale element with an **ENABLE** input. The **AND** and **OR** gates have active-high inputs—they require 1s on the input to assert their outputs. Likewise, the larger-scale element has an active-high **ENABLE** input, which must be 1 to enable the element to do its thing. In (b), we show the same logic elements with active-low input and output pins. Exactly the same logic functions are performed *inside* the symbolic outlines, but the inversion bubbles indicate that 0s must now be applied to the input pins to activate the logic functions, and that the outputs are 0 when they are "doing their thing."

Thus, active levels may be associated with the input and output pins of gates and larger-scale logic elements. We use an inversion bubble to indicate an active-low pin and the absence of a bubble to indicate an active-high pin. For example, the **AND** gate in Figure 5-6(a) performs the logical **AND** of two active-high inputs and produces an active-high output: if both inputs are asserted (1), the output is asserted (1). The **NAND** gate in (b) also performs the **AND** function, but it produces an active-low output. Even a **NOR** or **OR** gate can be construed to perform the **AND** function using active-low inputs and outputs, as shown in (c) and (d). All four gates in the figure can be said to perform the same function: the output of each gate is asserted if both of its inputs are asserted. Figure 5-7 shows the same idea for the **OR** function: The output of each gate is asserted if either of its inputs is asserted.

Figure 5-6 Four ways of obtaining an AND function: (a) AND gate (74x08);
(b) NAND gate (74x00); (c) NOR gate (74x02); (d) OR gate (74x32).

(a)   (b)   (c)   (d)

Figure 5-7 Four ways of obtaining an OR function: (a) OR gate (74x32); (b) NOR gate (74x02); (c) NAND gate (74x00); (d) AND gate

Figure 5-8 Alternate logic symbols: (a, b) inverters; (c, d) noninverting buffers.

Sometimes a noninverting buffer is used simply to boost the fanout of a logic signal without changing its function. Figure 5-8 shows the possible logic symbols for both inverters and noninverting buffers. In terms of active levels, all of the symbols perform exactly the same function: Each asserts its output signal if and only if its input is asserted.

5.1.5 Bubble-to-Bubble Logic Design

Experienced logic circuit designers formulate their circuits in terms of the logic functions performed *inside* the symbolic outlines. Whether you're designing with discrete gates or in an HDL like ABEL or VHDL, it's easiest to think of logic signals and their interactions using active-high names. However, once you're ready to realize your circuit, you may have to deal with active-low signals due to the requirements of the environment.

When you design with discrete gates, either at board or ASIC level, a key requirement is often speed. As we showed in Section 3.3.6, inverting gates are typically faster than noninverting ones, so there's often a significant performance payoff in carrying some signals in active-low form.

When you design with larger-scale elements, many of them may be off-the-shelf chips or other existing components that already have some inputs and outputs fixed in active-low form. The reasons that they use active-low signals may range from performance improvement to years of tradition, but in any case, you still have to deal with it.

NAME THAT SIGNAL!	Although it is absolutely necessary to name only a circuit's main inputs and outputs, most logic designers find it useful to name internal signals as well. During circuit debugging, it's nice to have a name to use when pointing to an internal signal that's behaving strangely. Most computer-aided design systems automatically generate labels for unnamed signals, but a user-chosen name is preferable to a computer-generated one like XSIG1057.

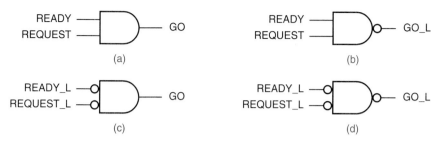

Figure 5-9 Many ways to GO: (a) active-high inputs and output;
(b) active-high inputs, active-low output; (c) active-low
inputs, active-high output; (d) active-low inputs and outputs.

Bubble-to-bubble logic design is the practice of choosing logic symbols *bubble-to-bubble logic*
and signal names, including active-level designators, that make the function of a *design*
logic circuit easier to understand. Usually, this means choosing signal names and
gate types and symbols so that most of the inversion bubbles "cancel out" and
the logic diagram can be analyzed as if all of the signals were active high.

For example, suppose we need to produce a signal that tells a device to
"GO" when we are "READY" and we get a "REQUEST." Clearly from the problem statement, an AND function is required; in switching algebra, we would
write GO = READY · REQUEST. However, we can use different gates to perform
the AND function, depending on the active level required for the GO signal and
the active levels of the available input signals.

Figure 5-9(a) shows the simplest case, where GO must be active-high and
the available input signals are also active-high; we use an AND gate. If, on the
other hand, the device that we're controlling requires an active-low GO_L signal,
we can use a NAND gate as shown in (b). If the available input signals are active-
low, we can use a NOR or OR gate as shown in (c) and (d).

The active levels of available signals don't always match the active levels
of available gates. For example, suppose we are given input signals READY_L
(active-low) and REQUEST (active-high). Figure 5-10 shows two different ways
to generate GO using an inverter to generate the active level needed for the AND
function. The second way is generally preferred, since inverting gates like NOR
are generally faster than noninverting ones like AND. We drew the inverter
differently in each case to make the output's active level match its signal name.

Figure 5-10 Two more ways to GO, with mixed input levels: (a) with an AND
gate; (b) with a NOR gate.

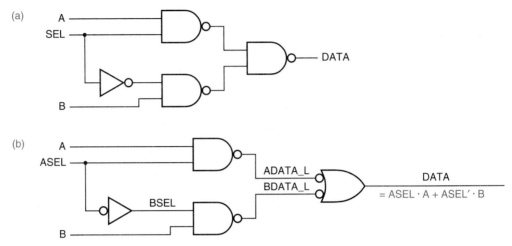

Figure 5-11 A 2-input multiplexer (you're not supposed to know what that is yet): (a) cryptic logic diagram; (b) proper logic diagram using active-level designators and alternate logic symbols.

To understand the benefits of bubble-to-bubble logic design, consider the circuit in Figure 5-11(a). What does it do? In Section 4.2 we showed several ways to analyze such a circuit, and we could certainly obtain a logic expression for the DATA output using these techniques. However, when the circuit is redrawn in Figure 5-11(b), the output function can be read directly from the logic diagram, as follows. The DATA output is asserted when either ADATA_L or BDATA_L is asserted. If ASEL is asserted, then ADATA_L is asserted if and only if A is asserted; that is, ADATA_L is a copy of A. If ASEL is negated, BSEL is asserted and BDATA_L is a copy of B. In other words, DATA is a copy of A if ASEL is asserted, and DATA is a copy of B if ASEL is negated. Even though there are five inversion bubbles in the logic diagram, we mentally had to perform only one negation to understand the circuit—that BSEL is asserted if ASEL is not asserted.

If we wish, we can write an algebraic expression for the DATA output. We use the technique of Section 4.2, simply propagating expressions through gates toward the output. In doing so, we can ignore pairs of inversion bubbles that cancel, and directly write the expression shown in color in the figure.

Figure 5-12 Another properly drawn logic diagram.

BUBBLE-TO-BUBBLE LOGIC DESIGN RULES	The following rules are useful for performing bubble-to-bubble logic design:

- The signal name on a device's output should have the same active level as the device's output pin—that is, active-low if the device symbol has an inversion bubble on the output pin, active-high if not.

- If the active level of an input signal is the same as that of the input pin to which it is connected, then the logic function inside the symbolic outline is activated when the signal is asserted. This is the most common case in a logic diagram.

- If the active level of an input signal is the opposite of that of the input pin to which it is connected, then the logic function inside the symbolic outline is activated when the signal is negated. This case should be avoided whenever possible because it forces us to keep track mentally of a logical negation to understand the circuit.

Another example is shown in Figure 5-12. Reading directly from the logic diagram, we see that ENABLE_L is asserted if READY_L and REQUEST_L are asserted or if TEST is asserted. The HALT output is asserted if READY_L and REQUEST_L are not both asserted or if LOCK_L is asserted. Once again, this example has only one place where a gate input's active level does not match the input signal level, and this is reflected in the verbal description of the circuit.

We can, if we wish, write algebraic equations for the ENABLE_L and HALT outputs. As we propagate expressions through gates toward the output, we obtain expressions like $READY\_L' \cdot REQUEST'$. However, we can use our active-level naming convention to simplify terms like $READY\_L'$. The circuit contains no signal with the name READY; but if it did, it would satisfy the relationship $READY = READY\_L'$ according to the naming convention. This allows us to write the ENABLE_L and HALT equations as shown. Complementing both sides of the ENABLE_L equation, we obtain an equation that describes a hypothetical active-high ENABLE output in terms of hypothetical active-high inputs.

We'll see more examples of bubble-to-bubble logic design in this and later chapters, especially as we begin to use larger-scale logic elements.

5.1.6 Drawing Layout

Logic diagrams and schematics should be drawn with gates in their "normal" orientation with inputs on the left and outputs on the right. The logic symbols for larger-scale logic elements are also normally drawn with inputs on the left and outputs on the right.

A complete schematic page should be drawn with system inputs on the left and outputs on the right, and the general flow of signals should be from left to right. If an input or output appears in the middle of a page, it should be extended to the left or right edge, respectively. In this way, a reader can find all inputs and outputs by looking at the edges of the page only. All signal paths on the page

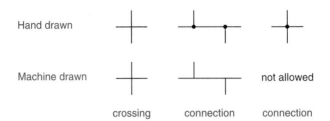

Figure 5-13
Line crossings and
connections.

should be connected when possible; paths may be broken if the drawing gets crowded, but breaks should be flagged in both directions, as described later.

Sometimes block diagrams are drawn without crossing lines for a neater appearance, but this is never done in logic diagrams. Instead, lines are allowed to cross and connections are indicated clearly with a dot. Still, some computer-aided design systems (and some designers) can't draw legible connection dots. To distinguish between crossing lines and connected lines, they adopt the convention that only "T"-type connections are allowed, as shown in Figure 5-13. This is a good convention to follow in any case.

Schematics that fit on a single page are the easiest to work with. The largest practical paper size for a schematic might be E-size (34"×44"). Although its drawing capacity is great, such a large paper size is unwieldy to work with. The best compromise of drawing capacity and practicality is B-size (11"×17"). It can be easily folded for storage and quick reference in standard 3-ring notebooks, and it can be copied on most office copiers. Regardless of paper size, schematics

Figure 5-14 Flat schematic structure.

come out best when the page is used in landscape format, that is, with its long dimension oriented from left to right, the direction of most signal flow.

Schematics that don't fit on a single page should be broken up into individual pages in a way that minimizes the connections (and confusion) between pages. They may also use a coordinate system, like that of a road map, to flag the sources and destinations of signals that travel from one page to another. An outgoing signal should have flags referring to all of the destinations of that signal, while an incoming signal should have a flag referring to the source only. That is, an incoming signal should be flagged to the place where it is generated, not to a place somewhere in the middle of a chain of destinations that use the signal.

signal flags

A multiple-page schematic usually has a "flat" structure. As shown in Figure 5-14, each page is carved out from the complete schematic and can connect to any other page as if all the pages were on one large sheet. However, much like programs, schematics can also be constructed hierarchically, as illustrated in Figure 5-15. In this approach, the "top-level" schematic is just a single page that may take the place of a block diagram. Typically, the top-level

flat schematic structure

hierarchical schematic structure

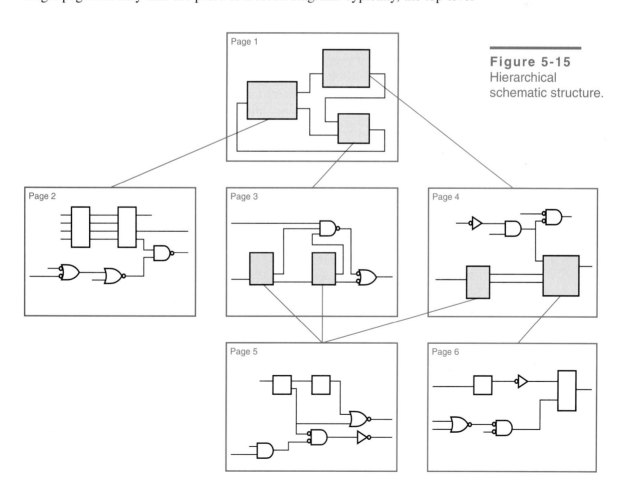

Figure 5-15
Hierarchical schematic structure.

schematic contains no gates or other logic elements; it only shows blocks corresponding to the major subsystems, and their interconnections. The blocks or subsystems are turn on lower-level pages, which may contain ordinary gate-level descriptions, or which may themselves use blocks defined in lower-level hierarchies. If a particular lower-level hierarchy needs to be used more than once, it may be reused (or "called," in the programming sense) multiple times by the higher-level pages.

Most computer-aided logic design systems support both flat and hierarchical schematics. Proper signal naming is very important in both styles, since there are a number of common errors that can occur:

- Like any other program, a schematic-entry program does what you say, not what you mean. If you use slightly different names for what you intend to be the same signal on different pages, they won't be connected.

- Conversely, if you inadvertently use the same name for different signals on different pages of a flat schematic, many programs will dutifully connect them together, even if you haven't connected them with an off-page flag. (In a hierarchical schematic, reusing a name at different places in the hierarchy is generally OK, because the program qualifies each name with its position in the hierarchy.)

- In a hierarchical schematic, you have to be careful in naming the external interface signals on pages in the lower levels of the hierarchy. These are the names that will appear inside the blocks corresponding to these pages when they are used at higher levels of the hierarchy. It's very easy to transpose signal names or use a name with the wrong active level, yielding incorrect results when the block is used.

- This is not usually a naming problem, but all schematic programs seem to have quirks in which signals that appear to be connected are not. Using the "T" convention in Figure 5-13 can help minimize this problem.

Fortunately, most schematic programs have error-checking facilities that can catch many of these errors, for example, by searching for signal names that have no inputs, no outputs, or multiple outputs associated with them. But most logic designers learn the importance of careful, manual schematic double-checking only through the bitter experience of building a printed-circuit board or an ASIC based on a schematic containing some dumb error.

5.1.7 Buses

As defined previously, a bus is a collection of two or more related signal lines. For example, a microprocessor system might have an address bus with 16 lines, ADDR0–ADDR15, and a data bus with 8 lines, DATA0–DATA7. The signal names in a bus are not necessarily related or ordered as in these first examples. For example, a microprocessor system might have a control bus containing five signals, ALE, MIO, RD_L, WR_L, and RDY.

Logic diagrams use special notation for buses in order to reduce the amount of drawing and to improve readability. As shown in Figure 5-16, a bus has its own descriptive name, such as ADDR[15:0], DATA[7:0], or CONTROL. A bus name may use brackets and a colon to denote a range. Buses are drawn with thicker lines than ordinary signals. Individual signals are put into or pulled out of the bus by connecting an ordinary signal line to the bus and writing the signal name. Often a special connection dot is also used, as in the example.

A computer-aided design system keeps track of the individual signals in a bus. When it actually comes time to build a circuit from the schematic, signal lines in a bus are treated just as though they had all been drawn individually.

The symbols at the righthand edge of Figure 5-16 are interpage signal flags. They indicate that LA goes out to page 2, DB is bidirectional and connects to page 2, and CONTROL is bidirectional and connects to pages 2 and 3.

Figure 5-16
Examples of buses.

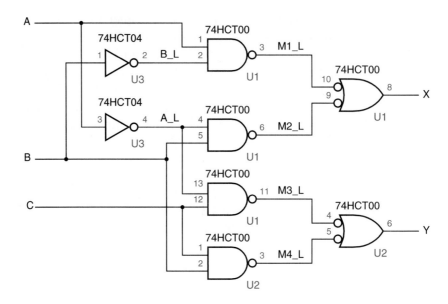

Figure 5-17
Schematic diagram
for a circuit using a
74HCT00.

5.1.8 Additional Schematic Information

Complete schematic diagrams indicate IC types, reference designators, and pin
numbers, as in Figure 5-17. The *IC type* is a part number identifying the integrated circuit that performs a given logic function. For example, a 2-input NAND
gate might be identified as a 74HCT00 or a 74LS00. In addition to the logic
function, the IC type identifies the device's logic family and speed.

IC type

The *reference designator* for an IC identifies a particular instance of that IC
type installed in the system. In conjunction with the system's mechanical
documentation, the reference designator allows a particular IC to be located
during assembly, test, and maintenance of the system. Traditionally, reference
designators for ICs begin with the letter U (for "unit").

reference designator

Once a particular IC is located, *pin numbers* are used to locate individual
logic signals on its pins. The pin numbers are written near the corresponding
inputs and outputs of the standard logic symbol, as shown in Figure 5-17.

pin number

In the rest of this book, just to make you comfortable with properly drawn
schematics, we'll include reference designators and pin numbers for all of the
logic-circuit examples that use SSI and MSI parts.

Figure 5-18 shows the pinouts of many different SSI ICs that are used in
examples throughout this book. Some special graphic elements appear in a few
of the symbols:

- Symbols for the 74x14 Schmitt-trigger inverter have a special element
  inside the symbol to indicate hysteresis.

- Symbols for the 74x03 quad NAND and the 74x266 quad Exclusive NOR
  have a special element to indicate an open-drain or open-collector output.

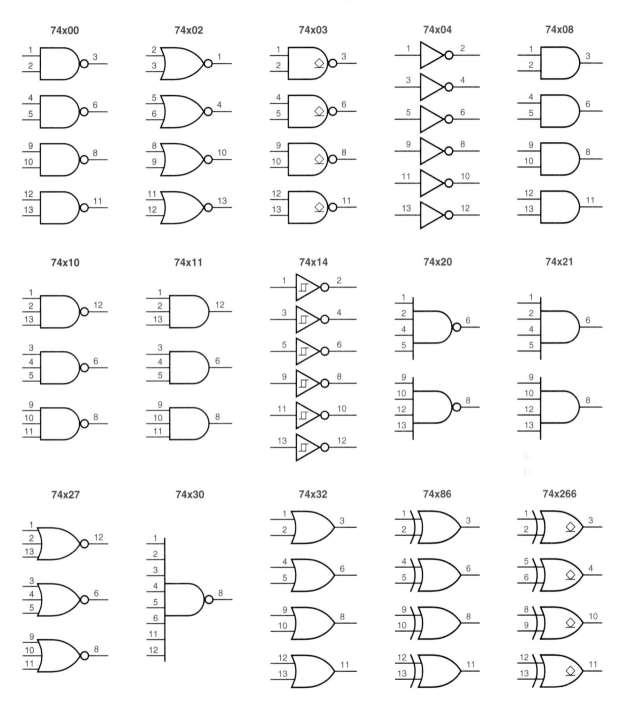

Figure 5-18 Pinouts for SSI ICs in standard dual-inline packages.

When you prepare a schematic diagram for a board-level design using a schematic drawing program, the program automatically provides the pin numbers for the devices that you select from its component library. Note that an IC's pin numbers may differ depending on package type, so you have to be careful to select the right version of the component from the library. Figure 5-18 shows the pin numbers that are used in a dual-inline-pin (DIP) package, the type of package that you would use in a digital design laboratory course or in a low-density, "thru-hole" commercial printed-circuit board.

5.2 Circuit Timing

"Timing is everything"—in investing, in comedy, and yes, in digital design. As we studied in Section 3.6, the outputs of real circuits take time to react to their inputs, and many of today's circuits and systems are so fast that even the speed-of-light delay in propagating an output signal to an input on the other side of a board or chip is significant.

Most digital systems are sequential circuits that operate step-by-step under the control of a periodic clock signal, and the speed of the clock is limited by the worst-case time that it takes for the operations in one step to complete. Thus, digital designers need to be keenly aware of timing behavior in order to build fast circuits that operate correctly under all conditions.

The last several years have seen great advances in the number and quality of CAD tools for analyzing circuit timing. Still, quite often the greatest challenge in completing a board-level or especially an ASIC design is achieving the required timing performance. In this section, we start with the basics, so you can understand what the tools are doing when you use them, and so you can figure out how to fix your circuits when their timing isn't quite making it.

5.2.1 Timing Diagrams

timing diagram

A *timing diagram* illustrates the logical behavior of signals in a digital circuit as a function of time. Timing diagrams are an important part of the documentation of any digital system. They can be used both to explain the timing relationships among signals within a system and to define the timing requirements of external signals that are applied to the system.

Figure 5-19(a) is the block diagram of a simple combinational circuit with two inputs and two outputs. Assuming that the ENB input is held at a constant value, (b) shows the delay of the two outputs with respect to the GO input. In each waveform, the upper line represents a logic 1 and the lower line a logic 0. Signal transitions are drawn as slanted lines to remind us that they do not occur in zero time in real circuits. (Also, slanted lines look nicer than vertical ones.)

causality

delay

Arrows are sometimes drawn, especially in complex timing diagrams, to show *causality*—which input transitions cause which output transitions. In any case, the most important information provided by a timing diagram is a specification of the *delay* between transitions.

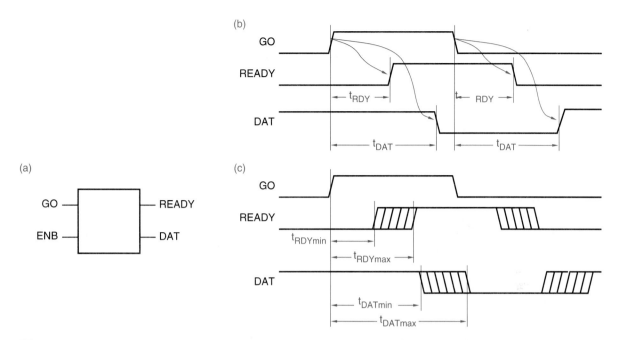

Figure 5-19 Timing diagrams for a combinational circuit: (a) block diagram of circuit; (b) causality and propagation delay; (c) minimum and maximum delays.

Different paths through a circuit may have different delays. For example, Figure 5-19(b) shows that the delay from GO to READY is shorter than the delay from GO to DAT. Similarly, the delays from the ENB input to the outputs may vary, and could be shown in another timing diagram. And, as we'll discuss later, the delay through any given path may vary depending on whether the output is changing from LOW to HIGH or from HIGH to LOW (this phenomenon is not shown in the figure).

Delay in a real circuit is normally measured between the centerpoints of transitions, so the delays in a timing diagram are marked this way. A single timing diagram may contain many different delay specifications. Each different delay is marked with a different identifier, such as t_{RDY} and t_{DAT} in the figure. In large timing diagrams, the delay identifiers are usually numbered for easier reference (e.g., t_1, t_2, ..., t_{42}). In either case, the timing diagram is normally accompanied by a *timing table* that specifies each delay amount and the conditions under which it applies.

timing table

Since the delays of real digital components can vary depending on voltage, temperature, and manufacturing parameters, delay is seldom specified as a single number. Instead, a timing table may specify a range of values by giving *minimum*, *typical*, and *maximum* values for each delay. The idea of a range of delays is sometimes carried over into the timing diagram itself by showing the transitions to occur at uncertain times, as in Figure 5-19(c).

Figure 5-20
Timing diagrams for "data" signals:
(a) certain and uncertain transitions;
(b) sequence of values on an 8-bit bus.

For some signals, the timing diagram needn't show whether the signal changes from 1 to 0 or from 0 to 1 at a particular time, only that a transition occurs then. Any signal that carries a bit of "data" has this characteristic—the actual value of the data bit varies according to circumstances but, regardless of value, the bit is transferred, stored, or processed at a particular time relative to "control" signals in the system. Figure 5-20(a) is a timing diagram that illustrates this concept. The "data" signal is normally at a steady 0 or 1 value, and transitions occur only at the times indicated. The idea of an uncertain delay time can also be used with "data" signals, as shown for the DATAOUT signal.

Quite often in digital systems, a group of data signals in a bus is processed by identical circuits. In this case, all signals in the bus have the same timing, and they can be represented by a single line in the timing diagram and corresponding specifications in the timing table. If the bus bits are known to take on a particular combination at a particular time, this is sometimes shown in the timing diagram using binary, octal, or hexadecimal numbers, as in Figure 5-20(b).

5.2.2 Propagation Delay

propagation delay

In Section 3.6.2 we formally defined the *propagation delay* of a signal path as the time that it takes for a change at the input of the path to produce a change at the output of the path. A combinational circuit with many inputs and outputs has many different paths, and each one may have a different propagation delay. Also, the propagation delay when the output changes from LOW to HIGH (t_{pLH}) may be different from the delay when it changes from HIGH to LOW (t_{pHL}).

The manufacturer of a combinational-logic IC normally specifies all of these different propagation delays, or at least the delays that would be of interest in typical applications. A logic designer who combines ICs in a larger circuit uses the individual device specifications to analyze the overall circuit timing. The delay of a path through the overall circuit is the sum of the delays through subpaths in the individual devices.

5.2.3 Timing Specifications

The timing specification for a device may give minimum, typical, and maximum values for each propagation-delay path and transition direction:

- *Maximum*. This specification is the one that is most often used by experi-   *maximum delay*
  enced designers, since a path "never" has a propagation delay longer than
  the maximum. However, the definition of "never" varies among logic
  families and manufacturers. For example, "maximum" propagation delays
  of 74LS and 74S TTL devices are specified with $V_{CC} = 5$ V, $T_A = 25°C$, and
  almost no capacitive load. If the voltage or temperature is different, or if
  the capacitive load is more than 15 pF, the delay may be longer. On the
  other hand, a "maximum" propagation delay is specified for 74AC and
  74ACT devices over the full operating voltage and temperature range and
  with a heavier capacitive load of 50 pF.

- *Typical*. This specification is the one most often used by designers who   *typical delay*
  don't expect to be around when their product leaves the friendly environment of the engineering lab and is shipped to customers. The "typical"
  delay is what you see from a device that was manufactured on a good day
  and is operating under near-ideal conditions.

- *Minimum*. This is the smallest propagation delay that a path will ever   *minimum delay*
  exhibit. Most well-designed circuits don't depend on this number; that is,
  they will work properly even if the delay is zero. That's good because
  manufacturers don't specify minimum delay in most moderate-speed logic
  families, including 74LS and 74S TTL. However, in high-speed families,
  including ECL and 74AC and 74ACT CMOS, a nonzero minimum delay
  is specified to help the designer ensure that timing requirements of latches
  and flip-flops, discussed in Section 7.2, are met.

Table 5-2 lists the typical and maximum delays of several 74-series CMOS and TTL gates. Table 5-3 does the same thing for most of the CMOS and TTL MSI parts that are introduced later in this chapter.

HOW TYPICAL IS   Most ICs, perhaps 99%, really are manufactured on "good" days and exhibit delays
TYPICAL?   near the "typical" specifications. However, if you design a system that works only if
all of its 100 ICs meet the "typical" timing specs, probability theory suggests that
63% $(1 - .99^{100})$ of the systems won't work. But see the next box ....

Table 5-2 Propagation delay in nanoseconds of selected 5-V CMOS and TTL SSI parts.

| | 74HCT | | 74AHCT | | | | 74LS | | | |
| | Typical | Maximum | Typical | | Maximum | | Typical | | Maximum | |
Part number	t_{pLH}, t_{pHL}	t_{pLH}, t_{pHL}	t_{pLH}	t_{pHL}	t_{pLH}	t_{pHL}	t_{pLH}	t_{pHL}	t_{pLH}	t_{pHL}
'00, '10	11	35	5.5	5.5	9.0	9.0	9	10	15	15
'02	9	29	3.4	4.5	8.5	8.5	10	10	15	15
'04	11	35	5.5	5.5	8.5	8.5	9	10	15	15
'08, '11	11	35	5.5	5.5	9.0	9.0	8	10	15	20
'14	16	48	5.5	5.5	9.0	9.0	15	15	22	22
'20	11	35					9	10	15	15
'21	11	35					8	10	15	20
'27	9	29	5.6	5.6	9.0	9.0	10	10	15	15
'30	11	35					8	13	15	20
'32	9	30	5.3	5.3	8.5	8.5	14	14	22	22
'86 (2 levels)	13	40	5.5	5.5	10	10	12	10	23	17
'86 (3 levels)	13	40	5.5	5.5	10	10	20	13	30	22

A COROLLARY OF MURPHY'S LAW

Murphy's law states, "If something can go wrong, it will." A corollary to this is, "If you want something to go wrong, it won't."

In the boxed example on the previous page, you might think that you have a 63% chance of detecting the potential timing problems in the engineering lab. The problems aren't spread out evenly, though, since all ICs from a given batch tend to behave about the same. Murphy's corollary says that *all* of the engineering proto-types will be built with ICs from the same, "good" batches. Therefore, everything works fine for a while, just long enough for the system to get into volume production and for everyone to become complacent and self-congratulatory.

Then, unbeknownst to the production department, a "slow" batch of some IC type arrives from a supplier and gets used in every system that is built, so that *nothing* works. The production engineers scurry around trying to analyze the problem (not easy, because the designer is long gone and didn't bother to write a circuit descrip-tion), and in the meantime the company loses big bucks because it is unable to ship its product.

Table 5-3 Propagation delay in nanoseconds of selected CMOS and TTL MSI parts.

Part	From	To	74HCT Typical $t_{\rm pLH}, t_{\rm pHL}$	74HCT Maximum $t_{\rm pLH}, t_{\rm pHL}$	74AHCT / FCT Typical $t_{\rm pLH}, t_{\rm pHL}$	74AHCT / FCT Maximum $t_{\rm pLH}, t_{\rm pHL}$	74LS Typical $t_{\rm pLH}$	74LS Typical $t_{\rm pHL}$	74LS Maximum $t_{\rm pLH}$	74LS Maximum $t_{\rm pHL}$
'138	any select	output (2)	23	45	8.1 / 5	13 / 9	11	18	20	41
	any select	output (3)	23	45	8.1 / 5	13 / 9	21	20	27	39
	$\overline{\rm G2A}$, $\overline{\rm G2B}$	output	22	42	7.5 / 4	12 / 8	12	20	18	32
	G1	output	22	42	7.1 / 4	11.5 / 8	14	13	26	38
'139	any select	output (2)	14	43	6.5 / 5	10.5 / 9	13	22	20	33
	any select	output (3)	14	43	6.5 / 5	10.5 / 9	18	25	29	38
	enable	output	11	43	5.9 / 5	9.5 / 9	16	21	24	32
'151	any select	Y	17	51	- / 5	- / 9	27	18	43	30
	any select	$\overline{\rm Y}$	18	54	- / 5	- / 9	14	20	23	32
	any data	Y	16	48	- / 4	- / 7	20	16	32	26
	any data	$\overline{\rm Y}$	15	45	- / 4	- / 7	13	12	21	20
	enable	Y	12	36	- / 4	- / 7	26	20	42	32
	enable	$\overline{\rm Y}$	15	45	- / 4	- / 7	15	18	24	30
'153	any select	output	14	43	- / 5	- / 9	19	25	29	38
	any data	output	12	43	- / 4	- / 7	10	17	15	26
	enable	output	11	34	- / 4	- / 7	16	21	24	32
'157	select	output	15	46	6.8 / 7	11.5 / 10.5	15	18	23	27
	any data	output	12	38	5.6 / 4	9.5 / 6	9	9	14	14
	enable	output	12	38	7.1 / 7	12.0 / 10.5	13	14	21	23
'182	any $\overline{\rm Gi}$, $\overline{\rm Pi}$	C1–3	13	41			4.5	4.5	7	7
	any $\overline{\rm Gi}$, $\overline{\rm Pi}$	$\overline{\rm G}$	13	41			5	7	7.5	10.5
	any $\overline{\rm Pi}$	$\overline{\rm P}$	11	35			4.5	6.5	6.5	10
	C0	C1–3	17	50			6.5	7	10	10.5
'280	any input	EVEN	18	53	- / 6	- / 10	33	29	50	45
	any input	ODD	19	56	- / 6	- / 10	23	31	35	50
'283	C0	any Si	22	66			16	15	24	24
	any Ai, Bi	any Si	21	61			15	15	24	24
	C0	C4	19	58			11	11	17	22
	any Ai, Bi	C4	20	60			11	12	17	17
'381	CIN	any Fi					18	14	27	21
	any Ai, Bi	$\overline{\rm G}$					20	21	30	33
	any Ai, Bi	$\overline{\rm P}$					21	33	23	33
	any Ai, Bi	any Fi					20	15	30	23
	any select	any Fi					35	34	53	51
	any select	$\overline{\rm G}$, $\overline{\rm P}$					31	32	47	48
'682	any Pi	$\overline{\rm PEQQ}$	26	69	- / 7	- / 11	13	15	25	25
	any Qi	$\overline{\rm PEQQ}$	26	69	- / 7	- / 11	14	15	25	25
	any Pi	$\overline{\rm PGTQ}$	26	69	- / 9	- / 14	20	15	30	30
	any Qi	$\overline{\rm PGTQ}$	26	69	- / 9	- / 14	21	19	30	30

All inputs of an SSI gate have the same propagation delay to the output. Note that TTL gates usually have different delays for LOW-to-HIGH and HIGH-to-LOW transitions (t_{pLH} and t_{pHL}), but CMOS gates usually do not. CMOS gates have a more symmetrical output driving capability, so any difference between the two cases is usually not worth noting.

The delay from an input transition to the corresponding output transition depends on the internal path taken by the changing signal, and in larger circuits the path may be different for different input combinations. For example, the 74LS86 2-input XOR gate is constructed from four NAND gates as shown in Figure 5-71 on page 412, and has two different-length paths from either input to the output. If one input is LOW, and the other is changed, the change propagates through two NAND gates, and we observe the first set of delays shown in Table 5-2. If one input is HIGH, and the other is changed, the change propagates through *three* NAND gates internally, and we observe the second set of delays. Similar behavior is exhibited by the 74LS138 and 74LS139 in Table 5-3. However, the corresponding CMOS parts do not show these differences; they are small enough to be ignored.

5.2.4 Timing Analysis

To accurately analyze the timing of a circuit containing multiple SSI and MSI devices, a designer may have to study its logical behavior in excruciating detail. For example, when TTL inverting gates (NAND, NOR, etc.) are placed in series, a LOW-to-HIGH change at one gate's output causes a HIGH-to-LOW change at the next one's, and so the differences between t_{pLH} and t_{pHL} tend to average out. On the other hand, when noninverting gates (AND, OR, etc.) are placed in series, a transition causes all outputs to change in the same direction, and so the gap between t_{pLH} and t_{pHL} tends to widen. As a reader, you'll have the privilege of carrying out this sort of analysis in Drills 5.8–5.13.

The analysis gets more complicated if there are MSI devices in the delay path, or if there are multiple paths from a given input signal to a given output signal. Thus, in large circuits, analysis of all of the different delay paths and transition directions can be very complex.

To permit a simplified "worst-case" analysis, designers often use a single *worst-case delay* specification that is the maximum of t_{pLH} and t_{pHL} specifications. The worst-case delay through a circuit is then computed as the sum of the worst-case delays through the individual components, independent of the transition direction and other circuit conditions. This may give an overly pessimistic view of the overall circuit delay, but it saves design time and it's guaranteed to work.

worst-case delay

5.2.5 Timing Analysis Tools

Sophisticated CAD tools for logic design make timing analysis even easier. Their component libraries typically contain not only the logic symbols and functional models for various logic elements, but also their timing models. A simulator allows you to apply input sequences and observe how and when outputs are produced in response. You typically can control whether minimum, typical, maximum, or some combination of delay values are used.

Even with a simulator, you're not completely off the hook. It's usually up the designer to supply the input sequences for which the simulator should produce outputs. Thus, you'll need to have a good feel for what to look for and how to stimulate your circuit to produce and observe the worst-case delays.

Some timing analysis programs can automatically find all possible delay paths in a circuit and print out a sorted list of them, starting with the slowest. These results may be overly pessimistic, however, as some paths may actually not be used in normal operations of the circuit; the designer must still use some intelligence to interpret the results properly.

5.3 Combinational PLDs

5.3.1 Programmable Logic Arrays

Historically, the first PLDs were *programmable logic arrays (PLAs)*. A PLA is a combinational, two-level AND-OR device that can be programmed to realize any sum-of-products logic expression, subject to the size limitations of the device. Limitations are

programmable logic array (PLA)

* the number of inputs (*n*),
* the number of outputs (*m*), and
* the number of product terms (*p*).

inputs

outputs

product terms

We might describe such a device as "an $n \times m$ PLA with p product terms." In general, *p* is far less than the number of *n*-variable minterms (2^n). Thus, a PLA cannot perform arbitrary *n*-input, *m*-output logic functions; its usefulness is limited to functions that can be expressed in sum-of-products form using *p* or fewer product terms.

An $n \times m$ PLA with *p* product terms contains *p* 2*n*-input AND gates and *m* *p*-input OR gates. Figure 5-21 shows a small PLA with four inputs, six AND

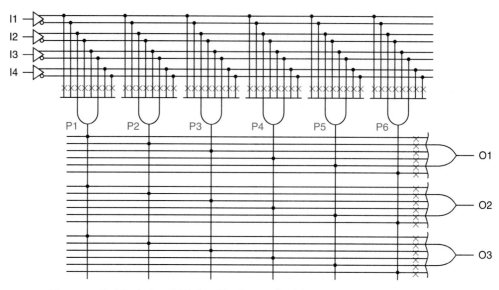

Figure 5-21 A 4 × 3 PLA with six product terms.

gates, and three **OR** gates and outputs. Each input is connected to a buffer that produces both a true and a complemented version of the signal for use within the array. Potential connections in the array are indicated by X's; the device is programmed by establishing only the connections that are actually needed. The needed connections are made by *fuses*, which are actual fusible links or non-volatile memory cells, depending on technology as we explain in Sections 5.3.4 and 5.3.5. Thus, each **AND** gate's inputs can be any subset of the primary input signals and their complements. Similarly, each **OR** gate's inputs can be any subset of the **AND**-gate outputs.

PLD fuses

PLA diagram

As shown in Figure 5-22, a more compact diagram can be used to represent a PLA. Moreover, the layout of this diagram more closely resembles the actual internal layout of a PLA chip (e.g., Figure 5-28 on page 346).

Figure 5-22
Compact representation of a 4 × 3 PLA with six product terms.

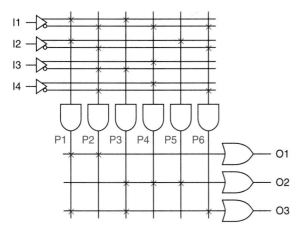

Figure 5-23
A 4 × 3 PLA programmed with a set of three logic equations.

The PLA in Figure 5-22 can perform any three 4-input combinational logic functions that can be written as sums of products using a total of six or fewer distinct product terms, for example:

$$O1 = I1 \cdot I2 + I1' \cdot I2' \cdot I3' \cdot I4'$$
$$O2 = I1 \cdot I3' + I1' \cdot I3 \cdot I4 + I2$$
$$O3 = I1 \cdot I2 + I1 \cdot I3' + I1' \cdot I2' \cdot I4'$$

These equations have a total of eight product terms, but the first two terms in the O3 equation are the same as the first terms in the O1 and O2 equations. The programmed connection pattern in Figure 5-23 matches these logic equations.

Sometimes a PLA output must be programmed to be a constant 1 or a constant 0. That's no problem, as shown in Figure 5-24. Product term P1 is always 1 because its product line is connected to no inputs and is therefore always pulled HIGH; this constant-1 term drives the O1 output. No product term drives the O2 output, which is therefore always 0. Another method of obtaining a constant-0 output is shown for O3. Product term P2 is connected to each input variable and its complement; therefore, it's always 0 ($X \cdot X' = 0$).

PLA constant outputs

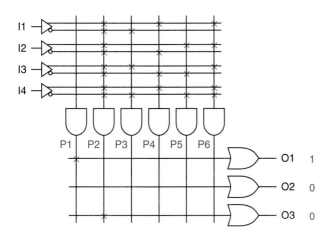

Figure 5-24
A 4 × 3 PLA programmed to produce constant 0 and 1 outputs.

Our example PLA has too few inputs, outputs, and AND gates (product terms) to be very useful. An n-input PLA could conceivably use as many as 2^n product terms, to realize all possible n-variable minterms. The actual number of product terms in typical commercial PLAs is far fewer, on the order of 4 to 16 per output, regardless of the value of n.

The Signetics 82S100 was a typical example of the PLAs that were introduced in the mid-1970s. It had 16 inputs, 48 AND gates, and 8 outputs. Thus, it had $2 \times 16 \times 48 = 1536$ fuses in the AND array and $8 \times 48 = 384$ in the OR array. Off-the-shelf PLAs like the 82S100 have since been supplanted by PALs, CPLDs, and FPGAs, but custom PLAs are often synthesized to perform complex combinational logic within a larger ASIC.

5.3.2 Programmable Array Logic Devices

programmable array logic (PAL) device

A special case of a PLA, and today's most commonly used type of PLD, is the *programmable array logic (PAL) device*. Unlike a PLA, in which both the AND and OR arrays are programmable, a PAL device has a *fixed* OR array.

The first PAL devices used TTL-compatible bipolar technology and were introduced in the late 1970s. Key innovations in the first PAL devices, besides the introduction of a catchy acronym, were the use of a fixed OR array and bidirectional input/output pins.

PAL16L8

These ideas are well illustrated by the *PAL16L8*, shown in Figures 5-25 and 5-26 and one of today's most commonly used combinational PLD structures. Its programmable AND array has 64 rows and 32 columns, identified for programming purposes by the small numbers in the figure, and $64 \times 32 = 2048$ fuses. Each of the 64 AND gates in the array has 32 inputs, accommodating 16 variables and their complements; hence, the "16" in "PAL16L8".

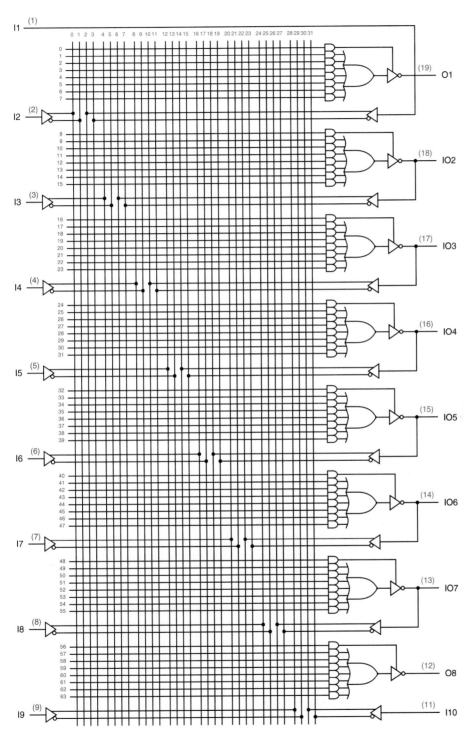

Figure 5-25 Logic diagram of the PAL16L8.

PAL16L8

Figure 5-26
Traditional logic symbol for
the PAL16L8.

Eight AND gates are associated with each output pin of the PAL16L8. Seven of them provide inputs to a fixed 7-input OR gate. The eighth, which we

output-enable gate

call the *output-enable gate*, is connected to the three-state enable input of the output buffer; the buffer is enabled only when the output-enable gate has a 1 output. Thus, an output of the PAL16L8 can perform only logic functions that can be written as sums of seven or fewer product terms. Each product term can be a function of any or all 16 inputs, but only seven such product terms are available.

Although the PAL16L8 has up to 16 inputs and up to 8 outputs, it is housed in a dual in-line package with only 20 pins, including two for power and ground (the corner pins, 10 and 20). This magic is the result of six bidirectional pins (13–18) that may be used as inputs or outputs or both. This and other differences between the PAL16L8 and a PLA structure are summarized below:

- The PAL16L8 has a fixed OR array, with seven AND gates permanently connected to each OR gate. AND-gate outputs cannot be shared; if a product term is needed by two OR gates, it must be generated twice.

- Each output of the PAL16L8 has an individual three-state output enable signal, controlled by a dedicated AND gate (the output-enable gate). Thus, outputs may be programmed as always enabled, always disabled, or enabled by a product term involving the device inputs.

**HOW USEFUL
ARE SEVEN
PRODUCT
TERMS?**

The worst-case logic function for two-level AND-OR design is an n-input XOR (parity) function, which requires 2^{n-1} product terms. However, less perverse functions with more than seven product terms of a PAL16L8 can often be built by decomposing them into a 4-level structure (AND-OR-AND-OR) that can be realized with two passes through the AND-OR array. Unfortunately, besides using up PLD outputs for the first-pass terms, this doubles the delay, since a first-pass input must pass through the PLD twice to propagate to the output.

COMBINATIONAL, NOT COMBINATORIAL!	A step *backward* in MMI's introduction of PAL devices was their popularization of the word "combinatorial" to describe combinational circuits. *Combinational* circuits have no memory—their output at any time depends on the current input *combination*. For well-rounded computer engineers, the word "combinatorial" should conjure up vivid images of binomial coefficients, problem-solving complexity, and computer-science-great Donald Knuth.

- There is an inverter between the output of each OR gate and the external pin of the device.

- Six of the output pins, called *I/O pins*, may also be used as inputs. This provides many possibilities for using each I/O pin, depending on how the device is programmed: *I/O pin*

  - If an I/O pin's output-control gate produces a constant 0, then the output is always disabled and the pin is used strictly as an input.

  - If the input signal on an I/O pin is not used by any gates in the AND array, then the pin may be used strictly as an output. Depending on the programming of the output-enable gate, the output may always be enabled, or it may be enabled only for certain input conditions.

  - If an I/O pin's output-control gate produces a constant 1, then the output is always enabled, but the pin may still be used as an input too. In this way, outputs can be used to generate first-pass "helper terms" for logic functions that cannot be performed in a single pass with the limited number of AND terms available for a single output. We'll show an example of this case on page 365.

  - In another case with an I/O pin always output enabled, the output may be used as an input to AND gates that affect the very same output. That is, we can embed a feedback sequential circuit in a PAL16L8. We'll discuss this case in Section 8.2.6.

The *PAL20L8* is another combinational PLD similar to the PAL16L8, except that its package has four more input-only pins and each of its AND gates has eight more inputs to accommodate them. Its output structure is the same as the PAL16L8's.   *PAL20L8*

5.3.3 Generic Array Logic Devices

In Section 8.3 we'll introduce sequential PLDs, programmable logic devices that provide flip-flops at some or all OR-gate outputs. These devices can be programmed to perform a variety of useful sequential-circuit functions.

COMBINATIONAL PLD SPEED	The speed of a combinational PLD is usually stated as a single number giving the propagation delay t_{PD} from any input to any output for either direction of transition. PLDs are available in a variety of speed grades; commonly used parts run at 10 ns. In 1998, the fastest available combinational PLDs included a bipolar PAL16L8 at 5 ns and a 3.3-V CMOS GAL22LV10 at 3.5 ns.

generic array logic
GAL device
GAL16V8

One type of sequential PLD, first introduced by Lattice Semiconductor, is called *generic array logic* or a *GAL device*, and is particularly popular. A single GAL device type, the *GAL16V8*, can be configured (via programming and a corresponding fuse pattern) to emulate the AND-OR, flip-flop, and output structure of any of a variety of combinational and sequential PAL devices, including the PAL16L8 introduced previously. What's more, the GAL device can be erased electrically and reprogrammed.

Figure 5-27 shows the logic diagram for a GAL16V8 when it has been configured as a strictly combinational device similar to the PAL16L8. This configuration is achieved by programming two "architecture-control" fuses, not shown. In this configuration, the device is called a *GAL16V8C*.

GAL16V8C

The most important thing to note about the GAL16V8C logic diagram, compared to that of a PAL16L8 on page 341, is that an XOR gate has been inserted between each OR output and the three-state output driver. One input of the XOR gate is "pulled up" to a logic 1 value but connected to ground (0) via a fuse. If this fuse is intact, the XOR gate simply passes the OR-gate's output unchanged, but if the fuse is blown, the XOR gate inverts the OR-gate's output. This fuse is said to control the *output polarity* of the corresponding output pin.

output polarity

Output-polarity control is a very important feature of modern PLDs, including the GAL16V8. As we discussed in Section 4.6.2, given a logic function to minimize, an ABEL compiler finds minimal sum-of-products expressions for both the function and its complement. If the complement yields fewer product terms, it can be used if the GAL16V8's output polarity fuse is set to invert. Unless overridden, the compiler automatically makes the best selection and sets up the fuse patterns appropriately.

PALCE16V8
GAL20V8
PALCE20V8

Several companies make a part that is equivalent to the GAL16V8, called the *PALCE16V8*. There is also a 24-pin GAL device, the *GAL20V8* or *PALCE20V8*, that can be configured to emulate the structure of the PAL20L8 or any of a variety of sequential PLDs, as described in Section 8.3.2.

LEGAL NOTICE	GAL is a trademark of Lattice Semiconductor, Hillsboro, OR 97124.

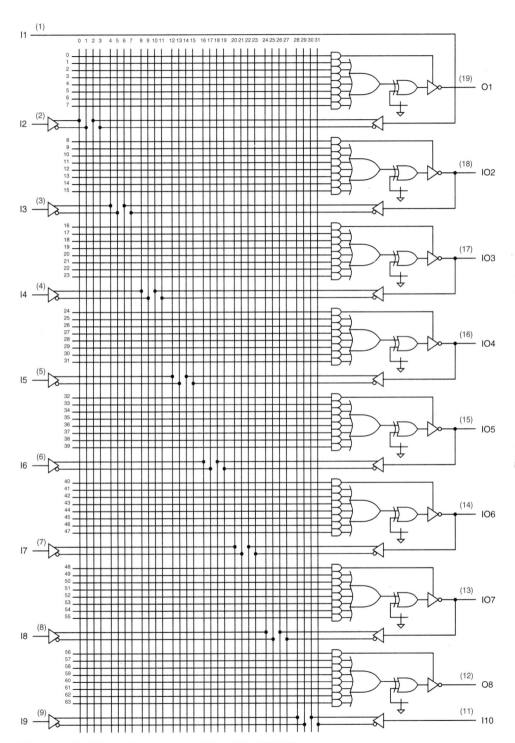

Figure 5-27 Logic diagram of the GAL16V8C.

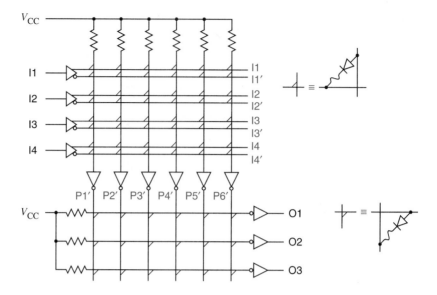

Figure 5-28
A 4 × 3 PLA built using
TTL-like open-collector
gates and diode logic.

*5.3.4 Bipolar PLD Circuits

There are several different circuit technologies for building and physically programming a PLD. Early commercial PLAs and PAL devices used bipolar circuits. For example, Figure 5-28 shows how the example 4 × 3 PLA circuit of the preceding section might be built in a bipolar, TTL-like technology. Each potential connection is made by a diode in series with a metal link that may be present or absent. If the link is present, then the diode connects its input into a diode-AND function. If the link is missing, then the corresponding input has no effect on that AND function.

A diode-AND function is performed because each and every horizontal "input" line that is connected via a diode to a particular vertical "AND" line must be HIGH in order for that AND line to be HIGH. If an input line is LOW, it pulls LOW all of the AND lines to which it is connected. This first matrix of circuit *AND plane*   elements that perform the AND function is called the *AND plane*.

Each AND line is followed by an inverting buffer, so overall a NAND function is obtained. The outputs of the first-level NAND functions are combined by another set of programmable diode AND functions, once again followed by inverters. The result is a two-level NAND-NAND structure that is functionally equivalent to the AND-OR PLA structure described in the preceding section. The matrix of circuit elements that perform the OR function (or the second NAND *OR plane*   function, depending on how you look at it) is called the *OR plane*.

A bipolar PLD chip is manufactured with all of its diodes present, but with *fusible link*   a tiny *fusible link* in series with each one (the little squiggles in Figure 5-28). By

---

* Throughout this book, optional sections are marked with an asterisk.

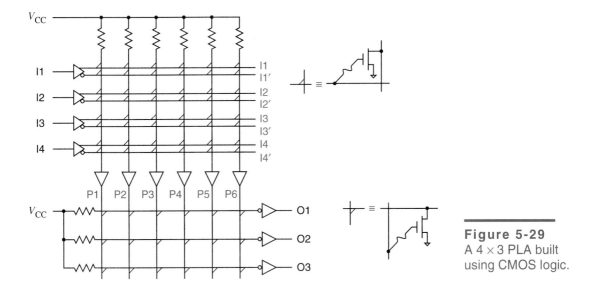

Figure 5-29
A 4 × 3 PLA built using CMOS logic.

applying special input patterns to the device, it is possible to select individual links, apply a high voltage (10–30 V), and thereby vaporize selected links.

Early bipolar PLDs had reliability problems. Sometimes the stored patterns changed because of incompletely vaporized links that would "grow back," and sometimes intermittent failures occurred because of floating shrapnel inside the IC package. However, these problems have been largely worked out, and reliable fusible-link technology is used in today's bipolar PLDs.

*5.3.5 CMOS PLD Circuits

Although they're still available, bipolar PLDs have been largely supplanted by CMOS PLDs with a number of advantages, including reduced power consumption and reprogrammability. Figure 5-29 shows a CMOS design for the 4 × 3 PLA circuit of Section 5.3.1.

Instead of a diode, an *n*-channel transistor with a programmable connection is placed at each intersection between an input line and a word line. If the input is LOW, then the transistor is "off," but if the input is HIGH, then the transistor is "on," which pulls the AND line LOW. Overall, an inverted-input AND (i.e., NOR) function is obtained. This is similar in structure and function to a normal CMOS *k*-input NOR gate, except that the usual series connection of *k* *p*-channel pull-up transistors has been replaced with a passive pull-up resistor (in practice, the pull-up is a single *p*-channel transistor with a constant bias).

As shown in color on Figure 5-29, the effects of using an *inverted*-input AND gate are canceled by using the opposite (complemented) input lines for each input, compared with Figure 5-28. Also notice that the connection between

Figure 5-30
AND plane of an
EPLD using floating-
gate MOS transistors.

the AND plane and the OR plane is noninverting, so the AND plane performs a
true AND function.

The outputs of the first-level AND functions are combined in the OR plane
by another set of NOR functions with programmable connections. The output of
each NOR function is followed by an inverter, so a true OR function is realized,
and overall the PLA performs an AND-OR function as desired.

In CMOS PLD technologies, the programmable links shown in
Figure 5-29 are not normally fuses. In non-field-programmable devices, such as
custom VLSI chips, the presence or absence of each link is simply established as
part of the metal mask pattern for the manufacture of the device. By far the most
common programming technology, however, is used in CMOS EPLDs, as
discussed next.

erasable programmable
logic device (EPLD)

An *erasable programmable logic device (EPLD)* can be programmed with
any desired link configuration, as well as "erased" to its original state, either
electronically or by exposing it to ultraviolet light. No, erasing does not cause
links to suddenly appear or disappear! Rather, EPLDs use a different technology,
called "floating-gate MOS."

floating-gate MOS
transistor

As shown in Figure 5-30, an EPLD uses *floating-gate MOS transistors*.
Such a transistor has two gates. The "floating" gate is unconnected and is
surrounded by extremely high-impedance insulating material. In the original,
manufactured state, the floating gate has no charge on it and has no effect on
circuit operation. In this state, all transistors are effectively "connected"; that is,
there is a logical link present at every crosspoint in the AND and OR planes.

To program an EPLD, the programmer applies a high voltage to the nonfloating gate at each location where a logical link is not wanted. This causes a

temporary breakdown in the insulating material and allows a negative charge to accumulate on the floating gate. When the high voltage is removed, the negative charge remains on the floating gate. During subsequent operations, the negative charge prevents the transistor from turning "on" when a HIGH signal is applied to the nonfloating gate; the transistor is effectively disconnected from the circuit.

EPLD manufacturers claim that a properly programmed bit will retain 70% of its charge for at least 10 years, even if the part is stored at 125°C, so for most applications the programming can be considered to be permanent. However, EPLDs can also be erased.

Although some early EPLDs were packaged with a transparent lid and used light for erasing, today's most popular devices are *electrically erasable PLDs*. The floating gates in an electrically erasable PLD are surrounded by an extremely thin insulating layer and can be erased by applying a voltage of the opposite polarity as the charging voltage to the nonfloating gate. Thus, the same piece of equipment that is normally used to program a PLD can also be used to erase an EPLD before programming it.

electrically erasable PLD

Larger-scale, "complex" PLDs (CPLDs), also use floating-gate programming technology. Even larger devices, often called *field-programmable gate arrays (FPGAs)*, use read/write memory cells to control the state of each connection. The read/write memory cells are volatile—they do not retain their state when power is removed. Therefore, when power is first applied to the FPGA, all of its read/write memory must be initialized to a state specified by a separate, external nonvolatile memory. This memory typically is either a programmable read-only memory (PROM) chip attached directly to the FPGA or is part of a microprocessor subsystem that initializes the FPGA as part of overall system initialization.

field-programmable gate array (FPGA)

*5.3.6 Device Programming and Testing

A special piece of equipment is used to vaporize fuses, charge up floating-gate transistors, or do whatever else is required to program a PLD. This piece of equipment, found nowadays in almost all digital design labs and production facilities, is called a *PLD programmer* or a *PROM programmer*. (It can be used

PLD programmer
PROM programmer

CHANGING HARDWARE ON THE FLY

PROMs are normally used to supply the connection pattern for a read/write-memory-based FPGA, but there are also applications where the pattern is actually read from a floppy disk. You just received a floppy with a new software version? Guess what, you just got a new hardware version too!

This concept leads us to the intriguing idea, already being applied in some applications, of "reconfigurable hardware," where a hardware subsystem is redefined, on the fly, to optimize its performance for the particular task at hand.

with programmable read-only memories, "PROMs," as well as for PLDs.) A typical PLD programmer includes a socket or sockets that physically accept the devices to be programmed, and a way to "download" desired programming patterns into the programmer, typically by connecting the programmer to a PC.

PLD programmers typically place a PLD into a special mode of operation in order to program it. For example, a PLD programmer typically programs the PLDs described in this chapter eight fuses at a time as follows:

1. Raise a certain pin to a predetermined, high voltage (such as 14 V) to put the device into programming mode.

2. Select a group of eight fuses by applying a binary "address" to certain inputs of the device. (For example, the 82S100 has 1920 fuses, and it would therefore require 8 inputs to select one of 240 groups of 8 fuses.)

3. Apply an 8-bit value to the *outputs* of the device to specify the desired programming for each fuse (the outputs are used as inputs in programming mode).

4. Raise a second predetermined pin to the high voltage for a predetermined length of time (such as 100 microseconds) to program the eight fuses.

5. Lower the second predetermined pin to a low voltage (such as 0 V) to read out and verify the programming of the eight fuses.

6. Repeat steps 1–5 for each group of eight fuses.

in-system
programmability

JTAG port

Many PLDs, especially larger CPLDs, feature *in-system programmability*. This means that the device can be programmed after it is already soldered into the system. In this case, the fuse patterns are applied to the device serially using four extra signals and pins, called the *JTAG port*, defined by IEEE standard 1149.1. These signals are defined so that multiple devices on the same printed-circuit board can be "daisy chained" and selected and programmed during the board manufacturing process using just one JTAG port on a special connector. No special high-voltage power supply is needed; each device uses a charge-pump circuit internally to generate the high voltage needed for programming.

As noted in step 5 above, fuse patterns are verified as they are programmed into a device. If a fuse fails to program properly the first time, the operation can be retried; if it fails to program properly after a few tries, the device is discarded (often with great prejudice and malice aforethought).

While verifying the fuse pattern of a programmed device proves that its fuses are programmed properly, it does not prove that the device will perform the logic function specified by those fuses. This is true because the device may have unrelated internal defects such as missing connections between the fuses and elements of the AND-OR array.

The only way to test for all defects is to put the device into its normal operational mode, apply a set of normal logic inputs, and observe the outputs. The required input and output patterns, called test vectors, can be specified by the designer as we showed in Section 4.6.7, or can be generated automatically by a special test-vector-generation program. Regardless of how the test vectors are generated, most PLD programmers have the ability to apply test-vector inputs to a PLD and to check its outputs against the expected results.

Most PLDs have a *security fuse* which, when programmed, disables the ability to read fuse patterns from the device. Manufacturers can program this fuse to prevent others from reading out the PLD fuse patterns in order to copy the product design. Even if the security fuse is programmed, test vectors still work, so the PLD can still be checked.

security fuse

5.4 Decoders

A *decoder* is a multiple-input, multiple-output logic circuit that converts coded inputs into coded outputs, where the input and output codes are different. The input code generally has fewer bits than the output code, and there is a one-to-one mapping from input code words into output code words. In a *one-to-one mapping*, each input code word produces a different output code word.

decoder

one-to-one mapping

The general structure of a decoder circuit is shown in Figure 5-31. The enable inputs, if present, must be asserted for the decoder to perform its normal mapping function. Otherwise, the decoder maps all input code words into a single, "disabled," output code word.

The most commonly used input code is an n-bit binary code, where an n-bit word represents one of 2^n different coded values, normally the integers from 0 through $2^n - 1$. Sometimes an n-bit binary code is truncated to represent fewer than 2^n values. For example, in the BCD code, the 4-bit combinations 0000 through 1001 represent the decimal digits 0–9, and combinations 1010 through 1111 are not used.

The most commonly used output code is a 1-out-of-m code, which contains m bits, where one bit is asserted at any time. Thus, in a 1-out-of-4 code with active-high outputs, the code words are 0001, 0010, 0100, and 1000. With active-low outputs, the code words are 1110, 1101, 1011, and 0111.

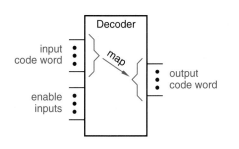

Figure 5-31
Decoder circuit structure.

Table 5-4
Truth table for a 2-to-4 binary decoder.

Inputs			Outputs			
EN	I1	I0	Y3	Y2	Y1	Y0
0	x	x	0	0	0	0
1	0	0	0	0	0	1
1	0	1	0	0	1	0
1	1	0	0	1	0	0
1	1	1	1	0	0	0

5.4.1 Binary Decoders

binary decoder

The most common decoder circuit is an n-to-2^n decoder or *binary decoder.* Such a decoder has an n-bit binary input code and a 1-out-of-2^n output code. A binary decoder is used when you need to activate exactly one of 2^n outputs based on an n-bit input value.

For example, Figure 5-32(a) shows the inputs and outputs and Table 5-4 is the truth table of a 2-to-4 decoder. The input code word I1,I0 represents an integer in the range 0–3. The output code word Y3,Y2,Y1,Y0 has Yi equal to 1 if and only if the input code word is the binary representation of i and the *enable input* EN is 1. If EN is 0, then all of the outputs are 0. A gate-level circuit for the 2-to-4 decoder is shown in Figure 5-32(b). Each AND gate *decodes* one combination of the input code word I1,I0.

enable input

decode

The binary decoder's truth table introduces a "don't-care" notation for input combinations. If one or more input values do not affect the output values for some combination of the remaining inputs, they are marked with an "x" for that input combination. This convention can greatly reduce the number of rows in the truth table, as well as make the functions of the inputs more clear.

The input code of an n-bit binary decoder need not represent the integers from 0 through 2^n-1. For example, Table 5-5 shows the 3-bit Gray-code output

Figure 5-32
A 2-to-4 decoder:
(a) inputs and outputs;
(b) logic diagram.

(a)   (b)

Disk Position	I2	I1	I0	Binary Decoder Output
0°	0	0	0	Y0
45°	0	0	1	Y1
90°	0	1	1	Y3
135°	0	1	0	Y2
180°	1	1	0	Y6
225°	1	1	1	Y7
270°	1	0	1	Y5
315°	1	0	0	Y4

Table 5-5
Position encoding for a 3-bit mechanical encoding disk.

of a mechanical encoding disk with eight positions. The eight disk positions can be decoded with a 3-bit binary decoder with the appropriate assignment of signals to the decoder outputs, as shown in Figure 5-33.

Also, it is not necessary to use all of the outputs of a decoder, or even to decode all possible input combinations. For example, a *decimal* or *BCD decoder* decodes only the first ten binary input combinations 0000–1001 to produce outputs Y0–Y9.

decimal decoder
BCD decoder

5.4.2 Logic Symbols for Larger-Scale Elements

Before describing some commercially available 74-series MSI decoders, we need to discuss general guidelines for drawing logic symbols for larger-scale logic elements.

The most basic rule is that logic symbols are drawn with inputs on the left and outputs on the right. The top and bottom edges of a logic symbol are not normally used for signal connections. However, explicit power and ground connections are sometimes shown at the top and bottom, especially if these connections are made on "nonstandard" pins. (Most MSI parts have power and ground connected to the corner pins, e.g., pins 8 and 16 of a 16-pin DIP package.)

Figure 5-33
Using a 3-to-8 binary decoder to decode a Gray code.

LOGIC FAMILIES   Most logic gates and larger-scale elements are available in a variety of CMOS and families, many of which we described in Sections 3.8 and 3.11. For example, the 74LS139, 74S139, 74ALS139, 74AS139, 74F139, 74HC139, 74HCT139, 74ACT139, 74AC139, 74FCT139 74AHC139, 74AHCT139, 74LC139, 74LVC139, and 74VHC139 are all dual 2-to-4 decoders with the same logic function, but in electrically different TTL and CMOS families and sometimes in different packages. In addition, "macro" logic elements with the same pin names and functions as the '139 and other popular 74-series devices are available as building blocks in most FPGA and ASIC design environments.

Throughout this text, we use "74x" as a generic prefix. And we'll sometimes omit the prefix and write, for example, '139. In a real schematic diagram for a circuit that you are going to build or simulate, you should include the full part number, since timing, loading, and packaging characteristics depend on the family.

Like gate symbols, the logic symbols for larger-scale elements associate an active level with each pin. With respect to active levels, it's important to use a consistent convention to naming the internal signals and external pins.

Larger-scale elements almost always have their signal names defined in terms of the functions performed *inside* their symbolic outline, as explained in Section 5.1.4. For example, Figure 5-34(a) shows the logic symbol for one section of a 74x139 dual 2-to-4 decoder, an MSI part that we'll fully describe in the next subsection. When the G input is asserted, one of the outputs Y0–Y3 is asserted, as selected by a 2-bit code applied to the A and B inputs. It is apparent from the symbol that the G input pin and all of the output pins are active low.

When the 74x139 symbol appears in the logic diagram for a real application, its inputs and outputs have signals connected to other devices, and each such signal has a name that indicates its function in the application. However, when we describe the 74x139 in isolation, we might still like to have a name for the signal on each external pin. Figure 5-34(b) shows our naming convention in this case. Active-high pins are given the same name as the internal signal, while active-low pins have the internal signal name followed by the suffix "_L".

Figure 5-34
Logic symbol for one-half of a 74x139 dual 2-to-4 decoder: (a) conventional symbol; (b) default signal names associated with external pins.

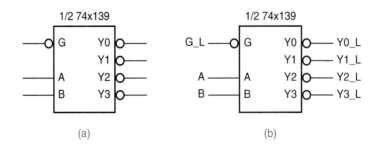

IEEE STANDARD LOGIC SYMBOLS Throughout this book, we use "traditional" symbols for larger-scale logic elements. The IEEE standard uses somewhat different symbols for larger-scale logic elements. IEEE standard symbols, as well as the pros and cons of IEEE versus traditional symbols, are discussed in a guide available on the web at www.ddpp.com.

5.4.3 The 74x139 Dual 2-to-4 Decoder

Two independent and identical 2-to-4 decoders are contained in a single MSI part, the *74x139*. The gate-level circuit diagram for this IC is shown in Figure 5-35(a). Notice that the outputs and the enable input of the '139 are active-low. Most MSI decoders were originally designed with active-low outputs, since TTL inverting gates are generally faster than noninverting ones. Also notice that the '139 has extra inverters on its select inputs. Without these inverters, each select input would present three AC or DC loads instead of one, consuming much more of the fanout budget of the device that drives it.

74x139

Figure 5-35 The 74x139 dual 2-to-4 decoder: (a) logic diagram, including pin numbers for a standard 16-pin dual in-line package; (b) traditional logic symbol; (c) logic symbol for one decoder.

Table 5-6
Truth table for one-half of a 74x139 dual 2-to-4 decoder.

Inputs			Outputs			
G_L	B	A	Y3_L	Y2_L	Y1_L	Y0_L
1	x	x	1	1	1	1
0	0	0	1	1	1	0
0	0	1	1	1	0	1
0	1	0	1	0	1	1
0	1	1	0	1	1	1

A logic symbol for the 74x139 is shown in Figure 5-35(b). Notice that all of the signal names inside the symbol outline are active-high (no "_L"), and that inversion bubbles indicate active-low inputs and outputs. Often a schematic may use a generic symbol for just one decoder, one-half of a '139, as shown in (c). In this case, the assignment of the generic function to one half or the other of a particular '139 package can be deferred until the schematic is completed.

Table 5-6 is the truth table for a 74x139-type decoder. The truth tables in some manufacturers' data books use L and H to denote the input and output signal voltage levels, so there can be no ambiguity about the electrical function of the device; a truth table written this way is sometimes called a *function table*.

function table

However, since we use positive logic throughout this book, we can use 0 and 1 without ambiguity. In any case, the truth table gives the logic function in terms of the *external pins* of the device. A truth table for the function performed *inside* the symbol outline would look just like Table 5-4, except that the input signal names would be G, B, A.

Some logic designers draw the symbol for 74x139s and other logic functions without inversion bubbles. Instead, they use an overbar on signal names inside the symbol outline to indicate negation, as shown in Figure 5-36(a). This notation is self-consistent, but it is inconsistent with our drawing standards for bubble-to-bubble logic design. The symbol shown in (b) is absolutely *incorrect*: according to this symbol, a logic 1, not 0, must be applied to the enable pin to enable the decoder.

Figure 5-36
More ways to symbolize a 74x139: (a) correct but to be avoided; (b) incorrect because of double negations.

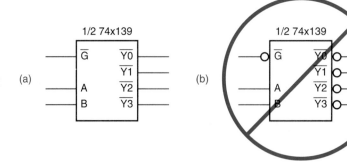

BAD NAMES   Some manufacturers' data sheets have inconsistencies similar to Figure 5-36(b). For example, Texas Instruments' data sheet for the 74AHC139 uses active-low names like $\overline{1G}$ for the enable inputs, with the overbar indicating an active-low pin, but active-high names like 1Y0 for all the active-low output pins. On the other hand, Motorola's data sheet for the 74VHC139 correctly uses overbars on the names for both the enable inputs and the outputs, but the overbars are barely visible in the device's function table due to a typographical problem.

I've also had the personal experience of building a printed-circuit board with many copies of a new device from a vendor whose documentation clearly indicated that a particular input was active low, only to find out upon the first power-on that the input was active high.

The moral of the story is that you have to study the description of each device to know what's really going on. And if it's a brand-new device, whether from a commercial vendor or your own company's ASIC group, you should double-check all of the signal polarities and pin assignments before committing to a PCB. Rest assured, however, that the signal names in this text *are* consistent and correct.

5.4.4 The 74x138 3-to-8 Decoder

The *74x138* is a commercially available MSI 3-to-8 decoder whose gate-level   *74x138*
circuit diagram and symbol are shown in Figure 5-37; its truth table is given in Table 5-7. Like the 74x139, the 74x138 has active-low outputs, and it has three enable inputs (G1, /G2A, /G2B), all of which must be asserted for the selected output to be asserted.

The logic function of the '138 is straightforward—an output is asserted if and only if the decoder is enabled and the output is selected. Thus, we can easily write logic equations for an internal output signal such as Y5 in terms of the internal input signals:

$$Y5 = \underbrace{G1 \cdot G2A \cdot G2B}_{\text{enable}} \cdot \underbrace{C \cdot B' \cdot A}_{\text{select}}$$

However, because of the inversion bubbles, we have the following relations between internal and external signals:

$$G2A = G2A\_L'$$
$$G2B = G2B\_L'$$
$$Y5 = Y5\_L'$$

Therefore, if we're interested, we can write the following equation for the external output signal Y5_L in terms of external input signals:

$$
\begin{aligned}
Y5\_L = Y5' &= (G1 \cdot G2A\_L' \cdot G2B\_L' \cdot C \cdot B' \cdot A)' \\
&= G1' + G2A\_L + G2B\_L + C' + B + A'
\end{aligned}
$$

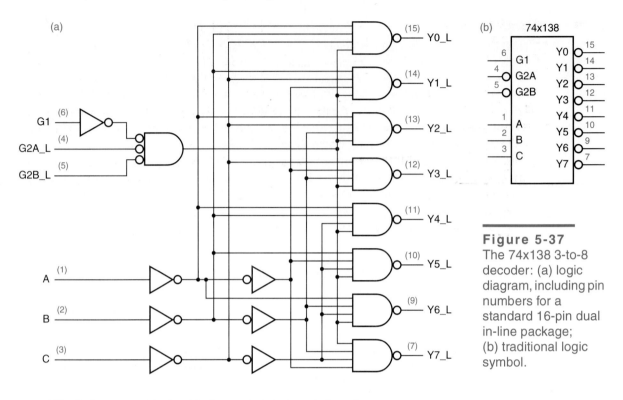

Figure 5-37
The 74x138 3-to-8 decoder: (a) logic diagram, including pin numbers for a standard 16-pin dual in-line package; (b) traditional logic symbol.

Table 5-7 Truth table for a 74x138 3-to-8 decoder.

Inputs						Outputs							
G1	G2A_L	G2B_L	C	B	A	Y7_L	Y6_L	Y5_L	Y4_L	Y3_L	Y2_L	Y1_L	Y0_L
0	x	x	x	x	x	1	1	1	1	1	1	1	1
x	1	x	x	x	x	1	1	1	1	1	1	1	1
x	x	1	x	x	x	1	1	1	1	1	1	1	1
1	0	0	0	0	0	1	1	1	1	1	1	1	0
1	0	0	0	0	1	1	1	1	1	1	1	0	1
1	0	0	0	1	0	1	1	1	1	1	0	1	1
1	0	0	0	1	1	1	1	1	1	0	1	1	1
1	0	0	1	0	0	1	1	1	0	1	1	1	1
1	0	0	1	0	1	1	1	0	1	1	1	1	1
1	0	0	1	1	0	1	0	1	1	1	1	1	1
1	0	0	1	1	1	0	1	1	1	1	1	1	1

On the surface, this equation doesn't resemble what you might expect for a decoder, since it is a logical sum rather than a product. However, if you practice bubble-to-bubble logic design, you don't have to worry about this; you just give the output signal an active-low name and remember that it's active low when you connect it to other inputs.

5.4.5 Cascading Binary Decoders

Multiple binary decoders can be used to decode larger code words. Figure 5-38 shows how two 3-to-8 decoders can be combined to make a 4-to-16 decoder. The availability of both active-high and active-low enable inputs on the 74x138 makes it possible to enable one or the other directly based on the state of the most significant input bit. The top decoder (U1) is enabled when N3 is 0, and the bottom one (U2) is enabled when N3 is 1.

To handle even larger code words, binary decoders can be cascaded hierarchically. Figure 5-39 shows how to use half of a 74x139 to decode the two high-order bits of a 5-bit code word, thereby enabling one of four 74x138s that decode the three low-order bits.

Figure 5-38
Design of a 4-to-16 decoder using 74x138s.

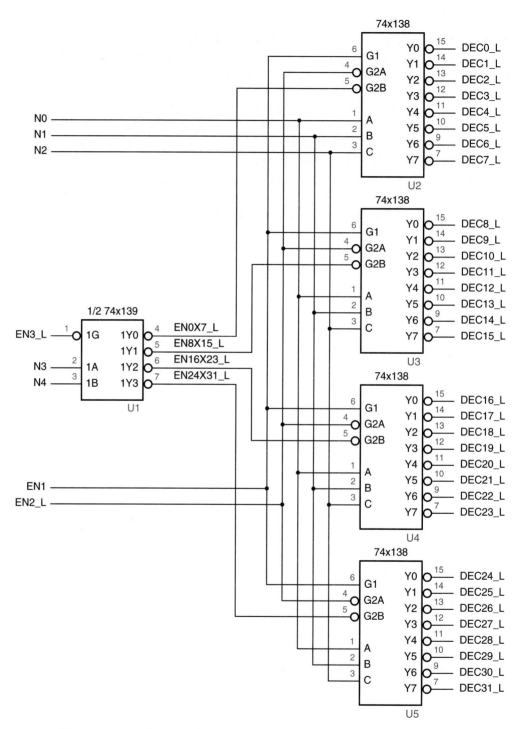

Figure 5-39 Design of a 5-to-32 decoder using 74x138s and a 74x139.

5.4.6 Decoders in ABEL and PLDs

Nothing in logic design is much easier than writing the PLD equations for a decoder. Since the logic expression for each output is typically just a single product term, decoders are very easily targeted to PLDs and use few product-term resources.

For example, Table 5-8 is an ABEL program for a 74x138-like 3-to-8 binary decoder as realized in a PAL16L8. Note that some of the input pins and all of the output pins have active-low names ("_L" suffix) in the pin declarations, corresponding to the logic diagram in Figure 5-37 on page 358. However, the program also defines a corresponding active-high name for each signal so that the equations can all be written "naturally," in terms of active-high signals. An alternate way to achieve the same effect is described in the box on page 363.

Table 5-8 An ABEL program for a 74x138-like 3-to-8 binary decoder.

```
module Z74X138
title '74x138 Decoder PLD
J. Wakerly, Stanford University'
Z74X138 device 'P16L8';

" Input and output pins
A, B, C, G2A_L, G2B_L, G1                      pin 1, 2, 3, 4, 5, 6;
Y0_L, Y1_L, Y2_L, Y3_L, Y4_L, Y5_L, Y6_L, Y7_L  pin 19..12 istype 'com';

" Active-high signal names for readability
G2A = !G2A_L;
G2B = !G2B_L;
Y0 = !Y0_L;
Y1 = !Y1_L;
Y2 = !Y2_L;
Y3 = !Y3_L;
Y4 = !Y4_L;
Y5 = !Y5_L;
Y6 = !Y6_L;
Y7 = !Y7_L;

" Constant expression
ENB = G1 & G2A & G2B;

equations
Y0 = ENB & !C & !B & !A;
Y1 = ENB & !C & !B &  A;
Y2 = ENB & !C &  B & !A;
Y3 = ENB & !C &  B &  A;
Y4 = ENB &  C & !B & !A;
Y5 = ENB &  C & !B &  A;
Y6 = ENB &  C &  B & !A;
Y7 = ENB &  C &  B &  A;

end Z74X138
```

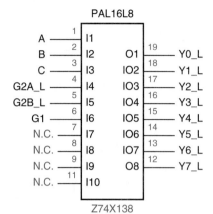

Figure 5-40
Logic diagram for
the PAL16L8 used as
a 74x138 decoder.

Also note that the ABEL program defines a constant expression for ENB. Here, ENB is not an input or output signal, but merely a user-defined name. In the equations section, the compiler substitutes the expression (G1 & G2A & G2B) everywhere that "ENB" appears. Assigning the constant expression to a user-defined name improves this program's readability and maintainability.

If all you needed was a '138, you'd be better off using a real '138 than a more expensive PLD. However, if you need nonstandard functionality, then the PLD can usually achieve it much more cheaply and easily than an MSI/SSI-based solution. For example, if you need the functionality of a '138 but with active-high outputs, you need only to change one line in the pin declarations of Table 5-8:

```
Y0, Y1, Y2, Y3, Y4, Y5, Y6, Y7            pin 19..12 istype 'com';
```

(Also, the original definitions of Y0–Y7 in Table 5-8 must be deleted.) Since each of the equations required a single product of six variables (including the three in the ENB expression), each complemented equation requires a *sum* of six product terms, less than the seven available in a PAL16L8. If you use a PAL16V8 or other device with output polarity selection, then the compiler can select non-inverted output polarity to use only one product term per output.

Table 5-9 Alternate declarations for a 74x138-like 3-to-8 binary decoder.

```
" Input and output pins
A, B, C, !G2A, !G2B, G1                    pin 1, 2, 3, 4, 5, 6;
!Y0, !Y1, !Y2, !Y3, !Y4, !Y5, !Y6, !Y7    pin 19..12 istype 'com';

" Constant expression
ENB = G1 & G2A & G2B;
```

ACTIVE-LOW PIN DEFINITIONS

ABEL allows you to use an inversion prefix (!) on signal names in the pin definitions of a program. When a pin name is defined with the inversion prefix, the compiler automatically prepends an inversion prefix to the signal name anywhere it appears elsewhere in the program. If it's already inverted, this results in a double inversion.

This feature can be used to define a different but consistent convention for defining active-low inputs and outputs—give each active-low signal an active-*high* name, but precede it with the inversion prefix in its pin definition. For the 3-to-8 decoder in Table 5-8, we replace the first part of the program with the code shown in Table 5-9; the equations section of the program stays exactly the same.

Which convention to use may be a matter of personal taste, but it can also depend on the capabilities of the CAD tools that you use to draw schematics. Many tools allow you to automatically create a schematic symbol from a logic block that is defined by an ABEL program. If the tool allows you to place inversion bubbles on selected inputs and outputs of the symbol, then the convention in Table 5-9 yields a symbol with active-high signal names inside the function outline. You can then specify external bubbles on the active-low signals to obtain a symbol that matches the conventions described in Section 5.4.2 and shown in Figure 5-41(a).

On the other hand, you may not be able or want to provide inversion bubbles on CAD-created symbols. In that case, you should use the convention in Table 5-8; this yields a CAD-created symbol in which the active level is indicated by the signal name inside the function outline; no external inversion bubbles are needed. This is shown in Figure 5-41(b). Note that unlike Figure 5-36(a) on page 356, we use a text-based convention (_L) rather than an overbar on the signal name to indicate active level. A properly chosen text-based convention provides portability among different CAD tools.

We'll somewhat arbitrarily select one convention or the other in each of the ABEL examples in the rest of this book, just to help you get comfortable with both approaches.

(a)

(b)

Figure 5-41
Possible CAD-created symbols for the PLD-based, 74x138-like decoder: (a) based on Table 5-9, after manual insertion of inversion bubbles; (b) based on Table 5-8.

Another easy change is to provide alternate enable inputs that are ORed with the main enable inputs. To do this, you need only define additional pins and modify the definition of ENB:

```
EN1, EN2_L                              pin 7, 8;
...
EN2 = !EN2_L;
...
ENB = G1 & G2A & G2B # EN1 # EN2;
```

This change expands the number of product terms per output to three, each having a form similar to

```
Y0 = G1 & G2A & G2B & !C & !B &!A
   # EN1 & !C & !B & !A
   # EN2 & !C & !B & !A;
```

(Remember that the PAL16L8 has a fixed inverter and the PAL16V8 has a selectable inverter between the AND-OR array and the output of the PLD, so the actual output is active low as desired.)

If you add the extra enables to the version of the program with active-high outputs, then the PLD must realize the complement of the sum-of-products expression above. It's not immediately obvious how many product terms this expression will have, and whether it will fit in a PAL16L8, but we can use the ABEL compiler to get the answer for us:

```
!Y0 = C # B # A # !G2B & !EN1 & !EN2
              # !G2A & !EN1 & !EN2
              # !G1  & !EN1 & !EN2;
```

The expression has a total of six product terms, so it fits in a PAL16L8.

As a final tweak, we can add an input to dynamically control whether the output is active high or active low, and modify all of the equations as follows:

```
POL                pin 9;
...
Y0 = POL $ (ENB & !C & !B & !A);
Y1 = POL $ (ENB & !C & !B &  A);
...
Y7 = POL $ (ENB &  C &  B &  A);
```

As a result of the XOR operation, the number of product terms needed per output increases to 9, in either output-pin polarity. Thus, even a PAL16V8 cannot implement the function as written.

helper output   The function can still be realized if we create a *helper output* to reduce the product-term explosion. As shown in Table 5-10, we allocate an output pin for

Table 5-10 ABEL program fragment showing two-pass logic.

```
...
" Output pins
Y0_L, Y1_L, Y2_L, Y3_L        pin 19, 18, 17, 16 istype 'com';
Y4_L, Y5_L, Y6_L, ENB         pin 15, 14, 13, 12 istype 'com';

equations
ENB = G1 & G2A & G2B # EN1 # EN2;
Y0 = POL $ (ENB & !C & !B & !A);
...
```

the ENB expression (losing the Y7_L output), and move the ENB equation into the equations section of the program. This reduces the product-term requirement to five in either polarity.

helper output

Besides sacrificing a pin for the helper output, this realization has the disadvantage of being slower. Any changes in the inputs to the helper expression must propagate through the PLD twice before reaching the final output. This is called *two-pass logic*. Many PLD and FPGA synthesis tools can automatically generate logic with two or more passes if a required expression cannot be realized in just one pass through the logic array.

two-pass logic

Decoders can be customized in other ways. A common customization is for a single output to decode more than one input combination. For example, suppose you needed to generate a set of enable signals according to Table 5-11. A 74x138 MSI decoder can be augmented as shown in Figure 5-42 to perform the required function. This approach, while potentially less expensive than a PLD, has the disadvantages that it requires extra components and delay to create the required outputs, and it is not easily modified.

Table 5-11
Truth table for a customized decoder function.

CS_L	RD_L	A2	A1	A0	Output(s) to Assert
1	x	x	x	x	none
x	1	x	x	x	none
0	0	0	0	0	BILL_L, MARY_L
0	0	0	0	1	MARY_L, KATE_L
0	0	0	1	0	JOAN_L
0	0	0	1	1	PAUL_L
0	0	1	0	0	ANNA_L
0	0	1	0	1	FRED_L
0	0	1	1	0	DAVE_L
0	0	1	1	1	KATE_L

Figure 5-42
Customized
decoder circuit.

A PLD solution to the same problem is shown in Table 5-12. Note that this program uses the pin active-low pin-naming convention described in the box on page 363 (you should be comfortable with either convention). Each of the last six equations uses a single AND gate in the PLD. The ABEL compiler will also minimize the MARY equation to use just one AND gate. Active-high output signals could be obtained just by changing two lines in the declaration section:

```
BILL, MARY, JOAN, PAUL      pin 19, 18, 17, 16 istype 'com';
ANNA, FRED, DAVE, KATE      pin 15, 14, 13, 12 istype 'com';
```

Table 5-12 ABEL equations for a customized decoder.

```
module CUSTMDEC
title 'Customized Decoder PLD
J. Wakerly, Stanford University'
CUSTMDEC device 'P16L8';

" Input pins
!CS, !RD, A0, A1, A2        pin 1, 2, 3, 4, 5;
" Output pins
!BILL, !MARY, !JOAN, !PAUL  pin 19, 18, 17, 16 istype 'com';
!ANNA, !FRED, !DAVE, !KATE  pin 15, 14, 13, 12 istype 'com';

equations
BILL = CS & RD & (!A2 & !A1 & !A0);
MARY = CS & RD & (!A2 & !A1 & !A0 # !A2 & !A1 &  A0);
KATE = CS & RD & (!A2 & !A1 &  A0 #  A2 &  A1 &  A0);
JOAN = CS & RD & (!A2 &  A1 & !A0);
PAUL = CS & RD & (!A2 &  A1 &  A0);
ANNA = CS & RD & ( A2 & !A1 & !A0);
FRED = CS & RD & ( A2 & !A1 &  A0);
DAVE = CS & RD & ( A2 &  A1 & !A0);

end CUSTMDEC
```

Another way of writing the equations is shown in Table 5-13. In most applications, this style is more clear, especially if the select inputs have numeric significance.

Table 5-13 Equivalent ABEL equations for a customized decoder.

```
ADDR = [A2,A1,A0];

equations
BILL = CS & RD & (ADDR == 0);
MARY = CS & RD & (ADDR == 0) # (ADDR == 1);
KATE = CS & RD & (ADDR == 1) # (ADDR == 7);
JOAN = CS & RD & (ADDR == 2);
PAUL = CS & RD & (ADDR == 3);
ANNA = CS & RD & (ADDR == 4);
FRED = CS & RD & (ADDR == 5);
DAVE = CS & RD & (ADDR == 6);
```

5.4.7 Decoders in VHDL

There are several ways to approach the design of decoders in VHDL. The most primitive approach would be to write a structural equivalent of a decoder logic circuit, as Table 5-14 does for the 2-to-4 binary decoder of Figure 5-32 on page 352. The components and3 and inv are assumed to already exist in the target technology. Of course, this mechanical conversion of an existing design into the equivalent of a netlist defeats the purpose of using VHDL in the first place.

Instead, we would like to write a program that uses VHDL to make our decoder design more understandable and maintainable. Table 5-15 shows one approach to writing code for a 3-to-8 binary decoder equivalent to the 74x138, using the dataflow style of VHDL. The address inputs A(2 downto 0) and the active-low decoded outputs Y_L(0 to 7) are declared using vectors to improve readability. A select statement enumerates the eight decoding cases and assigns the appropriate active-low output pattern to an 8-bit internal signal Y_L_i. This value is assigned to the actual circuit output Y_L only if all of the enable inputs are asserted.

This design is a good start, and it works, but it does have a potential pitfall. The adjustments that handle the fact that two inputs and all the outputs are active-low happen to be buried in the final assignment statement. While it's true that most VHDL programs are written almost entirely with active-high signals, if we're defining a device with active-low external pins, we really should handle them in a more systematic and easily maintainable way.

Table 5-14 VHDL structural program for the decoder in Figure 5-32.

```
library IEEE;
use IEEE.std_logic_1164.all;

entity V2to4dec is
  port (I0, I1, EN: in STD_LOGIC;
        Y0, Y1, Y2, Y3: out STD_LOGIC );
end V2to4dec;

architecture V2to4dec_s of V2to4dec is
  signal NOTI0, NOTI1: STD_LOGIC;
  component inv port (I: in STD_LOGIC; O: out STD_LOGIC ); end component;
  component and3 port (I0, I1, I2: in STD_LOGIC; O: out STD_LOGIC ); end component;
begin
  U1: inv port map (I0,NOTI0);
  U2: inv port map (I1,NOTI1);
  U3: and3 port map (NOTI0,NOTI1,EN,Y0);
  U4: and3 port map (   I0,NOTI1,EN,Y1);
  U5: and3 port map (NOTI0,   I1,EN,Y2);
  U6: and3 port map (   I0,   I1,EN,Y3);
end V2to4dec_s;
```

Table 5-15 Dataflow-style VHDL program for a 74x138-like 3-to-8 binary decoder.

```
library IEEE;
use IEEE.std_logic_1164.all;

entity V74x138 is
    port (G1, G2A_L, G2B_L: in STD_LOGIC;        -- enable inputs
          A: in STD_LOGIC_VECTOR (2 downto 0);   -- select inputs
          Y_L: out STD_LOGIC_VECTOR (0 to 7) );  -- decoded outputs
  end V74x138;

architecture V74x138_a of V74x138 is
  signal Y_L_i: STD_LOGIC_VECTOR (0 to 7);
begin
  with A select Y_L_i <=
    "01111111" when "000",
    "10111111" when "001",
    "11011111" when "010",
    "11101111" when "011",
    "11110111" when "100",
    "11111011" when "101",
    "11111101" when "110",
    "11111110" when "111",
    "11111111" when others;
  Y_L <= Y_L_i when (G1 and not G2A_L and not G2B_L)='1' else "11111111";
end V74x138_a;
```

Table 5-16 VHDL architecture with a maintainable approach to active-level handling.

```
architecture V74x138_b of V74x138 is
  signal G2A, G2B: STD_LOGIC;              -- active-high version of inputs
  signal Y: STD_LOGIC_VECTOR (0 to 7);    -- active-high version of outputs
  signal Y_s: STD_LOGIC_VECTOR (0 to 7);  -- internal signal
begin
  G2A <= not G2A_L; -- convert inputs
  G2B <= not G2B_L; -- convert inputs
  Y_L <= Y;         -- convert outputs
  with A select Y_s <=
    "10000000" when "000",
    "01000000" when "001",
    "00100000" when "010",
    "00010000" when "011",
    "00001000" when "100",
    "00000100" when "101",
    "00000010" when "110",
    "00000001" when "111",
    "00000000" when others;
  Y <= not Y_s when (G1 and G2A and G2B)='1' else "00000000";
end V74x138_b;
```

Table 5-16 shows such an approach. No changes are made to the `entity` declarations. However, active-high versions of the active-low external pins are defined within the V74x138_a architecture, and explicit assignment statements are used to convert between the active-high and active-low signals. The decoder function itself is defined in terms of only the active-high signals, probably the biggest advantage of this approach. Another advantage is that the design can be easily modified in just a few well-defined places if changes are required in the external active levels.

OUT-OF-ORDER EXECUTION   In Table 5-16, we've grouped all three of the active-level conversion statements together at the beginning of program, even a value isn't assigned to Y_L until *after* a value is assigned to Y, later in the program. Remember that this is OK because the assignment statements in the architecture body are executed concurrently. That is, an assignment to any signal causes all the other statements that use that signal to be reevaluated, regardless of their position in the architecture body.

You could put the "Y_L <= Y" statement at the end of the body if its current position bothers you, but the program is a bit more maintainable in its present form, with all the active-level conversions together.

Table 5-17 Hierarchical definition of 74x138-like decoder with active-level handling.

```
architecture V74x138_c of V74x138 is
  signal G2A, G2B: STD_LOGIC;              -- active-high version of inputs
  signal Y: STD_LOGIC_VECTOR (0 to 7);  -- active-high version of outputs
  component V3to8dec port (G1, G2, G3: in STD_LOGIC;
                             A: in STD_LOGIC_VECTOR (2 downto 0);
                             Y: out STD_LOGIC_VECTOR (0 to 7) ); end component;
begin
  G2A <= not G2A_L;   -- convert inputs
  G2B <= not G2B_L;   -- convert inputs
  Y_L <= not Y;       -- convert outputs
  U1: V3to8dec port map (G1, G2A, G2B, A, Y);
end V74x138_c;
```

Active levels can be handled in an even more structured way. As shown in Table 5-17, the V74x138 architecture can be defined hierarchically, using a fully active-high V3to8dec component that has its own dataflow-style definition in Table 5-18. Once again, no changes are required in the top-level definition of the V74x138 entity. Figure 5-43 shows the relationship between the entities.

Still another approach to decoder design is shown in Table 5-19, which can replace the V3to8dec_a architecture of Table 5-18. Instead of concurrent state-

Table 5-18
Dataflow definition of an active-high 3-to-8 decoder.

```
library IEEE;
use IEEE.std_logic_1164.all;

entity V3to8dec is
    port (G1, G2, G3: in STD_LOGIC;
          A: in STD_LOGIC_VECTOR (2 downto 0);
          Y: out STD_LOGIC_VECTOR (0 to 7) );
end V3to8dec;

architecture V3to8dec_a of V3to8dec is
  signal Y_s: STD_LOGIC_VECTOR (0 to 7);
begin
    with A select Y_s <=
      "10000000" when "000",
      "01000000" when "001",
      "00100000" when "010",
      "00010000" when "011",
      "00001000" when "100",
      "00000100" when "101",
      "00000010" when "110",
      "00000001" when "111",
      "00000000" when others;
    Y <= Y_s when (G1 and G2 and G3)='1'
              else "00000000";
end V3to8dec_a;
```

Figure 5-43 VHDL entity V74x138: (a) top level; (b) internal structure using architecture V74x138_c.

NAME MATCHING   In Figure 5-43, the port names of an entity are drawn inside the corresponding box. The names of the signals that are connected to the ports when the entity is used are drawn on the signal lines. Notice that the signal names may match, but they don't have to. The VHDL compiler keeps everything straight, associating a scope with each name. The situation is completely analogous to the way variable and parameter names are handled in structured, procedural programming languages like C.

ments, this architecture uses a process and sequential statements to define the decoder's operation in a behavioral style. However, a close comparison of the two architectures shows that they're really not that different except for syntax.

```
architecture V3to8dec_b of V3to8dec is
  signal Y_s: STD_LOGIC_VECTOR (0 to 7);
begin
process(A, G1, G2, G3, Y_s)
  begin
    case A is
      when "000" => Y_s <= "10000000";
      when "001" => Y_s <= "01000000";
      when "010" => Y_s <= "00100000";
      when "011" => Y_s <= "00010000";
      when "100" => Y_s <= "00001000";
      when "101" => Y_s <= "00000100";
      when "110" => Y_s <= "00000010";
      when "111" => Y_s <= "00000001";
      when others => Y_s <= "00000000";
    end case;
    if (G1 and G2 and G3)='1' then Y <= Y_s;
    else Y <= "00000000";
    end if;
  end process;
end V3to8dec_b;
```

Table 5-19
Behavioral-style architecture definition for a 3-to-8 decoder.

Table 5-20
Truly behavioral
architecture definition
for a 3-to-8 decoder.

```
architecture V3to8dec_c of V3to8dec is
begin
process (G1, G2, G3, A)
  variable i: INTEGER range 0 to 7;
  begin
    Y <= "00000000";
    if (G1 and G2 and G3) = '1' then
      for i in 0 to 7 loop
        if i=CONV_INTEGER(A) then Y(i) <= '1'; end if;
      end loop;
    end if;
  end process;
end V3to8dec_c;
```

As a final example, a more truly behavioral, process-based architecture for the 3-to-8 decoder is shown in Table 5-20. (Recall that the CONV_INTEGER function was defined in Section 4.7.4.) Of the examples we've given, this is the only one that describes the decoder function without essentially embedding a truth table in the VHDL program. In that respect, it is more flexible because it can be easily adapted to make a binary decoder of any size. In another respect, it is less flexible in that it does not have a truth table that can be easily modified to make custom decoders like the one we specified in Table 5-11 on page 365.

*5.4.8 Seven-Segment Decoders

seven-segment display

Look at your wrist and you'll probably see a *seven-segment display*. This type of display, which normally uses light-emitting diodes (LEDs) or liquid-crystal display (LCD) elements, is used in watches, calculators, and instruments to display decimal data. A digit is displayed by illuminating a subset of the seven line segments shown in Figure 5-44(a).

seven-segment decoder

A *seven-segment decoder* has 4-bit BCD as its input code and the "seven-segment code," which is graphically depicted in Figure 5-44(b), as its output code. Figure 5-45 and Table 5-21 are the logic diagram and truth table for a

74x49

74x49 seven-segment decoder. Except for the strange (clever?) connection of the "blanking input" BI_L, each output of the 74x49 is a minimal product-of-sums

Figure 5-44 Seven-segment display: (a) segment identification; (b) decimal digits.

(a)   (b)

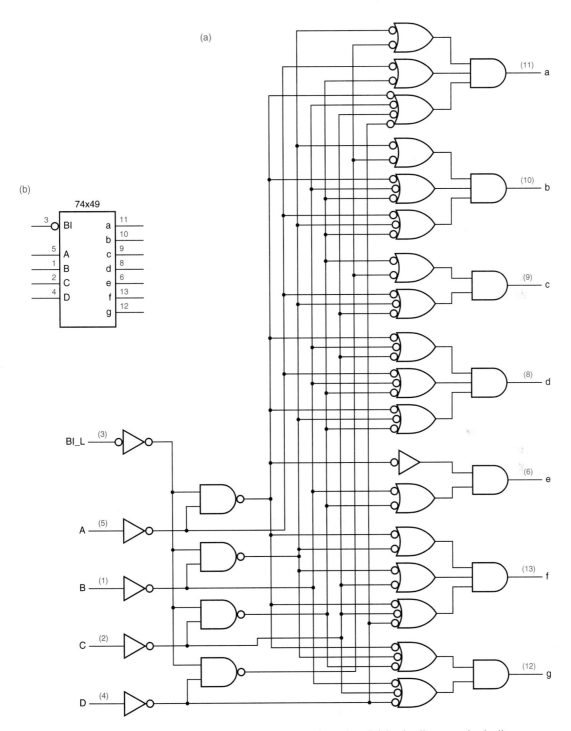

Figure 5-45 The 74x49 seven-segment decoder: (a) logic diagram, including pin numbers; (b) traditional logic symbol.

Table 5-21 Truth table for a 74x49 seven-segment decoder.

Inputs					Outputs						
BI_L	D	C	B	A	a	b	c	d	e	f	g
0	x	x	x	x	0	0	0	0	0	0	0
1	0	0	0	0	1	1	1	1	1	1	0
1	0	0	0	1	0	1	1	0	0	0	0
1	0	0	1	0	1	1	0	1	1	0	1
1	0	0	1	1	1	1	1	1	0	0	1
1	0	1	0	0	0	1	1	0	0	1	1
1	0	1	0	1	1	0	1	1	0	1	1
1	0	1	1	0	0	0	1	1	1	1	1
1	0	1	1	1	1	1	1	0	0	0	0
1	1	0	0	0	1	1	1	1	1	1	1
1	1	0	0	1	1	1	1	0	0	1	1
1	1	0	1	0	0	0	0	1	1	0	1
1	1	0	1	1	0	0	1	1	0	0	1
1	1	1	0	0	0	1	0	0	0	1	1
1	1	1	0	1	1	0	0	1	0	1	1
1	1	1	1	0	0	0	0	1	1	1	1
1	1	1	1	1	0	0	0	0	0	0	0

realization for the corresponding segment, assuming "don't-cares" for the non-decimal input combinations. The INVERT-OR-AND structure used for each output may seem a little strange, but it is equivalent under the generalized DeMorgan's theorem to an AND-OR-INVERT gate, which is a fairly fast and compact structure to build in CMOS or TTL.

Most modern seven-segment display elements have decoders built into them, so that a 4-bit BCD word can be applied directly to the device. Many of the older, discrete seven-segment decoders have special high-voltage or high-current outputs that are well suited for driving large, high-powered display elements.

Table 5-22 is an ABEL program for a seven-segment decoder. Sets are used to define the digit patterns to make the program more readable.

Table 5-22 ABEL program for a 74x49-like seven-segment decoder.

```
module Z74X49H
title 'Seven-Segment_Decoder
J. Wakerly, Micro Design Resources, Inc.'
Z74X49H device 'P16L8';

" Input pins
A, B, C, D                     pin 1, 2, 3, 4;
BI_L                           pin 5;
" Output pins
SEGA, SEGB, SEGC, SEGD         pin 19, 18, 17, 16 istype 'com';
SEGE, SEGF, SEGG               pin 15, 14, 13     istype 'com';

" Definitions
BI = !BI_L
DIGITIN = [D,C,B,A];
SEGOUT = [SEGA,SEGB,SEGC,SEGD,SEGE,SEGF,SEGG];

" Segment encodings for digits
DIG0 = [1,1,1,1,1,1,0];   " 0
DIG1 = [0,1,1,0,0,0,0];   " 1
DIG2 = [1,1,0,1,1,0,1];   " 2
DIG3 = [1,1,1,1,0,0,1];   " 3
DIG4 = [0,1,1,0,0,1,1];   " 4
DIG5 = [1,0,1,1,0,1,1];   " 5
DIG6 = [1,0,1,1,1,1,1];   " 6   'tail' included
DIG7 = [1,1,1,0,0,0,0];   " 7
DIG8 = [1,1,1,1,1,1,1];   " 8
DIG9 = [1,1,1,1,0,1,1];   " 9   'tail' included
DIGA = [1,1,1,0,1,1,1];   " A
DIGB = [0,0,1,1,1,1,1];   " b
DIGC = [1,0,0,1,1,1,0];   " C
DIGD = [0,1,1,1,1,0,1];   " d
DIGE = [1,0,0,1,1,1,1];   " E
DIGF = [1,0,0,0,1,1,1];   " F

equations

SEGOUT = !BI & ( (DIGITIN ==  0) & DIG0 # (DIGITIN ==  1) & DIG1
             # (DIGITIN ==  2) & DIG2 # (DIGITIN ==  3) & DIG3
             # (DIGITIN ==  4) & DIG4 # (DIGITIN ==  5) & DIG5
             # (DIGITIN ==  6) & DIG6 # (DIGITIN ==  7) & DIG7
             # (DIGITIN ==  8) & DIG8 # (DIGITIN ==  9) & DIG9
             # (DIGITIN == 10) & DIGA # (DIGITIN == 11) & DIGB
             # (DIGITIN == 12) & DIGC # (DIGITIN == 13) & DIGD
             # (DIGITIN == 14) & DIGE # (DIGITIN == 15) & DIGF );

end Z74X49H
```

5.5 Encoders

encoder

A decoder's output code normally has more bits than its input code. If the device's output code has *fewer* bits than the input code, the device is usually called an *encoder*. For example, consider a device with eight input bits representing an unsigned binary number, and two output bits indicating whether the number is prime or divisible by 7. We might call such a device a lucky/prime encoder.

2^n-to-n encoder

binary encoder

Probably the simplest encoder to build is a 2^n-to-n or *binary encoder*. As shown in Figure 5-46(a), it has just the opposite function as a binary *decoder*—its input code is the 1-out-of-2^n code and its output code is n-bit binary. The equations for an 8-to-3 encoder with inputs I0–I7 and outputs Y0–Y2 are given below:

$$Y0 = I1 + I3 + I5 + I7$$
$$Y1 = I2 + I3 + I6 + I7$$
$$Y2 = I4 + I5 + I6 + I7$$

The corresponding logic circuit is shown in (b). In general, a 2^n-to-n encoder can be built from n 2^{n-1}-input OR gates. Bit i of the input code is connected to OR gate j if bit j in the binary representation of i is 1.

Figure 5-46
Binary encoder:
(a) general structure;
(b) 8-to-3 encoder.

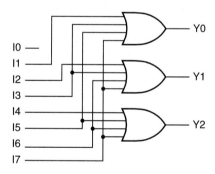

5.5.1 Priority Encoders

The 1-out-of-2^n coded outputs of an n-bit binary decoder are generally used to control a set of 2^n devices, where at most one device is supposed to be active at any time. Conversely, consider a system with 2^n *inputs*, each of which indicates a request for service, as in Figure 5-47. This structure is often found in microprocessor input/output subsystems, where the inputs might be interrupt requests.

In this situation, it may seem natural to use a binary encoder of the type shown in Figure 5-46 to observe the inputs and indicate which one is requesting service at any time. However, this encoder works properly only if the inputs are guaranteed to be asserted at most one at a time. If multiple requests can be made

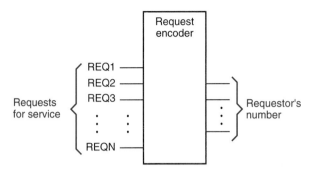

Figure 5-47
A system with 2^n requestors, and a "request encoder" that indicates which request signal is asserted at any time.

simultaneously, the encoder gives undesirable results. For example, suppose that inputs I2 and I4 of the 8-to-3 encoder are both 1; then the output is 110, the binary encoding of 6.

Either 2 or 4, not 6, would be a useful output in the preceding example, but how can the encoding device decide which? The solution is to assign *priority* to the input lines, so that when multiple requests are asserted, the encoding device produces the number of the highest-priority requestor. Such a device is called a *priority encoder*.

priority

priority encoder

The logic symbol for an 8-input priority encoder is shown in Figure 5-48. Input I7 has the highest priority. Outputs A2–A0 contain the number of the highest-priority asserted input, if any. The IDLE output is asserted if no inputs are asserted.

In order to write logic equations for the priority encoder's outputs, we first define eight intermediate variables H0–H7, such that Hi is 1 if and only if Ii is the highest priority 1 input:

$$H7 = I7$$
$$H6 = I6 \cdot I7'$$
$$H5 = I5 \cdot I6' \cdot I7'$$
$$\cdots$$
$$H0 = I0 \cdot I1' \cdot I2' \cdot I3' \cdot I4' \cdot I5' \cdot I6' \cdot I7'$$

Priority encoder

Figure 5-48
Logic symbol for a generic 8-input priority encoder.

Using these signals, the equations for the A2–A0 outputs are similar to the ones for a simple binary encoder:

$$A2 = H4 + H5 + H6 + H7$$
$$A1 = H2 + H3 + H6 + H7$$
$$A0 = H1 + H3 + H5 + H7$$

The IDLE output is 1 if no inputs are 1:

$$IDLE = (I0 + I1 + I2 + I3 + I4 + I5 + I6 + I7)'$$
$$= I0' \cdot I1' \cdot I2' \cdot I3' \cdot I4' \cdot I5' \cdot I6' \cdot I7'$$

74x148

5.5.2 The 74x148 Priority Encoder

The *74x148* is a commercially available, MSI 8-input priority encoder. Its logic symbol is shown in Figure 5-49 and its schematic is shown in Figure 5-50. The main difference between this IC and the "generic" priority encoder of Figure 5-48 is that its inputs and outputs are active low. Also, it has an enable input, EI_L, that must be asserted for any of its outputs to be asserted. The complete truth table is given in Table 5-23.

Instead of an IDLE output, the '148 has a GS_L output that is asserted when the device is enabled and one or more of the request inputs are asserted. The manufacturer calls this "Group Select," but it's easier to remember as "Got Something." The EO_L signal is an enable *output* designed to be connected to the EI_L input of another '148 that handles lower-priority requests. EO_L is asserted if EI_L is asserted but no request input is asserted; thus, a lower-priority '148 may be enabled.

Figure 5-51 shows how four 74x148s can be connected in this way to accept 32 request inputs and produce a 5-bit output, RA4–RA0, indicating the highest-priority requestor. Since the A2–A0 outputs of at most one '148 will be enabled at any time, the outputs of the individual '148s can be ORed to produce RA2–RA0. Likewise, the individual GS_L outputs can be combined in a 4-to-2 encoder to produce RA4 and RA3. The RGS output is asserted if any GS output is asserted.

Figure 5-49
Logic symbol for the 74x148 8-input priority encoder.

Table 5-23 Truth table for a 74x148 8-input priority encoder.

Inputs									Outputs				
EI_L	I0_L	I1_L	I2_L	I3_L	I4_L	I5_L	I6_L	I7_L	A2_L	A1_L	A0_L	GS_L	EO_L
1	x	x	x	x	x	x	x	x	1	1	1	1	1
0	x	x	x	x	x	x	x	0	0	0	0	0	1
0	x	x	x	x	x	x	0	1	0	0	1	0	1
0	x	x	x	x	x	0	1	1	0	1	0	0	1
0	x	x	x	x	0	1	1	1	0	1	1	0	1
0	x	x	x	0	1	1	1	1	1	0	0	0	1
0	x	x	0	1	1	1	1	1	1	0	1	0	1
0	x	0	1	1	1	1	1	1	1	1	0	0	1
0	0	1	1	1	1	1	1	1	1	1	1	0	1
0	1	1	1	1	1	1	1	1	1	1	1	1	0

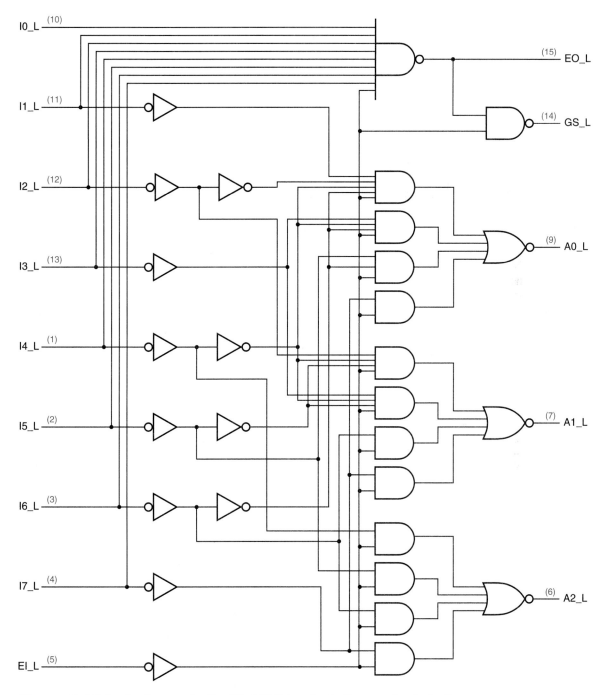

Figure 5-50 Logic diagram for the 74x148 8-input priority encoder, including pin numbers for a standard 16-pin dual in-line package.

Figure 5-51 Four 74x148s cascaded to handle 32 requests.

5.5.3 Encoders in ABEL and PLDs

Encoders can be designed in ABEL using an explicit equation for each input combinations, as in Table 5-8 on page 361, or using truth tables. However, since the number of inputs is usually large, the number of input combinations is very large, and this method often is not practical.

For example, how would we specify a 15-input priority encoder for inputs P0–P14? We obviously don't want to deal with all 2^{15} possible input combinations! One way to do it is to decompose the priority function into two parts. First, we write equations for 15 variables Hi ($0 \le i \le 14$) such that Hi is 1 if Pi is the highest-priority asserted input. Since by definition at most one Hi variable is 1 at any time, we can combine the Hi's in a binary encoder to obtain a 4-bit number identifying the highest-priority asserted input.

An ABEL program using this approach is shown in Table 5-24, and a logic diagram for the encoder using a single PAL20L8 or GAL20V8 is given in Figure 5-52. Inputs P0–P14 are asserted to indicate requests, with P14 having the highest priority. If EN_L (Enable) is asserted, then the Y3_L–Y0_L outputs give the number (active low) of the highest-priority request, and GS is asserted if any request is present. If EN_L is negated, then the Y3_L–Y0_L outputs are negated and GS is negated. ENOUT_L is asserted if EN_L is asserted and no request is present.

Notice that in the ABEL program, the equations for the Hi variables are written as "constant expressions," before the equations declaration. Thus, these signals will not be generated explicitly. Rather, they will be incorporated in the subsequent equations for Y0–Y3, which the compiler cranks on to obtain

Figure 5-52
Logic diagram for a
PLD-based 15-input
priority encoder

Table 5-24 An ABEL program for a 15-input priority encoder.

```
module PRIOR15
title '15-Input Priority Encoder '
PRIOR15 device 'P20L8';

" Input and output pins
P0..P14, EN_L                           pin 1..11, 13, 14, 23, 16, 17;
Y3_L, Y2_L, Y1_L, Y0_L, GS, ENOUT_L   pin 18..21, 15, 22 istype 'com';

" Active-level translation
EN = !EN_L; ENOUT = !ENOUT_L;
Y3 = !Y3_L; Y2 = !Y2_L; Y1 = !Y1_L; Y0 = !Y0_L;

" Constant expressions
H14 = EN&P14;
H13 = EN&!P14&P13;
H12 = EN&!P14&!P13&P12;
H11 = EN&!P14&!P13&!P12&P11;
H10 = EN&!P14&!P13&!P12&!P11&P10;
H9  = EN&!P14&!P13&!P12&!P11&!P10&P9;
H8  = EN&!P14&!P13&!P12&!P11&!P10&!P9&P8;
H7  = EN&!P14&!P13&!P12&!P11&!P10&!P9&!P8&P7;
H6  = EN&!P14&!P13&!P12&!P11&!P10&!P9&!P8&!P7&P6;
H5  = EN&!P14&!P13&!P12&!P11&!P10&!P9&!P8&!P7&!P6&P5;
H4  = EN&!P14&!P13&!P12&!P11&!P10&!P9&!P8&!P7&!P6&!P5&P4;
H3  = EN&!P14&!P13&!P12&!P11&!P10&!P9&!P8&!P7&!P6&!P5&!P4&P3;
H2  = EN&!P14&!P13&!P12&!P11&!P10&!P9&!P8&!P7&!P6&!P5&!P4&!P3&P2;
H1  = EN&!P14&!P13&!P12&!P11&!P10&!P9&!P8&!P7&!P6&!P5&!P4&!P3&!P2&P1;
H0  = EN&!P14&!P13&!P12&!P11&!P10&!P9&!P8&!P7&!P6&!P5&!P4&!P3&!P2&!P1&P0;

equations

Y3 = H8 # H9 # H10 # H11 # H12 # H13 # H14;
Y2 = H4 # H5 # H6 # H7 # H12 # H13 # H14;
Y1 = H2 # H3 # H6 # H7 # H10 # H11 # H14;
Y0 = H1 # H3 # H5 # H7 # H9 # H11 # H13;

GS    = EN&(P14#P13#P12#P11#P10#P9#P8#P7#P6#P5#P4#P3#P2#P1#P0);
ENOUT = EN&!P14&!P13&!P12&!P11&!P10&!P9&!P8&!P7&!P6&!P5&!P4&!P3&!P2&!P1&!P0;

end PRIOR15
```

minimal sum-of-products expressions. As it turns out, each Yi output has only seven product terms, as you can see from the structure of the equations.

The priority encoder can be designed even more intuitively using ABEL's WHEN statement. As shown in Table 5-25, a deeply nested series of WHEN statements expresses precisely the logical function of the priority encoder. This program yields exactly the same set of output equations as the previous program.

Table 5-25 Alternate ABEL program for the same 15-input priority encoder.

```
module PRIOR15W
title '15-Input Priority Encoder'
PRIOR15W device 'P20L8';

" Input and output pins
P0..P14, EN_L                       pin 1..11, 13, 14, 23, 16, 17;
Y3_L, Y2_L, Y1_L, Y0_L, GS, ENOUT_L  pin 18..21, 15, 22 istype 'com';

" Active-level translation
EN = !EN_L; ENOUT = !ENOUT_L;

" Set definition
Y_L = [Y3_L, Y2_L, Y1_L, Y0_L];

equations

WHEN !EN THEN !Y_L = 0;          " Note: !Y_L === active-high Y
ELSE WHEN P14 THEN !Y_L = 14;    " Can't define active-high Y set
  ELSE WHEN P13 THEN !Y_L = 13;  "       due to ABEL set quirks
    ELSE WHEN P12 THEN !Y_L = 12;
      ELSE WHEN P11 THEN !Y_L = 11;
        ELSE WHEN P10 THEN !Y_L = 10;
          ELSE WHEN P9 THEN !Y_L = 9;
            ELSE WHEN P8 THEN !Y_L = 8;
              ELSE WHEN P7 THEN !Y_L = 7;
                ELSE WHEN P6 THEN !Y_L = 6;
                  ELSE WHEN P5 THEN !Y_L = 5;
                    ELSE WHEN P4 THEN !Y_L = 4;
                      ELSE WHEN P3 THEN !Y_L = 3;
                        ELSE WHEN P2 THEN !Y_L = 2;
                          ELSE WHEN P1 THEN !Y_L = 1;
                            ELSE WHEN P0 THEN !Y_L = 0;
                              ELSE {!Y_L = 0; ENOUT = 1;};

GS = EN&(P14#P13#P12#P11#P10#P9#P8#P7#P6#P5#P4#P3#P2#P1#P0);

end PRIOR15W
```

5.5.4 Encoders in VHDL

The approach to specifying encoders in VHDL is similar to the ABEL approach. We could embed the equivalent of a truth table or explicit equations into the VHDL program, but a behavioral description is far more intuitive. Since VHDL's IF-THEN-ELSE construct best describes prioritization and is available only within a process, we use the process-based behavioral approach.

Table 5-26 is a behavioral VHDL program for a priority encoder whose function is equivalent to the 74x148. It uses a FOR loop to look for an asserted

Table 5-26 Behavioral VHDL program for a 74x148-like 8-input priority encoder.

```
library IEEE;
use IEEE.std_logic_1164.all;

entity V74x148 is
    port (
        EI_L: in STD_LOGIC;
        I_L: in STD_LOGIC_VECTOR (7 downto 0);
        A_L: out STD_LOGIC_VECTOR (2 downto 0);
        EO_L, GS_L: out STD_LOGIC
    );
end V74x148;

architecture V74x148p of V74x148 is
  signal EI: STD_LOGIC;                    -- active-high version of input
  signal I: STD_LOGIC_VECTOR (7 downto 0); -- active-high version of inputs
  signal EO, GS: STD_LOGIC;                -- active-high version of outputs
  signal A: STD_LOGIC_VECTOR (2 downto 0); -- active-high version of outputs
begin
  process (EI_L, I_L, EI, EO, GS, I, A)
  variable j: INTEGER range 7 downto 0;
  begin
    EI <= not EI_L; -- convert input
    I <= not I_L;   -- convert inputs
    EO <= '1'; GS <= '0'; A <= "000";
    if (EI)='0' then EO <= '0';
    else for j in 7 downto 0 loop
        if I(j)='1' then
          GS <= '1'; EO <= '0'; A <= CONV_STD_LOGIC_VECTOR(j,3);
          exit;
        end if;
      end loop;
    end if;
    EO_L <= not EO; -- convert output
    GS_L <= not GS; -- convert output
    A_L <= not A;   -- convert outputs
  end process;
end V74x148p;
```

input, starting with the highest-priority input. Like some of our previous programs, it performs explicit active-level conversion at the beginning and end. Also recall that the CONV_STD_LOGIC_VECTOR(j,n) function was defined in Section 4.7.4 to convert from an integer j to a STD_LOGIC_VECTOR of a specified length n. This program is easily modified to use a different priority order or a different number of inputs, or to add more functionality such as finding a second-highest-priority input, as explored in Section 6.2.3.

5.6 Three-State Devices

In Sections 3.7.3 and 3.10.5 we described the electrical design of CMOS and TTL devices whose outputs may be in one of three states—0, 1, or Hi-Z. In this section we'll show how to use them.

5.6.1 Three-State Buffers

The most basic three-state device is a *three-state buffer*, often called a *three-state driver*. The logic symbols for four physically different three-state buffers are shown in Figure 5-53. The basic symbol is that of a noninverting buffer [(a), (b)] or an inverter [(c), (d)]. The extra signal at the top of the symbol is a *three-state enable* input, which may be active high [(a), (c)] or active low [(b), (d)]. When the enable input is asserted, the device behaves like an ordinary buffer or inverter. When the enable input is negated, the device output "floats"; that is, it goes to a high-impedance (Hi-Z), disconnected state and functionally behaves as if it weren't even there.

three-state buffer
three-state driver

three-state enable

   Three-state devices allow multiple sources to share a single "party line," as long as only one device "talks" on the line at a time. Figure 5-54 gives an example of how this can be done. Three input bits, SSRC2–SSRC0, select one of eight sources of data that may drive a single line, SDATA. A 3-to-8 decoder, the 74x138, ensures that only one of the eight SEL lines is asserted at a time, enabling only one three-state buffer to drive SDATA. However, if not all of the EN lines are asserted, then none of the three-state buffers is enabled. The logic value on SDATA is undefined in this case.

Figure 5-53 Various three-state buffers: (a) noninverting, active-high enable; (b) noninverting, active-low enable; (c) inverting, active-high enable; (d) inverting, active-low enable.

(a)   (b)   (c)   (d)

DEFINING "UNDEFINED" The actual voltage level of a floating signal depends on circuit details, such as resistive and capacitive load, and may vary over time. Also, the interpretation of this level by other circuits depends on the input characteristics of those circuits, so it's best not to count on a floating signal as being anything other than "undefined." Sometimes a pull-up resistor is used on three-state party lines to ensure that a floating value is pulled to a HIGH voltage and interpreted as logic 1. This is especially important on party lines that drive CMOS devices, which may consume excessive current when their input voltage is halfway between logic 0 and 1.

Figure 5-54
Eight sources sharing
a three-state party line.

Typical three-state devices are designed so that they go into the Hi-Z state faster than they come out of the Hi-Z state. (In terms of the specifications in a data book, t_{pLZ} and t_{pHZ} are both less than t_{pZL} and t_{pZH}; also see Section 3.7.3.) This means that if the outputs of two three-state devices are connected to the same party line, and we simultaneously disable one and enable the other, the first device will get off the party line before the second one gets on. This is important because, if both devices were to drive the party line at the same time, and if both were trying to maintain opposite output values (0 and 1), then excessive current would flow and create noise in the system, as discussed in Section 3.7.7. This is

fighting

often called *fighting*.

Unfortunately, delays and timing skews in control circuits make it difficult to ensure that the enable inputs of different three-state devices change "simultaneously." Even when this is possible, a problem arises if three-state devices from different-speed logic families (or even different ICs manufactured on different days) are connected to the same party line. The turn-on time (t_{pZL} or t_{pZH}) of a "fast" device may be shorter than the turn-off time (t_{pLZ} or t_{pHZ}) of a "slow" one, and the outputs may still fight.

The only really safe way to use three-state devices is to design control logic

dead time

that guarantees a *dead time* on the party line during which no one is driving it.

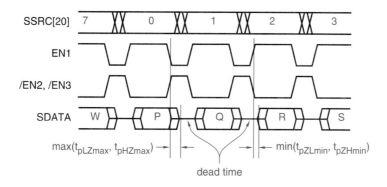

Figure 5-55
Timing diagram for the three-state party line.

The dead time must be long enough to account for the worst-case differences between turn-off and turn-on times of the devices and for skews in the three-state control signals. A timing diagram that illustrates this sort of operation for the party line of Figure 5-54 is shown in Figure 5-55. This timing diagram also illustrates a drawing convention for three-state signals—when in the Hi-Z state, they are shown at an "undefined" level halfway between 0 and 1.

5.6.2 Standard SSI and MSI Three-State Buffers

Like logic gates, several independent three-state buffers may be packaged in a single SSI IC. For example, Figure 5-56 shows the pinouts of *74x125* and *74x126*, each of which contains four independent noninverting three-state buffers in a 14-pin package. The three-state enable inputs in the '125 are active low, and in the '126 they are active high.

74x125
74x126

Most party-line applications use a bus with more than one bit of data. For example, in an 8-bit microprocessor system, the data bus is eight bits wide, and peripheral devices normally place data on the bus eight bits at a time. Thus, a peripheral device enables eight three-state drivers to drive the bus, all at the same time. Independent enable inputs, as in the '125 and '126, are not necessary.

Thus, to reduce the package size in wide-bus applications, most commonly used MSI parts contain multiple three-state buffers with common enable inputs. For example, Figure 5-57 shows the logic diagram and symbol for a *74x541* octal noninverting three-state buffer. *Octal* means that the part contains eight

74x541

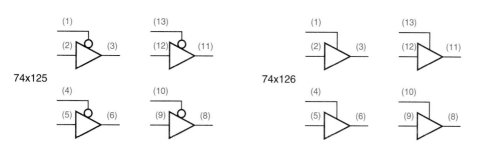

Figure 5-56
Pinouts of the 74x125 and 74x126 three-state buffers.

(b)

Figure 5-57
The 74x541 octal three-state buffer: (a) logic diagram, including pin numbers for a standard 20-pin dual in-line package; (b) traditional logic symbol.

(a)

Figure 5-58
Using a 74x541 as a microprocessor input port.

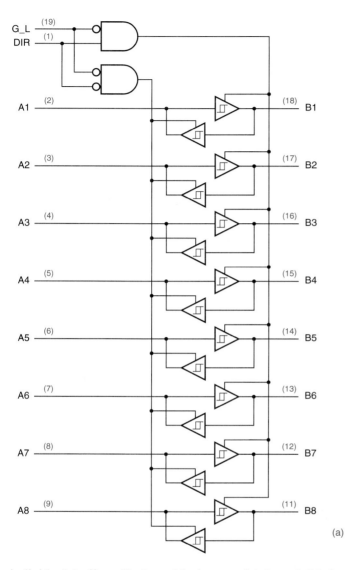

(b)

(a)

Figure 5-59
The 74x245 octal
three-state transceiver:
(a) logic diagram;
(b) traditional logic
symbol.

individual buffers. Both enable inputs, G1_L and G2_L, must be asserted to   *octal*
enable the device's three-state outputs. The little rectangular symbols inside the
buffer symbols indicate *hysteresis*, an electrical characteristic of the inputs that   *hysteresis*
improves noise immunity, as we explained in Section 3.7.2. The 74x541 inputs
typically have 0.4 volts of hysteresis.

Figure 5-58 shows part of a microprocessor system with an 8-bit data bus,
DB[0–7], and a 74x541 used as an input port. The microprocessor selects Input
Port 1 by asserting INSEL1 and requests a read operation by asserting READ.
The selected 74x541 responds by driving the microprocessor data bus with user-
supplied input data. Other input ports may be selected when a different INSEL
line is asserted along with READ.

Figure 5-60
Bidirectional buses
and transceiver
operation.

74x540

74x240

74x241

bus transceiver

74x245

Many other varieties of octal three-state buffers are commercially available. For example, the *74x540* is identical to the 74x541 except that it contains inverting buffers. The *74x240* and *74x241* are similar to the '540 and '541, except that they are split into two 4-bit sections, each with a single enable line.

A *bus transceiver* contains pairs of three-state buffers connected in opposite directions between each pair of pins, so that data can be transferred in either direction. For example, Figure 5-59 on the preceding page shows the logic diagram and symbol for a *74x245* octal three-state transceiver. The DIR input

Table 5-27 Modes of operation for a pair of bidirectional buses.

ENTFR_L	ATOB	*Operation*
0	0	Transfer data from a source on bus B to a destination on bus A.
0	1	Transfer data from a source on bus A to a destination on bus B.
1	x	Transfer data on buses A and B independently.

determines the direction of transfer, from A to B (DIR = 1) or from B to A (DIR = 0). The three-state buffer for the selected direction is enabled only if G_L is asserted.

A bus transceiver is typically used between two *bidirectional buses*, as shown in Figure 5-60. Three different modes of operation are possible, depending on the state of G_L and DIR, as shown in Table 5-27. As usual, it is the designer's responsibility to ensure that neither bus is ever driven simultaneously by two devices. However, independent transfers where both buses are driven at the same time may occur when the transceiver is disabled, as indicated in the last row of the table.

bidirectional bus

5.6.3 Three-State Outputs in ABEL and PLDs

The combinational-PLD applications in previous sections have used the bidirectional I/O pins (IO2–IO7 on a PAL16L8 or GAL16V8) statically, that is, always output-enabled or always output-disabled. In such applications, the compiler can take care of programming the output-enable gates appropriately—all fuses blown, or all fuses intact. By default in ABEL, a three-state output pin is programmed to be always enabled if its signal name appears on the lefthand side of an equation, and always disabled otherwise.

Three-state output pins can also be controlled dynamically, by a single input, by a product term, or, using two-pass logic, by a more complex logic expression. In ABEL, an *attribute suffix* .OE is attached to a signal name on the lefthand side of an equation to indicate that the equation applies to the output-enable for the signal. In a PAL16L8 or GAL16V8, the output enable is controlled by a single AND gate, so the righthand side of the enable equation must reduce to a single product term.

.OE attribute suffix

Table 5-28 shows a simple PLD program fragment with three-state control. Adapted from the program for a 74x138-like decoder on Table 5-8, this program includes a three-state output control OE for all eight decoder outputs. Notice that a set Y is defined to allow all eight output enables to be specified in a single equation; the .OE suffix is applied to each member of the set.

In the preceding example, the output pins Y0–Y7 are always either enabled or floating, and are used strictly as "output pins." I/O pins (IO2–IO7 in a 16L8 or 16V8) can be used as "bidirectional pins"; that is, they can be used dynamically

■ **Table 5-28** ABEL program for a 74x138-like 3-to-8 binary decoder with three-state output control.

```
module Z74X138T
title '74x138 Decoder with Three-State Output Enable'
Z74X138T device 'P16L8';

" Input pins
A, B, C, !G2A, !G2B, G1, !OE  pin 1, 2, 3, 4, 5, 6, 7;
" Output pins
!Y0, !Y1, !Y2, !Y3                pin 19, 18, 17, 16 istype 'com';
!Y4, !Y5, !Y6, !Y7                pin 15, 14, 13, 12 istype 'com';

" Constant expression
ENB = G1 & G2A & G2B;
Y = [Y0..Y7];

equations
Y.OE = OE;
Y0 = ENB & !C & !B & !A;
...
Y7 = ENB &  C &  B &  A;

end Z74X138T
```

as inputs or outputs depending on whether the output-enable gate is producing a 0 or a 1. An example application of I/O pins is a four-way, 2-bit bus transceiver with the following specifications:

- The transceiver handles four 2-bit bidirectional buses, A[1:2], B[1:2], C[1:2], and D[1:2].

- The source of data to drive the buses is selected by three select inputs, S[2:0], according to Table 5-29. If S2 is 0, the buses are driven with a constant value, otherwise they are driven with one of the other buses. However, when the selected source is a bus, the source bus is driven with 00.

Table 5-29
Bus-selection codes for a four-way bus transceiver.

S2	S1	S0	Source selected
0	0	0	00
0	0	1	01
0	1	0	10
0	1	1	11
1	0	0	A bus
1	0	1	B bus
1	1	0	C bus
1	1	1	D bus

Table 5-30 An ABEL program for four-way, 2-bit bus transceiver.

```
module XCVR4X2
title 'Four-way 2-bit Bus Transceiver'
XCVR4X2 device 'P16L8';

" Input pins
A1I, A2I                      pin 1, 11;
!AOE, !BOE, !COE, !DOE, !MOE  pin 2, 3, 4, 5, 6;
S0, S1, S2                    pin 7, 8, 9;
" Output and bidirectional pins
A1O, A2O                      pin 19, 12                       istype 'com';
B1, B2, C1, C2, D1, D2        pin 18, 17, 16, 15, 14, 13 istype 'com';

" Set definitions
ABUSO = [A1O,A2O];
ABUSI = [A1I,A2I];
BBUS  = [B1,B2];
CBUS  = [C1,C2];
DBUS  = [D1,D2];
SEL   = [S2,S1,S0];
CONST = [S1,S0];
" Constants
SELA = [1,0,0];
SELB = [1,0,1];
SELC = [1,1,0];
SELD = [1,1,1];

equations
ABUSO.OE = AOE & MOE;
BBUS.OE = BOE & MOE;
CBUS.OE = COE & MOE;
DBUS.OE = DOE & MOE;
ABUSO = !S2&CONST # (SEL==SELB)&BBUS # (SEL==SELC)&CBUS # (SEL==SELD)&DBUS;
BBUS  = !S2&CONST # (SEL==SELA)&ABUSI # (SEL==SELC)&CBUS # (SEL==SELD)&DBUS;
CBUS  = !S2&CONST # (SEL==SELA)&ABUSI # (SEL==SELB)&BBUS # (SEL==SELD)&DBUS;
DBUS  = !S2&CONST # (SEL==SELA)&ABUSI # (SEL==SELB)&BBUS # (SEL==SELC)&CBUS;

end XCVR4X2
```

- Each bus has its own output-enable signal, AOE_L, BOE_L, COE_L, or DOE_L. There is also a "master" output-enable signal, MOE_L. The transceiver drives a particular bus if and only if MOE_L and the output-enable signal for that bus are both asserted.

Table 5-30 is an ABEL program that performs the transceiver function. According to the enable (.OE) equations, each bus is output enabled if MOE and its own OE are asserted. Each bus is driven with S1 and S0 if S2 is 0, and with

Figure 5-61
PLD inputs and
outputs for a four-way,
2-bit bus transceiver.

the selected bus if a different bus is selected. If the bus itself is selected, the
output equation evaluates to 0, and the bus is driven with 00 as required.

Figure 5-61 is a logic diagram for a PAL16L8 (or GAL16V8) with the
required inputs and outputs. Since the device has only six bidirectional pins and
the specification requires eight, the A bus uses one pair of pins for input and
another for output. This is reflected in the program by the use of separate signals
and sets for the A-bus input and output.

Table 5-31 IEEE 1164 package declarations for STD_ULOGIC and STD_LOGIC.

```
PACKAGE std_logic_1164 IS
-- logic state system (unresolved)
    TYPE std_ulogic IS ( 'U',  -- Uninitialized
                         'X',  -- Forcing  Unknown
                         '0',  -- Forcing  0
                         '1',  -- Forcing  1
                         'Z',  -- High Impedance
                         'W',  -- Weak     Unknown
                         'L',  -- Weak     0
                         'H',  -- Weak     1
                         '-'   -- Don't care
                       );

-- unconstrained array of std_ulogic
    TYPE std_ulogic_vector IS ARRAY ( NATURAL RANGE <> ) OF std_ulogic;

-- resolution function
    FUNCTION resolved ( s : std_ulogic_vector ) RETURN std_ulogic;

-- *** industry standard logic type ***
    SUBTYPE std_logic IS resolved std_ulogic;
...
```

Table 5-32 IEEE 1164 package body for STD_ULOGIC and STD_LOGIC.

```
PACKAGE BODY std_logic_1164 IS
-- local type
    TYPE stdlogic_table IS ARRAY(std_ulogic, std_ulogic) OF std_ulogic;

-- resolution function
    CONSTANT resolution_table : stdlogic_table := (
    --      ---------------------------------------------------------
    --      | U    X    0    1    Z    W    L    H    -    |   |
    --      ---------------------------------------------------------
            ( 'U', 'U', 'U', 'U', 'U', 'U', 'U', 'U', 'U' ), -- | U |
            ( 'U', 'X', 'X', 'X', 'X', 'X', 'X', 'X', 'X' ), -- | X |
            ( 'U', 'X', '0', 'X', '0', '0', '0', '0', 'X' ), -- | 0 |
            ( 'U', 'X', 'X', '1', '1', '1', '1', '1', 'X' ), -- | 1 |
            ( 'U', 'X', '0', '1', 'Z', 'W', 'L', 'H', 'X' ), -- | Z |
            ( 'U', 'X', '0', '1', 'W', 'W', 'W', 'W', 'X' ), -- | W |
            ( 'U', 'X', '0', '1', 'L', 'W', 'L', 'W', 'X' ), -- | L |
            ( 'U', 'X', '0', '1', 'H', 'W', 'W', 'H', 'X' ), -- | H |
            ( 'U', 'X', 'X', 'X', 'X', 'X', 'X', 'X', 'X' )  -- | - |
        );

    FUNCTION resolved ( s : std_ulogic_vector ) RETURN std_ulogic IS
        VARIABLE result : std_ulogic := 'Z';  -- weakest state default
    BEGIN
        -- the test for a single driver is essential otherwise the
        -- loop would return 'X' for a single driver of '-' and that
        -- would conflict with the value of a single driver unresolved
        -- signal.
        IF    (s'LENGTH = 1) THEN    RETURN s(s'LOW);
        ELSE
            FOR i IN s'RANGE LOOP
                result := resolution_table(result, s(i));
            END LOOP;
        END IF;
        RETURN result;
    END resolved;
...
```

*5.6.4 Three-State Outputs in VHDL

VHDL itself does not have built-in types and operators for three-state outputs. However, it does have primitives which can be used to create signals and systems with three-state behavior; the IEEE 1164 package uses these primitives. As a start, the IEEE 1164 STD_LOGIC type defines 'Z' as one of its nine possible signal values; this value is used for the high-impedance state. You can assign this value to any STD_LOGIC signal, and the definitions of the standard logic functions account for the possibility of 'Z' inputs (generally a 'Z' input will cause a 'U' output).

Given the availability of three-state signals, how do we create three-state buses in VHDL? A three-state bus generally has two or more drivers, although the mechanisms we discuss work fine with just one driver. In VHDL, there is no explicit language construct for joining three-state outputs into a bus. Instead, the compiler automatically joins together signals that are driven in two or more different processes, that is, signals that appear on the lefthand side of a signal assignment statement in two or more processes. However, the signals must have the appropriate type, as explained below.

subtype
STD_ULOGIC
unresolved type
resolution function

The IEEE 1164 STD_LOGIC type is actually defined as a *subtype* of an unresolved type, *STD_ULOGIC*. In VHDL, an *unresolved type* is used for any signal that may be driven in two or more processes. The definition of an unresolved type includes a *resolution function* that is called every time an assignment is made to a signal having that type. As the name implies, this function resolves the value of the signal when it has multiple drivers.

Tables 5-31 and 5-32 on the preceding pages show the IEEE 1164 definitions of STD_ULOGIC, STD_LOGIC and the resolution function "resolved". This code uses a two-dimensional array resolution_table to determine the final STD_LOGIC value produced by *n* processes that drive a signal to *n* possibly different values passed in the input vector s. If, for example, a signal has four drivers, the VHDL compiler automatically constructs a 4-element vector containing the four driven values, and passes this vector to resolved every time that any one of those values changes. The result is passed back to the simulation.

Notice that the order in which the driven signal values appear in s does not affect the result produced by resolved, due to the strong ordering of "strengths" in the resolution_table: 'U'>'X'>'0','1'>'W'>'L','H'>'-'. That is, once a signal is partially resolved to a particular value, it never further resolves to a "weaker" value; and 0/1 and L/H conflicts always resolve to a stronger undefined value ('X' or 'W').

So, do you need to know all of this in order to use three-state outputs in VHDL? Well, usually not, but it can help if your simulations don't match up with reality. All that's normally required to use three-state outputs within VHDL is to declare the corresponding signals as type STD_ULOGIC.

For example, Table 5-33 describes a system that uses four 8-bit three-state drivers (in four processes) to select one of four 8-bit buses, A, B, C, and D, to drive onto a result bus X. Within each process, the IEEE 1164 standard function To_StdULogicVector is used to convert the input type of STD_LOGIC_VECTOR to STD_ULOGIC_VECTOR as required to make a legal assignment to result bus X. Recall that the construct "(others => 'Z')" denotes a vector of all Z's, with whatever length is needed to match the lefthand side of the assignment.

VHDL is flexible enough that you can use it to define other types of bus operation. For example, you could define a subtype and resolution function for open-drain outputs such that a wired-AND function is obtained. However, you'll

Table 5-33 VHDL program with four 8-bit three-state drivers.

```
library IEEE;
use IEEE.std_logic_1164.all;

entity V3statex is
    port (
        G_L: in STD_LOGIC;                          -- Global output enable
        SEL: in STD_LOGIC_VECTOR (1 downto 0);      -- Input select 0,1,2,3 ==> A,B,C,D
        A, B, C, D: in STD_LOGIC_VECTOR (1 to 8);   -- Input buses
        X: out STD_ULOGIC_VECTOR (1 to 8)           -- Output bus (three-state)
    );
end V3statex;

architecture V3states of V3statex is
begin
  process (G_L, SEL, A)
  begin
    if G_L='0' and SEL = "00" then X <= To_StdULogicVector(A);
    else X <= (others => 'Z');
    end if;
  end process;

  process (G_L, SEL, B)
  begin
    if G_L='0' and SEL = "01" then X <= To_StdULogicVector(B);
    else X <= (others => 'Z');
    end if;
  end process;

  process (G_L, SEL, C)
  begin
    if G_L='0' and SEL = "10" then X <= To_StdULogicVector(C);
    else X <= (others => 'Z');
    end if;
  end process;

  process (G_L, SEL, D)
  begin
    if G_L='0' and SEL = "11" then X <= To_StdULogicVector(D);
    else X <= (others => 'Z');
    end if;
  end process;

end V3states;
```

seldom have to use this feature; the definitions for specific output types in PLDs, FPGAs, and ASICs are usually already done for you in libraries provided by the component vendors.

5.7 Multiplexers

multiplexer

A *multiplexer* is a digital switch—it connects data from one of *n* sources to its output. Figure 5-62(a) shows the inputs and outputs of an *n*-input, *b*-bit multiplexer. There are *n* sources of data, each of which is *b* bits wide, and there are *b* output bits. In typical commercially available multiplexers, *n* = 1, 2, 4, 8, or 16, and *b* = 1, 2, or 4. There are *s* inputs that select among the *n* sources, so $s = \lceil \log_2 n \rceil$. An enable input EN allows the multiplexer to "do its thing"; when EN = 0, all of the outputs are 0. A multiplexer is often called a *mux* for short.

mux

Figure 5-62(b) shows a switch circuit that is roughly equivalent to the multiplexer. However, unlike a mechanical switch, a multiplexer is a unidirectional device: information flows only from inputs (on the left) to outputs (on the right).

We can write a general logic equation for a multiplexer output:

$$iY = \sum_{j=0}^{n-1} EN \cdot M_j \cdot iDj$$

Here, the summation symbol represents a logical sum of product terms. Variable iY is a particular output bit ($1 \leq i \leq b$), and variable iDj is input bit *i* of source *j* ($0 \leq j \leq n-1$). M_j represents minterm *j* of the *s* select inputs. Thus, when the multiplexer is enabled and the value on the select inputs is *j*, each output iY equals the corresponding bit of the selected input, iDj.

Multiplexers are obviously useful devices in any application in which data must be switched from multiple sources to a destination. A common application in computers is the multiplexer between the processor's registers and its arithmetic logic unit (ALU). For example, consider a 16-bit processor in which

Figure 5-62
Multiplexer structure:
(a) inputs and outputs;
(b) functional equivalent.

each instruction has a 3-bit field that specifies one of eight registers to use. This 3-bit field is connected to the select inputs of an 8-input, 16-bit multiplexer. The multiplexer's data inputs are connected to the eight registers, and its data outputs are connected to the ALU to execute the instruction using the selected register.

5.7.1 Standard MSI Multiplexers

The sizes of commercially available MSI multiplexers are limited by the number of pins available in an inexpensive IC package. Commonly used muxes come in 16-pin packages. At one extreme is the *74x151*, shown in Figure 5-63, which selects among eight 1-bit inputs. The select inputs are named C, B, and A, where C is most significant numerically. The enable input EN_L is active low; both active-high (Y) and active-low (Y_L) versions of the output are provided.

74x151

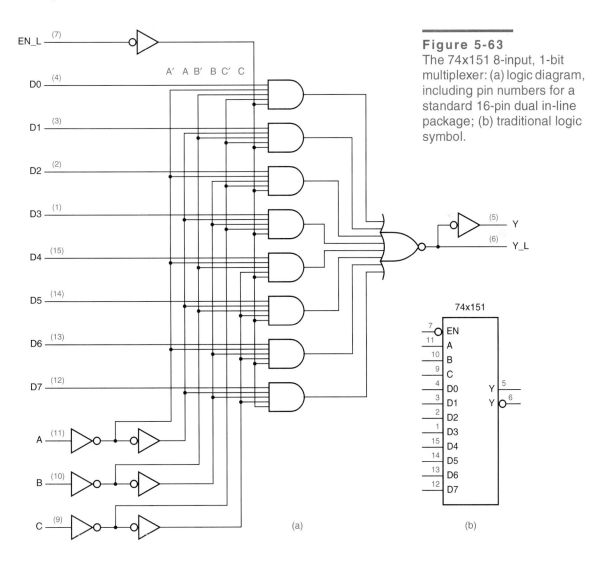

Figure 5-63
The 74x151 8-input, 1-bit multiplexer: (a) logic diagram, including pin numbers for a standard 16-pin dual in-line package; (b) traditional logic symbol.

Table 5-34
Truth table for a
74x151 8-input,
1-bit multiplexer.

Inputs				Outputs	
EN_L	C	B	A	Y	Y_L
1	x	x	x	0	1
0	0	0	0	D0	D0′
0	0	0	1	D1	D1′
0	0	1	0	D2	D2′
0	0	1	1	D3	D3′
0	1	0	0	D4	D4′
0	1	0	1	D5	D5′
0	1	1	0	D6	D6′
0	1	1	1	D7	D7′

Figure 5-64 The 74x157 2-input, 4-bit multiplexer: (a) logic diagram,
including pin numbers for a standard 16-pin dual in-line
package; (b) traditional logic symbol.

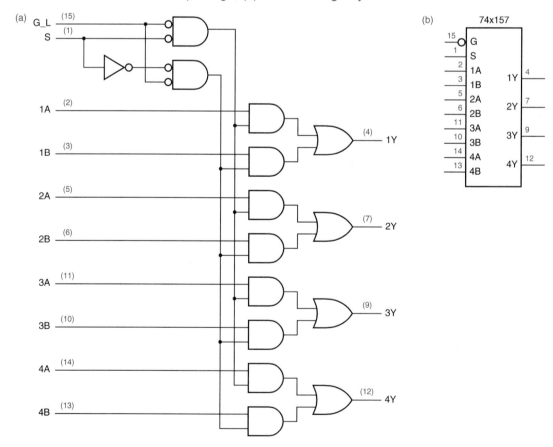

Inputs		Outputs			
G_L	S	1Y	2Y	3Y	4Y
1	x	0	0	0	0
0	0	1A	2A	3A	4A
0	1	1B	2B	3B	4B

Table 5-35
Truth table for a
74x157 2-input,
4-bit multiplexer.

The 74x151's truth table is shown in Table 5-34. Here we have once again extended our notation for truth tables. Up until now, our truth tables have specified an output of 0 or 1 for each input combination. In the 74x151's table, only a few of the inputs are listed under the "Inputs" heading. Each output is specified as 0, 1, or a simple logic function of the remaining inputs (e.g., D0 or D0′). This notation saves eight columns and eight rows in the table and presents the logic function more clearly than a larger table would.

At the other extreme of muxes in 16-pin packages, we have the *74x157*, *74x157* shown in Figure 5-64, which selects between two 4-bit inputs. Just to confuse things, the manufacturer has named the select input S and the active-low enable input G_L. Also note that the data sources are named A and B instead of D0 and D1 as in our generic example. Our extended truth-table notation makes the 74x157's description very compact, as shown in Table 5-35.

Intermediate between the 74x151 and 74x157 is the *74x153*, a 4-input, *74x153* 2-bit multiplexer. This device, whose logic symbol is shown in Figure 5-65, has separate enable inputs (1G, 2G) for each bit. As shown in Table 5-36, its function is very straightforward.

Inputs				Outputs	
1G_L	2G_L	B	A	1Y	2Y
0	0	0	0	1C0	2C0
0	0	0	1	1C1	2C1
0	0	1	0	1C2	2C2
0	0	1	1	1C3	2C3
0	1	0	0	1C0	0
0	1	0	1	1C1	0
0	1	1	0	1C2	0
0	1	1	1	1C3	0
1	0	0	0	0	2C0
1	0	0	1	0	2C1
1	0	1	0	0	2C2
1	0	1	1	0	2C3
1	1	x	x	0	0

Table 5-36
Truth table for a
74x153 4-input, 2-bit
multiplexer.

Figure 5-65
Traditional logic
symbol for the
74x153.

74x251

74x253
74x257

Some multiplexers have three-state outputs. The enable input of such a multiplexer, instead of forcing the outputs to zero, forces them to the Hi-Z state. For example, the *74x251* is identical to the '151 in its pinout and its internal logic design, except that Y and Y_L are three-state outputs. When the EN_L input is negated, instead of forcing the outputs to be negated, it forces the outputs into the Hi-Z state. Similarly, the *74x253* and *74x257* are three-state versions of the '153 and '157. The three-state outputs are especially useful when n-input muxes are combined to form larger muxes, as suggested in the next subsection.

5.7.2 Expanding Multiplexers

Seldom does the size of an MSI multiplexer match the characteristics of the problem at hand. For example, we suggested earlier that an 8-input, 16-bit multiplexer might be used in the design of a computer processor. This function could be performed by 16 74x151 8-input, 1-bit multiplexers or equivalent ASIC cells, each handling one bit of all the inputs and the output. The processor's 3-bit register-select field would be connected to the A, B, and C inputs of all 16 muxes, so they would all select the same register source at any given time.

The device that produces the 3-bit register-select field in this example must have enough fanout to drive 16 loads. With 74LS-series ICs this is possible because typical devices have a fanout of 20 LS-TTL loads.

Still, it is fortunate that the '151 was designed so that each of the A, B, and C inputs presents only one LS-TTL load to the circuit driving it. Theoretically, the '151 could have been designed without the first rank of three inverters shown on the select inputs in Figure 5-63, but then each select input would have presented five LS-TTL loads, and the drivers in the register-select application would need a fanout of 80.

Another dimension in which multiplexers can be expanded is the number of data sources. For example, suppose we needed a 32-input, 1-bit multiplexer. Figure 5-66 shows one way to build it. Five select bits are required. A 2-to-4 decoder (one-half of a 74x139) decodes the two high-order select bits to enable one of four 74x151 8-input multiplexers. Since only one '151 is enabled at a time, the '151 outputs can simply be ORed to obtain the final output.

CONTROL-SIGNAL FANOUT IN ASICS   Just the sort of fanout consideration that we described above occurs quite frequently in ASIC design. When a set of control signals, such as the register-select field in the example, controls a large number of bits, the required fanout can be enormous. In CMOS chips, the consideration is not DC loading but capacitive load, which slows down performance. In such an application, the designer must carefully partition the load and select points at which to buffer the control signals to reduce fanout. While inserting the extra buffers, the designer must be careful not increase the chip area significantly or to put so many buffers in series that their delay is unacceptable.

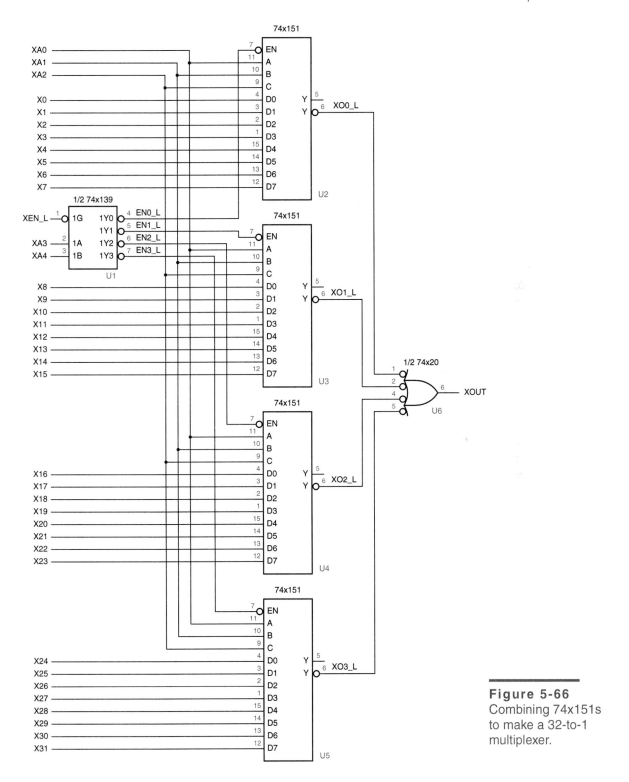

Figure 5-66
Combining 74x151s
to make a 32-to-1
multiplexer.

The 32-to-1 multiplexer can also be built using 74x251s. The circuit is identical to Figure 5-66, except that the output NAND gate is eliminated. Instead, the Y (and, if desired, Y_L) outputs of the four '251s are simply tied together. The '139 decoder ensures that at most one of the '251s has its three-state outputs enabled at any time. If the '139 is disabled (XEN_L is negated), then all of the '251s are disabled, and the XOUT and XOUT_L outputs are undefined. However, if desired, resistors may be connected from each of these signals to +5 volts to pull the output HIGH in this case.

Figure 5-67
A multiplexer driving a bus
and a demultiplexer
receiving the bus:
(a) switch equivalent;
(b) block-diagram symbols.

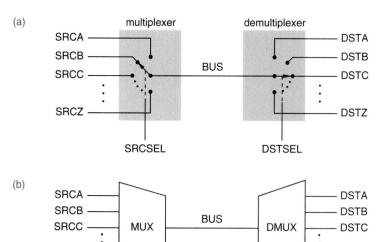

5.7.3 Multiplexers, Demultiplexers, and Buses

A multiplexer can be used to select one of n sources of data to transmit on a bus. At the far end of the bus, a *demultiplexer* can be used to route the bus data to one of m destinations. Such an application, using a 1-bit bus, is depicted in terms of our switch analogy in Figure 5-67(a). In fact, block diagrams for logic circuits often depict multiplexers and demultiplexers using the wedge-shaped symbols in (b), to suggest visually how a selected one of multiple data sources gets directed onto a bus and routed to a selected one of multiple destinations.

demultiplexer

The function of a demultiplexer is just the inverse of a multiplexer's. For example, a 1-bit, n-output demultiplexer has one data input and s inputs to select one of $n = 2^s$ data outputs. In normal operation, all outputs except the selected one are 0; the selected output equals the data input. This definition may be generalized for a b-bit, n-output demultiplexer; such a device has b data inputs, and its s select inputs choose one of $n = 2^s$ sets of b data outputs.

A binary decoder with an enable input can be used as a demultiplexer, as shown in Figure 5-68. The decoder's enable input is connected to the data line, and its select inputs determine which of its output lines is driven with the data bit. The remaining output lines are negated. Thus, the 74x139 can be used as a 2-bit, 4-output demultiplexer with active-low data inputs and outputs, and the 74x138 can be used as a 1-bit, 8-output demultiplexer. In fact, the manufacturer's catalog typically lists these ICs as "decoders/demultiplexers."

Figure 5-68 Using a 2-to-4 binary decoder as a 1-bit, 4-output demultiplexer: (a) generic decoder; (b) 74x139.

5.7.4 Multiplexers in ABEL and PLDs

Multiplexers are very easy to design using ABEL and combinational PLDs. For example, the function of a 74x153 4-input, 2-bit multiplexer can be duplicated in a PAL16L8 as shown in Figure 5-69 and Table 5-37. Several characteristics of the PLD-based design and program are worth noting:

- Signal names in the ABEL program are changed slightly from the signal names shown for a 74x153 in Figure 5-65 on page 401, since ABEL does not allow a number to be used as the first character of a signal name.

- A 74x153 has twelve inputs, while a PAL16L8 has only ten inputs. Therefore, two of the '153 inputs are assigned to 16L8 I/O pins, which are no longer usable as outputs.

Figure 5-69
Logic diagram for the
PAL16L8 used as a
74x153-like multiplexer.

- The '153 outputs (**1**Y and **2**Y) are assigned to pins 19 and 12 on the 16L8, which are usable *only* as outputs. This is preferred over assigning them to I/O pins; given a choice, it's better to leave I/O pins than output-only pins as spares.

- Although the multiplexer equations in the table are written quite naturally in sum-of-products form, they don't map directly onto the 16L8's structure

Table 5-37
ABEL program for a
74x153-like 4-input,
2-bit multiplexer.

```
module Z74X153
title '74x153-like multiplexer PLD '
Z74X153 device 'P16L8';

" Input and output pins
A, B, G1_L, G2_L          pin 17, 18, 1, 6;
C10, C11, C12, C13        pin 2, 3, 4, 5;
C20, C21, C22, C23        pin 7, 8, 9, 11;
Y1, Y2                    pin 19, 12 istype 'com';

" Active-level conversion
G1 = !G1_L;   G2 = !G2_L;

equations
Y1 = G1 & ( !B & !A & C10
        #  !B &  A & C11
        #   B & !A & C12
        #   B &  A & C13);

Y2 = G2 & ( !B & !A & C20
        #  !B &  A & C21
        #   B & !A & C22
        #   B &  A & C23);
end Z74X153
```

```
!Y1 = (!B &  !A & !C10
      # !B &   A & !C11
      #  B &  !A & !C12
      #  B &   A & !C13
      #  G1);
!Y2 = (!B &  !A & !C20
      # !B &   A & !C21
      #  B &  !A & !C22
      #  B &   A & !C23
      #  G2);
```

Table 5-38
Inverted, reduced
equations for 74x153-
like 4-input, 2-bit
multiplexer.

because of the inverter between the **AND-OR** array and the actual output
pins. Therefore, the ABEL compiler must complement the equations in the
table and then reduce the result to sum-of-products form. With a
GAL16V8, either version of the equations could be used.

Multiplexer functions are even easier to expression using ABEL's sets and
relations. For example, Table 5-39 shows the ABEL program for a 4-input, 8-bit
multiplexer. No device statement is included, because this function has too
many inputs and outputs to fit in any of the PLDs we've described so far.
However, it's quite obvious that a multiplexer of any size can be specified in just
a few lines of code in this way.

```
module mux4in8b
title '4-input, 8-bit wide multiplexer PLD'

" Input and output pins
!G                 pin;              " Output enable for Y bus
S1..S0             pin;              " Select inputs, 0-3 ==> A-D
A1..A8, B1..B8, C1..C8, D1..D8 pin; " 8-bit input buses A, B, C, D
Y1..Y8             pin istype 'com'; " 8-bit three-state output bus

" Sets
SEL = [S1..S0];
A = [A1..A8];
B = [B1..B8];
C = [C1..C8];
D = [D1..D8];
Y = [Y1..Y8];

equations
Y.OE = G;
WHEN (SEL == 0) THEN Y = A;
ELSE WHEN (SEL == 1) THEN Y = B;
ELSE WHEN (SEL == 2) THEN Y = C;
ELSE WHEN (SEL == 3) THEN Y = D;
end mux4in8b
```

Table 5-39
ABEL program for
a 4-input, 8-bit
multiplexer.

Table 5-40

Function table for
a specialized 4-input,
18-bit multiplexer.

S2	S1	S0	Input to Select
0	0	0	A
0	0	1	B
0	1	0	A
0	1	1	C
1	0	0	A
1	0	1	D
1	1	0	A
1	1	1	B

Likewise, it is easy to customize multiplexer functions using ABEL. For example, suppose that you needed a circuit that selects one of four 18-bit input buses, A, B, C, or D, to drive an 18-bit output bus F, as specified in Table 5-40 by three control bits. There are more control-bit combinations than multiplexer inputs, so a standard 4-input multiplexer doesn't quite fit the bill (but see Exercise 5.61). A 4-input, 3-bit multiplexer with the required behavior can be designed to fit into a single PAL16L8 or GAL16V8 as shown in Figure 5-70 and Table 5-41, and six copies of this device can be used to make the 18-bit mux. Alternatively, a single, larger PLD could be used. In any case, the ABEL program is very easily modified for different selection criteria.

Since this function uses all of the available pins on the PAL16L8, we had to make the pin assignment in Figure 5-70 carefully. In particular, we had to assign two output signals to the two output-only pins (O1 and O8), in order to maximize the number of input pins available.

Figure 5-70

Logic diagram for the
PAL16L8 used as a
specialized 4-input,
3-bit multiplexer.

■ **Table 5-41** ABEL program for a specialized 4-input, 3-bit multiplexer.

```
module mux4in3b
title 'Specialized 4-input, 3-bit Multiplexer'
mux4in3b device 'P16L8';

" Input and output pins
S2..S0                              pin 16..18;              " Select inputs
A0..A2, B0..B2, C0..C2, D0..D2  pin 1..9, 11, 13, 14;    " Bus inputs
F0..F2                              pin 19, 15, 12 istype 'com'; " Bus outputs

" Sets
SEL = [S2..S0];
A = [A0..A2];
B = [B0..B2];
C = [C0..C2];
D = [D0..D2];
F = [F0..F2];

equations
WHEN (SEL== 0) # (SEL== 2) # (SEL== 4) # (SEL== 6) THEN F = A;
ELSE WHEN (SEL== 1) # (SEL== 7) THEN F = B;
ELSE WHEN (SEL== 3) THEN F = C;
ELSE WHEN (SEL== 5) THEN F = D;

end mux4in3b
```

EASIEST, BUT NOT CHEAPEST As you've seen, it's very easy to program a PLD to perform decoder and multiplexer functions. Still, if you need the logic function of a standard decoder or multiplexer, it's usually less costly to use a standard MSI chip than it is to use a PLD. The PLD-based approach is best if the multiplexer has some nonstandard functional requirements, or if you think you may have to change its function as a result of debugging.

5.7.5 Multiplexers in VHDL

Multiplexers are very easy to describe in VHDL. In the dataflow style of architecture, the SELECT statement provides the required functionality, as shown in Table 5-42, the VHDL description of 4-input, 8-bit multiplexer.

In a behavioral architecture, a CASE statement is used. For example, Table 5-43 shows a process-based architecture for the same mux4in8b entity.

As in ABEL, it is very easy to customize the selection criteria in a VHDL multiplexer program. For example, Table 5-44 is a behavioral-style program for a specialized 4-input, 18-bit multiplexer with the selection criteria of Table 5-40.

In each example, if the select inputs are not valid (e.g., contain U's or X's), the output bus is set to "unknown" to help catch errors during simulation.

Table 5-42 Dataflow VHDL program for a 4-input, 8-bit multiplexer.

```
library IEEE;
use IEEE.std_logic_1164.all;

entity mux4in8b is
    port (
        S: in STD_LOGIC_VECTOR (1 downto 0);       -- Select inputs, 0-3 ==> A-D
        A, B, C, D: in STD_LOGIC_VECTOR (1 to 8); -- Data bus input
        Y: out STD_LOGIC_VECTOR (1 to 8)           -- Data bus output
    );
end mux4in8b;

architecture mux4in8b of mux4in8b is
begin
    with S select Y <=
        A when "00",
        B when "01",
        C when "10",
        D when "11",
        (others => 'U') when others; -- this creates an 8-bit vector of 'U'
end mux4in8b;
```

Table 5-43 Behavioral architecture for a 4-input, 8-bit multiplexer.

```
architecture mux4in8p of mux4in8b is
begin
process(S, A, B, C, D)
  begin
    case S is
      when "00" => Y <= A;
      when "01" => Y <= B;
      when "10" => Y <= C;
      when "11" => Y <= D;
      when others => Y <= (others => 'U');  -- 8-bit vector of 'U'
    end case;
  end process;
end mux4in8p;
```

5.8 Exclusive-OR Gates and Parity Circuits

5.8.1 Exclusive-OR and Exclusive-NOR Gates

Exclusive OR (XOR)
Exclusive NOR (XNOR)
Equivalence

An *Exclusive-OR (XOR)* gate is a 2-input gate whose output is 1 if exactly one of its inputs is 1. Stated another way, an XOR gate produces a 1 output if its inputs are different. An *Exclusive NOR (XNOR)* or *Equivalence* gate is just the opposite—it produces a 1 output if its inputs are the same. A truth table for these

Table 5-44 Behavioral VHDL program for a specialized 4-input, 3-bit multiplexer.

```
library IEEE;
use IEEE.std_logic_1164.all;

entity mux4in3b is
    port (
        S: in STD_LOGIC_VECTOR (2 downto 0);        -- Select inputs, 0-7 ==> ABACADAB
        A, B, C, D: in STD_LOGIC_VECTOR (1 to 18); -- Data bus inputs
        Y: out STD_LOGIC_VECTOR (1 to 18)           -- Data bus output
    );
end mux4in3b;

architecture mux4in3p of mux4in3b is
begin
process(S, A, B, C, D)
variable i: INTEGER;
  begin
    case S is
      when "000" | "010" | "100" | "110" => Y <= A;
      when "001" | "111" => Y <= B;
      when "011" => Y <= C;
      when "101" => Y <= D;
      when others => Y <= (others => 'U'); -- 18-bit vector of 'U'
    end case;
  end process;
end mux4in3p;
```

functions is shown in Table 5-45. The XOR operation is sometimes denoted by the symbol "⊕", that is,

$$X \oplus Y = X' \cdot Y + X \cdot Y'$$

Although EXCLUSIVE OR is not one of the basic functions of switching algebra, discrete XOR gates are fairly commonly used in practice. Most switching technologies cannot perform the XOR function directly; instead, they use multigate designs like the ones shown in Figure 5-71.

X	Y	$X \oplus Y$ (XOR)	$(X \oplus Y)'$ (XNOR)
0	0	0	1
0	1	1	0
1	0	1	0
1	1	0	1

Table 5-45
Truth table for XOR and XNOR functions.

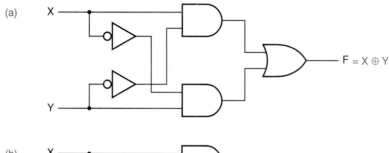

Figure 5-71
Multigate designs for
the 2-input XOR
function: (a) AND-OR;
(b) three-level NAND.

Figure 5-72 Equivalent symbols for (a) XOR gates; (b) XNOR gates.

The logic symbols for XOR and XNOR functions are shown in Figure 5-72. There are four equivalent symbols for each function. All of these alternatives are a consequence of a simple rule:

- Any two signals (inputs or output) of an XOR or XNOR gate may be complemented without changing the resulting logic function.

In bubble-to-bubble logic design, we choose the symbol that is most expressive of the logic function being performed.

74x86   Four XOR gates are provided in a single 14-pin SSI IC, the *74x86* shown in Figure 5-73. New SSI logic families do not offer XNOR gates, although they are readily available in FPGA and ASIC libraries and as primitives in HDLs.

Figure 5-73
Pinouts of the 74x86
quadruple 2-input
XOR gate.

(a)

(b)

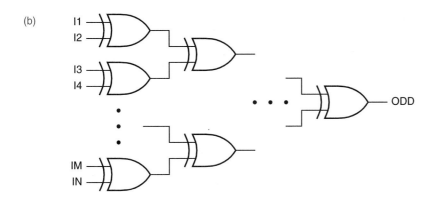

Figure 5-74
Cascading XOR
gates: (a) daisy-chain
connection; (b) tree
structure.

5.8.2 Parity Circuits

As shown in Figure 5-74(a), n XOR gates may be cascaded to form a circuit with
$n + 1$ inputs and a single output. This is called an *odd-parity circuit*, because its   *odd-parity circuit*
output is 1 if an odd number of its inputs are 1. The circuit in (b) is also an odd-
parity circuit, but it's faster because its gates are arranged in a treelike structure.
If the output of either circuit is inverted, we get an *even-parity circuit*, whose   *even-parity circuit*
output is 1 if an even number of its inputs are 1.

5.8.3 The 74x280 9-Bit Parity Generator

Rather than build a multibit parity circuit with discrete XOR gates, it is more
economical to put all of the XORs in a single MSI package with just the primary
inputs and outputs available at the external pins. The *74x280* 9-bit parity   *74x280*
generator, shown in Figure 5-75, is such a device. It has nine inputs and two
outputs that indicate whether an even or odd number of inputs are 1.

5.8.4 Parity-Checking Applications

In Section 2.15 we described error-detecting codes that use an extra bit, called a
parity bit, to detect errors in the transmission and storage of data. In an even-
parity code, the parity bit is chosen so that the total number of 1 bits in a code
word is even. Parity circuits like the 74x280 are used both to generate the correct
value of the parity bit when a code word is stored or transmitted and to check the
parity bit when a code word is retrieved or received.

Figure 5-75 The 74x280 9-bit odd/even parity generator: (a) logic diagram, including pin numbers for a standard 16-pin dual in-line package; (b) traditional logic symbol.

Figure 5-76 shows how a parity circuit might be used to detect errors in the memory of a microprocessor system. The memory stores 8-bit bytes, plus a parity bit for each byte. The microprocessor uses a bidirectional bus D[0:7] to transfer data to and from the memory. Two control lines, RD and WR, are used to indicate whether a read or write operation is desired, and an ERROR signal is asserted to indicate parity errors during read operations. Complete details of the memory chips, such as addressing inputs, are not shown; memory chips are described in detail in Chapter 10. For parity checking, we are concerned only with the data connections to the memory.

SPEEDING UP THE XOR TREE

If each XOR gate in Figure 5-75 were built using discrete NAND gates as in Figure 5-71(b), the 74x280 would be pretty slow, having a propagation delay equivalent to $4 \cdot 3 + 1$, or 13, NAND gates. Instead, a typical implementation of the 74x280 uses a 4-wide AND-OR-INVERT gate to perform the function of each shaded pair of XOR gates in the figure with about the same delay as a single NAND gate. The A–I inputs are buffered through two levels of inverters, so that each input presents just one unit load to the circuit driving it. Thus, the total propagation delay through this implementation of the 74x280 is about the same as five inverting gates, not 13.

Figure 5-76 Parity generation and checking for an 8-bit-wide memory system.

To store a byte into the memory chips, we specify an address (not shown), place the byte on D[0–7], generate its parity bit on PIN, and assert WR. The AND gate on the I input of the 74x280 ensures that I is 0 except during read operations, so that during writes the '280's output depends only on the parity of the D-bus data. The '280's ODD output is connected to PIN, so that the total number of 1s stored is even.

To retrieve a byte, we specify an address (not shown) and assert RD; the byte value appears on DOUT[0–7] and its parity appears on POUT. A 74x541 drives the byte onto the D bus, and the '280 checks its parity. If the parity of the 9-bit word DOUT[0–7],POUT is odd during a read, the ERROR signal is asserted.

Parity circuits are also used with error-correcting codes such as the Hamming codes described in Section 2.15.3. We showed the parity-check matrix for a 7-bit Hamming code in Figure 2-13 on page 63. We can correct errors in this code as shown in Figure 5-77. A 7-bit word, possibly containing an error, is presented on DU[1–7]. Three 74x280s compute the parity of the three bit-groups defined by the parity-check matrix. The outputs of the '280s form the syndrome, which is the number of the erroneous input bit, if any. A 74x138 is used to decode the syndrome. If the syndrome is zero, the NOERROR_L signal is asserted (this signal also could be named ERROR). Otherwise, the erroneous

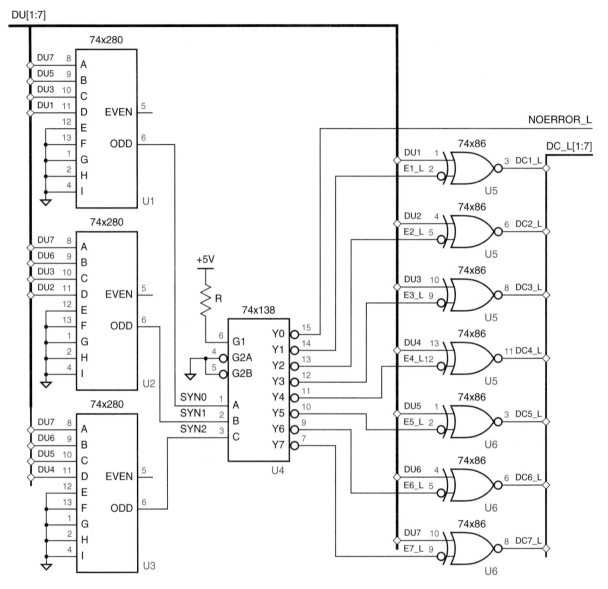

Figure 5-77 Error-correcting circuit for a 7-bit Hamming code.

bit is corrected by complementing it. The corrected code word appears on the DC_L bus.

Note that the active-low outputs of the '138 led us to use an active-low DC_L bus. If we required an active-high DC bus, we could have put a discrete inverter on each XOR input or output, or used a decoder with active-high outputs, or used XNOR gates.

5.8.5 Exclusive-OR Gates and Parity Circuits in ABEL and PLDs

The Exclusive-OR function is denoted in ABEL by the $ operator, and its complement, the Exclusive-NOR function, is denoted by !$. In principle, these operators may be used freely in ABEL expressions. For example, you could specify a PLD output equivalent to the 74x280's EVEN output using the following ABEL equation:

```
EVEN = !(A $ B $ C $ D $ E $ F $ G $ H $ I);
```

However, most PLDs realize expressions using two-level AND-OR logic and have little if any capability of realizing XOR functions directly. Unfortunately, the Karnaugh map of an n-input XOR function is a checkerboard with 2^{n-1} prime implicants. Thus, the sum-of-products realization of the simple equation above requires 256 product terms, well beyond the capability of any PLD.

So-called X-series PLDs can realize a two-input XOR function directly in a three-level structure that uses a single XOR gate ro combine two independent sums of products. This structure turns out to be useful in the design of counters. To create larger XOR functions, however, a board-level designer must normally use a specialized parity generator/checker component like the 74x280, and an ASIC designer must combine individual XOR gates in a multilevel parity tree similar to Figure 5-74(b) on page 413.

5.8.6 Exclusive-OR Gates and Parity Circuits in VHDL

Like ABEL, VHDL provides primitive operators, xor and xnor, for specifying XOR and XNOR functions (xnor is available only in VHDL-93). For example, Table 5-46 is a dataflow-style program for a 3-input XOR device that uses the xor primitive. It's also possible to specify XOR or parity functions behaviorally, as Table 5-47 does for a 9-input parity function similar to the 74x280.

Table 5-46 Dataflow-style VHDL program for a 3-input XOR device.

```
library IEEE;
use IEEE.std_logic_1164.all;

entity vxor3 is
    port (
        A, B, C: in STD_LOGIC;
        Y: out STD_LOGIC
    );
end vxor3;

architecture vxor3 of vxor3 is
begin
  Y <= A xor B xor C;
end vxor3;
```

Table 5-47 Behavioral VHDL program for a 9-input parity checker.

```
library IEEE;
use IEEE.std_logic_1164.all;

entity parity9 is
    port (
        I: in STD_LOGIC_VECTOR (1 to 9);
        EVEN, ODD: out STD_LOGIC
    );
end parity9;

architecture parity9p of parity9 is
begin
process (I)
  variable p : STD_LOGIC;
  begin
    p := I(1);
    for j in 2 to 9 loop
      if I(j) = '1' then p := not p; end if;
    end loop;
    ODD <= p;
    EVEN <= not p;
  end process;
end parity9p;
```

When a VHDL program containing large XOR functions is synthesized, the synthesis tool will do the best it can to realize the function in the targeted device technology. There's no magic—if we try to target the VHDL program in Table 5-47 to a 16V8 PLD, it still won't fit!

Typical ASIC and FPGA libraries contain two- and three-input XOR and XNOR functions as primitives. In CMOS ASICs, these primitives are usually realized very efficiently at the transistor level using transmission gates, as shown in Exercises 5.56 and 5.73. Fast and compact XOR trees can be built using these primitives. However, typical VHDL synthesis tools are not smart enough to create an efficient tree structure from a behavioral program like Table 5-47. Instead, we can use a structural program to get exactly what we want.

For example, Table 5-48 is a structural VHDL program for a 9-input XOR function that is equivalent to the 74x280 of Figure 5-75(a) in structure as well as function. In this example, we've used the previously defined vxor3 component as the basic building block of the XOR tree. In an ASIC, we would replace the vxor3 with a 3-input XOR primitive from the ASIC library. Also, if a 3-input XNOR were available, we could eliminate the explicit inversion for Y3N and instead use the XNOR for U5, using the noninverted Y3 signal as its last input.

Our final example is a VHDL version of the Hamming decoder circuit of Figure 5-77 and is shown in Table 5-49. A function syndrome(DU) is defined to

Table 5-48 Structural VHDL program for a 74x280-like parity checker.

```
library IEEE;
use IEEE.std_logic_1164.all;

entity V74x280 is
    port (
        I: in STD_LOGIC_VECTOR (1 to 9);
        EVEN, ODD: out STD_LOGIC
    );
end V74x280;

architecture V74x280s of V74x280 is
component vxor3
 port (A, B, C: in STD_LOGIC; Y: out STD_LOGIC);
end component;
signal Y1, Y2, Y3, Y3N: STD_LOGIC;
begin
  U1: vxor3 port map (I(1), I(2), I(3), Y1);
  U2: vxor3 port map (I(4), I(5), I(6), Y2);
  U3: vxor3 port map (I(7), I(8), I(9), Y3);
  Y3N <= not Y3;
  U4: vxor3 port map (Y1, Y2, Y3, ODD);
  U5: vxor3 port map (Y1, Y2, Y3N, EVEN);
end V74x280s;
```

return the 3-bit syndrome of a 7-bit uncorrected data input vector DU. In the "main" process, the corrected data output vector DC is initially set equal to DU. The CONV_INTEGER function, which we defined in Section 4.7.4, is used to convert the 3-bit syndrome to an integer. If the syndrome is nonzero, the corresponding bit of DC is complemented to correct the assumed 1-bit error. If the syndrome is zero, either no error or an undetectable error has occurred; the output NOERROR is set accordingly.

5.9 Comparators

Comparing two binary words for equality is a commonly used operation in computer systems and device interfaces. For example, in Figure 2-7(a) on page 56, we showed a system structure in which devices are enabled by comparing a "device select" word with a predetermined "device ID." A circuit that compares two binary words and indicates whether they are equal is called a *comparator*. *comparator* Some comparators interpret their input words as signed or unsigned numbers and also indicate an arithmetic relationship (greater or less than) between the words. These devices are often called *magnitude comparators*. *magnitude comparator*

Table 5-49 Behavioral VHDL program for Hamming error correction.

```
library IEEE;
use IEEE.std_logic_1164.all;
use IEEE.std_logic_unsigned.all;

entity hamcorr is
  port (
      DU: IN STD_LOGIC_VECTOR (1 to 7);
      DC: OUT STD_LOGIC_VECTOR (1 to 7);
      NOERROR: OUT STD_LOGIC
);
end hamcorr;

architecture hamcorr of hamcorr is

function syndrome (D: STD_LOGIC_VECTOR)
    return STD_LOGIC_VECTOR is
variable SYN: STD_LOGIC_VECTOR (2 downto 0);
begin
  SYN(0) := D(1) xor D(3) xor D(5) xor D(7);
  SYN(1) := D(2) xor D(3) xor D(6) xor D(7);
  SYN(2) := D(4) xor D(5) xor D(6) xor D(7);
  return(SYN);
end syndrome;

begin
process (DU)
variable i: INTEGER;
  begin
    DC <= DU;
    i := CONV_INTEGER(syndrome(DU));
    if i = 0 then NOERROR <= '1';
    else NOERROR <= '0'; DC(i) <= not DU(i); end if;
  end process;
end hamcorr;
```

5.9.1 Comparator Structure

Exclusive-OR and Exclusive-NOR gates may be viewed as 1-bit comparators. Figure 5-78(a) shows an interpretation of the 74x86 XOR gate as a 1-bit comparator. The active-high output, DIFF, is asserted if the inputs are different. The outputs of four XOR gates are ORed to create a 4-bit comparator in (b). The DIFF output is asserted if any of the input-bit pairs are different. Given enough XOR gates and wide enough OR gates, comparators with any number of input bits can be built.

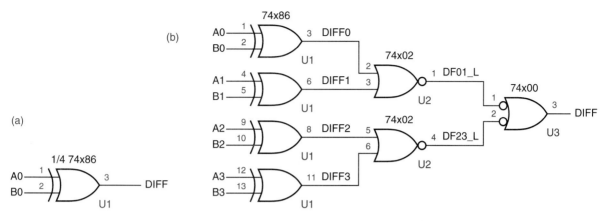

Figure 5-78 Comparators using the 74x86: (a) 1-bit comparator; (b) 4-bit comparator.

AN ITERATIVE COMPARATOR

The n-bit comparators in the preceding subsection might be called *parallel comparators* because they look at each pair of input bits simultaneously and deliver the 1-bit comparison results in parallel to an n-input OR or AND function. It is also possible to design an "iterative comparator" that looks at its bits one at a time using a small, fixed amount of logic per bit. Before looking at the iterative comparator design, you should understand the general class of "iterative circuits" described in the next subsection.

5.9.2 Iterative Circuits

An *iterative circuit* is a special type of combinational circuit, with the structure shown in Figure 5-79. The circuit contains n identical modules, each of which has both *primary inputs and outputs* and *cascading inputs and outputs*. The leftmost cascading inputs are called *boundary inputs* and are connected to fixed logic values in most iterative circuits. The rightmost cascading outputs are called *boundary outputs* and usually provide important information.

iterative circuit

primary inputs and outputs

cascading inputs and outputs

boundary outputs

Iterative circuits are well suited to problems that can be solved by a simple iterative algorithm:

1. Set C_0 to its initial value and set i to 0.
2. Use C_i and PI_i to determine the values of PO_i and C_{i+1}.
3. Increment i.
4. If $i < n$, go to step 2.

In an iterative circuit, the loop of steps 2–4 is "unwound" by providing a separate combinational circuit that performs step 2 for each value of i.

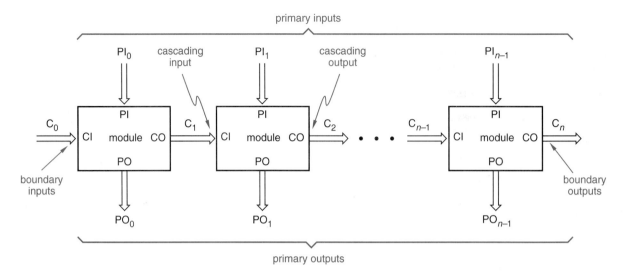

Figure 5-79 General structure of an iterative combinational circuit.

Examples of iterative circuits are the comparator circuit in the next subsection and the ripple adder in Section 5.10.2. The 74x85 4-bit comparator and the 74x283 4-bit adder are examples of MSI circuits that can be used as the individual modules in a larger iterative circuit. In Section 8.6 we'll explore the relationship between iterative circuits and corresponding sequential circuits that execute the 4-step algorithm above in discrete time steps.

5.9.3 An Iterative Comparator Circuit

Two n-bit values X and Y can be compared one bit at a time using a single bit EQ_i at each step to keep track of whether all of the bit-pairs have been equal so far:

1. Set EQ_0 to 1 and set i to 0.
2. If EQ_i is 1 and X_i and Y_i are equal, set EQ_{i+1} to 1. Else set EQ_{i+1} to 0.
3. Increment i.
4. If $i < n$, go to step 2.

Figure 5-80 shows a corresponding iterative circuit. Note that this circuit has no primary outputs; the boundary output is all that interests us. Other iterative circuits, such as the ripple adder of Section 5.10.2, have primary outputs of interest.

Given a choice between the iterative comparator circuit in this subsection and one of the parallel comparators shown previously, you would probably prefer the parallel comparator. The iterative comparator saves little if any cost, and it's very slow because the cascading signals need time to "ripple" from the leftmost to the rightmost module. Iterative circuits that process more than one bit

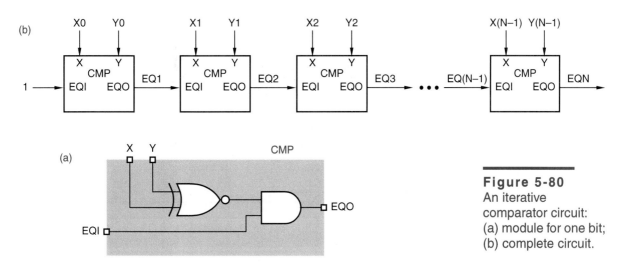

Figure 5-80
An iterative
comparator circuit:
(a) module for one bit;
(b) complete circuit.

at a time, using modules like the 74x85 4-bit comparator and 74x283 4-bit adder, are much more likely to be used in practical designs.

5.9.4 Standard MSI Comparators

Comparator applications are common enough that several MSI comparators have been developed commercially. The *74x85* is a 4-bit comparator with the logic symbol shown in Figure 5-81. It provides a greater-than output (AGTBOUT) and a less-than output (ALTBOUT) as well as an equal output (AEQBOUT). The '85 also has *cascading inputs* (AGTBIN, ALTBIN, AEQBIN) for combining multiple '85s to create comparators for more than four bits. Both the cascading inputs and the outputs are arranged in a 1-out-of-3 code, since in normal operation exactly one input and one output should be asserted.

    The cascading inputs are defined so the outputs of an '85 that compares less-significant bits are connected to the inputs of an '85 that compares more-

74x85

cascading inputs

Figure 5-81
Traditional logic symbol for
the 74x85 4-bit comparator.

```
        74x85
   2 ┌──────────────────┐ 7
 ────┤ ALTBIN   ALTBOUT ├────
   3 │                  │ 6
 ────┤ AEQBIN   AEQBOUT ├────
   4 │                  │ 5
 ────┤ AGTBIN   AGTBOUT ├────
  10 │                  │
 ────┤ A0               │
   9 │                  │
 ────┤ B0               │
  12 │                  │
 ────┤ A1               │
  11 │                  │
 ────┤ B1               │
  13 │                  │
 ────┤ A2               │
  14 │                  │
 ────┤ B2               │
  15 │                  │
 ────┤ A3               │
   1 │                  │
 ────┤ B3               │
     └──────────────────┘
```

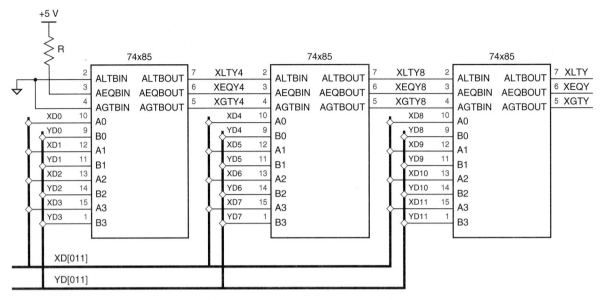

Figure 5-82 A 12-bit comparator using 74x85s.

significant bits, as shown in Figure 5-82 for a 12-bit comparator. This is an iter-ative circuit according to the definition in Section 5.9.2. Each '85 develops its cascading outputs roughly according to the following pseudo-logic equations:

$$\text{AGTBOUT} = (A > B) + (A = B) \cdot \text{AGTBIN}$$

$$\text{AEQBOUT} = (A = B) \cdot \text{AEQBIN}$$

$$\text{ALTBOUT} = (A < B) + (A = B) \cdot \text{ALTBIN}$$

The parenthesized subexpressions above are not normal logic expressions; rather, they indicate an arithmetic comparison that occurs between the A3–A0 and B3–B0 inputs. In other words, AGTBOUT is asserted if $A > B$ or if $A = B$ and AGTBIN is asserted (if the higher-order bits are equal, we have to look at the lower-order bits for the answer). We'll see this kind of expression again when we look at ABEL comparator design in Section 5.9.5. The arithmetic compari-sons can be expressed using normal logic expressions, for example,

$$(A > B) = A3 \cdot B3'+$$
$$(A3 \oplus B3)' \cdot A2 \cdot B2' +$$
$$(A3 \oplus B3)' \cdot (A2 \oplus B2)' \cdot A1 \cdot B1' +$$
$$(A3 \oplus B3)' \cdot (A2 \oplus B2)' \cdot (A1 \oplus B1)' \cdot A0 \cdot B0'$$

Figure 5-83
Traditional logic symbol for the 74x682 8-bit comparator.

Such expressions must be substituted into the pseudo-logic equations above to obtain genuine logic equations for the comparator outputs.

    Several 8-bit MSI comparators are also available. The simplest of these is the *74x682*, whose logic symbol is shown in Figure 5-83 and whose internal

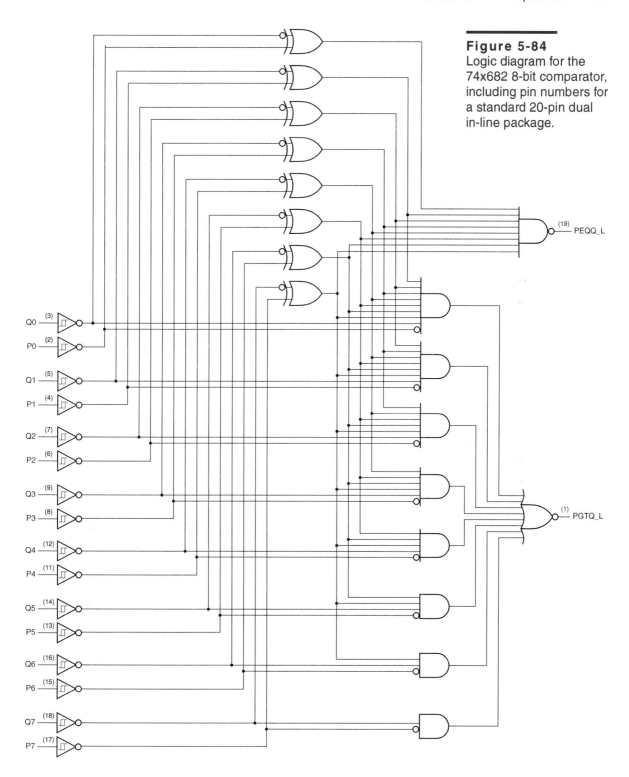

Figure 5-84
Logic diagram for the 74x682 8-bit comparator, including pin numbers for a standard 20-pin dual in-line package.

logic diagram is shown in Figure 5-84. The top half of the circuit checks the two 8-bit input words for equality. Each XNOR-gate output is asserted if its inputs are equal, and the PEQQ_L output is asserted if all eight input-bit pairs are equal. The bottom half of the circuit compares the input words arithmetically and asserts PGTQ_L if P[7–0] > Q[7–0].

Unlike the 74x85, the 74x682 does not have cascading inputs. Also unlike the '85, the '682 does not provide a "less than" output. However, any desired condition, including ≤ and ≥, can be formulated as a function of the PEQQ_L and PGTQ_L outputs, as shown in Figure 5-85.

Figure 5-85
Arithmetic conditions
derived from 74x682
outputs.

COMPARING COMPARATORS   The individual 1-bit comparators (XNOR gates) in the '682 are drawn in the opposite sense as the examples of the preceding subsection—outputs are asserted for *equal* inputs and then ANDed, rather than asserted for *different* inputs and then ORed. We can look at a comparator's function either way, as long as we're consistent.

5.9.5 Comparators in ABEL and PLDs

Comparing two sets for equality or inequality is very easy to do in ABEL using the "==" or "!=" operator in a relational expression. The only restriction is that the two sets must have an equal number of elements. Thus, given the relational expression "A!=B", where A and B are sets each with n elements, the compiler generates the logic expression

```
(A1 $ B1)  # (A2 $ B2)  # ... # (An $ Bn)
```

The logic expression for "A==B"is just the complement of the one above.

In the preceding logic expression, it takes one 2-input XOR function to compare each bit. Since a 2-input XOR function can be realized as a sum of two product terms, the complete expression can be realized in a PLD as a relatively modest sum of $2n$ product terms:

```
(A1&!B1 # !A1&B1) # (A2&!B2 # !A2&&B2) # ... # (An&!Bn # !An&&Bn)
```

Although ABEL has relational operators for less-than and greater-than comparisons, the resulting logic expressions are not so small or easy to derive. For example, consider the relational expression "A<B", where [An..A1] and [Bn..B1] are sets with n elements. To construct the corresponding logic expression, ABEL first constructs n equations of the form

```
Li = (!Ai & (Bi # Li-1) # (Ai & Bi & Li-1)
```

for $i = 1$ to n and $L0 = 0$ by definition. This is, in effect, an iterative definition of the less-than function, starting with the least-significant bit. Each Li equation says that, as of bit i, A is less than B if Ai is 0 and Bi is 1 or A was less than B as of the previous bit, or if Ai and Bi are both 1 and A was less than B as of the previous bit.

The logic expression for "A<B" is simply the equation for Ln. So, after creating the n equations above, ABEL collapses them into a single equation for Ln involving only elements of A and B. It does this by substituting the Ln-1 equation into the righthand side of the Ln equation, then substituting the Ln-2 equation into this result, and so on, until substituting 0 for L0. Finally, it derives a minimal sum-of-products expression from the result.

Collapsing an iterative circuit into a two-level sum-of-products realization usually creates an exponential expansion of product terms. The "<" comparison function follows this pattern, requiring $2^n - 1$ product terms for an n-bit comparator. Thus, comparators larger than a few bits cannot be realized practically in one pass through a PLD.

The results for ">" comparators are identical, of course, and logic expressions for ">=" and "<=" are at least as bad, being the complements of the expressions for "<" and ">". If we use a PLD with output polarity control, the inversion is free and the number of product terms is the same; otherwise, the minimal number of product terms after inverting is $2^n + 2^{n-1} - 1$.

5.9.6 Comparators in VHDL

VHDL has comparison operators for all of its built-in types. *Equality* (=) and *inequality* (/=) operators apply to all types; for array and record types, the operands must have equal size and structure, and the operands are compared component by component. We have used the equality operator to compare a signal or signal vector with a constant value in many examples in this chapter. If we compare two signals or variables, the synthesis engine generates equations similar to ABEL's in the preceding subsection.

equality, =
inequality, /=

VHDL's other comparison operators, >, <, >=, and <=, apply only to integer types, enumerated types (such as STD_LOGIC), and one-dimensional arrays of enumeration or integer types. Integer order from smallest to largest is the natural ordering, from minus infinity to plus infinity, and enumerated types use the ordering in which the elements of the type were defined, from first to last (unless you explicitly change the enumeration encoding using a command specific to the synthesis engine, in which case the ordering is that of your encoding).

The ordering for array types is defined iteratively, starting with the *leftmost* element in each array. Arrays are always compared from left to right, regardless of the order of their index range ("to" or "downto"). The order of the leftmost pair of unequal elements is the order of the array. If the arrays have unequal lengths and all the elements of the shorter array match the corresponding elements of the longer one, then the shorter array is considered to be the smaller.

The result of all this is that the built-in comparison operators compare equal-length arrays of type BIT_VECTOR or STD_LOGIC_VECTOR as if they represented unsigned integers. If the arrays have different lengths, then the operators do *not* yield a valid arithmetic comparison, what you'd get by extending the shorter array with zeroes on the left; more on this in a moment.

Table 5-50 is a VHDL program that produces all of the comparison outputs for comparing two 8-bit unsigned integers. Since the two input vectors A and B have equal lengths, the program produces the desired results.

Table 5-50
Behavioral VHDL program for comparing 8-bit unsigned integers.

```
library IEEE;
use IEEE.std_logic_1164.all;

entity vcompare is
    port (
        A, B: in STD_LOGIC_VECTOR (7 downto 0);
        EQ, NE, GT, GE, LT, LE: out STD_LOGIC
    );
end vcompare;

architecture vcompare_arch of vcompare is
begin
process (A, B)
  begin
    EQ <= '0'; NE <= '0'; GT <= '0'; GE <= '0'; LT <= '0'; LE <= '0';
    if A = B then EQ <= '1'; end if;
    if A /= B then NE <= '1'; end if;
    if A > B then GT <= '1'; end if;
    if A >= B then GE <= '1'; end if;
    if A < B then LT <= '1'; end if;
    if A <= B then LE <= '1'; end if;
  end process;
end vcompare_arch
```

To allow more flexible comparisons and arithmetic operations, IEEE standard 1076-3 created a standard package, `std_logic_arith`, which defines two important new types and a host of comparison and arithmetic functions that operate on them. The two new types are `SIGNED` and `UNSIGNED`:

```
type SIGNED is array (NATURAL range <>) of STD_LOGIC;
type UNSIGNED is array (NATURAL range <>) of STD_LOGIC;
```

As you can see, both types are defined just as indeterminate-length arrays of `STD_LOGIC`, no different from `STD_LOGIC_VECTOR`. The important thing is that the package also defines new comparison functions that are invoked when either or both comparison operands have one of the new types. For example, it defines eight new "less-than" functions with the following combinations of parameters:

```
function "<" (L: UNSIGNED; R: UNSIGNED) return BOOLEAN;
function "<" (L: SIGNED; R: SIGNED) return BOOLEAN;
function "<" (L: UNSIGNED; R: SIGNED) return BOOLEAN;
function "<" (L: SIGNED; R: UNSIGNED) return BOOLEAN;
function "<" (L: UNSIGNED; R: INTEGER) return BOOLEAN;
function "<" (L: INTEGER; R: UNSIGNED) return BOOLEAN;
function "<" (L: SIGNED; R: INTEGER) return BOOLEAN;
function "<" (L: INTEGER; R: SIGNED) return BOOLEAN;
```

Thus, the "<" operator can be used with any combination of `SIGNED`, `UNSIGNED`, and `INTEGER` operands; the compiler selects the function whose parameter types match the actual operands. Each of the functions is defined in the package to do the "right thing," including making the appropriate extensions and conversions when operands of different sizes or types are used. Similar functions are provided for the other five relational operators, =, /=, <=, >, and >=.

Using the `IEEE_std_logic_arith` package, you can write programs like the one in Table 5-51. Its 8-bit input vectors, A, B, C, and D, have three different types. In the comparisons involving A, B, and C, the compiler automatically selects the correct version of the comparison function; for example, for "A<B" it selects the first "<" function above, because both operands have type `UNSIGNED`.

In the comparisons involving D, explicit type conversions are used. The assumption is that the designer wants this particular `STD_LOGIC_VECTOR` to be interpreted as `UNSIGNED` in one case and `SIGNED` in another. The important thing to understand here is that the `std_logic_arith` package does not make any assumptions about how `STD_LOGIC_VECTORs` are to be interpreted; the user must specify the conversion.

Two other packages, `std_logic_signed` and `std_logic_unsigned`, do make assumptions and are useful if all `STD_LOGIC_VECTORs` are to be interpreted the same way. Each package contains three versions of each comparison function so that `STD_LOGIC_VECTORs` are interpreted as `SIGNED` or `UNSIGNED`, respectively, when compared with each other or with integers.

Table 5-51 Behavioral VHDL program for comparing 8-bit integers of various types.

```
library IEEE;
use IEEE.std_logic_1164.all;
use IEEE.std_logic_arith.all;

entity vcompa is
    port (
        A, B: in UNSIGNED (7 downto 0);
        C: in SIGNED (7 downto 0);
        D: in STD_LOGIC_VECTOR (7 downto 0);
        A_LT_B, B_GE_C, A_EQ_C, C_NEG, D_BIG, D_NEG: out STD_LOGIC
    );
end vcompa;

architecture vcompa_arch of vcompa is
begin
process (A, B, C, D)
  begin
    A_LT_B <= '0'; B_GE_C <= '0'; A_EQ_C <= '0'; C_NEG <= '0'; D_BIG <= '0'; D_NEG <= '0';
    if A < B then A_LT_B <= '1'; end if;
    if B >= C then B_GE_C <= '1'; end if;
    if A = C then A_EQ_C <= '1'; end if;
    if C < 0 then C_NEG <= '1'; end if;
    if UNSIGNED(D) > 200 then D_BIG <= '1'; end if;
    if SIGNED(D) < 0 then D_NEG <= '1'; end if;
  end process;
end vcompa_arch;
```

When a comparison function is specified in VHDL, realizing the function as a two-level sum of products takes just as many product terms as in ABEL. However, most VHDL synthesis engines will realize the comparator as an iterative circuit with far fewer gates, albeit more levels of logic. Also, better synthesis engines can detect opportunities to eliminate entire comparator circuits. For example, in the program of Table 5-50 on page 428, the NE, GE, and LE outputs could be realized with one inverter each, as the complements of the EQ, LT, and GT outputs, respectively.

*5.10 Adders, Subtractors, and ALUs

adder

subtractor

Addition is the most commonly performed arithmetic operation in digital systems. An *adder* combines two arithmetic operands using the addition rules described in Chapter 2. As we showed in Section 2.6, the same addition rules and therefore the same adders are used for both unsigned and two's-complement numbers. An adder can perform subtraction as the addition of the minuend and the complemented (negated) subtrahend, but you can also build *subtractor* circuits that perform subtraction directly. MSI devices called ALUs, described in Section 5.10.6, perform addition, subtraction, or any of several other operations according to an operation code supplied to the device.

*5.10.1 Half Adders and Full Adders

The simplest adder, called a *half adder*, adds two 1-bit operands X and Y, *half adder* producing a 2-bit sum. The sum can range from 0 to 2, which requires two bits to express. The low-order bit of the sum may be named HS (half sum), and the high-order bit may be named CO (carry out). We can write the following equations for HS and CO:

$$HS = X \oplus Y$$
$$= X \cdot Y' + X' \cdot Y$$
$$CO = X \cdot Y$$

To add operands with more than one bit, we must provide for carries between bit positions. The building block for this operation is called a *full adder*.   *full adder* Besides the addend-bit inputs X and Y, a full adder has a carry-bit input, CIN. The sum of the three inputs can range from 0 to 3, which can still be expressed with just two output bits, S and COUT, having the following equations:

$$S = X \oplus Y \oplus CIN$$
$$= X \cdot Y' \cdot CIN' + X' \cdot Y \cdot CIN' + X' \cdot Y' \cdot CIN + X \cdot Y \cdot CIN$$
$$COUT = X \cdot Y + X \cdot CIN + Y \cdot CIN$$

Here, S is 1 if an odd number of the inputs are 1, and COUT is 1 if two or more of the inputs are 1. These equations represent the same operation that was specified by the binary addition table in Table 2-3 on page 32.

One possible circuit that performs the full-adder equations is shown in Figure 5-86(a). The corresponding logic symbol is shown in (b). Sometimes the symbol is drawn as shown in (c), so that cascaded full adders can be drawn more neatly, as in the next subsection.

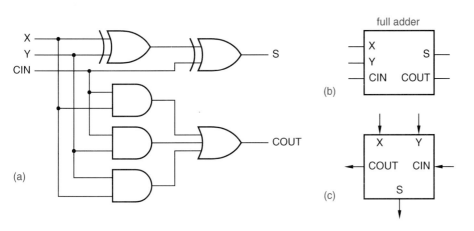

Figure 5-86
Full adder: (a) gate-level circuit diagram;
(b) logic symbol;
(c) alternate logic symbol suitable for cascading.

ripple adder

*5.10.2 Ripple Adders

Two binary words, each with n bits, can be added using a *ripple adder*—a cascade of n full-adder stages, each of which handles one bit. Figure 5-87 shows the circuit for a 4-bit ripple adder. The carry input to the least significant bit (c_0) is normally set to 0, and the carry output of each full adder is connected to the carry input of the next most significant full adder. The ripple adder is a classic example of an iterative circuit as defined in Section 5.9.2.

A ripple adder is slow, since in the worst case a carry must propagate from the least significant full adder to the most significant one. This occurs if, for example, one addend is 11 … 11 and the other is 00 … 01. Assuming that all of the addend bits are presented simultaneously, the total worst-case delay is

$$t_{ADD} = t_{XYCout} + (n - 2) \cdot t_{CinCout} + t_{CinS}$$

where t_{XYCout} is the delay from X or Y to COUT in the least significant stage, $t_{CinCout}$ is the delay from CIN to COUT in the middle stages, and t_{CinS} is the delay from CIN to S in the most significant stage.

A faster adder can be built by obtaining each sum output s_i with just two levels of logic. This can be accomplished by writing an equation for s_i in terms of x_0–x_i, y_0–y_i, and c_0, "multiplying out" or "adding out" to obtain a sum-of-products or product-of-sums expression, and building the corresponding AND-OR or OR-AND circuit. Unfortunately, beyond s_2 the resulting expressions have too many terms, requiring too many first-level gates and more inputs than typically possible on the second-level gate. For example, even assuming that $c_0 = 0$, a two-level AND-OR circuit for s_2 requires fourteen 4-input ANDs, four 5-input ANDs, and an 18-input OR gate; higher-order sum bits are even worse. Nevertheless, it is possible to build adders with just a few levels of delay using a more reasonable number of gates, as we'll see in Section 5.10.4.

*5.10.3 Subtractors

full subtractor

A binary subtraction operation analogous to binary addition was also specified in Table 2-3 on page 32. A *full subtractor* handles one bit of the binary subtraction algorithm, having input bits X (minuend), Y (subtrahend), and BIN (borrow

Figure 5-87
A 4-bit ripple adder.

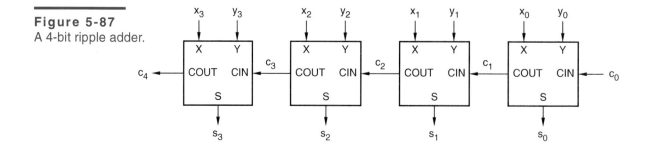

in), and output bits D (difference) and BOUT (borrow out). We can write logic equations corresponding to the binary subtraction table as follows:

$$D = X \oplus Y \oplus BIN$$

$$BOUT = X' \cdot Y + X' \cdot BIN + Y \cdot BIN$$

These equations are very similar to equations for a full adder, which should not be surprising. We showed in Section 2.6 that a two's-complement subtraction operation, $X - Y$, can be performed by an addition operation, namely, by adding the two's complement of Y to X. The two's complement of Y is $\overline{Y} + 1$, where $\overline{Y}$ is the bit-by-bit complement of Y. We also showed in Exercise 2.26 that a binary adder can be used to perform an unsigned subtraction operation $X - Y$ by performing the operation $X + \overline{Y} + 1$. We can now confirm that these statements are true by manipulating the logic equations above:

$BOUT = X' \cdot Y + X' \cdot BIN + Y \cdot BIN$

$BOUT' = (X + Y') \cdot (X + BIN') \cdot (Y' + BIN')$ (generalized DeMorgan's theorem)

$\quad = X \cdot Y' + X \cdot BIN' + Y' \cdot BIN'$ (multiply out)

$D = X \oplus Y \oplus BIN$

$\quad = X \oplus Y' \oplus BIN'$ (complementing XOR inputs)

For the last manipulation, recall that we can complement the two inputs of an XOR gate without changing the function performed.

Comparing with the equations for a full adder, the above equations tell us that we can build a full subtractor from a full adder as shown in Figure 5-88. Just to keep things straight, we've given the full-adder circuit in (a) a fictitious name, the "74x999." As shown in (c), we can interpret the function of this same physical circuit to be a full subtractor by giving it a new symbol with active-low borrow in, borrow out, and subtrahend signals.

Thus, to build a ripple subtractor for two n-bit active-high operands, we can use n 74x999s and inverters, as shown in (d). Note that for the subtraction operation, the borrow input of the least significant bit should be negated (no borrow), which for an active-low input means that the physical pin must be 1 or HIGH. This is just the opposite as in addition, where the same input pin is an active-high carry-in that is 0 or LOW.

By going back to the math in Chapter 2, we can show that this sort of manipulation works for all adder and subtractor circuits, not just ripple adders and subtractors. That is, any n-bit adder circuit can be made to function as a subtractor by complementing the subtrahend and treating the carry-in and carry-out signals as borrows with the opposite active level. The rest of this section discusses addition circuits only, with the understanding that they can easily be made to perform subtraction.

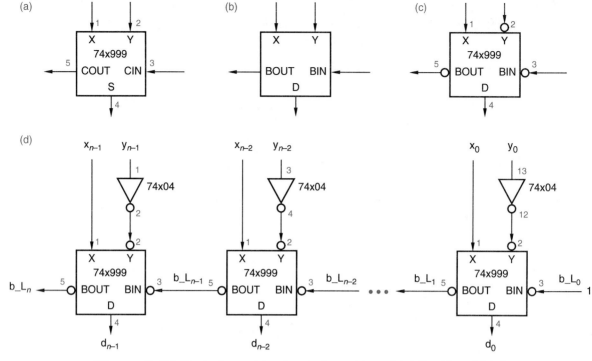

Figure 5-88 Designing subtractors using adders: (a) full adder; (b) full subtractor; (c) interpreting the device in (a) as a full subtractor; (d) ripple subtractor.

*5.10.4 Carry Lookahead Adders

The logic equation for sum bit i of a binary adder can actually be written quite simply:

$$s_i = x_i \oplus y_i \oplus c_i$$

More complexity is introduced when we expand c_i above in terms of $x_0 - x_{i-1}$, $y_0 - y_{i-1}$, and c_0, and we get a real mess expanding the XORs. However, if we're willing to forego the XOR expansion, we can at least streamline the design of c_i logic using ideas of *carry lookahead* discussed in this subsection.

carry lookahead

Figure 5-89 shows the basic idea. The block labeled "Carry Lookahead Logic" calculates c_i in a fixed, small number of logic levels for any reasonable value of i. Two definitions are the key to carry lookahead logic:

carry generate

- For a particular combination of inputs x_i and y_i, adder stage i is said to *generate* a carry if it produces a carry-out of 1 ($c_{i+1} = 1$) independent of the inputs on $x_0 - x_{i-1}$, $y_0 - y_{i-1}$, and c_0.

carry propagate

- For a particular combination of inputs x_i and y_i, adder stage i is said to *propagate* carries if it produces a carry-out of 1 ($c_{i+1} = 1$) in the presence of an input combination of $x_0 - x_{i-1}$, $y_0 - y_{i-1}$, and c_0 that causes a carry-in of 1 ($c_i = 1$).

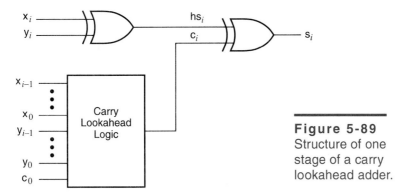

Figure 5-89
Structure of one
stage of a carry
lookahead adder.

Corresponding to these definitions, we can write logic equations for a carry-generate signal, g_i, and a carry-propagate signal, p_i, for each stage of a carry lookahead adder:

$$g_i = x_i \cdot y_i$$
$$p_i = x_i + y_i$$

That is, a stage unconditionally generates a carry if both of its addend bits are 1, and it propagates carries if at least one of its addend bits is 1. The carry output of a stage can now be written in terms of the generate and propagate signals:

$$c_{i+1} = g_i + p_i \cdot c_i$$

To eliminate carry ripple, we recursively expand the c_i term for each stage and multiply out to obtain a 2-level AND-OR expression. Using this technique, we can obtain the following carry equations for the first four adder stages:

$$c_1 = g_0 + p_0 \cdot c_0$$
$$c_2 = g_1 + p_1 \cdot c_1$$
$$= g_1 + p_1 \cdot (g_0 + p_0 \cdot c_0)$$
$$= g_1 + p_1 \cdot g_0 + p_1 \cdot p_0 \cdot c_0$$
$$c_3 = g_2 + p_2 \cdot c_2$$
$$= g_2 + p_2 \cdot (g_1 + p_1 \cdot g_0 + p_1 \cdot p_0 \cdot c_0)$$
$$= g_2 + p_2 \cdot g_1 + p_2 \cdot p_1 \cdot g_0 + p_2 \cdot p_1 \cdot p_0 \cdot c_0$$
$$c_4 = g_3 + p_3 \cdot c_3$$
$$= g_3 + p_3 \cdot (g_2 + p_2 \cdot g_1 + p_2 \cdot p_1 \cdot g_0 + p_2 \cdot p_1 \cdot p_0 \cdot c_0)$$
$$= g_3 + p_3 \cdot g_2 + p_3 \cdot p_2 \cdot g_1 + p_3 \cdot p_2 \cdot p_1 \cdot g_0 + p_3 \cdot p_2 \cdot p_1 \cdot p_0 \cdot c_0$$

Each equation corresponds to a circuit with just three levels of delay—one for the generate and propagate signals, and two for the sum of products shown. A *carry lookahead adder* uses three-level equations such as these in each adder stage for the block labeled "carry lookahead" in Figure 5-89. The sum output for a stage is produced by combining the carry bit with the two addend bits for the

carry lookahead adder

stage, as we showed in the figure. In the next subsection we'll study some commercial MSI adders and ALUs that use carry lookahead.

*5.10.5 MSI Adders

74x283

74x83

The *74x283* is a 4-bit binary adder that forms its sum and carry outputs with just a few levels of logic, using the carry lookahead technique. Figure 5-90 is a logic symbol for the 74x283. The older *74x83* is identical except for its pinout, which has nonstandard locations for power and ground.

Figure 5-90
Traditional logic symbol for the 74x283 4-bit binary adder.

The logic diagram for the '283, shown in Figure 5-91, has just a few differences from the general carry-lookahead design that we described in the preceding subsection. First of all, its addends are named A and B instead of X and Y; no big deal. Second, it produces active-low versions of the carry-generate (g_i') and carry-propagate (p_i') signals, since inverting gates are generally faster than noninverting ones. Third, it takes advantage of the fact that we can algebraically manipulate the half-sum equation as follows:

$$
\begin{aligned}
hs_i &= x_i \oplus y_i \\
&= x_i \cdot y_i' + x_i' \cdot y_i \\
&= x_i \cdot y_i' + x_i \cdot x_i' + x_i' \cdot y_i + y_i \cdot y_i' \\
&= (x_i + y_i) \cdot (x_i' + y_i') \\
&= (x_i + y_i) \cdot (x_i \cdot y_i)' \\
&= p_i \cdot g_i'
\end{aligned}
$$

Thus, an AND gate with an inverted input can be used instead of an XOR gate to create each half-sum bit.

Finally, the '283 creates the carry signals using an INVERT-OR-AND structure (the DeMorgan equivalent of an AND-OR-INVERT), which has about the same delay as a single CMOS or TTL inverting gate. This requires some explaining, since the carry equations that we derived in the preceding subsection are used in a slightly modified form. In particular, the c_{i+1} equation uses the term $p_i \cdot g_i$ instead of g_i. This has no effect on the output, since p_i is always 1 when g_i is 1. However, it allows the equation to be factored as follows:

$$
\begin{aligned}
c_{i+1} &= p_i \cdot g_i + p_i \cdot c_i \\
&= p_i \cdot (g_i + c_i)
\end{aligned}
$$

This leads to the following carry equations, which are used by the circuit :

$$
\begin{aligned}
c_1 &= p_0 \cdot (g_0 + c_0) \\
c_2 &= p_1 \cdot (g_1 + c_1) \\
&= p_1 \cdot (g_1 + p_0 \cdot (g_0 + c_0)) \\
&= p_1 \cdot (g_1 + p_0) \cdot (g_1 + g_0 + c_0)
\end{aligned}
$$

Figure 5-91
Logic diagram for
the 74x283 4-bit
binary adder.

$$c_3 = p_2 \cdot (g_2 + c_2)$$
$$= p_2 \cdot (g_2 + p_1 \cdot (g_1 + p_0) \cdot (g_1 + g_0 + c_0))$$
$$= p_2 \cdot (g_2 + p_1) \cdot (g_2 + g_1 + p_0) \cdot (g_2 + g_1 + g_0 + c_0)$$
$$c_4 = p_3 \cdot (g_3 + c_3)$$
$$= p_3 \cdot (g_3 + p_2 \cdot (g_2 + p_1) \cdot (g_2 + g_1 + p_0) \cdot (g_2 + g_1 + g_0 + c_0))$$
$$= p_3 \cdot (g_3 + p_2) \cdot (g_3 + g_2 + p_1) \cdot (g_3 + g_2 + g_1 + p_0) \cdot (g_3 + g_2 + g_1 + g_0 + c_0)$$

If you've followed the derivation of these equations and can obtain the same ones by reading the '283 logic diagram, then congratulations, you're up to speed on switching algebra! If not, you may want to review Sections 4.1 and 4.2.

group-ripple adder

The propagation delay from the C0 input to the C4 output of the '283 is very short, about the same as two inverting gates. As a result, fairly fast *group-ripple adders* with more than four bits can be made simply by cascading the carry outputs and inputs of '283s, as shown in Figure 5-92 for a 16-bit adder. The total propagation delay from C0 to C16 in this circuit is about the same as that of eight inverting gates.

Figure 5-92
A 16-bit group-ripple adder.

*5.10.6 MSI Arithmetic and Logic Units

An *arithmetic and logic unit (ALU)* is a combinational circuit that can perform any of a number of different arithmetic and logical operations on a pair of b-bit operands. The operation to be performed is specified by a set of function-select inputs. Typical MSI ALUs have 4-bit operands and three to five function-select inputs, allowing up to 32 different functions to be performed.

arithmetic and logic unit (ALU)

Figure 5-93 is a logic symbol for the *74x181* 4-bit ALU. The operation performed by the '181 is selected by the M and S3–S0 inputs, as detailed in Table 5-52. Note that the identifiers A, B, and F in the table refer to the 4-bit words A3–A0, B3–B0, and F3–F0; and the symbols · and + refer to logical AND and OR operations.

74x181

The 181's M input selects between arithmetic and logical operations. When M = 1, logical operations are selected, and each output Fi is a function only of the corresponding data inputs, Ai and Bi. No carries propagate between stages, and the CIN input is ignored. The S3–S0 inputs select a particular logical operation; any of the 16 different combinational logic functions on two variables may be selected.

Table 5-52 Functions performed by the 74x181 4-bit ALU.

| Inputs | | | | Function | |
S3	S2	S1	S0	M = 0 (arithmetic)	M = 1 (logic)
0	0	0	0	F = A minus 1 plus CIN	F = A'
0	0	0	1	F = A · B minus 1 plus CIN	F = A' + B'
0	0	1	0	F = A · B' minus 1 plus CIN	F = A' + B
0	0	1	1	F = 1111 plus CIN	F = 1111
0	1	0	0	F = A plus (A + B') plus CIN	F = A' · B'
0	1	0	1	F = A · B plus (A + B') plus CIN	F = B'
0	1	1	0	F = A minus B minus 1 plus CIN	F = A ⊕ B'
0	1	1	1	F = A + B' plus CIN	F = A + B'
1	0	0	0	F = A plus (A + B) plus CIN	F = A' · B
1	0	0	1	F = A plus B plus CIN	F = A ⊕ B
1	0	1	0	F = A · B' plus (A + B) plus CIN	F = B
1	0	1	1	F = A + B plus CIN	F = A + B
1	1	0	0	F = A plus A plus CIN	F = 0000
1	1	0	1	F = A · B plus A plus CIN	F = A · B'
1	1	1	0	F = A · B' plus A plus CIN	F = A · B
1	1	1	1	F = A plus CIN	F = A

Figure 5-93
Logic symbol for the 74x181 4-bit ALU.

When M = 0, arithmetic operations are selected, carries propagate between the stages, and CIN is used as a carry input to the least significant stage. For operations larger than four bits, multiple '181 ALUs may be cascaded like the group-ripple adder in Figure 5-92, with the carry-out (COUT) of each ALU connected to the carry-in (CIN) of the next most significant stage. The same function-select signals (M, S3–S0) are applied to all the '181s in the cascade.

To perform two's-complement addition, we use S3–S0 to select the operation "A plus B plus CIN." The CIN input of the least-significant ALU is normally set to 0 during addition operations. To perform two's-complement subtraction, we use S3–S0 to select the operation A minus B minus plus CIN. In this case, the CIN input of the least significant ALU is normally set to 1, since CIN acts as the complement of the borrow during subtraction.

The '181 provides other arithmetic operations, such as "A minus 1 plus CIN," that are useful in some applications (e.g., decrement by 1). It also provides a bunch of weird arithmetic operations, such as "A · B' plus (A + B) plus CIN," that are almost never used in practice, but that "fall out" of the circuit for free.

Notice that the operand inputs A3_L–A0_L and B3_L–B0_L, and the function outputs F3_L–F0_L of the '181 are active low. The '181 can also be used with active-high operand inputs and function outputs. In this case, a different version of the function table must be constructed. When M = 1, logical operations are still performed, but for a given input combination on S3–S0, the function obtained is precisely the dual of the one listed in Table 5-52. When M = 0, arithmetic operations are performed, but the function table is once again different. Refer to a '181 data sheet for more details.

74x381
74x382

Two other MSI ALUs, the *74x381* and *74x382* shown in Figure 5-94, encode their select inputs more compactly, and provide only eight different but useful functions, as detailed in Table 5-53. The only difference between the '381 and '382 is that one provides group-carry lookahead outputs (which we explain next), while the other provides ripple carry and overflow outputs.

Figure 5-94
Logic symbols for 4-bit ALUs: (a) 74x381; (b) 74x382.

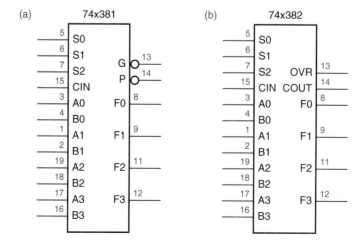

Inputs			
S2	S1	S0	*Function*
0	0	0	F = 0000
0	0	1	F = B minus A minus 1 plus CIN
0	1	0	F = A minus B minus 1 plus CIN
0	1	1	F = A plus B plus CIN
1	0	0	F = A ⊕ B
1	0	1	F = A + B
1	1	0	F = A · B
1	1	1	F = 1111

Table 5-53
Functions performed by the 74x381 and 74x382 4-bit ALUs.

*5.10.7 Group-Carry Lookahead

The '181 and '381 provide *group-carry lookahead* outputs that allow multiple ALUs to be cascaded without rippling carries between 4-bit groups. Like the 74x283, the ALUs use carry lookahead to produce carries internally. However, they also provide G_L and P_L outputs that are carry lookahead signals for the entire 4-bit group. The G_L output is asserted if the ALU generates a carry—that is, if it will produce a carry-out (COUT = 1) whether or not there is a carry-in (CIN = 1):

group-carry lookahead

$$G\_L = (g_3 + p_3 \cdot g_2 + p_3 \cdot p_2 \cdot g_1 + p_3 \cdot p_2 \cdot p_1 \cdot g_0)'$$

The P_L output is asserted if the ALU propagates a carry—that is, if it will produce a carry-out if there is a carry-in:

$$P\_L = (p_3 \cdot p_2 \cdot p_1 \cdot p_0)'$$

When ALUs are cascaded, the group-carry lookahead outputs may be combined in just two levels of logic to produce the carry input to each ALU. A *lookahead carry circuit*, the *74x182* shown in Figure 5-95, performs this operation. The '182 inputs are C0, the carry input to the least significant ALU ("ALU 0"), and G0–G3 and P0–P3, the generate and propagate outputs of ALUs 0–3. Using these inputs, the '182 produces carry inputs C1–C3 for ALUs 1–3. Figure 5-96 shows the connections for a 16-bit ALU using four '381s and a '182.

lookahead carry circuit
74x182

The 182's carry equations are obtained by "adding out" the basic carry lookahead equation of Section 5.10.4:

$$c_{i+1} = g_i + p_i \cdot c_i$$
$$= (g_i + p_i) \cdot (g_i + c_i)$$

Expanding for the first three values of i, we obtain the following equations:

C1 = (G0+P0) · (G0+C0)
C2 = (G1+P1) · (G1+G0+P0) · (G1+G0+C0)
C3 = (G2+P2) · (G2+G1+P1) · (G2+G1+G0+P0) · (G2+G1+G0+C0)

Figure 5-95
Logic symbol for the 74x182 lookahead carry circuit.

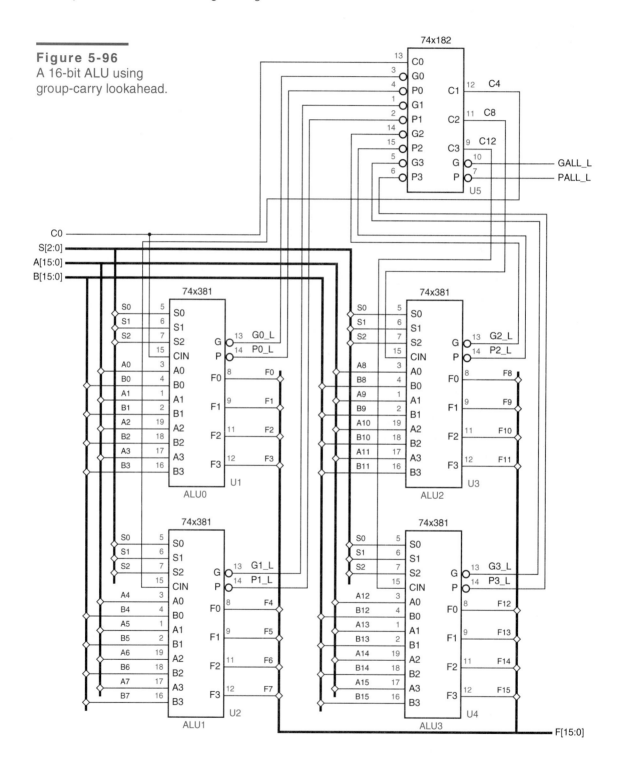

Figure 5-96
A 16-bit ALU using
group-carry lookahead.

The '182 realizes each of these equations with just one level of delay—an INVERT-OR-AND gate.

When more than four ALUs are cascaded, they may be partitioned into "supergroups," each with its own '182. For example, a 64-bit adder would have four supergroups, each containing four ALUs and a '182. The G_L and P_L outputs of each '182 can be combined in a next-level '182, since they indicate whether the supergroup generates or propagates carries:

$$G\_L = ((G3+P3) \cdot (G3+G2+P2) \cdot (G3+G2+G1+P1) \cdot (G3+G2+G1+G0))'$$
$$P\_L = (P0 \cdot P1 \cdot P2 \cdot P3)'$$

*5.10.8 Adders in ABEL and PLDs

ABEL supports addition (+) and subtraction (−) operators which can be applied to sets. Sets are interpreted as unsigned integers; for example, a set with n bits represents an integer in the range of 0 to 2^n-1. Subtraction is performed by negating the subtrahend and adding. Negation is performed in two's complement; that is, the operand is complemented bit-by-bit and then 1 is added.

Table 5-54 shows an example of addition in ABEL. The set definition for SUM was made one bit wider than the addends to accommodate the carry out of the MSB; otherwise this carry would be discarded. The set definitions for the addends were extended on the left with a 0 bit to match the size of SUM.

Even though the adder program is extremely small, it takes a long time to compile and it generates a huge number of terms in the minimal two-level sum of products. While SUM0 has only two product terms, subsequent terms SUMi have $5 \cdot 2^i - 4$ terms, or 636 terms for SUM7! And the carry out (SUM8) has $2^{80} - 1 = 255$ product terms. Obviously, adders with more than a few bits cannot be practically realized using two levels of logic.

```
module add
title 'Adder Exercise'

" Input and output pins
A7..A0, B7..B0          pin;
SUM8..SUM0              pin istype 'com';

" Set definitions
A = [0, A7..A0];
B = [0, B7..B0];
SUM = [SUM8..SUM0];

equations
SUM = A + B;

end add
```

Table 5-54
ABEL program for an 8-bit adder.

> **CARRYING ON**   The carry out (SUM8) in our adder example has the exactly same number of product terms (255) as the "less-than" or "greater-than" output of an 8-bit comparator. This is less surprising once you realize that the carry out from the addition A+B is functionally equivalent to the expression $A > \overline{B}$.

@CARRY directive

Recognizing that larger adders and comparators are still needed in PLDs from time to time, ABEL provides an *@CARRY directive* which tells the compiler to synthesize a group-ripple adder with *n* bits per group. For example, if the statement "@CARRY 1;" were included in Table 5-54, the compiler would create eight new signals for the carries out of bit positions 0 through 7. The equations for SUM1 through SUM8 would use these internal carries, essentially creating an 8-stage ripple adder with a worst-case delay of eight passes through the PLD.

If the statement "@CARRY 2;" were used, the compiler would compute carries two bits at a time, creating four new signals for carries out of bit positions 1, 3, 5, and 7. In this case, the maximum number of product terms needed for any output is still reasonable, only 7, and the worst-case delay path has just four passes through the PLD. With three bits per group (@CARRY 3;), the maximum number of product terms balloons to 28, which is impractical.

A special case that is often used in ABEL and PLDs is adding or subtracting a constant 1. This operation is used in the definition of counters, where the next state of the counter is just the current state plus 1 for an "up" counter or minus 1 for a "down" counter. The equation for bit *i* of an "up" counter can be stated very simply in words: "Complement bit *i* if counting is enabled and all of the bits lower than *i* are 1." This requires just $i + 2$ product terms for any value of *i*, and can be further reduced to just one product term and an XOR gate in some PLDs and CPLDs.

*5.10.9 Adders in VHDL

Although VHDL has addition (+) and subtraction (−) operators built in, they work only with the integer, real, and physical types. They specifically do *not* work with BIT_VECTOR types or the IEEE standard type STD_LOGIC_VECTOR. Instead, standard packages define these operators.

As we explained in Section 5.9.6, the IEEE_std_logic_arith package defines two new array types, SIGNED and UNSIGNED, and a set of comparison functions for operands of type INTEGER, SIGNED, or UNSIGNED. The package also defines addition and subtraction operations for the same kinds of operands as well as STD_LOGIC and STD_ULOGIC for 1-bit operands.

The large number of overlaid addition and subtraction functions may make it less than obvious what type an addition or subtraction result will have. Normally, if any of the operands is type SIGNED, the result is SIGNED, else the result is UNSIGNED. However, if the result is assigned to a signal or variable of

```
library IEEE;
use IEEE.std_logic_1164.all;
use IEEE.std_logic_arith.all;

entity vadd is
    port (
        A, B: in UNSIGNED (7 downto 0);
        C: in SIGNED (7 downto 0);
        D: in STD_LOGIC_VECTOR (7 downto 0);
        S: out UNSIGNED (8 downto 0);
        T: out SIGNED (8 downto 0);
        U: out SIGNED (7 downto 0);
        V: out STD_LOGIC_VECTOR (8 downto 0)
    );
end vadd;

architecture vadd_arch of vadd is
  begin
    S <= ('0' & A) + ('0' & B);
    T <= A + C;
    U <= C + SIGNED(D);
    V <= C - UNSIGNED(D);
end vadd_arch;
```

Table 5-55
VHDL program for adding and subtracting 8-bit integers of various types.

type STD_LOGIC_VECTOR, then the SIGNED or UNSIGNED result is converted to that type. The length of any result is normally the length of the longest operand. However, when an UNSIGNED operand is combined with a SIGNED or INTEGER operand, its length is increased by 1 to accommodate a sign bit of 0, and then the result's length is determined.

Incorporating these considerations, the VHDL program in Table 5-55 shows 8-bit additions for various operand and result types. The first result, S, is declared to be 9 bits long assuming the designer is interested in the carry from the 8-bit addition of UNSIGNED operands A and B. The concatenation operator & is used to extend A and B so that the addition function will return the carry bit in the MSB of the result.

The next result, T, is also 9 bits long, since the addition function extends the UNSIGNED operand A when combining it with the SIGNED operand C. In the third addition, an 8-bit STD_LOGIC_VECTOR D is type-converted to SIGNED and combined with C to obtain an 8-bit SIGNED result U. In the last statement, D is converted to UNSIGNED, automatically extended by one bit, and subtracted from C to produce a 9-bit result V.

Since addition and subtraction are fairly expensive in terms of the number of gates required, many VHDL synthesis engines will attempt to reuse adder blocks whenever possible. For example, Table 5-56 is a VHDL program that includes two different additions. Rather than building two adders and selecting

Table 5-56
VHDL program that
allows adder sharing.

```
library IEEE;
use IEEE.std_logic_1164.all;
use IEEE.std_logic_arith.all;

entity vaddshr is
    port (
        A, B, C, D: in SIGNED (7 downto 0);
        SEL: in STD_LOGIC;
        S: out SIGNED (7 downto 0)
    );
end vaddshr;

architecture vaddshr_arch of vaddshr is
begin
    S <= A + B when SEL = '1' else C + D;
end vaddshr_arch;
```

one's output with a multiplexer, the synthesis engine can build just one adder and select its inputs using multiplexers, potentially creating a smaller overall circuit.

*5.11 Combinational Multipliers

*5.11.1 Combinational Multiplier Structures

combinational multiplier

In Section 2.8 we outlined an algorithm that uses n shifts and adds to multiply n-bit binary numbers. Although the shift-and-add algorithm emulates the way that we do paper-and-pencil multiplication of decimal numbers, there is nothing inherently "sequential" or "time dependent" about multiplication. That is, given two n-bit input words X and Y, it is possible to write a truth table that expresses the $2n$-bit product $P = X \cdot Y$ as a *combinational* function of X and Y. A *combinational multiplier* is a logic circuit with such a truth table.

product component

Most approaches to combinational multiplication are based on the paper-and-pencil shift-and-add algorithm. Figure 5-97 illustrates the basic idea for an 8×8 multiplier for two unsigned integers, multiplicand $X = x_7x_6x_5x_4x_3x_2x_1x_0$ and multiplier $Y = y_7y_6y_5y_4y_3y_2y_1y_0$. We call each row a *product component*, a shifted

Figure 5-97
Partial products in an
8×8 multiplier.

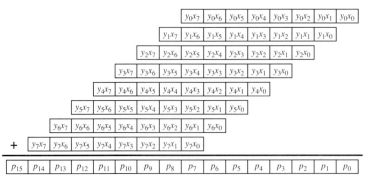

Figure 5-98
Interconnections
for an 8 × 8
combinational
multiplier.

multiplicand that is multiplied by 0 or 1 depending on the corresponding multiplier bit. Each small box represents one product-component bit $y_i x_j$, the logical AND of multiplier bit y_i and multiplicand bit x_j. The product $P = p_{15}p_{14}\cdots p_2 p_1 p_0$ has 16 bits and is obtained by adding together all the product components.

Figure 5-98 shows one way to add up the product components. Here, the product-component bits have been spread out to make space, and each "+" box is a full-adder equivalent to Figure 5-86(c) on page 431. The carries in each row of full adders are connected to make an 8-bit ripple adder. Thus, the first ripple adder combines the first two product components to product the first partial product, as defined in Section 2.8. Subsequent adders combine each partial product with the next product component.

It is interesting to study the propagation delay of the circuit in Figure 5-98. In the worst case, the inputs to the least significant adder ($y_0 x_1$ and $y_1 x_0$) can affect the MSB of the product (p_{15}). If we assume for simplicity that the delays from any input to any output of a full adder are equal, say t_{pd}, then the worst-case

Figure 5-99
Interconnections for a faster 8 × 8 combinational multiplier.

path goes through 20 adders and its delay is $20t_{pd}$. If the delays are different, then the answer depends on the relative delays; see Exercise 5.83.

sequential multiplier   *Sequential multipliers* use a single adder and a register to accumulate the partial products. The partial-product register is initialized to the first product component, and for an $n \times n$-bit multiplication, $n - 1$ steps are taken and the adder is used $n - 1$ times, once for each of the remaining $n - 1$ product components to be added to the partial-product register.

carry-save addition   Some sequential multipliers use a trick called *carry-save addition* to speed up multiplication. The idea is to break the carry chain of the ripple adder to shorten the delay of each addition. This is done by applying the carry output from bit i during step j to the carry input for bit $i + 1$ during the *next* step, $j + 1$. After the last product component is added, one more step is needed in which the carries are hooked up in the usual way and allowed to ripple from the least to the most significant bit.

The combinational equivalent of an $8{\times}8$ multiplier using carry-save addition is shown in Figure 5-99. Notice that the carry out of each full adder in the first seven rows is connected to an input of an adder *below* it. Carries in the eighth row of full adders are connected to create a conventional ripple adder. Although this adder uses exactly the same amount of logic as the previous one (64 2-input AND gates and 56 full adders), its propagation delay is substantially shorter. Its worst-case delay path goes through only 14 full adders. The delay can be further improved by using a carry lookahead adder for the last row.

The regular structure of combinational multipliers makes them ideal for VLSI and ASIC realization. The importance of fast multiplication in microprocessors, digital video, and many other applications has led to much study and development of even better structures and circuits for combinational multipliers; see the References.

*5.11.2 Multiplication in ABEL and PLDs

ABEL provides a multiplication operator $*$, but it can be used only with individual signals, numbers, or special constants, not with sets. Thus, ABEL cannot synthesize a multiplier circuit from a single equation like "P = X$*$Y."

Still, you can use ABEL to specify a combinational multiplier if you break it down into smaller pieces. For example, Table 5-57 shows the design of a $4{\times}4$ unsigned multiplier following the same general structure as Figure 5-97 on page page 446. Expressions are used to define the four product components, PC1, PC2, PC3, and PC4, which are then added in the equations section of the program. This does not generate an array of full adders as in Figure 5-98 or 5-99. Rather, the ABEL compiler will dutifully crunch the addition equation to

Table 5-57
ABEL program for a $4{\times}4$ combinational multiplier.

```
module mul4x4
title '4x4 Combinational Multiplier'

X3..X0, Y3..Y0 pin;  " multiplicand, multiplier
P7..P0         pin istype 'com';   " product

P = [P7..P0];
PC1 = Y0 & [0, 0, 0, 0,X3,X2,X1,X0];
PC2 = Y1 & [0, 0, 0,X3,X2,X1,X0, 0];
PC3 = Y2 & [0, 0,X3,X2,X1,X0, 0, 0];
PC4 = Y3 & [0,X3,X2,X1,X0, 0, 0, 0];

equations
P = PC1 + PC2 + PC3 + PC4;

end mul4x4
```

produce a minimal sum for each of the eight product output bits. Surprisingly, the worst-case output, P4, has only 36 product terms, a little high but certainly realizable in two passes through a PLD.

*5.11.3 Multiplication in VHDL

VHDL is rich enough to express multiplication in a number of different ways; we'll save the best for last.

Table 5-58 is a behavioral VHDL program that mimics the multiplier structure of Figure 5-99. In order to represent the internal signals in the figure, the program defines a new data type, `array8x8`, which is a two-dimensional array of STD_LOGIC (recall that STD_LOGIC_VECTOR is a one-dimensional array of STD_LOGIC). Variable PC is declared as such an array to hold the product-component bits, and variables PCS and PCC are similar arrays to hold the sum and carry outputs of the main array of full adders. One-dimensional arrays RAS and RAC hold the sum and carry outputs of the ripple adder. Figure 5-100 shows the variable naming and numbering scheme. Integer variables i and j are used as loop indices for rows and columns, respectively.

The program attempts to illustrate the logic gates that would be used in a faithful realization of Figure 5-99, even though a synthesizer could legitimately create quite a different structure from this behavioral program. If you want to control the structure, then you must use structural VHDL, as we'll show later.

In the program, the first, nested for statement performs 64 AND operations to obtain the product-component bits. The next for loop initializes boundary conditions at the top of the multiplier, using the notion of row-0 "virtual" full

Figure 5-100
VHDL variable names
for the 8 × 8 multiplier.

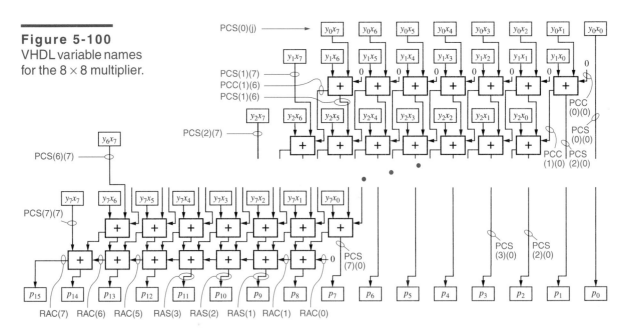

```vhdl
library IEEE;
use IEEE.std_logic_1164.all;

entity vmul8x8p is
    port ( X: in STD_LOGIC_VECTOR (7 downto 0);
           Y: in STD_LOGIC_VECTOR (7 downto 0);
           P: out STD_LOGIC_VECTOR (15 downto 0) );
end vmul8x8p;

architecture vmul8x8p_arch of vmul8x8p is
function MAJ (I1, I2, I3: STD_LOGIC) return STD_LOGIC is
  begin
    return ((I1 and I2) or (I1 and I3) or (I2 and I3));
  end MAJ;
begin
process (X, Y)
type array8x8 is array (0 to 7) of STD_LOGIC_VECTOR (7 downto 0);
variable PC: array8x8;      -- product component bits
variable PCS: array8x8;     -- full-adder sum bits
variable PCC: array8x8;     -- full-adder carry output bits
variable RAS, RAC: STD_LOGIC_VECTOR (7 downto 0); -- ripple adder sum
  begin                                            --   and carry bits
    for i in 0 to 7 loop for j in 0 to 7 loop
        PC(i)(j) := Y(i) and X(j); -- compute product component bits
    end loop;  end loop;
    for j in 0 to 7 loop
      PCS(0)(j) := PC(0)(j);  -- initialize first-row "virtual"
      PCC(0)(j) := '0';       --   adders (not shown in figure)
    end loop;
    for i in 1 to 7 loop      -- do all full adders except last row
      for j in 0 to 6 loop
        PCS(i)(j) := PC(i)(j) xor PCS(i-1)(j+1) xor PCC(i-1)(j);
        PCC(i)(j) := MAJ(PC(i)(j), PCS(i-1)(j+1), PCC(i-1)(j));
        PCS(i)(7) := PC(i)(7); -- leftmost "virtual" adder sum output
      end loop;
    end loop;
    RAC(0) := '0';
    for i in 0 to 6 loop  -- final ripple adder
      RAS(i) := PCS(7)(i+1) xor PCC(7)(i) xor RAC(i);
      RAC(i+1) := MAJ(PCS(7)(i+1), PCC(7)(i), RAC(i));
    end loop;
    for i in 0 to 7 loop
      P(i) <= PCS(i)(0);  -- first 8 product bits from full-adder sums
    end loop;
    for i in 8 to 14 loop
      P(i) <= RAS(i-8);   -- next 7 bits from ripple-adder sums
    end loop;
    P(15) <= RAC(7);      -- last bit from ripple-adder carry
  end process;
end vmul8x8p_arch;
```

Table 5-58
Behavioral VHDL
program for an 8×8
combinational
multiplier.

Table 5-59
Structural VHDL
architecture for an
8×8 combinational
multiplier.

```vhdl
architecture vmul8x8s_arch of vmul8x8s is
component AND2
  port( I0, I1: in STD_LOGIC;
        O: out STD_LOGIC );
end component;
component XOR3
  port( I0, I1, I2: in STD_LOGIC;
        O: out STD_LOGIC );
end component;
component MAJ   -- Majority function, O = I0*I1 + I0*I2 + I1*I2
  port( I0, I1, I2: in STD_LOGIC;
        O: out STD_LOGIC );
end component;

type array8x8 is array (0 to 7) of STD_LOGIC_VECTOR (7 downto 0);
signal PC: array8x8;      -- product-component bits
signal PCS: array8x8;     -- full-adder sum bits
signal PCC: array8x8;     -- full-adder carry output bits
signal RAS, RAC: STD_LOGIC_VECTOR (7 downto 0); -- sum, carry
begin
  g1: for i in 0 to 7 generate     -- product-component bits
    g2: for j in 0 to 7 generate
        U1: AND2 port map (Y(i), X(j), PC(i)(j));
    end generate;
  end generate;
  g3: for j in 0 to 7 generate
    PCS(0)(j) <= PC(0)(j);  -- initialize first-row "virtual" adders
    PCC(0)(j) <= '0';
  end generate;
  g4: for i in 1 to 7 generate   -- do full adders except the last row
    g5: for j in 0 to 6 generate
      U2: XOR3 port map (PC(i)(j),PCS(i-1)(j+1),PCC(i-1)(j),PCS(i)(j));
      U3: MAJ  port map (PC(i)(j),PCS(i-1)(j+1),PCC(i-1)(j),PCC(i)(j));
      PCS(i)(7) <= PC(i)(7); -- leftmost "virtual" adder sum output
    end generate;
  end generate;
  RAC(0) <= '0';
  g6: for i in 0 to 6 generate  -- final ripple adder
    U7: XOR3 port map (PCS(7)(i+1), PCC(7)(i), RAC(i), RAS(i));
    U3: MAJ  port map (PCS(7)(i+1), PCC(7)(i), RAC(i), RAC(i+1));
  end generate;
  g7: for i in 0 to 7 generate
    P(i) <= PCS(i)(0); -- get first 8 product bits from full-adder sums
  end generate;
  g8: for i in 8 to 14 generate
    P(i) <= RAS(i-8);  -- get next 7 bits from ripple-adder sums
  end generate;
  P(15) <= RAC(7);     -- get last bit from ripple-adder carry
end vmul8x8s_arch;
```

adders, not shown in the figure, whose sum outputs equal the first row of PC bits and whose carry outputs are 0. The third, nested for loop corresponds to the main array of adders in Figure 5-99, all except the last row, which is handled by the fourth for loop. The last two for loops assign the appropriate adder outputs to the multiplier output signals.

The program in Table 5-58 can be modified to use structural VHDL as shown in Table 5-59. This approach gives the designer complete control over the circuit structure that is synthesized, as might be desired in an ASIC realization. The program assumes that the architectures for AND2, XOR3, and MAJ3 have been defined elsewhere, for example, in an ASIC library.

This program makes good use of the *generate statement* to create the arrays of components used in the multiplier. The generate statement must have a label, and similar to a for-loop statement, it specifies an iteration scheme to control the repetition of the enclosed statements. Within for-generate, the enclosed statements can include any concurrent statements, IF-THEN-ELSE statements, and additional levels of looping constructs. Sometimes generate statements are combined with IF-THEN-ELSE to produce a kind of conditional compilation capability

generate statement

Well, we said we'd save the best for last, and here it is. The IEEE std_logic_arith library that we introduced in Section 5.9.6 defines multiplication functions for SIGNED and UNSIGNED types, and overlays these functions onto the "*" operator. Thus, the program in Table 5-60 can multiply unsigned numbers with a simple one-line assignment statement.

Within the IEEE std_logic_arith library, the multiplication function is defined behaviorally, using the shift-and-add algorithm. We could have showed you this approach at the beginning of this subsection, but then you wouldn't have read the rest of it, would you?

Table 5-60
Truly behavioral VHDL program for an 8×8 combinational multiplier.

```
library IEEE;
use IEEE.std_logic_1164.all;
use IEEE.std_logic_arith.all;

entity vmul8x8i is
    port (
        X: in UNSIGNED (7 downto 0);
        Y: in UNSIGNED (7 downto 0);
        P: out UNSIGNED (15 downto 0)
    );
end vmul8x8i;

architecture vmul8x8i_arch of vmul8x8i is
begin
  P <= X * Y;
end vmul8x8i_arch;
```

SIGNALS VS. VARIABLES

Variables are used rather than signals in the process in Table 5-58 to make the simulation run faster. Variables are faster because the simulator keeps track of their values only when the process is running. Because variable values are assigned sequentially, the process in Table 5-58 is carefully written to compute values in the proper order. That is, a variable cannot be used until a value has been assigned to it.

Signals, on the other hand, have a value at all times. When a signal value is changed in a process, the simulator schedules a future event in its event list for the value change. If the signal appears on the righthand side of an assignment statement in the process, then the signal must also be included in the process' sensitivity list. If a signal value changes, the process will then execute again, and keep repeating until all of the signals in the sensitivity list are stable.

In Table 5-58, if you wanted to observe internal values or timing during simulation, you could change all the variables (except i and j) to signals and include them in the sensitivity list. To make the program syntactically correct, you would also have to move the type and signal declarations to just after the architecture statement, and change all of the ":=" assignments to "<=".

The simulation will run a lot slower after you make these changes above. As we discussed in connection with Figure 5-99 on page 448, the worst-case signal propagation path through the adder array goes through 14 adders. This is also the number of simulation cycles that must occur for the signals in the changed VHDL program to stabilize, plus one more cycle for the simulator to notice that they've stabilized.

While the choice of signals vs. variables affects the speed of simulation, with most VHDL synthesis engines it does not affect the results of synthesis.

ON THE THRESHOLD OF A DREAM

A three-input "majority function," MAJ, is defined at the beginning of Table 5-58 and is subsequently used to compute carry outputs. An n-input *majority function* produces a 1 output if the majority of its inputs are 1, two out of three in the case of a 3-input majority function. (If n is even, $n/2 + 1$ inputs must be 1.)

Over thirty years ago, there was substantial academic interest in a more general class of n-input *threshold functions* which produce a 1 output if k or more of their inputs are 1. Besides providing full employment for logic theoreticians, threshold functions could realize many logic functions with a smaller number of elements than could a conventional AND/OR realization. For example, an adder's carry function requires three AND gates and one OR gate, but just one three-input threshold gate.

(Un)fortunately, an economical technology never emerged for threshold gates, and they remain, for now, an academic curiosity.

SYNTHESIS OF BEHAVIORAL DESIGNS

You've probably heard that compilers for high-level programming languages like C usually generate better code than people do writing in assembly language, even with "hand-tweaking." Most digital designers hope that compilers for behavioral HDLs will also some day produce results superior to a typical hand-tweaked design, be it a schematic or structural VHDL. Better compilers won't put the designers out of work, they will simply allow them to tackle bigger designs.

We're not quite there yet. However, the more advanced synthesis engines do include some nice optimizations for commonly used behavioral structures. For example, I have to admit that the FPGA synthesis engine that I used to test the VHDL programs in this subsection produced just as fast a multiplier from Table 5-60 as it did from any of the more detailed architectures!

References

Digital designers who want to improve4 their writing should start by reading the classic *Elements of Style,* third edition, by William Strunk, Jr., and E. B. White (Allyn & Bacon, 1979). Probably the most inexpensive and concise yet very useful guide to technical writing is *The Elements of Technical Writing*, by Gary Blake and Robert W. Bly (Macmillan, 1993). For a more encyclopedic treatment, see *Handbook of Technical Writing*, fifth edition, by Brusaw, Alred, and Oliu (St. Martin's Press, 1997).

The ANSI/IEEE standard for logic symbols is Std 91-1984, *IEEE Standard Graphic Symbols for Logic Functions*. Another standard of interest to logic designers is ANSI/IEEE 991-1986, *Logic Circuit Diagrams*. These two standards and ten others, including standard symbols for 10-inch gongs and maid's-signal plugs, can be found in one handy, five-pound reference, *Electrical and Electronics Graphic and Letter Symbols and Reference Designations Standards Collection*, published by the IEEE in 1996 (`www.ieee.org`).

Real logic devices are described in data sheets and data books published by the manufacturers. Updated editions of data books used to be published every few years, but recently the trend has been to minimize or eliminate the hardcopy editions and instead to publish up-to-date information on the Web. Two of the largest suppliers with the most comprehensive sites are Texas Instruments (`www.ti.com`) and Motorola (`www.mot.com`).

For a given logic family such as 74ALS, all manufacturers list generally equivalent specifications, so you can get by with just one data book per family. Some specifications, especially timing, may vary slightly between manufacturers, so when timing is tight it's best to check a couple of different sources and use the worst case. That's a *lot* easier than convincing your manufacturing department to buy a component only from a single supplier.

The first PAL devices were invented at Monolithic Memories, Inc. (MMI) in 1978 by John Birkner and H. T. Chua. The inventors earned U.S. patent number 4,124,899 for their invention, and MMI rewarded them by buying them a brand new Porsche and Mercedes, respectively! Seeing the value in this technology (PAL devices, not cars), Advanced Micro Devices (AMD) acquired MMI in the early 1980s and became a leading developer and supplier of new PLDs and CPLDs. In 1997, AMD spun off its PLD operations to a subsidiary, Vantis Corporation, which they sold in 1999 to former competitor Lattice Semiconductor.

Some of the best resources for learning about PLD-based design are provided by the PLD manufacturers. Xilinx Corporation, which started in the FPGA business, also makes CPLDs, and publishes a comprehensive *Xilinx Data Book* in paper (San Jose, CA 95124, 1999) and on the Web (www.xilinx.com). Similarly, GAL inventor Lattice Semiconductor has a comprehensive *Lattice Data Book* (Hillsboro, OR 97124, 1999) and Web site (www.latticesemi.com).

A much more detailed discussion of the internal operation of LSI and VLSI devices, including PLDs, ROMs, and RAMs, can be found in electronics texts such as *Microelectronics*, second edition, by J.Millman and A.Grabel (McGraw-Hill, 1987) and *VLSI Engineering* by Thomas E. Dillinger (Prentice Hall, 1988).

On the technical side of digital design, lots of textbooks cover digital design principles, but only a few cover practical aspects of design. An excellent short book focusing on digital design practices is *The Well-Tempered Digital Design* by Robert B. Seidensticker (Addison-Wesley, 1986). It contains hundreds of readily accessible digital-design "proverbs" in areas ranging from high-level design philosophy to manufacturability.

Drill Problems

5.1 Give three examples of combinational logic circuits that require *billions and billions* of rows to describe in a truth table. For each circuit, describe its inputs and output(s) and indicate exactly how many rows the truth table contains; you need not write out the truth table. (*Hint:* You can find several such circuits right in this chapter.)

5.2 Draw the DeMorgan equivalent symbol for a 74x30 8-input NAND gate.

5.3 Draw the DeMorgan equivalent symbol for a 74x27 3-input NOR gate.

5.4 What's wrong with the signal name "READY′"?

5.5 You may find it annoying to have to keep track of the active levels of all the signals in a logic circuit. Why not use only noninverting gates, so all signals are active high?

5.6 True or false: In bubble-to-bubble logic design, outputs with a bubble can be connected only to inputs with a bubble.

5.7 A digital communication system is being designed with twelve identical network ports. Which type of schematic structure is probably most appropriate for the design?

5.8   Determine the exact maximum propagation delay from IN to OUT of the circuit in Figure X5.8 for both LOW-to-HIGH and HIGH-to-LOW transitions, using the timing information given in Table 5-2. Repeat, using a single worst-case delay number for each gate, and compare and comment on your results.

Figure X5.8

5.9   Repeat Drill 5.8, substituting 74HCT00s for the 74LS00s.

5.10   Repeat Drill 5.8, substituting 74LS08s for the 74LS00s.

5.11   Repeat Drill 5.8, substituting 74AHCT02s for the 74LS00s, using constant 0 instead of constant 1 inputs, and using typical rather than maximum timing.

5.12   Estimate the minimum propagation delay from IN to OUT for the circuit shown in Figure X5.12. Justify your answer.

Figure X5.12

5.13   Determine the exact maximum propagation delay from IN to OUT of the circuit in Figure X5.12 for both LOW-to-HIGH and HIGH-to-LOW transitions, using the timing information given in Table 5-2. Repeat, using a single worst-case delay number for each gate, and compare and comment on your results.

5.14   Repeat Drill 5.13, substituting 74HCT86s for the 74LS86s.

5.15   Which would expect to be faster, a decoder with active-high outputs or one with active-low outputs?

5.16   Using the information in Table 5-3 for 74LS components, determine the maximum propagation delay from any input to any output in the 5-to-32 decoder circuit of Figure 5-39. You may use the "worst-case" analysis method.

5.17   Repeat Drill 5.16, performing a detailed analysis for each transition direction, and compare your results.

5.18   Draw the digits created by a 74x49 seven-segment decoder for the nondecimal inputs 1010 through 1111.

5.19   Show how to build each of the following single- or multiple-output logic functions using one or more 74x138 or 74x139 binary decoders and NAND gates. (*Hint:* Each realization should be equivalent to a sum of minterms.)

(a) $F = \Sigma_{X,Y,Z}(2,4,7)$
(b) $F = \Pi_{A,B,C}(3,4,5,6,7)$

(c) $F = \Sigma_{A,B,C,D}(2,4,6,14)$
(d) $F = \Sigma_{W,X,Y,Z}(0,1,2,3,5,7,11,13)$

(e) $F = \Sigma_{W,X,Y}(1,3,5,6)$
      $G = \Sigma_{W,X,Y}(2,3,4,7)$
(f) $F = \Sigma_{A,B,C}(0,4,6)$
      $G = \Sigma_{C,D,E}(1,2)$

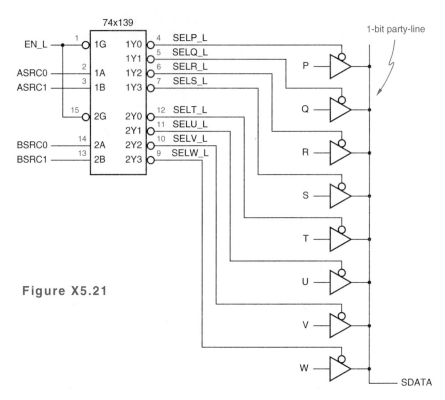

Figure X5.21

5.20 Starting with the logic diagram for the 74x148 priority encoder, write logic equations for its A2_L, A1_L, and A0_L outputs. How do they differ from the "generic" equations given in Section 5.5.1?

5.21 What's terribly wrong with the circuit in Figure X5.21? Suggest a change that eliminates the terrible problem.

5.22 Using the information in Tables 5-2 and 5-3 for 74LS components, determine the maximum propagation delay from any input to any output in the 32-to-1 multiplexer circuit of Figure 5-66. You may use the "worst-case" analysis method.

5.23 Repeat Exercise 5.22 using 74HCT components.

5.24 An n-input parity tree can be built with XOR gates in the style of Figure 5-74(a). Under what circumstances does a similar n-input parity tree built using XNOR gates perform exactly the same function?

5.25 Using the information in Tables 5-2 and 5-3 for 74LS components, determine the maximum propagation delay from the DU bus to the DC bus in the error-correction circuit of Figure 5-77. You may use the "worst-case" analysis method.

5.26 Repeat Exercise 5.25 using 74HCT components.

5.27 Starting with the equations given in Section 5.9.4, write a complete logic expression for the ALTBOUT output of the 74x85.

5.28 Write an algebraic expression for s_2, the third sum bit of a binary adder, as a function of inputs x_0, x_1, x_2, y_0, y_1, and y_2. Assume that $c_0 = 0$, and do not attempt to "multiply out" or minimize the expression.

5.29 Starting with the logic diagram for the 74x682, write a logic expression for the PGTQ_L output in terms of the inputs.

5.30 Using the information in Table 5-3 for 74LS components, determine the maximum propagation delay from any input to any output of the 16-bit group ripple adder of Figure 5-92. You may use the "worst-case" analysis method.

Exercises

5.31 A possible definition of a BUT gate (Exercise 4.50) is "Y1 is 1 if A1 and B1 are 1 *but* either A2 or B2 is 0; Y2 is defined symmetrically." Write the truth table and find minimal sum-of-products expressions for the BUT-gate outputs. Draw the logic diagram for a NAND-NAND circuit for the expressions, assuming that only uncomplemented inputs are available. You may use gates from 74x00, '04, '10, '20, and '30 packages.

5.32 Find a gate-level design for the BUT gate defined in Exercise 5.31 that uses a minimum number of transistors when realized in CMOS. You may use gates from 74x00, '02, '04, '10, '20, and '30 packages. Write the output expressions (which need not be two-level sums of products), and draw the logic diagram.

5.33 For each circuit in the two preceding exercises, compute the worst-case delay from input to output, using the delay numbers for 74HCT components in Table 5-2. Compare the cost (number of transistors), speed, and input loading of the two designs. Which is better?

5.34 Butify the function $F = \Sigma_{W,X,Y,Z}(3,7,11,12,13,14)$. That is, show how to perform F *butification* with a single BUT gate as defined in Exercise 5.31 and a single 2-input OR gate.

5.35 Suppose that a 74LS138 decoder is connected so that all enable inputs are asserted and C B A = 101. Using the information in Table 5-3 and the '138 internal logic diagram, determine the propagation delay from input to all relevant outputs for each possible single-input change. (*Hint:* There are a total of nine delay numbers, since a change on A, B, or C affects two outputs, and a change on any of the three enable inputs affects one output.)

5.36 Suppose that you are asked to design a new component, a decimal decoder that is optimized for applications in which only decimal input combinations are expected to occur. How can the cost of such a decoder be minimized compared to one that is simply a 4-to-16 decoder with six outputs removed? Write the logic equations for all ten outputs of the minimized decoder, assuming active-high inputs and outputs and no enable inputs.

5.37 How many Karnaugh maps would be required to work Exercise 5.36 using the formal multiple-output minimization procedure described in Section 4.3.8?

5.38 Suppose that a system requires a 5-to-32 binary decoder with a single active-low enable input, similar to Figure 5-39. With the EN1 input pulled HIGH, either the EN2_L or the EN3_L input in the figure could be used as the enable, with the other input grounded. Discuss the pros and cons of using EN2_L versus EN3_L.

5.39 Determine whether the a, b, and c output circuits in the 74x49 seven-segment decoder correspond to minimal product-of-sums expressions for these segments, assuming that the nondecimal input combinations are "don't cares" and BI = 1.

Figure X5.40

5.40 Redesign the MSI 74x49 seven-segment decoder so that the digits 6 and 9 have tails as shown in Figure X5.40. Are any of the digit patterns for nondecimal inputs 1010 through 1111 affected by your redesign?

5.41 Starting with the ABEL program in Table 5-22, write a program for a seven-segment decoder with the following enhancements:

- The outputs are all active low.
- Two new inputs, ENHEX and ERRDET, control segment-output decoding.
- If ENHEX = 0, the outputs match the behavior of a 74x49.
- If ENHEX = 1, then the outputs for digits 6 and 9 have tails, and the outputs for digits A–F are controlled by ERRDET.
- If ENHEX = 1 and ERRDET = 0, then the outputs for digits A–F look like the letters A–F, as in the original program.
- If ENHEX = 1 and ERRDET = 1, then digits A–F look like the letter S.

5.42 A famous logic designer decided to quit teaching and make a fortune by licensing the circuit design shown in Figure X5.42.

(a) Label the inputs and outputs of the circuit with appropriate signal names, including active-level indications.

(b) What does the circuit do? Be specific and account for all inputs and outputs.

(c) Draw the logic symbol that would go on the data sheet of this circuit.

(d) Write an ABEL or behavioral VHDL program for the circuit.

(e) With what standard building blocks does the new circuit compete? Do you think it would be successful as an MSI part?

5.43 An FCT three-state buffer drives ten FCT inputs and a 4.7-KΩ pull-up resistor to 5.0 V. When the output changes from LOW to Hi-Z, estimate how long it takes for the FCT inputs to see the output as HIGH. State any assumptions that you make.

5.44 On a three-state bus, ten FCT three-state buffers are driving ten FCT inputs and a 4.7-KΩ pull-up resistor to 5.0 V. Assuming that no other devices are driving the bus, estimate how long the bus signal remains at a valid logic level when an active output enters the Hi-Z state. State any assumptions that you make.

5.45 Design a 10-to-4 encoder with inputs in the 1-out-of-10 code and outputs in BCD.

5.46 Draw the logic diagram for a 16-to-4 encoder using just four 8-input NAND gates. What are the active levels of the inputs and outputs in your design?

5.47 Draw the logic diagram for a circuit that uses the 74x148 to resolve priority among eight active-high inputs, I0–I7, where I7 has the highest priority. The circuit should produce active-high address outputs A2–A0 to indicate the number of the highest-priority asserted input. If no input is asserted, then A2–A0 should be 111 and an IDLE output should be asserted. You may use discrete gates in addition to the '148. Be sure to name all signals with the proper active levels.

5.48 Draw the logic diagram for a circuit that resolves priority among eight active-low inputs, I0_L–I7_L, where I0_L has the highest priority. The circuit should produce active-high address outputs A2–A0 to indicate the number of the highest-priority asserted input. If at least one input is asserted, then an AVALID output should be asserted. Be sure to name all signals with the proper active levels. This circuit can be built with a single 74x148 and no other gates.

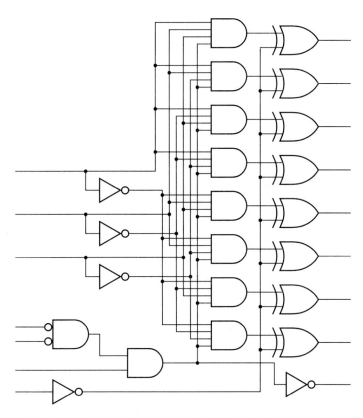

Figure X5.42

5.49 A purpose of Exercise 5.48 was to demonstrate that it is not always possible to maintain consistency in active-level notation unless you are willing to define alternate logic symbols for MSI parts that can be used in different ways. Define an alternate 74x148 symbol that provides this consistency in Exercise 5.48.

5.50 Design a combinational circuit with eight active-low request inputs, R0_L–R7_L, and eight outputs, A2–A0, AVALID, B2–B0, and BVALID. The R0_L–R7_L inputs and A2–A0 and AVALID outputs are defined as in Exercise 5.48. The B2–B0 and BVALID outputs identify the second-highest priority request input that is asserted. You should be able to design this circuit with no more than six SSI and MSI packages, but don't use more than 10 in any case.

5.51 Repeat Exercise 5.50 using ABEL. Does the design fit into a single GAL20V8?

5.52 Repeat Exercise 5.50 using VHDL.

5.53 Create a VHDL type, based on IEEE 1164, that models open-collector outputs, where typing outputs together creates a wired-AND function. You should also model a pull-up resistor element such that if there is no pull-up resistor and no device is driving the bus, then an "unknown" signal is produced. Test your definitions by modeling the circuit in Figure X3.92 for all input combinations, both with and without R1 present.

5.54 Design a 3-input, 5-bit multiplexer that fits in a 24-pin IC package. Write the truth table and draw a logic diagram and logic symbol for your multiplexer.

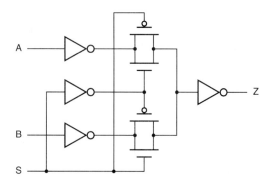

Figure X5.55

5.55   Write the truth table and a logic diagram for the logic function performed by the CMOS circuit in Figure X5.55. (The circuit contains transmission gates, which were introduced in Section 3.7.1.)

5.56   What logic function is performed by the CMOS circuit shown in Figure X5.56?

5.57   A famous logic designer decided to quit teaching and make a fortune by licensing the circuit design shown in Figure X5.57.

   (a) Label the inputs and outputs of the circuit with appropriate signal names, including active-level indications.

   (b) What does the circuit do? Be specific and account for all inputs and outputs.

   (c) Draw the logic symbol that would go on the data sheet of this circuit.

   (d) Write an ABEL or behavioral VHDL program for the circuit.

   (e) With what standard building blocks does the new circuit compete? Do you think it would be successful as an MSI part?

5.58   Write a VHDL program for 74x157 multiplexer with the function table shown in Table 5-35.

5.59   Write a VHDL program for 74x153 multiplexer with the function table shown in Table 5-36.

5.60   Show how to realize the 4-input, 18-bit multiplexer with the functionality described in Table 5-40 using 18 74x151s.

5.61   Show how to realize the 4-input, 18-bit multiplexer with the functionality of Table 5-40 using 9 74x153s and a "code converter" with inputs S2–S0 and outputs C1,C0 such that [C1,C0] = 00–11 when S2–S0 selects A–D, respectively.

Figure X5.56

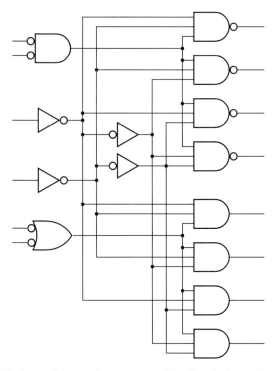

Figure X5.57

5.62 Design a 3-input, 2-output combinational circuit that performs the code conversion specified in the previous exercise, using discrete gates.

5.63 Add a three-state-output control input OE to the VHDL multiplexer program in Table 5-43. Your solution should have only one process.

5.64 A 16-bit *barrel shifter* is a combinational logic circuit with 16 data inputs, 16 data outputs, and 4 control inputs. The output word equals the input word, rotated by a number of bit positions specified by the control inputs. For example, if the input word equals ABCDEFGHIJKLMNOP (each letter represents one bit), and the control inputs are 0101 (5), then the output word is FGHIJKLMNOPABCDE. Design a 16-bit barrel shifter using combinational MSI parts discussed in this chapter. Your design should contain 20 or fewer ICs. Do not draw a complete schematic, but sketch and describe your design in general terms and indicate the types and total number of ICs required.

barrel shifter

5.65 Write an ABEL program for the barrel shifter in Exercise 5.64.

5.66 Write a VHDL program for the barrel shifter in Exercise 5.64.

5.67 A digital designer who built the circuit in Figure 5-76 accidentally used 74x00s instead of '08s in the circuit and found that the circuit still worked, except for a change in the active level of the ERROR signal. How was this possible?

5.68 An odd-parity circuit with 2^n inputs can be built with $2^n - 1$ XOR gates. Describe two different structures for this circuit, one of which gives a minimum worst-case input to output propagation delay and the other of which gives a maximum. For each structure, state the worst-case number of XOR-gate delays, and describe a situation where that structure might be preferred over the other.

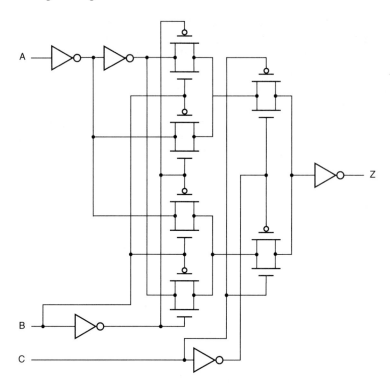

Figure X5.73

5.69   Write a 4-step iterative algorithm corresponding to the iterative comparator circuit of Figure 5-80.

5.70   Write a VHDL program for a 16-bit iterative camparator using the structure of Figure 5-80. Use VHDL's "generate" capability.

5.71   Design a 16-bit comparator using five 74x85s in a treelike structure, such that the maximum delay for a comparison equals twice the delay of one 74x85.

5.72   Write a VHDL program for a device with the functionality of a 74x85.

5.73   Write the truth table and a logic diagram for the logic function performed by the CMOS circuit in Figure X5.73.

5.74   Design a comparator similar to the 74x85 that uses the opposite cascading order. That is, to perform a 12-bit comparison, the cascading outputs of the high-order comparator would drive the cascading inputs of the mid-order comparator, and the mid-order outputs would drive the low-order inputs. You needn't do a complete logic design and schematic; a truth table and an application note showing the interconnection for a 12-bit comparison are sufficient.

5.75   Design a 24-bit comparator using three 74x682s and additional gates as required. Your circuit should compare two 24-bit unsigned numbers P and Q and produce two output bits that indicate whether $P = Q$ or $P > Q$.

5.76   Using the information in Table 5-3, determine the maximum propagation delay from any A or B bus input to any F bus output of the 16-bit carry lookahead adder of Figure 5-96. You may use the "worst-case" analysis method.

5.77 Starting with the logic diagram for the 74x283 in Figure 5-91, write a logic expression for the S2 output in terms of the inputs, and prove that it does indeed equal the third sum bit in a binary addition as advertised. You may assume that $c_0 = 0$ (i.e., ignore c_0).

5.78 Referring to the data sheet of a 74S182 carry lookahead circuit, determine whether or not its outputs match the equations given in Section 5.10.7.

5.79 Estimate the number of product terms in a minimal sum-of-products expression for the c_{32} output of a 32-bit binary adder. Be more specific than "billions and billions," and justify your answer.

5.80 Draw the logic diagram for a 64-bit ALU using sixteen 74x181s and five 74S182s for full carry lookahead (two levels of '182s). For the '181s, you need show only the CIN inputs and G_L and P_L outputs.

5.81 Write a VHDL model for a 74x181 ALU.

5.82 Show how to build all four of the following functions using one SSI package and one 74x138.

$$F1 = X' \cdot Y' \cdot Z' + X \cdot Y \cdot Z \qquad F2 = X' \cdot Y' \cdot Z + X \cdot Y \cdot Z'$$
$$F3 = X' \cdot Y \cdot Z' + X \cdot Y' \cdot Z \qquad F4 = X \cdot Y' \cdot Z' + X' \cdot Y \cdot Z$$

5.83 Determine the worst-case propagation delay of the multiplier in Figure 5-98, assuming that the propagation delay from any adder input to its sum output is twice as long as the delay to the carry output. Repeat, assuming the opposite relationship. If you were designing the adder cell from scratch, which path would you favor with the shortest delay? Is there an optimal balance?

5.84 Repeat the preceding exercise for the multiplier in Figure 5-99.

5.85 Design a customized decoder with the function table in Table X5.85 using MSI and SSI parts. Minimize the number of IC packages in your design.

Table X5.85

CS_L	A2	A1	A0	**Output to Assert**
1	x	x	x	none
0	0	0	x	BILL_L
0	0	x	0	MARY_L
0	0	1	x	JOAN_L
0	0	x	1	PAUL_L
0	1	0	x	ANNA_L
0	1	x	0	FRED_L
0	1	1	x	DAVE_L
0	1	x	1	KATE_L

5.86 Repeat Exercise 5.85 using ABEL and a single GAL16V8.

5.87 Repeat Exercise 5.85 using VHDL.

5.88 Based on the Hamming code used in the VHDL program in Table 5-77, write a VHDL program for a Hamming encoder entity with 4-bit data inputs and 7-bit encoded data outputs.

5.89   Using ABEL and a single GAL16V8, design a customized multiplexer with four
        3-bit input buses P, Q, R, T, and three select inputs S2–S0 that choose one of the
        buses to drive a 3-bit output bus Y according to Table X5.89.

Table X5.89

S2	S1	S0	Input to Select
0	0	0	P
0	0	1	P
0	1	0	P
0	1	1	Q
1	0	0	P
1	0	1	P
1	1	0	R
1	1	1	T

5.90   Design a customized multiplexer with four 8-bit input buses P, Q, R, and T, select-
        ing one of the buses to drive a 8-bit output bus Y according to Table X5.89. Use
        two 74x153s and a code converter that maps the eight possible values on S2–S0
        to four select codes for the '153. Choose a code that minimizes the size and
        propagation delay of the code converter.

5.91   Design a customized multiplexer with five 4-bit input buses A, B, C, D, and E,
        selecting one of the buses to drive a 4-bit output bus T according to Table X5.91.
        You may use no more than three MSI and SSI ICs.

Table X5.91

S2	S1	S0	Input to Select
0	0	0	A
0	0	1	B
0	1	0	A
0	1	1	C
1	0	0	A
1	0	1	D
1	1	0	A
1	1	1	E

5.92   Repeat Exercise 5.91 using ABEL and one or more PAL/GAL devices from this
        chapter. Minimize the number and size of the GAL devices.

5.93   Design a 3-bit equality checker with six inputs, SLOT[2–0] and GRANT[2–0], and
        one active-low output, MATCH_L. The SLOT inputs are connected to fixed values
        when the circuit is installed in the system, but the GRANT values are changed on
        a cycle-by-cycle basis during normal operation of the system. Using only SSI and
        MSI parts that appear in Tables 5-2 and 5-3, design a comparator with the shortest
        possible maximum propagation delay from GRANT[2–0] to MATCH_L. (*Note:*
        The author had to solve this problem "in real life" to shave 2 ns off the critical-
        path delay in a 25-MHz system design.)

Combinational-Circuit Design Examples

S
o far, we have looked at basic principles in several areas—number systems, digital circuits, and combinational logic—and we have described many of the basic building blocks of combinational design—decoders, multiplexers, and the like. All of that is a needed foundation, but the ultimate goal of studying digital design is eventually to be able to solve real problems by designing digital systems (well, duh...). That usually requires experience beyond what you can get by reading a textbook. We'll try to get you started by presenting a number of larger combinational design examples in this chapter.

The chapter is divided into three sections. The first gives design examples using combinational building blocks. While the examples are written in terms of MSI functions, the same functions are widely used in ASIC and schematic-based FPGA design. The idea of these examples is to show that you can often express a combinational function using a collection of smaller building blocks. This is important for a couple of reasons: a hierarchical approach usually simplifies the overall design task, and the smaller building blocks often have a more efficient, optimized realization in FPGA and ASIC cells than what you'd get if you wrote a larger, monolithic description in an HDL and then just hit the "synthesize" button.

The second section gives design examples using ABEL. These designs are all targeted to small PLDs such as 16V8s and 20V8s. Besides the general use of the ABEL language, some of the examples illustrate the partitioning decisions that a designer must make when an entire circuit does not fit into a single component.

A VHDL-based approach is especially appropriate for larger designs that will be realized in a single CPLD, FPGA, or ASIC, as described in the third section. You may notice that these examples do not target a specific CPLD or FPGA. Indeed, this is one of the benefits of HDL-based design; most or all of the design effort is "portable" and can be targeted to any of a variety of technologies.

The only prerequisites for this chapter are the chapters that precede it. The three sections are written to be pretty much independent of each other, so you don't have to read about ABEL if you're only interested in VHDL, or vice versa. Also, the rest of the book is written so that you can read this chapter now or skip it and come back later.

6.1 Building-Block Design Examples

6.1.1 Barrel Shifter

barrel shifter

A *barrel shifter* is a combinational logic circuit with n data inputs, n data outputs, and a set of control inputs that specify how to shift the data between input and output. A barrel shifter that is part of a microprocessor CPU can typically specify the direction of shift (left or right), the type of shift (circular, arithmetic, or logical), and the amount of shift (typically 0 to $n-1$ bits, but sometimes 1 to n bits).

In this subsection we'll look at the design of a simple 16-bit barrel shifter that does left circular shifts only, using a 4-bit control input S[3:0] to specify the amount of shift. For example, if the input word is ABCDEFGHGIHKLMNOP (where each letter represents one bit), and the control input is 0101 (5), then the output word is FGHGIHKLMNOPABCDE.

From one point of view, this problem is deceptively simple. Each output bit can be obtained from a 16-input multiplexer controlled by the shift-control inputs, which each multiplexer data input connected to the appropriate On the other hand, when you look at the details of the design, you'll see that there are tradeoffs in the speed and size of the circuit.

Let us first consider a design that uses off-the-shelf MSI multiplexers. A 16-input, one-bit multiplexer can be built using two 74x151s by applying S3 and its complement to the EN_L inputs and combining the Y_L data outputs with a NAND gate, as we showed in Figure 5-66 for a 32-input multiplexer. The low-order shift-control inputs, S2–S0, connect to the like-named select inputs of the '151s.

We complete the design by replicating this 16-input multiplexer 16 times and hooking up the data inputs appropriately, as shown in Figure 6-1. The top '151 of each pair is enabled by S3_L, and the bottom one by S3; the remaining select bits are connected to all 32 '151s. Data inputs D0–D7 of each '151 are connected to the DIN inputs in the listed order from left to right.

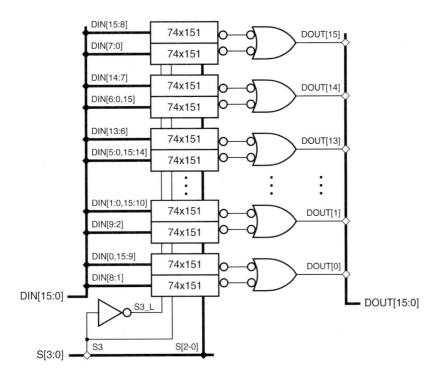

Figure 6-1
One approach to building a 16-bit barrel shifter.

The first row of Table 6-1 shows the characteristics of this first approach. About 36 chips (32 74x151s, 4 74x00s, and 1/6 74x04) are used in the MSI/SSI realization. We can reduce this to 32 chips by replacing the 74x151s with 74x251s and tying their three-state Y outputs together, as tabulated in the second row. Both of these designs have very heavy loading on the control inputs; each of the control bits S[2:0] must be connected to the like-named select input of all 32 multiplexers. The data inputs are also fairly heavily loaded; each data bit must connect to 16 different multiplexer data inputs, corresponding to the 16 possible shift amounts. However, assuming that the heavy control and data loads don't slow things down too much, the 74x251-based approach yields the shortest data delay, with each data bit passing through just one multiplexer.

Alternatively, we could build the barrel shifter using 16 74x157 2-input, 4-bit multiplexers, as tabulated in the last row of the table. We start by using four 74x157s to make a 2-input, 16-bit multiplexer. Then, we can hook up a first set

Multiplexer Component	Data Loading	Data Delay	Control Loading	Total ICs
74x151	16	2	32	36
74x251	16	1	32	32
74x153	4	2	8	16
74x157	2	4	4	16

Table 6-1
Properties of four different barrel-shifter design approaches.

Figure 6-2 A second approach to building a 16-bit barrel shifter.

of four '157s controlled by S0 to shift the input word left by 0 or 1 bit. The data outputs of this set are connected to the inputs of a second set, controlled by S1, which shifts its input word left by 0 or 2 bits. Continuing the cascade, a third and fourth set are controlled by S2 and S3 to shift selectively by 4 and 8 bits, as shown in Figure 6-2. Here, the 1A-4A and 1B-4B inputs and the 1Y-4Y outputs of each '157 are connected to the indicated signals in the listed order from left to right.

The '157-based approach requires only half as many MSI packages and has far less loading on the control and data inputs. On the other hand, it has the longest data-path delay, since each data bit must pass through four 74x157s.

Halfway between the two approaches, we can use eight 74x153 4-input, 2-bit multiplexers to build a 4-input, 16-bit multiplexer. Cascading two sets of these, we can use S[3:2] to shift selectively by 0, 4, 8, or 12 bits, and S[1:0] to shift by 0–3 bits. This approach has the performance characteristics shown in the third row of Table 6-1 and would appear to be the best compromise if you don't need to have the absolutely shortest possible data delay.

The same kind of considerations would apply if you were building the barrel shifter out of ASIC cells instead of MSI parts, except you'd be counting chip area instead of MSI/SSI packages.

Typical ASIC cell libraries have 1-bit-wide multiplexers, usually realized with CMOS transmission gates, with 2 to 8 inputs. To build a larger multiplexer, you have to put together the appropriate combination of smaller cells. Besides the kind of choices we encountered in the MSI example, you have the further complication that CMOS delays are highly dependent on loading. Thus, depending on the approach, you must decide where to add buffers to the control lines, the data lines, or both to minimize loading-related delays. An approach that looks good on paper, before analyzing these delays and adding buffers, may actually turn out to have poorer delay or more chip area than another approach.

6.1.2 Simple Floating-Point Encoder

The previous example used multiple copies of a single building block, a multiplexer, and it was pretty obvious from the problem statement that a multiplexer was the appropriate building block. The next example shows that you sometimes have to look a little harder to see the solution in terms of known building blocks.

Now let's look at a design problem whose MSI solution is not quite so obvious, a "fixed-point to floating-point encoder." An unsigned binary integer B in the range $0 \leq B < 2^{11}$ can be represented by 11 bits in "fixed-point" format, $B = b_{10}b_9...b_1b_0$. We can represent numbers in the same range with less precision using only 7 bits in a floating-point notation, $F = M \cdot 2^E$, where M is a 4-bit mantissa $m_3m_2m_1m_0$ and E is a 3-bit exponent $e_2e_1e_0$. The smallest integer in this format is $0 \cdot 2^0$ and the largest is $(2^4-1) \cdot 2^7$.

Given an 11-bit fixed-point integer B, we can convert it to our 7-bit floating-point notation by "picking off" four high-order bits beginning with the most significant 1, for example,

$$11010110100 = 1101 \cdot 2^7 + 0110100$$
$$00100101111 = 1001 \cdot 2^5 + 01111$$
$$00000111110 = 1111 \cdot 2^2 + 10$$
$$00000001011 = 1011 \cdot 2^0 + 0$$
$$00000000010 = 0010 \cdot 2^0 + 0$$

The last term in each equation is a truncation error that results from the loss of precision in the conversion. Corresponding to this conversion operation, we can write the specification for a fixed-point to floating-point encoder circuit:

- A combinational circuit is to convert an 11-bit unsigned binary integer B into a 7-bit floating-point number M,E, where M and E have 4 and 3 bits, respectively. The numbers have the relationship $B = M \cdot 2^E + T$, where T is the truncation error, $0 \leq T < 2^E$.

Starting with a problem statement like the one above, it takes some creativity to come up with an efficient circuit design—the specification gives no clue. However, we can get some ideas by looking at how we converted numbers by hand earlier. We basically scanned each input number from left to right to find the first position containing a 1, stopping at the b_3 position if no 1 was found. We picked off four bits starting at that position to use as the mantissa, and the starting position number determined the exponent. These operations are beginning to sound like MSI building blocks.

"Scanning for the first 1" is what a generic priority encoder does. The output of the priority encoder is a number that tells us the position of the first 1. The position number determines the exponent; first-1 positions of $b_{10} - b_3$ imply exponents of 7–0, and positions of $b_2 - b_0$ or no-1-found imply an exponent of 0. Therefore, we can scan for the first 1 with an 8-input priority encoder with inputs

I7 (highest priority) through I0 connected to $b_{10} - b_3$. We can use the priority encoder's A2 – A0 outputs directly as the exponent, as long as the no-1-found case produces A2 – A0 = 000.

"Picking off four bits" sounds like a "selecting" or multiplexing operation. The 3-bit exponent determines which four bits of B we pick off, so we can use the exponent bits to control an 8-input, 4-bit multiplexer that selects the appropriate four bits of B to form M.

An MSI circuit that results from these ideas is shown in Figure 6-3. It contains several optimizations:

- Since the available MSI priority encoder, the 74x148, has active-low inputs, the input number B is assumed to be available on an active-low bus B_L[10:0]. If only an active-high version of B is available, then eight inverters can be used to obtain the active-low version.

- If you think about the conversion operation a while, you'll realize that the most significant bit of the mantissa, m_3, is always 1, except in the no-1-found case. The '148 has a GS_L output that indicates this case, allowing us to eliminate the multiplexer for m_3.

- The '148 has active-low outputs, so the exponent bits (E0_L–E2_L) are produced in active-low form. Naturally, three inverters could be used to produce an active-high version.

- Since everything else is active-low, active-low mantissa bits are used, too. Active-high bits are also readily available on the '148 EO_L and the '151 Y_L outputs.

Strictly speaking, the multiplexers in Figure 6-3 are drawn incorrectly. The 74x151 symbol can be drawn alternatively as shown in Figure 6-4. In words, if the multiplexer's data inputs are active low, then the data outputs have an active level opposite that shown in the original symbol. The "active-low-data" symbol

Figure 6-4
Alternate logic symbol
for the 74x151 8-input
multiplexer.

should be preferred in Figure 6-3, since the active levels of the '151 inputs and outputs would then match their signal names. However, in data transfer and storage applications, designers (and the book) don't always "go by the book." It is usually clear from the context that a multiplexer (or a multibit register, in Section 8.2.5) does not alter the active level of its data.

6.1.3 Dual-Priority Encoder

Quite often MSI building blocks need a little help from their friends—ordinary gates—to get the job done. In this example, we'd like to build a priority encoder that identifies not only the highest but also the second-highest-priority asserted signal among a set of eight request inputs.

We'll assume for this example that the request inputs are active low and are named R0_L–R7_L, where R0_L has the highest priority. We'll use A2–A0 and AVALID to identify the highest-priority request, where AVALID is asserted only if at least one request input is asserted. We'll use B2–B0 and BVALID to identify the second-highest-priority request, where BVALID is asserted only if at least two request inputs are asserted.

Finding the highest-priority request is easy enough, we can just use a 74x148. To find the second-highest-priority request, we can use another '148, but only if we first "knock out" the highest-priority request before applying the request inputs. This can be done using a decoder to select a signal to knock out, based on A2–A0 and AVALID from the first '148. These ideas are combined in the solution shown in Figure 6-6. A 74x138 decoder asserts at most one of its eight outputs, corresponding to the highest-priority request input. The outputs are fed to a rank of NAND gates to "turn off" the highest-priority request.

A trick is used in this solution is to get active-high outputs from the '148s, as shown in Figure 6-5. We can rename the address outputs A2_L–A0_L to be active high if we also change the name of the request input that is associated with each output combination. In particular, we complement the bits of the request number. In the redrawn symbol, request input I0 has the highest priority.

Figure 6-5
Alternate logic symbols
for the 74x148 8-input
priority encoder.

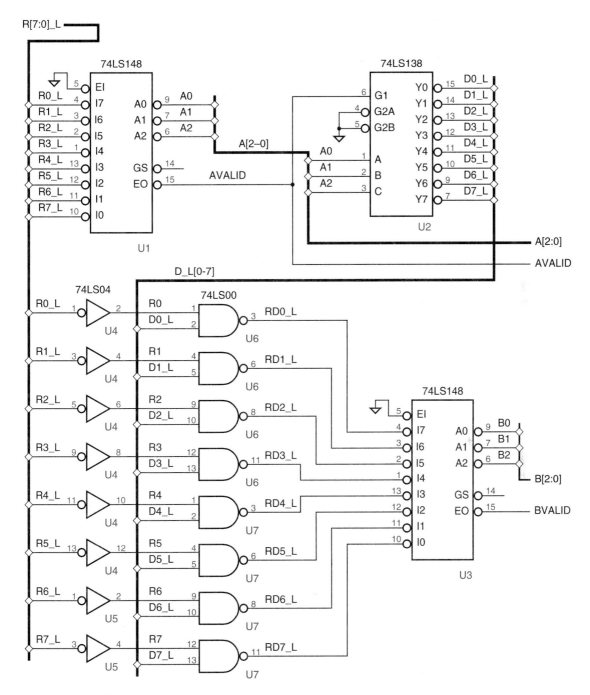

Figure 6-6 First- and second-highest-priority encoder circuit.

6.1.4 Cascading Comparators

In Section 5.9.4, we showed how 74x85 4-bit comparators can be cascaded to create larger comparators. Since the 74x85 uses a serial cascading scheme, it can be used to build arbitrarily large comparators. The 74x682 8-bit comparator, on the other hand, doesn't have any cascading inputs and outputs at all. Thus, at first glance, you might think that it can't be used to build larger comparators. But that's not true.

If you think about the nature of a large comparison, it is clear that two wide inputs, say 32 bits (four bytes) each, are equal only if their corresponding bytes are equal. If we're trying to do a greater-than or less-than comparison, then the corresponding most-significant that are not equal determine the result of the comparison.

Using these ideas, Figure 6-7 uses three 74x682 8-bit comparators to do equality and greater-than comparison on two 24-bit operands. The 24-bit results are derived from the individual 8-bit results using combinational logic for the following equations:

$$PEQQ = EQ2 \cdot EQ1 \cdot EQ0$$
$$PGTQ = GT2 + EQ2 \cdot GT1 + EQ2 \cdot EQ1 \cdot GT0$$

This "parallel" expansion approach is actually faster than the 74x85's serial cascading scheme, because it does not suffer the delay of propagating the cascading signals through a cascade of comparators. The parallel approach can be used to build very wide comparators using two-level AND-OR logic to combine the 8-bit results, limited only by the fan-in constraints of the AND-OR logic. Arbitrary large comparators can be made if you use additional levels of logic to do the combining.

6.1.5 Mode-Dependent Comparator

Quite often, the requirements for a digital-circuit application are specified in a way that makes an MSI or other building-block solution obvious. For example, consider the following problem:

- Design a combinational circuit whose inputs are two 8-bit unsigned binary integers, X and Y, and a control signal MIN/MAX. The output of the circuit is an 8-bit unsigned binary integer Z such that Z = min(X,Y) if MIN/MAX = 1, and Z = max(X,Y) if MIN/MAX = 0.

This circuit is fairly easy to visualize in terms of MSI functions. Clearly, we can use a comparator to determine whether X > Y. We can use the comparator's output to control multiplexers that produce min(X,Y) and max(X,Y), and we can use another multiplexer to select one of these results depending on MIN/MAX. Figure 6-8(a) is the block diagram of a circuit corresponding to this approach.

Our first solution approach works, but it's more expensive than it needs to be. Although it has three two-input multiplexers, there are only two input words,

Figure 6-7
24-bit comparator circuit.

X and Y, that may ultimately be selected and produced at the output of the circuit. Therefore, we should be able to use just a single two-input mux, and use some other logic to figure out which input to tell it to select. This approach is shown in Figure 6-8(b) and (c). The "other logic" is very simple indeed, just a single XOR gate.

Figure 6-8 Mode-dependent comparator circuit: (a) block diagram of a
"first-cut" solution; (b) block diagram of a more cost-effective
solution; (c) logic diagram for the second solution.

DON'T BE A BLOCKHEAD   The wastefulness of our original design approach in Figure 6-8(a) may have been obvious to you from the beginning, but it demonstrates an important approach to designing with building blocks:

- Use standard building blocks to handle data, and look for ways that a single block can perform different functions at different times or in different modes. Design control circuits to select the appropriate functions as needed, to reduce the total parts count of the design.

As Figure 6-8(c) dramatically shows, this approach can save a lot of chips. When designing with IC chips, you should *not* heed the slogan, "Have all you want, we'll make more"!

6.2 Design Examples Using ABEL and PLDs

6.2.1 Barrel Shifter

A barrel shifter, defined on page 468, is good example of something *not* to design using PLDs. However, we can put ABEL to good use to describe a barrel shifter's function, and we'll also see why a typical barrel shifter is not a good fit for a PLD.

Table 6-2 shows the equations for a 16-bit barrel shifter with the same functionality as the example on page 468—it does left circular shifts only, using a 4-bit control input S[3..0] to specify the amount of shift. ABEL makes it easy to specify the functionality of the overall circuit without worrying about how the circuit might be partitioned into multiple chips. Also, ABEL dutifully generates a minimal sum-of-products expression for each output bit. In this case, each output requires 16 product terms.

Partitioning the 16-bit barrel shifter into multiple PLDs is a difficult task in two different ways. First, it should be obvious that the nature of the function is such that every output bit depends on every input bit. A PLD that produces, say, the DOUT0 output must have all 16 DIN inputs and all four S inputs available to it. So, a GAL16V8 definitely cannot be used; it has only 16 inputs.

The GAL20V8 is similar to the GAL16V8, with the addition of four input-only pins. If we use all 20 available inputs, we are left with two output-only pins (corresponding to the top and bottom outputs in Figure 5-27 on page 345). Thus, it seems possible that we could realize the barrel shifter using eight 20V8 chips, producing two output bits per chip.

No, we still have a problem. The second dimension of difficulty in a PLD-based barrel shifter is the number of product terms per output. The barrel shifter requires 16, and the 20V8 provides only 7. We're stuck—any realization of the barrel shifter in 20V8s is going to require multiple-pass logic. At this point, we would be best advised to look at partitioning options along the lines that we did in Section 6.1.1.

Table 6-2
ABEL program for a
16-bit barrel shifter.

```
module barrel16
title '16-bit Barrel Shifter'

" Inputs and Outputs
DIN15..DIN0, S3..S0                    pin;
DOUT15..DOUT0                          pin istype 'com';

S = [S3..S0];

equations

[DOUT15..DOUT0] = (S==0)  & [DIN15..DIN0]
              #   (S==1)  & [DIN14..DIN0,DIN15]
              #   (S==2)  & [DIN13..DIN0,DIN15..DIN14]
              #   (S==3)  & [DIN12..DIN0,DIN15..DIN13]
                  . . .
              #   (S==12) & [DIN3..DIN0,DIN15..DIN4]
              #   (S==13) & [DIN2..DIN0,DIN15..DIN3]
              #   (S==14) & [DIN1..DIN0,DIN15..DIN2]
              #   (S==15) & [DIN0,DIN15..DIN1];

end barrel16
```

The 16-bit barrel shifter can be realized without much difficulty in a larger programmable device, that is, in a CPLD or an FPGA with enough I/O pins. However, imagine that we were trying to design a 32-bit or 64-bit barrel shifter. Clearly, we would need to use a device with even more I/O pins, but that's not all. The number of product terms and the large amount of connectivity (all the inputs connect to all the outputs) would still be challenging.

Indeed, a typical CPLD or FPGA fitter could have difficulty realizing a large barrel shifter with small delay or even at all. There is a critical resource that we took for granted in the partitioned, building-block barrel-shifter designs of Section 6.1.1—wires! An FPGA is somewhat limited in its internal connectivity, and a CPLD is even more so. Thus, even with modern FPGA and CPLD design tools, you may still have to "use your head" to partition the design in a way that helps the tools do their job.

Barrel shifters can be even more complex than what we've shown so far. Just for fun, Table 6-3 shows the design for a barrel shifter that supports six different kinds of shifting. This requires even more product terms, up to 40 per output! Although you'd never build this device in a PLD, CPLD, or small FPGA, the minimized ABEL equations are useful because they can help you understand the effects of some of your design choices. For example, by changing the coding of SLA and SRA to [1,.X.,0] and [1,.X.,1], you can reduce the total number of product terms in the design from 624 to 608. You can save more product terms by changing the coding of the shift amount for some shifts (see Exercise 6.3). The savings from these changes may carry over to other design approaches.

Table 6-3 ABEL program for a multi-mode 16-bit barrel shifter.

```
module barrl16f
Title 'Multi-mode 16-bit Barrel Shifter'

" Inputs and Outputs
DIN15..DIN0, S3..S0, C2..C0          pin;
DOUT15..DOUT0                        pin istype 'com';

S = [S3..S0]; C = [C2..C0]; " Shift amount and mode
L = DIN15; R = DIN0;        " MSB and LSB

ROL = (C == [0,0,0]); " Rotate (circular shift) left
ROR = (C == [0,0,1]); " Rotate (circular shift) right
SLL = (C == [0,1,0]); " Shift logical left (shift in 0s)
SRL = (C == [0,1,1]); " Shift logical right (shift in 0s)
SLA = (C == [1,0,0]); " Shift left arithmetic (replicate LSB)
SRA = (C == [1,0,1]); " Shift right arithmetic (replicate MSB)

equations

[DOUT15..DOUT0] = ROL & (S==0) & [DIN15..DIN0]
              # ROL & (S==1) & [DIN14..DIN0,DIN15]
              # ROL & (S==2) & [DIN13..DIN0,DIN15..DIN14]
              ...
              # ROL & (S==15) & [DIN0,DIN15..DIN1]
              # ROR & (S==0) & [DIN15..DIN0]
              # ROR & (S==1) & [DIN0,DIN15..DIN1]
              ...
              # ROR & (S==14) & [DIN13..DIN0,DIN15..DIN14]
              # ROR & (S==15) & [DIN14..DIN0,DIN15]
              # SLL & (S==0) & [DIN15..DIN0]
              # SLL & (S==1) & [DIN14..DIN0,0]
              ...
              # SLL & (S==14) & [DIN1..DIN0,0,0,0,0,0,0,0,0,0,0,0,0,0]
              # SLL & (S==15) & [DIN0,0,0,0,0,0,0,0,0,0,0,0,0,0,0,0]
              # SRL & (S==0) & [DIN15..DIN0]
              # SRL & (S==1) & [0,DIN15..DIN1]
              ...
              # SRL & (S==14) & [0,0,0,0,0,0,0,0,0,0,0,0,0,0,DIN15..DIN14]
              # SRL & (S==15) & [0,0,0,0,0,0,0,0,0,0,0,0,0,0,0,DIN15]
              # SLA & (S==0) & [DIN15..DIN0]
              # SLA & (S==1) & [DIN14..DIN0,R]
              ...
              # SLA & (S==14) & [DIN1..DIN0,R,R,R,R,R,R,R,R,R,R,R,R,R,R]
              # SLA & (S==15) & [DIN0,R,R,R,R,R,R,R,R,R,R,R,R,R,R,R]
              # SRA & (S==0) & [DIN15..DIN0]
              # SRA & (S==1) & [L,DIN15..DIN1]
              ...
              # SRA & (S==14) & [L,L,L,L,L,L,L,L,L,L,L,L,L,L,DIN15..DIN14]
              # SRA & (S==15) & [L,L,L,L,L,L,L,L,L,L,L,L,L,L,L,DIN15];
end barrl16f
```

6.2.2 Simple Floating-Point Encoder

We defined a simple floating-point number format on page 471 and posed the design problem of converting a number from fixed-point to this floating-point format. The I/O-pin requirements of this design are limited—11 inputs and 7 outputs—so we can potentially use a single PLD to replace the four parts that were used in the MSI solution.

An ABEL program for the fixed-to-floating-point converter is given in Table 6-4. The WHEN statement expresses the operation of determining the exponent value E in a very natural way. Then E is used to select the appropriate bits of B to use as the mantissa M.

Despite the deep nesting of the WHEN statement, only four product terms are needed in the minimal sum for each bit of E. The equations for the M bits are not too bad either, requiring only eight product terms each. Unfortunately, the

Table 6-4
An ABEL program for the fixed-point to floating-point PLD.

```
module fpenc
title 'Fixed-point to Floating-point Encoder'
FPENC device 'P20L8';

" Input and output pins
B10..B0                 pin 1..11;
E2..E0, M3..M0          pin 21..15 istype 'com';

" Constant expressions
B = [B10..B0];
E = [E2..E0];
M = [M3..M0];

equations

WHEN B < 16 THEN E = 0;
ELSE WHEN B < 32 THEN E = 1;
ELSE WHEN B < 64 THEN E = 2;
ELSE WHEN B < 128 THEN E = 3;
ELSE WHEN B < 256 THEN E = 4;
ELSE WHEN B < 512 THEN E = 5;
ELSE WHEN B < 1024 THEN E = 6;
ELSE E = 7;

M = (E==0) & [B3..B0]
  # (E==1) & [B4..B1]
  # (E==2) & [B5..B2]
  # (E==3) & [B6..B3]
  # (E==4) & [B7..B4]
  # (E==5) & [B8..B5]
  # (E==6) & [B9..B6]
  # (E==7) & [B10..B7];

end fpenc
```

GAL20V8 has available only seven product terms per output. However, the GAL22V10 (Figure 8-22 on page 688) has more product terms available, so we can use that if we like.

One drawback of the design in Table 6-4 is that the [M3..M0] outputs are slow; since they use [E2..E0], they take two passes through the PLD. A faster approach, if it fits, would be to rewrite the "select" terms (E==0, etc.) as intermediate equations before the equations section, and let ABEL expand the resulting M equations in a single level of logic. Unfortunately, ABEL does not allow WHEN statements outside of the equations section, so we'll have to roll up our sleeves and write our own logic expressions in the intermediate equations.

Table 6-5 shows the modified approach. The expressions for S7–S0 are just mutually-exclusive AND-terms that indicate exponent values of 7–0 depending on the location of the most significant 1 bit in the fixed-point input number. The exponent [E2..E0] is a binary encoding of the select terms, and the mantissa bits [M3..M0] are generated using a select term for each case. It turns out that these M equations still require 8 product terms per output bit, but at least they're a lot faster, since they use just one level of logic.

```
module fpence
title 'Fixed-point to Floating-point Encoder'
FPENCE device 'P20L8';

" Input and output pins
B10..B0                     pin 1..11;
E2..E0, M3..M0              pin 21..15 istype 'com';

" Intermediate equations
S7 = B10;
S6 = !B10 & B9;
S5 = !B10 & !B9 & B8;
S4 = !B10 & !B9 & !B8 & B7;
S3 = !B10 & !B9 & !B8 & !B7 & B6;
S2 = !B10 & !B9 & !B8 & !B7 & !B6 & B5;
S1 = !B10 & !B9 & !B8 & !B7 & !B6 & !B5 & B4;
S0 = !B10 & !B9 & !B8 & !B7 & !B6 & !B5 & !B4;


equations

E2 = S7 # S6 # S5 # S4;
E1 = S7 # S6 # S3 # S2;
E0 = S7 # S5 # S3 # S1;

[M3..M0] = S0 & [B3..B0] # S1 & [B4..B1] # S2 & [B5..B2]
           # S3 & [B6..B3] # S4 & [B7..B4] # S5 & [B8..B5]
           # S6 & [B9..B6] # S7 & [B10..B7];

end fpenc
```

Table 6-5
Alternative ABEL program for the fixed-point to floating-point PLD.

6.2.3 Dual-Priority Encoder

In this example, we'll design a PLD-based priority encoder that identifies both the highest-priority and the second-highest-priority asserted signal among a set of eight active-high request inputs named [R0..R7], where R0 has the highest priority. We'll use [A2..A0] and AVALID to identify the highest-priority request, asserting AVALID only if a highest-priority request is present. Similarly, we'll use [B2:B0] and BVALID to identify the second-highest-priority request.

Table 6-6 shows an ABEL program for the priority encoder. As usual, a nested WHEN statement is perfect for expressing priority behavior. To find the

Table 6-6
ABEL program for a dual-priority encoder.

```
title 'Dual Priority Encoder'
PRIORTWO device 'P16V8';

" Input and output pins
R7..R0                             pin 1..8;
AVALID, A2..A0, BVALID, B2..B0     pin 19..12 istype 'com';

" Set definitions
A = [A2..A0]; B = [B2..B0];

equations

WHEN R0==1 THEN A=0;
ELSE WHEN R1==1 THEN A=1;
ELSE WHEN R2==1 THEN A=2;
ELSE WHEN R3==1 THEN A=3;
ELSE WHEN R4==1 THEN A=4;
ELSE WHEN R5==1 THEN A=5;
ELSE WHEN R6==1 THEN A=6;
ELSE WHEN R7==1 THEN A=7;

AVALID = ([R7..R0] != 0);

WHEN (R0==1) & (A!=0) THEN B=0;
ELSE WHEN (R1==1) & (A!=1) THEN B=1;
ELSE WHEN (R2==1) & (A!=2) THEN B=2;
ELSE WHEN (R3==1) & (A!=3) THEN B=3;
ELSE WHEN (R4==1) & (A!=4) THEN B=4;
ELSE WHEN (R5==1) & (A!=5) THEN B=5;
ELSE WHEN (R6==1) & (A!=6) THEN B=6;
ELSE WHEN (R7==1) & (A!=7) THEN B=7;

BVALID = (R0==1) & (A!=0) # (R1==1) & (A!=1)
       # (R2==1) & (A!=2) # (R3==1) & (A!=3)
       # (R4==1) & (A!=4) # (R5==1) & (A!=5)
       # (R6==1) & (A!=6) # (R7==1) & (A!=7);

end priortwo
```

P-Terms	Fan-in	Fan-out	Type	Name
8/1	8	1	Pin	AVALID
4/5	8	1	Pin	A2
4/5	8	1	Pin	A1
4/5	8	1	Pin	A0
24/8	11	1	Pin	BVALID
24/17	11	1	Pin	B2
20/21	11	1	Pin	B1
18/22	11	1	Pin	B0
106/84		Best P-Term Total: 76		
		Total Pins: 16		
	Average P-Term/Output: 9			

Table 6-7
Product-term usage
in the dual priority
encoder PLD.

second-highest priority input, we exclude an input if its input number matches
the highest-priority input number, which is A. Thus, we're using two-pass logic
to compute the B outputs. The equation for AVALID is easy; AVALID is 1 if the
request inputs are not all 0. To compute BVALID, we OR all of the conditions that
set B in the WHEN statement.

Even with two-pass logic, the B outputs use too many product terms to fit in
a 16V8; Table 6-7 shows the product-term usage. The B outputs use too many
terms even for a 22V10, which has 16 terms for two of its output pins and 8–14
for the others. Sometimes you just have to work harder to make things fit!

So, how can we save some product terms? One important thing to notice is
that R0 can never be the second-highest-priority asserted input, and therefore B
can never be valid and 0. Thus, we can eliminate the WHEN clause for the R0==1
case. Making this change reduces the minimum number of terms for B2–B0 to
14, 17, and 15, respectively. We can almost fit the design in a 22V10, if we can
just know one term out of the B1 equation.

Well, let's try something else. The second WHEN clause, for the R0==2 case,
also fails to make use of everything we know. We don't need the full generality

**SUMS OF
PRODUCTS AND
PRODUCTS OF
SUMS
(SAY THAT 5
TIMES FAST)**   You may recall from Section 4.3.6 that the minimal sum-of-products expression for
the complement of a function can be manipulated through DeMorgan's theorem to
obtain a minimal product-of-sums expression for the original function. You may also
recall that the number of product terms in the minimal sum of products may differ
from the number of sum terms in the minimal product of sums. The "P-Terms"
column in Table 6-7 lists the number of terms in both minimal forms (product/sum
terms). If *either* minimal form has less than or equal to the number of product terms
available in a 22V10's AND-OR array, then the function can be made to fit.

of A!=0; this case is only important when R0 is 1. So, let us replace the first two lines of the original WHEN statement with

WHEN (R1==1) & (R0==1) THEN B=1;

This subtle change reduces the minimum number of terms for B2–B0 to 12, 16, and 13, respectively. We made it! Can the number of product terms be reduced further, enough to fit into a 16V8 while maintaining the same functionality? It's not likely, but we'll leave that as an exercise (6.4) for the reader!

6.2.4 Cascading Comparators

We showed in Section 5.9.5 that equality comparisons are easy to realize in PLDs but that magnitude comparisons (greater-than or less-than) of more than a few bits are not good candidates for PLD realization, owing to the large number of product terms required. Thus, comparators are best realized using discrete MSI comparator components or as specialized cells within an FPGA or ASIC library. However, PLDs are quite suitable for realizing the combinational logic used in "parallel expansion" schemes that construct wider comparators from smaller ones, as we'll show here.

In Section 5.9.4 we showed how to connect 74x85 4-bit comparators in series to create larger comparators. Although a serial cascading scheme requires no extra logic to build arbitrarily large comparators, it has the major drawback that the delay increases linearly with the length of the cascade.

In Section 6.1.4, on the other hand, we showed how multiple copies of the 74x682 8-bit comparator could be used in parallel along with combinational logic to perform a 24-bit comparison. This scheme can be generalized for comparisons of arbitrary width.

Table 6-8 is an ABEL program that uses a GAL22V10 to perform a 64-bit comparison, using eight 74x682s to combine the equal (EQ) and greater-than (GT) outputs from the individual byte to produce all six possible relations of the two 64-bit input values ($=, \neq, >, \geq, <, \leq$).

In this program, the PEQQ and PNEQ outputs can be realized with one product term each. The remaining eight outputs use eight product terms each. As we've mentioned previously, the 22V10 provides 8–16 product terms per output, so the design fits.

HAVE IT YOUR WAY   Early PLDs such as the PAL16L8s did not have output-polarity control. Designers who used these devices were forced to choose a particular polarity, active high or active low, for some outputs in order to obtain reduced equations that would fit. When a 16V8, 20V8, 22V10, or any of a plethora of modern CPLDs is used, no such restriction exists. If an equation *or its complement* can be reduced to the number of product terms available, then the corresponding output can be made active high or active low by programming the output-polarity fuse appropriately.

```
module compexp
title 'Expansion logic for 64-bit comparator'
COMPEXP device 'P22V10';

" Inputs from the individual comparators, active-low, 7 = MSByte
EQ_L7..EQ_L0, GT_L7..GT_L0                pin 1..11, 13..14, 21..23;

" Comparison outputs
PEQQ, PNEQ, PGTQ, PGEQ, PLTQ, PLEQ        pin 15..20 istype 'com';

" Active-level conversions
EQ7 = !EQ_L7; EQ6 = !EQ_L6; EQ5 = !EQ_L5; EQ4 = !EQ_L4;
EQ3 = !EQ_L3; EQ2 = !EQ_L2; EQ1 = !EQ_L1; EQ0 = !EQ_L0;
GT7 = !GT_L7; GT6 = !GT_L6; GT5 = !GT_L5; GT4 = !GT_L4;
GT3 = !GT_L3; GT2 = !GT_L2; GT1 = !GT_L1; GT0 = !GT_L0;

" Less-than terms
LT7 = !(EQ7 # GT7); LT6 = !(EQ6 # GT6); LT5 = !(EQ5 # GT5);
LT4 = !(EQ4 # GT4); LT3 = !(EQ3 # GT3); LT2 = !(EQ2 # GT2);
LT1 = !(EQ1 # GT1); LT0 = !(EQ0 # GT0);


equations

PEQQ = EQ7 & EQ6 & EQ5 & EQ4 & EQ3 & EQ2 & EQ1 & EQ0;

PNEQ = !(EQ7 & EQ6 & EQ5 & EQ4 & EQ3 & EQ2 & EQ1 & EQ0);

PGTQ = GT7  #  EQ7 & GT6  #  EQ7 & EQ6 & GT5
     # EQ7 & EQ6 & EQ5 & GT4  #  EQ7 & EQ6 & EQ5 & EQ4 & GT3
     # EQ7 & EQ6 & EQ5 & EQ4 & EQ3 & GT2
     # EQ7 & EQ6 & EQ5 & EQ4 & EQ3 & EQ2 & GT1
     # EQ7 & EQ6 & EQ5 & EQ4 & EQ3 & EQ2 & EQ1 & GT0;

PLEQ = !(GT7  #  EQ7 & GT6  #  EQ7 & EQ6 & GT5
      # EQ7 & EQ6 & EQ5 & GT4  #  EQ7 & EQ6 & EQ5 & EQ4 & GT3
      # EQ7 & EQ6 & EQ5 & EQ4 & EQ3 & GT2
      # EQ7 & EQ6 & EQ5 & EQ4 & EQ3 & EQ2 & GT1
      # EQ7 & EQ6 & EQ5 & EQ4 & EQ3 & EQ2 & EQ1 & GT0);

PLTQ = LT7  #  EQ7 & LT6  #  EQ7 & EQ6 & LT5
     # EQ7 & EQ6 & EQ5 & LT4  #  EQ7 & EQ6 & EQ5 & EQ4 & LT3
     # EQ7 & EQ6 & EQ5 & EQ4 & EQ3 & LT2
     # EQ7 & EQ6 & EQ5 & EQ4 & EQ3 & EQ2 & LT1
     # EQ7 & EQ6 & EQ5 & EQ4 & EQ3 & EQ2 & EQ1 & LT0;

PGEQ = !(LT7  #  EQ7 & LT6  #  EQ7 & EQ6 & LT5
      # EQ7 & EQ6 & EQ5 & LT4  #  EQ7 & EQ6 & EQ5 & EQ4 & LT3
      # EQ7 & EQ6 & EQ5 & EQ4 & EQ3 & LT2
      # EQ7 & EQ6 & EQ5 & EQ4 & EQ3 & EQ2 & LT1
      # EQ7 & EQ6 & EQ5 & EQ4 & EQ3 & EQ2 & EQ1 & LT0);

end compexp
```

Table 6-8
ABEL program for combining eight 74x682s into a 64-bit comparator.

Table 6-9
Mode-control bits for
the mode-dependent
comparator.

M1	M0	Comparison
0	0	32-bit
0	1	31-bit
1	0	30-bit
1	1	not used

6.2.5 Mode-Dependent Comparator

For the next example, let us suppose we have a system in which we need to compare two 32-bit words under normal circumstances, but where we must sometimes ignore one or two low-order bits of the input words. The operating mode is specified by two mode-control bits, M1 and M0, as shown in Table 6-9.

As we've noted previously, comparing, adding, and other "iterative" operations are usually poor candidates for PLD-based design, because an equivalent two-level sum-of-products expression has far too many product terms. In Section 5.9.5 we calculated how many product terms are needed for an n-bit comparator. Based on these results, we certainly wouldn't be able to build the 32-bit mode-dependent comparator or even an 8-bit slice of it with a PLD; the 74x682 8-bit comparator is just about the most efficient possible single chip we can use to perform an 8-bit comparison. However, a PLD-based design is quite reasonable for handling the mode-control logic and the part of the comparison that is dependent on mode (the two low-order bits).

Figure 6-9 shows a complete circuit design resulting from this idea, and Table 6-11 is the ABEL program for a 16V8 MODECOMP PLD that handles the "random logic." Four '682s are used to compare most of the bits, and the 16V8 combines the '682 outputs and handles the two low-order bits as a function of the mode. Intermediate expressions EQ30 and GT30 are defined to save typing in the equations section of the program.

As shown in Table 6-10, the XEQY and XGTY outputs use 7 and 11 product terms, respectively. Thus, XGTY does not fit into the 7 product terms available on a 16V8 output. However, this is another example where we have some flexibility in our coding choices. By changing the coding of MODE30 to [1,.X.], we can reduce the product-term requirements for XGTY to 7/12 and thereby fit the design into a 16V8.

Table 6-10
Product-term usage
for the MODECOMP
PLD.

```
P-Terms   Fan-in  Fan-out   Type   Name
---------  ------  -------   ----   --------

  7/9        10       1      Pin    XEQY
 11/13       14       1      Pin    XGTY
=========
 18/22            Best P-Term Total:  18
                      Total Pins:  16
              Average P-Term/Output:  9
```

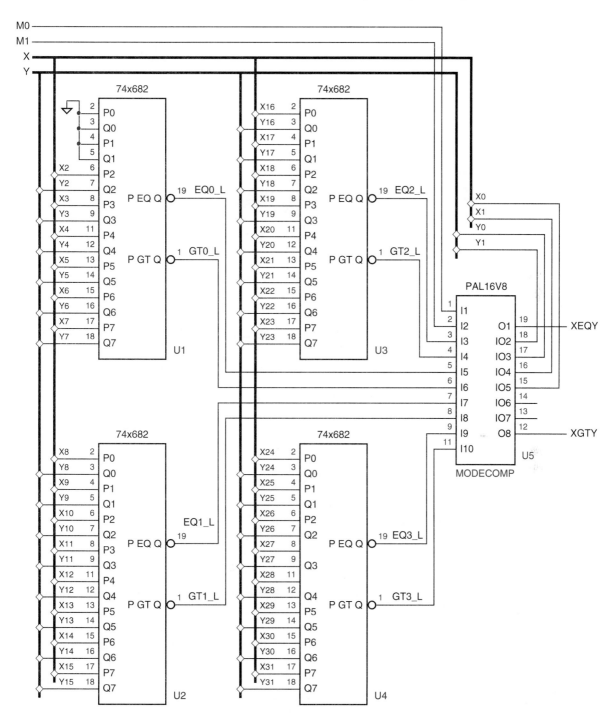

Figure 6-9 A 32-bit mode-dependent comparator.

Table 6-11
ABEL program for
combining eight
74x682s into a
64-bit comparator.

```
module modecomp
title 'Control PLD for Mode-Dependent Comparator'
MODECOMP device 'P16V8';

" Input and output pins
M0, M1, EQ2_L, GT2_L, EQ0_L, GT0_L              pin 1..6;
EQ1_L, GT1_L, EQ3_L, GT3_L, X0, X1, Y0, Y1   pin 7..9, 10, 15..18;
XEQY, XGTY                                      pin 19, 12 istype 'com';

" Active-level conversions
EQ3 = !EQ3_L; EQ2 = !EQ2_L; EQ1 = !EQ1_L; EQ0 = !EQ0_L;
GT3 = !GT3_L; GT2 = !GT2_L; GT1 = !GT1_L; GT0 = !GT0_L;

" Mode definitions
MODE32 = ([M1,M0] == [0,0]); " 32-bit comparison
MODE31 = ([M1,M0] == [0,1]); " 31-bit comparison
MODE30 = ([M1,M0] == [1,0]); " 30-bit comparison
MODEXX = ([M1,M0] == [1,1]); " Unused

" Expressions for 30-bit equal and greater-than
EQ30 = EQ3 & EQ2 & EQ1 & EQ0;
GT30 = GT3 # (EQ3 & GT2) # (EQ3 & EQ2 & GT1) # (EQ3 & EQ2 & EQ1 & GT0);


equations

WHEN MODE32 THEN {
  XEQY = EQ30 & (X1==Y1) & (X0==Y0);
  XGTY = GT30 # (EQ30 & (X1>Y1)) # (EQ30 & (X1==Y1) & (X0>Y0));
  }
ELSE WHEN MODE31 THEN {
  XEQY = EQ30 & (X1==Y1);
  XGTY = GT30 # (EQ30 & (X1>Y1));
  }
ELSE WHEN MODE30 THEN {
  XEQY = EQ30;
  XGTY = GT30;
  }

end modecomp
```

6.2.6 Ones Counter

There are several important algorithms that include the step of counting the number of "1" bits in a data word. In fact, some microprocessor instruction sets have been extended recently to include ones counting as a basic instruction.

Counting the ones in a data word can be done easily as an iterative process, where you scan the word from one end to the other and increment a counter each time a "1" is encountered. However, this operation must be done more quickly inside the arithmetic and logic unit of a microprocessor. Ideally, we would like

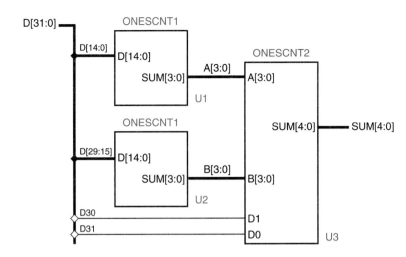

Figure 6-10
Possible partitioning
for the ones-counting
circuit.

ones counting to run as fast as any other arithmetic operation, such as adding two words. Therefore, a combinational circuit is required.

In this example, let us suppose that we have a requirement to build a 32-bit ones counter as part of a larger system. Based on the number of inputs and outputs required, we obviously can't fit the design into a single 22V10-class PLD, but we might be able to partition the design into a reasonably small number of PLDs.

Figure 6-10 shows such a partition. Two copies of a first 22V10, ONESCNT1, are used to count the ones in two 15-bit chunks of the 32-bit input word D[31:0], each producing a 4-bit sum output. A second 22V10, ONESCNT2, is used to add the two 4-bit sums and the last 2 input bits.

The program for ONESCNT1 is deceptively simple, as shown in Table 6-12. The statement "@CARRY 1" is included to limit the carry chain to one stage; as explained in Section 5.10.8, this reduces product-term requirements at the expense of helper outputs and increased delay.

Table 6-12
ABEL program for
counting the 1 bits
in a 15-bit word.

```
module onescnt1
title 'Count the ones in a 15-bit word'
ONESCNT1 device 'P22V10';

" Input and output pins
D14..D0              pin 1..11, 13..15, 23;
SUM3..SUM0           pin 17..20 istype 'com';

equations

@CARRY 1;
[SUM3..SUM0] = D0 + D1 + D2 + D3 + D4 + D5 + D6 + D7
            + D8 + D9 + D10 + D11 + D12 + D13 + D14;

end onescnt1
```

Unfortunately, when I compiled this program, my computer just sat there, CPU-bound, for an hour without producing any results. That gave me time to use my brain, a good exercise for those of us who have become too dependent on CAD tools. I then realized that I could write the logic function for the SUM0 output by hand in just a few seconds,

$$SUM0 = D0 \oplus D1 \oplus D2 \oplus D3 \oplus D4 \oplus D5 \oplus D6 \oplus D7 \oplus \ldots \oplus D13 \oplus D14$$

The Karnaugh map for this function is a checkerboard, and the minimal sum-of-products expression has 2^{14} product terms. Obviously this is not going to fit in one or a few passes through a 22V10! So, anyway, I killed the ABEL compiler process and rebooted Windows just in case the compiler had gone awry.

Obviously, a partitioning into smaller chunks is required to design the ones-counting circuit. Although we could pursue this further using ABEL and PLDs, it's more interesting to do a structural design using VHDL, as we will in Section 6.3.6. The ABEL and PLD version is left as an exercise (6.6).

6.2.7 Tic-Tac-Toe

In this example, we'll design a combinational circuit that picks a player's next move in the game of Tic-Tac-Toe. The first thing we'll do is decide on a strategy for picking the next move. Let us try to emulate the typical human's strategy by following the decision steps below:

1. Look for a row, column, or diagonal that has two of my marks (X or O, depending on which player I am) and one empty cell. If one exists, place my mark in the empty cell; I win!

2. Else, look for a row, column, or diagonal that has two of my opponent's marks and one empty cell. If one exists, place my mark in the empty cell to block a potential win by my opponent.

3. Else, pick a cell based on experience. For example, if the middle cell is open, it's usually a good bet to take it. Otherwise, the corner cells are good bets. Intelligent players can also notice and block a developing pattern by the opponent or "look ahead" to pick a good move.

TIC-TAC-TOE, IN CASE YOU DIDN'T KNOW

The game of Tic-Tac-Toe is played by two players on a 3×3 grid of cells that are initially empty. One player is "X" and the other is "O". The players alternate in placing their mark in an empty cell; "X" always goes first. The first player to get three of his or her own marks in the same row, column, or diagonal wins. Although the first player to move (X) has a slight advantage, it can be shown that a game between two intelligent players will always end in a draw; neither player will get three in a row before the grid fills up.

	1	2	3	column
row				
1	X11,Y11	X13,Y12	X13,Y13	
2	X21,Y21	X23,Y22	X23,Y23	
3	X21,Y21	X23,Y22	X23,Y23	

Figure 6-11
Tic-Tac-Toe grid and
ABEL signal names.

Planning ahead, we'll call the second player "Y" to avoid confusion between "O" and "0" in our programs. The next thing to think about is how we might encode the inputs and outputs of the circuit. There are only nine possible moves that a player can make, so the output can be encoded in just four bits. The circuit's input is the current state of the playing grid. There are nine cells, and each cell has one of three possible states (empty, occupied by X, occupied by Y).

There are several choices of how to code the state of one cell. Because the game is symmetric, we'll choose a symmetric encoding that may help us later:

00 Cell is empty.

10 Cell is occupied by X.

01 Cell is occupied by Y.

So, we can encode the 3×3 grid's state in 18 bits. As shown in Figure 6-11, we'll number the grid with row and column numbers and use ABEL signals Xij and Yij to denote the presence of X or Y in cell i,j. We'll look at the output coding later.

With a total of 18 inputs and 4 outputs, the Tic-Tac-Toe circuit could conceivably fit in just one 22V10. However, experience suggests that there's just no way. We're going to have to find a partitioning of the function, and partitioning along the lines of the decision steps on the preceding page seems like a good idea.

In fact, steps 1 and 2 are very similar; they differ only in reversing the roles of the player and the opponent. Here's where our symmetric encoding can pay

COMPACT ENCODING   Since each cell in the Tic-Tac-Toe grid can have only three states, not four, the total number of board configurations is 3^9, or 19,683. This is less than 2^{15}, so the board state can be encoded in only 15 bits. However, such an encoding would lead to much larger circuits for picking a move, unless the move-picking circuit was a read-only memory (see Exercise 10.26).

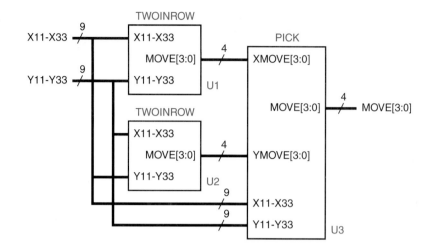

Figure 6-12
Preliminary PLD partitioning for the Tic-Tac-Toe game.

off. A PLD that finds me two of my marks in a row along with one empty cell for a winning move (step 1) can find two of my opponent's marks in a row plus an empty for a blocking move (step 2). All we have to do is swap the encodings for X and Y. With our selected coding, that doesn't require any logic, just physically swapping the Xij and Yij signals for each cell. With this in mind, we can use two copies of the same PLD, TWOINROW, to perform steps 1 and 2 as shown in Figure 6-12. Notice that the X11–X33 signals are connected to the top inputs of the first TWOINROW PLD but to the bottom inputs of the second.

The moves from the two TWOINROW PLDs can be examined in another PLD, PICK. This device picks a move from the first two PLDs if either found one; else it performs step 3. It looks like PICK has too many inputs and outputs to fit in a 22V10, but we'll come back to that later.

Table 6-13 is a program for the TWOINROW PLD. It looks at the grid's state from the point of view of X; that is, it looks for a move where X can get three in a row. The program makes extensive use of intermediate equations to

Table 6-13
ABEL program to find two in a row in Tic-Tac-Toe.

```
module twoinrow
Title 'Find Two Xs and an empty cell in a row, column, or diagonal'
TWOINROW device 'P22V10';

" Inputs and Outputs
X11, X12, X13, X21, X22, X23, X31, X32, X33 pin 1..9;
Y11, Y12, Y13, Y21, Y22, Y23, Y31, Y32, Y33 pin 10,11,13..15,20..23;
MOVE3..MOVE0                                pin 16..19 istype 'com';

" MOVE output encodings
MOVE   = [MOVE3..MOVE0];
MOVE11 = [1,0,0,0]; MOVE12 = [0,1,0,0]; MOVE13 = [0,0,1,0];
MOVE21 = [0,0,0,1]; MOVE22 = [1,1,0,0]; MOVE23 = [0,1,1,1];
MOVE31 = [1,0,1,1]; MOVE32 = [1,1,0,1]; MOVE33 = [1,1,1,0];
NONE   = [0,0,0,0];
```

Table 6-13
(continued)

```
" Find moves in rows.  Rxy ==> a move exists in cell xy
R11 = X12 & X13 & !X11 & !Y11;
R12 = X11 & X13 & !X12 & !Y12;
R13 = X11 & X12 & !X13 & !Y13;
R21 = X22 & X23 & !X21 & !Y21;
R22 = X21 & X23 & !X22 & !Y22;
R23 = X21 & X22 & !X23 & !Y23;
R31 = X32 & X33 & !X31 & !Y31;
R32 = X31 & X33 & !X32 & !Y32;
R33 = X31 & X32 & !X33 & !Y33;

" Find moves in columns.  Cxy ==> a move exists in cell xy
C11 = X21 & X31 & !X11 & !Y11;
C12 = X22 & X32 & !X12 & !Y12;
C13 = X23 & X33 & !X13 & !Y13;
C21 = X11 & X31 & !X21 & !Y21;
C22 = X12 & X32 & !X22 & !Y22;
C23 = X13 & X33 & !X23 & !Y23;
C31 = X11 & X21 & !X31 & !Y31;
C32 = X12 & X22 & !X32 & !Y32;
C33 = X13 & X23 & !X33 & !Y33;

" Find moves in diagonals.  Dxy or Exy ==> a move exists in cell xy
D11 = X22 & X33 & !X11 & !Y11;
D22 = X11 & X33 & !X22 & !Y22;
D33 = X11 & X22 & !X33 & !Y33;
E13 = X22 & X31 & !X13 & !Y13;
E22 = X13 & X31 & !X22 & !Y22;
E31 = X13 & X22 & !X31 & !Y31;

" Combine moves for each cell.  Gxy ==> a move exists in cell xy
G11 = R11 # C11 # D11;
G12 = R12 # C12;
G13 = R13 # C13 # E13;
G21 = R21 # C21;
G22 = R22 # C22 # D22 # E22;
G23 = R23 # C23;
G31 = R31 # C31 # E31;
G32 = R32 # C32;
G33 = R33 # C33 # D33;

equations

WHEN G22 THEN MOVE= MOVE22;
ELSE WHEN G11 THEN MOVE = MOVE11;
ELSE WHEN G13 THEN MOVE = MOVE13;
ELSE WHEN G31 THEN MOVE = MOVE31;
ELSE WHEN G33 THEN MOVE = MOVE33;
ELSE WHEN G12 THEN MOVE = MOVE12;
ELSE WHEN G21 THEN MOVE = MOVE21;
ELSE WHEN G23 THEN MOVE = MOVE23;
ELSE WHEN G32 THEN MOVE = MOVE32;
ELSE MOVE = NONE;

end twoinrow
```

define all possible row, column, and diagonal moves. It combines all of the moves for a cell i,j in an expression for Gij, and finally the equations section uses a WHEN statement to select a move.

Note that a nested WHEN statement must be used rather than nine parallel WHEN statements or assignments, because we can select only one move even if multiple moves are available. Also note that G22, the center cell, is checked first, followed by the corners. This was done hoping that we could minimize the number of terms by putting the most common moves early in the nested WHEN. Alas, the design still requires a ton of product terms, as shown in Table 6-14.

By the way, we still haven't explained why we chose the output coding that we did (as defined by MOVE11, MOVE22, etc. in the program). It's pretty clear that changing the encoding is never going to save us enough product terms to fit the design into a 22V10. But there's still method to this madness, as we'll now show.

Clearly we'll have to split TWOINROW into two or more pieces. As in any design problem, several different strategies are possible. The first strategy I tried was to use two different PLDs, one to find moves in all the rows and one of the diagonals, and the other to work on all the columns and the remaining diagonal. That helped, but not nearly enough to fit each half into a 22V10.

With the second strategy, I tried slicing the problem a different way. The first PLD finds all the moves in cells 11, 12, 13, 21, and 22, and the second PLD finds all the moves in the remaining cells. That worked! The first PLD, named TWOINHAF, is obtained from Table 6-13 simply by commenting out the four lines of the WHEN statement for the moves to cells 23, 31, 32, and 33.

We could obtain the second PLD from TWOINROW in a similar way, but let's wait a minute. In the manufacture of real digital systems, it is always desirable to minimize the number of distinct parts that are used; this saves on inventory costs and complexity. With programmable parts, it is desirable to minimize the number of distinct programs that are used. Even though the physical parts are identical, a different set of test vectors must be devised at some cost for each different program. Also, it's possible that the product will be successful enough for us to save money by converting the PLDs into hard-coded devices, a different one for each program, again encouraging us to minimize programs.

Table 6-14
Product-term usage for the TWOINROW PLD.

P-Terms	Fan-in	Fan-out	Type	Name
61/142	18	1	Pin	MOVE3
107/129	18	1	Pin	MOVE2
77/88	17	1	Pin	MOVE1
133/87	18	1	Pin	MOVE0

378/446	Best P-Term Total: 332
	Total Pins: 22
	Average P-Term/Output: 83

The Tic-Tac-Toe game is the same game even if we rotate the grid 90° or 180°. Thus, the TWOINHAF PLD can find moves for cells 33, 32, 31, 23, and 22 if we rotate the grid 180°. Because of the way we defined the grid state, with a separate pair of inputs for each cell, we can "rotate the grid" simply by shuffling the input pairs appropriately. That is, we swap 33↔11, 32↔12, 31↔13, and 23↔21.

Of course, once we rearrange inputs, TWOINHAF will still produce output move codes corresponding to cells in the top half of the grid. To keep things straight, we should transform these into codes for the proper cells in the bottom half of the grid. We would like this transformation to take a minimum of logic. This is where our choice of output code comes in. If you look carefully at the MOVE coding defined at the beginning of Table 6-13, you'll see that the code for a given position in the 180°-rotated grid is obtained by complementing and reversing the order of the code bits for the same position in the unrotated grid. In other words, the code transformation can be done with four inverters and a rearrangement of wires. This can be done "for free" in the PLD that looks at the TWOINHAF outputs.

You probably never thought that Tic-Tac-Toe could be so tricky. Well, we're halfway there. Figure 6-13 shows the partitioning of the design as we'll now continue it. Each TWOINROW PLD from our original partition is replaced

Figure 6-13
Final PLD partitioning for the Tic-Tac-Toe game.

by a pair of **TWOINHALF** PLDs. The bottom PLD of each pair is preceded by a box labeled "P", which permutes the inputs to rotate the grid 180°, as discussed previously. Likewise, it is followed by a box labeled "T", which compensates for the rotation by transforming the output code; this box will actually be absorbed into the PLD that follows it, PICK1.

The function of PICK1 is pretty straightforward. As shown in Table 6-15, it is simply picks a winning move or a blocking move if one is available. Since there are two extra input pins available on the 22V10, we use them to input the state of the center cell. In this way, we can perform the first part of step 3 of the "human" algorithm on page 492, to pick the center cell if no winning or blocking move is available. The PICK1 PLD uses at most 9 product terms per output.

Table 6-15 ABEL program to pick one move based on four inputs.

```
module pick1
Title 'Pick One Move from Four Possible'
PICK1 device 'P22V10';

" Inputs from TWOINHAF PLDs
WINA3..WINA0        pin 1..4;        "Winning moves in cells 11,12,13,21,22
WINB3..WINB0        pin 5..8;        "Winning moves in cells 11,12,13,21,22 of rotated grid
BLKA3..BLKA0        pin 9..11, 13;   "Blocking moves in cells 11,12,13,21,22
BLKB3..BLKB0        pin 14..16, 21;  "Blocking moves in cells 11,12,13,21,22 of rotated grid
" Inputs from grid
X22, Y22            pin 22..23;      "Center cell; pick if no other moves
" Move outputs to PICK2 PLD
MOVE3..MOVE0        pin 17..20 istype 'com';

" Sets
WINA = [WINA3..WINA0];  WINB = [WINB3..WINB0];
BLKA = [BLKA3..BLKA0];  BLKB = [BLKB3..BLKB0];
MOVE = [MOVE3..MOVE0];

" Non-rotated move input and output encoding
MOVE11 = [1,0,0,0]; MOVE12 = [0,1,0,0]; MOVE13 = [0,0,1,0];
MOVE21 = [0,0,0,1]; MOVE22 = [1,1,0,0]; MOVE23 = [0,1,1,1];
MOVE31 = [1,0,1,1]; MOVE32 = [1,1,0,1]; MOVE33 = [1,1,1,0];
NONE   = [0,0,0,0];

equations

WHEN WINA != NONE THEN MOVE = WINA;
ELSE WHEN WINB != NONE THEN MOVE = ![WINB0..WINB3];  " Map rotated coding
ELSE WHEN BLKA != NONE THEN MOVE = BLKA;
ELSE WHEN BLKB != NONE THEN MOVE = ![BLKB0..BLKB3];  " Map rotated coding
ELSE WHEN !X22 & !Y22 THEN MOVE = MOVE22;            " Pick center cell if it's empty
ELSE MOVE = NONE;

end pick1
```

The final part of the design in Figure 6-13 is the PICK2 PLD. This PLD must provide most of the "experience" in step 3 of the human algorithm if PICK1 does not find a move.

We have a little problem with PICK2 in that a 22V10 does not have enough pins to accommodate the 4-bit input from PICK1, its own 4-bit output, and all 18 bits of grid state; it has only 22 I/O pins. Actually, we don't need to connect X22 and Y22, since they were already examined in PICK1, but that still leaves us two pins short. So, the purpose of the "other logic" block in Figure 6-13 is to encode

▌ **Table 6-16** ABEL program to pick one move using "experience."

```
module pick2
Title 'Pick a move using experience'
PICK2 device 'P22V10';

" Inputs from PICK1 PLD
PICK3..PICK0                            pin 1..4;   " Move, if any, from PICK1 PLD
" Inputs from Tic-Tac-Toe grid corners
X11, Y11, X13, Y13, X31, Y31, X33, Y33    pin 5..11, 13;
" Combined inputs from external NOR gates; 1 ==> corresponding cell is empty
E12, E21, E23, E32                      pin 14..15, 22..23;
" Move output
MOVE3..MOVE0                            pin 17..20 istype 'com';

PICK = [PICK3..PICK0];   " Set definition
" Non-rotated move input and output encoding
MOVE = [MOVE3..MOVE0];
MOVE11 = [1,0,0,0]; MOVE12 = [0,1,0,0]; MOVE13 = [0,0,1,0];
MOVE21 = [0,0,0,1]; MOVE22 = [1,1,0,0]; MOVE23 = [0,1,1,1];
MOVE31 = [1,0,1,1]; MOVE32 = [1,1,0,1]; MOVE33 = [1,1,1,0];
NONE   = [0,0,0,0];

" Intermediate equations for empty corner cells
E11 = !X11 & !Y11;  E13 = !X13 & !Y13;  E31 = !X31 & !Y31;  E33 = !X33 & !Y33;

equations

"Simplest approach -- pick corner if available, else side
WHEN PICK != NONE THEN MOVE = PICK;
ELSE WHEN E11 THEN MOVE = MOVE11;
ELSE WHEN E13 THEN MOVE = MOVE13;
ELSE WHEN E31 THEN MOVE = MOVE31;
ELSE WHEN E33 THEN MOVE = MOVE33;
ELSE WHEN E12 THEN MOVE = MOVE12;
ELSE WHEN E21 THEN MOVE = MOVE21;
ELSE WHEN E23 THEN MOVE = MOVE23;
ELSE WHEN E32 THEN MOVE = MOVE32;
ELSE MOVE = NONE;

end pick2
```

some of the information to save two pins. The method that we'll use here is to combine the signals for the middle edge cells 12, 21, 23, and 32 to produce four signals E12, E21, E23, and E32 that are asserted if and only if the corresponding cells are empty. This can be done with four 2-input NOR gates and actually leaves two spare inputs or outputs on the 22V10.

Assuming the four NOR gates as "other logic," Table 6-16 on the preceding page gives a program for the PICK2 PLD. When it must pick a move, this program uses the simplest heuristic possible—it picks a corner cell if one is empty, else it picks a middle edge cell. This program could use some improvement, because it will sometimes lose (see Exercise 6.8). Luckily, the equations resulting from Table 6-16 require only 8 to 10 terms per output, so it's possible to put in more intelligence (see Exercises 6.9 and 6.10).

6.3 Design Examples Using VHDL

6.3.1 Barrel Shifter

On page 468 we defined a barrel shifter as a combinational logic circuit with n data inputs, n data outputs, and a set of control inputs that specify how to shift the data between input and output. We showed in Section 6.1.1 how to build a simple barrel shifter that performs only left circular shifts using MSI building blocks. Later, in Section 6.2.1, we showed how to define a more capable barrel shifter using ABEL, but we also pointed out that PLDs are normally unsuitable for realizing barrel shifters. In this subsection we'll show how VHDL can be used to describe both the behavior and structure of barrel shifters for FPGA or ASIC realization.

Table 6-17 is a behavioral VHDL program for a 16-bit barrel shifter that performs any of six different combinations of shift type and direction. The shift types are circular, logical, and arithmetic, as defined previously in Table 6-3, and the directions are, of course, left and right. As shown in the entity declaration, a 4-bit control input S gives the shift amount, and a 3-bit control input C gives the shift mode (type and direction). We used the IEEE std_logic_arith package and defined the shift amount S to be type UNSIGNED so we could later use the CONV_INTEGER function in that package.

Notice that the entity declaration includes six constant definitions that establish the correspondence between shift modes and the value of C. Although we didn't discuss it in Section 4.7, VHDL allows you to put constant, type, signal, and other declarations within an entity declaration. It makes sense to define such items within the entity declaration only if they must be the same in any architecture. In this case, we are pinning down the shift-mode encodings, so they should go here. Other items should go in the architecture definition.

In the architecture part of the program we define six functions, one for each kind of shift on a 16-bit STD_LOGIC_VECTOR. We defined the subtype DATAWORD to save typing in the function definitions.

Table 6-17 VHDL behavioral description of a 6-function barrel shifter.

```vhdl
library IEEE;
use IEEE.std_logic_1164.all;
use IEEE.std_logic_arith.all;

entity barrel16 is
    port (
        DIN: in STD_LOGIC_VECTOR (15 downto 0);  -- Data inputs
        S: in UNSIGNED (3 downto 0);             -- Shift amount, 0-15
        C: in STD_LOGIC_VECTOR (2 downto 0);     -- Mode control
        DOUT: out STD_LOGIC_VECTOR (15 downto 0) -- Data bus output
    );
    constant Lrotate:  STD_LOGIC_VECTOR := "000";  -- Define the coding of
    constant Rrotate:  STD_LOGIC_VECTOR := "001";  -- the different shift modes
    constant Llogical: STD_LOGIC_VECTOR := "010";
    constant Rlogical: STD_LOGIC_VECTOR := "011";
    constant Larith:   STD_LOGIC_VECTOR := "100";
    constant Rarith:   STD_LOGIC_VECTOR := "101";
end barrel16;

architecture barrel16_behavioral of barrel16 is
subtype DATAWORD is STD_LOGIC_VECTOR(15 downto 0);

function Vrol (D: DATAWORD; S: UNSIGNED)
     return DATAWORD is
  variable N: INTEGER;
  variable TMPD: DATAWORD;
  begin
    N := CONV_INTEGER(S); TMPD := D;
    for i in 1 to N loop
      TMPD := TMPD(14 downto 0) & TMPD(15);
    end loop;
    return TMPD;
  end Vrol;

...

begin
process(DIN, S, C)
  begin
    case C is
      when Lrotate  => DOUT <= Vrol(DIN,S);
      when Rrotate  => DOUT <= Vror(DIN,S);
      when Llogical => DOUT <= Vsll(DIN,S);
      when Rlogical => DOUT <= Vsrl(DIN,S);
      when Larith   => DOUT <= Vsla(DIN,S);
      when Rarith   => DOUT <= Vsra(DIN,S);
      when others   => null;
    end case;
  end process;
end barrel16_behavioral;
```

ROLLING
YOUR OWN
VHDL-93 actually has built-in array operators, rol, ror, sll, srl, sla, and sra, corresponding to the shift operations that we defined in Table 6-3. Since these operations are not provided in VHDL-87, we've defined our own functions in Table 6-17. Well, actually we've defined only one of them (Vrol); the rest are left as an exercise for the reader (Exercise 6.11).

Table 6-17 shows the details of only the first function (Vrol); the rest are similar with only a one-line change. We define a variable N for converting the shift amount S into an integer for the for loop. We also assign the input vector D to a local variable TMPD, which is shifted N times in the for loop. In the body of the for loop, a single assignment statement takes a 15-bit slice of the data word (TMPD 14 downto 0)) and uses concatenation (&) to put it back together with the bit that "falls off" the left end (TMPD(15)). Other shift types can be described with similar operations. Note that the shift functions might not be defined in other, nonbehavioral descriptions of the barrel16 entity, for example in structural architectures.

The "concurrent statements" part of the architecture is a single process that has all of the entity's inputs in its sensitivity list. Within this process, a case statement assigns a result to DOUT by calling the appropriate function based on the value of the mode-control input C.

The process in Table 6-17 is a nice behavioral description of the barrel shifter, but most synthesis tools cannot synthesize a circuit from it. The problem is that most tools require the range of a for loop to be static at the time it is analyzed. The range of the for loop in the Vrol function is dynamic; it depends on the value of input signal S when the circuit is operating.

Well, that's OK; it's hard to predict what kind of circuit the synthesis tool would come up with even if it could handle a dynamic for range. This is an example where as designers we should take a little more control over the circuit structure to obtain a reasonably fast, efficient synthesis result.

In Figure 6-2 on page 470 we showed how to design a 16-bit barrel shifter for left circular shifts using MSI building blocks. We used a cascade of four 16-bit, 2-input multiplexers to shift their inputs by 0 or 1, 2, 4, or 8 positions, depending on the values of S0 through S3, respectively. We can express the same kind of behavior and structure using the VHDL program shown in Table 6-18. Even though the program uses a process and is "behavioral" in style, we can be pretty sure that most synthesis engines will generate a 1-input multiplexer for each "if" statement in the program, thereby creating a cascade similar to Figure 6-2.

Table 6-18 VHDL program for a 16-bit barrel shifter for left circular shifts only.

```
library IEEE;
use IEEE.std_logic_1164.all;

entity rol16 is
    port (
        DIN: in STD_LOGIC_VECTOR(15 downto 0);  -- Data inputs
        S: in STD_LOGIC_VECTOR (3 downto 0);     -- Shift amount, 0-15
        DOUT: out STD_LOGIC_VECTOR(15 downto 0)  -- Data bus output
    );
end rol16;

architecture rol16_arch of rol16 is
begin
process(DIN, S)
  variable X, Y, Z: STD_LOGIC_VECTOR(15 downto 0);
  begin
    if S(0)='1' then X := DIN(14 downto 0) & DIN(15); else X := DIN; end if;
    if S(1)='1' then Y := X(13 downto 0) & X(15 downto 14); else Y := X; end if;
    if S(2)='1' then Z := Y(11 downto 0) & Y(15 downto 12); else Z := Y; end if;
    if S(3)='1' then DOUT <= Z(7 downto 0) & Z(15 downto 8); else DOUT <= Z; end if;
  end process;
end rol16_arch;
```

Of course, our problem statement requires a barrel shifter that can shift both left and right. Table 6-19 revises the previous program to do circular shifts in either direction. An additional input, DIR, specifies the shift direction, 0 for left, 1 for right. Each rank of shifting is specified by a case statement that picks one of four possibilities based on the values of DIR and the bit of S that controls that rank. Notice that we created local 2-bit variables CTRLi to hold the pair of values DIR and S(i); each case statement is controlled by one of these variables. You might like to eliminate these variables and simply control each case statement with a concatenation "DIR & S(i)", but VHDL syntax doesn't allow that because the type of this concatenation would be unknown.

A typical VHDL synthesis tool will generate a 3- or 4-input multiplexer for each of the case statements in Table 6-19. A good synthesis tool will generate only a 2-input multiplexer for the last case statement.

So, now we have a barrel shifter that will do left or right circular shifts, but we're not done yet—we need to take care of the logical and arithmetic shifts in both directions. Figure 6-14 shows our strategy for completing the design. We start out with the ROLR16 component that we just completed, and we use other logic to control the shift direction as a function of C.

Table 6-19 VHDL program for a 16-bit barrel shifter for left and right circular shifts.

```
library IEEE;
use IEEE.std_logic_1164.all;

entity rolr16 is
    port (
        DIN:  in STD_LOGIC_VECTOR(15 downto 0);   -- Data inputs
        S:    in STD_LOGIC_VECTOR (3 downto 0);   -- Shift amount, 0-15
        DIR:  in STD_LOGIC;                       -- Shift direction, 0=>L, 1=>R
        DOUT: out STD_LOGIC_VECTOR(15 downto 0)   -- Data bus output
    );
end rolr16;

architecture rol16r_arch of rolr16 is
begin
process(DIN, S, DIR)
  variable X, Y, Z: STD_LOGIC_VECTOR(15 downto 0);
  variable CTRL0, CTRL1, CTRL2, CTRL3: STD_LOGIC_VECTOR(1 downto 0);
  begin
    CTRL0 := S(0) & DIR; CTRL1 := S(1) & DIR; CTRL2 := S(2) & DIR; CTRL3 := S(3) & DIR;
    case CTRL0 is
      when "00" | "01" => X := DIN;
      when "10" => X := DIN(14 downto 0) & DIN(15);
      when "11" => X := DIN(0) & DIN(15 downto 1);
      when others => null;  end case;
    case CTRL1 is
      when "00" | "01" => Y := X;
      when "10" => Y := X(13 downto 0) & X(15 downto 14);
      when "11" => Y := X(1 downto 0) & X(15 downto 2);
      when others => null;  end case;
    case CTRL2 is
      when "00" | "01" => Z := Y;
      when "10" => Z := Y(11 downto 0) & Y(15 downto 12);
      when "11" => Z := Y(3 downto 0) & Y(15 downto 4);
      when others => null;  end case;
    case CTRL3 is
      when "00" | "01" => DOUT <= Z;
      when "10" | "11" => DOUT <= Z(7 downto 0) & Z(15 downto 8);
      when others => null;  end case;
  end process;
end rol16r_arch;
```

Next we must "fix up" some of the result bits if we are doing a logical or arithmetic shift. For a left logical or arithmetic n-bit shift, we must set the rightmost $n-1$ bits to 0 or the original rightmost bit value, respectively. For a right logical or arithmetic n-bit shift, we must set the leftmost $n-1$ bits to 0 or the original leftmost bit value, respectively.

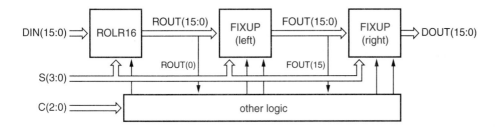

Figure 6-14
Barrel-shifter
components.

As shown in Figure 6-14, our strategy is to follow the circular shifter (ROLR16) with a fix-up circuit (FIXUP) that plugs in appropriate low-order bits for a left logical or arithmetic shift, and follow that with another fix-up circuit that plugs in high-order bits for a right logical or arithmetic shift.

Table 6-20 is a behavioral VHDL program for the left-shift fix-up circuit. The circuit has 16 bits of data input and output, DIN and DOUT. Its control inputs are the shift amount S, an enable input FEN, and the new value FDAT to be plugged into the fixed-up data bits. For each output bit DOUT(i), the circuit puts out the fixed-up bit value if i is less than S and the circuit is enabled; else it puts out the unmodified data input DIN(i).

The for loop in Table 6-20 is readily synthesizable, but you can't really be sure what kind of logic the synthesis tool will generate. In particular, the ">"

Table 6-20 Behavioral VHDL program for left-shift fix-ups.

```
library IEEE;
use IEEE.std_logic_1164.all;
use IEEE.std_logic_arith.all;

entity fixup is
    port (
        DIN:  in STD_LOGIC_VECTOR(15 downto 0); -- Data inputs
        S:    in UNSIGNED(3 downto 0);          -- Shift amount, 0-15
        FEN:  in STD_LOGIC;                     -- Fixup enable
        FDAT: in STD_LOGIC;                     -- Fixup data
        DOUT: out STD_LOGIC_VECTOR(15 downto 0) -- Data bus output
    );
end fixup;

architecture fixup_arch of fixup is
begin
process(DIN, S, FEN, FDAT)
  begin
    for i in 0 to 15 loop
      if (i < CONV_INTEGER(S)) and (FEN = '1') then DOUT(i) <= FDAT;
      else DOUT(i) <= DIN(i);  end if;
    end loop;
  end process;
end fixup_arch;
```

A SERIAL FIX-UP STRUCTURE

A structural architecture for the fix-up logic is shown in Table 6-21. Here, we have defined what is in effect an iterative circuit to create a 16-bit vector FSEL, where FSEL(i) is 1 if bit i needs fixing up. We start by setting FSEL(15) to 0, since that bit never needs fixing up. Then we note that for the remaining values of i, FSEL(i) should be 1 if S equals i+1 or if FSEL(i+1) is already asserted. Thus, the FSEL assignment within the generate statement creates a serial chain of 2-input OR gates, where one input is asserted if S=i (decoded with a 4-input AND gate), and the other input is connected to the previous OR gate's output. The DOUT(i) assignment statement creates 16 2-input multiplexers that select either DIN(i) or the fix-up data (FDAT), depending on the value of FSEL(i).

Although the serial realization is compact, it is very slow compared to one that realizes each FSEL output as a 2-level sum-of-products circuit. However, the long delay may not matter, because the fix-up circuit appears near the end of the data path. If speed is still a problem, there is a zero-cost trick that cuts the delay in half (see Exercise 6.12).

operation in each step of the loop may cause the synthesis of a general-purpose magnitude comparator, even though one of the operands is a constant and each output could therefore be generated with no more than a handful of gates. (In fact, the logic for "7 < CONV_INTEGER(S)" is just a wire, S(3)!) For a structural version of this function, see the box on this page.

For right shifts, fix-ups start from the opposite end of the data word, so it would appear that we need a second version of the fix-up circuit. However, we can use the original version if we just reverse the order of its input and output bits, as we'll soon see.

Table 6-22 puts together a structural architecture for the complete, 16-bit, 6-function barrel shifter using the design approach of Figure 6-14 on page 505. The entity declaration for barrel16 is unchanged from the original in Table 6-17 on page 501. The architecture declares two components, rolr16 and fixup; these use our previous entity definitions. The statement part of the archi-

Table 6-21 Structural VHDL architecture for left-shift fix-ups.

```
architecture fixup_struc of fixup is
signal FSEL: STD_LOGIC_VECTOR(15 downto 0);         -- Fixup select
begin
  FSEL(15) <= '0'; DOUT(15) <= DIN(15);
  U1: for i in 14 downto 0 generate
    FSEL(i) <= '1' when CONV_INTEGER(S) = i+1 else FSEL(i+1);
    DOUT(i) <= FDAT when (FSEL(i) = '1' and FEN = '1') else DIN(i);
  end generate;
end fixup_struc;
```

tecture instantiates `rolr16` and `fixup` and has several assignment statements that create needed control signals (the "other logic" in Figure 6-14).

For example, the first assignment statement asserts `DIR_RIGHT` if C specifies one of the right shifts. The enable inputs for the left and right fix-up circuits are `FIX_LEFT` and `FIX_RIGHT`, asserted for left and right logical and arithmetic shifts. The fix-up data values are `FIX_LEFT_DAT` and `FIX_RIGHT_DAT`.

While all the statements in the architecture execute concurrently, they are listed in Table 6-22 in the order of the actual dataflow to improve readability. First, `rolr16` is instantiated to perform the basic left or right circular shift as specified. Its outputs are hooked up to the inputs of the first `fixup` component (U2) to handle fix-ups for left logical and arithmetic shifts. Next comes U3, a `generate` statement that reverses the order of the data inputs for the next `fixup`

■ **Table 6-22** VHDL structural architecture for the 6-function barrel shifter.

```
architecture barrel16_struc of barrel16 is

component rolr16 port (
        DIN:  in STD_LOGIC_VECTOR(15 downto 0); -- Data inputs
        S:    in UNSIGNED(3 downto 0);          -- Shift amount, 0-15
        DIR:  in STD_LOGIC;                     -- Shift direction, 0=>L, 1=>R
        DOUT: out STD_LOGIC_VECTOR(15 downto 0) -- Data bus output
    ); end component;

component fixup port (
        DIN:  in STD_LOGIC_VECTOR(15 downto 0); -- Data inputs
        S:    in UNSIGNED(3 downto 0);          -- Shift amount, 0-15
        FEN:  in STD_LOGIC;                     -- Fixup enable
        FDAT: in STD_LOGIC;                     -- Fixup data
        DOUT: out STD_LOGIC_VECTOR(15 downto 0) -- Data bus output
    ); end component;

signal DIR_RIGHT, FIX_RIGHT, FIX_RIGHT_DAT, FIX_LEFT, FIX_LEFT_DAT: STD_LOGIC;
signal ROUT, FOUT, RFIXIN, RFIXOUT: STD_LOGIC_VECTOR(15 downto 0);

begin
  DIR_RIGHT <= '1' when C = Rrotate or C = Rlogical or C = Rarith else '0';
  FIX_LEFT <= '1' when DIR_RIGHT='0' and (C = Llogical or C = Larith) else '0';
  FIX_RIGHT <= '1' when DIR_RIGHT='1' and (C = Rlogical or C = Rarith) else '0';
  FIX_LEFT_DAT <= DIN(0) when C = Larith else '0';
  FIX_RIGHT_DAT <= DIN(15) when C = Rarith else '0';
  U1: rolr16 port map (DIN, S, DIR_RIGHT, ROUT);
  U2: fixup port map (ROUT, S, FIX_LEFT, FIX_LEFT_DAT, FOUT);
  U3: for i in 0 to 15 generate RFIXIN(i) <= FOUT(15-i); end generate;
  U4: fixup port map (RFIXIN, S, FIX_RIGHT, FIX_RIGHT_DAT, RFIXOUT);
  U5: for i in 0 to 15 generate DOUT(i) <= RFIXOUT(15-i); end generate;
end barrel16_struc;
```

component (U4), which handles fix-ups for right logical and arithmetic shifts. Finally U5, another generate statement, undoes the bit reversing of U3. Note that in synthesis, U3 and U5 are merely permutations of wires.

Many other architectures are possible for the original barrel16 entity. In Exercise 6.14, we suggest an architecture that enables the circular shifting to be done by the rol16 entity, which uses only 2-input multiplexers, rather than the more expensive rolr16.

6.3.2 Simple Floating-Point Encoder

We defined a simple floating-point number format on page 471 and posed the design problem of converting a number from fixed-point to this floating-point format. The problem of determining the exponent of the floating-point number mapped nicely into an MSI priority encoder. In an HDL, the same problem maps into nested "if" statements.

Table 6-23 is a behavioral VHDL program for the floating-point encoder. Within the fpenc_arch architecture, a nested "if" statement checks the range of the input B and sets M and E appropriately. Notice that the program uses the IEEE std_logic_arith package; this is done to get the UNSIGNED type and the comparison operations that go along with it, as we described in Section 5.9.6. Just to save typing, a variable BU is defined to hold the value of B as converted to the UNSIGNED type; alternatively, we could have written "UNSIGNED(B)" in each nested "if" clause.

Although the code in Table 6-23 is fully synthesizable, some synthesis tools may not be smart enough to recognize that the nested comparisons require just one bit to be checked at each level, and might instead generate a full 11-bit comparator at each level. Such logic would be a lot bigger and slower than what otherwise would be possible. If faced with this problem, we can always write the architecture a little differently and more explicitly to help out the tool, as shown in Table 6-24.

```vhdl
library IEEE;
use IEEE.std_logic_1164.all;
use IEEE.std_logic_arith.all;

entity fpenc is
  port (
    B: in STD_LOGIC_VECTOR(10 downto 0); -- fixed-point number
    M: out STD_LOGIC_VECTOR(3 downto 0); -- floating-point mantissa
    E: out STD_LOGIC_VECTOR(2 downto 0)  -- floating-point exponent
  );
end fpenc;

architecture fpenc_arch of fpenc is
begin
  process(B)
  variable BU: UNSIGNED(10 downto 0);
  begin
    BU := UNSIGNED(B);
    if    BU < 16   then M <= B( 3 downto 0); E <= "000";
    elsif BU < 32   then M <= B( 4 downto 1); E <= "001";
    elsif BU < 64   then M <= B( 5 downto 2); E <= "010";
    elsif BU < 128  then M <= B( 6 downto 3); E <= "011";
    elsif BU < 256  then M <= B( 7 downto 4); E <= "100";
    elsif BU < 512  then M <= B( 8 downto 5); E <= "101";
    elsif BU < 1024 then M <= B( 9 downto 6); E <= "110";
    else                 M <= B(10 downto 7); E <= "111";
    end if;
  end process;
end fpenc_arch;
```

Table 6-23
Behavioral VHDL program for fixed-point to floating-point conversion.

```vhdl
architecture fpence_arch of fpenc is
begin
  process(B)
  begin
    if    B(10) = '1' then M <= B(10 downto 7); E <= "111";
    elsif B(9)  = '1' then M <= B( 9 downto 6); E <= "110";
    elsif B(8)  = '1' then M <= B( 8 downto 5); E <= "101";
    elsif B(7)  = '1' then M <= B( 7 downto 4); E <= "100";
    elsif B(6)  = '1' then M <= B( 6 downto 3); E <= "011";
    elsif B(5)  = '1' then M <= B( 5 downto 2); E <= "010";
    elsif B(4)  = '1' then M <= B( 4 downto 1); E <= "001";
    else                   M <= B( 3 downto 0); E <= "000";
    end if;
  end process;
end fpence_arch;
```

Table 6-24
Alternative VHDL architecture for fixed-point to floating-point conversion.

On the other hand, we might like to use the real comparators and spend even more gates to improve the functionality of our design. In particular, the present design performs truncation rather than rounding when generating the mantissa bits. A more accurate result is achieved with rounding, but this is a much more complicated design. First, we will need an adder to add 1 to the selected mantissa bits when we round up. However, adding 1 when the mantissa is already 1111 will bump us into the next exponent range, so we need to watch out for this case. Finally, we can never round up if the unrounded mantissa and exponent are 1111 and 111, because there's no higher value in our floating-point representation to round to.

The program in Table 6-25 performs rounding as desired. The function round takes a selected 5-bit slice from the fixed-point number and returns the four high-order bits, adding 1 if the LSB is 1. Thus, if we think of the binary point as being just to the left of the LSB, rounding occurs if the truncated part of the mantissa is 1/2 or more. In each clause in the nested "if" statement in the process, the comparison value is selected so that rounding up will occur only if it does not "overflow," pushing the result into the next exponent range. Otherwise, conversion and rounding occurs in the next clause. In the last clause, we ensure that we do not round up when we're at the end of the floating-point range.

```
architecture fpencr_arch of fpenc is
function round (BSLICE: STD_LOGIC_VECTOR(4 downto 0))
    return STD_LOGIC_VECTOR is
  variable BSU: UNSIGNED(3 downto 0);
  begin
    if BSLICE(0) = '0' then return BSLICE(4 downto 1);
    else null;
      BSU := UNSIGNED(BSLICE(4 downto 1)) + 1;
      return STD_LOGIC_VECTOR(BSU);
    end if;
  end;
begin
  process(B)
  variable BU: UNSIGNED(10 downto 0);
  begin
    BU := UNSIGNED(B);
    if     BU < 16      then M <= B( 3 downto 0); E <= "000";
    elsif BU < 32-1     then M <= round(B( 4 downto 0)); E <= "001";
    elsif BU < 64-2     then M <= round(B( 5 downto 1)); E <= "010";
    elsif BU < 128-4    then M <= round(B( 6 downto 2)); E <= "011";
    elsif BU < 256-8    then M <= round(B( 7 downto 3)); E <= "100";
    elsif BU < 512-16   then M <= round(B( 8 downto 4)); E <= "101";
    elsif BU < 1024-32  then M <= round(B( 9 downto 5)); E <= "110";
    elsif BU < 2048-64  then M <= round(B(10 downto 6)); E <= "111";
    else                     M <= "1111";               E <= "111";
    end if;
  end process;
end fpencr_arch;
```

Table 6-25
Behavioral VHDL architecture for fixed-point to floating-point conversion with rounding.

Once again, synthesis results for this behavioral program may or may not be efficient. Besides the multiple comparison statements, we now must worry about the multiple 4-bit adders that might be synthesized as a result of the multiple calls to the round function. Restructuring the architecture so that only a single adder is synthesized is left as an exercise (6.15).

6.3.3 Dual-Priority Encoder

In this example we'll use VHDL to create a behavioral description of a PLD priority encoder that identifies both the highest-priority and the second-highest-priority asserted signal among a set of request inputs R(0 to 7), where R(0) has the highest priority. We'll use A(2 downto 0) and AVALID to identify the highest-priority request, asserting AVALID only if a highest-priority request is present. Similarly, we'll use B(2 downto 0) and BVALID to identify the second-highest-priority request.

Table 6-26
Behavioral VHDL program for a dual-priority encoder.

```vhdl
library IEEE;
use IEEE.std_logic_1164.all;
use IEEE.std_logic_arith.all;

entity Vprior2 is
    port (
        R: in STD_LOGIC_VECTOR (0 to 7);
        A, B: out STD_LOGIC_VECTOR (2 downto 0);
        AVALID, BVALID: buffer STD_LOGIC
    );
end Vprior2;

architecture Vprior2_arch of Vprior2 is
begin
  process(R, AVALID, BVALID)
  begin
    AVALID <= '0'; BVALID <= '0'; A <= "000"; B <= "000";
    for i in 0 to 7 loop
      if R(i) = '1' and AVALID = '0' then
        A <= CONV_STD_LOGIC_VECTOR(i,3); AVALID <= '1';
      elsif R(i) = '1' and BVALID = '0' then
        B <= CONV_STD_LOGIC_VECTOR(i,3); BVALID <= '1';
      end if;
    end loop;
  end process;
end Vprior2_arch;
```

Table 6-26 shows a behavioral VHDL program for the priority encoder. Instead of the nested "if" approach of the previous example, we've used a "for" loop. This approach allows us to take care of both the first and the second priorities within the same loop, working our way from highest to lowest priority. Besides std_logic_1164, the program uses the IEEE std_logic_arith package in order to get the CONV_STD_LOGIC_VECTOR function. We also wrote this function explicitly in Table 4-39 on page 279.

Notice in the table that ports AVALID and BVALID are declared as mode buffer, because they are read within the architecture. If you were stuck with an entity definition that declared AVALID and BVALID as mode out, you could still use the same architecture approach, but you would have to declare local variables corresponding to AVALID and BVALID within the process. Notice also that we included AVALID and BVALID in the process sensitivity list. Although this is not strictly necessary, it prevents warnings that the compiler otherwise would give about using the value of a signal that is not on the sensitivity list.

The nested "if" approach can also be used for the dual-priority encoder, but it yields a longer program with more accidents waiting to happen, as shown in Table 6-27. On the other hand, it may yield a better synthesis result; the only

```
architecture Vprior2i_arch of Vprior2 is
begin
  process(R, A, AVALID, BVALID)
  begin
    if    R(0) = '1' then A <= "000"; AVALID <= '1';
    elsif R(1) = '1' then A <= "001"; AVALID <= '1';
    elsif R(2) = '1' then A <= "010"; AVALID <= '1';
    elsif R(3) = '1' then A <= "011"; AVALID <= '1';
    elsif R(4) = '1' then A <= "100"; AVALID <= '1';
    elsif R(5) = '1' then A <= "101"; AVALID <= '1';
    elsif R(6) = '1' then A <= "110"; AVALID <= '1';
    elsif R(7) = '1' then A <= "111"; AVALID <= '1';
    else                  A <= "000"; AVALID <= '0';
    end if;
    if    R(1) = '1' and A /= "001" then B <= "001"; BVALID <= '1';
    elsif R(2) = '1' and A /= "010" then B <= "010"; BVALID <= '1';
    elsif R(3) = '1' and A /= "011" then B <= "011"; BVALID <= '1';
    elsif R(4) = '1' and A /= "100" then B <= "100"; BVALID <= '1';
    elsif R(5) = '1' and A /= "101" then B <= "101"; BVALID <= '1';
    elsif R(6) = '1' and A /= "110" then B <= "110"; BVALID <= '1';
    elsif R(7) = '1' and A /= "111" then B <= "111"; BVALID <= '1';
    else                                  B <= "000"; BVALID <= '0';
    end if;
  end process;
end Vprior2i_arch;
```

Table 6-27
Alternative VHDL architecture for a dual-priority encoder.

way to know with a particular tool is to synthesize the circuit and analyze the results in terms of delay and cell or gate count.

Both nested "if" statements and "for" statements may lead to long delay chains in synthesis. To guarantee that you get a faster dual-priority encoder, you must follow a structural or semistructural design approach. For example, you can start by writing a dataflow model of a fast 8-input priority encoder using the ideas found in the 74x148 logic diagram (Figure 5-50 on page 379) or in a related ABEL program (Table 5-24 on page 382). Then you can put two of these together in a structure that "knocks out" the highest-priority input in order to find the second, as we showed in Figure 6-6 on page 475.

6.3.4 Cascading Comparators

Cascading comparators is something we typically would not do in a VHDL behavioral model, because the language and the IEEE std_logic_arith package let us define comparators of any desired length directly. However, we may indeed need to write structural or semistructural VHDL programs that hook up smaller comparator components in a specific way to obtain high performance.

Table 6-28
Behavioral VHDL
program for a 64-bit
comparator.

```
library IEEE;
use IEEE.std_logic_1164.all;
use IEEE.std_logic_unsigned.all;

entity comp64 is
    port ( A, B: in STD_LOGIC_VECTOR (63 downto 0);
            EQ, GT: out STD_LOGIC );
end comp64;

architecture comp64_arch of comp64 is
begin
  EQ <= '1' when A = B else '0';
  GT <= '1' when A > B else '0';
end comp64_arch;
```

Table 6-28 is a simple behavioral model of a 64-bit comparator with equals and greater-than outputs. This program uses the IEEE std_logic_unsigned package, whose built-in comparison functions automatically treat all signals of type STD_LOGIC_VECTOR as unsigned integers. Although the program is fully synthesizable, the speed and size of the result depends on the "intelligence" of the particular tool that is used.

An alternative is to build the comparator by cascading smaller components, such as 8-bit comparators. Table 6-29 is the behavioral model of an 8-bit comparator. A particular tool may or may not synthesize a very fast comparator from this program, but it's sure to be significantly faster than a 64-bit comparator in any case.

Next, we can write a structural program that instantiates eight of these 8-bit comparators and hooks up their individual outputs through additional logic to calculate the overall comparison result. One way to do this is shown Table 6-30. A generate statement creates not only the individual 8-bit comparators but also

Table 6-29
VHDL program for an
8-bit comparator.

```
library IEEE;
use IEEE.std_logic_1164.all;
use IEEE.std_logic_unsigned.all;

entity comp8 is
    port ( A, B: in STD_LOGIC_VECTOR (7 downto 0);
            EQ, GT: out STD_LOGIC );
end comp8;

architecture comp8_arch of comp8 is
begin
  EQ <= '1' when A = B else '0';
  GT <= '1' when A > B else '0';
end comp8_arch;
```

Table 6-30 VHDL structural architecture for a 64-bit comparator.

```
architecture comp64s_arch of comp64 is
component comp8
  port ( A, B: in STD_LOGIC_VECTOR (7 downto 0);
         EQ, GT: out STD_LOGIC);
end component;
signal EQ8, GT8: STD_LOGIC_VECTOR (7 downto 0); -- =, > for 8-bit slice
signal SEQ, SGT: STD_LOGIC_VECTOR (8 downto 0); -- serial chain of slice results
begin
  SEQ(8) <= '1'; SGT(8) <= '0';
  U1: for i in 7 downto 0 generate
        U2: comp8 port map (A(7+i*8 downto i*8), B(7+i*8 downto i*8), EQ8(i), GT8(i));
        SEQ(i) <= SEQ(i+1) and EQ8(i);
        SGT(i) <= SGT(i+1) or (SEQ(i+1) and GT8(i));
  end generate;
  EQ <= SEQ(0); GT <= SGT(0);
end comp64s_arch;
```

cascading logic that serially builds up the overall result from most significant to least significant stage.

An unsophisticated tool could synthesize a slow iterative comparator circuit for our original 64-bit comparator architecture in Table 6-28. In this situation, the architecture in Table 6-30 yields a faster synthesized circuit because it explicitly "pulls" out the cascading information for each 8-bit slice and combines it in a faster combinational circuit (just 8 levels of AND-OR logic, not 64). A more sophisticated tool might flatten the 8-bit comparator into faster, non-iterative structure similar to the 74x682 MSI comparator (Figure 5-84 on page 425), and it might flatten our iterative cascading logic in Table 6-30 into two-level sum-of-products equations similar to the ones in the ABEL solution on page 487.

6.3.5 Mode-Dependent Comparator

For the next example, let us suppose we have a system in which we need to compare two 32-bit words under normal circumstances, but where we must sometimes ignore one or two low-order bits of the input words. The operating mode is specified by two mode-control bits, M1 and M0, as shown in Table 6-9 on page 488.

The desired functionality can be obtained very easily in VHDL using a case statement to select the behavior by mode, as shown in the program in Table 6-31. This is a perfectly good behavioral description that is also fully synthesizable. However, it has one major drawback in synthesis—it will, in all likelihood, cause the creation of three separate equality and magnitude comparators (32-, 31-, and 30-bit), one for each case in the case statement. The

Table 6-31 VHDL behavioral architecture of a 32-bit mode-dependent comparator.

```
library IEEE;
use IEEE.std_logic_1164.all;
use IEEE.std_logic_unsigned.all;

entity Vmodecmp is
    port ( M: in STD_LOGIC_VECTOR (1 downto 0);      -- mode
           A, B: in STD_LOGIC_VECTOR (31 downto 0); -- unsigned integers
           EQ, GT: out STD_LOGIC );                  -- comparison results
end Vmodecmp;

architecture Vmodecmp_arch of Vmodecmp is
begin
  process (M, A, B)
  begin
    case M is
      when "00" =>
        if A = B then EQ <= '1'; else EQ <= '0'; end if;
        if A > B then GT <= '1'; else GT <= '0'; end if;
      when "01" =>
        if A(31 downto 1) = B(31 downto 1) then EQ <= '1'; else EQ <= '0'; end if;
        if A(31 downto 1) > B(31 downto 1) then GT <= '1'; else GT <= '0'; end if;
      when "10" =>
        if A(31 downto 2) = B(31 downto 2) then EQ <= '1'; else EQ <= '0'; end if;
        if A(31 downto 2) > B(31 downto 2) then GT <= '1'; else GT <= '0'; end if;
      when others => EQ <= '0';  GT <= '0';
    end case;
  end process;
end Vmodecmp_arch;
```

individual comparators may or may not be fast, as discussed in the previous subsection, but we won't worry about speed for this example.

A more efficient alternative is to perform just one comparison for the 30 high-order bits of the inputs, using additional logic that is dependent on mode to give a final result using the low-order bits as necessary. This approach is shown in Table 6-32. Two variables, EQ30 and GT30, are used within the process to hold the results of the comparison of the 30 high-order bits. A case statement similar to the previous architecture's is then used to obtain the final results as a function of the mode. If desired, the speed of the 30-bit comparison can be optimized using the methods discussed in the preceding subsection.

6.3.6 Ones Counter

Several important algorithms include the step of counting the number of "1" bits in a data word. In fact, some microprocessor instruction sets have been extended recently to include ones counting as a basic instruction. In this example, let us suppose that we have a requirement to design a combinational circuit that counts ones in a 32-bit word as part of the arithmetic and logic unit of a microprocessor.

Table 6-32 More efficient architecture for a 32-bit mode-dependent comparator.

```vhdl
architecture Vmodecpe_arch of Vmodecmp is
begin
  process (M, A, B)
  variable EQ30, GT30: STD_LOGIC; -- 30-bit comparison results
  begin
    if A(31 downto 2) = B(31 downto 2) then EQ30 := '1'; else EQ30 := '0'; end if;
    if A(31 downto 2) > B(31 downto 2) then GT30 := '1'; else GT30 := '0'; end if;
    case M is
      when "00" =>
        if EQ30='1' and A(1 downto 0) = B(1 downto 0) then
          EQ <= '1'; else EQ <= '0'; end if;
        if GT30='1' or (EQ30='1' and A(1 downto 0) > B(1 downto 0)) then
          GT <= '1'; else GT <= '0'; end if;
      when "01" =>
        if EQ30='1' and A(1) = B(1) then EQ <= '1'; else EQ <= '0'; end if;
        if GT30='1' or (EQ30='1' and A(1) > B(1)) then
          GT <= '1'; else GT <= '0'; end if;
      when "10" => EQ <= EQ30;  GT <= GT30;
      when others => EQ <= '0';  GT <= '0';
    end case;
  end process;
end Vmodecpe_arch;
```

Ones counting can be described very easily by a behavioral VHDL program, as shown in Table 6-33. This program is fully synthesizable, but it may generate a very slow, inefficient realization with 32 5-bit adders in series.

```vhdl
library IEEE;
use IEEE.std_logic_1164.all;
use IEEE.std_logic_unsigned.all;

entity Vcnt1s is
    port ( D: in STD_LOGIC_VECTOR (31 downto 0);
           SUM: out STD_LOGIC_VECTOR (4 downto 0) );
end Vcnt1s;

architecture Vcnt1s_arch of Vcnt1s is
begin
  process (D)
  variable S: STD_LOGIC_VECTOR(4 downto 0);
  begin
    S := "00000";
    for i in 0 to 31 loop
      if D(i) = '1' then S := S + "00001"; end if;
    end loop;
    SUM <= S;
  end process;
end Vcnt1s_arch;
```

Table 6-33
Behavioral VHDL program for a 32-bit ones counter.

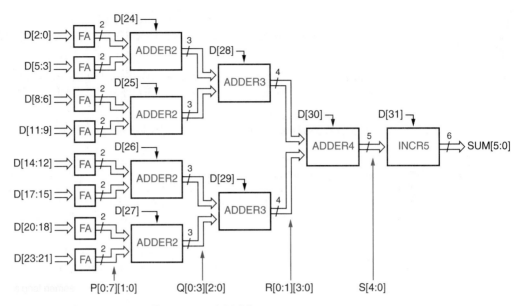

Figure 6-15 Structure of 32-bit ones counter.

To synthesize a more efficient realization of the ones counter, we must come up with an efficient structure and then write an architecture that describes it. Such a structure is the adder tree shown in Figure 6-15. A full adder (FA) adds three input bits to produce a 2-bit sum. Pairs of 2-bit numbers are added by 2-bit adders (ADDER2), each of which also has a carry input that can include add another 1-bit input to its sum. The resulting 3-bit sums are combined by 3-bit adders (ADDER3), and the final pair of 4-bit sums are combined in a 4-bit adder (ADDER4). By making use of the available carry inputs, this tree structure can combine 31 bits. A separate 5-bit incrementer is used at the end to handle the one remaining input bit.

The structure of Figure 6-15 can be created nicely by a structural VHDL architecture, as shown in Table 6-34. The program begins by declaring all of the components that will be used in the design, corresponding to the blocks in the figure.

The letter under each column of signals in Figure 6-15 corresponds to the name used for that signal in the program. Each of signals P, Q, and R is an array with one STD_LOGIC_VECTOR per connection in the corresponding column. The program defines a corresponding type for each of these, followed by the actual signal declaration.

The program in Table 6-34 makes good use of generate statements to create the multiple adder components on the lefthand side of the figure—eight FAs, four ADDER2s, and two ADDER3s. Finally, it instantiates one each of ADDER4 and INCR5.

```
architecture Vcnt1str_arch of Vcnt1str is

component FA port ( A, B, CI: in  STD_LOGIC;
                    S, CO:    out STD_LOGIC );
end component;

component ADDER2 port ( A, B: in  STD_LOGIC_VECTOR(1 downto 0);
                        CI:  in  STD_LOGIC;
                        S:   out STD_LOGIC_VECTOR(2 downto 0) );
end component;

component ADDER3 port ( A, B: in  STD_LOGIC_VECTOR(2 downto 0);
                        CI:  in  STD_LOGIC;
                        S:   out STD_LOGIC_VECTOR(3 downto 0) );
end component;

component ADDER4 port ( A, B: in  STD_LOGIC_VECTOR(3 downto 0);
                        CI:  in  STD_LOGIC;
                        S:   out STD_LOGIC_VECTOR(4 downto 0) );
end component;

component INCR5 port ( A:  in  STD_LOGIC_VECTOR(4 downto 0);
                       CI: in  STD_LOGIC;
                       S:  out STD_LOGIC_VECTOR(5 downto 0) );
end component;

type Ptype is array (0 to 7) of STD_LOGIC_VECTOR(1 downto 0);
type Qtype is array (0 to 3) of STD_LOGIC_VECTOR(2 downto 0);
type Rtype is array (0 to 1) of STD_LOGIC_VECTOR(3 downto 0);
signal P: Ptype;  signal Q: Qtype;  signal R: Rtype;
signal S: STD_LOGIC_VECTOR(4 downto 0);

begin
  U1: for i in 0 to 7 generate
    U1C: FA port map (D(3*i), D(3*i+1), D(3*i+2), P(i)(0), P(i)(1));
  end generate;
  U2: for i in 0 to 3 generate
    U2C: ADDER2 port map (P(2*i), P(2*i+1), D(24+i), Q(i));
  end generate;
  U3: for i in 0 to 1 generate
    U3C: ADDER3 port map (Q(2*i), Q(2*i+1), D(28+i), R(i));
  end generate;
  U4: ADDER4 port map (R(0), R(1), D(30), S);
  U5: INCR5 port map (S, D(31), SUM);
end Vcnt1str_arch;
```

Table 6-34
VHDL structural architecture for a 32-bit ones counter.

The definitions of the ones counter's individual component entities and architectures, from FA to INCR, can be made in separate structural or behavioral programs. For example, Table 6-35 is a structural program for FA. The rest of the components are left as exercises (6.20–6.22).

Table 6-35
Structural VHDL program for a full adder.

```
library IEEE;
use IEEE.std_logic_1164.all;

entity FA is
    port ( A, B, CI: in  STD_LOGIC;
           S, CO:    out STD_LOGIC );
end FA;

architecture FA_arch of FA is
begin
  S <= A xor B xor CI;
  CO <= (A and B) or (A and CI) or (B and CI);
end FA_arch;
```

6.3.7 Tic-Tac-Toe

Our last example is the design of a combinational circuit that picks a player's next move in the game of Tic-Tac-Toe. In case you're not familiar with the game, the rules are explained in the box on page 492. We'll repeat here our strategy for playing and winning the game:

1. Look for a row, column, or diagonal that has two of my marks (X or O, depending on which player I am) and one empty cell. If one exists, place my mark in the empty cell; I win!

2. Else, look for a row, column, or diagonal that has two of my opponent's marks and one empty cell. If one exists, place my mark in the empty cell to block a potential win by my opponent.

3. Else, pick a cell based on experience. For example, if the middle cell is open, it's usually a good bet to take it. Otherwise, the corner cells are good bets. Intelligent players can also notice and block a developing pattern by the opponent or "look ahead" to pick a good move.

To avoid confusion between "O" and "0" in our programs, we'll call the second player "Y". Now we can think about how to encode the inputs and outputs of the circuit. The inputs represent the current state of the playing grid. There are nine cells, and each cell has one of three possible states (empty, occupied by X, occupied by Y). The outputs represent the move to make, assuming that it is X's turn. There are only nine possible moves that a player can make, so the output can be encoded in just four bits.

There are several choices of how to code the state of one cell. Because the game is symmetric, we choose a symmetric encoding that can help us later:

00 Cell is empty.

10 Cell is occupied by X.

01 Cell is occupied by Y.

```
library IEEE;
use IEEE.std_logic_1164.all;

package TTTdefs is

type TTTgrid is array (1 to 3) of STD_LOGIC_VECTOR(1 to 3);
subtype TTTmove is STD_LOGIC_VECTOR (3 downto 0);

constant MOVE11: TTTmove := "1000";
constant MOVE12: TTTmove := "0100";
constant MOVE13: TTTmove := "0010";
constant MOVE21: TTTmove := "0001";
constant MOVE22: TTTmove := "1100";
constant MOVE23: TTTmove := "0111";
constant MOVE31: TTTmove := "1011";
constant MOVE32: TTTmove := "1101";
constant MOVE33: TTTmove := "1110";
constant NONE:   TTTmove := "0000";

end TTTdefs;
```

Table 6-36
VHDL package with definitions for the Tic-Tac-Toe project.

So, we can encode the 3 × 3 grid's state in 18 bits. Since VHDL supports arrays, it is useful to define an array type, TTTgrid, that contains elements corresponding to the cells in the grid. Since this type will be used throughout our Tic-Tac-Toe project, it is convenient to put this definition, along with several others that we'll come to, in a VHDL package, as shown in Table 6-36.

It would be natural to define TTTgrid as a two-dimensional array of STD_LOGIC, but not all VHDL tools support two-dimensional arrays. Instead, we define it as an array of 3-bit STD_LOGIC_VECTORs, which is almost the same thing. To represent the Tic-Tac-Toe grid, we'll use two signals X and Y of this type, where an element of a variable is 1 if the like-named player has a mark in the corresponding cell. Figure 6-16 shows the correspondence between signal names and cells in the grid.

Figure 6-16
Tic-Tac-Toe grid and VHDL signal names.

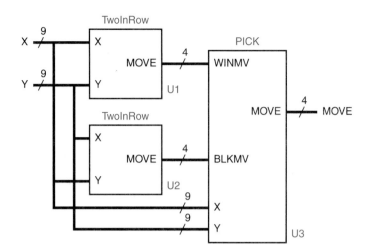

Figure 6-17
Entity partitioning for
the Tic-Tac-Toe game.

The package in Table 6-36 also defines a 4-bit type TTTmove for encoded moves. A player has nine possible moves, and one more code is used for the case where no move is possible. This particular coding was chosen and used in the package for no other reason than that it's the same coding that was used in the ABEL version of this example in Section 6.2.7. By defining the coding in the package, we can easily change the definition later without having to change the entities that use it (for example, see Exercise 6.23).

Rather than try to design the Tic-Tac-Toe move-finding circuit as a single monolithic entity, it makes sense for us to try to partition it into smaller pieces. In fact, partitioning it along the lines of the three-step strategy at the beginning of this section seems like a good idea.

We note that steps 1 and 2 of our strategy are very similar; they differ only in reversing the roles of the player and the opponent. An entity that finds a winning move for me can also find a blocking move for my opponent. Looking at this characteristic from another point of view, an entity that finds a winning move for me can find a blocking move for me if the encodings for me and my opponent are swapped. Here's where our symmetric encoding pays off—we can swap players merely by swapping signals X and Y.

With this in mind, we can use two copies of the same entity, TwoInRow, to perform steps 1 and 2 as shown in Figure 6-17. Notice that signal X is connected to the top input of the first TwoInRow entity, but to the bottom input of the second. A third entity, PICK, picks a winning move if one is available from U1, else it picks a blocking move if available from U2, else it uses "experience" (step 3) to pick a move.

Table 6-37 is a structural VHDL program for the top-level entity, GETMOVE. In addition to the IEEE std_logic_1164 package, it uses our TTTdefs package. Notice that the "use" clause for the TTTdefs packages specifies that it is stored in the "work" library, which is automatically created for our project.

```
library IEEE;
use IEEE.std_logic_1164.all;
use work.TTTdefs.all;

entity GETMOVE is
    port ( X, Y: in  TTTgrid;
           MOVE: out TTTmove );
end GETMOVE;

architecture GETMOVE_arch of GETMOVE is

component TwoInRow port ( X, Y: in  TTTgrid;
                          MOVE: out STD_LOGIC_VECTOR(3 downto 0) );
end component;

component PICK port ( X, Y:        in  TTTgrid;
                      WINMV, BLKMV: in  STD_LOGIC_VECTOR(3 downto 0);
                      MOVE:         out STD_LOGIC_VECTOR(3 downto 0) );
end component;

signal WIN, BLK: STD_LOGIC_VECTOR(3 downto 0);

begin
  U1: TwoInRow port map (X, Y, WIN);
  U2: TwoInRow port map (Y, X, BLK);
  U3: PICK port map (X, Y, WIN, BLK, MOVE);
end GETMOVE_arch;
```

Table 6-37
Top-level structural
VHDL entity for
picking a move in
Tic-Tac-Toe.

The architecture in Table 6-37 declares and uses just two components, TwoInRow and PICK, which will be defined shortly. The only internal signals are WIN and BLK, which pass winning and blocking moves from the two instances of TwoInRow to PICK, as in Figure 6-17. The statement part of the architecture has just three statements to instantiate the three blocks in the figure.

Now comes the interesting part, the design of the individual entities in Figure 6-17. We'll start with TwoInRow, since it accounts for two-thirds of the design. Its entity definition is very simple, as shown in Table 6-38. But there's plenty to discuss about its architecture, shown in Table 6-39.

```
library IEEE;
use IEEE.std_logic_1164.all;
use work.TTTdefs.all;

entity TwoInRow is
    port ( X, Y: in  TTTgrid;
           MOVE: out TTTmove );
end TwoInRow;
```

Table 6-38
Declaration of
TwoInRow entity.

◼ **Table 6-39** Architecture of TwoInRow entity.

```
architecture TwoInRow_arch of TwoInRow is

function R(X, Y: TTTgrid; i, j: INTEGER) return BOOLEAN is
  variable result: BOOLEAN;
  begin                                 -- Find 2-in-row with empty cell i,j
    result := TRUE;
    for jj in 1 to 3 loop
      if jj = j then result := result and X(i)(jj)='0' and Y(i)(jj)='0';
      else result := result and X(i)(jj)='1'; end if;
    end loop;
    return result;
  end R;

function C(X, Y: TTTgrid; i, j: INTEGER) return BOOLEAN is
  variable result: BOOLEAN;
  begin                                 -- Find 2-in-column with empty cell i,j
    result := TRUE;
    for ii in 1 to 3 loop
      if ii = i then result := result and X(ii)(j)='0' and Y(ii)(j)='0';
      else result := result and X(ii)(j)='1'; end if;
    end loop;
    return result;
  end C;

function D(X, Y: TTTgrid; i, j: INTEGER) return BOOLEAN is
  variable result: BOOLEAN;          -- Find 2-in-diagonal with empty cell i,j.
  begin                              -- This is for 11, 22, 33 diagonal.
    result := TRUE;
    for ii in 1 to 3 loop
      if ii = i then result := result and X(ii)(ii)='0' and Y(ii)(ii)='0';
      else result := result and X(ii)(ii)='1'; end if;
    end loop;
    return result;
  end D;

function E(X, Y: TTTgrid; i, j: INTEGER) return BOOLEAN is
  variable result: BOOLEAN;          -- Find 2-in-diagonal with empty cell i,j.
  begin                              -- This is for 13, 22, 31 diagonal.
    result := TRUE;
    for ii in 1 to 3 loop
      if ii = i then result := result and X(ii)(4-ii)='0' and Y(ii)(4-ii)='0';
      else result := result and X(ii)(4-ii)='1'; end if;
    end loop;
    return result;
  end E;
```

Table 6-39
(continued)

```
begin
  process (X, Y)
  variable G11, G12, G13, G21, G22, G23, G31, G32, G33: BOOLEAN;
  begin
    G11 := R(X,Y,1,1) or C(X,Y,1,1) or D(X,Y,1,1);
    G12 := R(X,Y,1,2) or C(X,Y,1,2);
    G13 := R(X,Y,1,3) or C(X,Y,1,3) or E(X,Y,1,3);
    G21 := R(X,Y,2,1) or C(X,Y,2,1);
    G22 := R(X,Y,2,2) or C(X,Y,2,2) or D(X,Y,2,2) or E(X,Y,2,2);
    G23 := R(X,Y,2,3) or C(X,Y,2,3);
    G31 := R(X,Y,3,1) or C(X,Y,3,1) or E(X,Y,3,1);
    G32 := R(X,Y,3,2) or C(X,Y,3,2);
    G33 := R(X,Y,3,3) or C(X,Y,3,3) or D(X,Y,3,3);
    if    G11 then MOVE <= MOVE11;
    elsif G12 then MOVE <= MOVE12;
    elsif G13 then MOVE <= MOVE13;
    elsif G21 then MOVE <= MOVE21;
    elsif G22 then MOVE <= MOVE22;
    elsif G23 then MOVE <= MOVE23;
    elsif G31 then MOVE <= MOVE31;
    elsif G32 then MOVE <= MOVE32;
    elsif G33 then MOVE <= MOVE33;
    else           MOVE <= NONE;
    end if;
  end process;
end TwoInRow_arch;
```

The architecture defines several functions, each of which determines whether there is a winning move (from X's point of view) in a particular cell i,j. A winning move exists if cell i,j is empty and the other two cells in the same row, column, or diagonal contain an X. Functions R and C look for winning moves in cell i,j's row and column, respectively. Functions D and E look in the two diagonals.

Within the architecture's single process, nine BOOLEAN variables G11–G33 are declared to indicate whether each of the cells has a winning move possible. Assignment statements at the beginning of the process set each variable to TRUE if there is such a move, calling and combining all of the appropriate functions for cell i,j.

The rest of the process is a deeply nested "if" statement that looks for a winning move in all possible cells. Although it typically results in slower synthesized logic nested "if" is required rather than some form of "case" statement, because multiple moves may be possible. If no winning move is possible, the value "NONE" is assigned.

EXPLICIT IMPURITY In addition to a cell index i,j, the functions R, C, D, and E in Table 6-39 are passed the grid state X and Y. This is necessary because VHDL functions are by default *pure*, which means that signals and variables declared in the function's parents are *not* directly visible within the function. However, you can relax this restriction by explicitly declaring a function to be *impure* by placing the keyword impure before the keyword function in its definition.

The PICK entity combines the results of two TwoInRow entities according to the program in Table 6-40. First priority is given to a winning move, followed by a blocking move. Otherwise, function MT is called for each cell, starting with the middle and ending with the side cells, to find an available move. This completes the design of the Tic-Tac-Toe circuit.

Table 6-40
VHDL program to pick a winning or blocking Tic-Tac-Toe move or else use "experience."

```
library IEEE;
use IEEE.std_logic_1164.all;
use work.TTTdefs.all;

entity PICK is
  port ( X, Y:        in  TTTgrid;
         WINMV, BLKMV: in  STD_LOGIC_VECTOR(3 downto 0);
         MOVE:        out STD_LOGIC_VECTOR(3 downto 0) );
end PICK;

architecture PICK_arch of PICK is
function MT(X, Y: TTTgrid; i, j: INTEGER) return BOOLEAN is
  begin                          -- Determine if cell i,j is empty
    return X(i)(j)='0' and Y(i)(j)='0';
  end MT;
begin
  process (X, Y, WINMV, BLKMV)
  begin                                    -- If available, pick:
    if    WINMV /= NONE then MOVE <= WINMV;  -- winning move
    elsif BLKMV /= NONE then MOVE <= BLKMV;  -- else blocking move
    elsif MT(X,Y,2,2)   then MOVE <= MOVE22; -- else center cell
    elsif MT(X,Y,1,1)   then MOVE <= MOVE11; -- else corner cells
    elsif MT(X,Y,1,3)   then MOVE <= MOVE13;
    elsif MT(X,Y,3,1)   then MOVE <= MOVE31;
    elsif MT(X,Y,3,3)   then MOVE <= MOVE33;
    elsif MT(X,Y,1,2)   then MOVE <= MOVE12; -- else side cells
    elsif MT(X,Y,2,1)   then MOVE <= MOVE21;
    elsif MT(X,Y,2,3)   then MOVE <= MOVE23;
    elsif MT(X,Y,3,2)   then MOVE <= MOVE32;
    else                     MOVE <= NONE;   -- else grid is full
    end if;
  end process;
end PICK_arch;
```

Exercises

6.1   Explain how the 16-bit barrel shifter of Section 6.1.1 can be realized with a combination of 74x157s and 74x151s. How does this approach compare with the others in delay and parts count?

6.2   Show how the 16-bit barrel shifter of Section 6.1.1 can be realized in eight identical GAL22V10s.

6.3   Find a coding of the shift amounts (S[3:0]) and modes (C[2:0]) in the barrel shifter of Table 6-3 that further reduces the total number of product terms used by the design.

6.4   Make changes to the dual-priority encoder program of Table 6-6 to further reduce the number of product terms required. State whether your changes increase the delay of the circuit when realized in a GAL22V10. Can you reduce the product terms enough to fit the design into a GAL16V8?

6.5   Here's an exercise where you can use your brain, like the author had to when figuring out the equation for the SUM0 output in Table 6-12. Do each of the SUM1–SUM3 outputs require more terms or fewer terms than SUM0?

6.6   Complete the design of the ABEL and PLD-based ones-counting circuit that was started in Section 6.2.6. Use 22V10 or smaller PLDs and try to minimize the total number of PLDs required. State the total delay of your design in terms of the worst-case number of PLD delays in a signal path from input to output.

6.7   Find another code for the Tic-Tac-Toe moves in Table 6-13 that has the same rotation properties as the original code. That is, it should be possible to compensate for a 180° rotation of the grid using just inverters and wire rearrangement. Determine whether the TWOINHAF equations will still fit in a single 22V10 using the new code.

6.8   Using a simulator, demonstrate a sequence of moves in which the PICK2 PLD in Table 6-16 will lose a Tic-Tac-Toe game, even if X goes first.

6.9   Modify the program in Table 6-16 to give the program a better chance of winning, or at least not losing. Can your new program still lose?

6.10   Modify both the "other logic" in Figure 6-13 and the program in Table 6-16 to give the program a better chance of winning, or at least not losing. Can your new program still lose?

6.11   Write the VHDL functions for Vror, Vsll, Vsrl, Vsla, and Vsra in Table 6-17 using the ror, sll, srl, sla, and sra operations as defined in Table 6-3.

6.12   The iterative-circuit version of fixup in Table 6-20 has a worst-case delay path of 15 OR gates from the first decoded value of i (14) to the FSEL(0) signal. Figure out a trick that cuts this delay path almost in half with no cost (or negative cost) in gates. How can this trick be extended further to save gates or gate inputs?

6.13   Rewrite the barrel16 entity definition in Table 6-17 and the architecture in Table 6-22 so that a single direction-control bit is made explicitly available to the architecture.

Figure X6.14

6.14 Rewrite the `barrel16` architecture definition in Table 6-22 to use the structure shown in Figure X6.14. Use the existing ROL16 and FIXUP entities; it's up to you to come up with MAGIC and the other logic.

6.15 Write a semibehavioral or structural version of the `fpencr_arch` architecture of Table 6-25 that generates only one adder in synthesis and that does not generate multiple 10-bit comparators for the nested "`if`" statement.

6.16 Repeat Exercise 6.15, including a structural definition of an efficient rounding circuit that performs the `round` function. Your circuit should require significantly fewer gates than a 4-bit adder.

6.17 Redesign the VHDL dual-priority encoder of Section 6.3.3 to get better, known performance, as suggested in the last paragraph of the section.

6.18 Write a structural VHDL architecture for a 64-bit comparator that is similar to Table 6-30 except that it builds up the comparison result serially from least to most significant stage.

6.19 What significant change occurs in the synthesis of the VHDL program in Table 6-31 if we change the statements in the "`when others`" case to "`null`"?

6.20 Write behavioral VHDL programs for the "ADDERx" components used in Table 6-34.

6.21 Write a structural VHDL programs for the "ADDERx" components in Table 6-34. Use a generic definition so that the same entity can be instantiated for ADDER2, ADDER3, and ADDER5, and show what changes must be made in Table 6-34 to do this.

6.22 Write a structural VHDL program for the "INCR5" component in Table 6-34.

6.23 Using an available VHDL synthesis tool, synthesize the Tic-Tac-Toe design of Section 6.3.7, fit it into an available FPGA, and determine how many internal resources it uses. Then try to reduce the resource requirements by specifying a different encoding of the moves in the TTTdefs package.

6.24 The Tic-Tac-Toe program in Section 6.3.7 eventually loses against an intelligent opponent if applied to the grid state shown in Figure X6.24. Use an available VHDL simulator to prove that this is true. Then modify the PICK entity to win in this and similar situations and verify your design using the simulator.

Figure X6.24

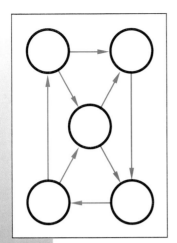

Sequential
Logic Design Principles

L
ogic circuits are classified into two types, "combinational" and "sequential." A *combinational* logic circuit is one whose outputs depend only on its current inputs. The rotary channel selector knob on an old-fashioned TV is like a combinational circuit—its "output" selects a channel based only on its current "input"—the position of the knob.

A *sequential* logic circuit is one whose outputs depend not only on its current inputs, but also on the past sequence of inputs, possibly arbitrarily far back in time. The circuit controlled by the channel-up and channel-down pushbuttons on a TV or VCR is a sequential circuit—the channel selection depends on the past sequence of up/down pushes, at least since when you started viewing 10 hours before, and perhaps as far back as when you first plugged the device into the wall.

So it is inconvenient, and often impossible, to describe the behavior of a sequential circuit by means of a table that lists outputs as a function of the input sequence that has been received up until the current time. To know where you're going next, you need to know where you are now. With the TV channel selector, it is impossible to determine what channel is currently selected by looking only at the preceding sequence of presses on the up and down pushbuttons, whether we look at the preceding 10 presses or the preceding 1,000. More information, the current "state" of the channel selector, is needed. Probably the best definition of "state" that I've seen appeared in Herbert Hellerman's book on *Digital Computer System Principles* (McGraw-Hill, 1967):

state
state variable

The *state* of a sequential circuit is a collection of *state variables* whose values at any one time contain all the information about the past necessary to account for the circuit's future behavior.

In the channel-selector example, the current channel number is the current state. Inside the TV, this state might be stored as seven binary state variables representing a decimal number between 0 and 127. Given the current state (channel number), we can always predict the next state as a function of the inputs (presses of the up/down pushbuttons). In this example, one highly visible output of the sequential circuit is an encoding of the state itself—the channel-number display. Other outputs, internal to the TV, may be combinational functions of the state alone (e.g., VHF/UHF/cable tuner selection) or of both state and input (e.g., turning off the TV if the current state is 0 and the "down" button is pressed).

State variables need not have direct physical significance, and there are often many ways to choose them to describe a particular sequential circuit. For example, in the TV channel selector, the state might be stored as three BCD digits or 12 bits, with many of the bit combinations (4,096 possible) going unused.

In a digital logic circuit, state variables are binary values, corresponding to certain logic signals in the circuit, as we'll see in later sections. A circuit with n binary state variables has 2^n possible states. As large as it might be, 2^n is always finite, never infinite, so sequential circuits are sometimes called *finite-state machines*.

finite-state machine

The state changes of most sequential circuits occur at times specified by a free-running *clock* signal. Figure 7-1 gives timing diagrams and nomenclature for typical clock signals. By convention, a clock signal is active high if state changes occur at the clock's rising edge or when the clock is HIGH, and active low in the complementary case. The *clock period* is the time between successive transitions in the same direction, and the *clock frequency* is the reciprocal of the period. The first edge or pulse in a clock period or sometimes the period itself is called a *clock tick*. The *duty cycle* is the percentage of time that the clock signal

clock

clock period
clock frequency
clock tick
duty cycle

NON-FINITE-STATE MACHINES

A group of mathematicians recently proposed a non-finite-state machine, but they're still busy listing its states. . . . Sorry, that's just a joke. There *are* mathematical models for infinite-state machines, such as Turing machines. They typically contain a small finite-state-machine control unit, and an infinite amount of auxiliary memory, such as an endless tape.

(a)

state changes occur here

CLK

t_H t_L

t_{per}

period = t_{per}
frequency = 1 / t_{per}
duty cycle = t_H / t_{per}

(b)

state changes occur here

CLK_L

t_L t_H

t_{per}

duty cycle = t_L / t_{per}

Figure 7-1
Clock signals:
(a) active high;
(b) active low.

is at its asserted level. Typical digital systems, from digital watches to supercomputers, use a quartz-crystal oscillator to generate a free-running clock signal. Clock frequencies might range from 32.768 kHz (for a watch) to 500 MHz (for a CMOS RISC microprocessor with a cycle time of 2 ns); "typical" systems using TTL and CMOS parts have clock frequencies in the 5–150 MHz range.

In this chapter we'll discuss two types of sequential circuits that account for the majority of practical discrete designs. A *feedback sequential circuit* uses ordinary gates and feedback loops to obtain memory in a logic circuit, thereby creating sequential-circuit building blocks such as latches and flip-flops that are used in higher-level designs. A *clocked synchronous state machine* uses these building blocks, in particular edge-triggered D flip-flops, to create circuits whose inputs are examined and whose outputs change in accordance with a controlling clock signal. Other sequential circuit types, such as general fundamental mode, multiple-pulse mode, and multiphase circuits, are sometimes useful in high-performance systems and VLSI and are discussed in advanced texts.

feedback sequential circuit

clocked synchronous state machine

7.1 Bistable Elements

The simplest sequential circuit consists of a pair of inverters forming a feedback loop, as shown in Figure 7-2. It has *no* inputs and two outputs, Q and Q_L.

7.1.1 Digital Analysis

The circuit of Figure 7-2 is often called a *bistable*, since a strictly digital analysis shows that it has two stable states. If Q is HIGH, then the bottom inverter has a HIGH input and a LOW output, which forces the top inverter's output HIGH as we assumed in the first place. But if Q is LOW, then the bottom inverter has a LOW input and a HIGH output, which forces Q LOW, another stable situation. We could use a single state variable, the state of signal Q, to describe the state of the circuit; there are two possible states, Q = 0 and Q = 1.

bistable

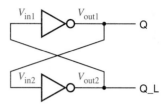

Figure 7-2
A pair of inverters forming
a bistable element.

The bistable element is so simple that it has no inputs and therefore no way of controlling or changing its state. When power is first applied to the circuit, it randomly comes up in one state or the other and stays there forever. Still, it serves our illustrative purposes very well, and we *will* actually show a couple of applications for it in Sections 8.2.3 and 8.2.4.

7.1.2 Analog Analysis

The analysis of the bistable has more to reveal if we consider its operation from an analog point of view. The dark line in Figure 7-3 shows the steady-state (DC) transfer function T for a single inverter; the output voltage is a function of input voltage, $V_{out} = T(V_{in})$. With two inverters connected in a feedback loop as in Figure 7-2, we know that $V_{in1} = V_{out2}$ and $V_{in2} = V_{out1}$; therefore, we can plot the transfer functions for both inverters on the same graph with an appropriate labeling of the axes. Thus, the dark line is the transfer function for the top inverter in Figure 7-2, and the colored line is the transfer function for the bottom one.

Considering only the steady-state behavior of the bistable's feedback loop, and not dynamic effects, the loop is in equilibrium if the input and output voltages of both inverters are constant DC values consistent with the loop connection and the inverters' DC transfer function. That is, we must have

$$
\begin{aligned}
V_{in1} &= V_{out2} \\
&= T(V_{in2}) \\
&= T(V_{out1}) \\
&= T(T(V_{in1}))
\end{aligned}
$$

Figure 7-3
Transfer functions for
inverters in a bistable
feedback loop.

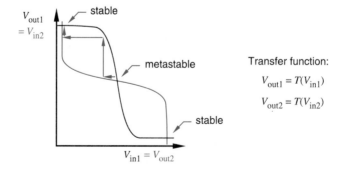

Likewise, we must have

$$V_{in2} = T(T(V_{in2}))$$

We can find these equilibrium points graphically from Figure 7-3; they are the points at which the two transfer curves meet. Surprisingly, we find that there are not two but *three* equilibrium points. Two of them, labeled *stable*, correspond to the two states that our "strictly digital" analysis identified earlier, with Q either 0 (LOW) or 1 (HIGH).

stable

The third equilibrium point, labeled *metastable*, occurs with V_{out1} and V_{out2} about halfway between a valid logic 1 voltage and a valid logic 0 voltage; so Q and Q_L are not valid logic signals at this point. Yet the loop equations are satisfied; if we can get the circuit to operate at the metastable point, it could theoretically stay there indefinitely.

metastable

7.1.3 Metastable Behavior

Closer analysis of the situation at the metastable point shows that it is aptly named. It is not truly stable, because random noise will tend to drive a circuit that is operating at the metastable point toward one of the stable operating points, as we'll now demonstrate.

Suppose the bistable is operating precisely at the metastable point in Figure 7-3. Now let us assume that a small amount of circuit noise reduces V_{in1} by a tiny amount. This tiny change causes V_{out1} to *increase* by a small amount. But since V_{out1} produces V_{in2}, we can follow the first horizontal arrow from near the metastable point to the second transfer characteristic, which now demands a lower voltage for V_{out2}, which is V_{in1}. Now we're back where we started, except we have a much larger change in voltage at V_{in1} than the original noise produced, and the operating point is still changing. This "regenerative" process continues until we reach the stable operating point at the upper lefthand corner of Figure 7-3. However, if we perform a "noise" analysis for either of the stable operating points, we find that feedback brings the circuit back toward the stable operating point, rather than away from it.

Metastable behavior of a bistable can be compared to the behavior of a ball dropped onto a hill, as shown in Figure 7-4. If we drop a ball from overhead, it will probably roll down immediately to one side of the hill or the other. But if it lands right at the top, it may precariously sit there for a while before random

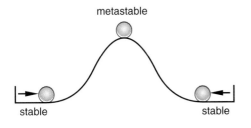

Figure 7-4
Ball and hill analogy for metastable behavior.

forces (wind, rodents, earthquakes) start it rolling down the hill. Like the ball at the top of the hill, the bistable may stay in the metastable state for an unpredictable length of time before nondeterministically settling into one stable state or the other.

If the *simplest* sequential circuit is susceptible to metastable behavior, you can be sure that *all* sequential circuits are susceptible. And this behavior is not something that only occurs at power-up.

Returning to the ball-and-hill analogy, consider what happens if we try to kick the ball from one side of the hill to the other. Apply a strong force (Arnold Schwarzenegger), and the ball goes right over the top and lands in a stable resting place on the other side. Apply a weak force (Mr. Rogers), and the ball falls back to its original starting place. But apply a wishy-washy force (Charlie Brown), and the ball goes to the top of the hill, teeters, and eventually falls back to one side or the other.

This behavior is completely analogous to what happens to flip-flops under marginal triggering conditions. For example, we'll soon study S-R flip-flops, where a pulse on the S input forces the flip-flop from the 0 state to the 1 state. A minimum pulse width is specified for the S input. Apply a pulse of this width or longer, and the flip-flop immediately goes to the 1 state. Apply a very short pulse, and the flip-flop stays in the 0 state. Apply a pulse just under the minimum width, and the flip-flop may go into the metastable state. Once the flip-flop is in the metastable state, its operation depends on "the shape of its hill." Flip-flops built from high-gain, fast technologies tend to come out of metastability faster than ones built from low-performance technologies.

We'll say more about metastability in the next section in connection with specific flip-flop types, and in Section 8.9 with respect to synchronous design methodology and synchronizer failure.

7.2 Latches and Flip-Flops

Latches and flip-flops are the basic building blocks of most sequential circuits. Typical digital systems use latches and flip-flops that are prepackaged, functionally specified devices in a standard integrated circuit. In ASIC design environments, latches and flip-flops are typically predefined cells specified by the ASIC vendor. However, within a standard IC or an ASIC, each latch or flip-flop cell is typically designed as a feedback sequential circuit using individual logic gates and feedback loops. We'll study these discrete designs for two reasons—to understand the behavior of the prepackaged elements better, and to gain the capability of building a latch or flip-flop "from scratch," as is required occasionally in digital-design practice and often in digital-design exams.

flip-flop   All digital designers use the name *flip-flop* for a sequential device that normally samples its inputs and changes its outputs only at times determined by a

S	R	Q	QN
0	0	last Q	last QN
0	1	0	1
1	0	1	0
1	1	0	0

(a)   (b)

Figure 7-5
S-R latch: (a) circuit design using NOR gates; (b) function table.

clocking signal. On the other hand, most digital designers use the name *latch* for   *latch*
a sequential device that watches all of its inputs continuously and changes its
outputs at any time, independent of a clocking signal. We follow this standard
convention in this text. However, some textbooks and digital designers may
(incorrectly) use the name "flip-flop" for a device that we call a "latch."

In any case, because the functional behaviors of latches and flip-flops are
quite different, it is important for the logic designer to know which type is being
used in a design, either from the device's part number (e.g., 74x374 vs. 74x373)
or from other contextual information. We discuss the most commonly used types
of latches and flip-flops in the following subsections.

7.2.1 S-R Latch

An *S-R (set-reset) latch* based on NOR gates is shown in Figure 7-5(a). The   *S-R latch*
circuit has two inputs, S and R, and two outputs, labeled Q and QN, where QN
is normally the complement of Q. Signal QN is sometimes labeled $\overline{Q}$ or Q_L.

If S and R are both 0, the circuit behaves like the bistable element—we
have a feedback loop that retains one of two logic states, Q = 0 or Q = 1. As
shown in Figure 7-5(b), either S or R may be asserted to force the feedback loop
to a desired state. S *sets* or *presets* the Q output to 1; R *resets* or *clears* the Q   *set*
output to 0. After the S or R input is negated, the latch remains in the state that it   *preset*
was forced into. Figure 7-6(a) shows the functional behavior of an S-R latch for   *reset*
a typical sequence of inputs. Colored arrows indicate causality, that is, which   *clear*
input transitions cause which output transitions.

Figure 7-6 Typical operation of an S-R latch: (a) "normal" inputs; (b) S and R asserted simultaneously.

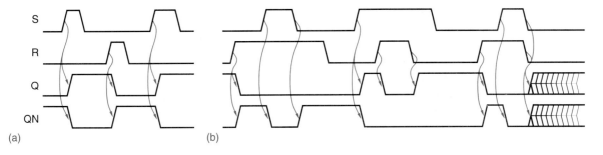

(a)   (b)

Q VERSUS QN   In most applications of an S-R latch, the QN (a.k.a. $\overline{Q}$) output is always the complement of the Q output. However, the $\overline{Q}$ name is not quite correct, because there is one
case where this output is not the complement of Q. If both S and R are 1, as they are
in several places in Figure 7-6(b), then both outputs are forced to 0. Once we negate
either input, the outputs return to complementary operation as usual. However, if we
negate both inputs simultaneously, the latch goes to an unpredictable next state, and it
may in fact oscillate or enter the metastable state. Metastability may also occur if a 1
pulse that is too short is applied to S or R.

Three different logic symbols for the same S-R latch circuit are shown in
Figure 7-7. The symbols differ in the treatment of the complemented output.
Historically, the first symbol was used, showing the active-low or complemented
signal name inside the function rectangle. However, in bubble-to-bubble logic
design the second form of the symbol is preferred, showing an inversion bubble
outside the function rectangle. The last form of the symbol is obviously wrong.

propagation delay   Figure 7-8 defines timing parameters for an S-R latch. The *propagation
delay* is the time it takes for a transition on an input signal to produce a transition
on an output signal. A given latch or flip-flop may have several different
propagation-delay specifications, one for each pair of input and output signals.
Also, the propagation delay may be different depending on whether the output
makes a LOW-to-HIGH or HIGH-to-LOW transition. With an S-R latch, a
LOW-to-HIGH transition on S can cause a LOW-to-HIGH transition on Q, so a
propagation delay $t_{\text{pLH(SQ)}}$ occurs, as shown in transition 1 in the figure.
Similarly, a LOW-to-HIGH transition on R can cause a HIGH-to-LOW transition
on Q, with propagation delay $t_{\text{pHL(RQ)}}$ as shown in transition 2. Not shown in the
figure are the corresponding transitions on QN, which would have propagation
delays $t_{\text{pHL(SQN)}}$ and $t_{\text{pLH(RQN)}}$.

minimum pulse width   Minimum-pulse-width specifications are usually given for the S and R
inputs. As shown in Figure 7-8, the latch may go into the metastable state and
remain there for a random length of time if a pulse shorter than the minimum
width $t_{\text{pw(min)}}$ is applied to S or R. The latch can be deterministically brought out
of the metastable state only by applying a pulse to S or R that meets or exceeds
the minimum-pulse-width requirement.

Figure 7-7
Symbols for an S-R latch:
(a) without bubble;
(b) preferred for bubble-
to-bubble design;
(c) incorrect because
of double negation.

(a)   (b)   (c)

HOW CLOSE IS CLOSE?	As mentioned in the previous note, an S-R latch may go into the metastable state if S and R are negated simultaneously. Often, but not always, a commercial latch's specifications define "simultaneously" (e.g., S and R negated within 20 ns of each other). This parameter is sometimes called the *recovery time*, t_{rec}. It is the minimum delay between negating S and R for them to be considered nonsimultaneous and it is closely related to the minimum-pulse-width specification. Both specifications are measures of how long it takes for the latch's feedback loop to stabilize during a change of state.

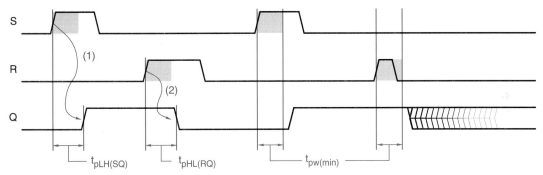

Figure 7-8 Timing parameters for an S-R latch.

7.2.2 $\overline{S}$-$\overline{R}$ Latch

An $\overline{S}$-$\overline{R}$ *latch* (read "S-bar-R-bar latch") with active-low set and reset inputs may be built from NAND gates as shown in Figure 7-9(a). In TTL and CMOS logic families, $\overline{S}$-$\overline{R}$ latches are used much more often than S-R latches because NAND gates are preferred over NOR gates.

S-R latch

As shown by the function table, Figure 7-9(b), operation of the $\overline{S}$-$\overline{R}$ latch is similar to that of the S-R, with two major differences. First, $\overline{S}$ and $\overline{R}$ are active low, so the latch remembers its previous state when $\overline{S} = \overline{R} = 1$; the active-low inputs are clearly indicated in the symbols in (c). Second, when both $\overline{S}$ and $\overline{R}$ are asserted simultaneously, both latch outputs go to 1, not 0 as in the S-R latch. Except for these differences, operation of the $\overline{S}$-$\overline{R}$ is the same as the S-R, including timing and metastability considerations.

Figure 7-9 $\overline{S}$-$\overline{R}$ latch: (a) circuit design using NAND gates; (b) function table; (c) logic symbol.

(a)

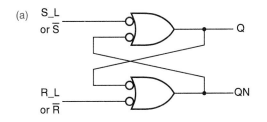

(b)

S_L	R_L	Q	QN
0	0	1	1
0	1	1	0
1	0	0	1
1	1	last Q	last QN

(c)

S	R	C	Q	QN
0	0	1	last Q	last QN
0	1	1	0	1
1	0	1	1	0
1	1	1	1	1
x	x	0	last Q	last QN

Figure 7-10 S-R latch with enable: (a) circuit using NAND gates; (b) function table; (c) logic symbol.

7.2.3 S-R Latch with Enable

S-R latch with enable

An S-R or $\overline{\text{S}}$-$\overline{\text{R}}$ latch is sensitive to its S and R inputs at all times. However, it may easily be modified to create a device that is sensitive to these inputs only when an enabling input C is asserted. Such an S-R *latch with enable* is shown in Figure 7-10. As shown by the function table, the circuit behaves like an S-R latch when C is 1, and retains its previous state when C is 0. The latch's behavior for a typical set of inputs is shown in Figure 7-11. If both S and R are 1 when C changes from 1 to 0, the circuit behaves like an S-R latch in which S and R are negated simultaneously—the next state is unpredictable and the output may become metastable.

7.2.4 D Latch

D latch

S-R latches are useful in control applications, where we often think in terms of setting a flag in response to some condition and resetting it when conditions change. So, we control the set and reset inputs somewhat independently. However, we often need latches simply to store bits of information—each bit is presented on a signal line, and we'd like to store it somewhere. A *D latch* may be used in such an application.

Figure 7-12 shows a D latch. Its logic diagram is recognizable as that of an S-R latch with enable, with an inverter added to generate S and R inputs from the

Figure 7-11 Typical operation of an S-R latch with enable.

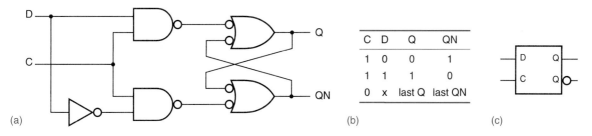

Figure 7-12 D latch: (a) circuit design using NAND gates; (b) function table; (c) logic symbol.

C	D	Q	QN
1	0	0	1
1	1	1	0
0	x	last Q	last QN

Figure 7-13 Functional behavior of a D latch for various inputs.

single D (data) input. This eliminates the troublesome situation in S-R latches, where S and R may be asserted simultaneously. The control input of a D latch, labeled C in (c), is sometimes named ENABLE, CLK, or G and is active low in some D-latch designs.

An example of a D latch's functional behavior is given in Figure 7-13. When the C input is asserted, the Q output follows the D input. In this situation, the latch is said to be "open" and the path from D input to Q output is "transparent"; the circuit is often called a *transparent latch* for this reason. When the C input is negated, the latch "closes"; the Q output retains its last value and no longer changes in response to D, as long as C remains negated.

transparent latch

More detailed timing behavior of the D latch is shown in Figure 7-14. Four different delay parameters are shown for signals that propagate from the C or D input to the Q output. For example, at transitions 1 and 4 the latch is initially

Figure 7-14 Timing parameters for a D latch.

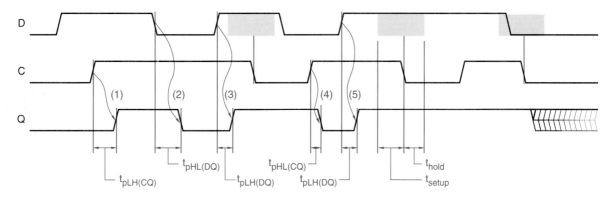

"closed" and the D input is the opposite of Q output, so that when C goes to 1 the latch "opens up" and the Q output changes after delay $t_{pLH(CQ)}$ or $t_{pHL(CQ)}$. At transitions 2 and 3 the C input is already 1 and the latch is already open, so that Q transparently follows the transitions on D with delay $t_{pHL(DQ)}$ and $t_{pLH(DQ)}$. Four more parameters specify the delay to the QN output, not shown.

Although the D latch eliminates the S = R = 1 problem of the S-R latch, it does not eliminate the metastability problem. As shown in Figure 7-14, there is a (shaded) window of time around the falling edge of C when the D input must not change. This window begins at time t_{setup} before the falling (latching) edge

setup time

of C; t_{setup} is called the *setup time*. The window ends at time t_{hold} afterward; t_{hold}

hold time

is called the *hold time*. If D changes at any time during the setup- and hold-time window, the output of the latch is unpredictable and may become metastable, as shown for the last latching edge in the figure.

7.2.5 Edge-Triggered D Flip-Flop

positive-edge-triggered
D flip-flop

A *positive-edge-triggered D flip-flop* combines a pair of D latches, as shown in Figure 7-15, to create a circuit that samples its D input and changes its Q and QN outputs only at the rising edge of a controlling CLK signal. The first latch is

master

called the *master*; it is open and follows the input when CLK is 0. When CLK goes to 1, the master latch is closed and its output is transferred to the second

slave

latch, called the *slave*. The slave latch is open all the while that CLK is 1, but changes only at the beginning of this interval, because the master is closed and unchanging during the rest of the interval.

dynamic-input
indicator

The triangle on the D flip-flop's CLK input indicates edge-triggered behavior and is called a *dynamic-input indicator*. Examples of the flip-flop's functional behavior for several input transitions are presented in Figure 7-16. The QM signal shown is the output of the master latch. Notice that QM changes only when CLK is 0. When CLK goes to 1, the current value of QM is transferred to Q, and QM is prevented from changing until CLK goes to 0 again.

Figure 7-17 shows more detailed timing behavior for the D flip-flop. All propagation delays are measured from the rising edge of CLK, since that's the only event that causes an output change. Different delays may be specified for LOW-to-HIGH and HIGH-to-LOW output changes.

Figure 7-15 Positive-edge-triggered D flip-flop: (a) circuit design using D latches; (b) function table; (c) logic symbol.

(b)

D	CLK	Q	QN
0	⌐_	0	1
1	_⌐	1	0
x	0	last Q	last QN
x	1	last Q	last QN

Figure 7-16 Functional behavior of a positive-edge-triggered D flip-flop.

Like a D latch, the edge-triggered D flip-flop has a setup and hold time window during which the D inputs must not change. This window occurs around the triggering edge of CLK, and is indicated by shaded color in Figure 7-17. If the setup and hold times are not met, the flip-flop output will usually go to a stable, though unpredictable, 0 or 1 state. In some cases, however, the output will oscillate or go to a metastable state halfway between 0 and 1, as shown at the second-to-last clock tick in the figure. If the flip-flop goes into the metastable state, it will return to a stable state on its own only after a probabilistic delay, as explained in Section 8.9. It can also be forced into a stable state by applying another triggering clock edge with a D input that meets the setup- and hold-time requirements, as shown at the last clock tick in the figure.

A *negative-edge-triggered* D *flip-flop* simply inverts the clock input, so that all the action takes place on the falling edge of CLK_L; by convention, a falling-edge trigger is considered to be active low. The flip-flop's function table and logic symbol are shown in Figure 7-18.

Some D flip-flops have *asynchronous inputs* that may be used to force the flip-flop to a particular state independent of the CLK and D inputs. These inputs, typically labeled PR (*preset*) and CLR (*clear*), behave like the set and reset

*negative-edge-triggered
 D flip-flop*

*asynchronous inputs
 preset
 clear*

Figure 7-17 Timing behavior of a positive-edge-triggered D flip-flop.

Figure 7-18 Negative-edge triggered D flip-flop: (a) circuit design using D latches; (b) function table; (c) logic symbol.

inputs on an SR latch. The logic symbol and NAND circuit for an edge-triggered D flip-flop with these inputs is shown in Figure 7-19. Although asynchronous inputs are used by some logic designers to perform tricky sequential functions, they are best reserved for initialization and testing purposes, to force a sequential circuit into a known starting state; more on this when we discuss synchronous design methodology in Section 8.7.

7.2.6 Edge-Triggered D Flip-Flop with Enable

A commonly desired function in D flip-flops is the ability to hold the last value stored, rather than load a new value, at the clock edge. This is accomplished by *enable input* adding an *enable input*, called EN or CE (*clock enable*). While the name "clock *clock-enable input* enable" is descriptive, the extra input's function is not obtained by controlling

Figure 7-19 Positive-edge-triggered D flip-flop with preset and clear: (a) logic symbol; (b) circuit design using NAND gates.

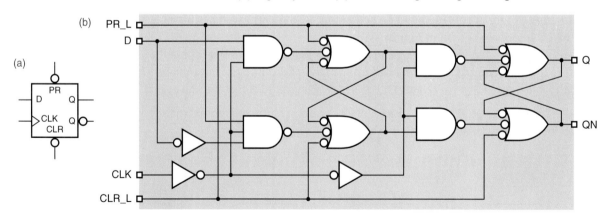

TIME FOR A COMMERCIAL Commercial TTL positive-edge-triggered D flip-flops do not use the master-slave latch design of Figure 7-15 or Figure 7-19. Instead, flip-flops like the 74LS74 use the six-gate design of Figure 7-20, which is smaller and faster. We'll show how to formally analyze the next-state behavior of both designs in Section 7.9.

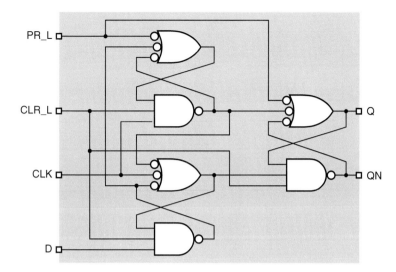

Figure 7-20
Commercial circuit for a positive-edge-triggered D flip-flop such as 74LS74.

the clock in any way whatsoever. Rather, as shown in Figure 7-21(a), a 2-input multiplexer controls the value applied to the internal flip-flop's D input. If EN is asserted, the external D input is selected; if EN is negated, the flip-flop's current output is used. The resulting function table is shown in (b). The flip-flop symbol is shown in (c); in some flip-flops, the enable input is active low, denoted by an inversion bubble on this input.

7.2.7 Scan Flip-Flop

An important flip-flop function for ASIC testing is so-called *scan capability*.   *scan capability*
The idea is to be able to drive the flip-flop's D input with an alternate source of data during device testing. When all of the flip-flops are put into testing mode, a test pattern can be "scanned in" to the ASIC using the flip-flops' alternate data inputs. After the test pattern is loaded, the flip-flops are put back into "normal" mode, and all of the flip-flops are clocked normally. After one or more clock ticks, the flip-flops are put back into test mode, and the test results are "scanned out."

Figure 7-21 Positive-edge-triggered D flip-flop with enable: (a) circuit design; (b) function table; (c) logic symbol.

(a)

(b)

D	EN	CLK	Q	QN
0	1	↑	0	1
1	1	↑	1	0
x	0	↑	last Q	last QN
x	x	0	last Q	last QN
x	x	1	last Q	last QN

(c)

TE	TI	D	CLK	Q	QN
0	x	0	⌐_⌐	0	1
0	x	1	⌐_⌐	1	0
1	0	x	⌐_⌐	0	1
1	1	x	⌐_⌐	1	0
x	x	x	0	last Q	last QN
x	x	x	1	last Q	last QN

(a)   (b)   (c)

Figure 7-22 Positive-edge-triggered D flip-flop with scan: (a) circuit design; (b) function table; (c) logic symbol.

Figure 7-22(a) shows the design of a typical scan flip-flop. It is nothing more than a D flip-flop with a 2-input multiplexer on the D input. When the TE (*test enable*) input is negated, the circuit behaves like an ordinary D flip-flop. When TE is asserted, it takes its data from TI (*test input*) instead of from D. This functional behavior is shown in (b), and a symbol for the device is given in (c).

test-enable input, TE
test input, TI

scan chain

The extra inputs are used to connect all of an ASIC's flip-flops in a *scan chain* for testing purposes. Figure 7-23 is a simple example with four flip-flops in the scan chain. The TE inputs of all the flip-flops are connected to a global TE input, while each flip-flop's Q output is connected to another's TI input in serial (daisy-chain) fashion. The TI, TE, and TO (test output) connections are strictly for testing purposes; the additional logic connected to the D inputs and Q outputs needed to make the circuit do something useful is not shown.

To test the circuit, including the additional logic, the global TE input is asserted while *n* clock ticks occur and *n* test-vector bits are applied to the global TI input and are thereby scanned (shifted) into the *n* flip-flops; *n* equals 4 in Figure 7-23. Then TE is negated, and the circuit is allowed to run for one or more additional clock ticks. The new state of the circuit, represented by the new values in the *n* flip-flops, can be observed (scanned out) at TO by asserting TE while *n* more clock ticks occur. To make the testing process more efficient, another test vector can be scanned in while the previous result is being scanned out.

Figure 7-23 A scan chain with four flip-flops.

There are many different types of scan flip-flops, corresponding to different types of basic flip-flop functionality. For example, scan capability could be added to the D flip-flop with enable in Figure 7-21 by replacing its internal 2-input multiplexer with a 3-input one. At each clock tick the flip-flop would load D, TI, or its current state, depending on the values of EN and TE. Scan capability can also be added to other flip-flop types, such as J-K and T introduced later in this section.

*7.2.8 Master/Slave S-R Flip-Flop

We indicated earlier that S-R latches are useful in "control" applications, where we may have independent conditions for setting and resetting a control bit. If the control bit is supposed to be changed only at certain times with respect to a clock signal, then we need an S-R flip-flop that, like a D flip-flop, changes its outputs only on a certain edge of the clock signal. This subsection and the next two describe flip-flops that are useful for such applications.

If we substitute S-R latches for the D latches in the negative-edge-triggered D flip-flop of Figure 7-18(a), we get a *master/slave S-R flip-flop*, shown in Figure 7-24. Like a D flip-flop, the S-R flip-flop changes its outputs only at the falling edge of a control signal C. However, the new output value depends on input values not just at the falling edge, but during the entire interval in which C is 1 prior to the falling edge. As shown in Figure 7-25, a short pulse on S any time during this interval can set the master latch; likewise, a pulse on R can reset it. The value transferred to the flip-flop output on the falling edge of C depends on whether the master latch was last set or cleared while C was 1.

master/slave S-R flip-flop

Shown in Figure 7-24(c), the logic symbol for the master/slave S-R flip-flop does not use a dynamic-input indicator, because the flip-flop is not truly edge triggered. It is more like a latch that follows its input during the entire interval that C is 1 but changes its output to reflect the final latched value only when C goes to 0. In the symbol, a *postponed-output indicator* indicates that the output signal does not change until enable input C is negated. Flip-flops with this kind of behavior are sometimes called *pulse-triggered flip-flops*.

postponed-output indicator

pulse-triggered flip-flop

Figure 7-24 Master/slave S-R flip-flop: (a) circuit using S-R latches;
(b) function table; (c) logic symbol.

(a)

(b)

S	R	C	Q	QN
x	x	0	last Q	last QN
0	0	⎍	last Q	last QN
0	1	⎍	0	1
1	0	⎍	1	0
1	1	⎍	undef.	undef.

(c)

*Throughout this book, optional sections are marked with an asterisk.

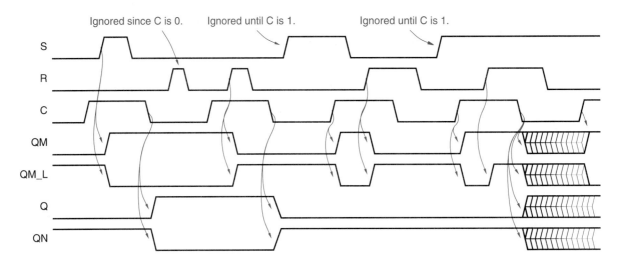

Figure 7-25 Internal and functional behavior of a master/slave S-R flip-flop.

The operation of the master/slave S-R flip-flop is unpredictable if both S and R are asserted at the falling edge of C. In this case, just before the falling edge, both the Q and QN outputs of the master latch are 1. When C goes to 0, the master latch's outputs change unpredictably and may even become metastable. At the same time, the slave latch opens up and propagates this garbage to the flip-flop output.

*7.2.9 Master/Slave J-K Flip-Flop

master/slave J-K flip-flop

The problem of what to do when S and R are asserted simultaneously is solved in a *master/slave J-K flip-flop*. The J and K inputs are analogous to S and R. However, as shown in Figure 7-26, asserting J asserts the master's S input only if the flip-flop's QN output is currently 1 (i.e., Q is 0), and asserting K asserts the master's R input only if Q is currently 1. Thus, if J and K are asserted simultaneously, the flip-flop goes to the opposite of its current state.

Figure 7-26 Master/slave J-K flip-flop: (a) circuit design using S-R latches; (b) function table; (c) logic symbol.

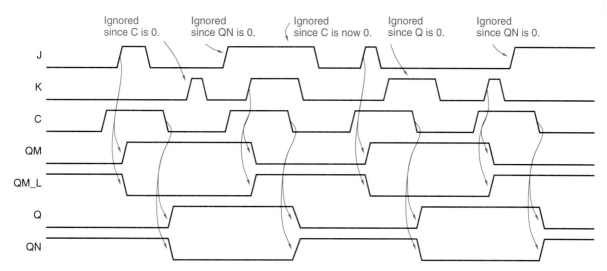

Figure 7-27 Internal and functional behavior of a master/slave J-K flip-flop.

Figure 7-27 shows the functional behavior of a J-K master/slave flip-flop for a typical set of inputs. Note that the J and K inputs need not be asserted at the end of the triggering pulse for the flip-flop output to change at that time. In fact, because of the gating on the master latch's S and R inputs, it is possible for the flip-flop output to change to 1 even though K and not J is asserted at the end of the triggering pulse. This behavior, known as *1s catching*, is illustrated in the second-to-last triggering pulse in the figure. An analogous behavior known as *0s catching* is illustrated in the last triggering pulse. Because of this behavior, the J and K inputs of a J-K master/slave flip-flop must be held valid during the entire interval that C is 1.

1s catching

0s catching

7.2.10 Edge-Triggered J-K Flip-Flop

The problem of 1s and 0s catching is solved in an *edge-triggered* J-K *flip-flop*, whose functional equivalent is shown in Figure 7-28. Using an edge-triggered D flip-flop internally, the edge-triggered J-K flip-flop samples its inputs at the

edge-triggered J-K flip-flop

Figure 7-28 Edge-triggered J-K flip-flop: (a) equivalent function using an edge-triggered D flip-flop; (b) function table; (c) logic symbol.

J	K	CLK	Q	QN
x	x	0	last Q	last QN
x	x	1	last Q	last QN
0	0	⌐	last Q	last QN
0	1	⌐	0	1
1	0	⌐	1	0
1	1	⌐	last QN	last Q

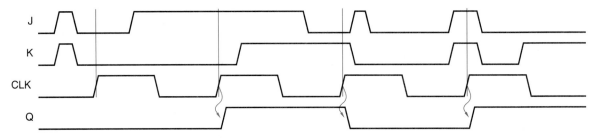

Figure 7-29 Functional behavior of a positive-edge-triggered J-K flip-flop.

rising edge of the clock and produces its next output according to the "characteristic equation" $Q^* = J \cdot Q' + K' \cdot Q$ (see Section 7.3.3).

Typical functional behavior of an edge-triggered J-K flip-flop is shown in Figure 7-29. Like the D input of an edge-triggered D flip-flop, the J and K inputs of a J-K flip-flop must meet published setup- and hold-time specifications with respect to the triggering clock edge for proper operation.

Because they eliminate the problems of 1s and 0s catching and of simultaneously asserting both control inputs, edge-triggered J-K flip-flops have largely obsoleted the older pulse-triggered types. The *74x109* is a TTL positive-edge-triggered J-$\overline{\text{K}}$ flip-flop with an active-low K input (named $\overline{\text{K}}$ or K_L).

74x109

ANOTHER COMMERCIAL (FLIP-FLOP, THAT IS)	The internal design of the 74LS109 is very similar to that of the 74LS74, which we showed in Figure 7-20. As shown in Figure 7-30, the '109 simply replaces the bottom-left gate of the '74, which realizes the characteristic equation $Q^* = D$, with an AND-OR structure that realizes the J-$\overline{\text{K}}$ characteristic equation, $Q^* = J \cdot Q' + K\_L \cdot Q$.

Figure 7-30
Internal logic diagram
for the 74LS109
positive-edge-triggered
J-$\overline{\text{K}}$ flip-flop.

Figure 7-31 Positive-edge-triggered T flip-flop: (a) logic symbol; (b) functional behavior.

The most common application of J-K flip-flops is in clocked synchronous state machines. As we'll explain in Section 7.4.5, the next-state logic for J-K flip-flops is sometimes simpler than for D flip-flops. However, most state machines are still designed using D flip-flops, because the design methodology is a bit simpler and because most sequential programmable logic devices contain D, not J-K, flip-flops. Therefore, we'll give most of our attention to D flip-flops.

7.2.11 T Flip-Flop

A *T (toggle) flip-flop* changes state on every tick of the clock. Figure 7-31 shows   *T flip-flop*
the symbol and illustrates the behavior of a positive-edge-triggered T flip-flop. Notice that the signal on the flip-flop's Q output has precisely half the frequency of the T input. Figure 7-32 shows how to obtain a T flip-flop from a D or J-K flip-flop. T flip-flops are most often used in counters and frequency dividers, as we'll show in Section 8.4.

In many applications of T flip-flops, the flip-flop need not be toggled on every clock tick. Such applications can use a *T flip-flop with enable*. As shown   *T flip-flop with enable*
in Figure 7-33, the flip-flop changes state at the triggering edge of the clock only if the enable signal EN is asserted. Like the D, J, and K inputs on other edge-triggered flip-flops, the EN input must meet specified setup and hold times with respect to the triggering clock edge. The circuits of Figure 7-32 are easily modified to provide an EN input, as shown in Figure 7-34.

Figure 7-32
Possible circuit designs for a T flip-flop: (a) using a D flip-flop; (b) using a J-K flip-flop.

Figure 7-33 Positive-edge-triggered T flip-flop with enable: (a) logic symbol; (b) functional behavior.

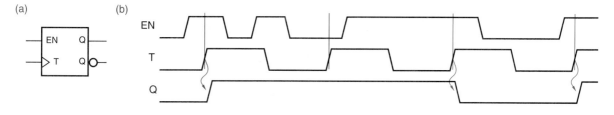

Figure 7-34
Possible circuits for a
T flip-flop with enable:
(a) using a D flip-flop;
(b) using a J-K flip-flop.

(a)   (b)

7.3 Clocked Synchronous State-Machine Analysis

Although latches and flip-flops, the basic building blocks of sequential circuits, are themselves feedback sequential circuits that can be formally analyzed, we'll first study the operation of *clocked synchronous state machines*, since they are the easiest to understand. "State machine" is a generic name given to these sequential circuits; "clocked" refers to the fact that their storage elements (flip-flops) employ a clock input; and "synchronous" means that all of the flip-flops use the same clock signal. Such a state machine changes state only when a triggering edge or "tick" occurs on the clock signal.

clocked synchronous state machine

7.3.1 State-Machine Structure

Figure 7-35 shows the general structure of a clocked synchronous state machine. The *state memory* is a set of n flip-flops that store the current state of the machine, and it has 2^n distinct states. The flip-flops are all connected to a common clock signal that causes them to change state at each *tick* of the clock. What constitutes a tick depends on the flip-flop type (edge triggered, pulse triggered, etc.). For the positive-edge-triggered D and J-K flip-flops considered in this section, a tick is the rising edge of the clock signal.

state memory

tick

The next state of the state machine in Figure 7-35 is determined by the *next-state logic F* as a function of the current state and input. The *output logic G*

next-state logic

output logic

Figure 7-35 Clocked synchronous state-machine structure (Mealy machine).

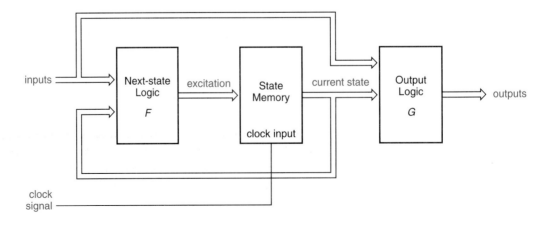

determines the output as a function of the current state and input. Both *F* and *G* are strictly combinational logic circuits. We can write

$$\text{Next state} = F(\text{current state, input})$$

$$\text{Output} = G(\text{current state, input})$$

State machines may use positive-edge-triggered D flip-flops for their state memory, in which case a tick occurs at each rising edge of the clock signal. It is also possible for the state memory to use negative-edge-triggered D flip-flops, D latches, or J-K flip-flops. However, inasmuch as most state machines are designed nowadays using programmable logic devices with positive-edge-triggered D flip-flops, that's what we'll concentrate on.

7.3.2 Output Logic

A sequential circuit whose output depends on both state and input as shown in Figure 7-35 is called a *Mealy machine*. In some sequential circuits the output depends on the state alone:

$$\text{Output} = G(\text{current state})$$

Such a circuit is called a *Moore machine*, and its general structure is shown in Figure 7-36.

Obviously, the only difference between the two state-machine models is in how outputs are generated. In practice, many state machines must be categorized as Mealy machines, because they have one or more *Mealy-type outputs* that depend on input as well as state. However, many of these same machines also have one or more *Moore-type outputs* that depend only on state.

In the design of high-speed circuits, it is often necessary to ensure that state-machine outputs are available as early as possible and do not change during each clock period. One way to get this behavior is to encode the state so that the state variables themselves serve as outputs. We call this an *output-coded state assignment*; it produces a Moore machine in which the output logic of Figure 7-36 is null, consisting of just wires.

output-coded state assignment

Figure 7-36 Clocked synchronous state-machine structure (Moore machine).

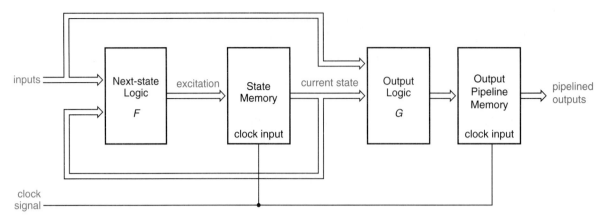

Figure 7-37 Mealy machine with pipelined outputs.

pipelined outputs

Another approach is to design the state machine so that the outputs during one clock period depend on the state and inputs during the *previous* clock period. We call these *pipelined outputs*, and they are obtained by attaching another stage of memory (flip-flops) to a Mealy machine's outputs, as in Figure 7-37.

With appropriate circuit or drawing manipulations, you can map one state-machine model into another. For example, you could declare the flip-flops that produce pipelined outputs from a Mealy machine to be part of its state memory and thereby obtain a Moore machine with an output-coded state assignment.

The exact classification of a state machine into one style or another is not so important. What's important is how you think about output structure and how it satisfies your overall design objectives, including timing and flexibility. For example, pipelined outputs are great for timing, but you can use them only in situations where you can figure out the desired next output value one clock period in advance. In any given application you may use different styles for different output signals. For example, we'll see in Section 7.11.5 that different statements can be used to specify different output styles in ABEL.

7.3.3 Characteristic Equations

The functional behavior of a latch or flip-flop can be described formally by a *characteristic equation*

characteristic equation

that specifies the flip-flop's next state as a function of its current state and inputs.

* suffix

The characteristic equations of the flip-flops in Section 7.2 are listed in Table 7-1. By convention, the * suffix in $Q*$ means "the next value of Q." Notice that the characteristic equation does not describe detailed timing behavior of the device (latching vs. edge-triggered, etc.), only the functional response to the control inputs. This simplified description is useful in the analysis of state machines, as we'll soon show.

Device Type	Characteristic Equation
S-R latch	$Q* = S + R' \cdot Q$
D latch	$Q* = D$
Edge-triggered D flip-flop	$Q* = D$
D flip-flop with enable	$Q* = EN \cdot D + EN' \cdot Q$
Master/slave S-R flip-flop	$Q* = S + R' \cdot Q$
Master/slave J-K flip-flop	$Q* = J \cdot Q' + K' \cdot Q$
Edge-triggered J-K flip-flop	$Q* = J \cdot Q' + K' \cdot Q$
T flip-flop	$Q* = Q'$
T flip-flop with enable	$Q* = EN \cdot Q' + EN' \cdot Q$

Table 7-1
Latch and flip-flop characteristic equations.

7.3.4 Analysis of State Machines with D Flip-Flops

Consider the formal definition of a state machine that we gave previously:

$$\text{Next state} = F(\text{current state, input})$$

$$\text{Output} = G(\text{current state, input})$$

Recalling our notion that "state" embodies all we need to know about the past history of the circuit, the first equation tells us that what we next need to know can be determined from what we currently know and the current input. The second equation tells us that the current output can be determined from the same information. The goal of sequential circuit analysis is to determine the next-state and output functions so that the behavior of a circuit can be predicted.

The analysis of a clocked synchronous state machine has three basic steps:

1. Determine the next-state and output functions F and G.

2. Use F and G to construct a *state/output table* that completely specifies the next state and output of the circuit for every possible combination of current state and input. *state/output table*

3. (Optional) Draw a *state diagram* that presents the information from the previous step in graphical form. *state diagram*

Figure 7-38 shows a simple state machine with two positive-edge-triggered D flip-flops. To determine the next-state function F, we must first consider the behavior of the state memory. At the rising edge of the clock signal, each D flip-flop samples its D input and transfers this value to its Q output; the characteristic equation of a D flip-flop is $Q* = D$. Therefore, to determine the next value of Q (i.e., Q^*), we must first determine the current value of D.

In Figure 7-38 there are two D flip-flops, and we have named the signals on their outputs Q0 and Q1. These two outputs are the state variables; their value is

Figure 7-38 Clocked synchronous state machine using positive-edge-triggered D flip-flops.

excitation

excitation equation

the current state of the machine. We have named the signals on the corresponding D inputs D0 and D1. These signals provide the *excitation* for the D flip-flops at each clock tick. Logic equations that express the excitation signals as functions of the current state and input are called *excitation equations* and can be derived from the circuit diagram:

$$D0 \; = \; Q0 \cdot EN' + Q0' \cdot EN$$
$$D1 \; = \; Q1 \cdot EN' + Q1' \cdot Q0 \cdot EN + Q1 \cdot Q0' \cdot EN$$

* suffix

By convention, the next value of a state variable after a clock tick is denoted by appending a star to the state-variable name, for example, $Q0*$ or $Q1*$. Using the characteristic equation of D flip-flops, $Q* = D$, we can describe the next-state function of the example machine with equations for the next value of the state variables:

$$Q0* \; = \; D0$$
$$Q1* \; = \; D1$$

Substituting the excitation equations for D0 and D1, we can write

$$Q0* \; = \; Q0 \cdot EN' + Q0' \cdot EN$$
$$Q1* \; = \; Q1 \cdot EN' + Q1' \cdot Q0 \cdot EN + Q1 \cdot Q0' \cdot EN$$

(a)

Q1 Q0	EN 0	1
00	00	01
01	01	10
10	10	11
11	11	00
	Q1* Q0*	

(b)

S	EN 0	1
A	A	B
B	B	C
C	C	D
D	D	A
	S*	

(c)

S	EN 0	1
A	A, 0	B, 0
B	B, 0	C, 0
C	C, 0	D, 0
D	D, 0	A, 1
	S*, MAX	

Table 7-2
Transition, state, and state/output tables for the state machine in Figure 7-38.

These equations, which express the next value of the state variables as a function of current state and input, are called *transition equations*.

transition equation

For each combination of current state and input value, the transition equations predict the next state. Each state is described by two bits, the current values of Q0 and Q1: (Q1 Q0) = 00, 01, 10, or 11. [The reason for "arbitrarily" picking the order (Q1 Q0) instead of (Q0 Q1) will become apparent shortly.] For each state, our example machine has just two possible input values, EN = 0 or EN = 1, so there are a total of 8 state/input combinations. (In general, a machine with s state bits and i inputs has 2^{s+i} state/input combinations.)

Table 7-2(a) shows a *transition table* that is created by evaluating the transition equations for every possible state/input combination. Traditionally, a transition table lists the states along the left and the input combinations along the top of the table, as shown in the example.

transition table

The function of our example machine is apparent from its transition table—it is a 2-bit binary counter with an enable input EN. When EN = 0, the machine maintains its current count, but when EN = 1, the count advances by 1 at each clock tick, rolling over to 00 when it reaches a maximum value of 11.

If we wish, we may assign alphanumeric *state names* to each state. The simplest naming is 00 = A, 01 = B, 10 = C, and 11 = D. Substituting the state names for combinations of Q1 and Q0 (and Q1* and Q0*) in Table 7-2(a) produces the *state table* in (b). Here "S" denotes the current state and "S*" denotes the next state of the machine. A state table is usually easier to understand than a transition table, because in complex machines we can use state names that have meaning. However, a state table contains less information than a transition table because it does not indicate the binary values assumed by the state variables in each named state.

state names

state table

Once a state table is produced, we have only the output logic of the machine left to analyze. In the example machine there is only a single output signal, and it is a function of both current state and input (this is a Mealy machine). So we can write a single *output equation*:

output equation

$$MAX = Q1 \cdot Q0 \cdot EN$$

The output behavior predicted by this equation can be combined with the next-state information to produce a *state/output table* as shown in Table 7-2(c).

state/output table

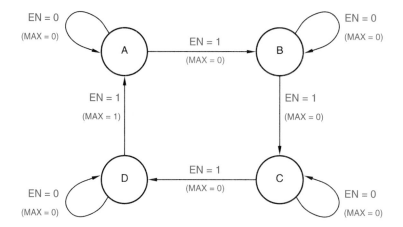

Figure 7-39
State diagram corresponding to the state machine of Table 7-2.

State/output tables for Moore machines are slightly simpler. For example, in the circuit of Figure 7-38 suppose we removed the EN signal from the AND gate that produces the MAX output, producing a Moore-type output MAXS. Then MAXS is a function of the state only, and the state/output table can list MAXS in a single column, independent of the input values. This is shown in Table 7-3.

state diagram
node
directed arc

A *state diagram* presents the information from the state/output table in a graphical format. It has one circle (or *node*) for each state and an arrow (or *directed arc*) for each transition. Figure 7-39 shows the state diagram for our example state machine. The letter inside each circle is a state name. Each arrow leaving a given state points to the next state for a given input combination; it also shows the output value produced in the given state for that input combination.

The state diagram for a Moore machine can be somewhat simpler. In this case, the output values can be shown inside each state circle, since they are

Table 7-3
State/output table for a Moore machine.

S	EN 0	1	MAXS
A	A	B	0
B	B	C	0
C	C	D	0
D	D	A	1
	S*		

A CLARIFICATION   The state-diagram notation for output values in Mealy machines is a little misleading. You should remember that the listed output value is produced continuously when the machine is in the indicated state and has the indicated input, not just during the transition to the next state.

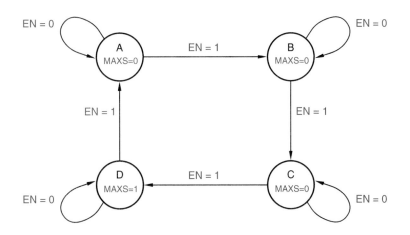

Figure 7-40
State diagram
corresponding to the
state machine of
Table 7-3.

functions of state only. The state diagram for a Moore machine using this
convention is shown in Figure 7-40.

The original logic diagram of our example state machine, Figure 7-38, was
laid out to match our conceptual model of a Mealy machine. However, nothing
requires us to group the next-state logic, state memory, and output logic in this
way. Figure 7-41 shows another logic diagram for the same state machine. To
analyze this circuit, the designer (or analyzer, in this case) can still extract the
required information from the diagram as drawn. The only circuit difference in

Figure 7-41 Redrawn logic diagram for a clocked synchronous state machine.

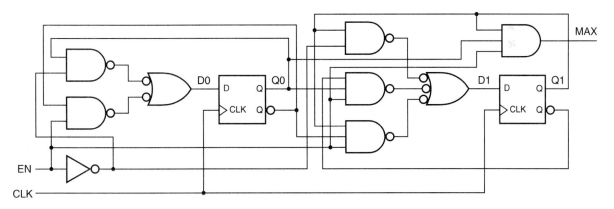

LITTLE ARROWS, LITTLE ARROWS EVERYWHERE	Since there is only one input in our example machine, there are only two possible input combinations, and two arrows leaving each state. In a machine with n inputs, we would have 2^n arrows leaving each state. This is messy if n is large. Later, in Figure 7-44, we'll describe a convention whereby a state needn't have one arrow leaving it for each input combination, only one arrow for each different next state.

SUGGESTIVE DRAWINGS   Using the transition, state, and output tables, we can construct a timing diagram that shows the behavior of a state machine for any desired starting state and input sequence. For example, Figure 7-42 shows the behavior of our example machine with a starting state of 00 (**A**) and a particular pattern on the **EN** input.

Notice that the value of the **EN** input affects the next state only at the rising edge of the **CLOCK** input; that is, the counter counts only if **EN** = 1 at the rising edge of **CLOCK**. On the other hand, since **MAX** is a Mealy-type output, its value is affected by **EN** at all times. If we also provide a Moore-type output **MAXS** as suggested in the text, its value depends only on state as shown in the figure.

The timing diagram is drawn in a way that shows changes in the **MAX** and **MAXS** outputs occurring slightly later than the state and input changes that cause them, reflecting the combinational-logic delay of the output circuits. Naturally, the drawings are merely suggestive; precise timing is normally indicated by a timing table of the type suggested in Section 5.2.1.

the new diagram is that we have used the flip-flops' **QN** outputs (which are normally the complement of **Q**) to save a couple of inverters.

In summary, the detailed steps for analyzing a clocked synchronous state machine are as follows:

excitation equations   1. Determine the excitation equations for the flip-flop control inputs.

transition equations   2. Substitute the excitation equations into the flip-flop characteristic equations to obtain transition equations.

transition table   3. Use the transition equations to construct a transition table.

output equations   4. Determine the output equations.

Figure 7-42 Timing diagram for example state machine.

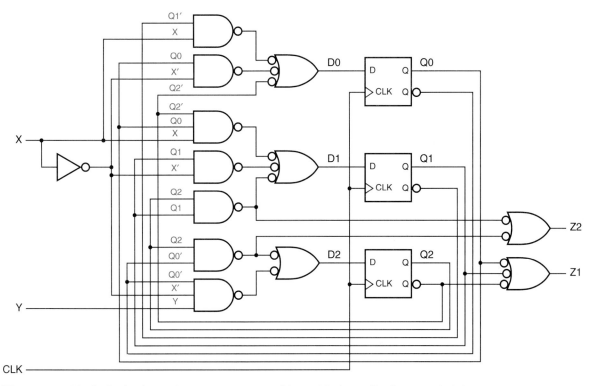

Figure 7-43 A clocked synchronous state machine with three flip-flops and eight states.

5. Add output values to the transition table for each state (Moore) or state/input combination (Mealy) to create a transition/output table.   *transition/output table*

6. Name the states and substitute state names for state-variable combinations in the transition/output table to obtain a state/output table.   *state names*   *state/output table*

7. (Optional) Draw a state diagram corresponding to the state/output table.   *state diagram*

We'll go through this complete sequence of steps to analyze another clocked synchronous state machine, shown in Figure 7-43. Reading the logic diagram, we find that the excitation equations are as follows:

$$D0 = Q1' \cdot X + Q0 \cdot X' + Q2$$
$$D1 = Q2' \cdot Q0 \cdot X + Q1 \cdot X' + Q2 \cdot Q1$$
$$D2 = Q2 \cdot Q0' + Q0' \cdot X' \cdot Y$$

Substituting into the characteristic equation for D flip-flops, we obtain the transition equations:

$$Q0* = Q1' \cdot X + Q0 \cdot X' + Q2$$
$$Q1* = Q2' \cdot Q0 \cdot X + Q1 \cdot X' + Q2 \cdot Q1$$
$$Q2* = Q2 \cdot Q0' + Q0' \cdot X' \cdot Y$$

Table 7-4
Transition/output
and state/output
tables for the
state machine
in Figure 7-43.

(a)

Q2 Q1 Q0	XY 00	01	10	11	Z1 Z2
000	000	100	001	001	10
001	001	001	011	011	10
010	010	110	000	000	10
011	011	011	010	010	00
100	101	101	101	101	11
101	001	001	001	001	10
110	111	111	111	111	11
111	011	011	011	011	11
	Q2* Q1* Q0*				

(b)

S	XY 00	01	10	11	Z1 Z2
A	A	E	B	B	10
B	B	B	D	D	10
C	C	G	A	A	10
D	D	D	C	C	00
E	F	F	F	F	11
F	B	B	B	B	10
G	H	H	H	H	11
H	D	D	D	D	11
	S*				

A transition table based on these equations is shown in Table 7-4(a). Reading the logic diagram, we can write two output equations:

$$Z1 = Q2 + Q1' + Q0'$$
$$Z2 = Q2 \cdot Q1 + Q2 \cdot Q0'$$

The resulting output values are shown in the last column of (a). Assigning state names A–H, we obtain the state/output table shown in (b).

A state diagram for the example machine is shown in Figure 7-44. Since our example is a Moore machine, the output values are written with each state. Each arc is labeled with a *transition expression*; a transition is taken for input combinations for which the transition expression is 1. Transitions labeled "1" are always taken.

transition expression

Figure 7-44 State diagram corresponding to Table 7-4.

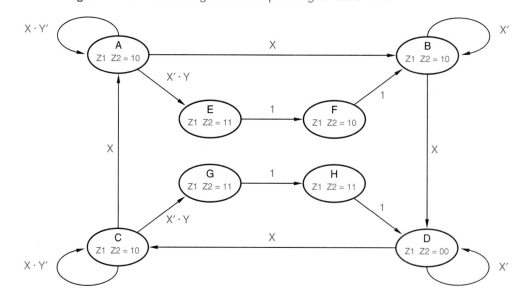

The transition expressions on arcs leaving a particular state must be mutually exclusive and all inclusive, as explained below:

- No two transition expressions can equal 1 for the same input combination, *mutual exclusion* since a machine can't have two next states for one input combination.

- For every possible input combination, some transition expression must *all inclusion* equal 1, so that all next states are defined.

Starting with the state table, a transition expression for a particular current state and next state can be written as a sum of minterms for the input combinations that cause that transition. If desired, the expression can then be minimized to give the information in a more compact form. Transition expressions are most useful in the *design* of state machines, where the expressions may be developed from the word description of the problem, as we'll show in Section 7.5.

*7.3.5 Analysis of State Machines with J-K Flip-Flops

Clocked synchronous state machines built from J-K flip-flops can also be analyzed by the basic procedure in the preceding subsection. The only difference is that there are two excitation equations for each flip-flop—one for J and the other for K. To obtain the transition equations, both of these must be substituted into the J-K's characteristic equation, $Q* = J \cdot Q' + K' \cdot Q$.

Figure 7-45 is an example state machine using J-K flip-flops. Reading the logic diagram, we can derive the following excitation equations:

$$J0 = X \cdot Y'$$
$$K0 = X \cdot Y' + Y \cdot Q1$$
$$J1 = X \cdot Q0 + Y$$
$$K1 = Y \cdot Q0' + X \cdot Y' \cdot Q0$$

Figure 7-45
Clocked synchronous state machine using J-K flip-flops.

Table 7-5
Transition/output
and state/output
tables for the
state machine
in Figure 7-45.

(a)

Q1 Q0	XY 00	01	10	11
00	00, 0 0,1	1 1,0	0 0,1	1
01	01, 0 1,1	0 0,1	0 1,1	0
10	10, 0 0,0	0 1,1	0 0,0	0
11	11, 0 0,1	0 0,0	1 0,1	1

Q1*Q0*, Z

(b)

S	XY 00	01	10	11
A	A, 0	C, 1	B, 0	C, 1
B	B, 0	D, 0	C, 0	D, 0
C	C, 0	A, 0	D, 0	A, 0
D	D, 0	C, 0	A, 1	C, 1

S   *, Z

Substituting into the characteristic equation for J-K flip-flops, we obtain the transition equations:

$$Q0* = J0 \cdot Q0' + K0' \cdot Q0$$
$$= X \cdot Y' \cdot Q0' + (X \cdot Y' + Y \cdot Q1)' \cdot Q0$$
$$= X \cdot Y' \cdot Q0' + X' \cdot Y' \cdot Q0 + X' \cdot Q1' \cdot Q0 + Y \cdot Q1' \cdot Q0$$
$$Q1* = J1 \cdot Q1' + K1' \cdot Q1$$
$$= (X \cdot Q0 + Y) \cdot Q1' + (Y \cdot Q0' + X \cdot Y' \cdot Q0)' \cdot Q1$$
$$= X \cdot Q1' \cdot Q0 + Y \cdot Q1' + X' \cdot Y' \cdot Q1 + Y' \cdot Q1 \cdot Q0' + X' \cdot Q1 \cdot Q0 + Y \cdot Q1 \cdot C$$

A transition table based on these equations is shown in Table 7-5(a). Reading the logic diagram, we can write the output equation:

$$Z = X \cdot Q1 \cdot Q0 + Y \cdot Q1' \cdot Q0'$$

The resulting output values are shown in each column of (a) along with the next state. Assigning state names A–D, we obtain the state/output table shown in (b). A corresponding state diagram that uses transition expressions is shown in Figure 7-46.

Figure 7-46
State diagram
corresponding to the
state machine of
Table 7-5.

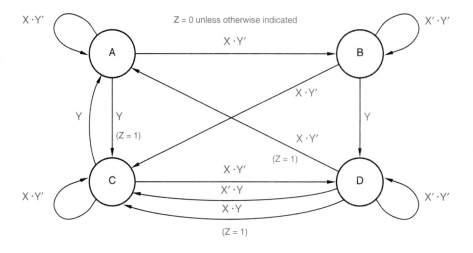

7.4 Clocked Synchronous State-Machine Design

The steps for designing a clocked synchronous state machine, starting from a word description or specification, are just about the reverse of the analysis steps that we used in the preceding section:

1. Construct a state/output table corresponding to the word description or specification, using mnemonic names for the states. (It's also possible to start with a state diagram; this method is discussed in Section 7.5.)   *state/output table*

2. (Optional) Minimize the number of states in the state/output table.   *state minimization*

3. Choose a set of state variables and assign state-variable combinations to the named states.   *state assignment*

4. Substitute the state-variable combinations into the state/output table to create a transition/output table that shows the desired next state-variable combination and output for each state/input combination.   *transition/output table*

5. Choose a flip-flop type (e.g., D or J-K) for the state memory. In most cases you'll already have a choice in mind at the outset of the design, but this step is your last chance to change your mind.

6. Construct an excitation table that shows the excitation values required to obtain the desired next state for each state/input combination.   *excitation table*

7. Derive excitation equations from the excitation table.   *excitation equations*

8. Derive output equations from the transition/output table.   *output equations*

9. Draw a logic diagram that shows the state-variable storage elements and realizes the required excitation and output equations. (Or realize the equations directly in a programmable logic device.)   *logic diagram*

In this section we'll describe each of these basic steps in state-machine design. Step 1 is the most important, since it is here that the designer really *designs*, going through the creative process of translating a (perhaps ambiguous) English-language description of the state machine into a formal tabular description. Step 2 is hardly ever performed by experienced digital designers, but designers bring much of their experience to bear in step 3.   *design*

Once the first three steps are completed, all of the remaining steps can be completed by "turning the crank," that is, by following a well-defined synthesis procedure. Steps 4 and 6–9 are the most tedious, but they are easily automated. For example, when you design a state machine that will be realized in a programmable logic device, you can use an ABEL compiler to do the cranking, as shown in Section 7.11.2. Still, it's important for you to understand the details of the synthesis procedure, both to give you an appreciation of the compiler's function and to give you a chance of figuring out what's really going on when the compiler produces unexpected results. Therefore, all nine steps of the state-machine design procedure are discussed in the remainder of this section.

STATE-TABLE DESIGN AS A KIND OF PROGRAMMING

Designing a state table (or equivalently, a state diagram) is a creative process that is like writing a computer program in many ways:

- You start with a fairly precise description of inputs and outputs, but a possibly ambiguous description of the desired relationship between them, and usually no clue about how to actually obtain the desired outputs from the inputs.

- During the design you may have to identify and choose among different ways of doing things, sometimes using common sense, and sometimes arbitrarily.

- You may have to identify and handle special cases that weren't included in the original description.

- You will probably have to keep track of several ideas in your head during the design process.

- Since the design process is not an algorithm, there's no guarantee that you can complete the state table or program using a finite number of states or lines of code. However, unless you work for the government, you must try to do so.

- When you finally run the state machine or program, it will do exactly what you told it to do—no more, no less.

- There's no guarantee that the thing will work the first time; you may have to debug and iterate on the whole process.

Although state-table design is a challenge, there's no need to be intimidated. If you've made it this far in your education, then you've probably written a few programs that worked, and you can become just as good at designing state tables.

7.4.1 State-Table Design Example

There are several different ways to describe a state machine's state table. Later, we'll see how ABEL and VHDL can specify state tables indirectly. In this section, however, we deal only with state tables that are specified directly, in the same tabular format that we used in the previous section for analysis.

We'll present the state-table design process as well as the synthesis procedure in later subsections, using the simple design problem below:

Design a clocked synchronous state machine with two inputs, A and B, and a single output Z that is 1 if:

- A had the same value at each of the two previous clock ticks, *or*

- B has been 1 since the last time that the first condition was true.

Otherwise, the output should be 0.

If the meaning of this specification isn't crystal clear to you at this point, don't worry. Part of your job as a designer is to convert such a specification into a state table that is absolutely unambiguous; even if it doesn't match what was originally intended, it at least forms a basis for further discussion and refinement.

REALIZING RELIABLE RESET

For proper system operation, the hardware design of a state machine should ensure that it enters a known initial state on power-up, such as the INIT state in our design example. Most systems have a RESET signal that is asserted during power-up.

The RESET signal is typically generated by an analog circuit. Such a reset circuit typically detects a voltage (say, 4.5 V) close to the power supply's full voltage, and follows that with a delay (say, 100 ms) to ensure that all components (including oscillators) have had time to stabilize before it "unresets" the system. The Texas Instruments TL7705 is such an analog reset IC; it has an internal 4.5-V reference for the detector and uses an external resistor and capacitor to determine the "unreset" time constant.

If a state machine is built using discrete flip-flops with asynchronous preset and clear inputs, the RESET signal can be applied to these inputs to force the machine into the desired initial state. If preset and clear inputs are not available, or if reset must be synchronous (as in systems using high-speed microprocessors), then the RESET signal may be used as another input to the state machine, with all of the next-state entries going to the desired initial state when RESET is asserted.

As an additional "hint" or requirement, state-table design problems often include timing diagrams that show the state machine's expected behavior for one or more sequences of inputs. Such a timing diagram is unlikely to specify unambiguously the machine's behavior for all possible sequences of inputs, but, again, it's a good starting point for discussion and a benchmark against which proposed designs can be checked. Figure 7-47 is such a timing diagram for our example state-table design problem.

The first step in the state-table design is to construct a template. From the word description, we know that our example is a Moore machine—its output depends only on the current state, that is, what happened in previous clock periods. Thus, as shown in Figure 7-48(a), we provide one next-state column for each possible input combination and a single column for the output values. The order in which the input combinations are written doesn't affect this part of the

Figure 7-47 Timing diagram for example state machine.

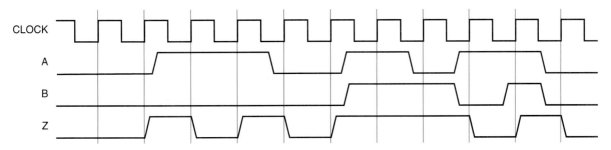

(a)

Meaning	S	A B 00	01	11	10	Z
Initial state	INIT					0
. . .						
. . .						
. . .						
			S*			

(b)

Meaning	S	A B 00	01	11	10	Z
Initial state	INIT	A0	A0	A1	A1	0
Got a 0 on A	A0					0
Got a 1 on A	A1					0
			S*			

(c)

Meaning	S	A B 00	01	11	10	Z
Initial state	INIT	A0	A0	A1	A1	0
Got a 0 on A	A0	OK	OK	A1	A1	0
Got a 1 on A	A1					0
Got two equal A inputs	OK					1
			S*			

(d)

Meaning	S	A B 00	01	11	10	Z
Initial state	INIT	A0	A0	A1	A1	0
Got a 0 on A	A0	OK	OK	A1	A1	0
Got a 1 on A	A1	A0	A0	OK	OK	0
Got two equal A inputs	OK					1
			S*			

Figure 7-48 Evolution of a state table.

process, but we've written them in Karnaugh-map order to simplify the derivation of excitation equations later. In a Mealy machine we would omit the output column and write the output values along with the next-state values under each input combination. The leftmost column is simply an English-language reminder of the meaning of each state or the "history" associated with it.

The word description isn't specific about what happens when this machine is first started, so we'll just have to improvise. We'll assume that when power is first applied to the system, the machine enters an *initial state*, called INIT in this example. We write the name of the initial state (INIT) in the first row and leave room for enough rows (states) to complete the design. We can also fill in the value of Z for the INIT state; common sense says it should be 0 because there were *no* inputs beforehand.

initial state

Next, we must fill in the next-state entries for the INIT row. The Z output can't be 1 until we've seen at least two inputs on A, so we'll provide two states, A0 and A1, that "remember" the value of A on the previous clock tick, as shown in Figure 7-48(b). In both of these states Z is 0, since we haven't satisfied the conditions for a 1 output yet. The precise meaning of state A0 is "Got A = 0 on the previous tick, A ≠ 0 on the tick before that, and B ≠ 1 at some time since the previous pair of equal A inputs." State A1 is defined similarly.

At this point we know that our state machine has at least three states, and we have created two more blank rows to fill in. Hmmmm, this isn't such a good trend! In order to fill in the next-state entries for *one* state (INIT), we had to create *two* new states A0 and A1. If we kept going this way, we could end up with 65,535 states by bedtime! Instead, we should be on the lookout for existing states

(a)

Meaning	S	A B				Z
		00	01	11	10	
Initial state	INIT	A0	A0	A1	A1	0
Got a 0 on A	A0	OK	OK	A1	A1	0
Got a 1 on A	A1	A0	A0	OK	OK	0
Got two equal A inputs	OK	?	OK	OK	?	1
			S*			

(b)

Meaning	S	A B				Z
		00	01	11	10	
Initial state	INIT	A0	A0	A1	A1	0
Got a 0 on A	A0	OK0	OK0	A1	A1	0
Got a 1 on A	A1	A0	A0	OK1	OK1	0
Two equal, A=0 last	OK0					1
Two equal, A=1 last	OK1					1
			S*			

(c)

Meaning	S	A B				Z
		00	01	11	10	
Initial state	INIT	A0	A0	A1	A1	0
Got a 0 on A	A0	OK0	OK0	A1	A1	0
Got a 1 on A	A1	A0	A0	OK1	OK1	0
Two equal, A=0 last	OK0	OK0	OK0	OK1	A1	1
Two equal, A=1 last	OK1					1
			S*			

(d)

Meaning	S	A B				Z
		00	01	11	10	
Initial state	INIT	A0	A0	A1	A1	0
Got a 0 on A	A0	OK0	OK0	A1	A1	0
Got a 1 on A	A1	A0	A0	OK1	OK1	0
Two equal, A=0 last	OK0	OK0	OK0	OK1	A1	1
Two equal, A=1 last	OK1	A0	OK0	OK1	OK1	1
			S*			

Figure 7-49 Continued evolution of a state table.

that have the same meaning as new ones that we might otherwise create. Let's see how it goes.

In state A0, we know that input A was 0 at the previous clock tick. Therefore, if A is 0 again, we go to a new state OK with Z = 1, as shown in Figure 7-48(c). If A is 1, then we don't have two equal inputs in a row, so we go to state A1 to remember that we just got a 1. Likewise in state A1, shown in (d), we go to OK if we get a second 1 input in a row, or to A0 if we get a 0.

Once we get into the OK state, the machine description tells us we can stay there as long as B = 1, irrespective of the A input, as shown in Figure 7-49(a). If B = 0, we have to look for two 1s or two 0s in a row on A again. However, we've got a little problem in this case. The current A input may or may not be the second equal input in a row, so we may still be "OK" or we may have to go back to A0 or A1. We defined the OK state too broadly—it doesn't "remember" enough to tell us which way to go.

The problem is solved in Figure 7-49(b) by splitting OK into two states, OK0 and OK1, that "remember" the previous A input. All of the next states for OK0 and OK1 can be selected from existing states, as shown in (c) and (d). For example, if we get A = 0 in OK0, we can just stay in OK0; we don't have to create a new state that "remembers" three 0s in a row, because the machine's description doesn't require us to distinguish that case. Thus, we have achieved "closure" of the state table, which now describes a *finite*-state machine. As a sanity check, Figure 7-50 repeats the timing diagram of Figure 7-47, listing the states that should be visited according to our final state table.

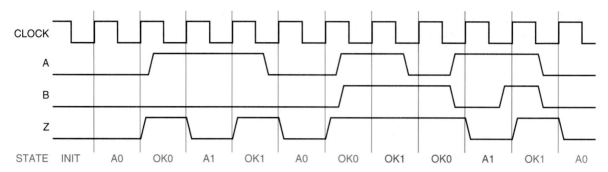

Figure 7-50 Timing diagram and state sequence for example state machine.

7.4.2 State Minimization

Figure 7-49(d) is a "minimal" state table for our original word description, in the sense that it contains the fewest possible states. However, Figure 7-51 shows other state tables, with more states, that also do the job. Formal procedures can be used to minimize the number of states in such tables.

equivalent states

The basic idea of formal minimization procedures is to identify *equivalent states*, where two states are equivalent if it is impossible to distinguish them by observing only the current and future *outputs* of the machine (and *not* the internal state variables). A pair of equivalent states can be replaced by a single state.

Two states S1 and S2 are equivalent if two conditions are true. First, S1 and S2 must produce the same values at the state-machine output(s); in a Mealy machine, this must be true for all input combinations. Second, for each input combination, S1 and S2 must have either the same next state or equivalent next states.

Thus, a formal state-minimization procedure shows that states OK00 and OKA0 in Figure 7-51(a) are equivalent because they produce the same output and their next-state entries are identical. Since the states are equivalent, state

Figure 7-51 Nonminimal state tables equivalent to Figure 7-49(d).

(a)

Meaning	S	00	01	11	10	Z
Initial state	INIT	A0	A0	A1	A1	0
Got a 0 on A	A0	OK00	OK00	A1	A1	0
Got a 1 on A	A1	A0	A0	OK11	OK11	0
Got 00 on A	OK00	OK00	OK00	OKA1	A1	1
Got 11 on A	OK11	A0	OKA0	OK11	OK11	1
OK, got a 0 on A	OKA0	OK00	OK00	OKA1	A1	1
OK, got a 1 on A	OKA1	A0	OKA0	OK11	OK11	1

S*

(b)

Meaning	S	00	01	11	10	Z
Initial state	INIT	A0	A0	A1	A1	0
Got a 0 on A	A0	OK00	OK00	A1	A1	0
Got a 1 on A	A1	A0	A0	OK11	OK11	0
Got 00 on A	OK00	OK00	OK00	A001	A1	1
Got 11 on A	OK11	A0	A110	OK11	OK11	1
Got 001 on A, B=1	A001	A0	AE10	OK11	OK11	1
Got 110 on A, B=1	A110	OK00	OK00	AE01	A1	1
Got bb...10 on A, B=1	AE10	OK00	OK00	AE01	A1	1
Got bb...01 on A, B=1	AE01	A0	AE10	OK11	OK11	1

S*

In the header of both tables: A B spans columns 00, 01, 11, 10.

OK00 may be eliminated and its occurrences in the table replaced by OKA0, or vice versa. Likewise, states OK11 and OKA1 are equivalent.

To minimize the state table in Figure 7-51(b), a formal procedure must use a bit of circular reasoning. States OK00, A110, and AE10 all produce the same output and have almost identical next-state entries, so they might be equivalent. They are equivalent only if A001 and AE01 are equivalent. Similarly, OK11, A001, and AE01 are equivalent only if A110 and AE10 are equivalent. In other words, the states in the first set are equivalent if the states in the second set are, and vice versa. So, let's just go ahead and say they're equivalent.

IS THIS REALLY ALL NECESSARY?	Details of formal state-minimization procedures are discussed in advanced textbooks, cited in the References. However, these procedures are seldom used by most digital designers. By carefully matching state meanings to the requirements of the problem, experienced digital designers produce state tables for small problems with a minimal or near-minimal number of states, without using a formal minimization procedure. Also, there are situations where *increasing* the number of states may simplify the design or reduce its cost, so even an automated state-minimization procedure doesn't necessarily help. A designer can do more to simplify a state machine during the state-assignment phase of the design, discussed in the next subsection.

7.4.3 State Assignment

The next step in the design process is to determine how many binary variables are required to represent the states in the state table, and to assign a specific combination to each named state. We'll call the binary combination assigned to a particular state a *coded state*. The *total number of states* in a machine with n flip-flops is 2^n, so the number of flip-flops needed to code s states is $\lceil \log_2 s \rceil$, the smallest integer greater than or equal to $\log_2 s$.

coded state
total number of states

For reference, the state/output table of our example machine is repeated in Table 7-6. It has five states, so it requires three flip-flops. Of course, three flip-flops provide a total of eight states, so there will be $8 - 5 = 3$ *unused states*. We'll discuss alternatives for handling the unused states at the end of this subsection. Right now, we have to deal with lots of choices for the five coded states.

unused states

INITIAL VERSUS IDLE STATES	The example state machine in this subsection visits its initial state only during reset. Many machines are designed instead with an "idle" state that is entered both at reset and whenever the machine has nothing in particular to do.

Table 7-6
State and output table
for example problem.

S	00	01	11	10	Z
INIT	A0	A0	A1	A1	0
A0	OK0	OK0	A1	A1	0
A1	A0	A0	OK1	OK1	0
OK0	OK0	OK0	OK1	A1	1
OK1	A0	OK0	OK1	OK1	1
			S*		

The table above has the heading **A B** spanning columns 00, 01, 11, 10, and the bottom row label is S*.

The simplest assignment of s coded states to 2^n possible states is to use the first s binary integers in binary counting order, as shown in the first assignment column of Table 7-7. However, the simplest state assignment does not always lead to the simplest excitation equations, output equations, and resulting logic circuit. In fact, the state assignment often has a major effect on circuit cost, and it may interact with other factors, such as the choice of storage elements (e.g., D vs. J-K flip-flops) and the realization approach for excitation and output logic (e.g., a sum of products, a product of sums, or an ad hoc design).

So, how do we choose the best state assignment for a given problem? In general, the only formal way to find the *best* assignment is to try *all* the assignments. That's too much work, even for students. Instead, most digital designers rely on experience and several practical guidelines for making reasonable state assignments:

- Choose an initial coded state into which the machine can easily be forced at reset (00...00 or 11...11 in typical circuits).

- Minimize the number of state variables that change on each transition.

CAUTION: MATH   The number of different ways to choose m coded states out of a set of n possible states is given by a *binomial coefficient*, denoted $\binom{n}{m}$, whose value is $\frac{n!}{m! \cdot (n-m)}$. (We used binomial coefficients previously in Section 2.10 in the context of decimal coding.) In our example, there are $\binom{8}{5}$ different ways to choose five coded states out of eight possible states, and 5! ways to assign the five named states to each different choice. So there are $\frac{8!}{5! \cdot 3!} \cdot 5!$ or 6,720 different ways to assign the five states of our example machine to combinations of three binary state variables. We don't have time to look at all of them.

| | | Assignment | | |
State Name	Simplest Q1–Q3	Decomposed Q1–Q3	One-hot Q1–Q5	Almost One-hot Q1–Q4
INIT	000	000	00001	0000
A0	001	100	00010	0001
A1	010	101	00100	0010
OK0	011	110	01000	0100
OK1	100	111	10000	1000

Table 7-7
Possible state assignments for the state machine in Table 7-6.

- Maximize the number of state variables that don't change in a group of related states (i.e., a group of states in which most of transitions stay in the group).

- Exploit symmetries in the problem specification and the corresponding symmetries in the state table. That is, suppose that one state or group of states means almost the same thing as another. Once an assignment has been established for the first, a similar assignment, differing only in one bit, should be used for the second.

- If there are unused states (i.e., if $s < 2^n$ where $n = \lceil \log_2 s \rceil$), then choose the "best" of the available state-variable combinations to achieve the foregoing goals. That is, don't limit the choice of coded states to the first s n-bit integers.

- Decompose the set of state variables into individual bits or fields, where each bit or field has a well-defined meaning with respect to the input effects or output behavior of the machine.

- Consider using more than the minimum number of state variables to make a decomposed assignment possible.

Some of these ideas are incorporated in the "decomposed" state assignment in Table 7-7. As before, the initial state is 000, which is easy to force either asynchronously (applying the RESET signal to the flip-flop CLR inputs) or synchronously (by ANDing RESET' with all of the D flip-flop inputs). After this point, the assignment takes advantage of the fact that there are only four states in addition to INIT, which is a fairly "special" state that is never reentered once the machine gets going. Therefore, Q1 can be used to indicate whether or not the machine is in the INIT state, and Q2 and Q3 can be used to distinguish among the four non-INIT states.

The non-INIT states in the "decomposed" column of Table 7-7 appear to have been assigned in binary counting order, but that's just a coincidence. State bits Q2 and Q3 actually have individual meanings in the context of the state machine's inputs and output. Q3 gives the previous value of A, and Q2 indicates that the conditions for a 1 output are satisfied in the current state. By decomposing the state-bit meanings in this way, we can expect the next-state and output logic to be simpler than in a "random" assignment of Q2,Q3 combinations to the non-INIT states. We'll continue the state-machine design based on this assignment in later subsections.

one-hot assignment

Another useful state assignment, one that can be adapted to any state machine, is the *one-hot assignment* shown in Table 7-7. This assignment uses more than the minimum number of state variables—it uses one bit per state. In addition to being simple, a one-hot assignment has the advantage of usually leading to small excitation equations, since each flip-flop must be set to 1 for transitions into only one state. An obvious disadvantage of a one-hot assignment, especially for machines with many states, is that it requires (many) more than the minimum number of flip-flops. However, the one-hot encoding is ideal for a machine with s states that is required to have a set of 1-out-of-s coded outputs indicating its current state. The one-hot-coded flip-flop outputs can be used directly for this purpose, with no additional combinational output logic.

The last column of Table 7-7 is an "almost one-hot assignment" that uses the "no-hot" combination for the initial state. This makes a lot of sense for two reasons: it's easy to initialize most storage devices to the all-0s state, and the initial state in this machine is never revisited once the machine gets going. Completing the state-machine design using this state assignment is considered in Exercises 7.37 and 7.40.

unused states

We promised earlier to consider the disposition of *unused states* when the number of states available with n flip-flops, 2^n, is greater than the number of states required, s. There are two approaches that make sense, depending on the application requirements:

- *Minimal risk.* This approach assumes that it is possible for the state machine somehow to get into one of the unused (or "illegal") states, perhaps because of a hardware failure, an unexpected input, or a design error. Therefore, all of the unused state-variable combinations are identified and explicit next-state entries are made so that, for any input combination, the unused states go to the "initial" state, the "idle" state, or some other "safe" state. This is an automatic consequence of some design methodologies if the initial state is coded 00. . . 00.

- *Minimal cost.* This approach assumes that the machine will never enter an unused state. Therefore, in the transition and excitation tables, the next-state entries of the unused states can be marked as "don't-cares." In most cases this simplifies the excitation logic. However, the machine's behavior if it ever does enter an unused state may be pretty weird.

We'll look at both of these approaches as we complete the design of our example state machine.

7.4.4 Synthesis Using D Flip-Flops

Once we've assigned coded states to the named states of a machine, the rest of the design process is pretty much "turning the crank." In fact, in Section 7.11.2 we'll describe software tools that can turn the crank for you. Just so that you'll appreciate those tools, however, we'll go through the process by hand in this subsection.

Coded states are substituted for named states in the (possibly minimized) state table to obtain a *transition table*. The transition table shows the next coded state for each combination of current coded state and input. Table 7-8 shows the transition and output table that is obtained from the example state machine of Table 7-6 on page 570 using the "decomposed" assignment of Table 7-7 on page 571.

transition table

The next step is to write an *excitation table* that shows, for each combination of coded state and input, the flip-flop excitation input values needed to make the machine go to the desired next coded state. This structure and content of this table depend on the type of flip-flops that are used (D, J-K, T, etc.). We *usually* have a particular flip-flop type in mind at the beginning of a design—and we *certainly* do in this subsection, given its title. In fact, most state-machine designs nowadays use D flip-flops, because of their availability in both discrete packages and programmable logic devices, and because of their ease of use (compare with J-K flip-flops in the next subsection).

excitation table

Of all flip-flop types, a D flip-flop has the simplest characteristic equation, $Q* = D$. Each D flip-flop in a state machine has a single excitation input, D, and the excitation table must show the value required at each flip-flop's D input for each coded-state/input combination. Table 7-9 shows the excitation table for our example problem. Since $D = Q*$, the excitation table is identical to the transition table, except for labeling of its entries. Thus, with D flip-flops, you don't really need to write a separate excitation table; you can just call the first table a *transition/excitation table*.

transition/excitation table

The excitation table is like a truth table for three combinational logic functions (D1, D2, D3) of five variables (A, B, Q1, Q2, Q3). Accordingly, we can

		A B			
Q1 Q2 Q3	00	01	11	10	Z
000	100	100	101	101	0
100	110	110	101	101	0
101	100	100	111	111	0
110	110	110	111	101	1
111	100	110	111	111	1
		Q1* Q2* Q3*			

Table 7-8
Transition and output table for example problem.

Table 7-9
Excitation and output table for Table 7-8 using D flip-flops.

Q1 Q2 Q3	A B				Z
	00	**01**	**11**	**10**	
000	100	100	101	101	0
100	110	110	101	101	0
101	100	100	111	111	0
110	110	110	111	101	1
111	100	110	111	111	1
	D1 D2 D3				

excitation maps

design circuits to realize these functions using any of the combinational design methods at our disposal. In particular, we can transfer the information in the excitation table to Karnaugh maps, which we may call *excitation maps*, and find a minimal sum-of-products or product-of-sums expression for each function.

5-variable Karnaugh map

Excitation maps for our example state machine are shown in Figure 7-52. Each function, such as D1, has five variables and therefore uses a *5-variable Karnaugh map*. A 5-variable map is drawn as a pair of 4-variable maps, where cells in the same position in the two maps are considered to be adjacent. These maps are a bit unwieldy, but if you want to design by hand any but the most trivial state machines, you're going to get stuck with 5-variable maps and worse. At least we had the foresight to label the input combinations of the original state table in Karnaugh-map order, which makes it easier to transfer information to the maps in this step. However, note that the *states* were not assigned in Karnaugh-map order; in particular, the rows for states 110 and 111 are in the opposite order in the map as in the excitation table.

Figure 7-52
Excitation maps for D1, D2, and D3 assuming that unused states go to state 000.

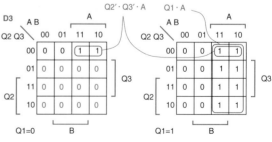

It is in this step, transferring the excitation table to excitation maps, that we discover why the excitation table is not quite a truth table—it does not specify functional values for *all* input combinations. In particular, the next-state information for the unused states, 001, 010, and 011, is not specified. Here we must make a choice, discussed in the preceding subsection, between a minimal-risk and a minimal-cost strategy for handling the unused states. Figure 7-52 has taken the minimal-risk approach: The next state for each unused state and input combination is 000, the INIT state. The three rows of colored 0s in each Karnaugh map are the result of this choice. With the maps completely filled in, we can now obtain minimal sum-of-products expressions for the flip-flop excitation inputs:

$$D1 \; = \; Q1 + Q2' \cdot Q3'$$
$$D2 \; = \; Q1 \cdot Q3' \cdot A' + Q1 \cdot Q3 \cdot A + Q1 \cdot Q2 \cdot B$$
$$D3 \; = \; Q1 \cdot A + Q2' \cdot Q3' \cdot A$$

An output equation can easily be developed directly from the information in Table 7-9. The output equation is simpler than the excitation equations, because the output is a function of state only. We could use a Karnaugh map, but it's easy to find a minimal-risk output function algebraically, by writing it as the sum of the two coded states (110 and 111) in which Z is 1:

$$Z \; = \; Q1 \cdot Q2 \cdot Q3' + Q1 \cdot Q2 \cdot Q3$$
$$= \; Q1 \cdot Q2$$

At this point, we're just about done with the state-machine design. If the state machine is going to be built with discrete flip-flops and gates, then the final step is to draw a logic diagram. On the other hand, if we are using a programmable logic device, then we only have to enter the excitation and output equations into a computer file that specifies how to program the device, as an example shows in Section 7.11.1. Or, if we were thinking ahead, we specified the machine using a state-machine description language like ABEL in the first place (Section 7.11.2), and the computer did all the work in this subsection for us!

MINIMAL-COST SOLUTION   If we choose in our example to derive minimal-cost excitation equations, we write "don't-cares" in the next-state entries for the unused states. The colored d's in Figure 7-53 are the result of this choice. The excitation equations obtained from this map are somewhat simpler than before:

$$D1 \; = \; 1$$
$$D2 \; = \; Q1 \cdot Q3' \cdot A' + Q3 \cdot A + Q2 \cdot B$$
$$D3 \; = \; A$$

For a minimal-cost output function, the value of Z is a "don't-care" for the unused states. This leads to an even simpler output function, $Z = Q2$. The logic diagram for the minimal-cost solution is shown in Figure 7-54.

Figure 7-53
Excitation maps for D1, D2, and D3 assuming that next states of unused states are "don't-cares."

Figure 7-54
Logic diagram resulting from Figure 7-53.

*7.4.5 Synthesis Using J-K Flip-Flops

At one time, J-K flip-flops were popular for discrete SSI state-machine designs, since a J-K flip-flop embeds more functionality than a D flip-flop in the same size SSI package. By "more functionality" we mean that the combination of J and K inputs yields more possibilities for controlling the flip-flop than a single D input does. As a result, a state machine's excitation logic may be simpler using J-K flip-flops than using D flip-flops, which reduced package count when SSI gates were used for the excitation logic.

JUST FOR FUN   While minimizing excitation logic was a big deal in the days of SSI-based design, the name of the game has changed with PLDs and ASICs. As you might guess from your knowledge of the AND-OR structure of combinational PLDs, the need to provide separate AND-OR arrays for the J and K inputs of a J-K flip-flop would be a distinct disadvantage in a sequential PLD.

In ASIC technologies, J-K flip-flops aren't so desirable either. For example, in LSI Logic Corp.'s LCA500K series of CMOS gate arrays, an FD1QP D flip-flop macrocell uses 7 "gate cells," while an FJK1QP J-K flip-flop macrocell uses 9 gate cells, over 25% more chip area. Therefore, a more cost-effective design usually results from sticking with D flip-flops and using the extra chip area for more complex excitation logic in just the cases where it's really needed.

Still, this subsection describes the J-K synthesis process "just for fun."

Up through the state-assignment step, the design procedure with J-K flip-flops is basically the same as with D flip-flops. The only difference is that a designer might select a slightly different state assignment, knowing the sort of behavior that can easily be obtained from J-K flip-flops (e.g., "toggling" by setting J and K to 1).

The big difference occurs in the derivation of an excitation table from the transition table. With D flip-flops, the two tables are identical; using the D's characteristic equation, $Q* = D$, we simply substitute $D = Q*$ for each entry. With J-K flip-flops, each entry in the excitation table has twice as many bits as in the transition table, since there are two excitation inputs per flip-flop.

A J-K flip-flop's characteristic equation, $Q* = J \cdot Q' + K' \cdot Q$, cannot be rearranged to obtain independent equations for J and K. Instead, the required values for J and K are expressed as functions of Q and Q* in a J-K *application table*, Table 7-10. According to the first row, if Q is currently 0, all that is required to obtain 0 as the next value of Q is to set J to 0; the value of K doesn't matter. Similarly, according to the third row, if Q is currently 1, the next value of Q will be 0 if K is 1, regardless of J's value. Each desired transition can be obtained by either of two different combinations on the J and K inputs, so we get a "don't-care" entry in each row of the application table.

J-K application table

To obtain a J-K excitation table, the designer must look at both the current and desired next value of each state bit in the transition table and substitute the

Q	Q*	J	K
0	0	0	d
0	1	1	d
1	0	d	1
1	1	d	0

Table 7-10
Application table for
J-K flip-flops.

corresponding pair of J and K values from the application table. For the transition table in Table 7-8 on page 573, these substitutions produce the excitation table in Table 7-11. For example, in state 100 under input combination 00, Q1 is 1 and the required Q1* is 1; therefore, "d0" is entered for J1 K1. For the same state/input combination, Q2 is 0 and Q2* is 1, so "1d" is entered for J2 K2. Obviously, it takes quite a bit of patience and care to fill in the entire excitation table (a job best left to a computer).

As in the D synthesis example of the preceding subsection, the excitation table is *almost* a truth table for the excitation functions. This information is transferred to Karnaugh maps in Figure 7-55.

The excitation table does not specify next states for the unused states, so once again we must choose between the minimal-risk and minimal-cost approaches. The colored entries in the Karnaugh maps result from taking the minimal-risk approach.

Figure 7-55 Excitation maps for J1, K1, J2, K2, J3, and K3, assuming that unused states go to state 000.

Q1 Q2 Q3	AB				Z
	00	**01**	**11**	**10**	
000	1d, 0d, 0d	1d, 0d, 0d	1d, 0d, 1d	1d, 0d, 1d	0
100	d0, 1d, 0d	d0, 1d, 0d	d0, 0d, 1d	d0, 0d, 1d	0
101	d0, 0d, d1	d0, 0d, d1	d0, 1d, d0	d0, 1d, d0	0
110	d0, d0, 0d	d0, d0, 0d	d0, d0, 1d	d0, d1, 1d	1
111	d0, d1, d1	d0, d0, d1	d0, d0, d0	d0, d0, d0	1
	J1 K1, J2 K2, J3 K3				

Table 7-11
Excitation and output table for the state machine of Table 7-8, using J-K flip-flops.

Note that even though the "safe" next state for unused states is 000, we didn't just put 0s in the corresponding map cells, as we were able to do in the D case. Instead, we still had to work with the application table to determine the proper combination of J and K needed to get $Q* = 0$ for each unused state entry, once again a tedious and error-prone process.

Using the maps in Figure 7-55, we can derive sum-of-products excitation equations:

$$J1 = Q2' \cdot Q3' \qquad\qquad K1 = 0$$
$$J2 = Q1 \cdot Q3' \cdot A' + Q1 \cdot Q3 \cdot A \qquad K2 = Q1' + Q3' \cdot A \cdot B' + Q3 \cdot A' \cdot B'$$
$$J3 = Q2' \cdot A + Q1 \cdot A \qquad\qquad K3 = Q1' + A'$$

These equations take two more gates to realize than do the preceding subsection's minimal-risk equations using D flip-flops, so J-K flip-flops didn't save us anything in this example, least of all design time.

MINIMAL-COST SOLUTION

In the preceding design example, excitation maps for the minimal-cost approach would have been somewhat easier to construct, since we could have just put d's in all of the unused state entries. Sum-of-products excitation equations obtained from the minimal-cost maps (not shown) are as follows:

$$J1 = 1 \qquad\qquad K1 = 0$$
$$J2 = Q1 \cdot Q3' \cdot A' + Q3 \cdot A \qquad K2 = Q3' \cdot A \cdot B' + Q3 \cdot A' \cdot B'$$
$$J3 = A \qquad\qquad K3 = A'$$

The state encoding for the J-K circuit is the same as in the D circuit, so the output equation is the same, $Z = Q1 \cdot Q2$ for minimal risk, $Z = Q2$ for minimal cost.

A logic diagram corresponding to the minimal-cost equations is shown in Figure 7-56. This circuit has two more gates than the minimal-cost D circuit in Figure 7-54, so J-K flip-flops still didn't save us anything.

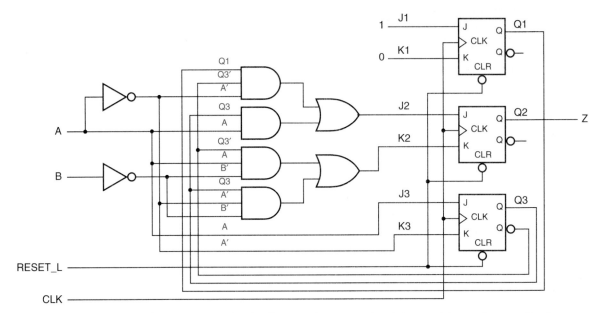

Figure 7-56 Logic diagram for example state machine using J-K flip-flops and minimal-cost excitation logic.

7.4.6 More Design Examples Using D Flip-Flops

We'll conclude this section with two more state-machine design examples using D flip-flops. The first example is a "1s-counting machine":

> Design a clocked synchronous state machine with two inputs, X and Y, and one output, Z. The output should be 1 if the number of 1 inputs on X and Y since reset is a multiple of 4, and 0 otherwise.

At first glance, you might think the machine needs an infinite number of states, since it counts 1 inputs over an arbitrarily long time. However, since the output indicates the number of inputs received *modulo 4*, four states are sufficient. We'll name them S0–S3, where S0 is the initial state and the total number of 1s received in S*i* is *i* modulo 4. Table 7-12 is the resulting state and output table.

Table 7-12
State and output
table for 1s-counting
machine.

Meaning	S	00	01	11	10	Z
		XY				
Got zero 1s (modulo 4)	S0	S0	S1	S2	S1	1
Got one 1 (modulo 4)	S1	S1	S2	S3	S2	0
Got two 1s (modulo 4)	S2	S2	S3	S0	S3	0
Got three 1s (modulo 4)	S3	S3	S0	S1	S0	0
		S*				

	XY				
Q1 Q2	**00**	**01**	**11**	**10**	**Z**
00	00	01	11	01	1
01	01	11	10	11	0
11	11	10	00	10	0
10	10	00	01	00	0
	Q1∗ Q2∗ or D1 D2				

Table 7-13
Transition/excitation
and output table for
1s-counting machine.

The 1s-counting machine can use two state variables to code its four states, with no unused states. In this case, there are only 4! possible assignments of coded states to named states. Still, we'll try only one of them. We'll assign coded states to the named states in Karnaugh-map order (00, 01, 11, 10) for two reasons: in this state table, it minimizes the number of state variables that change for most transitions, potentially simplifying the excitation equations; and it simplifies the mechanical transfer of information to excitation maps.

A transition/excitation table based on our chosen state assignment is shown in Table 7-13. Since we're using D flip-flops, the transition and excitation tables are the same. Corresponding Karnaugh maps for D1 and D2 are shown in Figure 7-57. Since there are no unused states, all of the information we need is in the excitation table; no choice is required between minimal-risk and minimal-cost approaches. The excitation equations can be read from the maps, and the output equation can be read directly from the transition/excitation table:

$$D1 = Q2 \cdot X' \cdot Y + Q1' \cdot X \cdot Y + Q1 \cdot X' \cdot Y' + Q2 \cdot X \cdot Y'$$
$$D2 = Q1' \cdot X' \cdot Y + Q1' \cdot X \cdot Y' + Q2 \cdot X' \cdot Y' + Q2' \cdot X \cdot Y$$
$$Z = Q1' \cdot Q2'$$

A logic diagram using D flip-flops and AND-OR or NAND-NAND excitation logic can be drawn from these equations.

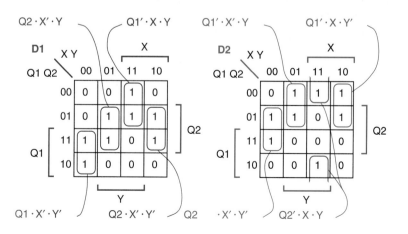

Figure 7-57
Excitation maps for
D1 and D2 inputs in
1s-counting machine.

The second example is a "combination lock" state machine that activates an "unlock" output when a certain binary input sequence is received:

Design a clocked synchronous state machine with one input, X, and two outputs, UNLK and HINT. The UNLK output should be 1 if and only if X is 0 and the sequence of inputs received on X at the preceding seven clock ticks was 0110111. The HINT output should be 1 if and only if the current value of X is the correct one to move the machine closer to being in the "unlocked" state (with UNLK = 1).

It should be apparent from the word description that this is a Mealy machine. The UNLK output depends on both the past history of inputs and X's current value, and HINT depends on both the state and the current X (indeed, if the current X produces HINT = 0, then the clued-in user will want to change X before the clock tick).

A state and output table for the combination lock is presented in Table 7-14. In the initial state, A, we assume that we have received no inputs in the required sequence; we're looking for the first 0 in the sequence. Therefore, as long as we get 1 inputs, we stay in state A, and we move to state B when we receive a 0. In state B, we're looking for a 1. If we get it, we move on to C; if we don't, we can stay in B, since the 0 we just received might still turn out to be the first 0 in the required sequence. In each successive state, we move on to the next state if we get the correct input, and we go back to A or B if we get the wrong one. An exception occurs in state G; if we get the wrong input (a 0) there, the previous three inputs might still turn out to be the first three inputs of the required sequence, so we go back to state E instead of B. In state H, we've received the required sequence, so we set UNLK to 1 if X is 0. In each state, we set HINT to 1 for the value of X that moves us closer to state H.

Table 7-14
State and output table for combination-lock machine.

Meaning	S	X 0	X 1
Got zip	A	B, 01	A, 00
Got 0	B	B, 00	C, 01
Got 01	C	B, 00	D, 01
Got 011	D	E, 01	A, 00
Got 0110	E	B, 00	F, 01
Got 01101	F	B, 00	G, 01
Got 011011	G	E, 00	H, 01
Got 0110111	H	B, 11	A, 00
		S*, UNLK HINT	

Table 7-15
Transition/excitation table for combination-lock machine.

Q1 Q2 Q3	X	
	0	1
000	001, 01	000, 00
001	001, 00	010, 01
010	001, 00	011, 01
011	100, 01	000, 00
100	001, 00	101, 01
101	001, 00	110, 01
110	100, 00	111, 01
111	001, 11	000, 00
	Q1* Q2* Q3*, UNLK HINT	

The combination lock's eight states can be coded with three state variables, leaving no unused states. There are 8! state assignments to choose from. To keep things simple, we'll use the simplest, and assign the states in binary counting order, yielding the transition/excitation table in Table 7-15. Corresponding Karnaugh maps for D1, D2, and D3 are shown in Figure 7-58. The excitation equations can be read from the maps:

$$D1 = Q1 \cdot Q2' \cdot X + Q1' \cdot Q2 \cdot Q3 \cdot X' + Q1 \cdot Q2 \cdot Q3'$$
$$D2 = Q2' \cdot Q3 \cdot X + Q2 \cdot Q3' \cdot X$$
$$D3 = Q1 \cdot Q2' \cdot Q3' + Q1 \cdot Q3 \cdot X' + Q2' \cdot X' + Q3' \cdot Q1' \cdot X' + Q2 \cdot Q3' \cdot X$$

Figure 7-58 Excitation maps for D1, D2, and D3 in combination-lock machine.

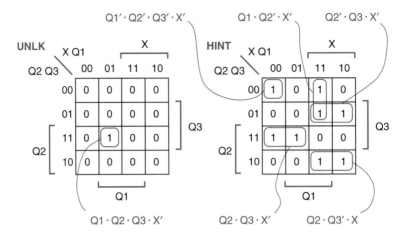

Figure 7-59
Karnaugh maps for output functions UNLK and HINT in combination-lock machine.

The output values are transferred from the transition/excitation and output table to another set of maps in Figure 7-59. The corresponding output equations are:

$$UNLK = Q1 \cdot Q2 \cdot Q3 \cdot X'$$
$$HINT = Q1' \cdot Q2' \cdot Q3' \cdot X' + Q1 \cdot Q2' \cdot X + Q2' \cdot Q3 \cdot X + Q2 \cdot Q3 \cdot X' + Q2 \cdot Q3' \cdot X$$

Note that some product terms are repeated in the excitation and output equations, yielding a slight savings in the cost of the AND-OR realization. If we went through the trouble of performing a formal multiple-output minimization of all five excitation and output functions, we could save two more gates (see Exercise 7.55).

7.5 Designing State Machines Using State Diagrams

Aside from planning the overall architecture of a digital system, designing state machines is probably the most creative task of a digital designer. Most people like to take a graphical approach to design—you've probably solved many problems just by doodling. For that reason, state diagrams are often used to design small- to medium-sized state machines. In this section we'll give examples of state-diagram design and describe a simple procedure for synthesizing circuits from the state diagrams. This procedure is the basis of the method used by CAD tools that can synthesize logic from graphical or even text-based "state diagrams."

Designing a state diagram is much like designing a state table, which, as we showed in Section 7.4.1, is much like writing a program. However, there is one fundamental difference between a state diagram and a state table, a difference that makes state-diagram design simpler but also more error prone:

- A state table is an exhaustive listing of the next states for each state/input combination. No ambiguity is possible.

- A state diagram contains a set of arcs labeled with transition expressions. Even when there are many inputs, only one transition expression is required per arc. However, when a state diagram is constructed, there is no guarantee that the transition expressions written on the arcs leaving a particular state cover all the input combinations exactly once.

In an improperly constructed (*ambiguous*) state diagram, the next state for some input combinations may be unspecified, which is generally undesirable, while multiple next states may be specified for others, which is just plain wrong. Thus, considerable care must be taken in the design of state diagrams; we'll give several examples.

ambiguous state diagram

Our first example is a state machine that controls the tail lights of a 1965 Ford Thunderbird, shown in Figure 7-60. There are three lights on each side, and for turns they operate in sequence to show the turning direction, as illustrated in Figure 7-61. The state machine has two input signals, LEFT and RIGHT, that carry the driver's request for a left turn or a right turn. It also has an emergency-flasher input, HAZ, that requests the tail lights to be operated in hazard mode— all six lights flashing on and off in unison. We also assume the existence of a free-running clock signal whose frequency equals the desired flashing rate for the lights.

Figure 7-60
T-bird tail lights.

LC   LB   LA   RA   RB   RC

WHOSE REAR END?   Actually, Figure 7-60 looks more like the rear end of a Mercury Capri, which also had sequential tail lights.

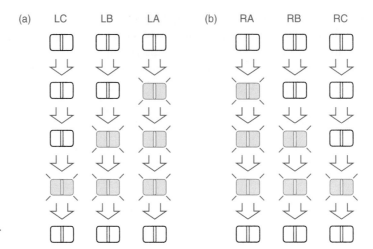

Figure 7-61
Flashing sequence for
T-bird tail lights:
(a) left turn; (b) right turn.

Figure 7-62
Initial state diagram
and output table for
T-bird tail lights.

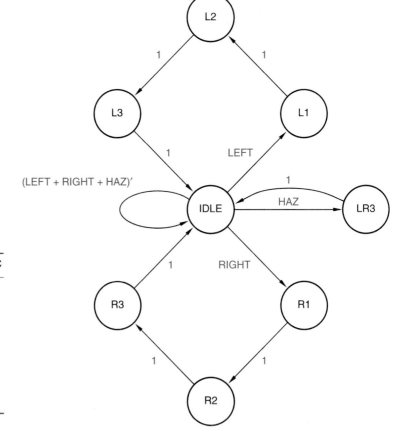

Output Table

State	LC	LB	LA	RA	RB	RC
IDLE	0	0	0	0	0	0
L1	0	0	1	0	0	0
L2	0	1	1	0	0	0
L3	1	1	1	0	0	0
R1	0	0	0	1	0	0
R2	0	0	0	1	1	0
R3	0	0	0	1	1	1
LR3	1	1	1	1	1	1

Given the foregoing requirements, we can design a clocked synchronous state machine to control the T-bird tail lights. We will design a Moore machine, so that the state alone determines which lights are on and which are off. For a left turn, the machine should cycle through four states in which the righthand lights are off and 0, 1, 2, or 3 of the lefthand lights are on. Likewise, for a right turn, it should cycle through four states in which the lefthand lights are off and 0, 1, 2, or 3 of the righthand lights are on. In hazard mode, only two states are required— all lights on and all lights off.

Figure 7-62 shows our first cut at a state diagram for the machine. A common IDLE state is defined in which all of the lights are off. When a left turn is requested, the machine goes through three states in which 1, 2, and 3 of the lefthand lights are on, and then back to IDLE; right turns work similarly. In the hazard mode, the machine cycles back and forth between the IDLE state and a state in which all six lights are on. Since there are so many outputs, we've included a separate output table rather than writing output values on the state diagram. Even without assigning coded states to the named states, we can write output equations from the output table, if we let each state name represent a logic expression that is 1 only in that state:

$$LA = L1 + L2 + L3 + LR3 \qquad RA = R1 + R2 + R3 + LR3$$
$$LB = L2 + L3 + LR3 \qquad RB = R2 + R3 + LR3$$
$$LC = L3 + LR3 \qquad RC = R3 + LR3$$

There's one big problem with the state diagram of Figure 7-62—it doesn't properly handle multiple inputs asserted simultaneously. For example, what happens in the IDLE state if both LEFT and HAZ are asserted? According to the state diagram, the machine goes to two states, L1 and LR3, which is impossible. In reality, the machine would have only one next state, which could be L1, LR3, or a totally unrelated (and possibly unused) third state, depending on details of the state machine's realization (e.g., see Exercise 7.58).

The problem is fixed in Figure 7-63, where we have given the HAZ input priority. Also, we treat LEFT and RIGHT asserted simultaneously as a hazard request, since the driver is clearly confused and needs help.

The new state diagram is unambiguous because the transition expressions on the arcs leaving each state are mutually exclusive and all-inclusive. That is, for each state, no two expressions are 1 for the same input combination, and some expression is 1 for every input combination. This can be confirmed algebraically for this or any other state diagram by performing two steps:

1. *Mutual exclusion.* For each state, show that the logical product of each possible pair of transition expressions on arcs leaving that state is 0. If there are n arcs, then there are $n(n-1)/2$ logical products to evaluate. *mutual exclusion*

2. *All inclusion.* For each state, show that the logical sum of the transition expressions on all arcs leaving that state is 1. *all inclusion*

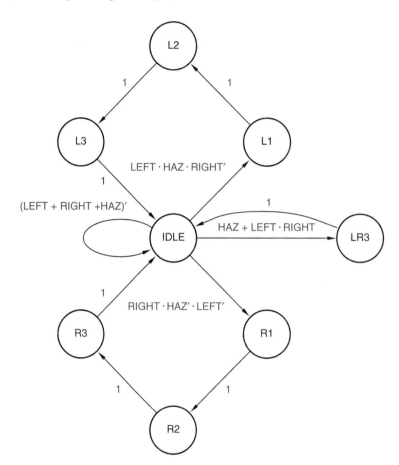

Figure 7-63
Corrected state
diagram for T-bird
tail lights.

If there are many transitions leaving each state, these steps, especially the first one, are very difficult to perform. However, typical state machines, even ones with lots of states and inputs, don't have many transitions leaving each state, since most designers can't dream up such complex machines in the first place. This is where the tradeoff between state-table and state-diagram design occurs. In state-table design, the foregoing steps are not required, because the structure of a state table guarantees mutual exclusion and all inclusion. But if there are a lot of inputs, the state table has *lots* of columns.

Verifying that a state diagram is unambiguous may be difficult in principle, but it's not too bad in practice for small state diagrams. In Figure 7-63, most of the states have a single arc with a transition expression of 1, so verification is trivial. Real work is needed only to verify the IDLE state, which has four transitions leaving it. This can be done on a sheet of scratch paper by listing the eight combinations of the three inputs and checking off the combinations covered by each transition expression. Each combination should have exactly one check. As another example, consider the state diagrams in Figures 7-44 and 7-46 on pages 560 and 562; both can be verified mentally.

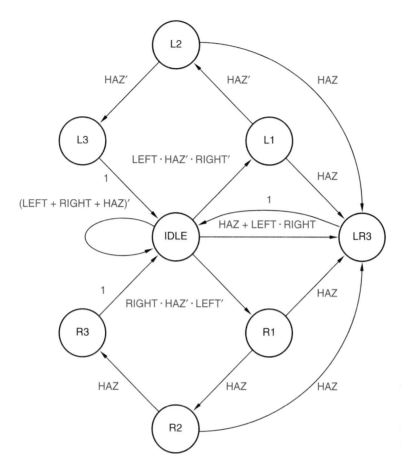

Figure 7-64
Enhanced state
diagram for T-bird
tail lights.

Returning to the T-bird tail-lights machine, we can now synthesize a circuit from the state diagram if we wish. However, if we want to change the machine's behavior, now is the time to do it, before we do all the work of synthesizing a circuit. In particular, notice that once a left- or right-turn cycle has begun, the state diagram in Figure 7-63 allows the cycle to run to completion, even if HAZ is asserted. While this may have a certain aesthetic appeal, it would be safer for the car's occupants to have the machine go into hazard mode as soon as possible. The state diagram is modified to provide this behavior in Figure 7-64.

Now we're finally ready to synthesize a circuit for the T-bird machine. The state diagram has eight states, so we'll need a minimum of three flip-flops to code the states. Obviously, many state assignments are possible (8! to be exact); we'll use the one in Table 7-16 for the following reasons:

1. An initial (idle) state of 000 is compatible with most flip-flops and registers, which are easily initialized to the 0 state.

2. Two state variables, Q1 and Q0, are used to "count" in Gray-code sequence for the left-turn cycle (IDLE→L1→L2→L3→IDLE). This minimizes the

Table 7-16
State assignment
for T-bird tail-lights
state machine.

State	Q2	Q1	Q0
IDLE	0	0	0
L1	0	0	1
L2	0	1	1
L3	0	1	0
R1	1	0	1
R2	1	1	1
R3	1	1	0
LR3	1	0	0

number of state-variable changes per state transition, which can often simplify the excitation logic.

3. Because of the symmetry in the state diagram, the same sequence on Q1 and Q0 is used to "count" during a right-turn cycle, while Q2 is used to distinguish between left and right.

4. The remaining state-variable combination is used for the LR3 state.

The next step is to write a sort of transition table. However, we must use a format different from the transition tables of Section 7.4.4, because the transitions in a state diagram are specified by expressions rather than by an exhaustive tabulation of next states. We'll call the new format a *transition list*, because it has one row for each transition or arc in the state diagram.

transition list

Table 7-17 is the transition list for the state diagram of Figure 7-64 and the state assignment of Table 7-16. Each row contains the current state, next state, and transition expression for one arc in the state diagram. Both the named and coded versions of the current state and next state are shown. The named states are useful for reference purposes, while the coded states are used to develop transition equations.

Once we have a transition list, the rest of the synthesis steps are pretty much "turning the crank." Synthesis procedures are described in Section 7.6. Although these procedures can be applied manually, they are usually embedded in a CAD software package; thus, Section 7.6 can help you understand what's going on (or going wrong) in your favorite CAD package.

We also encountered one "turn-the-crank" step in this section—finding the ambiguities in state diagrams. Even though the procedure we discussed can be easily automated, few if any CAD programs perform this step in this way. For example, one "state diagram entry" tool silently removes duplicated transitions and goes to the state coded "00...00" for missing transitions, without warning the user. Thus, in most design environments, the designer is responsible for writing a state-machine description that is unambiguous. The state-machine description languages at the end of this chapter provide a good way to do this.

S	Q2	Q1	Q0	Transition Expression	S*	Q2*	Q1*	Q0*
IDLE	0	0	0	(LEFT + RIGHT + HAZ)′	IDLE	0	0	0
IDLE	0	0	0	LEFT · HAZ′ · RIGHT′	L1	0	0	1
IDLE	0	0	0	HAZ + LEFT · RIGHT	LR3	1	0	0
IDLE	0	0	0	RIGHT · HAZ′ · LEFT′	R1	1	0	1
L1	0	0	1	HAZ′	L2	0	1	1
L1	0	0	1	HAZ	LR3	1	0	0
L2	0	1	1	HAZ′	L3	0	1	0
L2	0	1	1	HAZ	LR3	1	0	0
L3	0	1	0	1	IDLE	0	0	0
R1	1	0	1	HAZ′	R2	1	1	1
R1	1	0	1	HAZ	LR3	1	0	0
R2	1	1	1	HAZ′	R3	1	1	0
R2	1	1	1	HAZ	LR3	1	0	0
R3	1	1	0	1	IDLE	0	0	0
LR3	1	0	0	1	IDLE	0	0	0

Table 7-17
Transition list for T-bird tail-lights state machine.

*7.6 State-Machine Synthesis Using Transition Lists

Once a machine's state diagram has been designed and a state assignment has been made, the creative part of the design process is pretty much over. The rest of the synthesis procedure can be carried out by CAD programs.

As we showed in the preceding section, a transition list can be constructed from a machine's state diagram and state assignment. This section shows how to synthesize a state machine from its transition list. It also delves into some of the options and nuances of state-machine design using transition lists. Although this material is useful for synthesizing machines by hand, its main purpose is to help you understand the internal operation and the external quirks of CAD programs and languages that deal with state machines.

*7.6.1 Transition Equations

The first step in synthesizing a state machine from a transition list is to develop a set of transition equations that define each next-state variable V* in terms of the current state and input. The transition list can be viewed as a sort of hybrid

---

* This section and all of its subsections are optional.

truth table in which the state-variable combinations for the current state are list-ed explicitly and input combinations are listed algebraically. Reading down a V* column in a transition list, we find a sequence of 0s and 1s, indicating the value of V* for various (if we've done it right, all) state/input combinations.

A transition equation for a next-state variable V* can be written using a sort of hybrid canonical sum:

$$V* = \sum_{\text{transition-list rows where } V* = 1} \text{(transition p-term)}$$

transition p-term

That is, the transition equation has one "transition p-term" for each row of the transition list that contains a 1 in the V* column. A row's *transition p-term* is the product of the current state's minterm and the transition expression.

Based on the transition list in Table 7-17, the transition equation for Q2* in the T-bird machine can be written as the sum of eight p-terms:

$$
\begin{aligned}
Q2* = \ & Q2' \cdot Q1' \cdot Q0' \cdot (HAZ + LEFT \cdot RIGHT) \\
& + Q2' \cdot Q1' \cdot Q0' \cdot (RIGHT \cdot HAZ' \cdot LEFT') \\
& + Q2' \cdot Q1' \cdot Q0 \cdot (HAZ) \\
& + Q2' \cdot Q1 \cdot Q0 \cdot (HAZ) \\
& + Q2 \cdot Q1' \cdot Q0 \cdot (HAZ') \\
& + Q2 \cdot Q1' \cdot Q0 \cdot (HAZ) \\
& + Q2 \cdot Q1 \cdot Q0 \cdot (HAZ') \\
& + Q2 \cdot Q1 \cdot Q0 \cdot (HAZ)
\end{aligned}
$$

Some straightforward algebraic manipulations lead to a simplified transition equation that combines the first two, second two, and last four p-terms above:

$$
\begin{aligned}
Q2* = \ & Q2' \cdot Q1' \cdot Q0' \cdot (HAZ + RIGHT) \\
& + Q2' \cdot Q0 \cdot (HAZ) \\
& + Q2 \cdot Q0
\end{aligned}
$$

Transition equations for Q1* and Q0* may be obtained in a similar manner:

$$
\begin{aligned}
Q1* = \ & Q2' \cdot Q1' \cdot Q0 \cdot (HAZ') \\
& + Q2' \cdot Q1 \cdot Q0 \cdot (HAZ') \\
& + Q2 \cdot Q1' \cdot Q0 \cdot (HAZ') \\
& + Q2 \cdot Q1 \cdot Q0 \cdot (HAZ') \\
= \ & Q0 \cdot HAZ' \\
Q0* = \ & Q2' \cdot Q1' \cdot Q0' \cdot (LEFT \cdot HAZ' \cdot RIGHT') \\
& + Q2' \cdot Q1' \cdot Q0' \cdot (RIGHT \cdot HAZ' \cdot LEFT') \\
& + Q2' \cdot Q1' \cdot Q0 \cdot (HAZ') \\
& + Q2 \cdot Q1' \cdot Q0 \cdot (HAZ') \\
= \ & Q2' \cdot Q1' \cdot HAZ' \cdot (LEFT \oplus RIGHT) + Q1' \cdot Q0 \cdot HAZ'
\end{aligned}
$$

Except for Q1*, there's no guarantee that the transition equations above are in any sense minimal—in fact, the expressions for Q2* and Q0* aren't even in standard sum-of-products or product-of-sums form. The simplified equations, or the original unsimplified ones, merely provide an unambiguous starting point for whatever combinational design method you might choose to synthesize the excitation logic for the state machine—ad hoc, NAND-NAND, MSI-based, or whatever. In a PLD-based design, you could simply plug the equations into an ABEL program and let the compiler calculate the minimal sum-of-products expressions for the PLD's AND-OR array.

*7.6.2 Excitation Equations

While we're on the subject of excitation logic, note that so far we have derived only *transition equations*, not *excitation equations*. However, if we use D flip-flops as the memory elements in our state machines, then the excitation equations are trivial to derive from the transition equations, since the characteristic equation of a D flip-flop is $Q* = D$. Therefore, if the transition equation for a state variable $Qi*$ is

$$Qi* \ = \ \text{expression}$$

then the excitation equation for the corresponding D flip-flop input is

$$Di \ = \ \text{expression}$$

Efficient excitation equations for other flip-flop types, especially J-K, are not so easy to derive (see Exercise 7.63). For that reason, the vast majority of discrete, PLD-based, and ASIC-based state-machine designs employ D flip-flops.

*7.6.3 Variations on the Scheme

There are other ways to obtain transition and excitation equations from a transition list. If the column for a particular next-state variable contains fewer 0s than 1s, it may be advantageous to write that variable's transition equation in terms of the 0s in its column. That is, we write

$$V*' \ = \ \sum_{\text{transition-list rows where } V* = 0} (\text{transition p-term})$$

That is, $V*'$ is 1 for all of the p-terms for which $V*$ is 0. Thus, a transition equation for $Q2*'$ may be written as the sum of seven p-terms:

$$
\begin{aligned}
Q2*' \ = \ & Q2' \cdot Q1' \cdot Q0' \cdot ((\text{LEFT}+ \text{RIGHT}+ \text{HAZ})') \\
& + Q2' \cdot Q1' \cdot Q0' \cdot (\text{LEFT} \cdot \text{HAZ}' \cdot \text{RIGHT}') \\
& + Q2' \cdot Q1' \cdot Q0 \cdot (\text{HAZ}') \\
& + Q2' \cdot Q1 \cdot Q0 \cdot (\text{HAZ}') \\
& + Q2' \cdot Q1 \cdot Q0' \cdot (1) \\
& + Q2 \cdot Q1 \cdot Q0' \cdot (1) \\
& + Q2 \cdot Q1' \cdot Q0' \cdot (1) \\
= \ & Q2' \cdot Q1' \cdot Q0' \cdot \text{HAZ}' \cdot \text{RIGHT}' + Q2' \cdot Q0 \cdot \text{HAZ}' + Q1 \cdot Q0' + Q2 \cdot Q0'
\end{aligned}
$$

To obtain an equation for Q2∗, we simply complement both sides of the reduced equation.

To obtain an expression for a next-state variable V∗ directly, using the 0s in the transition list, we can complement the righthand side of the general V∗′ equation using DeMorgan's theorem, obtaining a sort of hybrid canonical product:

$$V* = \prod_{\text{transition-list rows where } V* = 0} (\text{transition s-term})$$

transition s-term

Here, a row's *transition s-term* is the sum of the maxterm for the current state and the complement of the transition expression. If the transition expression is a simple product term, then its complement is a sum, and the transition equation expresses V∗ in product-of-sums form.

*7.6.4 Realizing the State Machine

Once you have the excitation equations for a state machine, all you're left with is a multiple-output combinational logic design problem. Of course, there are many ways to realize combinational logic from equations, but the easiest way is just to type them into an ABEL or VHDL program and use the compiler to synthesize a PLD, FPGA, or ASIC realization.

Combinational PLDs, such as the PAL16L8 and GAL16V8 that we studied in Section 5.3, can be used to realize excitation equations up to a certain number of inputs, outputs, and product terms. Better yet, in Section 8.3 we'll introduce sequential PLDs that include D flip-flops on the same chip with the combinational AND-OR array. For a given number of PLD input and output pins, these sequential PLDs can realize larger state machines than their combinational counterparts, because the excitation signals never have to go off the chip. In Section 9.1.3, we'll show how to realize the T-bird tail-lights machine in a sequential PLD.

*7.7 Another State-Machine Design Example

This section gives one more example of state-machine design using a state diagram. The example provides a basis for further discussion of a few topics: unused states, output-coded state assignments, and "don't-care" state codings.

*7.7.1 The Guessing Game

Our final state-machine example is a "guessing game" that can be built as an amusing lab project:

Design a clocked synchronous state machine with four inputs, G1–G4, that are connected to pushbuttons. The machine has four outputs, L1–L4, connected to lamps or LEDs located near the like-numbered pushbuttons. There is also an ERR output connected to a red lamp. In normal operation the L1–L4 outputs display a 1-out-of-4 pattern. At each clock tick, the pattern is rotated by one position; the clock frequency is about 4 Hz.

Guesses are made by pressing a pushbutton, which asserts an input Gi. When any Gi input is asserted, the ERR output is asserted if the "wrong" pushbutton was pressed, that is, if the Gi input detected at the clock tick does not have the same number as the lamp output that was asserted before the clock tick. Once a guess has been made, play stops and the ERR output maintains the same value for one or more clock ticks until the Gi input is negated, then play resumes.

Clearly, we will have to provide four states, one for each position of the rotating pattern, and we'll need at least one state to indicate that play has stopped. A possible state diagram is shown in Figure 7-65. The machine cycles through states S1–S4 as long as no Gi input is asserted, and it goes to the STOP state when a guess is made. Each Li output is asserted in the like-numbered state.

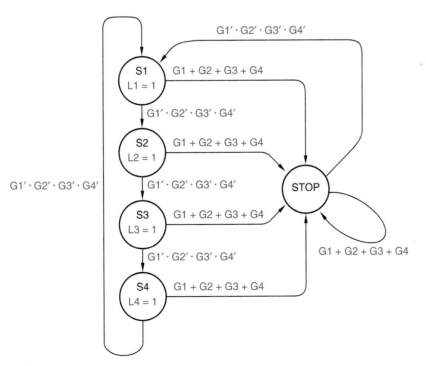

Figure 7-65
First try at a state diagram for the guessing game.

The only problem with this state diagram is that it doesn't "remember" in the STOP state whether the guess was correct, so it has no way to control the ERR output. This problem is fixed in Figure 7-66, which has two "stopped" states, SOK and SERR. On an incorrect guess, the machine goes to SERR, where ERR is asserted; otherwise, it goes to SOK. Although the machine's word description doesn't require it, the state diagram is designed to go to SERR even if the user tries to fool it by pressing two or more pushbuttons simultaneously, or by changing guesses while stopped.

A transition list corresponding to the state diagram is shown in Table 7-18, using a simple 3-bit binary state encoding with Gray-code order for the S1–S4 cycle. Transition equations for Q1* and Q0* can be obtained from the table as follows:

$$
\begin{aligned}
Q1* &= Q2' \cdot Q1' \cdot Q0 \cdot (G1' \cdot G2' \cdot G3' \cdot G4') \\
&\quad + Q2' \cdot Q1 \ \cdot Q0 \cdot (G1' \cdot G2' \cdot G3' \cdot G4') \\
&= Q2' \ \cdot Q0 \cdot G1' \cdot G2' \cdot G3' \cdot G4' \\
Q0* &= Q2' \cdot Q1' \cdot Q0' \cdot (G1' \cdot G2' \cdot G3' \cdot G4') \\
&\quad + Q2' \cdot Q1' \cdot Q0' \cdot (G2 + G3 + G4) \\
&\quad + Q2' \cdot Q1' \cdot Q0 \ \cdot (G1' \cdot G2' \cdot G3' \cdot G4') \\
&\quad + Q2' \cdot Q1' \cdot Q0 \ \cdot (G1 + G3 + G4) \\
&\quad + Q2' \cdot Q1 \ \cdot Q0 \ \cdot (G1 + G2 + G4) \\
&\quad + Q2' \cdot Q1 \ \cdot Q0' \cdot (G1 + G2 + G3) \\
&\quad + Q2 \ \cdot Q1' \cdot Q0 \ \cdot (G1 + G2 + G3 + G4)
\end{aligned}
$$

Using a logic-minimization program, the Q0* expression can be reduced to 11 product terms in two-level sum-of-products form. An expression for Q2* is best formulated in terms of the 0s in the Q2* column of the transition list:

$$
\begin{aligned}
Q2*' &= Q2' \cdot Q1' \cdot Q0' \cdot (G1' \cdot G2' \cdot G3' \cdot G4') \\
&\quad + Q2' \cdot Q1' \cdot Q0 \ \cdot (G1' \cdot G2' \cdot G3' \cdot G4') \\
&\quad + Q2' \cdot Q1 \ \cdot Q0 \ \cdot (G1' \cdot G2' \cdot G3' \cdot G4') \\
&\quad + Q2' \cdot Q1 \ \cdot Q0' \cdot (G1' \cdot G2' \cdot G3' \cdot G4') \\
&\quad + Q2 \ \cdot Q1' \cdot Q0' \cdot (G1' \cdot G2' \cdot G3' \cdot G4') \\
&\quad + Q2 \ \cdot Q1' \cdot Q0 \ \cdot (G1' \cdot G2' \cdot G3' \cdot G4') \\
&= (Q2' + Q1') \cdot (G1' \cdot G2' \cdot G3' \cdot G4')
\end{aligned}
$$

The last five columns of Table 7-18 show output values. Thus, output equations can be developed in much the same way as transition equations. However, since this example is a Moore machine, outputs are independent of the transition expressions; only one row of the transition list must be considered for each current state. The output equations are

$L1 = Q2' \cdot Q1' \cdot Q0'$	$L3 = Q2' \cdot Q1 \cdot Q0$	$ERR = Q2 \cdot Q1' \cdot Q0$
$L2 = Q2' \cdot Q1' \cdot Q0$	$L4 = Q2' \cdot Q1 \cdot Q0'$	

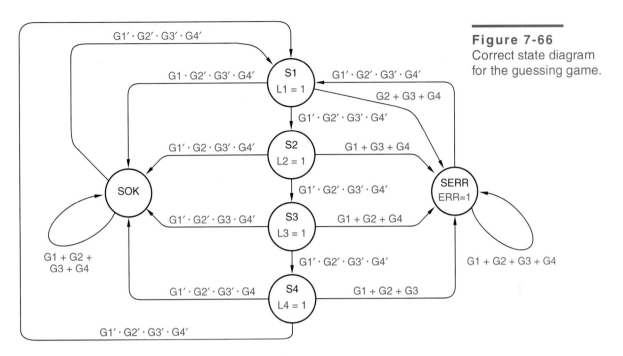

Figure 7-66
Correct state diagram for the guessing game.

Table 7-18 Transition list for guessing-game machine.

Current State					Next State				Output				
S	Q2	Q1	Q0	Transition Expression	S*	Q2*	Q1*	Q0*	L1	L2	L3	L4	ERR
S1	0	0	0	G1′ · G2′ · G3′ · G4′	S2	0	0	1	1	0	0	0	0
S1	0	0	0	G1 · G2′ · G3′ · G4′	SOK	1	0	0	1	0	0	0	0
S1	0	0	0	G2 + G3 + G4	SERR	1	0	1	1	0	0	0	0
S2	0	0	1	G1′ · G2′ · G3′ · G4′	S3	0	1	1	0	1	0	0	0
S2	0	0	1	G1′ · G2 · G3′ · G4′	SOK	1	0	0	0	1	0	0	0
S2	0	0	1	G1 + G3 + G4	SERR	1	0	1	0	1	0	0	0
S3	0	1	1	G1′ · G2′ · G3′ · G4′	S4	0	1	0	0	0	1	0	0
S3	0	1	1	G1′ · G2′ · G3 · G4′	SOK	1	0	0	0	0	1	0	0
S3	0	1	1	G1 + G2 + G4	SERR	1	0	1	0	0	1	0	0
S4	0	1	0	G1′ · G2′ · G3′ · G4′	S1	0	0	0	0	0	0	1	0
S4	0	1	0	G1′ · G2′ · G3′ · G4	SOK	1	0	0	0	0	0	1	0
S4	0	1	0	G1 + G2 + G3	SERR	1	0	1	0	0	0	1	0
SOK	1	0	0	G1 + G2 + G3 + G4	SOK	1	0	0	0	0	0	0	0
SOK	1	0	0	G1′ · G2′ · G3′ · G4′	S1	0	0	0	0	0	0	0	0
SERR	1	0	1	G1 + G2 + G3 + G4	SERR	1	0	1	0	0	0	0	1
SERR	1	0	1	G1′ · G2′ · G3′ · G4′	S1	0	0	0	0	0	0	0	1

*7.7.2 Unused States

Our state diagram for the guessing game has six states, but the actual state machine, built from three flip-flops, has eight. By omitting the unused states from the transition list, we treated them as "don't-cares" in a very limited sense:

- When we wrote equations for Q1∗ and Q0∗, we formed a sum of transition p-terms for state/input combinations that had an explicit 1 in the corresponding columns of the transition list. Although we didn't consider the unused states, our procedure implicitly treated them as if they had 0s in the Q1∗ and Q0∗ columns.

- Conversely, we wrote the Q2∗′ equation as a sum of transition p-terms for state/input combinations that had an explicit 0 in the corresponding columns of the transition list. Unused states were implicitly treated as if they had 1s in the Q2∗ column.

As a consequence of these choices, all of the unused states in the guessing-game machine have a coded next state of 100 for all input combinations. That's safe, acceptable behavior should the machine stray into an unused state, since 100 is the coding for one of the normal states (SOK).

To treat the unused states as true "don't-cares," we would have to allow them to go to any next state under any input combination. This is simple in principle but may be difficult in practice.

At the end of Section 7.4.4 we showed how to handle unused states as "don't-cares" in the Karnaugh-map method for developing transition/excitation equations. Unfortunately, for all but the smallest problems, Karnaugh maps are unwieldy. Commercially available logic-minimization programs can easily handle larger problems, but many of them don't handle "don't-cares" or require the designer to insert special code to handle them. In ABEL state machines, don't-care next states can be handled fairly easily using the @DCSET directive, as we discuss in the box on page 632. In VHDL the process is a bit unwieldy.

*7.7.3 Output-Coded State Assignment

Let's look at another realization of the guessing-game machine. The machine's outputs are a function of state only; furthermore, a *different* output combination is produced in each named state. Therefore, we can use the outputs as state variables and assign each named state to the required output combination. This sort of *output-coded state assignment* can sometimes result in excitation equations that are simpler than the set of excitation and output equations obtained with a state assignment using a minimum number of state variables.

output-coded state assignment

Table 7-19 is the guessing-game transition list that results from an output-coded state assignment. Each transition/excitation equation has very few transition p-terms because the transition list has so few 1s in the next-state columns:

Table 7-19 Transition list for guessing-game machine using outputs as state variables.

Current State						Transition Expression	Next State					
S	L1	L2	L3	L4	ERR		S*	L1*	L2*	L3*	L4*	ERR*
S1	1	0	0	0	0	$G1' \cdot G2' \cdot G3' \cdot G4'$	S2	0	1	0	0	0
S1	1	0	0	0	0	$G1 \cdot G2' \cdot G3' \cdot G4'$	SOK	0	0	0	0	0
S1	1	0	0	0	0	$G2 + G3 + G4$	SERR	0	0	0	0	1
S2	0	1	0	0	0	$G1' \cdot G2' \cdot G3' \cdot G4'$	S3	0	0	1	0	0
S2	0	1	0	0	0	$G1' \cdot G2 \cdot G3' \cdot G4'$	SOK	0	0	0	0	0
S2	0	1	0	0	0	$G1 + G3 + G4$	SERR	0	0	0	0	1
S3	0	0	1	0	0	$G1' \cdot G2' \cdot G3' \cdot G4'$	S4	0	0	0	1	0
S3	0	0	1	0	0	$G1' \cdot G2' \cdot G3 \cdot G4'$	SOK	0	0	0	0	0
S3	0	0	1	0	0	$G1 + G2 + G4$	SERR	0	0	0	0	1
S4	0	0	0	1	0	$G1' \cdot G2' \cdot G3' \cdot G4'$	S1	1	0	0	0	0
S4	0	0	0	1	0	$G1' \cdot G2' \cdot G3' \cdot G4$	SOK	0	0	0	0	0
S4	0	0	0	1	0	$G1 + G2 + G3$	SERR	0	0	0	0	1
SOK	0	0	0	0	0	$G1 + G2 + G3 + G4$	SOK	0	0	0	0	0
SOK	0	0	0	0	0	$G1' \cdot G2' \cdot G3' \cdot G4'$	S1	1	0	0	0	0
SERR	0	0	0	0	1	$G1 + G2 + G3 + G4$	SERR	0	0	0	0	1
SERR	0	0	0	0	1	$G1' \cdot G2' \cdot G3' \cdot G4'$	S1	1	0	0	0	0

$$L1* = L1' \cdot L2' \cdot L3' \cdot L4 \cdot ERR' \cdot (G1' \cdot G2' \cdot G3' \cdot G4')$$
$$+ L1' \cdot L2' \cdot L3' \cdot L4' \cdot ERR' \cdot (G1' \cdot G2' \cdot G3' \cdot G4')$$
$$+ L1' \cdot L2' \cdot L3' \cdot L4' \cdot ERR \cdot (G1' \cdot G2' \cdot G3' \cdot G4')$$
$$L2* = L1 \cdot L2' \cdot L3' \cdot L4' \cdot ERR' \cdot (G1' \cdot \underline{G2'} \cdot G3' \cdot G4')$$
$$L3* = L1' \cdot L2 \cdot L3' \cdot L4' \cdot ERR' \cdot (G1' \cdot G2' \cdot G3' \cdot G4')$$
$$L4* = L1' \cdot L2' \cdot L3 \cdot L4' \cdot ERR' \cdot (G1' \cdot G2' \cdot G3' \cdot G4')$$
$$ERR* = L1 \cdot L2' \cdot L3' \cdot L4' \cdot ERR' \cdot (G2 + G3 + G4)$$
$$+ L1' \cdot L2 \cdot L3' \cdot L4' \cdot ERR' \cdot (G1 + G3 + G4)$$
$$+ L1' \cdot L2' \cdot L3 \cdot L4' \cdot ERR' \cdot (G1 + G2 + G4)$$
$$+ L1' \cdot L2' \cdot L3' \cdot L4 \cdot ERR' \cdot (G1 + G2 + G3)$$
$$+ L1' \cdot L2' \cdot L3' \cdot L4' \cdot ERR \cdot (G1 + G2 + G3 + G4)$$

There are no output equations, of course. The ERR* equation above is the worst
in the group, requiring 16 terms to express in either minimal sum-of-products or
product-of-sums form.

As a group, the equations developed above have just about the same complexity as the transition and output equations that we developed from Table 7-18. Even though the output-coded assignment does not produce a simpler set of equations in this example, it can still save cost in a PLD-based design, since fewer PLD macrocells or outputs are needed overall.

*7.7.4 "Don't-Care" State Codings

Out of the 32 possible coded states using five variables, only six are used in Table 7-19. The rest of the states are unused and have a next state of 00000 if the machine is built using the equations in the preceding subsection. Another possible disposition for unused states, one that we haven't discussed before, is obtained by careful use of "don't-cares" in the assignment of coded states to current states.

Table 7-20 shows one such state assignment for the guessing-game machine, derived from the output-coded state assignment of the preceding subsection. In this example, every possible combination of current-state variables corresponds to one of the coded states (e.g., 10111 = S1, 00101 = S3). However, *next states* are coded using the same unique combinations as in the preceding subsection. Table 7-21 shows the resulting transition list.

In this approach, each unused current state behaves like a nearby "normal" state; Figure 7-67 illustrates the concept. The machine is well behaved and goes to a "normal state" if it inadvertently enters an unused state. Yet the approach still allows some simplification of the excitation and output logic by introducing

Table 7-20
Current-state assignment for the guessing-game machine using don't-cares.

State	L1	L2	L3	L4	ERR
S1	1	x	x	x	x
S2	0	1	x	x	x
S3	0	0	1	x	x
S4	0	0	0	1	x
SOK	0	0	0	0	0
SERR	0	0	0	0	1

Figure 7-67
State assignment using don't-cares for current states.

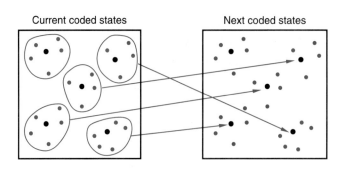

Current coded states   Next coded states

Table 7-21 Transition list for guessing-game machine using don't-care state codings.

Current State						Transition Expression	Next State					
S	L1	L2	L3	L4	ERR		S*	L1*	L2*	L3*	L4*	ERR*
S	1	x	x	x	x	$G1' \cdot G2' \cdot G3' \cdot G4'$	S2	0	1	0	0	0
S	1	x	x	x	x	$G1 \cdot G2' \cdot G3' \cdot G4'$	SOK	0	0	0	0	0
S	1	x	x	x	x	$G2 + G3 + G4$	SERR	0	0	0	0	1
S2	0	1	x	x	x	$G1' \cdot G2' \cdot G3' \cdot G4'$	S3	0	0	1	0	0
S2	0	1	x	x	x	$G1' \cdot G2 \cdot G3' \cdot G4'$	SOK	0	0	0	0	0
S2	0	1	x	x	x	$G1 + G3 + G4$	SERR	0	0	0	0	1
S3	0	0	1	x	x	$G1' \cdot G2' \cdot G3' \cdot G4'$	S4	0	0	0	1	0
S3	0	0	1	x	x	$G1' \cdot G2' \cdot G3 \cdot G4'$	SOK	0	0	0	0	0
S3	0	0	1	x	x	$G1 + G2 + G4$	SERR	0	0	0	0	1
S4	0	0	0	1	x	$G1' \cdot G2' \cdot G3' \cdot G4'$	S1	1	0	0	0	0
S4	0	0	0	1	x	$G1' \cdot G2' \cdot G3' \cdot G4$	SOK	0	0	0	0	0
S4	0	0	0	1	x	$G1 + G2 + G3$	SERR	0	0	0	0	1
SOK	0	0	0	0	0	$G1 + G2 + G3 + G4$	SOK	0	0	0	0	0
SOK	0	0	0	0	0	$G1' \cdot G2' \cdot G3' \cdot G4'$	S1	1	0	0	0	0
SERR	0	0	0	0	1	$G1 + G2 + G3 + G4$	SERR	0	0	0	0	1
SERR	0	0	0	0	1	$G1' \cdot G2' \cdot G3' \cdot G4'$	S1	1	0	0	0	0

don't-cares in the transition list. When a row's transition p-term is written, current-state variables that are don't-cares in that row are omitted; for example,

$$
\begin{aligned}
ERR* \; = \; & L1 \cdot (G2 + G3 + G4) \\
& + L1' \cdot L2 \cdot (G1 + G3 + G4) \\
& + L1' \cdot L2' \cdot L3 \cdot (G1 + G2 + G4) \\
& + L1' \cdot L2' \cdot L3' \cdot L4 \cdot (G1 + G2 + G3) \\
& + L1' \cdot L2' \cdot L3' \cdot L4' \cdot ERR \cdot (G1 + G2 + G3 + G4)
\end{aligned}
$$

Compared with the ERR* equation in the preceding subsection, the one above still requires 16 terms to express as a sum of products. However, it requires only five terms in minimal product-of-sums form, which makes its complement more suitable for realization in a PLD.

*7.8 Decomposing State Machines

Just like large procedures or subroutines in a programming language, large state machines are difficult to conceptualize, design, and debug. Therefore, when faced with a large state-machine problem, digital designers often look for opportunities to solve it with a collection of smaller state machines.

state-machine
decomposition

There's a well-developed theory of *state-machine decomposition* that you can use to analyze any given, monolithic state machine to determine whether it can be realized as a collection of smaller ones. However, decomposition theory is not too useful for designers who want to avoid designing large state machines in the first place. Rather, a practical designer tries to cast the original design problem into a natural, hierarchical structure, so that the uses and functions of submachines are obvious, making it unnecessary ever to write a state table for the equivalent monolithic machine.

main machine
submachines

The simplest and most commonly used type of decomposition is illustrated in Figure 7-68. A *main machine* provides the primary inputs and outputs and executes the top-level control algorithm. *Submachines* perform low-level steps under the control of the main machine and may optionally handle some of the primary inputs and outputs.

Perhaps the most commonly used submachine is a counter. The main machine starts the counter when it wishes to stay in a particular main state for n clock ticks, and the counter asserts a DONE signal when n ticks have occurred. The main machine is designed to wait in the same state until DONE is asserted. This adds an extra output and input to the main machine (START and DONE), but it saves $n - 1$ states.

An example decomposed state machine designed along these lines is based on the guessing game of Section 7.7. The original guessing game is easy to win after a minute of practice because the lamps cycle at a very consistent rate of

Figure 7-68
A typical, hierarchical
state-machine
structure.

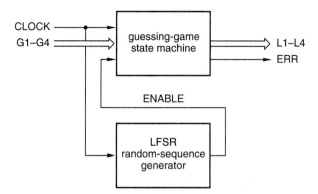

Figure 7-69
Block diagram of
guessing game with
random delay.

4 Hz. To make the game more challenging, we can double or triple the clock speed but allow the lamps to stay in each state for a random length of time. Then the user truly must guess whether a given lamp will stay on long enough for the corresponding pushbutton to be pressed.

A block diagram for the enhanced guessing game is shown in Figure 7-69. The main machine is basically the same as before, except that it advances from one lamp state to the next only if the enable input EN is asserted, as shown by the state diagram in Figure 7-70. The enable input is driven by the output of a pseudorandom sequence generator, a linear feedback shift register (LFSR).

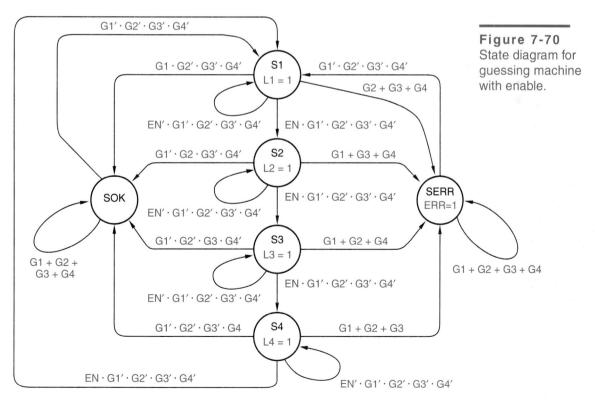

Figure 7-70
State diagram for
guessing machine
with enable.

A SHIFTY CIRCUIT   LFSR circuits are described in Section 8.5.8. In Figure 7-69, the low-order bit of an n-bit LFSR counter is used as the enable signal. Thus, the length of time that a particular lamp stays on depends on the counting sequence of the LFSR.

  In the best case for the user, the LFSR contains 10...00; in this case the lamp is on for $n-1$ clock ticks, because it takes that long for the single 1 to shift into the low-order bit position. In the worst case, the LFSR contains 11...11, and shifting occurs for n consecutive clock ticks. At other times, shifting stops for a time determined by the number of consecutive 0s starting with the low-order bit of the LFSR.

  All of these cases are quite unpredictable, unless the user has memorized the shifting cycle ($2^n - 1$ states) or is very fast at Galois-field arithmetic. Obviously, a large value of n (≥ 16) provides the most fun.

  Another obvious candidate for decomposition is a state machine that performs binary multiplication using the shift-and-add algorithm, or binary division using the shift-and-subtract algorithm. To perform an n-bit operation, these algorithms require an initialization step, n computation steps, and possible cleanup steps. The main machine for such an algorithm contains states for initialization, generic computation, and cleanup steps, and a modulo-n counter can be used as a submachine to control the number of generic computation steps executed.

*7.9 Feedback Sequential Circuits

The simple bistable and the various latches and flip-flops that we studied earlier in this chapter are all feedback sequential circuits. Each has one or more feedback loops that, ignoring their behavior during state transitions, store a 0 or a 1 at all times. The feedback loops are memory elements, and the circuits' behavior depends on both the current inputs and the values stored in the loops.

*7.9.1 Analysis

fundamental-mode circuit

Feedback sequential circuits are the most common example of *fundamental-mode circuits*. In such circuits, inputs are not normally allowed to change simultaneously. The analysis procedure assumes that inputs change one at a time, allowing enough time between successive changes for the circuit to settle into a stable internal state. This differs from clocked circuits, in which multiple inputs can change at almost arbitrary times without affecting the state, and all input values are sampled and state changes occur with respect to a clock signal.

---

*This section and all of its subsections are optional.

KEEP YOUR FEEDBACK TO YOURSELF	Only rarely does a logic designer encounter a situation where a feedback sequential circuit must be analyzed or designed. The most commonly used feedback sequential circuits are the flip-flops and latches that are used as the building blocks in larger sequential-circuit designs. Their internal design and operating specifications are supplied by an IC manufacturer.
	Even an ASIC designer typically does not design gate-level flip-flop or latch circuits, since these elements are supplied in a "library" of commonly used functions in the particular ASIC technology. Still, you may be curious about how off-the-shelf flip-flops and latches "do their thing"; this section will show you how to analyze such circuits.

Like clocked synchronous state machines, feedback sequential circuits may be structured as Mealy or Moore circuits, as shown in Figure 7-71. A circuit with n feedback loops has n binary state variables and 2^n states.

To analyze a feedback sequential circuit, we must break the feedback loops in Figure 7-71 so that the next value stored in each loop can be predicted as a function of the circuit inputs and the current value stored in all loops. Figure 7-72 shows how to do this for the NAND circuit for a D latch, which has only one feedback loop. We conceptually break the loop by inserting a fictional buffer in the loop as shown. The output of the buffer, named Y, is the single state variable for this example.

Let us assume that the propagation delay of the fictional buffer is 10 ns (but any nonzero number will do) and that all of the other circuit components have zero delay. If we know the circuit's current state (Y) and inputs (D and C), then we can predict the value Y will have in 10 ns. The next value of Y, denoted Y∗, is a combinational function of the current state and inputs. Thus, reading the circuit diagram, we can write an *excitation equation* for Y∗:

$$Y* = (C \cdot D) + (C \cdot D' + Y')'$$
$$= C \cdot D + C' \cdot Y + D \cdot Y$$

excitation equation

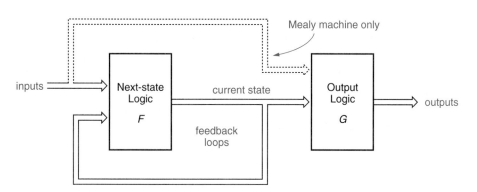

inputs

Mealy machine only

Next-state Logic

F

current state

feedback loops

Output Logic

G

outputs

Figure 7-71
Feedback sequential circuit structure for Mealy and Moore machines.

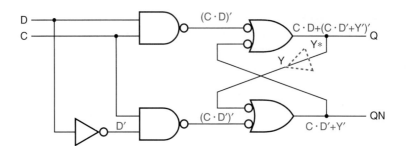

Figure 7-72
Feedback analysis
of a D latch.

transition table

Now the state of the feedback loop (and the circuit) can be written as a function of the current state and input, and enumerated by a *transition table* as shown in Figure 7-73. Each cell in the transition table shows the fictional-buffer output value that will occur 10 ns (or whatever delay you've assumed) after the corresponding state and input combination occurs.

Figure 7-73
Transition table for the
D latch in Figure 7-72.

Y	00	01	11	10
			C D	
0	0	0	1	0
1	1	1	1	0
		Y*		

A transition table has one row for each possible combination of the state variables, so a circuit with n feedback loops has 2^n rows in its transition table. The table has one column for each possible input combination, so a circuit with m inputs has 2^m columns in its transition table.

By definition, a fundamental-mode circuit such as a feedback sequential circuit does not have a clock to tell it when to sample its inputs. Instead, we can imagine that the circuit is evaluating its current state and input *continuously* (or every 10 ns, if you prefer). As the result of each evaluation, it goes to a next state predicted by the transition table. Most of the time, the next state is the same as the current state; this is the essence of fundamental-mode operation. We make some definitions below that will help us study this behavior in more detail.

JUST ONE LOOP   The way the circuit in Figure 7-72 is drawn, it may look like there are two feedback loops. However, once we make one break as shown, there are no more loops. That is, each signal can be written as a combinational function of the other signals, not including itself.

	C D			
S	00	01	11	10
S0	ⓈO	ⓈO	S1	ⓈO
S1	Ⓢ1	Ⓢ1	Ⓢ1	S0
		S*		

Figure 7-74
State table for the D
latch in Figure 7-72.
showing stable total
states.

In a fundamental-mode circuit, a *total state* is a particular combination of *internal state* (the values stored in the feedback loops) and *input state* (the current value of the circuit inputs). A *stable total state* is a combination of internal state and input state such that the next internal state predicted by the transition table is the same as the current internal state. If the next internal state is different, then the combination is an *unstable total state*. We have rewritten the transition table for the D latch in Figure 7-74 as a *state table*, giving the names S0 and S1 to the states and drawing a circle around the stable total states.

total state
internal state
input state
stable total state

unstable total state
state table

To complete the analysis of the circuit, we must also determine how the outputs behave as functions of the internal state and inputs. There are two outputs and hence two *output equations*:

output equation

$$Q = C \cdot D + C' \cdot Y + D \cdot Y$$
$$QN = C \cdot D' + Y'$$

Note that Q and QN are *outputs*, not state variables. Even though the circuit has two outputs, which can theoretically take on four combinations, it has only one state variable Y, and hence only two states.

The output values predicted by the Q and QN equations can be incorporated in a combined state and output table that completely describes the operation of the circuit, as shown in Figure 7-75. Although Q and QN are normally complementary, it is possible for them to have the same value (1) momentarily, during the transition from S0 to S1 under the C D = 11 column of the table.

We can now predict the behavior of the circuit from the transition and output table. First of all, notice that we have written the column labels in our state tables in "Karnaugh map" order, so that only a single input bit changes

	C D			
S	00	01	11	10
S0	ⓈO, 01	ⓈO, 01	S1, 11	ⓈO, 01
S1	Ⓢ1, 10	Ⓢ1, 10	Ⓢ1, 10	S0, 01
		S*, Q QN		

Figure 7-75
State and output table
for the D latch.

		C D		
S	00	01	11	10
S0	(S0), 01	(S0), 01	S1, 11	(S0), 01
S1	(S1), 10	(S1), 10	(S1), 10	S0, 01
		S*, Q QN		

Figure 7-76
Analysis of the D latch
for a few transitions.

between adjacent columns of the table. This layout helps our analysis because we assume that only one input changes at a time and that the circuit always reaches a stable total state before another input changes.

At any time, the circuit is in a particular internal state and a particular input is applied to it; we called this combination the total state of the circuit. Let us start at the stable total state "S0/00" (S = S0, C D = 00), as shown in Figure 7-76. Now suppose that we change D to 1. The total state moves to one cell to the right; we have a new stable total state, S0/01. The D input is different, but the internal state and output are the same as before. Next, let us change C to 1. The total state moves one cell to the right to S0/11, which is unstable. The next-state entry in this cell sends the circuit to internal state S1, so the total state moves down one cell, to S1/11. Examining the next-state entry in the new cell, we find that we have reached a stable total state. We can trace the behavior of the circuit for any desired sequence of single input changes in this way.

Now we can revisit the question of simultaneous input changes. Even though "almost simultaneous" input changes may occur in practice, we must assume that nothing happens simultaneously in order to analyze the behavior of sequential circuits. The impossibility of simultaneous events is supported by the varying delays of circuit components themselves, which depend on voltage, temperature, and fabrication parameters. What this tells us is that a set of n inputs that appear to us to change "simultaneously" may actually change in any of $n!$ different orders from the point of view of the circuit operation.

For example, consider the operation of the D latch as shown in Figure 7-77. Let us assume that it starts in stable total state S1/11. Now suppose that C and D

Figure 7-77
Multiple input changes
with the D latch.

		C D		
S	00	01	11	10
S0	(S0), 01	(S0), 01	S1, 11	(S0), 01
S1	(S1), 10	(S1), 10	(S1), 10	S0, 01
		S*, Q QN		

are both "simultaneously" set to 0. In reality, the circuit behaves as if one or the other input went to 0 first. Suppose that C changes first. Then the sequence of two left-pointing arrows in the table tells us that the circuit goes to stable total state S1/00. However, if D changes first, then the other sequence of arrows tells us that the circuit goes to stable total state S0/00. So the final state of the circuit is unpredictable, a clue that the feedback loop may actually become metastable if we set C and D to 0 simultaneously. The time span over which this view of simultaneity is relevant is the setup- and hold-time window of the D latch.

Simultaneous input changes don't always cause unpredictable behavior. However, we must analyze the effects of all possible orderings of signal changes to determine this; if all orderings give the same result, then the circuit output is predictable. For example, consider the behavior of the D latch starting in total state S0/00 with C and D simultaneously changing from 0 to 1; it always ends up in total state S1/11.

*7.9.2 Analyzing Circuits with Multiple Feedback Loops

In circuits with multiple feedback loops, we must break all of the loops, creating one fictional buffer and state variable for each loop that we break. There are many possible ways, which mathematicians call *cut sets*, to break the loops in a given circuit, so how do we know which one is best? The answer is that any *minimal cut set*—a cut set with a minimum number of cuts—is fine. Mathematicians can give you an algorithm for finding a minimal cut set, but as a digital designer working on small circuits, you can just eyeball the circuit to find one.

cut set

minimal cut set

Different cut sets for a circuit lead to different excitation equations, transition tables, and state/output tables. However, the stable total states derived from one minimal cut set correspond one-to-one to the stable total states derived from any other minimal cut set for the same circuit. That is, state/output tables derived from different minimal cut sets display the same input/output behavior, with only the names and coding of the states changed.

If you use more than the minimal number of cuts to analyze a feedback sequential circuit, the resulting state/output table will still describe the circuit correctly. However, it will use 2^m times as many states as necessary, where m is the number of extra cuts. Formal state-minimization procedures can be used to reduce this larger table to the proper size, but it's a much better idea to select a minimal cut set in the first place.

A good example of a sequential circuit with multiple feedback loops is the commercial circuit design for a positive edge-triggered TTL D flip-flop that we showed in Figure 7-20. The circuit is redrawn in simplified form in Figure 7-78, assuming that the original circuit's PR_L and CLR_L inputs are never asserted, and also showing fictional buffers to break the three feedback loops. These three loops give rise to eight states, compared with the minimum of four states used by the two-loop design in Figure 7-19. We'll address this curious difference later.

Figure 7-78
Simplified positive
edge-triggered
D flip-flop for analysis.

The following excitation and output equations can be derived from the logic diagram in Figure 7-78:

$$Y1* = Y2 \cdot D + Y1 \cdot CLK$$
$$Y2* = Y1 + CLK' + Y2 \cdot D$$
$$Y3* = Y1 \cdot CLK + Y1 \cdot Y3 + Y3 \cdot CLK' + Y2 \cdot Y3 \cdot D$$
$$Q = Y1 \cdot CLK + Y1 \cdot Y3 + Y3 \cdot CLK' + Y2 \cdot Y3 \cdot D$$
$$QN = Y3' + Y1' \cdot Y2' \cdot CLK + Y1' \cdot CLK \cdot D'$$

Figure 7-79
Transition table
for the D flip-flop
in Figure 7-78.

| Y1 Y2 Y3 | CLK D | | | |
	00	01	11	10
000	010	010	(000)	(000)
001	011	011	000	000
010	(010)	110	110	000
011	(011)	111	111	000
100	010	010	111	111
101	011	011	111	111
110	010	(110)	111	111
111	011	(111)	(111)	(111)
	Y1* Y2* Y3*			

Y1 Y2 Y3	CLK D			
	00	01	11	10
000	010	010	(000)	(000)
001	011	011	000	000
010	(010)	110	110	000
011	(011)	111	111	000
		Y1* Y2* Y3*		

Figure 7-80
Portion of the D flip-flop transition table showing a noncritical race.

The corresponding transition table is shown in Figure 7-79, with the stable total states circled. Before going further, we must introduce the concept of "races."

*7.9.3 Races

In a feedback sequential circuit, a *race* is said to occur when multiple internal variables change state as a result of a single input changing state. In the example of Figure 7-79, a race occurs in stable total state 011/00 when CLK is changed from 0 to 1. The table indicates that the next internal state is 000, a 2-variable change from 011.

    As we've discussed previously, logic signals never really change "simultaneously." Thus, the internal state may make the change 011→000 as either 011→001→000 or 011→010→000. Figure 7-80 indicates that the example circuit, starting in total state 011/00, should go to total state 000/10 when CLK changes from 0 to 1. However, it may temporarily visit total state 001/10 or 010/10 along the way. That's OK, because the next internal state for both of these temporary states is 000; therefore, even in the temporary states, the excitation logic continues to drive the feedback loops toward the same stable total state, 000/10. Since the final state does not depend on the order in which the state variables change, this is called a *noncritical race*.

    Now suppose that the next-state entry for total state 010/10 is changed to 110, as shown in Figure 7-81, and consider the case that we just analyzed. Starting in stable total state 011/00 and changing CLK to 1, the circuit may end up in internal state 000 or 111, depending on the order and speed of the internal variable changes. This is called a *critical race*.

race

noncritical race

critical race

WATCH OUT FOR CRITICAL RACES!	When you design a feedback-based sequential circuit, you must ensure that its transition table does not contain any critical races. Otherwise, the circuit may operate unpredictably, with the next state for racy transitions depending on factors like temperature, voltage, and the phase of the moon.

Y1 Y2 Y3	CLK D			
	00	01	11	10
000	010	010	(000)	(000)
001	011	011	000	000
010	(010)	110	110	110
011	(011)	111	111	000
100	010	010	111	111
101	011	011	111	111
110	010	(110)	111	111
111	011	(111)	(111)	(111)
		Y1* Y2* Y3*		

Figure 7-81
A transition table
containing a critical race.

*7.9.4 State Tables and Flow Tables

Analysis of the real transition table, Figure 7-79, for our example D flip-flip circuit, shows that it does not have any critical races; in fact, it has no races except the noncritical one we identified earlier. Once we've determined this fact, we no longer need to refer to state variables. Instead, we can name the state-variable combinations and determine the output values for each state/input combination to obtain a state/output table such as Figure 7-82.

Figure 7-82
State/output table
for the D flip-flop in
Figure 7-78.

S	CLK D			
	00	01	11	10
S0	S2 , 01	S2 , 01	(S0) , 01	(S0) , 01
S1	S3 , 10	S3 , 10	S0 , 10	S0 , 10
S2	(S2) , 01	S6 , 01	S6 , 01	S0 , 01
S3	(S3) , 10	S7 , 10	S7 , 10	S0 , 01
S4	S2 , 01	S2 , 01	S7 , 11	S7 , 11
S5	S3 , 10	S3 , 10	S7 , 10	S7 , 10
S6	S2 , 01	(S6) , 01	S7 , 11	S7 , 11
S7	S3 , 10	(S7) , 10	(S7) , 10	(S7) , 10
		S* , Q QN		

		CLK D		
S	00	01	11	10
S0	S2 , 01	S6 , 01	(S0) , 01	(S0) , 01
S2	(S2) , 01	S6 , 01	— , –	S0 , 10
S3	(S3) , 10	S7 , 10	— , –	S0 , 01
S6	S2 , 01	(S6) , 01	S7 , 11	— , –
S7	S3 , 10	(S7) , 10	(S7) , 10	(S7) , 10
		S* , Q QN		

Figure 7-83
Flow and output table
for the D flip-flop in
Figure 7-78.

The state table shows that for some single input changes, the circuit takes multiple "hops" to get to a new stable total state. For example, in state S0/11, an input change to 01 sends the circuit first to state S2 and then to stable total state S6/01. A *flow table* eliminates multiple hops and shows only the ultimate destination for each transition. The flow table also eliminates the rows for unused internal states—ones that are stable for no input combination—and eliminates the next-state entries for total states that cannot be reached from a stable total state as the result of a single input change. Using these rules, Figure 7-83 shows the flow table for our D flip-flop example.

flow table

The flip-flop's edge-triggered behavior can be observed in the series of state transitions shown in Figure 7-84. Let us assume that the flip-flop starts in internal state S0/10. That is, the flip-flop is storing a 0 (since Q = 0), CLK is 1, and D is 0. Now suppose that D changes to 1; the flow table shows that we move one cell to the left, still a stable total state with the same output value. We can change D between 0 and 1 as much as we want, and just bounce back and forth between these two cells. However, once we change CLK to 0, we move to

		CLK D		
S	00	01	11	10
S0	S2 , 01	S6 , 01	(S0) , 01	(S0) , 01
S2	(S2) , 01	S6 , 01	— ,	– S0 , 10
S3	(S3) , 10	S7 , 10	— ,	– S0 , 01
S6	S2 , 01	(S6) , 01	S7 , 11	— , –
S7	S3 , 10	(S7) , 10	(S7) , 10	(S7) , 10
		S* , Q QN		

Figure 7-84
Flow and output table
showing the D flip-flop's
edge-triggered behavior.

	CLK D			
S	00	01	11	10
SB	(SB) , 01	S6 , 01	(SB) , 01	(SB) , 01
S3	(S3) , 10	S7 , 10	— , —	SB , 01
S6	SB , 01	(S6) , 01	S7 , 11	— , —
S7	S3 , 10	(S7) , 10	(S7) , 10	(S7) , 10
	S* , Q QN			

Figure 7-85
Reduced flow and output table for a positive edge-triggered D flip-flop.

internal state S2 or S6, depending on whether D was 0 or 1 at the time; but still the output is unchanged. Once again, we can change D between 0 and 1 as much as we want, this time bouncing between S2 and S6 without changing the output.

The moment of truth finally comes when CLK changes to 1. Depending on whether we are in S2 or S6, we go back to S0 (leaving Q at 0) or to S7 (setting Q to 1). Similar behavior involving S3 and S7 can be observed on a rising clock edge that causes Q to change from 1 to 0.

In Figure 7-19 we showed a circuit for a positive edge-triggered D flip-flop with only two feedback loops and hence four states. The circuit that we just analyzed has three loops and eight states. Even after eliminating unused states, the flow table has five states. However, a formal state-minimization procedure can be used to show that states S0 and S2 are "compatible," so that they can be merged into a single state SB that handles the transitions for both original states, as shown in Figure 7-85. Thus, the job really could have been done by a four-state circuit. In fact, in Exercise 7.66 you'll show that the circuit in Figure 7-19 does the job specified by the reduced flow table.

*7.9.5 CMOS D Flip-Flop Analysis

CMOS flip-flops typically use transmission gates in their feedback loops. For example, Figure 7-86 shows the circuit design of the "FD1Q" positive-edge-triggered D flip-flop in LSI Logic's LCA500K series of CMOS gate arrays. Such a flip-flop can be analyzed in the same way as a purely logic-gate-based design, once you recognize the feedback loops. Figure 7-86 has two feedback loops, each of which has a pair of transmission gates in a mux-like configuration controlled by CLK and CLK', yielding the following loop equations:

$$Y1* = CLK' \cdot D' + CLK \cdot Y1$$
$$Y2* = CLK \cdot Y1' + CLK' \cdot Y2$$

Except for the double inversion of the data as it goes from D to Y2* (once in the Y1* equation and again in the Y2* equation), these equations are very reminiscent of the master/slave-latch structure of the D flip-flop in Figure 7-15.

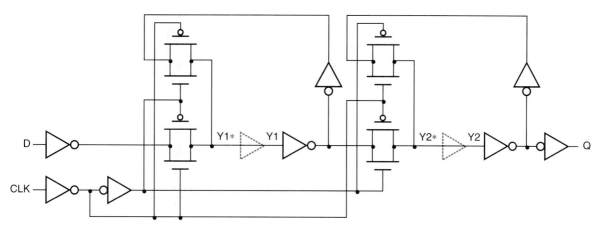

Figure 7-86 Positive edge-triggered CMOS D flip-flop for analysis.

Completing the formal analysis of the circuit is left as an exercise (7.73). Note, however, that since there are just two feedback loops, the resulting state and flow tables will have the minimum of just four states.

FEEDBACK CIRCUIT DESIGN The feedback sequential circuits that we've analyzed in this section exhibit quite reasonable behavior, since, after all, they are latch and flip-flop circuits that have been used for years. However, if we throw together a "random" collection of gates and feedback loops, we won't necessarily get "reasonable" sequential circuit behavior. In a few rare cases we may not get a sequential circuit at all (see Exercise 7.67), and in many cases the circuit may be unstable for some or all input combinations (see Exercise 7.71). Thus, the design of feedback sequential circuits continues to be something of a black art and is practiced only by a small fraction of digital designers. Still, the next section introduces basic concepts that help you do simple designs.

*7.10 Feedback Sequential-Circuit Design

It's sometimes useful to design a small feedback sequential circuit, such as a specialized latch or a pulse catcher; this section will show you how. It's even possible that you might go on to be an IC designer and be responsible for designing high-performance latches and flip-flops from scratch. This section will serve as an introduction to the basic concepts you'll need, but you'll still need considerably more study, experience, and finesse to do it right.

*7.10.1 Latches

Although the design of feedback sequential circuits is generally a hard problem, some circuits can be designed pretty easily. Any circuit with one feedback loop

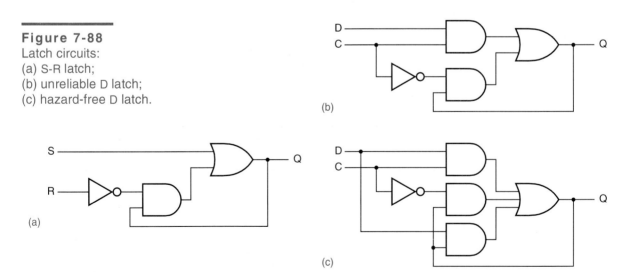

Figure 7-87
General structure
of a latch.

is just a variation of an S-R or D latch. It has the general structure shown in Figure 7-87 and an excitation equation with the following format:

$$Q* = (\text{forcing term}) + (\text{holding term}) \cdot Q$$

For example, the excitation equations for S-R and D latches are

$$Q* = S + R' \cdot Q$$
$$Q* = C \cdot D + C' \cdot Q$$

Corresponding circuits are shown in Figure 7-88(a) and (b).

hazard-free excitation logic

In general, *the excitation logic in a feedback sequential circuit must be hazard free;* we'll demonstrate this fact by way of an example. Figure 7-89(a) is a Karnaugh map for the D-latch excitation circuit of Figure 7-88(b). The map exhibits a static-1 hazard when D and Q are 1 and C is changing. Unfortunately, the latch's feedback loop may not hold its value if a hazard-induced glitch occurs. For example, consider what happens when D and Q are 1 and C changes from 1 to 0; the circuit should latch a 1. However, unless the inverter is very fast, the output of the top AND gate goes to 0 before the output of the bottom one goes to 1, the OR-gate output goes to 0, and the feedback loop stores a 0.

Figure 7-88
Latch circuits:
(a) S-R latch;
(b) unreliable D latch;
(c) hazard-free D latch.

Figure 7-89
Karnaugh maps for
D-latch excitation
functions: (a) original,
containing a static-1
hazard; (b) hazard
eliminated.

Hazards can be eliminated using the methods described in Section 4.5. In the D latch, we simply include the consensus term in the excitation equation:

$$Q* \ = \ C \cdot D + C' \cdot Q + D \cdot Q$$

Figure 7-88(c) shows the corresponding hazard-free, correct D-latch circuit.

Now, suppose we need a specialized "D" latch with three data inputs, D1–D3, that stores a 1 only if D1–D3 = 010. We can convert this word description into an excitation equation that mimics the equation for a simple D latch:

$$Q* \ = \ C \cdot (D1' \cdot D2 \cdot D3') + C' \cdot Q$$

Eliminating hazards, we get

$$Q* \ = \ C \cdot D1' \cdot D2 \cdot D3' + C' \cdot Q + D1' \cdot D2 \cdot D3' \cdot Q$$

The hazard-free excitation equation can be realized with discrete gates or in a PLD, as we'll show in Section 8.2.6.

*7.10.2 Designing Fundamental-Mode Flow Table

To design feedback sequential circuits more complex than latches, we must first convert the word description into a flow table. Once we have a flow table, we can turn the crank (with some effort) to obtain a circuit.

When we construct the flow table for a feedback sequential circuit, we give each state a meaning in the context of the problem, much as we did in the design

---

PRODUCT-TERM EXPLOSION   In some cases, the need to cover hazards can cause an explosion in the number of product terms in a two-level realization of the excitation logic. For example, suppose we need a specialized latch with two control inputs, C1 and C2, and three data inputs as before. The latch is to be "open" only if both control inputs are 1, and is to store a 1 if any data input is 1. The minimal excitation equation is

$$Q* \ = \ C1 \cdot C2 \cdot (D1 + D2 + D3) + (C1 \cdot C2)' \cdot Q$$
$$= \ C1 \cdot C2 \cdot D1 + C1 \cdot C2 \cdot D2 + C1 \cdot C2 \cdot D3 + C1' \cdot Q + C2' \cdot Q$$

However, it takes six consensus terms to eliminate hazards (see Exercise 7.76).

Figure 7-90 Typical functional behavior of a pulse-catching circuit.

of clocked state machines. However, it's easier to get confused when constructing the flow table for a feedback sequential circuit, because not every total state is stable. Therefore, the recommended procedure is to construct a *primitive flow table*—one that has only one stable total state in each row. Since there is only one stable state per row, the output may be shown as a function of state only.

primitive flow table

In a primitive flow table, each state has a more precise "meaning" than it might otherwise have, and the table's structure clearly displays the underlying fundamental-mode assumption: inputs change one at a time, with enough time between changes for the circuit to settle into a new stable state. A primitive flow table usually has extra states, but we can "turn the crank" later to minimize the number of states, once we have a flow table that we believe to be correct.

We'll use the following problem, a "pulse-catching" circuit, to demonstrate flow-table design:

Design a feedback sequential circuit with two inputs, P (pulse) and R (reset), and a single output Z that is normally 0. The output should be set to 1 whenever a 0-to-1 transition occurs on P, and should be reset to 0 whenever R is 1. Typical functional behavior is shown in Figure 7-90.

Figure 7-91 is a primitive flow table for the pulse catcher. Let's walk through how this table was developed.

We assume that the pulse catcher is initially idle, with P and R both 0; this is the IDLE state, with Z = 0. In this state, if reset occurs (R = 1), we could probably stay in the same state, but since this is supposed to be a *primitive* flow table, we create a new state, RES1, so as not to have two stable total states in the same row. On the other hand, if a pulse occurs (P = 1) in the IDLE state, we definitely want to go to a different state, which we've named PLS1, since we've caught a pulse and we must set the output to 1. Input combination 11 is not considered in the IDLE state, because of the fundamental-mode assumption that only one input changes at a time; we assume the circuit always makes it to another stable state before input combination 11 can occur.

Next, we fill in the next-state entries for the newly created RES1 state. If reset goes away, we can go back to the IDLE state. If a pulse occurs, we must remain in a "reset" state, since, according to the timing diagram, a 0-to-1 transition that occurs while R is 1 is ignored. Again, to keep the flow table in primitive form, we must create a new state for this case, RES2.

			P R			
Meaning	S	00	01	11	10	Z
Idle, waiting for pulse	IDLE	(IDLE)	RES1	—	PLS1	0
Reset, no pulse	RES1	IDLE	(RES1)	RES2	—	0
Got pulse, output on	PLS1	PLS2	—	RES2	(PLS1)	1
Reset, got pulse	RES2	—	RES1	(RES2)	PLSN	0
Pulse gone, output on	PLS2	(PLS2)	RES1	—	PLS1	1
Got pulse, but output off	PLSN	IDLE	—	RES2	(PLSN)	0
				S*		

Figure 7-91 Primitive flow table for pulse-catching circuit.

Now that we have one stable total state in each column, we may be able to go to existing states for more transitions, instead of always defining new states. Sure enough, starting in stable total state PLS1/10, for R = 1 we can go to RES2, which fits the requirement of producing a 0 output. On the other hand, where should we go for P = 0? IDLE is a stable total state in the 00 column, but it produces the wrong output value. In PLS1, we've gotten a pulse and haven't seen a reset, so if the pulse goes away, we should go to a state that still has Z = 1. Thus, we must create a new state PLS2 for this case.

In RES2, we can safely go to RES1 if the pulse goes away. However, we've got to be careful if reset goes away, as shown in the timing diagram. Since we've already passed the pulse's 0-to-1 transition, we can't go to the PLS1 state, since that would give us a 1 output. Instead, we create a new state PLSN with a 0 output.

Finally, we can complete the next-state entries for PLS2 and PLSN without creating any new states. Notice that, starting in PLS2, we bounce back and forth between PLS2 and PLS1 and maintain a continuous 1 output if we get a series of pulses without an intervening reset input.

*7.10.3 Flow-Table Minimization

As we mentioned earlier, a primitive flow table usually has more states than required. However, there exists a formal procedure, discussed in the References, for minimizing the number of states in a flow table. This procedure is often complicated by the existence of don't-care entries in the flow table.

Fortunately, our example flow table is small and simple enough to minimize by inspection. States IDLE and RES1 produce the same output, and they have the same next-state entry for input combinations where they are both specified. Therefore, they are compatible and may be replaced by a single state

		P R			
S	00	01	11	10	Z
IDLE	(IDLE)	(IDLE)	RES	PLS	0
PLS	(PLS)	IDLE	RES	(PLS)	1
RES	IDLE	IDLE	(RES)	(RES)	0
			S*		

Figure 7-92
Reduced flow table
for pulse-catching
circuit.

(IDLE) in a reduced flow table. The same can be said for states PLS1 and PLS2 (replaced by PLS) and for RES2 and PLSN (replaced by RES). The resulting reduced flow table, which has only three states, is shown in Figure 7-92.

*7.10.4 Race-Free State Assignment

The next somewhat creative (read "difficult") step in feedback sequential circuit design is to find a race-free assignment of coded states to the named states in the reduced flow table. Recall from Section 7.9.3 that a race occurs when multiple internal variables change state as a result of a single input change. A feedback-based sequential circuit must not contain any critical races; otherwise, the circuit may operate unpredictably. As we'll see, eliminating races often necessitates *increasing* the number of states in the circuit.

state adjacency
diagram

A circuit's potential for having races in its transition table can be analyzed by means of a *state adjacency diagram* for its flow table. The adjacency diagram is a simplified state diagram that omits self-loops and does not show the direction of other transitions (A→B is drawn the same as B→A) or the input combinations that cause them. Figure 7-93 is an example fundamental-mode flow table and Figure 7-94(a) is the corresponding adjacency diagram.

adjacent states

Two states are said to be *adjacent* if there is an arc between them in the state adjacency diagram. For race-free transitions, adjacent coded states must differ in only one bit. If two states A and B are adjacent, it doesn't matter whether

Figure 7-93
Example flow table for
the state-assignment
problem.

		X Y		
S	00	01	11	10
A	(A)	B	(A)	B
B	(B)	(B)	D	(B)
C	(C)	A	A	(C)
D	(D)	B	(D)	C
			S*	

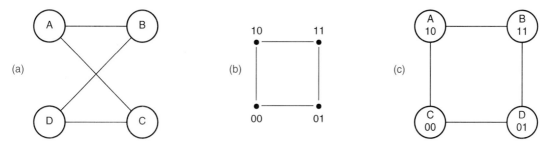

Figure 7-94 State-assignment example: (a) adjacency diagram; (b) a 2-cube; (c) one of eight possible race-free state assignments.

the original flow table had transitions from A to B, from B to A, or both. Any one of these transitions is a race if A and B differ in more than one variable. That's why we don't need to show the direction of transitions in an adjacency diagram.

The problem of finding a race-free assignment of states to *n* state variables is equivalent to the problem of mapping the nodes and arcs of the adjacency diagram onto the nodes and arcs of an *n*-cube. In Figure 7-94, the problem is to map the adjacency diagram (a) onto a 2-cube (b). You can visually identify eight ways to do this (four rotations times two flips), one of which produces the state assignment shown in (c).

Figure 7-95(a) is an adjacency diagram for our pulse-catching circuit, based on the reduced flow table in Figure 7-92. Clearly, there's no way to map this "triangle" of states onto a 2-cube. At this point, we can only go back and modify the original flow table. In particular, the flow table tells us the destination state that we *eventually* must reach for each transition, but it doesn't prevent us from going through other states on the way. As shown in Figure 7-96, we can create a new state RESA and make the transition from PLS to RES by going through RESA. The modified state table has the new adjacency diagram shown in Figure 7-95(b), which has many race-free assignments possible. A transition table based on the assignment in (c) is shown in Figure 7-98. Note that the PLS→RESA→RES transition will be slower than the other transitions in the

Figure 7-95 Adjacency diagrams for the pulse catcher: (a) using original flow table; (b) after adding a state; (c) showing one of eight possible race-free state assignments.

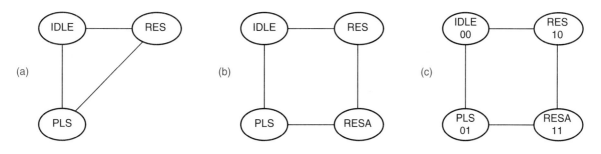

S	00	01	11	10	Z
		P R			
IDLE	(IDLE)	(IDLE)	RES	PLS	0
PLS	(PLS)	IDLE	RESA	(PLS)	1
RESA	—	—	RES	—	—
RES	IDLE	IDLE	(RES)	(RES)	0
			S*		

Figure 7-96
State table allowing a race-free assignment for the pulse-catching circuit.

original flow table because it requires two internal state changes, with two propagation delays through the feedback loops.

Even though we added a state in the previous example, we still got by with just two state variables. However, we may sometimes have to add one or more state variables to make a race-free assignment. Figure 7-97(a) shows the worst possible adjacency diagram for four states—every state is adjacent to every other state. Clearly, this adjacency diagram cannot be mapped onto a 2-cube. However, there is a race-free assignment of states to a 3-cube, shown in (b), where each state in the original flow table is represented by two equivalent states in the final state table. Both states in a pair, such as A1 and A2, are equivalent and produce the same output. Each state is adjacent to one of the states in every other pair, so a race-free transition may be selected for each next-state entry.

HANDLING THE GENERAL ASSIGNMENT CASE	In the general case of a flow table with 2^n rows, it can be shown that a race-free assignment can be obtained using $2n - 1$ state variables (see References). However, there aren't many applications for fundamental-mode circuits with more than a few states, so the general case is of little more than academic interest.

Figure 7-97
A worst-case scenario: (a) 4-state adjacency diagram; (b) assignment using pairs of equivalent states.

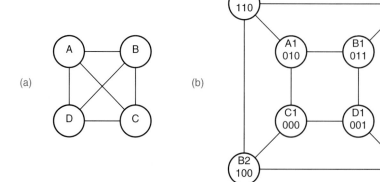

Y1 Y2	00	01	11	10	Z
			P R		
00	(00)	(00)	10	01	0
01	(01)	00	11	(01)	1
11	—	—	10	—	–
10	00	00	(10)	(10)	0
		Y1* Y2*			

Figure 7-98
Race-free transition table for the pulse-catching circuit.

*7.10.5 Excitation Equations

Once we have a race-free transition table for a circuit, we can just "turn the crank" to obtain excitation equations for the feedback loops. Figure 7-99 shows Karnaugh maps derived from Figure 7-98, the pulse catcher's transition table. Notice that "don't-care" next-state and output entries give rise to corresponding entries in the maps, simplifying the excitation and output logic. The resulting minimal sum-of-products excitation and output equations are as follows:

$$Y1* = P \cdot R + P \cdot Y1$$
$$Y2* = Y2 \cdot R' + Y1' \cdot Y2 \cdot P + Y1' \cdot P \cdot R'$$
$$Z = Y2$$

Recall that the excitation logic in a feedback sequential circuit must be hazard free. The sum-of-products expressions we derived happen to be hazard free as well as minimal. The logic diagram of Figure 7-100 uses these expressions to build the pulse-catching circuit.

Figure 7-99 Karnaugh maps for pulse-catcher excitation and output logic.

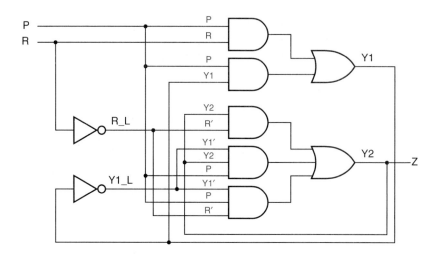

Figure 7-100
Pulse-catching
circuit.

*7.10.6 Essential Hazards

After all this effort, you'd think that we'd have a pulse-catching circuit that would operate reliably all of the time. Unfortunately, we're not quite there yet. A fundamental-mode circuit must generally satisfy five requirements for proper operation:

1. Only one input signal may change at a time, with a minimum bound between successive input changes.

2. There must be a maximum propagation delay through the excitation logic and feedback paths; this maximum must be less than the time between successive input changes.

3. The state assignment (transition table) must be free of critical races.

4. The excitation logic must be hazard free.

5. The minimum propagation delay through the excitation logic and feedback paths must be greater than the maximum timing skew through the "input logic."

Without the first requirement, it would be impossible to satisfy the major premise of fundamental-mode operation—that the circuit has time to settle into a stable total state between successive input changes. The second requirement says that the excitation logic is fast enough to do just that. The third requirement ensures that the proper state changes are made, even if the excitation circuits for different state variables have different delays. The fourth requirement ensures that state variables that aren't supposed to change on a particular transition don't.

The last requirement deals with subtle timing-dependent errors that can occur in fundamental-mode circuits, even ones that satisfy the first four

essential hazard   requirements. An *essential hazard* is the possibility of a circuit's going to an

Figure 7-101 Physical conditions in pulse-catching circuit for exhibiting an essential hazard.

erroneous next state as the result of a single input change; the error occurs if the input change is not seen by all of the excitation circuits before the resulting state-variable transition(s) propagate back to the inputs of the excitation circuits. In a world where "faster is better" is the usual rule, a designer may sometimes have to *slow down* excitation logic to mask these hazards.

Essential hazards are best explained in terms of an example, our pulse-catching circuit. Suppose we built our circuit on a PCB or a chip, and we (or, more likely, our CAD system) inadvertently connected input signal P through a long, slow path at the point shown in Figure 7-101. Let's assume that this delay is longer than the propagation delay of the AND-OR excitation logic.

Now consider what can happen if P R = 10, the circuit is in internal state 10, and P changes from 1 to 0. According to the transition table, repeated in Figure 7-102, the circuit should go to internal state 00, and that's that. But let's look at the actual operation of the circuit, as traced in Figure 7-101:

- (Changes shown with "→") The first thing that happens after P changes is that Y1 changes from 1 to 0. Now the circuit is in internal state 00.
- (Changes shown with "↠") Y1_L changes from 0 to 1. The change in Y1_L at AND gate A causes its output to go to 1, which in turn forces Y2 to 1. Whoops, now the circuit is in internal state 01.
- (Changes shown with "⇒") The change in Y2 at AND gates B and C causes their outputs to go to 1, reinforcing the 1 output at Y2. All this time, we've been waiting for the 1-to-0 change in P to appear at point PD.
- (Changes shown with "⇛") Finally, PD changes from 1 to 0, forcing the outputs of AND gates A and B to 0. However, AND gate C still has a 1 output, and the circuit remains in state 01—*the wrong state*.

| | P R | | | | |
Y1 Y2	00	01	11	10	Z
00	00	00	10	01	0
01	01	00	11	01	1
11	—	—	10	—	—
10	00	00	10	10	0
			Y1* Y2*		

Figure 7-102
Transition table for the pulse-catching circuit, exhibiting an essential hazard.

The only way to avoid this erroneous behavior in general is to ensure that changes in P arrive at the inputs of all the excitation circuits before any changes in state variables do. Thus, the inevitable difference in input arrival times, called *timing skew*, must be less than the propagation delay of the excitation circuits and feedback loops. This timing requirement can generally be satisfied only by careful design *at the electrical circuit level.*

timing skew

In the example circuit, it would appear that the hazard is easily masked, even by non-electrical engineers, since the designer need only ensure that a straight wire has shorter propagation delay than an AND-OR structure—easy in most technologies.

Still, many feedback sequential circuits, such as the TTL edge-triggered D flip-flop in Figure 7-19, have essential hazards in which the input skew paths include inverters. In such cases, the input inverters must be guaranteed to be faster than the excitation logic; that's not so trivial in either board-level or IC design. For example, if the excitation circuit in Figure 7-101 were physically built using AND-OR-INVERT gates, the delay from input changes to Y1_L could be very short indeed, as short as the delay through a single inverter.

Essential hazards can be found in most but not all fundamental-mode circuits. There's an easy rule for detecting them; in fact, this is the definition of "essential hazard" in some texts:.

- A fundamental-mode flow table contains an essential hazard for a stable total state S and an input variable X if, starting in state S, the stable total state reached after three successive transitions in X is different from the stable total state reached after one transition in X.

THESE HAZARDS ARE, WELL, ESSENTIAL! Essential hazards are called "essential" because they are inherent in the flow table for a particular sequential function and will appear in any circuit realization of that function. They can be masked only by controlling the delays in the circuit. Compare with static hazards in combinational logic, where we could eliminate hazards by adding consensus terms to a logic expression.

Thus, the essential hazard in the pulse catcher is detected by the arrows in Figure 7-102, starting in internal state 10 with $P R = 10$.

A fundamental-mode circuit must have at least three states to have an essential hazard, so latches don't have them. On the other hand, all flip-flops (circuits that sample inputs on a clock edge) do.

*7.10.7 Summary

In summary, you use the following steps to design a feedback sequential circuit:

1. Construct a primitive flow table from the circuit's word description.

2. Minimize the number of states in the flow table.

3. Find a race-free assignment of coded states to named states, adding auxiliary states or splitting states as required.

4. Construct the transition table.

5. Construct excitation maps and find a hazard-free realization of the excitation equations.

6. Check for essential hazards. Modify the circuit if necessary to ensure that minimum excitation and feedback delays are greater than maximum inverter or other input-logic delays.

7. Draw the logic diagram.

Also note that some circuits routinely violate the basic fundamental-mode assumption that inputs change one at a time. For example, in a positive-edge-triggered D flip-flop, the D input may change at the same time that CLK changes from 1 to 0, and the flip-flop still operates properly. The same thing certainly cannot be said at the 0-to-1 transition of CLK. Such situations require analysis of the transition table and circuit on a case-by-case basis if proper operation in "special cases" is to be guaranteed.

A FINAL QUESTION   Given the difficulty of designing fundamental-mode circuits that work properly, let alone ones that are fast or compact, how did anyone ever come up with the 6-gate, 8-state, commercial D flip-flop design in Figure 7-20? Don't ask me, I don't know!

7.11 ABEL Sequential-Circuit Design Features

7.11.1 Registered Outputs

ABEL has several features that support the design of sequential circuits. As we'll show in Section 8.3, most PLD outputs can be configured by the user to be *registered outputs* that provide a D flip-flop following the AND-OR logic, as in Figure 7-103. To configure one or more outputs to be registered, an ABEL

registered output

Figure 7-103
PLD registered output.

reg

program's pin declarations normally must contain an istype clause using the keyword "reg" (rather than "com") for each registered output. Table 7-22 is an example program that has three registered outputs and two combinational outputs.

As suggested by Figure 7-103, a registered output has at least two other attributes associated with it. The three-state buffer driving the output pin has an output-enable input OE, and the flip-flop itself has a clock input CLK. As shown in Table 7-22, the signals that drive these inputs are specified in the equations section of the program. Each input signal is specified as the corresponding main output signal name followed by an attribute suffix, *.CLK* or *.OE*. Some PLDs have flip-flops with additional controllable inputs. For example, asynchronous reset and preset inputs have attribute suffixes *.AR* and *.AP*; synchronous reset and preset inputs use suffixes *.SR* and *.SP*. And some PLDs provide flip-flop types other than D; their inputs are specified with suffixes like *.J* and *.K*.

.CLK
.OE
.AR
.AP
.SR
.SP

Within the equations section of the ABEL program, the logic values for registered outputs are established using the *clocked assignment operator, :=*. When the PLD is compiled, the expression on the righthand side will be applied to the D input of the output flip-flop. All of the same rules as for combinational outputs apply to output polarity control, on-set generation, don't-cares, and so on. In Table 7-22, the state bits Q1–Q3 are registered outputs, so they use clocked assignment, ":=". The UNLK and HINT signals are Mealy outputs, combinational functions of current state and input, so they use unclocked assignment, "=". A machine with pipelined outputs (Figure 7-37 on page 552), would instead use clocked assignment for such outputs.

clocked assignment operator, :=

ABEL's truth-table syntax (Table 4-16 on page 257) can also be used with registered outputs. The only difference is that the "->" operator between input and output items is changed to ":>".

clocked truth-table operator, :>

IS istype ESSENTIAL?   Older devices, such as the PAL16Rx family, contain a fixed, preassigned mix of combinational and registered outputs and are not configurable. With these devices, the compiler can deduce each output's type from its pin number and the istype statement is not necessary. Even with configurable outputs, some compilers can correctly deduce the output type from the equations. Still, it's a good idea to include the istype information, both as a double check and to enhance design portability.

```
module CombLock
Title 'Combination-Lock State Machine'

" Input and Outputs
X, CLOCK        pin;
UNLK, HINT      pin istype 'com';
Q1, Q2, Q3      pin istype 'reg';

Q = [Q1..Q3];

Equations

Q.CLK = CLOCK; Q.OE = 1;

" State variables
Q1 := Q1 & !Q2 & X  #  !Q1 & Q2 & Q3 & !X  #  Q1 & Q2 & !Q3;
Q2 := !Q2 & Q3 & X  #   Q2 & !Q3 & X;
Q3 := Q1 & !Q2 & !Q3 #   Q1 & Q3 & !X  #  !Q2 & !X
      #  !Q1 & !Q3 & !X  #   Q2 & !Q3 & X;

" Mealy outputs
UNLK = Q1 & Q2 & Q3 & !X;
HINT = !Q1 & !Q2 & !Q3 & !X  #   Q1 & !Q2 & X  #  !Q2 & Q3 & X
      #  Q2 & Q3 & !X  #   Q2 & !Q3 & X;

end CombLock
```

Table 7-22
ABEL program using registered outputs.

You can also design feedback sequential circuits in ABEL without using any of the language's sequential-circuit features. For example, in Section 8.2.6 we show how to specify latches using ABEL.

7.11.2 State Diagrams

The state-machine example in the previous subsection is just a transcription of the combination-lock machine that we synthesized by hand in Section 7.4.6 beginning on page 582. However, most PLD programming languages have a notation for defining, documenting, and synthesizing state machines directly, without ever writing a state, transition, or excitation table or deriving excitation equations by hand. Such a notation is called a *state-machine description language*. In ABEL, this notation is called a "state diagram," and the ABEL compiler does all the work of generating excitation equations that realize the specified machine.

In ABEL, the keyword `state_diagram` indicates the beginning of a state-machine definition. Table 7-23 shows the textual structure of an ABEL "state diagram." Here *state-vector* is an ABEL set that lists the state variables of the machine. If there are n variables in the set, then the machine has 2^n possible states corresponding to the 2^n different assignments of constant values to

state-machine description language

state_diagram

state-vector

Table 7-23
Structure of a "state diagram" in ABEL.

```
state_diagram state-vector
state state-value 1 : transition statement;
state state-value 2 : transition statement;
...
state state-value 2ⁿ : transition statement;
```

variables in the set. States are usually given symbolic names in an ABEL program; this makes it easy to try different assignments simply by changing the constant definitions.

state

The on-set for each state variable is created according to the information in the "state diagram." If a state variable also appears on the lefthand side of an equation in the `equations` section, the effects are cumulative (see the box on page 739 for a useful application of this behavior). The keyword `state` indicates that the next states and current outputs for a particular current state are about to be defined; a *state-value* is a constant that defines state-variable values for the current state. A *transition statement* defines the possible next states for the current state.

state-value
transition statement

GOTO statement
IF statement

ABEL has two commonly used transition statements. The *GOTO statement* unconditionally specifies the next state, for example "GOTO INIT". The *IF statement* defines the possible next states as a function of an arbitrary logic expressions. (There's also a seldom-used `CASE` statement that we don't cover.)

Table 7-24 shows the syntax of the ABEL IF statement. Here *TrueState* and *FalseState* are state values that the machine will go to if *LogicExpression* is true or false, respectively. These statements can be nested: *FalseState* can itself be another IF statement, and *TrueState* can be an IF statement if it is enclosed in braces. When multiple next states are possible, a nested IF-THEN-ELSE structure eliminates the ambiguities that can occur in hand-drawn state diagrams, where the transition conditions leaving a state can overlap (Section 7.5).

Our first example using ABEL's "state diagram" capability is based on our first state-machine design example from Section 7.4.1 on page 564. A state table for this machine was developed in Figure 7-49 on page 567. It is adapted to ABEL in Table 7-25. Several characteristics of this program should be noted:

- The definition of QSTATE uses three variables to encode state.
- The definitions of INIT–XTRA3 determine the individual state encodings.
- IF-THEN-ELSE statements are nested. A particular next state may appear in multiple places in one set of nested IF-THEN-ELSE clauses (e.g., see states OK0 and OK1).

Table 7-24
Structure of an ABEL IF statement.

```
IF LogicExpression THEN
    TrueState;
ELSE
    FalseState;
```

Table 7-25
An example of ABEL's
state-diagram
notation.

```
module SMEX1
title 'PLD Version of Example State Machine'

" Input and output pins
CLOCK, RESET_L, A, B            pin;
Q1..Q3                         pin istype 'reg';
Z                              pin istype 'com';

" Definitions
QSTATE = [Q1,Q2,Q3];               " State variables
INIT   = [ 0, 0, 0];
A0     = [ 0, 0, 1];
A1     = [ 0, 1, 0];
OK0    = [ 0, 1, 1];
OK1    = [ 1, 0, 0];
XTRA1  = [ 1, 0, 1];
XTRA2  = [ 1, 1, 0];
XTRA3  = [ 1, 1, 1];
RESET = !RESET_L;

state_diagram QSTATE

state INIT: IF RESET THEN INIT
            ELSE IF (A==0) THEN A0
            ELSE A1;

state A0:   IF RESET THEN INIT
            ELSE IF (A==0) THEN OK0
            ELSE A1;

state A1:   IF RESET THEN INIT
            ELSE IF (A==0) THEN A0
            ELSE OK1;

state OK0:  IF RESET THEN INIT
            ELSE IF (B==1)&(A==0) THEN OK0
            ELSE IF (B==1)&(A==1) THEN OK1
            ELSE IF (A==0) THEN OK0
            ELSE IF (A==1) THEN A1;

state OK1:  IF RESET THEN INIT
            ELSE IF (B==1)&(A==0) THEN OK0
            ELSE IF (B==1)&(A==1) THEN OK1
            ELSE IF (A==0) THEN A0
            ELSE IF (A==1) THEN OK1;

state XTRA1: GOTO INIT;
state XTRA2: GOTO INIT;
state XTRA3: GOTO INIT;

equations

QSTATE.CLK = CLOCK; QSTATE.OE = 1;
Z = (QSTATE == OK0) # (QSTATE == OK1);

END SMEX1
```

Table 7-26
Reduced equations
for SMEX1 PLD.

```
Q1 := (!Q2.FB & !Q3.FB & RESET_L
    # Q1.FB & RESET_L);
Q1.C = (CLOCK);
Q1.OE = (1);

Q2 := (Q1.FB & !Q3.FB & RESET_L & !A
    # Q1.FB & Q3.FB & RESET_L & A
    # Q1.FB & Q2.FB & RESET_L & B);
Q2.C = (CLOCK);
Q2.OE = (1);

Q3 := (!Q2.FB & !Q3.FB & RESET_L & A
    # Q1.FB & RESET_L & A);
Q3.C = (CLOCK);
Q3.OE = (1);

Z = (Q2 & Q1);
```

- Expressions like "(B==1)*(A==0)" were used instead of equivalents like "B*!A" only because the former are a bit more readable.

- The first IF statement in each of states INIT–OK1 ensures that the machine goes to the INIT state if RESET is asserted.

- Next-state equations are given for XTRA1–XTRA3 to ensure that the machine goes to a "safe" state if it somehow gets into an unused state.

- The single equation in the "equations" section of the program determines the behavior of the Moore-type output.

Table 7-26 shows the resulting excitation and output equations produced by the ABEL compiler (the reverse-polarity equations are not shown). Notice the use of variable names like "Q1.FB" in the righthand sides of the equations. Here, the ".FB" attribute suffix refers to the "feedback" signal into the AND-OR array coming from the flip-flop's Q output. This is done to make it clear that the signal is coming from the flip-flop, not from the corresponding PLD output pin, which can be selected in some complex PLDs. As shown in Figure 7-104, ABEL

USE IT OR ELSE

ABEL's IF-THEN-ELSE structure eliminates the transition ambiguity that can occur in state diagrams. However, the ELSE clause of an IF statement is optional. If it is omitted, the next state for some input combinations will be unspecified. Usually this is not the designer's intention.

Nevertheless, if you can guarantee that the unspecified input combinations will never occur, you may be able to reduce the size of the transition logic. If the @DCSET directive is given, the ABEL compiler treats the transition outputs for the unspecified state/input combinations as "don't-cares." In addition, it treats *all* transitions out of unused states as "don't-cares."

Figure 7-104
Output selection capability in a complex PLD.

actually allows you to select among three possible values on the righthand side of an equation using an attribute suffix on the signal name:

.Q   The actual flip-flop output pin before any programmable inversion.   *.Q*

.FB   A value equal to the value that the output pin would have if enabled.   *.FB*

.PIN   The actual signal at the PLD output pin. This signal is floating or driven   *.PIN*
by another device if the three-state driver is not enabled.

Obviously, the .PIN value should not be used in a state-machine excitation equation, since it is not guaranteed always to equal the state variable.

Despite the use of "high-level language," the program's author still had to refer to the original, hand-constructed state table in Figure 7-49 to come up with ABEL version in Table 7-25. A different approach is shown in Table 7-27. This program was developed directly from the word description of the state machine, which is repeated below:

Design a clocked synchronous state machine with two inputs, A and B, and a single output Z that is 1 if:

— A had the same value at each of the two previous clock ticks, *or*

— B has been 1 since the last time that the first condition was true.

Otherwise, the output should be 0.

A key idea in the new approach is to remove the last value of A from the state definitions, and instead to have a separate flip-flop that keeps track of it (LASTA). Then only two non-INIT states must be defined: LOOKING ("still looking for a match") and OK ("got a match or B has been 1 since last match"). The Z output is a simple combinational decode of the OK state.

**PHANTOM
(OF THE)
OPERAND**
Real CPLDs typically have only a two-input output-select multiplexer and omit the .FB input shown in Figure 7-104. When an equation calls for a signal with the .FB attribute, the ABEL compiler uses the corresponding .Q signal and simply adjusts it with the appropriate inversion (or not).

Table 7-27
A more "natural"
ABEL program for
the example state
machine.

```
module SMEX2
title 'Alternate Version of Example State Machine'

" Input and output pins
CLOCK, RESET_L, A, B            pin;
LASTA, Q1, Q2                   pin istype 'reg';
Z                               pin istype 'com';

" Definitions
QSTATE  = [Q1,Q2];                  " State variables
INIT    = [ 0, 0];                  " State encodings
LOOKING = [ 0, 1];
OK      = [ 1, 0];
XTRA    = [ 1, 1];
RESET = !RESET_L;

state_diagram QSTATE

state INIT:    IF RESET THEN INIT ELSE LOOKING;

state LOOKING: IF RESET THEN INIT
                    ELSE IF (A == LASTA) THEN OK
                    ELSE LOOKING;

state OK:      IF RESET THEN INIT
                    ELSE IF B THEN OK
                    ELSE IF (A == LASTA) THEN OK
                    ELSE LOOKING;

state XTRA:    GOTO INIT;

equations
LASTA.CLK = CLOCK; QSTATE.CLK = CLOCK; QSTATE.OE = 1;

LASTA := A;
Z = (QSTATE == OK);

END SMEX2
```

*7.11.3 External State Memory

In some situations, the state memory of a PLD-based state machine may be kept in flip-flops external to the PLD. ABEL provides a special version of the state_diagram statement to handle this situation:

> state_diagram *current-state-variables* -> *next-state variables*

current-state-variables
next-state-variables

Here *current-state-variables* is an ABEL set that lists the input signals which represent the current state of the machine, and *next-state-variables* is a set that lists the corresponding output signals which are the excitation for external D flip-flops holding the state of the machine, for example,

> state_diagram [CURQ1, CURQ2] -> [NEXTQ1, NEXTQ2]

```
state_diagram state-variables
state state-value 1 :
      optional equation ;
      optional equation ;
      . . .
      transition statement ;
state state-value 2 :
      optional equation ;
      optional equation ;
      . . .
      transition statement ;
. . .
state state-value 2ⁿ :
      optional equation ;
      optional equation ;
      . . .
      transition statement ;
```

Table 7-28
Structure of an ABEL
state diagram with
Moore outputs
defined.

*7.11.4 Specifying Moore Outputs

The output Z in our example state machine is a Moore output, a function of state only, and we defined this output in Tables 7-25 and 7-27 using an appropriate equation in the equations section of the program. Alternatively, ABEL allows Moore outputs to be specified along with the state definitions themselves. The transition statement in a state definition may be preceded by one or more optional equations, as shown in Table 7-28. To use this capability with the machine in Table 7-27, for example, we would eliminate the Z equation in the equations section and rewrite the state diagram as shown in Table 7-29.

As in other ABEL equations, when a variable such as Z appears on the left-hand side of multiple equations, the righthand sides are OR'ed together to form the final result (as discussed in Section 4.6.3). Also notice that Z is still specified

```
state_diagram QSTATE
state INIT:    Z = 0;
               IF RESET THEN INIT ELSE LOOKING;
state LOOKING: Z = 0;
               IF RESET THEN INIT
               ELSE IF (A == LASTA) THEN OK
               ELSE LOOKING;
state OK:      Z = 1;
               IF RESET THEN INIT
               ELSE IF B THEN OK
               ELSE IF (A == LASTA) THEN OK
               ELSE LOOKING;
state XTRA:    Z = 0;
               GOTO INIT;
```

Table 7-29
State machine with
embedded Moore
output definitions.

as a combinational, not a registered, output. If Z were a registered output, the desired output value would occur one clock tick after the machine visited the corresponding state.

*7.11.5 Specifying Mealy and Pipelined Outputs with WITH

Some state-machine outputs are functions of the inputs as well as state. In Section 7.3.2 we called them Mealy outputs or pipelined outputs, depending on whether they occurred immediately upon an input change or only after a clock edge. ABEL's *WITH statement* provides a way to specify these outputs side by side with the next states, rather than separately in the equations section of the program.

WITH statement

As shown in Table 7-30, the syntax of the WITH statement is very simple. Any next-state value which is part of a transition statement can be followed by the keyword WITH and a bracketed list of equations that are "executed" for the specified transition. Formally, let "E" an excitation expression that is true only when the specified transition is to be taken. Then for each equation in the WITH's bracketed list, the righthand side is AND'ed with E and assigned to the lefthand side. The equations can use either unclocked or clocked assignment to create Mealy or pipelined outputs, respectively.

Table 7-30
Structure of ABEL
WITH statement.

```
next-state WITH {
    equation;
    equation;
    . . .
}
```

We developed an example "combination lock" state machine with Mealy outputs in Table 7-14 on page 582. The same state machine is specified by the ABEL program in Table 7-31, using WITH statements for the Mealy outputs. Note that closing brackets take the place of the semicolons that normally end the transition statements for the states.

Based on the combination lock's word description, it is not possible to realize UNLK and HINT as pipelined outputs, since they depend on the current value of X. However, if we redefine UNLK to be asserted for the entire "unlocked" state, and HINT to be the actual recommended next value of X, we can create a new machine with pipelined outputs, as shown in Table 7-32. Notice that we used the clocked assignment operator for the outputs. More importantly, notice that the values of UNLK and HINT are different than in the Mealy example, since they have to "look ahead" one clock tick.

Because of "lookahead," pipelined outputs can be more difficult than Mealy outputs to design and understand. In the example above, we even had to modify the problem statement to accommodate them. The advantage of

```
module SMEX4
title 'Combination-Lock State Machine'

" Input and output pins
CLOCK, X                        pin;
Q1..Q3                          pin istype 'reg';
UNLK, HINT                      pin istype 'com';

" Definitions
S        = [Q1,Q2,Q3];          " State variables
ZIP      = [ 0, 0, 0];          " State encodings
X0       = [ 0, 0, 1];
X01      = [ 0, 1, 0];
X011     = [ 0, 1, 1];
X0110    = [ 1, 0, 0];
X01101   = [ 1, 0, 1];
X011011  = [ 1, 1, 0];
X0110111 = [ 1, 1, 1];

state_diagram S

state ZIP:       IF X==0 THEN X0    WITH {UNLK = 0; HINT = 1}
                 ELSE ZIP           WITH {UNLK = 0; HINT = 0}

state X0:        IF X==0 THEN X0    WITH {UNLK = 0; HINT = 0}
                 ELSE X01           WITH {UNLK = 0; HINT = 1}

state X01:       IF X==0 THEN X0    WITH {UNLK = 0; HINT = 0}
                 ELSE X011          WITH {UNLK = 0; HINT = 1}

tate X011:       IF X==0 THEN X0110 WITH {UNLK = 0; HINT = 1}
                 ELSE ZIP           WITH {UNLK = 0; HINT = 0}

state X0110:     IF X==0 THEN X0    WITH {UNLK = 0; HINT = 0}
                 ELSE X01101        WITH {UNLK = 0; HINT = 1}

state X01101:    IF X==0 THEN X0    WITH {UNLK = 0; HINT = 0}
                 ELSE X011011       WITH {UNLK = 0; HINT = 1}

state X011011:   IF X==0 THEN X0110 WITH {UNLK = 0; HINT = 0}
                 ELSE X0110111      WITH {UNLK = 0; HINT = 1}

state X0110111:  IF X==0 THEN X0    WITH {UNLK = 1; HINT = 1}
                 ELSE ZIP           WITH {UNLK = 0; HINT = 0}

equations
S.CLK = CLOCK;

END SMEX4
```

Table 7-31
State machine with embedded Mealy output definitions.

pipelined outputs is that, since they are connected directly to register outputs, they are valid a few gate-delays sooner after a state change than Moore or Mealy outputs, which normally include additional combinational logic. In the combination-lock example, it's probably not that important to open your lock or see your hint a few nanoseconds earlier. However, shaving off a few gate delays can be quite important in high-speed applications.

Table 7-32
State machine with
embedded pipelined
output definitions.

```
module SMEX5
title 'Combination-Lock State Machine'

" Input and output pins
CLOCK, X                         pin;
Q1..Q3, UNLK, HINT               pin istype 'reg';

" Definitions
S        = [Q1,Q2,Q3];           " State variables
ZIP      = [ 0, 0, 0];           " State encodings
X0       = [ 0, 0, 1];
X01      = [ 0, 1, 0];
X011     = [ 0, 1, 1];
X0110    = [ 1, 0, 0];
X01101   = [ 1, 0, 1];
X011011  = [ 1, 1, 0];
X0110111 = [ 1, 1, 1];

state_diagram S

state ZIP:      IF X==0 THEN X0    WITH {UNLK := 0; HINT := 1}
                ELSE ZIP           WITH {UNLK := 0; HINT := 0}

state X0:       IF X==0 THEN X0    WITH {UNLK := 0; HINT := 1}
                ELSE X01           WITH {UNLK := 0; HINT := 1}

state X01:      IF X==0 THEN X0    WITH {UNLK := 0; HINT := 1}
                ELSE X011          WITH {UNLK := 0; HINT := 0}

state X011:     IF X==0 THEN X0110 WITH {UNLK := 0; HINT := 1}
                ELSE ZIP           WITH {UNLK := 0; HINT := 0}

state X0110:    IF X==0 THEN X0    WITH {UNLK := 0; HINT := 1}
                ELSE X01101        WITH {UNLK := 0; HINT := 1}

state X01101:   IF X==0 THEN X0    WITH {UNLK := 0; HINT := 1}
                ELSE X011011       WITH {UNLK := 0; HINT := 1}

state X011011:  IF X==0 THEN X0110 WITH {UNLK := 0; HINT := 1}
                ELSE X0110111      WITH {UNLK := 1; HINT := 0}

state X0110111: IF X==0 THEN X0    WITH {UNLK := 0; HINT := 1}
                ELSE ZIP           WITH {UNLK := 0; HINT := 0}

equations
S.CLK = CLOCK; UNLK.CLK = CLOCK; HINT.CLK = CLOCK;

END SMEX5
```

7.11.6 Test Vectors

.C., clock edge

Test vectors for sequential circuits in ABEL have the same uses and limitations as test vectors for combinational circuits, as described in Section 4.6.7. One important addition to their syntax is the use of the constant ".C." to denote a clock edge, $0 \rightarrow 1 \rightarrow 0$. Thus, Table 7-33 is an ABEL program, with test vectors, for a simple 8-bit register with a clock-enable input. A variety of vectors are used to test loading and holding different input values.

```
module REG8EN
title '8-bit register with clock enable'

" Input and output pins
CLK, EN, D1..D8          pin;
Q1..Q8                   pin istype 'reg';

" Sets
D = [D1..D8];
Q = [Q1..Q8];

equations

Q.CLK = CLK;

WHEN EN == 1 THEN Q := D ELSE Q := Q;

test_vectors ([CLK, EN,  D  ] -> [ Q  ])
             [.C.,  1, ^h00] -> [^h00]; " 0s in every bit
             [.C.,  0, ^hFF] -> [^h00]; " Hold capability, EN=0
             [.C.,  1, ^hFF] -> [^hFF]; " 1s in every bit
             [.C.,  0, ^h00] -> [^hFF]; " Hold capability
             [.C.,  1, ^h55] -> [^h55]; " Adjacent bits shorted
             [.C.,  0, ^hAA] -> [^h55]; " Hold capability
             [.C.,  1, ^hAA] -> [^hAA]; " Adjacent bits shorted
             [.C.,  1, ^h55] -> [^h55]; " Load with quick setup
             [.C.,  1, ^hAA] -> [^hAA]; " Again
END REG8EN
```

Table 7-33
ABEL program with test vectors for a simple 8-bit register.

A typical approach to testing state machines is to write vectors that cause the machine not only to visit every state, but also to exercise every transition from every state. A key difference and challenge compared to combinational-circuit test vectors is that the vectors must first drive the machine into the desired state before testing a transition, and then come back again for each different transition from that state.

Thus, Table 7-34 shows test vectors for the state machine in Table 7-27. It's important to understand that, unlike combinational vectors, these vectors work only if applied in exactly the order they are written. Notice that the vectors were written to be independent of the state encoding. As a result, they don't have to be modified if the state encoding is changed.

We encounter another challenge if we attempt to create test vectors for the combination-lock state machine of Table 7-31 on page 637. This machine has a major problem when it comes to testing—it has no reset input. Its starting state at power-up may be different in PLD devices and technologies—the individual flip-flops may be all set, all reset, or all in random states. In the machine's actual application, we didn't necessarily need a reset input, but for testing purposes we somehow have to get to a known starting state.

Table 7-34 Test vectors for the state machine in Table 7-27.

```
test_vectors
([RESET_L, CLOCK, A, B] -> [QSTATE , LASTA, Z])
  [  0   , .C. , 0, 0] -> [INIT   ,  0  , 0]; " Check -->INIT (RESET)
  [  0   , .C. , 1, 0] -> [INIT   ,  1  , 0]; "  and LASTA flip-flop
  [  1   , .C. , 0, 0] -> [LOOKING,  0  , 0]; " Come out of initialization
  [  0   , .C. , 0, 0] -> [INIT   ,  0  , 0]; " Check LOOKING-->INIT (RESET)
  [  1   , .C. , 0, 0] -> [LOOKING,  0  , 0]; " Come out of initialization
  [  1   , .C. , 1, 0] -> [LOOKING,  1  , 0]; " --> LOOKING since 0!=1
  [  1   , .C. , 1, 0] -> [OK     ,  1  , 1]; " --> OK since 1==1
  [  0   , .C. , 0, 0] -> [INIT   ,  0  , 0]; " Check OK-->INIT (RESET)
  [  1   , .C. , 0, 0] -> [LOOKING,  0  , 0]; " Go back towards OK ...
  [  1   , .C. , 0, 0] -> [OK     ,  0  , 1]; " --> OK since 0==0
  [  1   , .C. , 1, 1] -> [OK     ,  1  , 1]; " --> OK since B, even though 1!=0
  [  1   , .C. , 1, 0] -> [OK     ,  1  , 1]; " --> OK since 1==1
  [  1   , .C. , 0, 0] -> [LOOKING,  0  , 0]; " --> LOOKING since 0!=1
```

synchronizing sequence   Luckily, the combination-lock machine has a *synchronizing sequence*—a fixed sequence of one or more input values that will always drive it to a certain known state. In particular, starting from any state, if we apply X=1 to the machine for four ticks, we will always be in state ZIP by the fourth tick. This is the approach taken by the first four vectors in Table 7-35. Until we get to the known state, we indicate the next state on the righthand side of the vector as being "don't care," so the simulator or the physical device tester will not flag a random state as an error.

Once we get going, we encounter something else that's new—Mealy outputs that must be tested. As shown by the fourth and fifth vectors, we don't have to transition the clock in every test vector. Instead, we can keep the clock fixed at 0, where the last transition left it, and observe the Mealy output values produced by the two input values of X. Then we can test the next state transition.

For the state transitions, we list the expected next state but we show the output values as don't-cares. For a correct test vector, the outputs must show the values attained *after* the transition, a function of the *next* state. Although it's possible to figure them out and include them, the complexity is enough to give you a headache, and they will be tested by the next CLOCK=0 vectors anyway.

Creating test vectors for a state machine by hand is a painstaking process, and no matter how careful you are, there's no guarantee that you've tested all its functions and potential hardware faults. For example, the vectors in Table 7-34

SYNCHRONIZING SEQUENCES AND RESET INPUTS	We lucked out with the combination lock; not all state machines have synchronizing sequences. This is why most state machines are designed with a reset input, which in effect allows a synchronizing sequence of length one.

Table 7-35 Test vectors for the combination-lock state machine of Table 7-31.

```
test_vectors
([CLOCK, X] -> [  S    , UNLK, HINT])
 [ .C. , 1] -> [.X.    , .X. , .X. ]; " Since no reset input, apply
 [ .C. , 1] -> [.X.    , .X. , .X. ]; "   a 'synchronizing sequence'
 [ .C. , 1] -> [.X.    , .X. , .X. ]; "   to reach a known starting
 [ .C. , 1] -> [ZIP    , .X. , .X. ]; "   state
 [ 0   , 0] -> [ZIP    , 0   , 1   ]; " Test Mealy outputs for both
 [ 0   , 1] -> [ZIP    , 0   , 0   ]; "   values of X
 [ .C. , 1] -> [ZIP    , .X. , .X. ]; " Test ZIP-->ZIP (X==1)
 [ .C. , 0] -> [X0     , .X. , .X. ]; "   and ZIP-->X0 (X==0)
 [ 0   , 0] -> [X0     , 0   , 0   ]; " Test Mealy outputs for both
 [ 0   , 1] -> [X0     , 0   , 1   ]; "   values of X
 [ .C. , 0] -> [X0     , .X. , .X. ]; " Test X0-->X0 (X==0)
 [ .C. , 1] -> [X01    , .X. , .X. ]; "   and X0-->X01 (X==1)
 [ 0   , 0] -> [X01    , 0   , 0   ]; " Test Mealy outputs for both
 [ 0   , 1] -> [X01    , 0   , 1   ]; "   values of X
 [ .C. , 0] -> [X0     , .X. , .X. ]; " Test X01-->X0 (X==0)
 [ .C. , 1] -> [X01    , .X. , .X. ]; " Get back to X01
 [ .C. , 1] -> [X011   , .X. , .X. ]; " Test X01-->X011 (X==1)
```

do not test (A LASTA) = 10 in state LOOKING, or (A B LASTA) = 100 in state OK. Thus, generating a complete set of test vectors for fault-detection purposes is a process best left to an automatic test-generation program. In Table 7-35 we petered out after writing vectors for the first few states; completing the test vectors is left as an exercise (7.92). Still, on the functional testing side, writing a few vectors to exercise the machine's most basic functions can weed out obvious design errors early in the process. More subtle design errors are best detected by a thorough system-level simulation.

7.12 VHDL Sequential-Circuit Design Features

Most of the VHDL features that are needed to support sequential-circuit design, in particular, processes, were already introduced in Section 4.7 and were used in the VHDL sections in Chapter 5. This section introduces just a couple more features and gives simple examples of how they are used. Larger examples appear in the VHDL sections of Chapter 8.

7.12.1 Feedback Sequential Circuits

A VHDL process and the simulator's event-list mechanism for tracking signal changes form the fundamental basis for handling feedback sequential circuits in VHDL. Remember that feedback sequential circuits may change state in response to input changes, and these state changes are manifested by changes propagating in a feedback loop until the feedback loop stabilizes. In simulation,

Table 7-36
Dataflow VHDL for
an S-R latch.

```
library IEEE;
use IEEE.std_logic_1164.all;

entity Vsrlatch is
  port (S, R: in STD_LOGIC;
        Q, QN: buffer STD_LOGIC );
end Vsrlatch;

architecture Vsrlatch_arch of Vsrlatch is
begin
  QN <= S nor Q;
  Q  <= R nor QN;
end Vsrlatch_arch;
```

this is manifested by the simulator putting signal changes on the event list and scheduling processes to rerun in "delta time" and propagate these signal changes until no more signal changes are scheduled.

Table 7-36 is a VHDL program for an S-R latch. The architecture contains two concurrent assignment statements, each of which gives rise to a process, as discussed in Section 4.7.9. These processes interact to create the simple latching behavior of an S-R latch.

The VHDL simulation is faithful enough to handle the case where both S and R are asserted simultaneously. The most interesting result in simulation occurs if you negate S and R simultaneously. Recall from the box on page 536 that a real S-R latch may oscillate or go into a metastable state in this situation. The simulation will potentially loop forever as each execution of one assignment statement triggers another execution of the other. After some number of repetitions, a well-designed simulator will discover the problem—delta time keeps advancing while simulated time does not—and halt the simulation.

**WHAT DO 'U'
WANT?**

It would be nice if the S-R-latch model in Table 7-36 produced a 'U' output whenever S and R were negated simultaneously, but it's not *that* good. However, the language is powerful enough that experienced VHDL designers can easily write a model with that behavior. Such a model would make use of VHDL's time-modeling facilities, which we haven't discussed, to model the latch's "recovery time" (see box on page 537) and force a 'U' output if a second input change occurred too soon. It's even possible to model a maximum assumed metastability resolution time in this way.

Note that if a circuit has the possibility of entering a metastable state, there's no guarantee that the simulation will detect it, especially in larger designs. The best way to avoid metastability problems in a system design is to clearly identify and protect its asynchronous boundaries, as discussed in Section 8.9.

```
library IEEE;
use IEEE.std_logic_1164.all;

entity VposDff is
  port (CLK, CLR, D: in STD_LOGIC;
        Q, QN: out STD_LOGIC );
end VposDff;

architecture VposDff_arch of VposDff is
begin
  process (CLK, CLR)
  begin
    if CLR='1' then Q <= '0'; QN <= '1';
    elsif CLK'event and CLK='1' then Q <= D; QN <= not D;
    end if;
  end process;
end VposDff_arch;
```

Table 7-37
Behavioral VHDL
for a positive-edge-
triggered D flip-flop.

7.12.2 Clocked Circuits

In practice, the majority of digital designs that are modeled using VHDL are clocked, synchronous systems using edge-triggered flip-flops. In addition to what we've already learned about VHDL, there's just one more feature needed to describe edge-triggered behavior. The *event attribute* can be attached to a signal name to yield a value of type boolean that is true if an event on the signal caused the encompassing process to run in the current simulation cycle, and false otherwise.

event attribute

Using the event attribute, we can model the behavior of a positive-edge-triggered D flip-flop with asynchronous clear as shown in Table 7-37. Here, the asynchronous clear input CLR overrides any behavior on the clock input CLK and is therefore checked first, in the "if" clause. If CLR is negated, then the "elsif" clause is checked, and its statements are executed on the rising edge of CLK. Note that "CLK'event" is true for any change on CLK, and "CLK='1'" is checked to limit triggering to just the rising edge of CLK. There are many other ways to construct processes or statements with edge-triggered behavior; Table 7-38 shows two more ways to describe a D flip-flop (without a CLR input).

In the test bench for a clocked circuit, one of things you need to do is to generate a system clock signal. This can be done quite easily with a loop inside a process, as shown in Table 7-39 for a 100-MHz clock with a 60% duty cycle.

```
process
  wait until CLK'event and CLK='1';
  Q <= D;
end process;

Q <= D when CLK'event and CLK='1' else Q;
```

Table 7-38
Two more ways to
describe a positive-
edge-triggered
D flip-flop.

SYNTHESIS STUFF

You may be wondering, how does a synthesis tool convert the edge-triggered behavior described in Table 7-37 or 7-38 into an efficient flip-flop realization? Most tools recognize only a few predetermined ways of expressing edge-triggered behavior and map those into predetermined flip-flop components in the target technology.

The Synopsis synthesis engine used in the Xilinx Foundation Series 1.5 software recognizes the "CLK'event and CLK='1'" expression that we use in this book. Even with that as a given, VHDL has many different ways of expressing the same functionality, as we showed in Table 7-38. Peter Ashenden, author of *The Designer's Guide to VHDL* (Morgan Kaufmann, 1996), ran these statements and one other, with some modification, through several different synthesis tools. Only one of them was able to synthesize three out of the four forms; most could handle only two. So, you need to follow the method that is prescribed by the tool you use.

Table 7-39
Clock process within a test bench.

```
architecture TB_arch of TB is
signal MCLK: STD_LOGIC;
signal ... -- Declare other input and output signals

process -- Clock generator
begin
  MCLK <= 1;  -- Start at 1 at time 0
  loop
    MCLK <= 0 after 6 ns;
    MCLK <= 1 after 4 ns;
  end loop;
end process;

process -- Generate the rest of the input stimuli , check outputs
begin
  ...
end;
```

References

The problem of metastability has been around for a long time. Greek philosophers wrote about the problem of indecision thousands of years ago. A group of modern philosophers named Devo sang about metastability in the title song of their *Freedom of Choice* album. And the U.S. Congress still can't decide how to "save" Social Security.

Scan capability was first deployed in latches, not flip-flops, in IBM IC designs decades ago. Edward J. McCluskey has a very good discussion of this and other scan methods in *Logic Design Principles* (Prentice Hall, 1986).

Most ASICs and MSI-, PLD- and FPGA-based designs use the sequential-circuit types described in this chapter. However, there are other types that are used in both older discrete designs (going all the way back to vacuum-tube logic) and as well as in modern, custom VLSI designs.

For example, clocked synchronous state machines are a special case of a more general class of *pulse-mode circuits*. Such circuits have one or more *pulse inputs* such that (a) only one pulse occurs at a time; (b) nonpulse inputs are stable when a pulse occurs; (c) only pulses can cause state changes; and (d) a pulse causes at most one state change. In clocked synchronous state machines, the clock is the single pulse input, and a "pulse" is the triggering edge of the clock. However, it is also possible to build circuits with multiple pulse inputs, and it is possible to use storage elements other than the familiar edge-triggered flip-flops. These possibilities are discussed very thoroughly in McCluskey's *Logic Design Principles* (Prentice Hall, 1986).

pulse-mode circuit
pulse input

A particularly important type of pulse-mode circuit that is discussed by McCluskey and others is the *two-phase latch machine*. The rationale for a two-phase clocking approach in VLSI circuits is discussed by Carver Mead and Lynn Conway in *Introduction to VLSI Systems* (Addison-Wesley, 1980). These machines essentially eliminate the essential hazards present in edge-triggered flip-flops by using pairs of latches that are enabled by nonoverlapping clocks.

two-phase latch machine

Methods for reducing both completely and incompletely specified state tables are described in advanced logic design texts, including McCluskey's 1986 book. A more mathematical discussion of these methods and other "theoretical" topics in sequential machine design appears in *Switching and Finite Automata Theory*, second edition, by Zvi Kohavi (McGraw-Hill, 1978).

As we showed in this chapter, improperly constructed state diagrams may yield an ambiguous description of next-state behavior. The "IF–THEN–ELSE" structures in HDLs like ABEL and VHDL can eliminate these ambiguities, but they were not the first to do so. *Algorithmic-state-machine (ASM)* notation, a flowchart-like equivalent of nested IF–THEN–ELSE statements, has been around for over 25 years. So-called *ASM charts* were pioneered at Hewlett-Packard Laboratories by Thomas E. Osborne and were further developed by Osborne's colleague Christopher R. Clare in a book, *Designing Logic Systems Using State Machines* (McGraw-Hill, 1973). Design and synthesis methods using ASM charts subsequently found a home in many digital design texts, including *The Art of Digital Design* by F. P. Prosser and D. E. Winkel (Prentice-Hall, 1987, second edition) and *Digital Design* by M. Morris Mano (Prentice-Hall, 1984), as well as in the first two editions of the book you're reading.

algorithmic state machine (ASM)
ASM chart

Another notation for describing state machines is an extension of "traditional" state-diagram notation called the mnemonic documented state (MDS) diagram; it was developed by William I. Fletcher in *An Engineering Approach to Digital Design* (Prentice-Hall, 1980). ASM charts and MDS diagrams have now been largely replaced by HDLs and their compilers.

Many CAD environments for digital design include a graphical state-diagram entry tool. Unfortunately, these typically support only traditional state diagrams, making it very easy for a designer to create an ambiguous description of next-state behavior. As a result, my personal recommendation is that you stay away from state-diagram editors and instead use an HDL to describe your state machines.

We mentioned the importance of synchronizing sequences in connection with state-machine test vectors. There's actually a very well developed but almost forgotten theory and practice of synchronizing sequences and somewhat less powerful "homing experiments," described by Frederick C. Hennie in *Finite-State Models for Logical Machines* (Wiley, 1968). Unless you've got this old classic on your bookshelf and know how to apply its teachings, please just remember to provide a reset input in every state machine that you design!

Drill Problems

7.1   Give three examples of metastability that occur in everyday life, other than ones discussed in this chapter.

7.2   Sketch the outputs of an S-R latch of the type shown in Figure 7-5 for the input waveforms shown in Figure X7.2. Assume that input and output rise and fall times are zero, that the propagation delay of a NOR gate is 10 ns, and that each time division below is 10 ns.

Figure X7.2

7.3   Repeat Drill 7.2 using the input waveforms shown in Figure X7.3. Although you may find the result unbelievable, this behavior can actually occur in real devices whose transition times are short compared to their propagation delay.

Figure X7.3

7.4   Figure 7-34 showed how to build a T flip-flop with enable using a D flip-flop and combinational logic. Show how to build a D flip-flop using a T flip-flop with enable and combinational logic.

7.5   Show how to build a J-K flip-flop using a T flip-flop with enable and combinational logic.

7.6   Show how to build an S-R latch using a single 74x74 positive-edge-triggered D flip-flop and *no* other components.

7.7   Show how to build a flip-flop equivalent to the 74x109 positive-edge-triggered J-$\overline{\text{K}}$ flip-flop using a 74x74 positive-edge-triggered D flip-flop and one or more gates from a 74x00 package.

7.8   Show how to build a flip-flop equivalent to the 74x74 positive-edge-triggered D flip-flop using a 74x109 positive-edge-triggered J-$\overline{\text{K}}$ flip-flop and *no* other components.

7.9   Analyze the clocked synchronous state machine in Figure X7.9. Write excitation equations, excitation/transition table, and state/output table (use state names A–D for Q1 Q2 = 00–11).

Figure X7.9

7.10   Repeat Drill 7.9, swapping AND and OR gates in the logic diagram. Is the new state/output table the "dual" of the original one? Explain.

7.11   Draw a state diagram for the state machine described by Table 7-6.

7.12   Draw a state diagram for the state machine described by Table 7-12.

7.13   Draw a state diagram for the state machine described by Table 7-14.

7.14   Construct a state and output table equivalent to the state diagram in Figure X7.14. Note that the diagram is drawn with the convention that the state does not change except for input conditions that are explicitly shown.

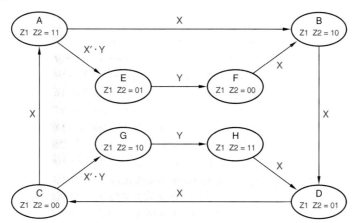

Figure X7.14

7.15   Analyze the clocked synchronous state machine in Figure 7.15. Write excitation equations, excitation/transition table, and state table (use state names A–H for Q2 Q1 Q0 = 000–111).

Figure X7.15

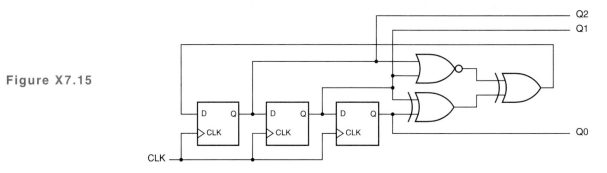

7.16   Analyze the clocked synchronous state machine in Figure X7.16. Write excitation equations, excitation/transition table, and state/output table (use state names A–H for Q1 Q2 Q3 = 000–111).

Figure X7.16

7.17   Analyze the clocked synchronous state machine in Figure X7.17. Write excitation equations, transition equations, transition table, and state/output table (use state names A–D for Q1 Q2 = 00–11). Draw a state diagram, and draw a timing diagram for CLK, X, Q1, and Q2 for 10 clock ticks, assuming that the machine starts in state 00 and X is continuously 1.

Figure X7.17

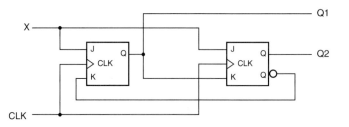

7.18   Analyze the clocked synchronous state machine in Figure X7.18. Write excitation equations, transition equations, transition table, and state/output table (use state names A–D for Q1 Q0 = 00–11). Draw a state diagram, and draw a timing diagram for CLK, EN, Q1, and Q0 for 10 clock ticks, assuming that the machine starts in state 00 and EN is continuously 1.

Figure X7.18

7.19   Analyze the clocked synchronous state machine in Figure X7.19. Write excitation equations, excitation/transition table, and state/output table (use state names A–D for Q1 Q2 = 00–11).

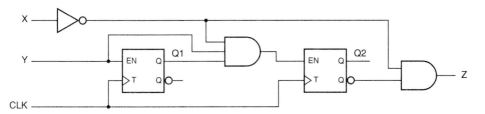

Figure X7.19

7.20   All of the state diagrams in Figure X7.20 are ambiguous. List all of the ambiguities in these state diagrams. (*Hint:* Use Karnaugh maps where necessary to find uncovered and double-covered input combinations.)

Figure X7.20

(a)

(b)

(c)

(d)

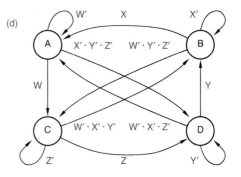

7.21   Synthesize a circuit for the state diagram of Figure 7-64 using six variables to encode the state, where the LA–LC and RA–RC outputs equal the state variables themselves. Write a transition list, a transition equation for each state variable as a sum of p-terms, and simplified transition/excitation equations for a realization using D flip-flops. Draw a circuit diagram using SSI and MSI components.

7.22   Starting with the transition list in Table 7-18, find a minimal sum-of-products expression for Q2*, assuming that the next states for the unused states are true don't-cares.

7.23   Modify the state diagram of Figure 7-64 so that the machine goes into hazard mode immediately if LEFT and RIGHT are asserted simultaneously during a turn. Write the corresponding transition list.

Exercises

7.24   Explain how metastability occurs in a D latch when the setup and hold times are not met, analyzing the behavior of the feedback loop inside the latch.

7.25   What is the minimum setup time of a pulse-triggered flip-flop such as a master/ slave J-K or S-R flip-flop? (*Hint:* It depends on certain characteristics of the clock.)

7.26   Describe a situation, other than the metastable state, in which the Q and QN outputs of a 74x74 edge-triggered D flip-flop may be noncomplementary for an arbitrarily long time.

7.27   Compare the circuit in Figure X7.27 with the D latch in Figure 7-12. Prove that the circuits function identically. In what way is Figure X7.27, which is used in some commercial D latches, better?

Figure X7.27

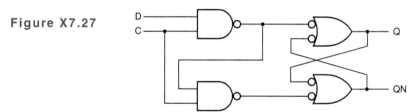

7.28   Suppose that a clocked synchronous state machine with the structure of Figure 7-35 is designed using D latches with active-high C inputs as storage elements. For proper next-state operation, what relationships must be satisfied among the following timing parameters?

t_{Fmin}, t_{Fmax} Minimum and maximum propagation delay of the next-state logic.
t_{CQmin}, t_{CQmax} Minimum and maximum clock-to-output delay of a D latch.
t_{DQmin}, t_{DQmax} Minimum and maximum data-to-output delay of a D latch.
t_{setup}, t_{hold} Setup and hold times of a D latch.
t_H, t_L Clock HIGH and LOW times.

7.29   Redesign the state machine in Drill 7.9 using just three inverting gates—NAND or NOR—and no inverters.

7.30 Draw a state diagram for a clocked synchronous state machine with two inputs, INIT and X, and one Moore-type output Z. As long as INIT is asserted, Z is continuously 0. Once INIT is negated, Z should remain 0 until X has been 0 for two successive ticks and 1 for two successive ticks, regardless of the order of occurrence. Then Z should go to 1 and remain 1 until INIT is asserted again. Your state diagram should be neatly drawn and planar (no crossed lines). (*Hint:* No more than ten states are required.)

7.31 Repeat Exercise 7.30, but write the state diagram in ABEL.

7.32 Design a clocked synchronous state machine that checks a serial data line for even parity. The circuit should have two inputs, SYNC and DATA, in addition to CLOCK, and one Moore-type output, ERROR. Devise a state/output table that does the job using just four states and include a description of each state's meaning in the table. Choose a 2-bit state assignment, write transition and excitation equations, and draw the logic diagram. Your circuit may use D flip-flops, J-K flip-flops, or one of each.

7.33 Repeat Exercise 7.32, but do the design using ABEL and a GAL16V8 PLD.

7.34 Design a clocked synchronous state machine with the state/output table shown in Table X7.34, using D flip-flops. Use two state variables, Q1 Q2, with the state assignment A = 00, B = 01, C = 11, D = 10.

Table X7.34

	X		
S	0	1	Z
A	B	D	0
B	C	B	0
C	B	A	1
D	B	C	0
	S*		

7.35 Repeat Exercise 7.34 using J-K flip-flops.

7.36 Write a new transition table and derive minimal-cost excitation and output equations for the state table in Table 7-6 using the "simplest" state assignment in Table 7-7 and D flip-flops. Compare the cost of your excitation and output logic (when realized with a two-level AND-OR circuit) with the circuit in Figure 7-54.

7.37 Repeat Exercise 7.36 using the "almost one-hot" state assignment in Table 7-7.

7.38 Suppose that the state machine in Figure 7-54 is to be built using 74LS74 D flip-flops. What signals should be applied to the flip-flop preset and clear inputs?

7.39 Write new transition and excitation tables and derive minimal-cost excitation and output equations for the state table in Table 7-6 using the "simplest" state assignment in Table 7-7 and J-K flip-flops. Compare the cost of your excitation and output logic (when realized with a two-level AND-OR circuit) with the circuit in Figure 7-56.

7.40 Repeat Exercise 7.39 using the "almost one-hot" state assignment in Table 7-7.

7.41   Construct an application table similar to Table 7-10 for each of the following flip-flop types: (a) S-R; (b) T with enable; (c) D with enable. Discuss the unique problem that you encounter when trying to make the most efficient use of don't-cares with one of these flip-flops.

7.42   Construct a new excitation table and derive minimal-cost excitation and output equations for the state machine of Table 7-8 using T flip-flops with enable inputs (Figure 7-33). Compare the cost of your excitation and output logic (when realized with a two-level AND-OR circuit) with the circuit in Figure 7-54.

7.43   Determine the full 8-state table of the circuit in Figure 7-54. Use the names U1, U2, and U3 for the unused states (001, 010, and 011). Draw a state diagram and explain the behavior of the unused states.

7.44   Repeat Exercise 7.43 for the circuit of Figure 7-56.

7.45   Write a transition table for the nonminimal state table in Figure 7-51(a) that results from assigning the states in binary counting order, INIT–OKA1 = 000–110. Write corresponding excitation equations for D flip-flops, assuming a minimal-cost disposition of the unused state 111. Compare the cost of your equations with the minimal-cost equations for the minimal state table presented in the text.

7.46   Write the application table for a T flip-flop with enable.

7.47   In many applications, the outputs produced by a state machine during or shortly after reset are irrelevant, as long as the machine begins to behave correctly a short time after the reset signal is removed. If this idea is applied to Table 7-6, the INIT state can be removed and only two state variables are needed to code the remaining four states. Redesign the state machine using this idea. Write a new state table, transition table, excitation table for D flip-flops, minimal-cost excitation and output equations, and logic diagram. Compare the cost of the new circuit with that of Figure 7-54.

7.48   Repeat Exercise 7.47 using J-K flip-flops, and use Figure 7-56 to compare cost.

7.49   Redesign the 1s-counting machine of Table 7-12, assigning the states in binary counting order (S0–S3 = 00, 01, 10, 11). Compare the cost of the resulting sum-of-products excitation equations with the ones derived in the text.

7.50   Repeat Exercise 7.49 using J-K flip-flops.

7.51   Repeat Exercise 7.49 using T flip-flops with enable.

7.52   Redesign the 1s-counting machine of Table 7-12 as an ABEL state diagram. Try to find a state assignment that minimizes the total number of product terms, assuming that you can use either polarity of output equations. How many different state assignments must you examine?

7.53   Redesign the combination-lock machine of Table 7-14, assigning coded states in Gray-code order (A–H = 000, 001, 011, 010, 110, 111, 101, 100). Compare the cost of the resulting sum-of-products excitation equations with the ones derived in the text.

7.54   Find a 3-bit state assignment for the combination-lock machine of Table 7-14 that results in less costly excitation equations than the ones derived in the text. (*Hint:* Use the fact that inputs 1–3 are the same as inputs 4–6 in the required input sequence.)

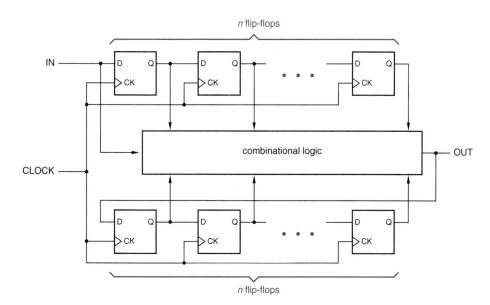

Figure X7.56

7.55 What changes would be made to the excitation and output equations for the combination-lock machine in Section 7.4.6 as the result of performing a formal multiple-output minimization procedure (Section 4.3.8) on the five functions? You need not construct 31 product maps and go through the whole procedure; you should be able to "eyeball" the excitation and output maps in Section 7.4.6 to see what savings are possible.

7.56 The output of a *finite-memory machine* is completely determined by its current input and its inputs and outputs during the previous n clock ticks, where n is a finite, bounded integer. Any machine that can be realized as shown in Figure X7.56 is a finite-memory machine. Note that a finite-*state* machine need not be a finite-*memory* machine; for example, a modulo-n counter with an enable input and a "MAX" output has only n states, but its output may depend on the value of the enable input at every clock tick since initialization. Show how to realize the combination-lock machine of Table 7-14 as a finite-memory machine.

finite-memory machine

7.57 Synthesize a circuit for the ambiguous state diagram in Figure 7-62. Use the state assignment in Table 7-16. Write a transition list, a transition equation for each state variable as a sum of p-terms, and simplified transition/excitation equations for a realization using D flip-flops. Determine the actual next state of the circuit, starting from the IDLE state, for each of the following input combinations on (LEFT, RIGHT, HAZ): (1,0,1), (0,1,1), (1,1,0), (1,1,1). Comment on the machine's behavior in these cases.

7.58 Suppose that for a state SA and an input combination I, an ambiguous state diagram indicates that there are two next states, SB and SC. The actual next state SD for this transition depends on the state machine's realization. If the state machine is synthesized using the method ($V* = \Sigma$ p-terms where $V* = 1$) to obtain transition/excitation equations for D flip-flops, what is the relationship between the coded states for SB, SC, and SD? Explain.

7.59   Repeat Exercise 7.58, assuming that the machine is synthesized using the method ($V*' = \Sigma$ p-terms where $V* = 0$).

7.60   Suppose that for a state SA and an input combination I, an ambiguous state diagram does not define a next state. The actual next state SD for this transition depends on the state machine's realization. Suppose that the state machine is synthesized using the method ($V* = \Sigma$ p-terms where $V* = 1$) to obtain transition/excitation equations for D flip-flops. What coded state is SD? Explain.

7.61   Repeat Exercise 7.60, assuming that the machine is synthesized using the method ($V*' = \Sigma$ p-terms where $V* = 0$).

7.62   Given the transition equations for a clocked synchronous state machine that is to be built using master/slave S-R flip-flops, how can the excitation equations for the S and R inputs be derived? (*Hint:* Show that any transition equation, $Q_i* = expr$, can be written in the form $Q_i* = Q_i \cdot expr1 + Q_i' \cdot expr2$, and see where that leads.)

7.63   Repeat Exercise 7.62 for J-K flip-flops. How can the "don't-cares" that are possible in a J-K design be specified?

7.64   Draw a logic diagram for the output logic of the guessing-game machine in Table 7-18 using a single 74x139 dual 2-to-4 decoder. (*Hint:* Use active-low outputs.)

7.65   What does the personalized license plate in Figure 7-60 stand for? (*Hint:* It's the author's old plate, a computer engineer's version of OTTFFSS.)

7.66   Analyze the feedback sequential circuit in Figure 7-19, assuming that the PR_L and CLR_L inputs are always 1. Derive excitation equations, construct a transition table, and analyze the transition table for critical and noncritical races. Name the states, and write a state/output table and a flow/output table. Show that the flow table performs the same function as Figure 7-85.

7.67   Draw the logic diagram for a circuit that has one feedback loop but is *not* a sequential circuit. That is, the circuit's output should be a function of its current input only. In order to prove your case, break the loop and analyze the circuit as if it were a feedback sequential circuit, and demonstrate that the outputs for each input combination do not depend on the "state."

BUT flop
NBUT gate

7.68   A BUT *flop* may be constructed from an NBUT gate as shown in Figure X7.68. (An NBUT *gate* is simply a BUT gate with inverted outputs; see Exercise 5.31 for the definition of a BUT gate.) Analyze the BUT flop as a feedback sequential circuit and obtain excitation equations, transition table, and flow table. Is this circuit good for anything, or is it a flop?

7.69   Repeat Exercise 7.68 for the BUT flop in Figure X7.69.

Figure X7.68   **Figure X7.69**

Figure X7.70

7.70 A clever student designed the circuit in Figure X7.70 to create a BUT gate. But the circuit didn't always work correctly. Analyze the circuit and explain why.

7.71 Analyze the feedback sequential circuit in Figure X7.71. Break the feedback loops, write excitation equations, and construct a transition and output table, showing the stable total states. What application might this circuit have?

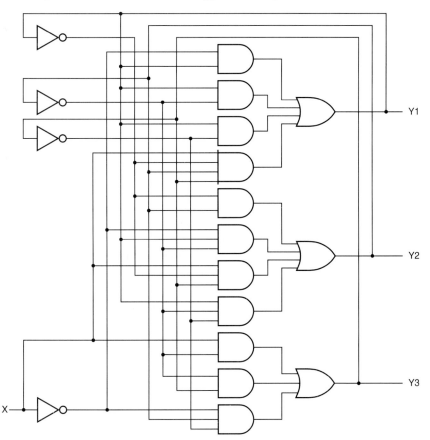

Figure X7.71

7.72 Show that a 4-bit ones'-complement adder with end-around carry is a feedback sequential circuit.

7.73 Complete the analysis of the positive-edge-triggered D flip-flop in Figure 7-86, including transition/output, state/output, and flow/output tables. Show that its behavior is equivalent to that of the D flip-flop in Figure 7-78.

7.74 We claimed in Section 7.10.1 that all single-loop feedback sequential circuits have an excitation equation of the form

$$Q* = (\text{forcing term}) + (\text{holding term}) \cdot Q$$

Why aren't there any practical circuits whose excitation equation substitutes Q' for Q above?

7.75 Simulate the latch circuit of Figure 7-88(b) under the conditions described in the text on page 616, either using a simulator in unit-delay mode or by hand assuming that each gate has a delay of 1 ns. Does the circuit behave as claimed in the text? Replace the inverter in the circuit with three inverters, repeat the simulation, and explain the results. What would you expect to happen in the real circuit?

7.76 Design a latch with two control inputs, C1 and C2, and three data inputs, D1, D2, and D3. The latch is to be "open" only if both control inputs are 1, and it is to store a 1 if any of the data inputs is 1. Use hazard-free two-level sum-of-products circuits for the excitation functions.

7.77 Repeat Exercise 7.76, but minimize the number of gates required; the excitation circuits may have multiple levels of logic.

7.78 Redraw the timing diagram in Figure 7-90, showing the internal state variables of the pulse-catching circuit of Figure 7-100, assuming that it starts in state 00.

7.79 The general solution for obtaining a race-free state assignment of 2^n states using 2^{n-1} state variables yields the adjacency diagram shown in Figure X7.79 for the $n = 2$ case. Compare this diagram with Figure 7-97. Which is better, and why?

7.80 Design a fundamental-mode flow table for a pulse-catching circuit similar to the one described in Section 7.10.2, except that the circuit should detect both 0-to-1 and 1-to-0 transitions on P.

Figure X7.79

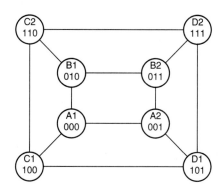

S	00	01	11	10
A	B	C	—	(A)
B	(B)	E	—	(B)
C	F	(C)	—	E
D	(D)	F	—	B
E	D	(E)	—	(E)
F	(F)	(F)	—	A

X Y

S*

Table X7.85

7.81 Design a fundamental-mode flow table for a double-edge-triggered D flip-flop, one that samples its inputs and changes its outputs on both edges of the clock signal.

7.82 Design a fundamental-mode flow table for a circuit with two inputs, EN and CLKIN, and a single output, CLKOUT, with the following behavior. A clock period is defined to be the interval between successive rising edges of CLKIN. If EN is asserted during an entire given clock period, then CLKOUT should be "on" during the next clock period; that is, it should be identical to CLKIN. If EN is negated during an entire given clock period, then CLKOUT should be "off" (constant 1) during the next clock period. If EN is both asserted and negated during a given clock period, then CLKOUT should be on in the next period if it had been off, and off if it had been on. After writing the fundamental-mode flow table, reduce it by combining "compatible" states if possible.

7.83 Design a circuit that meets the specifications of Exercise 7.82 using edge-triggered D flip-flops (74x74) or J-K flip-flops (74x109) and NAND and NOR gates without feedback loops. Give a complete circuit diagram and word description of how your circuit achieves the desired behavior.

7.84 Which of the circuits of the two preceding exercises is (are) subject to meta-stability, and under what conditions?

7.85 For the flow table in Table X7.85, find an assignment of state variables that avoids all critical races. You may add additional states as necessary, but use as few state variables as possible. Assign the all-0s combination to state A. Draw an adjacency diagram for the original flow table, and write the modified flow table and another adjacency diagram to support your final state-variable assignment.

7.86 Prove that the fundamental-mode flow table of any flip-flop that samples input(s) and change(s) outputs on the rising edge only of a clock signal CLK contains an essential hazard.

7.87   Locate the essential hazard(s) in the flow table for a positive-edge-triggered D flip-flop, Figure 7-85.

7.88   Identify the essential hazards, if any, in the flow table developed in Exercise 7.81.

7.89   Identify the essential hazards, if any, in the flow table developed in Exercise 7.82.

7.90   Build a verbal flip-flop—a logical word puzzle that can be answered correctly in either of two ways depending on state. How might such a device be adapted to the political arena?

7.91   Modify the ABEL program in Table 7-27 to use an output-coded state assignment, thereby reducing the total number of PLD outputs required by one.

7.92   Finish writing the test vectors, started in Table 7-35, for the combination-lock state machine of Table 7-31. The complete set of vectors should test all of the state transitions and all of the output values for every state and input combination.

74x163

2 CLK	
1 CLR	
9 LD	
7 ENP	
10 ENT	
3 A	QA 14
4 B	QB 13
5 C	QC 12
6 D	QD 11
	RCO 15

Sequential Logic Design Practices

The purpose of this chapter is to familiarize you with the most commonly used and dependable sequential-circuit design methods. Therefore, we will emphasize *synchronous systems*—that is, systems in which all flip-flops are clocked by the same common clock signal. Although all the world does not march to the tick of a common clock, within the confines of a digital system or subsystem we can make it so. When we interconnect digital systems or subsystems that use different clocks, we can usually identify a limited number of asynchronous signals that need special treatment, as we'll show later.

We begin this chapter with a quick summary of sequential-circuit documentation standards. After revisiting the most basic building blocks of sequential-circuit design—latches and flip-flops—we describe some of the most flexible building blocks—sequential PLDs. Next we show how counters and shift registers are realized in both MSI devices and PLDs and show some of their applications. Finally, we show how these elements come together in synchronous systems and how the inevitable asynchronous inputs are handled.

8.1 Sequential-Circuit Documentation Standards

8.1.1 General Requirements

Basic documentation standards in areas like signal naming, logic symbols, and schematic layout, which we introduced in Chapter 5, apply to digital systems as a whole and therefore to sequential circuits in particular. We highlight the following ideas, however, for system elements that are specifically "sequential":

- *State-machine layout.* Within a logic diagram, a collection of flip-flops and combinational logic that forms a state machine should be drawn together in a logical format on the same page, so the fact that it is a state machine is obvious. (You shouldn't have to flip pages to find the feedback path!)

- *Cascaded elements.* In a similar way, registers, counters, and shift registers that use multiple ICs should have the ICs grouped together in the schematic so that the cascading structure is obvious.

- *Flip-flops.* The symbols for individual sequential-circuit elements, especially flip-flops, should rigorously follow the appropriate drawing standards, so that the type, function, and clocking behavior of the elements are clear.

- *State-machine descriptions.* State machines should be described by state tables, state diagrams, transition lists, or text files in a state-machine description language such as ABEL or VHDL.

- *Timing diagrams.* The documentation package for sequential circuits should include timing diagrams that show the general timing assumptions and timing behavior of the circuit.

- *Timing specifications.* A sequential circuit should be accompanied by a specification of the timing requirements for proper internal operation (e.g., maximum clock frequency), as well as the requirements for any externally supplied inputs (e.g., setup- and hold-time requirements with respect to the system clock, minimum pulse widths, etc.).

8.1.2 Logic Symbols

We introduced traditional symbols for flip-flops in Section 7.2. Flip-flops are always drawn as rectangular-shaped symbols and follow the same general guidelines as other rectangular-shaped symbols—inputs on the left, outputs on the right, bubbles for active levels, and so on. In addition, some specific guidelines apply to flip-flop symbols:

- A dynamic indicator is placed on edge-triggered clock inputs.

- A postponed-output indicator is placed on master/slave outputs that change at the end interval during which the clock is asserted.

- Asynchronous preset and clear inputs may be shown at the top and bottom of a flip-flop symbol—preset at the top and clear at the bottom.

The logic symbols for larger-scale sequential elements, such as the counters and shift register described later in this chapter, are generally drawn with all inputs, including presets and clears, on the left, and all outputs on the right. Bidirectional signals may be drawn on the left or the right, whichever is convenient.

Like individual flip-flops, larger-scale sequential elements use a dynamic indicator to indicate edge-triggered clock inputs. In "traditional" symbols, the names of the inputs and outputs give a clue of their function, but they are sometimes ambiguous. For example, two elements described later in this chapter, the 74x161 and 74x163 4-bit counters, have exactly the same traditional symbol, even though the behavior of their CLR inputs is completely different.

IEEE STANDARD SYMBOLS   IEEE standard symbols have a rich set of notation that can provide an unambiguous definition of every signal's function. A guide to IEEE symbols, including all of the sequential elements in this chapter, can be found on the web at www.ddpp.com.

8.1.3 State-Machine Descriptions

So far we have dealt with six different representations of state machines:

- Word descriptions
- State tables
- State diagrams
- Transition lists
- ABEL programs
- VHDL programs

You might think that having all these different ways to represent state machines is a problem—too much for you to learn! Well, they're not all that difficult to learn, but there *is* a subtle problem here.

Consider a similar problem in programming, where high-level "pseudocode" or perhaps a flowchart might be used to document how a program works. The pseudocode may express the programmer's intentions very well, but errors, misinterpretations, and typos can occur when it is translated into real code. In any creative process, inconsistencies can occur when there are multiple representations of how things work.

The same kind of inconsistencies can occur in state-machine design. A logic designer may document a machine's desired behavior with a 100%-correct hand-drawn state diagram, but you can make mistakes translating the diagram into a program, and there are *lots* of opportunities to mess up if you have to "turn the crank" manually to translate the state diagram into a state table, transition table, excitation equations, and logic diagram.

inconsistent state-machine representations

The solution to this problem is similar to the one adopted by programmers who write self-documenting code using a high-level language. The key is to select a representation that is both expressive of the designer's intentions *and* translatable into a physical realization using an error-free, automated process. (You don't hear many programmers screaming "Compiler bug!" when their programs don't work the first time.)

The best solution (for now, at least) is to write state-machine "programs" directly in a high-level state-machine description language like ABEL or VHDL and to avoid alternate representations, other than general, summary word descriptions. Languages like ABEL and VHDL are easily readable and allow automatic conversion of the description into a PLD-, FPGA-, or ASIC-based realization. Some CAD tools allow state machines to be specified and synthesized using state diagrams, or even using sample timing diagrams, but these can often lead to ambiguities and unanticipated results. Thus, we'll use ABEL/VHDL approach exclusively for the rest of this book.

8.1.4 Timing Diagrams and Specifications

We showed many examples of timing diagrams in Chapters 5 and 7. In the design of synchronous systems, most timing diagrams show the relationship between the clock and various input, output, and internal signals.

Figure 8-1 shows a fairly typical timing diagram that specifies the requirements and characteristics of input and output signals in a synchronous circuit. The first line shows the system clock and its nominal timing parameters. The remaining lines show a range of delays for other signals.

For example, the second line shows that flip-flops change their outputs at some time between the rising edge of **CLOCK** and time t_{ffpd} afterward. External circuits that sample these signals should not do so while they are changing. The timing diagram is drawn as if the minimum value of t_{ffpd} were zero; a complete

Figure 8-1
A detailed timing diagram showing propagation delays and setup and hold times with respect to the clock.

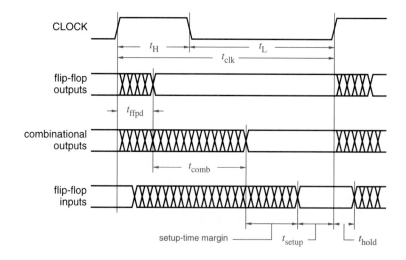

documentation package would include a timing table indicating the actual minimum, typical, and maximum values of t_{ffpd} and all other timing parameters.

The third line of the timing diagram shows the additional time, t_{comb}, required for the flip-flop output changes to propagate through combinational logic elements, such as flip-flop excitation logic. The excitation inputs of flip-flops and other clocked devices require a setup time of t_{setup}, as shown in the fourth line. For proper circuit operation we must have $t_{clk} - t_{ffpd} - t_{comb} > t_{setup}$.

Timing margins indicate how much "worse than worst-case" the individual components of a circuit can be without causing the circuit to fail. Well-designed systems have positive, nonzero timing margins to allow for unexpected circumstances (marginal components, brownouts, engineering errors, etc.) and clock skew (Section 8.8.1). *timing margin*

The value $t_{clk} - t_{ffpd(max)} - t_{comb(max)} - t_{setup}$ is called the *setup-time margin;* if this is negative, the circuit won't work. Note that *maximum* propagation delays are used to calculate setup-time margin. Another timing margin involves the hold-time requirement t_{hold}; the sum of the *minimum* values of t_{ffpd} and t_{comb} must be greater than t_{hold}, and the *hold-time margin* is $t_{ffpd(min)} + t_{comb(min)} - t_{hold}$. *setup-time margin*

hold-time margin

The timing diagram in Figure 8-1 does not show the timing differences between different flip-flop inputs or combinational-logic signals, even though such differences exist in most circuits. For example, one flip-flop's Q output may be connected directly to another flip-flop's D input, so that t_{comb} for that path is zero, while another's may go the ripple-carry path of a 32-bit adder before reaching a flip-flop input. When proper synchronous design methodology is used, these relative timings are not critical, since none of these signals affect the state of the circuit until a clock edge occurs. You merely have to find the longest delay path in one clock period to determine whether the circuit will work. However, you may have to analyze several different paths in order to find the worst-case one.

Another, perhaps more common, type of timing diagram shows only functional behavior and is not concerned with actual delay amounts; an example is shown in Figure 8-2. Here, the clock is "perfect." Whether to show signal changes as vertical or slanted lines is strictly a matter of personal taste in this and all other timing diagrams, unless rise and fall times must be explicitly indicated. Clock transitions are shown as vertical lines in this and other figures in keeping with the idea that the clock is a "perfect" reference signal.

Figure 8-2
Functional timing of a synchronous circuit.

The other signals in Figure 8-2 may be flip-flop outputs, combinational outputs, or flip-flop inputs. Shading is used to indicate "don't-care" signal values; crosshatching as in Figure 8-1 could be used instead. All of the signals are shown to change immediately after the clock edge. In reality, the outputs change sometime later, and inputs may change just barely before the next clock edge. However, "lining up" everything on the clock edge allows the timing diagram to display more clearly which functions are performed during each clock period. Signals that are lined up with the clock are simply understood to change sometime *after* the clock edge, with timing that meets the setup- and hold-time requirements of the circuit. Many timing diagrams of this type appear in this chapter.

Table 8-1 shows manufacturer's timing parameters for commonly used flip-flops, registers, and latches in CMOS and TTL. "Typical" values are for

Table 8-1 Propagation delay in ns of selected CMOS flip-flops, registers, and latches.

Part	Parameter	74HCT		74AHCT		74FCT		74LS	
		Typ.	Max.	Typ.	Max.	Min.	Max.	Typ.	Max.
'74	t_{pd}, CLK↑ to Q or $\overline{Q}$	35	44	6.3	10			25	40
	t_{pd}, $\overline{PR}$↓ or $\overline{CLR}$↓ to Q or $\overline{Q}$	40	50	8.1	13			25	40
	t_s, D to CLK↑	12	15		5				20
	t_h, D from CLK↑	3	3		0				5
	t_{rec}, CLK↑ from $\overline{PR}$↑ or $\overline{CLR}$↑	6	8		3.5				
	t_w, CLK low or high	18	23		5				25
	t_w, $\overline{PR}$ or $\overline{CLR}$ low	16	20		5				25
'174	t_{pd}, CLK↑ to Q	40	50	6.3	10			21	30
	t_{pd}, $\overline{CLR}$↓ to Q	44	55	8.1	13			23	35
	t_s, D to CLK↑	16	20		5				20
	t_h, D from CLK↑	5	5		0				5
	t_{rec}, CLK↑ from $\overline{CLR}$↑	12	15		3.5				25
	t_w, CLK low or high	20	25		5				20
	t_w, $\overline{CLR}$ low	25	31		5				20

Table 8-1 (continued) Propagation delay in ns of selected CMOS flip-flops, registers, and latches.

Part	Parameter	74HCT Typ.	74HCT Max.	74AHCT Typ.	74AHCT Max.	74FCT Min.	74FCT Max.	74LS Typ.	74LS Max.
'175	t_{pd}, CLK↑ to Q or $\overline{Q}$	33	41					21	30
	t_{pd}, $\overline{CLR}$↓ to Q or $\overline{Q}$	35	44					23	35
	t_s, D to CLK↑	20	25						20
	t_h, D from CLK↑	5	5						5
	t_{rec}, CLK↑ from $\overline{CLR}$↑	5	5						25
	t_w, CLK low or high	20	25						20
	t_w, $\overline{CLR}$ low	20	25						20
'273	t_{pd}, CLK↑ to Q	30	38	6.8	11	2	7.2	18	27
	t_{pd}, $\overline{CLR}$↓ to Q	32	40	8.5	12.6	2	7.2	18	27
	t_s, D to CLK↑	12	15		5		2		20
	t_h, D from CLK↑	3	3		0		1.5		5
	t_{rec}, CLK↑ from $\overline{CLR}$↑	10	13		2.5		2		25
	t_w, CLK low or high	20	25		6.5		4		20
	t_w, $\overline{CLR}$ low	12	15		6		5		20
'373	t_{pd}, C↑ to Q	35	44	8.5	14.5	2	8.5	24	36
	t_{pd}, D to Q	32	40	5.9	10.5	1.5	5.2	18	27
	t_s, D to C↓	10	13		1.5		2		0
	t_h, D from C↓	5	5		3.5		1.5		10
	t_{pHZ}, $\overline{OE}$ to Q	35	44		12	1.5	6.5	12	20
	t_{pLZ}, $\overline{OE}$ to Q	35	44		12	1.5	6.5	16	25
	t_{pZH}, $\overline{OE}$ to Q	35	44		13.5	1.5	5.5	16	28
	t_{pZL}, $\overline{OE}$ to Q	35	44		13.5	1.5	5.5	22	36
	t_w, C low or high	16	20		6.5		5		15
'374	t_{pd}, CLK↑ to Q	33	41	6.4	11.5	2	6.5	22	34
	t_s, D to CLK↑	12	15		2.5		2		20
	t_h, D from CLK↑	5	5		2.5		1.5		0
	t_{pHZ}, $\overline{OE}$ to Q	28	35		12	1.5	6.5		18
	t_{pLZ}, $\overline{OE}$ to Q	28	35		12	1.5	6.5		24
	t_{pZH}, $\overline{OE}$ to Q	30	38		12.5	1.5	5.5		28
	t_{pZL}, $\overline{OE}$ to Q	30	38		12.5	1.5	5.5		36
	t_w, CLK low or high	16	20		6.5		5		15
'377	t_{pd}, CLK↑ to Q	38	48			2	7.2	18	27
	t_s, D to CLK↑	12	15				2		20
	t_h, D from CLK↑	3	3				1.5		5
	t_s, $\overline{EN}$ to CLK↑	12	15				2		25
	t_h, $\overline{EN}$ from CLK↑	5	5				1.5		5
	t_w, CLK low or high	20	25				6		20

devices operating at 25°C, but, depending on the logic family, they could be for a typical part and nominal power-supply voltage, or they could be for a worst-case part at worst-case supply voltage. "Maximum" values are generally valid over the commercial operating range of voltage and temperature, except TTL values, which are specified at 25°C. Also note that the "maximum" values of t_s, t_h, t_{rec}, or t_w are the maximum values of the *minimum* setup time, hold time, recovery time, or pulse width that the specified part will exhibit.

Different manufacturers may use slightly different definitions for the same timing parameters, and they may specify different numbers for the same part. A given manufacturer may even use different definitions for different families or part numbers in the same family. Thus, all of the specifications in Table 8-1 are merely representative; for exact numbers *and* their definitions, you must consult the data sheet for the particular part and manufacturer.

8.2 Latches and Flip-Flops

8.2.1 SSI Latches and Flip-Flops

Several different types of discrete latches and flip-flops are available as SSI parts. These devices are sometimes used in the design of state machines and "unstructured" sequential circuits that don't fall into the categories of shift registers, counters, and other sequential MSI functions presented later in this chapter. However, SSI latches and flip-flops have been eliminated to a large extent in modern designs as their functions are embedded in PLDs and FPGAs. Nevertheless, a handful of these discrete building blocks still appear in many digital systems, so it's important to be familiar with them.

74x375

Figure 8-3 shows the pinouts for several SSI sequential devices. The only latch in the figure is the *74x375,* which contains four D latches, similar in function to the "generic" D latches described in Section 7.2.4. Because of pin limitations, the latches are arranged in pairs with a common C control line for each pair.

74x74

Among the devices in Figure 8-3, the most important is the *74x74,* which contains two independent positive-edge-triggered D flip-flops with preset and clear inputs. We described the functional operation, timing, and internal structure of edge-triggered D flip-flops in general, and the 74x74 in particular, in Section 7.2.5. Besides the use of the 74x74 in "random" sequential circuits, fast

74F74
74ACT74

versions of the part, such as the *74F74* and *74ACT74,* find application in synchronizers for asynchronous input signals, as discussed in Section 8.9.

74x109

The *74x109* is a positive-edge-triggered J-$\overline{\text{K}}$ flip-flop with an active-low K input (named $\overline{\text{K}}$ or K_L). We discussed the internal structure of the '109 in Section 7.2.10. Another J-K flip-flop is the *74x112,* which has an active-low clock input.

74x112

Figure 8-3
Pinouts for SSI latches and flip-flops.

*8.2.2 Switch Debouncing

A common application of simple bistables and latches is switch debouncing. We're all familiar with electrical switches from experience with lights, garbage disposals, and other appliances. Switches connected to sources of constant logic 0 and 1 are often used in digital systems to supply "user inputs." However, in digital logic applications we must consider another aspect of switch operation, the time dimension. A simple make or break operation, which occurs instantly as far as we slow-moving humans are concerned, actually has several phases that are discernible by high-speed digital logic.

Figure 8-4(a) shows how a single-pole, single-throw (SPST) switch might be used to generate a single logic input. A pull-up resistor provides a logic-1 value when the switch is opened, and the switch contact is tied to ground to provide a logic-1 value when the switch is pushed.

As shown in (b), it takes a while after a push for the wiper to hit the bottom contact. Once it hits, it doesn't stay there for long; it bounces a few times before finally settling. The result is that several transitions are seen on the SW_L and DSW logic signals for each single switch push. This behavior is called *contact bounce*. Typical switches bounce for 10–20 ms, a very long time compared to the switching speeds of logic gates.

contact bounce

Contact bounce may or may not be a problem, depending on the switch application. For example, some computers have configuration information specified by small switches, called *DIP switches* because they are the same size as a dual in-line package (DIP). Since DIP switches are normally changed only when the computer is inactive, there's no problem. Contact bounce *is* a problem

DIP switch

---

* Throughout this book, optional sections are marked with an asterisk.

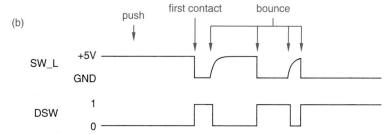

Figure 8-4
Switch input without
debouncing.

if a switch is being used to count or signal some event (e.g., laps in a race). Then we must provide a circuit (or, in microprocessor-based systems, software) to

debounce

debounce the switch—to provide just one signal change or pulse for each external event.

*8.2.3 The Simplest Switch Debouncer

Switch debouncing is a good application for the simplest sequential circuit, the bistable element of Section 7.1, which can be used as shown in Figure 8-5. This circuit uses a single-pole, double-throw (SPDT) switch. The switch contacts and wiper have a "break before make" behavior, so the wiper terminal is "floating" at some time halfway through the switch depression.

Before the button is pushed, the top contact holds SW at 0 V, a valid logic 0, and the top inverter produces a logic 1 on SW_L and on the bottom contact. When the button is pushed and contact is broken, feedback in the bistable holds SW at V_{OL} (≤ 0.5 V for LS-TTL), still a valid logic 0.

Next, when the wiper hits the bottom contact, the circuit operates quite unconventionally for a moment. The top inverter in the bistable is trying to maintain a logic 1 on the SW_L signal; the top transistor in its totem-pole output is "on" and connecting SW_L through a small resistance to +5 V. Suddenly, the switch contact makes a metallic connection of SW_L to ground, 0.0 V. Not surprisingly, the switch contact wins.

A short time later (30 ns for the 74LS04), the forced logic 0 on SW_L propagates through the two inverters of the bistable, so that the top inverter gives up its vain attempt to drive a 1, and instead drives a logic 0 onto SW_L. At this point, the top inverter output is no longer shorted to ground, and feedback in the bistable maintains the logic 0 on SW_L even if the wiper bounces off the bottom contact, as it does. (It does not bounce far enough to touch the top contact again.)

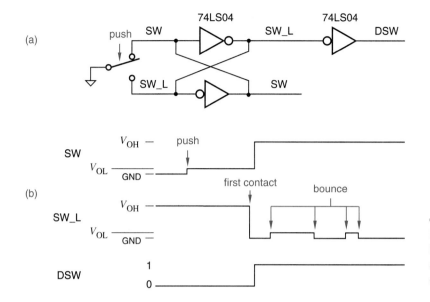

Figure 8-5
Switch input using a bistable for debouncing

Advantages of this circuit compared to other debouncing approaches are that it has a low chip count (one-third of a 74LS04), no pull-up resistors are required, and both polarities of the input signal (active-high and active-low) are produced. In situations where momentarily shorting gate outputs must be avoided, a similar circuit can be designed using an $\overline{S}\text{-}\overline{R}$ latch and pull-up resistors, as suggested in Figure 8-6.

Figure 8-6
Switch input using an $\overline{S}\text{-}\overline{R}$ latch for debouncing.

WHERE WIMPY WORKS WELL   The circuit in Figure 8-5, while elegant, should not be used with high-speed CMOS devices, like the 74ACT04, whose outputs are capable of sourcing large amounts of current in the HIGH state. While shorting such outputs to ground momentarily will not cause any damage, it will generate a noise pulse on power and ground signals that may trigger improper operation of the circuit elsewhere. The debouncing circuit in the figure works well with wimpy logic families like HCT and LS-TTL.

*8.2.4 Bus Holder Circuit

In Sections 3.7.3 and 5.6 we described three-state outputs and how they are tied together to create three-state buses. At any time, at most one output can drive the bus; sometimes, no output is driving the bus, and the bus is "floating." When high-speed CMOS inputs are connected to a bus that is left floating for a long time (in the fastest circuits, more than a clock tick or two), bad things can happen. In particular, noise, crosstalk, and other effects can drive the high-impedance floating bus signals to a voltage level near the CMOS devices' input switching threshold, which in turn allows excessive current to flow in the device outputs. For this reason, it is desirable and customary to provide pull-up resistors that quickly pull a floating bus to a valid HIGH logic level.

Pull-up resistors aren't all goodness—they cost money and they occupy valuable printed-circuit-board real estate. A big problem they have in very high-speed circuits is the choice of resistance value. If the resistance is too high, when a bus goes from LOW to floating, the transition from LOW to pulled-up (HIGH) will be slow due to the high RC time constant, and input levels may spend too much time near the switching threshold. If the pull-up resistance is too low, devices trying to pull the bus LOW will have to sink too much current.

bus holder circuit

The solution to this problem is to eliminate pull-up resistors in favor of an active *bus holder circuit*, shown in Figure 8-7. This is nothing but a bistable with a resistor R in one leg of the feedback loop. The bus holder's INOUT signal is connected to the three-state bus line which is to be held. When the three-state output currently driving the line LOW or HIGH changes to floating, the bus holder's righthand inverter holds the line in its current state. When a three-state output tries to change the line from LOW to HIGH or vice versa, it must source or sink a small amount of additional current through R to overcome the bus holder. This additional current flow persists only for the short time that it takes for the bistable to flip into its other stable state.

The choice of the value of R in the bus holder is a compromise between low override current (high R) and good noise immunity on the held bus line (low R). A typical example, bus holder circuits in the 3.3-V CMOS LVC family specify a maximum override current of 500 μA, implying $R \approx 3.3 / 0.0005 = 6.6 \text{ K}\Omega$.

Bus holder circuits are often built into another MSI device, such as an octal CMOS bus driver or transceiver. They require no extra pins and very little chip area, so they are essentially free. And there's no real problem in having multiple (n) bus holders on the same signal line, as long as the bus drivers can provide n times the override current for a few nanoseconds during switching. Note that bus holders normally are not effective on buses that have TTL inputs attached to them (see Exercise 8.25).

Figure 8-7
Bus holder circuit.

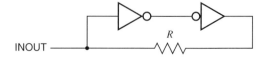

INOUT

8.2.5 Multibit Registers and Latches

A collection of two or more D flip-flops with a common clock input is called a *register*. Registers are often used to store a collection of related bits, such as a byte of data in a computer. However, a single register can also be used to store unrelated bits of data or control information; the only real constraint is that all of the bits are stored using the same clock signal.

register

Figure 8-8 shows the logic diagram and logic symbol for a commonly used MSI register, the *74x175*. The 74x175 contains four edge-triggered D flip-flops with a common clock and asynchronous clear inputs. It provides both active-high and active-low outputs at the external pins of the device.

74x175

The individual flip-flops in a '175 are negative-edge triggered, as indicated by the inversion bubbles on their CLK inputs. However, the circuit also contains an inverter that makes the flip-flops positive-edge triggered with respect to the device's external CLK input pin. The common, active-low, clear signal (CLR_L) is connected to the asynchronous clear inputs of all four flip-flops. Both CLK and CLR_L are buffered before fanning out to the four flip-flops, so that a device driving one of these inputs sees only one unit load instead of four. This is especially important if a common clock or clear signal must drive many such registers.

Figure 8-8
The 74x175 4-bit register:
(a) logic diagram, including
pin numbers for a standard
16-pin dual in-line package;
(b) traditional logic symbol.

Figure 8-9
Logic symbol for the
74x174 6-bit register.

The logic symbol for the *74x174*, 6-bit register is shown in Figure 8-9. The internal structure of this device is similar to the 74x175's, except that it eliminates the active-low outputs and provides two more flip-flops instead.

Many digital systems, including computers, telecommunications devices, and stereo equipment, process information 8, 16, or 32 bits at a time; as a result, ICs that handle 8 bits are very popular. One such MSI IC is the *74x374* octal edge-triggered D flip-flop, also known simply as an 8-bit register. (Once again, "octal" means that the device has eight sections.)

As shown in Figure 8-10, the 74x374 contains eight edge-triggered D flip-flops that all sample their inputs and change their outputs on the rising edge of a common CLK input. Each flip-flop output drives a three-state buffer that in turn drives an active-high output. All of the three-state buffers are enabled by a

Figure 8-10
The 74x374 8-bit register:
(a) logic diagram, including pin
numbers for a standard 20-pin
dual in-line package;
(b) traditional logic symbol.

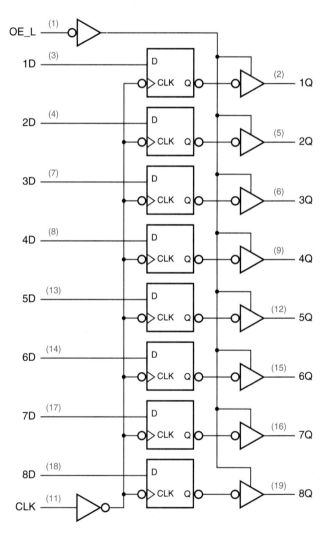

common, active-low OE_L (output enable) input. As in the other registers that we've studied, the control inputs (CLK and OE_L) are buffered so that they present only one unit load to a device that drives them.

One variation of the 74x374 is the *74x373*, whose symbol is shown in Figure 8-11. The '373 uses D latches instead of edge-triggered flip-flops. Therefore, its outputs follow the corresponding inputs whenever C is asserted, and they latch the last input values when C is negated. Another variation is the *74x273*, shown in Figure 8-12. This octal register has non-three-state outputs and no OE_L input; instead it uses pin 1 for an asynchronous clear input CLR_L.

The *74x377*, whose symbol is shown in Figure 8-13(a), is an edge-triggered register like the '374, but it does not have three-state outputs. Instead, pin 1 is used as an active-low clock-enable input EN_L. If EN_L is asserted (LOW) at the rising edge of the clock, then the flip-flops are loaded from the data inputs; otherwise, they retain their present values, as shown logically in (b).

Figure 8-11
Logic symbol for the
74x373 8-bit latch.

Figure 8-12
Logic symbol for the
74x273 8-bit register.

Figure 8-13 The 74x377 8-bit register with gated clock:
(a) logic symbol; (b) logical behavior of one bit.

High pin-count surface-mount packaging supports even wider registers, drivers, and transceivers. Most common are 16-bit devices, but there are also devices with 18 bits (for byte parity) and 32 bits. Also, the larger packages can offer more control functions, such as clear, clock enable, multiple output enables, and even a choice of latching vs. registered behavior all in one device.

8.2.6 Registers and Latches in ABEL and PLDs

As we showed in Section 7.11, registers are very easy to specify in ABEL. For example, Table 7-33 on page 639 showed an ABEL program for an 8-bit register with enable. Obviously, ABEL allows the functions performed at the D inputs of register to be customized in almost any way desired, limited only by the number of inputs and product terms in the targeted PLD. We describe sequential PLDs in Section 8.3.

With most sequential PLDs, few if any customizations can be applied to a register's clock input (e.g, polarity choice) or to the asynchronous inputs (e.g., different preset conditions for different bits). However, ABEL does provide appropriate syntax to apply these customizations in devices that support them, as described in Section 7.11.1.

Very few PLDs have latches built in; edge-triggered registers are much more common and generally more useful. However, you can also synthesize a latch using combinational logic and feedback. For example, the excitation equation for an S-R latch is

$$Q* = S + R' \cdot Q$$

Thus, you could build an S-R latch using one combinational output of a PLD, using the ABEL equation "Q = S # !R & Q." Furthermore, the S and R signals above could be replaced with more complex logic functions of the PLD's inputs, limited only by the availability of product terms (seven per output in a 16V8C or 16L8) to realize the final excitation equation. The feedback loop can be created only when Q is assigned to a bidirectional pin (in a 16V8C or 16L8, pins IO2–IO7, not O1 or O8). Also, the output pin must be continuously output-enabled; otherwise, the feedback loop would be broken and the latch's state lost.

Probably the handiest latch to build out of a combinational PLD is a D latch. The basic excitation equation for a D latch is

$$Q* = C \cdot D + C' \cdot Q$$

However, we showed in Section 7.10.1 that this equation contains a static hazard, and the corresponding circuit does not latch data reliably. To build a reliable D latch, we must include a consensus term in the excitation equation:

$$Q* = C \cdot D + C' \cdot Q + D \cdot Q$$

The D input in this equation may be replaced with a more complicated expression, but the equation's structure remains the same:

$$Q* = C \cdot \textit{expression} + C' \cdot Q + \textit{expression} \cdot Q$$

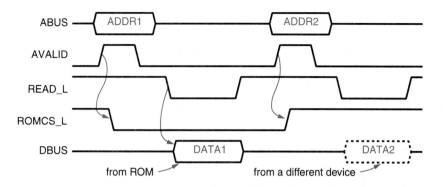

Figure 8-14
Timing diagram for a
microprocessor read
operation.

It is also possible to use a more complex expression for the **C** input, as we showed in Section 7.10.1. In any case, it is very important for the consensus term to be included in the PLD realization. The compiler can work against you in this case, since its minimization step will find that the consensus term is redundant and remove it.

Some versions of the ABEL compiler let you prevent elimination of consensus terms by including a keyword "`retain`" in the property list of the `istype` declaration for any output which is not to be minimized. In other versions, your only choice is to turn off minimization for the entire design.

retain property

Probably the most common use of a PLD-based latch is to simultaneously decode and latch addresses in order to select memory and I/O devices in microprocessor systems. Figure 8-14 is a timing diagram for this function in a typical system. The microprocessor selects a device and a location within the device by placing an address on its address bus (**ABUS**) and asserting an "address valid" signal (**AVALID**). A short time later, it asserts a read signal (**READ_L**), and the selected device responds by placing data on the data bus (**DBUS**).

Notice that the address does not stay valid on **ABUS** for the entire operation. The microprocessor bus protocol expects the address to be latched using **AVALID** as an enable, then decoded, as shown in Figure 8-15. The decoder selects different devices to be enabled or "chip-selected" according to the high-order bits of the address (the 12 high-order bits in this example). The low-order bits are used to address individual locations of a selected device.

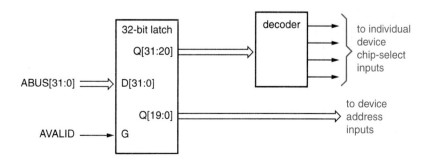

Figure 8-15
Microprocessor
address latching and
decoding circuit.

WHY A LATCH?   The microprocessor bus protocol in Figure 8-14 raises several questions:

- Why not keep the address valid on ABUS for the entire operation? In a real system using this protocol, the functions of ABUS and DBUS are combined (multiplexed) onto one three-state bus to save pins and wires.

- Why not use AVALID as the clock input to a positive-edge-triggered register to capture the address? There isn't enough setup time; in a real system, the address may first be valid at or slightly after the rising edge of AVALID.

- OK, so why not use AVALID to clock a negative-edge-triggered register? This works, but the latched outputs are available sooner; valid values on ABUS flow through a latch immediately, without waiting for the falling clock edge. This relaxes the access-time requirements of memories and other devices driven by the latched outputs.

Using a PLD, the latching and decoding functions for the high-order bits can be combined into a single device, yielding the block diagram in Figure 8-16. Compared with Figure 8-15, the "latching decoder" saves devices and pins, and it may produce a valid chip-select output more quickly (see Exercise 8.9).

Table 8-2 is an ABEL program for the latching decoder. Since it operates on only the high-order bits ABUS[31..20], it can decode addresses only in 1-Mbyte or larger chunks ($2^{20} = 1$ M). A read-only memory (ROM) is located in the highest 1-Mbyte chunk, addresses 0xfff00000–0xffffffff, and is selected by ROMCS. Three 16-Mbyte banks of read/write memory (RAM) are located at lower addresses, starting at addresses 0x00000000, 0x00100000, and 0x00200000, respectively. Notice how don't-cares are used in the definitions of the RAM bank address ranges to decode a chunk larger than 1 Mbyte. Other approaches to these definitions are also possible (e.g., see Exercise 8.1).

The equations in Table 8-2 for the chip-select outputs follow the D-latch template that we gave on page 674. The expressions that select a device, such as "ABUS==ROM," each generate a single product term, and each equation generates three product terms. Notice the use of the "retain" property in the pin declarations to prevent the compiler from optimizing away the consensus terms.

Figure 8-16
Using a combined
address latching and
decoding circuit.

```
module latchdec
title 'Latching Microprocessor Address Decoder'

" Inputs
AVALID, ABUS31..ABUS20          pin;
" Latched and decoded outputs
ROMCS, RAMCS0, RAMCS1, RAMCS2   pin istype 'com,retain';

ABUS = [ABUS31..ABUS20];
ROM = ^hFFF;
RAMBANK0 = [0,0,0,0,0,0,0,0,.X.,.X.,.X.,.X.];
RAMBANK1 = [0,0,0,0,0,0,0,1,.X.,.X.,.X.,.X.];
RAMBANK2 = [0,0,0,0,0,0,1,0,.X.,.X.,.X.,.X.];

equations

ROMCS  = AVALID & (ABUS==ROM) # !AVALID & ROMCS
       # (ABUS==ROM) & ROMCS;
RAMCS0 = AVALID & (ABUS==RAMBANK0) # !AVALID & RAMCS0
       # (ABUS==RAMBANK0) & RAMCS0;
RAMCS1 = AVALID & (ABUS==RAMBANK1) # !AVALID & RAMCS1
       # (ABUS==RAMBANK1) & RAMCS1;
RAMCS2 = AVALID & (ABUS==RAMBANK2) # !AVALID & RAMCS2
       # (ABUS==RAMBANK2) & RAMCS2;

end latchdec
```

Table 8-2
ABEL program
for a latching
address decoder.

After seeing how easy it is to build S-R and D latches using combinational PLDs, you might be tempted to go further and try to build an edge-triggered D flip-flop. Although this is possible, it is expensive because an edge-triggered flip-flop has four internal states and thus two feedback loops, consuming two PLD outputs. Furthermore, the setup and hold times and propagation delays of such a flip-flop would be quite poor compared to those of a discrete flip-flop in the same technology. Finally, as we discussed in Section 7.10.6, the flow tables of all edge-triggered flip-flops contain essential hazards, which can be masked only by controlling path delays, difficult in a PLD-based design.

8.2.7 Registers and Latches in VHDL

Register and latch circuits can be specified using structural VHDL. For example, Table 8-3 is a structural VHDL program corresponding to the D latch circuit of Figure 7-12 on page 539. However, writing structural programs is not really our motivation for using VHDL; our goal is to use behavioral programs to model the operation of circuits more intuitively.

Table 8-4 is a process-based behavioral architecture for the D latch that requires, in effect, just one line of code to describe the latch's behavior. Note that the VHDL compiler "infers" a latch from this description—since the code

Table 8-3 VHDL structural program for the D latch in Figure 7-12.

```
library IEEE;
use IEEE.std_logic_1164.all;

entity Vdlatch is
  port (D, C: in STD_LOGIC;
        Q, QN: buffer STD_LOGIC );
end Vdlatch;

architecture Vdlatch_s of Vdlatch is
  signal DN, SN, RN: STD_LOGIC;
  component inv port (I: in STD_LOGIC; O: out STD_LOGIC ); end component;
  component nand2b port (I0, I1: in STD_LOGIC; O: buffer STD_LOGIC ); end component;
begin
  U1: inv port map (D,DN);
  U2: nand2b port map (D,C,SN);
  U3: nand2b port map (C,DN,RN);
  U4: nand2b port map (SN,QN,Q);
  U5: nand2b port map (Q,RN,QN);
end Vdlatch_s;
```

Table 8-4 VHDL behavioral architecture for a D latch.

```
architecture Vdlatch_b of Vdlatch is
begin
process(C, D, Q)
  begin
    if (C='1') then Q <= D; end if;
    QN <= not Q;
  end process;
end Vdlatch_b;
```

inferred latch

doesn't say what to do if C is not 1, the compiler creates an *inferred latch* to retain the value of Q between process invocations. In general, a VHDL compiler infers a latch for a signal that is assigned a value in an `if` or `case` statement if not all input combinations are accounted for.

event attribute

In order to describe edge-triggered behavior of flip-flops, we need to use one of VHDL's predefined signal attributes, the *event* attribute. If "SIG" is a signal name, then the construction "SIG'event" returns the value `true` at any delta time when SIG changes from one value to another, and `false` otherwise.

Using the event attribute, we can model a positive-edge triggered flip-flop as shown in Table 8-6. In the IF statement, "CLK'event" returns true on any clock edge, and "CLK='1'" ensures that D is assigned to Q only on a *rising* edge. Note that the process sensitivity list includes only CLK; changes on D at other times are not relevant in this functional model.

BUFFS 'N' STUFF   Note that in Table 8-3 we defined the type of Q and QN to be buffer rather than out, since these signals are used as inputs as well as outputs in the architecture definition. Then we had to define a special 2-input NAND gate nand2b with output type buffer to avoid having a type mismatch (out vs. buffer) in the component instantiations for U4 and U5. Alternatively, we could have used internal signals to get around the problem, as shown in Table 8-5. As you know by now, VHDL has many different ways to express the same thing.

Table 8-5 Alternative VHDL structural program for the D latch in Figure 7-12.

```
library IEEE;
use IEEE.std_logic_1164.all;

entity Vdlatch is
  port (D, C: in STD_LOGIC;
        Q, QN: out STD_LOGIC );
end Vdlatch;

architecture Vdlatch_s2 of Vdlatch is
  signal DN, SN, RN, IQ, IQN: STD_LOGIC;
  component inv port (I: in STD_LOGIC; O: out STD_LOGIC ); end component;
  component nand2 port (I0, I1: in STD_LOGIC; O: out STD_LOGIC ); end component;
begin
  U1: inv port map (D,DN);
  U2: nand2 port map (D,C,SN);
  U3: nand2 port map (C,DN,RN);
  U4: nand2 port map (SN,IQN,IQ);
  U5: nand2 port map (IQ,RN,IQN);
  Q <= IQ; QN <= IQN;
end Vdlatch_s2;
```

Table 8-6 VHDL behavioral model of an edge-triggered D flip-flop.

```
library IEEE;
use IEEE.std_logic_1164.all;

entity Vdff is
  port (D, CLK: in STD_LOGIC;
        Q: out STD_LOGIC );
end Vdff;

architecture Vdff_b of Vdff is
begin
process(CLK)
  begin
    if (CLK'event and CLK='1') then Q <= D; end if;
  end process;
end Vdff_b;
```

▌ **Table 8-7** VHDL model of a 74x74-like D flip-flop with preset and clear.

```vhdl
library IEEE;
use IEEE.std_logic_1164.all;

entity Vdff74 is
  port (D, CLK, PR_L, CLR_L: in STD_LOGIC;
        Q, QN: out STD_LOGIC );
end Vdff74;

architecture Vdff74_b of Vdff74 is
signal PR, CLR: STD_LOGIC;
begin
process(CLR_L, CLR, PR_L, PR, CLK)
  begin
    PR <= not PR_L; CLR <= not CLR_L;
    if (CLR and PR) = '1' then Q <= '0'; QN <= '0';
    elsif CLR = '1' then Q <= '0'; QN <= '1';
    elsif PR = '1' then Q <= '1'; QN <= '0';
    elsif (CLK'event and CLK='1') then Q <= D; QN <= not D;
    end if;
  end process;
end Vdff74_b;
```

▌ **Table 8-8** VHDL model of a 16-bit register with many features.

```vhdl
library IEEE;
use IEEE.std_logic_1164.all;

entity Vreg16 is
  port (CLK, CLKEN, OE_L, CLR_L: in STD_LOGIC;
        D: in STD_LOGIC_VECTOR(1 to 16);      -- Input bus
        Q: out STD_ULOGIC_VECTOR (1 to 16) ); -- Output bus (three-state)
end Vreg16;

architecture Vreg16 of Vreg16 is
signal CLR, OE: STD_LOGIC;  -- active-high versions of signals
signal IQ: STD_LOGIC_VECTOR(1 to 16); -- internal Q signals
begin
process(CLK, CLR_L, CLR, OE_L, OE, IQ)
  begin
    CLR <= not CLR_L;  OE <= not OE_L;
    if (CLR = '1') then IQ <= (others => '0');
    elsif (CLK'event and CLK='1') then
      if (CLKEN='1') then IQ <= D; end if;
    end if;
    if OE = '1' then Q <= To_StdULogicVector(IQ);
    else Q <= (others => 'Z'); end if;
  end process;
end Vreg16;
```

The D-flip-flop model can be augmented to include asynchronous inputs and a complemented output as in the 74x74 discrete flip-flop, as shown in Table 8-7. This more detailed functional model shows the non-complementary behavior of the Q and QN outputs when preset and clear are asserted simultaneously. However, it does not include timing behavior such as propagation delay and setup and hold times, which are beyond the scope of the VHDL coverage in this book.

Larger registers can be modeled, of course, by defining the data inputs and outputs to be vectors, and additional functions can also be included. For example, Table 8-8 models a 16-bit register with three-state outputs and clock-enable, output-enable, and clear inputs. An internal signal vector IQ is used to hold the flip-flop outputs, and three-state outputs are defined and enabled as in Section 5.6.4.

**SYNTHESIS
RESTRICTIONS**
In Table 8-8, the first elsif statement theoretically could have included all of the conditions needed to assign D to IQ. That is, it could have read "elsif (CLK'event) and (CLK='1') and (CLKEN='1') then . . ." instead of using a nested if statement to check CLKEN. However, it was written as shown for a very pragmatic reason.

Only a subset of the VHDL language can be synthesized by the VHDL compiler that was used to prepare this chapter; this is true of any VHDL compiler today. In particular, use of the "event" attribute is limited to the form shown in the example, and a few others, for detecting simple edge-triggered behavior. This gets mapped into edge-triggered D flip-flops during synthesis. The nested IF statement that checks CLKEN in the example leads to the synthesis of multiplexer logic on the D inputs of these flip-flops.

8.3 Sequential PLDs

8.3.1 Bipolar Sequential PLDs

The *PAL16R8*, shown in Figure 8-17, is representative of the first generation of sequential PLDs, which used bipolar (TTL) technology. This device has eight primary inputs, eight outputs, and common clock and output-enable inputs, and it fits in a 20-pin package.

PAL16R8

The PAL16R8's AND-OR array is exactly the same as the one found in the PAL16L8 combinational PLD. However, the PAL16R8 has edge-triggered D flip-flops between the AND-OR array and its eight outputs, O1–O8. All of the flip-flops are connected to a common clock input, CLK, and change state on the rising edge of the clock. Each flip-flop drives an output pin through a 3-state buffer; the buffers have a common output-enable signal, OE_L. Notice that, like the combinational output pins of a PAL16L8, the registered output pins of the PAL16R8 contain the complement of the signal produced by the AND-OR array.

The possible inputs to the PAL16R8's AND-OR array are eight primary inputs (I1–I8) and the eight D flip-flop outputs. The connection from the D flip-flop outputs into the AND-OR array makes it easy to design shift registers, counters, and general state machines. Unlike the PAL16L8's combinational outputs, the PAL16R8's D-flip-flop outputs are available to the AND-OR array whether or not the O1–O8 three-state drivers are enabled. Thus, the internal flip-flops can go to a next state that is a function of the current state even when the O1–O8 outputs are disabled.

Many applications require combinational as well as sequential PLD outputs. The manufacturers of bipolar PLDs addressed this need by providing a few variants of the PAL16R8 that omitted the D flip-flops on some output pins, and instead provided input and output capability identical to that of the PAL16L8's bidirectional pins. For example, Figure 8-18 is the logic diagram of the *PAL16R6*, which has only six registered outputs. Two pins, IO1 and IO8, are bidirectional, serving both as inputs and as combinational outputs with separate 3-state enables, just like the PAL16L8's bidirectional pins. Thus, the possible inputs to the PAL16R6's AND-OR array are the eight primary inputs (I1–I8), the six D-flip-flop outputs, and the two bidirectional pins (IO1, IO8).

PAL16R6

PAL16L8
PAL16R4
PAL16R8
PAL20L8
PAL20R4

Table 8-9 shows eight standard bipolar PLDs with differing numbers and types of inputs and outputs. All of the PAL16xx parts in the table use the same AND-OR array, where each output has eight AND gates, each with 16 variables and their complements as possible inputs. The PAL20xx parts use a similar

Table 8-9 Characteristics of standard bipolar PLDs.

Part number	Package pins	AND-gate inputs	Inputs to AND array			
			Primary inputs	Bidirectional combinational outputs	Registered outputs	Combinational outputs
PAL16L8	20	16	10	6	0	2
PAL16R4	20	16	8	4	4	0
PAL16R6	20	16	8	2	6	0
PAL16R8	20	16	8	0	8	0
PAL20L8	24	20	14	6	0	2
PAL20R4	24	20	12	4	4	0
PAL20R6	24	20	12	2	6	0
PAL20R8	24	20	12	0	8	0

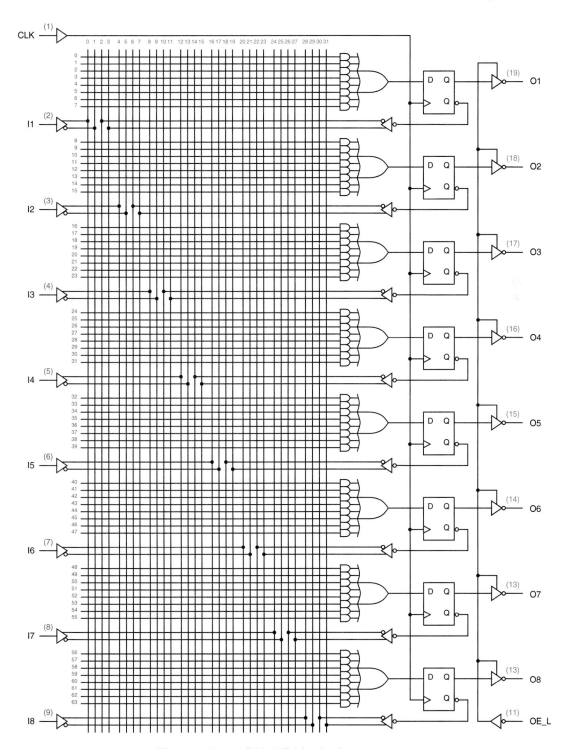

Figure 8-17 PAL16R8 logic diagram.

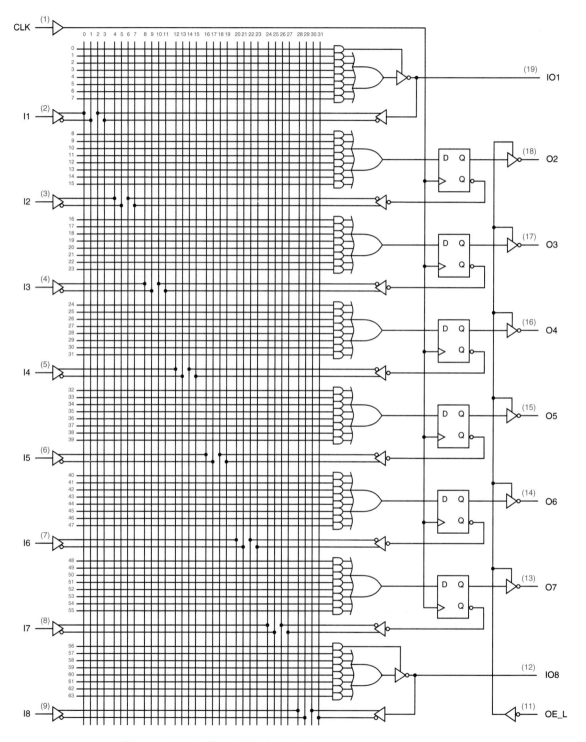

Figure 8-18 PAL16R6 logic diagram.

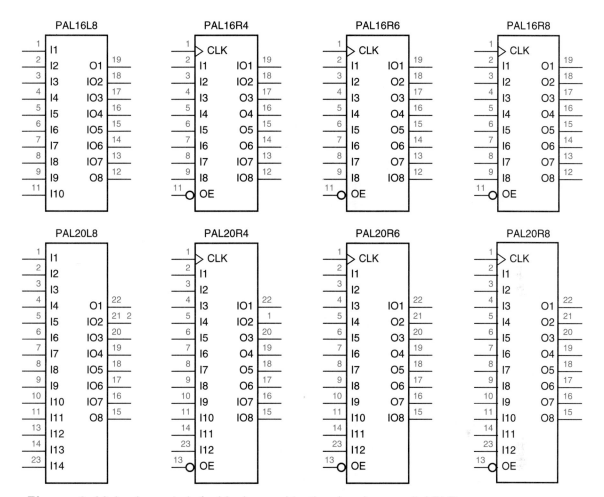

Figure 8-19 Logic symbols for bipolar combinational and sequential PLDs.

AND-OR array with 20 variables and their complements as possible inputs. *PAL20R6*
Figure 8-19 shows logic symbols for all of the PLDs in the table. *PAL20R8*

8.3.2 Sequential GAL Devices

The GAL16V8 electrically erasable PLD was introduced in Section 5.3.3. Two
"architecture-control" fuses are used to select among three basic configurations
of this device. Section 5.3.3 described the *16V8C* ("complex") configuration, *16V8C*
shown in Figure 5-27 on page 345, a structure similar to that of a bipolar
combinational PAL device, the PAL16L8. The *16V8S* ("simple") configuration *16V8S*
provides a slightly different combinational logic capability (see box on
page 687).

The third configuration, called the *16V8R*, allows a flip-flop to be provided *16V8R*
on any or all of the outputs. Figure 8-20 shows the structure of the device when

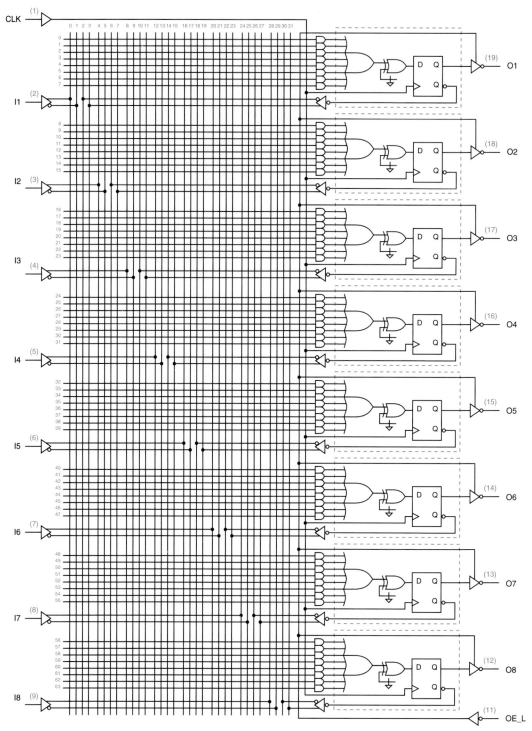

Figure 8-20 Logic diagram for the 16V8 in the "registered" configuration.

Figure 8-21 Output logic macrocells for the 16V8R: (a) registered; (b) combinational.

flip-flops are provided on all outputs. Notice that all of the flip-flops are controlled by a common clock signal on pin 1, as in the bipolar devices of the preceding subsection. Likewise, all of the output buffers are controlled by a common output-enable signal on pin 11.

The circuitry inside each dotted box in Figure 8-20 is called an *output logic macrocell*. The 16V8R is much more flexible than a PAL16R8 because each macrocell may be individually configured to bypass the flip-flop—that is, to produce a combinational output. Figure 8-21 shows the two macrocell configurations that are possible in the 16V8R; (a) is registered and (b) is combinational. Thus, it is possible to program the device to have any set of registered and combinational outputs, up to eight total.

output logic macrocell

The *20V8* is similar to the 16V8 but comes in a 24-pin package with four extra input-only pins. Each AND gate in the 20V8 has 20 inputs, hence the "20" in "20V8."

20V8

THE "SIMPLE" 16V8S   The "simple" 16V8S configuration of the GAL16V8 is not often used, because its capabilities are mostly a subset of the 16V8C's. Instead of an AND term, the 16V8S uses one fuse per output to control whether the output buffers are enabled. That is, each output pin may be programmed either to be always enabled or to be always disabled (except pins 15 and 16, which are always enabled). All of the output pins (except 15 and 16) are available as inputs to the AND array regardless of whether the output buffer is enabled.

The only advantage of a 16V8S compared to a 16V8C is that it has eight, not seven, AND terms as inputs to the OR gate on each output. The 16V8S architecture was designed mainly for emulation of certain now-obsolete bipolar PAL devices, some of which either had eight product terms per output or had inputs on pins 12 and 19, which are not inputs in the 16V8C configuration. With appropriate programming, the 16V8S can be used as a pin-for-pin compatible replacement for these devices, which included the PAL10H8, PAL12H6, PAL14H4, PAL16H2, PAL10L8, PAL12L6, PAL14L4, and PAL16L2.

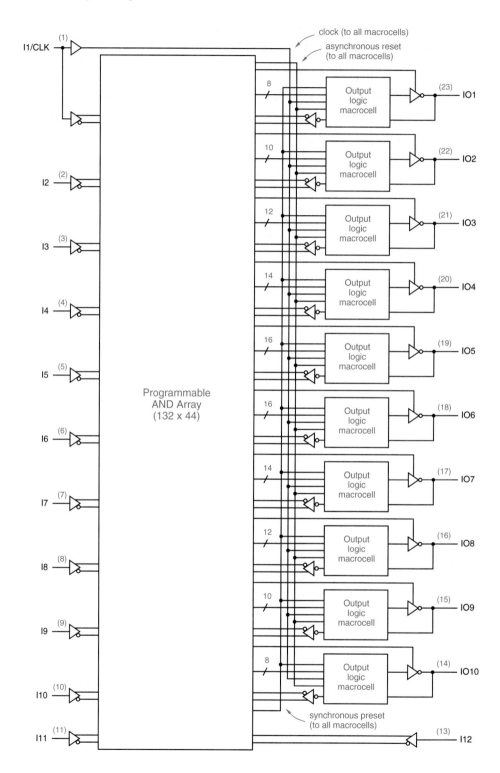

Figure 8-22
Logic diagram for
the 22V10.

The *22V10*, whose basic structure is shown in Figure 8-22, also comes in a 24-pin package but is somewhat more flexible than the 20V8. The 22V10 does not have "architecture control" bits like the 16V8's and 20V8's, but it can realize any function that is realizable with a 20V8, and more. It has more product terms, two more general-purpose inputs, and better output-enable control than the 20V8. Key differences are summarized below:

22V10

- Each output logic macrocell is configurable to have a register or not, as in the 20V8R architecture. However, the macrocells are different from the 16V8's and 20V8's, as shown in Figure 8-23.

- A single product term controls the output buffer, regardless of whether the registered or the combinational configuration is selected for a macrocell.

- Every output has at least eight product terms available, regardless of whether the registered or the combinational configuration is selected. Even more product terms are available on the inner pins, with 16 available on each of the two innermost pins. ("Innermost" is with respect to the right-hand side of the Figure 8-22, which also matches the arrangement of these pins on a 24-pin dual-inline package.)

- The clock signal on pin 1 is also available as a combinational input to any product term.

- A single product term is available to generate a global, asynchronous reset signal that resets all internal flip-flops to 0.

- A single product term is available to generate a global, synchronous preset signal that sets all internal flip-flops to 1 on the rising edge of the clock.

- Like the 16V8 and 20V8, the 22V10 has programmable output polarity. However, in the registered configuration, the polarity change is made at the output, rather than the input, of the D flip-flop. This affects the details of programming when the polarity is changed but does not affect the overall capability of the 22V10 (i.e., whether a given function can be realized). In fact, the difference in polarity-change location is transparent when you use a PLD programming language such as ABEL.

Figure 8-23 Output logic macrocells for the 22V10: (a) registered; (b) combinational.

Figure 8-24 Logic symbols for popular GAL devices.

For most of the 1990s, the 16V8, 20V8, and 22V10 were the most popular and cost-effective PLDs (but see the box on page 693). Figure 8-24 shows generic logic symbols for these three devices. Most of the examples in the rest of this chapter can fit into the smallest of the three devices, the 16V8.

PALS? GALS?   Lattice Semiconductor introduced GAL devices including the GAL16V8 and GAL20V8 in the mid-1980s. Advanced Micro Devices later followed up with a pin-compatible device which they call the PALCE16V8 ("C" is for CMOS, "E" is for erasable). Several other manufacturers make differently numbered but compatible devices as well. Rather than get caught up in the details of different manufacturers' names, in this chapter we usually refer to commonly used GAL devices with their generic names, 16V8, 20V8, and 22V10.

8.3.3 PLD Timing Specifications

Several timing parameters are specified for combinational and sequential PLDs. The most important ones are illustrated in Figure 8-25 and are explained below:

t_{PD}

feedback input

t_{PD} This parameter applies to combinational outputs. It is the propagation delay from a primary input pin, bidirectional pin, or "feedback" input to the combinational output. A *feedback input* is an internal input of the AND-OR array that is driven by the registered output of an internal macrocell.

t_{CO} This parameter applies to registered outputs. It is the propagation delay t_{CO}
from the rising edge of **CLK** to a primary output.

t_{CF} This parameter also applies to registered outputs. It is the propagation t_{CF}
delay from the rising edge of **CLK** to a macrocell's registered output that
connects back to a feedback input. If specified, t_{CF} is normally less than
t_{CO}. However, some manufacturers do not specify t_{CF}, in which case you
must assume that $t_{CF} = t_{CO}$.

t_{SU} This parameter applies to primary, bidirectional, and feedback inputs t_{SU}
that propagate to the D inputs of flip-flops. It is the setup time during
which the input signal must be stable before the rising edge of **CLK**.

t_H This parameter also applies to signals that propagate to the D inputs of t_H
flip-flops. It is the hold time during which the input signal must be stable
after the rising edge of **CLK**.

f_{max} This parameter applies to clocked operation. It is the highest frequency f_{max}
at which the PLD can operate reliably and is the reciprocal of the
minimum clock period. Two versions of this parameter can be derived
from the previous specifications, depending on whether the device is
operating with external feedback or internal feedback.

External feedback refers to a circuit in which a registered PLD output is   *external feedback*
connected to the input of another registered PLD with similar timing; for
proper operation, the sum of t_{CO} for the first PLD and t_{SU} for the second
must not exceed the clock period.

Internal feedback refers to a circuit in which a registered PLD output is fed   *internal feedback*
back to a register in the same PLD; in this case, the sum of t_{CF} and t_{SU} must
not exceed the clock period.

Each of the PLDs that we described in previous sections is available in
several different speed grades. The speed grade is usually indicated by a suffix
on the part number, such as "16V8-10"; the suffix usually refers to the t_{PD}

Figure 8-25 PLD timing parameters.

Table 8-10 Timing specifications, in nanoseconds, of popular bipolar and CMOS PLDs.

Part numbers	Suffix	t_{PD}	t_{CO}	t_{CF}	t_{SU}	t_H
PAL16L8, PAL16Rx, PAL20L8, PAL20Rx	-5	5	4	–	4.5	0
PAL16L8, PAL16Rx, PAL20L8, PAL20Rx	-7	7.5	6.5	–	7	0
PAL16L8, PAL16Rx, PAL20L8, PAL20Rx	-10	10	8	–	10	0
PAL16L8, PAL16Rx, PAL20L8, PAL20Rx	B	15	12	–	15	0
PAL16L8, PAL16Rx, PAL20L8, PAL20Rx	B-2	25	15	–	25	0
PAL16L8, PAL16Rx, PAL20L8, PAL20Rx	A	25	15	–	25	0
PALCE16V8, PALCE20V8	-5	5	4	–	3	0
GAL16V8, GAL20V8	-7	7.5	5	3	5	0
GAL16V8, GAL20V8	-10	10	7.5	6	7.5	0
GAL16V8, GAL20V8	-15	15	10	8	12	0
GAL16V8, GAL20V8	-25	25	12	10	15	0
PALCE22V10	-5	5	4	–	3	0
PALCE22V10	-7	7.5	4.5	–	4.5	0
GAL22V10	-10	10	7	2.5	7	0
GAL22V10	-15	15	8	2.5	10	0
GAL22V10	-25	25	15	13	15	0

specification, in nanoseconds. Table 8-10 shows the timing of several popular bipolar and CMOS PLDs. Note that only the t_{PD} column applies to the combinational outputs of a device, while the last four columns apply to registered outputs. All of the timing specifications are worst-case numbers over the commercial operating range.

When sequential PLDs are used in applications with critical timing, it's important to remember that they normally have longer setup times than discrete edge-triggered registers in the same technology, owing to the delay of the AND-OR array on each D input. Conversely, under typical conditions, a PLD actually has a negative hold-time requirement because of the delay through AND-OR array. However, you can't count on it having a negative hold time—the worst-case specification is normally zero.

8.4 Counters

The name *counter* is generally used for any clocked sequential circuit whose state diagram contains a single cycle, as in Figure 8-26. The *modulus* of a counter is the number of states in the cycle. A counter with m states is called a *modulo-m counter* or, sometimes, a *divide-by-m counter*. A counter with a non-power-of-2 modulus has extra states that are not used in normal operation.

counter
modulus

modulo-m counter
divide-by-m counter

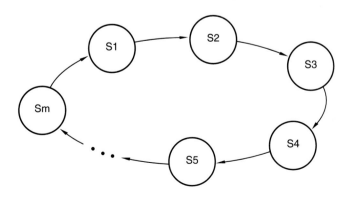

Figure 8-26
General structure of a counter's state diagram—a single cycle.

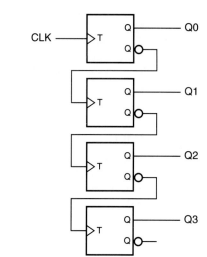

Figure 8-27
A 4-bit binary
ripple counter.

n-bit binary counter

Probably the most commonly used counter type is an *n-bit binary counter*. Such a counter has n flip-flops and has 2^n states, which are visited in the sequence 0, 1, 2, ... , $2^n - 1$, 0, 1, ... Each of these states is encoded as the corresponding n-bit binary integer.

8.4.1 Ripple Counters

An n-bit binary counter can be constructed with just n flip-flops and no other components, for any value of n. Figure 8-27 shows such a counter for $n = 4$. Recall that a T flip-flop changes state (toggles) on every rising edge of its clock input. Thus, each bit of the counter toggles if and only if the immediately preceding bit changes from 1 to 0. This corresponds to a normal binary counting sequence—when a particular bit changes from 1 to 0, it generates a carry to the next most significant bit. The counter is called a *ripple counter* because the carry information ripples from the less significant bits to the more significant bits, one bit at a time.

ripple counter

8.4.2 Synchronous Counters

Although a ripple counter requires fewer components than any other type of binary counter, it does so at a price—it is slower than any other type of binary counter. In the worst case, when the most significant bit must change, the output is not valid until time $n \cdot t_{TQ}$ after the rising edge of CLK, where t_{TQ} is the propagation delay from input to output of a T flip-flop.

synchronous counter

A *synchronous counter* connects all of its flip-flop clock inputs to the same common CLK signal, so that all of the flip-flop outputs change at the same time, after only t_{TQ} ns of delay. As shown in Figure 8-28, this requires the use of T flip-flops with enable inputs; the output toggles on the rising edge of T if and only if EN is asserted. Combinational logic on the EN inputs determines which, if any, flip-flops toggle on each rising edge of T.

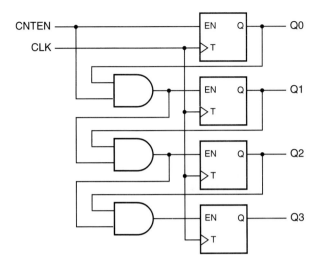

Figure 8-28
A synchronous 4-bit
binary counter with
serial enable logic.

As shown in Figure 8-28, it is also possible to provide a master count-enable signal CNTEN. Each T flip-flop toggles if and only if CNTEN is asserted and all of the lower-order counter bits are 1. Like the binary ripple counter, a synchronous n-bit binary counter can be built with a fixed amount of logic per bit—in this case, a T flip-flop with enable and a 2-input AND gate.

The counter structure in Figure 8-28 is sometimes called a *synchronous serial counter* because the combinational enable signals propagate serially from the least significant to the most significant bits. If the clock period is too short, there may not be enough time for a change in the counter's LSB to propagate to the MSB. This problem is eliminated in Figure 8-29 by driving each EN input with a dedicated AND gate, just a single level of logic. Called a *synchronous parallel counter,* this is the fastest binary counter structure.

synchronous serial counter

synchronous parallel counter

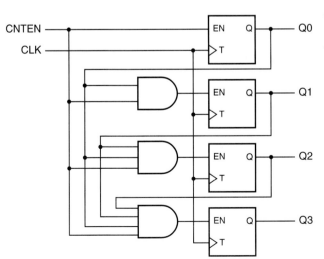

Figure 8-29
A synchronous 4-bit
binary counter with
parallel enable logic

8.4.3 MSI Counters and Applications

The most popular MSI counter is the *74x163*, a synchronous 4-bit binary counter with active-low load and clear inputs, with the traditional logic symbol shown in Figure 8-30. Its function is summarized by the state table in Table 8-11, and its internal logic diagram is shown in Figure 8-31.

The '163 uses D flip-flops rather than T flip-flops internally to facilitate the load and clear functions. Each D input is driven by a 2-input multiplexer consisting of an OR gate and two AND gates. The multiplexer output is 0 if the CLR_L input is asserted. Otherwise, the top AND gate passes the data input (A, B, C, or D) to the output if LD_L is asserted. If neither CLR_L nor LD_L is asserted, the bottom AND gate passes the output of an XNOR gate to the multiplexer output.

Figure 8-30
Traditional logic symbol for the 74x163.

Table 8-11 State table for a 74x163 4-bit binary counter.

Inputs				Current State				Next State			
CLR_L	LD_L	ENT	ENP	QD	QC	QB	QA	QD*	QC*	QB*	QA*
0	x	x	x	x	x	x	x	0	0	0	0
1	0	x	x	x	x	x	x	D	C	B	A
1	1	0	x	x	x	x	x	QD	QC	QB	QA
1	1	x	0	x	x	x	x	QD	QC	QB	QA
1	1	1	1	0	0	0	0	0	0	0	1
1	1	1	1	0	0	0	1	0	0	1	0
1	1	1	1	0	0	1	0	0	0	1	1
1	1	1	1	0	0	1	1	0	1	0	0
1	1	1	1	0	1	0	0	0	1	0	1
1	1	1	1	0	1	0	1	0	1	1	0
1	1	1	1	0	1	1	0	0	1	1	1
1	1	1	1	0	1	1	1	1	0	0	0
1	1	1	1	1	0	0	0	1	0	0	1
1	1	1	1	1	0	0	1	1	0	1	0
1	1	1	1	1	0	1	0	1	0	1	1
1	1	1	1	1	0	1	1	1	1	0	0
1	1	1	1	1	1	0	0	1	1	0	1
1	1	1	1	1	1	0	1	1	1	1	0
1	1	1	1	1	1	1	0	1	1	1	1
1	1	1	1	1	1	1	1	0	0	0	0

Figure 8-31 Logic diagram for the 74x163 synchronous 4-bit binary counter, including pin numbers for a standard 16-pin dual in-line package.

Figure 8-32
Connections for the
74x163 to operate in
a free-running mode.

The XNOR gates perform the counting function in the '163. One input of each XNOR is the corresponding count bit (QA, QB, QC, or QD); the other input is 1, which complements the count bit, if and only if both enables ENP and ENT are asserted and all of the lower-order count bits are 1. The RCO ("ripple carry out") signal indicates a carry from the most significant bit position and is 1 when all of the count bits are 1 and ENT is asserted.

free-running counter

Even though most MSI counters have enable inputs, they are often used in a *free-running* mode in which they are enabled continuously. Figure 8-32 shows the connections to make a '163 operate in this way, and Figure 8-33 shows the resulting output waveforms. Notice that, starting with QA, each signal has half the frequency of the preceding one. Thus, a free-running '163 can be used as a divide-by-2, -4, -8, or -16 counter, by ignoring any unnecessary high-order output bits.

Note that the '163 is fully synchronous; that is, its outputs change only on the rising edge of CLK. Some applications need an asynchronous clear function,

Figure 8-33 Clock and output waveforms for a free-running divide-by-16 counter.

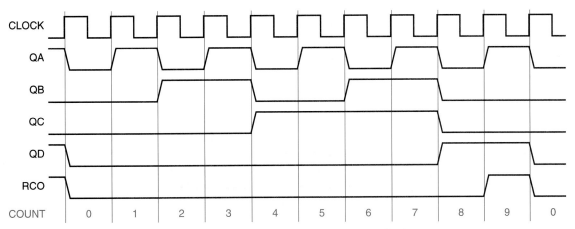

Figure 8-34 Clock and output waveforms for a free-running divide-by-10 counter.

as provided by the *74x161*. The '161 has the same pinout as the '163, but its CLR_L input is connected to the asynchronous clear inputs of its flip-flops.

74x161

The *74x160* and *74x162* are more variations with the same pinouts and general functions as the '161 and '163, except that the counting sequence is modified to go to state 0 after state 9. In other words, these are modulo-10 counters, sometimes called *decade counters*. Figure 8-34 shows the output waveforms for a free-running '160 or '162. Notice that although the QD and QC outputs have one-tenth of the CLK frequency, they do not have a 50% duty cycle, and the QC output, with one-fifth of the input frequency, does not have a constant duty cycle. We'll show the design of a divide-by-10 counter with a 50% duty-cycle output later in this subsection.

74x160
74x162

decade counter

Although the '163 is a modulo-16 counter, it can be made to count in a modulus less than 16 by using the CLR_L or LD_L input to shorten the normal counting sequence. For example, Figure 8-35 shows one way of using the '163 as a modulo-11 counter. The RCO output, which detects state 15, is used to force

Figure 8-35
Using the 74x163 as a modulo-11 counter with the counting sequence 5, 6, …, 15, 5, 6, ….

Figure 8-36
Using the 74x163 as
a modulo-11 counter
with the counting
sequence 0, 1, 2, …,
10, 0, 1, ….

the next state to 5, so that the circuit will count from 5 to 15 and then start at 5 again, for a total of 11 states per counting cycle.

A different approach for modulo-11 counting with the '163 is shown in Figure 8-36. This circuit uses a NAND gate to detect state 10 and force the next state to 0. Notice that only a 2-input gate is used to detect state 10 (binary 1010). Although a 4-input gate would normally be used to detect the condition CNT10 = $Q3 \cdot Q2' \cdot Q1 \cdot Q0'$, the 2-input gate takes advantage of the fact that no other state in the normal counting sequence of 0–10 has $Q3 = 1$ and $Q1 = 1$. In general, to detect state N in a binary counter that counts from 0 to N, we need to AND only the state bits that are 1 in the binary encoding of N.

There are many other ways to make a modulo-11 counter using a '163. The choice of approach—one of the preceding or a combination of them (as in Exercise 8.36)—depends on the application. As another example, in Section 2.10 we promised to show you how to build a circuit that counts in the

Figure 8-37
A 74x163 used as an
excess-3 decimal
counter.

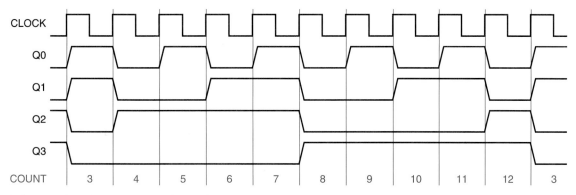

Figure 8-38 Timing waveforms for the '163 used as an excess-3 decimal counter.

excess-3 decimal code, shown in Table 2-9 on page 49. Figure 8-37 shows the connections for a '163 to count in the excess-3 sequence. A NAND gate detects state 1100 and forces 0011 to be loaded as the next state. Figure 8-38 shows the resulting timing waveforms. Notice that the Q3 output has a 50% duty cycle, which may be desirable for some applications.

A binary counter with a modulus greater than 16 can be built by cascading 74x163s. Figure 8-39 shows the general connections for such a counter. The CLK, CLR_L, and LD_L inputs of all the '163s are connected in parallel, so that all of them count or are cleared or loaded at the same time. A master count-enable (CNTEN) signal is connected to the low-order '163. The RCO4 output is asserted if and only if the low-order '163 is in state 15 *and* CNTEN is asserted; RCO4 is connected to the enable inputs of the high-order '163. Thus, both the carry information and the master count-enable ripple from the output of one

Figure 8-39 General cascading connections for 74x163-based counters.

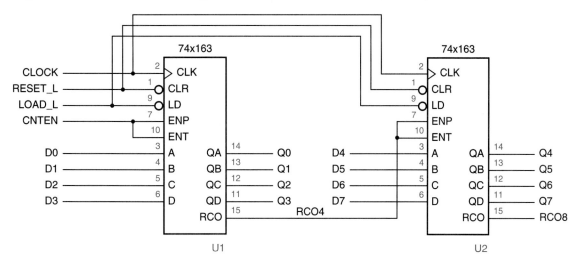

4-bit counter stage to the next. Like the synchronous serial counter of Figure 8-28, this scheme can be extended to build a counter with any desired number of bits; the maximum counting speed is limited by the propagation delay of the ripple carry signal through all of the stages (but see Exercise 8.38).

Even experienced digital designers are sometimes confused about the difference between the ENP and ENT enable inputs of the '163 and similar counters, since both must be asserted for the counter to count. However, a glance at the 163's internal logic diagram, Figure 8-31 on page 697, shows the difference quite clearly—ENT goes to the ripple carry output as well. In many applications this distinction is important.

For example, Figure 8-40 shows an application that uses two '163s as a modulo-193 counter that counts from 63 to 255. The MAXCNT output detects state 255 and stops the counter until GO_L is asserted. When GO_L is asserted, the counter is reloaded with 63 and counts up to 255 again. (Note that the value of GO_L is relevant only when the counter is in state 255.) To keep the counter

Figure 8-40
Using 74x163s as a modulo-193 counter with the counting sequence 63, 64, …, 255, 63, 64, ….

stopped, MAXCNT must be asserted in state 255 even while the counter is stopped. Therefore, the low-order counter's ENT input is always asserted, its RCO output is connected to the high-order ENT input, and MAXCNT detects state 255 even if CNTEN is not asserted (compare with the behavior of RCO8 in Figure 8-39). To enable counting, CNTEN is connected to the ENP inputs in parallel. A NAND gate asserts RELOAD_L to go back to state 63 only if GO_L is asserted and the counter is in state 255.

Another counter with functions similar to 74x163's is the *74x169,* whose logic symbol is shown in Figure 8-41. One difference in the '169 is that its carry output and enable inputs are active low. More importantly, the '169 is an *up/down counter;* it counts in ascending or descending binary order depending on the value of an input signal, UP/DN. The '169 counts up when UP/DN is 1 and down when UP/DN is 0.

74x169

up/down counter

Figure 8-41
Logic symbol for the
74x169 up/down
counter.

8.4.4 Decoding Binary-Counter States

A binary counter may be combined with a decoder to obtain a set of 1-out-of-*m*-coded signals, where one signal is asserted in each counter state. This is useful when counters are used to control a set of devices, where a different device is enabled in each counter state. In this approach, each output of the decoder enables a different device.

Figure 8-42 shows how a 74x163 wired as a modulo-8 counter can be combined with a 74x138 3-to-8 decoder to provide eight signals, each one representing a counter state. Figure 8-43 shows typical timing for this circuit. Each decoder output is asserted during a corresponding clock period.

Notice that the decoder outputs may contain "glitches" on state transitions where two or more counter bits change, even though the '163 outputs are glitch free and the '138 does not have any static hazards. In a synchronous counter like the '163, the outputs don't change at exactly the same time. More important,

decoding glitches

Figure 8-42
A modulo-8 binary
counter and decoder.

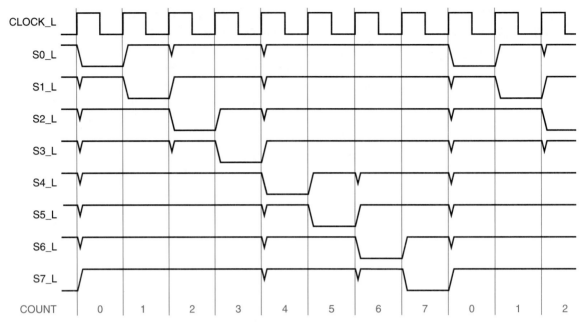

Figure 8-43 Timing diagram for a modulo-8 binary counter and decoder, showing decoding glitches.

multiple signal paths in a decoder like the '138 have different delays; for example, the path from B to Y1_L is faster than the path from A to Y1_L. Thus, even if the input changes simultaneously from 011 to 100, the decoder may behave as if the input were temporarily 001, and the Y1_L output may have a glitch. In the present example, it can be shown that the glitches can occur in *any* realization of the binary decoder function; this problem is an example of a *function hazard*.

function hazard

In most applications, the decoder output signals portrayed in Figure 8-43 would be used as control inputs to registers, counters, and other edge-triggered devices (e.g., EN_L in a 74x377, LD_L in a 74x163, or ENP_L in a 74x169). In such a case, the decoding glitches in the figure are not a problem, since they occur *after* the clock tick. They are long gone before the next tick comes along, when the decoder outputs are sampled by other edge-triggered devices. However, the glitches *would* be a problem if they were applied to something like the S_L or R_L inputs of an $\overline{\text{S}}$-$\overline{\text{R}}$ latch. Likewise, using such potentially glitchy signals as clocks for edge-triggered devices is a definite no-no.

If necessary, one way to "clean up" the glitches in Figure 8-43 is to connect the '138 outputs to another register that samples the stable decoded outputs on the next clock tick, as shown in Figure 8-44. Notice that the decoded outputs have been renamed to account for the 1-tick delay through the register. However, once you decide to pay for an 8-bit register, a less costly solution is to use an 8-bit "ring counter," which provides glitch-free decoded outputs directly, as we'll show in Section 8.5.6.

Figure 8-44 A modulo-8 binary counter and decoder with glitch-free outputs.

8.4.5 Counters in ABEL and PLDs

Binary counters are good candidates for ABEL- and PLD-based design, for several reasons:

- A large state machine can often be decomposed into two or more smaller state machines, where one of the smaller machines is a binary counter that keeps track of how long the other machine should stay in a particular state. This may simplify both the conceptual design and the circuit design of the machine.

- Many applications require almost-binary-modulus counters with special requirements for initialization, state detection, or state skipping. For example, a counter in an elevator controller may skip state 13. Instead of using an off-the-shelf binary counter and extra logic for the special requirements, a designer can specify exactly the required functions in an ABEL program.

- Most standard MSI counters have only 4 bits, while a single 24-pin PLD can be used to create a binary counter with up to 10 bits.

The most popular MSI counter is the 74x163 4-bit binary counter, shown in Figure 8-31 on page 697. A glance at this figure shows that the excitation logic for this counter isn't exactly simple, especially considering its use of XNOR gates. Nevertheless, ABEL provides a very simple way of defining counter behavior, which we describe next.

Recall that ABEL uses the "+" symbol to specify integer addition. When two sets are "added" with this operator, each is interpreted as a binary number; the rightmost set element corresponds to the least significant bit of the number. Thus, the function of a 74x163 can be specified by the ABEL program in Table 8-12. When the counter is enabled, 1 is added to the current state.

Table 8-12 ABEL program for a 74x163-like 4-bit binary counter.

```
module Z74X163
title '4-bit Binary Counter'

" Input pins
CLK, !LD, !CLR, ENP, ENT        pin;
A, B, C, D                      pin;

" Output pins
QA, QB, QC, QD                  pin istype 'reg';
RCO                             pin istype 'com';

" Set definitions
INPUT = [D, C, B, A];
COUNT = [QD, QC, QB, QA];


equations

COUNT.CLK = CLK;

COUNT := !CLR & ( LD & INPUT
                # !LD & (ENT & ENP) & (COUNT + 1)
                # !LD & !(ENT & ENP) & COUNT);

RCO = (COUNT == [1,1,1,1]) & ENT;

end Z74X163
```

Table 8-13 MInimized equations for the 4-bit binary counter in Table 8-12.

```
QA := (CLR & LD & ENT & ENP & !QA          QD := (CLR & LD & ENT & ENP & !QD & QC & QB & QA
    # CLR & LD & !ENP & QA                     # CLR & !LD & D
    # CLR & LD & !ENT & QA                     # CLR & LD & QD & !QB
    # CLR & !LD & A);                          # CLR & LD & QD & !QC
                                               # CLR & LD & !ENP & QD
QB := (CLR & LD & ENT & ENP & !QB & QA         # CLR & LD & !ENT & QD
    # CLR & LD & QB & !QA                       # CLR & LD & QD & !QA);
    # CLR & LD & !ENP & QB
    # CLR & LD & !ENT & QB                  RCO = (ENT & QD & QC & QB & QA);
    # CLR & !LD & B);

QC := (CLR & LD & ENT & ENP & !QC & QB & QA
    # CLR & LD & QC & !QA
    # CLR & LD & QC & !QB
    # CLR & LD & !ENP & QC
    # CLR & LD & !ENT & QC
    # CLR & !LD & C);
```

Table 8-13 shows the minimized logic equations that ABEL generates for the 4-bit counter. Notice that each more significant output bit requires one more product term. As a result, the size of counters that can be realized in a 16V8 or even a 20V8 is generally limited to five or six bits. Other devices, including the X-series PLDs and some CPLDs, contain an XOR structure that can realize larger counters without increasing product-term requirements.

Designing a specialized counting sequence in ABEL is much simpler than adapting a standard binary counter. For example, the ABEL program in Table 8-12 can be adapted to count in excess-3 sequence (Figure 8-38 on page 701) by changing the equations as follows:

```
COUNT := !CLR & ( LD & INPUT
       # !LD & (ENT & ENP) &
          ((COUNT==12) & 3) # ((COUNT!=12) & (COUNT + 1))
       # !LD & !(ENT & ENP) & COUNT);

RCO = (COUNT == 12) & ENT;
```

PLDs can be cascaded to obtain wider counters by providing each counter stage with a carry output that indicates when it is about to roll over. There are two basic approaches to generating the carry output:

- *Combinational*. The carry equation indicates that the counter is enabled and is currently in its last state before rollover. For a 5-bit binary up counter, we have

  combinational carry output

  ```
  COUT = CNTEN & Q4 & Q3 & Q2 & Q1 & Q0;
  ```

  Since CNTEN is included, this approach allows carries to be rippled through cascaded counters by connecting each COUT to the next CNTEN.

- *Registered*. The carry equation indicates that the counter is about to enter its last state before rollover. Thus, at the next clock tick, the counter enters this last state and the carry output is asserted. For a 5-bit binary up counter with load and clear inputs, we have

  registered carry output

  ```
  COUT := !CLR & !LD & CNTEN
          & Q4 & Q3 & Q2 & Q1 & !Q0
       # !CLR * !LD * !CNTEN
          & Q4 & Q3 & Q2 & Q1 & Q0
       # !CLR & LD
          & D4 & D3 & D2 & D1 & D0;
  ```

The second approach has the advantage of producing COUT with less delay than the combinational approach. However, external gates are now required between stages, since the CNTEN signal for each stage should be the logical AND of the master count-enable signal and the COUT outputs of all lower-order counters. These external gates can be avoided if the higher-order counters have multiple enable inputs.

8.4.6 Counters in VHDL

Like ABEL, VHDL allows counters to be specified fairly easily. The biggest challenge in VHDL, with its strong type checking, is to get all of the signal types defined correctly and consistently.

Table 8-14 is a VHDL program for a 74x163-like binary counter. Notice that the program uses the IEEE.std_logic_arith.all library, which includes the UNSIGNED type, as we described in Section 5.9.6 on page 427. This library includes definitions of "+" and "−" operators that perform unsigned addition and subtraction on UNSIGNED operands. The counter program declares the counter input and output as UNSIGNED vectors and uses "+" to increment the counter value as required.

In the program, we defined an internal signal IQ to hold the counter value. We could have used Q directly, but then we'd have to declare its port type as buffer rather than out. Also, we could have defined the type of ports D and Q to be STD_LOGIC_VECTOR, but then we would have to perform type conversions inside the body of the process (see Exercise 8.51).

Table 8-14 VHDL program for a 74x163-like 4-bit binary counter.

```
library IEEE;
use IEEE.std_logic_1164.all;
use IEEE.std_logic_arith.all;

entity V74x163 is
    port ( CLK, CLR_L, LD_L, ENP, ENT: in STD_LOGIC;
            D: in UNSIGNED (3 downto 0);
            Q: out UNSIGNED (3 downto 0);
            RCO: out STD_LOGIC );
end V74x163;

architecture V74x163_arch of V74x163 is
signal IQ: UNSIGNED (3 downto 0);
begin
process (CLK, ENT, IQ)
  begin
    if (CLK'event and CLK='1') then
      if CLR_L='0' then IQ <= (others => '0');
      elsif LD_L='0' then IQ <= D;
      elsif (ENT and ENP)='1' then IQ <= IQ + 1;
      end if;
    end if;
    if (IQ=15) and (ENT='1') then RCO <= '1';
    else RCO <= '0';
    end if;
    Q <= IQ;
  end process;
end V74x163_arch;
```

Table 8-15 VHDL architecture for counting in excess-3 order.

```
architecture V74xs3_arch of V74x163 is
signal IQ: UNSIGNED (3 downto 0);
begin
process (CLK, ENT, IQ)
  begin
    if CLK'event and CLK='1' then
      if CLR_L='0' then IQ <= (others => '0');
      elsif LD_L='0' then IQ <= D;
      elsif (ENT and ENP)='1' and (IQ=12) then IQ <= ('0','0','1','1');
      elsif (ENT and ENP)='1' then IQ <= IQ + 1;
      end if;
    end if;
    if (IQ=12) and (ENT='1') then RCO <= '1';
    else RCO <= '0';
    end if;
    Q <= IQ;
  end process;
end V74xs3_arch;
```

As in ABEL, specialized counting sequences can be specified very easily using behavioral VHDL code. For example, Table 8-15 modifies the 74x163-like counter to count in excess-3 sequence (3, … , 12, 3, …).

Unfortunately, some VHDL synthesis engines do not synthesize counters particularly well. In particular, they tend to synthesize the counting step using a binary adder with the counter value and a constant 1 as operands. This approach requires much more combinational logic than what we've shown for discrete counters; it is particularly wasteful in CPLDs and FPGAs containing T flip-flops, XOR gates, or other structures optimized for counters. In this case, a useful alternative is to write structural VHDL that is targeted to the cells available in a particular CPLD, FPGA, or ASIC technology.

For example, we can construct one bit-cell for a 74x163-like counter using the circuit in Figure 8-45. This circuit is designed to use serial propagation for the carry bits, so the same circuit can be used at any stage of an arbitrarily large counter, subject to fanout constraints on the common signals that drive all of the stages. The signals in the bit-cell have the following definitions:

CLK (common) The clock input for all stages.

LDNOCLR (common) Asserted if the counter's LD input is asserted and CLR is negated.

NOCLRORLD (common) Asserted if the counter's CLR and LD inputs are both negated.

CNTENP (common) Asserted if the counter's ENP input is asserted.

Di (per cell) Load data input for cell i.

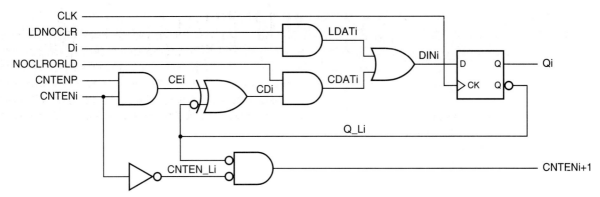

Figure 8-45 One bit-cell of a synchronous serial, 74x163-like counter.

CNTENi   (per cell) Serial count enable input for cell *i*.

CNTENi+1   (per cell) Serial count enable output for cell *i*.

Qi   (per cell) Counter output for cell *i*.

Table 8-16 is a VHDL program corresponding to the bit-cell in the figure. In the program, the D flip-flop component Vdffqqn is assumed to be already defined; it is similar to the D flip-flop in Table 8-6 with the addition of a QN (complemented) output. In an FPGA or ASIC design, a flip-flop component type would be chosen from the manufacturer's standard cell library.

Table 8-16 VHDL program for counter cell of Figure 8-45.

```
library IEEE;
use IEEE.std_logic_1164.all;

entity syncsercell is
  port( CLK, LDNOCLR, NOCLRORLD, CNTENP, D, CNTEN: in STD_LOGIC;
        CNTENO, Q: out STD_LOGIC );
end syncsercell;

architecture syncsercell_arch of syncsercell is
component Vdffqqn
  port( CLK, D: in STD_LOGIC;
        Q, QN: out STD_LOGIC );
end component;
signal LDAT, CDAT, DIN, Q_L: STD_LOGIC;
begin
  LDAT <= LDNOCLR and D;
  CDAT <= NOCLRORLD and ((CNTENP and CNTEN) xor not Q_L);
  DIN  <= LDAT or CDAT;
  CNTENO <= (not Q_L) and CNTEN;
  U1: Vdffqqn port map (CLK, DIN, Q, Q_L);
end syncsercell_arch;
```

> **A MATTER OF STYLE** Note that Table 8-16 uses a combination of dataflow and structural VHDL styles. It could have been written completely structurally, for example using an ASIC manufacturer's gate component definitions, to guarantee that the synthesized circuit conforms exactly to Figure 8-45. However, most synthesis engines can do a good job of picking the best gate realization for the simple signal assignments used here.

Table 8-17 shows how to create an 8-bit synchronous serial counter using the cell defined previously. The first two assignments in the architecture body synthesize the common LDNOCLR and NOCLRORLD signals. The next two statements handle boundary conditions for the serial count-enable chain. Finally, the generate statement (introduced on page 453) instantiates eight 1-bit counter cells and hooks up the count-enable chain as required.

It should be clear that a larger or smaller counter can be created simply by changing a few definitions in the program. You can put VHDL's generic statement to good use here to allow you to change the counter's size with a one-line change (see Exercise 8.53).

Table 8-17 VHDL program for an 8-bit 74x163-like synchronous serial counter.

```vhdl
library IEEE;
use IEEE.std_logic_1164.all;

entity V74x163s is
  port( CLK, CLR_L, LD_L, ENP, ENT: in STD_LOGIC;
        D: in STD_LOGIC_VECTOR (7 downto 0);
        Q: out STD_LOGIC_VECTOR (7 downto 0);
        RCO: out STD_LOGIC );
end V74x163s;

architecture V74x163s_arch of V74x163s is
component syncsercell
  port( CLK, LDNOCLR, NOCLRORLD, CNTENP, D, CNTEN: in STD_LOGIC;
        CNTENO, Q: out STD_LOGIC );
end component;
signal LDNOCLR, NOCLRORLD: STD_LOGIC; -- common signals
signal SCNTEN: STD_LOGIC_VECTOR (8 downto 0); -- serial count-enable inputs
begin
  LDNOCLR <= (not LD_L) and CLR_L; -- create common load and clear controls
  NOCLRORLD <= LD_L and CLR_L;
  SCNTEN(0) <= ENT; -- serial count-enable into the first stage
  RCO <= SCNTEN(8);  -- RCO is equivalent to final count-enable output
  g1: for i in 0 to 7 generate    -- generate the eight syncsercell stages
      U1: syncsercell port map ( CLK, LDNOCLR, NOCLRORLD, ENP, D(i), SCNTEN(i),
                                 SCNTEN(i+1), Q(i));
  end generate;
end V74x163s_arch;
```

8.5 Shift Registers

8.5.1 Shift-Register Structure

A *shift register* is an *n*-bit register with a provision for shifting its stored data by one bit position at each tick of the clock. Figure 8-46 shows the structure of a serial-in, serial-out shift register. The *serial input,* SERIN, specifies a new bit to
be shifted into one end at each clock tick. This bit appears at the *serial output,* SEROUT, after *n* clock ticks, and is lost one tick later. Thus, an *n*-bit serial-in, serial-out shift register can be used to delay a signal by *n* clock ticks.

A *serial-in, parallel-out shift register,* shown in Figure 8-47, has outputs for all of its stored bits, making them available to other circuits. Such a shift
register can be used to perform *serial-to-parallel conversion,* as explained later in this section.

Conversely, it is possible to build a *parallel-in, serial-out shift register.* Figure 8-48 shows the general structure of such a device. At each clock tick the register either loads new data from inputs 1D–ND or it shifts its current contents, depending on the value of the LOAD/SHIFT control input (which could be named LOAD or SHIFT_L). Internally, the device uses a 2-input multiplexer on each flip-flop's D input to select between the two cases. A parallel-in, serial-out
shift register can be used to perform *parallel-to-serial conversion,* as explained later in this section.

By providing outputs for all of the stored bits in a parallel-in shift register, we obtain the p*arallel-in, parallel-out shift register* shown in Figure 8-49. Such a device is general enough to be used in any of the applications of the previous shift registers.

Figure 8-46 Structure of a serial-in, serial-out shift register.

Figure 8-47 Structure of a serial-in, parallel-out shift register.

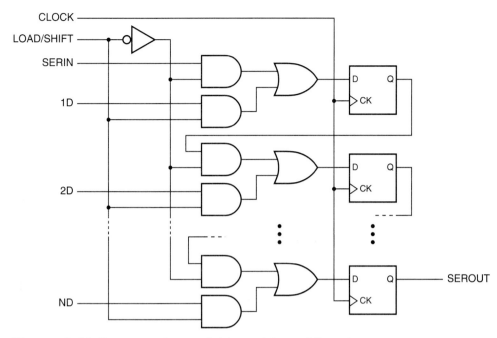

Figure 8-48 Structure of a parallel-in, serial-out shift register.

Figure 8-49 Structure of a parallel-in, parallel-out shift register.

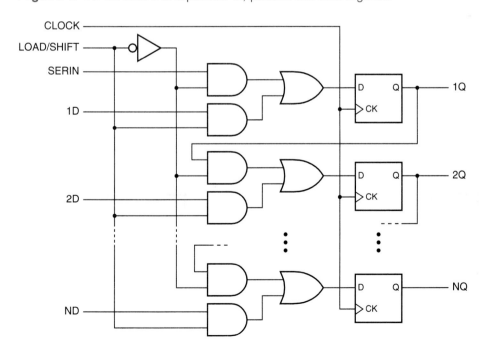

8.5.2 MSI Shift Registers

74x164

Figure 8-50 shows logic symbols for three popular MSI 8-bit shift registers. The *74x164* is a serial-in, parallel-out device with an asynchronous clear input (CLR_L). It has two serial inputs that are ANDed internally. That is, both SERA and SERB must be 1 for a 1 to be shifted into the first bit of the register.

74x166

The *74x166* is a parallel-in, serial-out shift register, also with an asynchronous clear input. This device shifts when SH/LD is 1 and loads new data otherwise. The '166 has an unusual clocking arrangement called a "gated clock" (see also Section 8.8.2); it has two clock inputs that are connected to the internal flip-flops as shown in Figure 8-50(c). The designers of the '166 intended for CLK to be connected to a free-running system clock, and for CLKINH to be asserted to inhibit CLK, so that neither shifting nor loading occurs on the next clock tick, and the current register contents are held. However, for this to work, CLKINH must be changed only when CLK is 1; otherwise, undesired clock edges occur on the internal flip-flops. A much safer way of obtaining a "hold" function is employed in the next devices that we discuss.

74x194

unidirectional shift register

bidirectional shift register

The *74x194* is an MSI 4-bit bidirectional, parallel-in, parallel-out shift register. Its logic diagram is shown in Figure 8-51. The shift registers that we've studied previously are called *unidirectional shift registers* because they shift in only one direction. The '194 is a *bidirectional shift register* because its contents may be shifted in either of two directions, depending on a control input. The two

Figure 8-50

Traditional logic symbols for MSI shift registers:
(a) 74x164 8-bit serial-in, parallel-out shift register;
(b) 74x166 8-bit parallel-in, serial-out shift register;
(c) equivalent circuit for 74x166 clock inputs;
(d) 74x194 universal shift register.

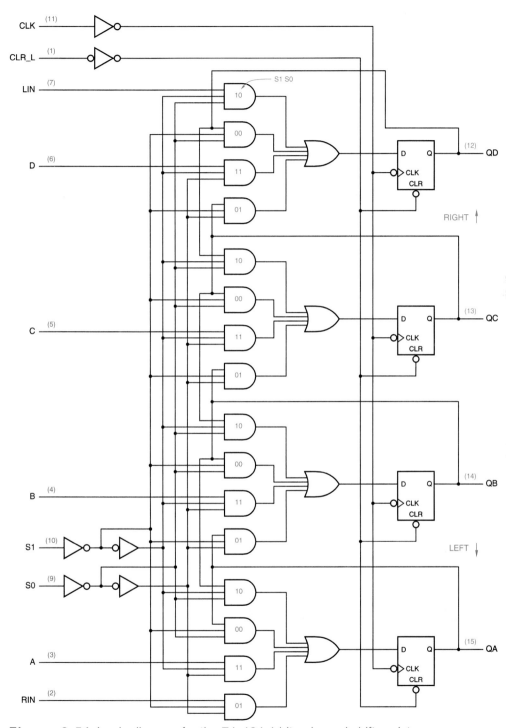

Figure 8-51 Logic diagram for the 74x194 4-bit universal shift register, including pin numbers for a standard 16-pin dual in-line package.

Table 8-18
Function table for the
74x194 4-bit universal
shift register.

	Inputs		Next state			
Function	S1	S0	QA*	QB*	QC*	QD*
Hold	0	0	QA	QB	QC	QD
Shift right	0	1	RIN	QA	QB	QC
Shift left	1	0	QB	QC	QD	LIN
Load	1	1	A	B	C	D

directions are called "left" and "right," even though the logic diagram and the logic symbol aren't necessarily drawn that way. In the '194, *left* means "in the direction from QD to QA," and *right* means "in the direction from QA to QD." Our logic diagram and symbol for the '194 are consistent with these names if you rotate them 90° clockwise.

left

right

Table 8-18 is a function table for the 74x194. This function table is highly compressed, since it does not contain columns for most of the inputs (A–D, RIN, LIN) or the current state QA–QD. Still, by expressing each next-state value as a function of these implicit variables, it completely defines the operation of the '194 for all 2^{12} possible combinations of current state and input, and it sure beats a 4,096-row table!

Figure 8-52
Traditional logic
symbol for
the 74x299.

Note that the '194's LIN (left-in) input is conceptually located on the "righthand" side of the chip, but it is the serial input for *left* shifts. Likewise, RIN is the serial input for right shifts.

The '194 is sometimes called a *universal* shift register because it can be made to function like any of the less general shift-register types that we've discussed (e.g., unidirectional; serial-in, parallel-out; parallel-in, serial-out). In fact, many of our design examples in the next few subsections contain '194s configured to use just a subset of their available functions.

The *74x299* is an 8-bit universal shift register in a 20-pin package; its symbol and logic diagram are given in Figures 8-52 and 8-53. The '299's functions and function table are similar to the '194's, as shown in Table 8-19. To save pins, the '299 uses bidirectional three-state lines for input and output, as shown in the logic diagram. During load operations (S1 S0 = 11), the three-state

Table 8-19 Function table for a 74x299 8-bit universal shift register.

	Inputs		Next state							
Function	S1	S0	QA*	QB*	QC*	QD*	QE*	QF*	QG*	QH*
Hold	0	0	QA	QB	QC	QD	QE	QF	QG	QH
Shift right	0	1	RIN	QA	QB	QC	QD	QE	QF	QG
Shift left	1	0	QB	QC	QD	QE	QF	QG	QH	LIN
Load	1	1	AQA	BQB	CQC	DQD	EQE	FQF	GQG	HQH

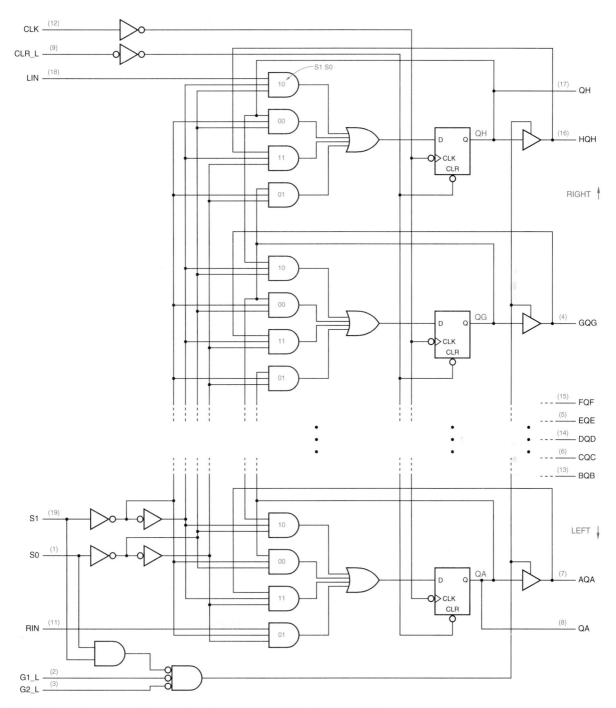

Figure 8-53 Logic diagram for the 74x299 8-bit universal shift register, including pin numbers for a standard 20-pin dual in-line package.

drivers are disabled and data is loaded through the AQA–HQH pins. At other times the stored bits are driven onto these same pins if G1_L and G2_L are asserted. The leftmost and rightmost stored bits are available at all times on separate output-only pins, QA and QH.

8.5.3 The World's Biggest Shift-Register Application

The most common application of shift registers is to convert parallel data into serial format for transmission or storage, and to convert serial data back to parallel format for processing or display (see Section 2.16.1). The most common example of serial data transmission, one that you almost certainly take part in every day, is in *digital telephony.*

digital telephony

CO

For years, TPCs (The Phone Companies) have been installing digital switching equipment in their central offices (*COs*). Most home phones have a two-wire analog connection to the central office. However, an analog-to-digital converter samples the analog voice signal 8,000 times per second (once every 125 μs) when it enters the CO and produces a corresponding sequence of 8,000 8-bit bytes representing the sign and amplitude of the analog signal at each sampling point. Subsequently, your voice is transmitted digitally on 64-Kbps *serial channels* throughout the phone network, until it is converted back to an analog signal by a digital-to-analog converter at the far-end CO.

serial channel

The 64-Kbps bandwidth required by a single digital voice signal is far *less* than can be obtained on a single digital signal line or switched by digital ICs. Therefore most digital telephone equipment *multiplexes* many 64-Kbps channels onto a single wire, saving both wires and digital ICs for switching. In the next subsection, we show how 32 channels can be processed by a handful of MSI chips; and these chips could be easily integrated into a single CPLD. This is a classic example of a *space/time tradeoff* in digital design—by running the chips faster, you can accomplish a larger task with fewer chips. Indeed, this is the main reason that the telephone network has "gone digital."

multiplex

space/time tradeoff

I STILL DON'T KNOW   ISDN (Integrated Services Digital Network) technology was developed in the late 1980s to extend full-duplex 144-Kbps serial digital channels to home phones. The idea was to carry two 64-Kbps voice conversations plus a 16-Kbps control channel on a single pair of wires, thus increasing the capacity of installed wiring.

In the first edition of this book, we noted that delays in ISDN deployment had led some people in the industry to rename it "Imaginary Services Delivered Nowhere." In the mid-1990s, ISDN finally took off in the U.S., but it was deployed not so much to carry voice as to provide "high-speed" connections to the Internet.

Unfortunately for TPCs, the growth in ISDN was cut short first by the deployment of inexpensive 56-Kbps analog modems and later by the growing availability of very high-speed connections (160 Kbps to 2 Mbps or higher) using newer DSL (Digital Subscriber Line) and cable-modem technologies.

8.5.4 Serial/Parallel Conversion

A typical application of serial data transfer between two modules (possibly in a piece of CO switching equipment) is shown in Figure 8-54. Three signals are normally connected between a source module and a destination module to accomplish the transfer:

*   *Clock.* The clock signal provides the timing reference for transfers, _defining the time to transfer one bit. In systems with just two modules, the clock may be part of control circuits located on the source module as shown. In larger systems, the clock may be generated at a common point and distributed to all of the modules.

*   *Serial data.* The data itself is transmitted on a single line.

*   *Synchronization.* A *synchronization pulse* (or *sync pulse*) provides a reference point for defining the data format, such as the beginning of a byte or word in the serial data stream. Some systems omit this signal and instead use a unique pattern on the serial data line for synchronization.

synchronization pulse
sync pulse

The general timing characteristics of these signals in a typical digital telephony application are shown in Figure 8-55(a). The CLOCK signal has a frequency of 2.048 MHz to allow the transmission of $32 \times 8,000$ 8-bit bytes per

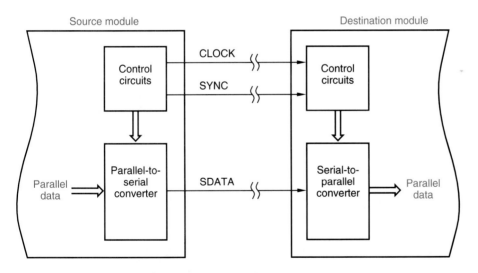

Figure 8-54
A system that transmits data serially between modules.

THE NATION'S CLOCK   Believe it or not, in the phone network, a very precise 8-KHz clock is generated in St. Louis and distributed throughout the U.S.! The clock signal that is distributed in a particular piece of local CO equipment is normally derived from the national clock. For example, the 2.048-MHz clock in this section's example could be derived by a phase-locked loop circuit that multiplies the national clock frequency by 256.

Figure 8-55 Timing diagram for parallel-to-serial conversion: (a) a complete frame; (b) one byte at the beginning of the frame.

frame
timeslot

second. The 1-bit-wide pulse on the SYNC signal identifies the beginning of a 125-μs interval called a *frame*. A total of 256 bits are transmitted on SDATA during this interval, which is divided into 32 *timeslots* containing eight bits each. Each timeslot carries one digitally encoded voice signal. Both timeslot numbers and bit positions within a timeslot are located relative to the SYNC pulse.

Figure 8-56 shows a circuit that converts parallel data to the serial format of Figure 8-55(a), with detailed timing shown in (b). Two 74x163 counters are wired as a free-running modulo-256 counter to define the frame. The five high-order and three low-order counter bits are the timeslot number and bit number, respectively.

WHICH BIT FIRST?   Most real serial links for digitized voice actually transmit bit 7 first, because this is the first bit generated by the analog-to-digital converter that digitizes the voice signal. However, to simplify our examples, we transmit bit 0 first so that counter state equals bit number.

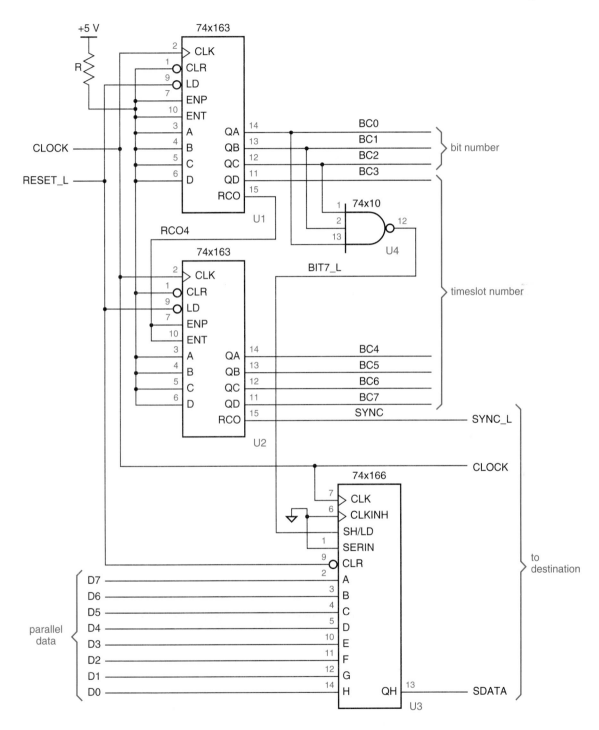

Figure 8-56 Parallel-to-serial conversion using a parallel-in shift register.

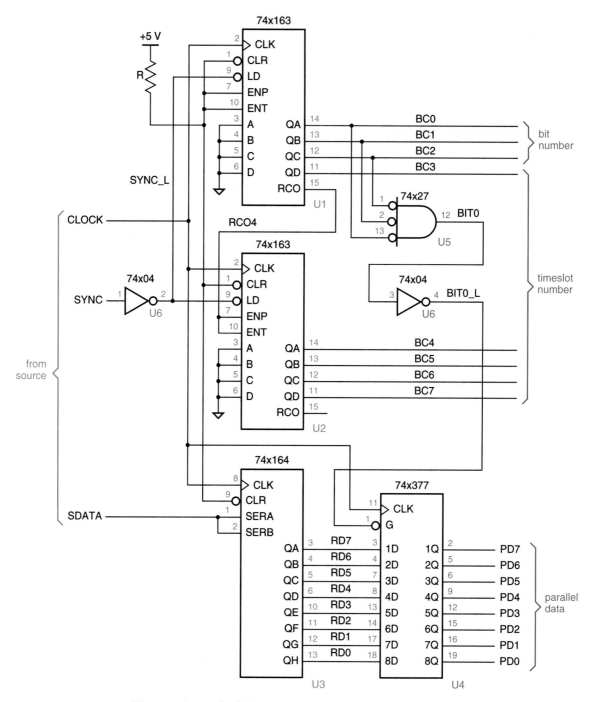

Figure 8-57 Serial-to-parallel conversion using a parallel-out shift register.

A 74x166 parallel-in shift register performs the parallel-to-serial conversion. Bit 0 of the parallel data (D0–D7) is connected to the '166 input closest to the SDATA output, so bits are transmitted serially in the order 0 through 7.

During bit 7 of each timeslot, the BIT7_L signal is asserted, which causes the '166 to be loaded with D0–D7. The value of D0–D7 is irrelevant except during the setup- and hold-time window around the clock edge on which the '166 is loaded, as shown by shading in the timing diagram. This leaves open the possibility that the parallel data bus could be used for other things at other times (see Exercise 8.54).

A destination module can convert the serial data back into parallel format using the circuit of Figure 8-57. A modulo-256 counter built from a pair of '163s is used to reconstruct the timeslot and bit numbers. Although SYNC is asserted during state 255 of the counter on the source module, SYNC loads the destination module's counter with 0 so that both counters go to 0 on the same clock edge. The counter's high-order bits (timeslot number) are not used in the figure, but they may be used by other circuits in the destination module to identify the byte from a particular timeslot on the parallel data bus (PD0–PD7).

Figure 8-58 shows detailed timing for the serial-to-parallel conversion circuit. A complete received byte is available at parallel output of the 74x164 shift register during the clock period following the reception of the last bit (7) of the byte. The parallel data in this example is *double-buffered*—once it is fully received, it is transferred into a 74x377 register, where it is available on PD0-PD7 for eight full clock periods until the next byte is fully received. The BIT0_L signal enables the '377 to be loaded at the proper time. Additional registers and decoding could be provided to load the byte from each timeslot into a different register, making each byte available for 125 μs (see Section 8.57).

double-buffered data

Figure 8-58 Timing diagram for serial-to-parallel conversion.

Once the received data is in parallel format, it can easily be stored or modified by other digital circuits; we'll give examples in Section 10.1.6. In digital telephony, the received parallel data is converted back into an analog voltage that is filtered and transmitted to an earpiece or speaker for 125 μs, until the next voice sample arrives.

8.5.5 Shift-Register Counters

Serial/parallel conversion is a "data" application, but shift registers have "non-data" applications as well. A shift register can be combined with combinational logic to form a state machine whose state diagram is cyclic. Such a circuit is called a *shift-register counter.* Unlike a binary counter, a shift-register counter does not count in an ascending or descending binary sequence, but it is useful in many "control" applications nonetheless.

shift-register counter

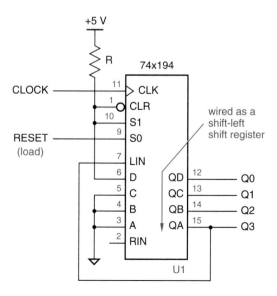

Figure 8-59
Simplest design for a 4-bit, 4-state ring counters with a single circulating 1.

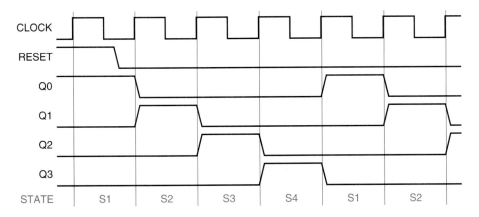

Figure 8-60
Timing diagram for a
4-bit ring counter.

8.5.6 Ring Counters

The simplest shift-register counter uses an n-bit shift register to obtain a counter
with n states, and is called a *ring counter*. Figure 8-59 is the logic diagram for a   *ring counter*
4-bit ring counter. The 74x194 universal shift register is wired so that it normally
performs a left shift. However, when RESET is asserted, it loads 0001 (refer to
the '194's function table, Table 8-18 on page 716). Once RESET is negated, the
'194 shifts left on each clock tick. The LIN serial input is connected to the
"leftmost" output, so the next states are 0010, 0100, 1000, 0001, 0010, ... Thus,
the counter visits four unique states before repeating. A timing diagram is shown
in Figure 8-60. In general, an n-bit ring counter visits n states in a cycle.

  The ring counter in Figure 8-59 has one major problem—it is not robust. If
its single 1 output is lost due to a temporary hardware problem (e.g., noise), the
counter goes to state 0000 and stays there forever. Likewise, if an extra 1 output
is set (i.e., state 0101 is created), the counter will go through an incorrect cycle
of states and stay in that cycle forever. These problems are quite evident if we
draw the *complete* state diagram for the counter circuit, which has 16 states. As
shown in Figure 8-61, there are 12 states that are not part of the normal counting
cycle. If the counter somehow gets off the normal cycle, it stays off it.

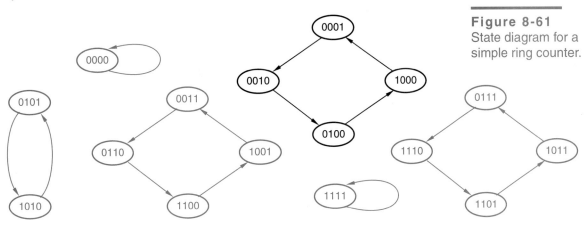

Figure 8-61
State diagram for a
simple ring counter.

Figure 8-62
Self-correcting
4-bit, 4-state ring
counter with a single
circulating 1.

self-correcting counter   A *self-correcting counter* is designed so that all abnormal states have transitions leading to normal states. Self-correcting counters are desirable for the same reason that we use a minimal-risk approach to state assignment in Section 7.4.3: If something unexpected happens, a counter or state machine should go to a "safe" state.

self-correcting ring   A *self-correcting ring counter* circuit is shown in Figure 8-62. The circuit
counter   uses a NOR gate to shift a 1 into LIN only when the three least significant bits are 0. This results in the state diagram in Figure 8-63; all abnormal states lead back

Figure 8-63
State diagram for a
self-correcting ring
counter.

Figure 8-64
Self-correcting
4-bit, 4-state ring
counter with a single
circulating 0.

into the normal cycle. Notice that, in this circuit, an explicit RESET signal is not necessarily required. Regardless of the initial state of the shift register on power-up, it reaches state 0001 within four clock ticks. Therefore, an explicit reset signal is required only if it is necessary to ensure that the counter starts up synchronously with other devices in the system or to provide a known starting point in simulation.

In the general case, an n-bit self-correcting ring counter uses an $(n-1)$-input NOR gate and corrects an abnormal state within $n-1$ clock ticks.

In CMOS and TTL logic families, wide NAND gates are generally easier to come by than NORs, so it may be more convenient to design a self-correcting ring counter as shown in Figure 8-64. States in this counter's normal cycle have a single circulating 0.

The major appeal of a ring counter for control applications is that its states appear in 1-out-of-n decoded form directly on the flip-flop outputs. That is, exactly one flip-flop output is asserted in each state. Furthermore, these outputs are "glitch free"; compare with the binary counter and decoder approach of Figure 8-42 on page 703.

*8.5.7 Johnson Counters

An n-bit shift register with the complement of the serial output fed back into the serial input is a counter with *2n* states and is called a *twisted-ring*, *Moebius*, or *Johnson counter*. Figure 8-65 is the basic circuit for a Johnson counter and Figure 8-66 is its timing diagram. The normal states of this counter are listed in

twisted-ring counter
Moebius counter
Johnson counter

Figure 8-65
Basic 8-bit, 8-state
Johnson counter.

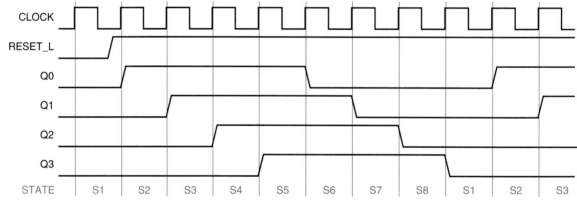

Figure 8-66 Timing diagram for a 4-bit Johnson counter.

Table 8-20
States of a 4-bit
Johnson counter.

State Name	Q3	Q2	Q1	Q0	Decoding
S1	0	0	0	0	Q3′ · Q0′
S2	0	0	0	1	Q1′ · Q0
S3	0	0	1	1	Q2′ · Q1
S4	0	1	1	1	Q3′ · Q2
S5	1	1	1	1	Q3 · Q0
S6	1	1	1	0	Q1 · Q0′
S7	1	1	0	0	Q2 · Q1′
S8	1	0	0	0	Q3 · Q2′

Figure 8-67
Self-correcting 4-bit,
8-state Johnson
counter.

Table 8-20. If both the true and complemented outputs of each flip-flop are available, each normal state of the counter can be decoded with a 2-input AND or NAND gate, as shown in the table. The decoded outputs are glitch free.

An n-bit Johnson counter has $2^n - 2n$ abnormal states and is therefore subject to the same robustness problems as a ring counter. A 4-bit *self-correcting Johnson counter* can be designed as shown in Figure 8-67. This circuit loads 0001 as the next state whenever the current state is 0xx0. A similar circuit using a single 2-input NOR gate can perform correction for a Johnson counter with any number of bits. The correction circuit must load 00...01 as the next state whenever the current state is 0x...x0.

self-correcting Johnson counter

| THE SELF-CORRECTION CIRCUIT IS ITSELF CORRECT! | We can prove that the Johnson-counter self-correction circuit corrects any abnormal state as follows. An abnormal state can always be written in the form x...x10x...x, since the only states that can't be written in this form are normal states (00...00, 11...11, 01...1, 0...01...1, and 0...01). Therefore, within $n-2$ clock ticks, the shift register will contain 10x...x. One tick later it will contain 0x...x0, and one tick after that the normal state 00...01 will be loaded. |

*maximum-length
sequence generator*

finite fields

*maximum-length
sequence*

*8.5.8 Linear Feedback Shift-Register Counters

The *n*-bit shift register counters that we've shown so far have far less than the maximum of 2^n normal states. An *n*-bit *linear feedback shift-register (LFSR) counter* can have $2^n - 1$ states, almost the maximum. Such a counter is often called a *maximum-length sequence generator.*

The design of LFSR counters is based on the theory of *finite fields,* which was developed by French mathematician Évariste Galois (1811–1832) shortly before he was killed in a duel with a political opponent. The operation of an LFSR counter corresponds to operations in a finite field with 2^n elements.

Figure 8-68 shows the structure of an *n*-bit LFSR counter. The shift register's serial input is connected to the sum modulo 2 of a certain set of output bits. These feedback connections determine the state sequence of the counter. By convention, outputs are always numbered and shifted in the direction shown.

Using finite-field theory, it can be shown that for any value of *n*, there exists at least one feedback equation that makes the counter go through all $2^n - 1$ nonzero states before repeating. This is called a *maximum-length sequence.*

Table 8-21 lists feedback equations that result in maximum-length sequences for selected values of *n*. For each value of *n* greater than 3, there are many other feedback equations that result in maximum-length sequences, all different.

An LFSR counter designed according to Figure 8-68 can never cycle through all 2^n possible states. Regardless of the connection pattern, the next state for the all-0s state is the same—all 0s.

Figure 8-68 General structure of a linear feedback shift-register counter.

n	Feedback Equation
2	X2 = X1 ⊕ X0
3	X3 = X1 ⊕ X0
4	X4 = X1 ⊕ X0
5	X5 = X2 ⊕ X0
6	X6 = X1 ⊕ X0
7	X7 = X3 ⊕ X0
8	X8 = X4 ⊕ X3 ⊕ X2 ⊕ X0
12	X12 = X6 ⊕ X4 ⊕ X1 ⊕ X0
16	X16 = X5 ⊕ X4 ⊕ X3 ⊕ X0
20	X20 = X3 ⊕ X0
24	X24 = X7 ⊕ X2 ⊕ X1 ⊕ X0
28	X28 = X3 ⊕ X0
32	X32 = X22 ⊕ X2 ⊕ X1 ⊕ X0

Table 8-21
Feedback equations
for linear feedback
shift-register counters.

WORKING IN THE FIELD

A finite field has a finite number of elements and two operators, addition and multiplication, that satisfy certain properties. An example of a finite field with P elements, where P is any prime, is the set of integers modulo P. The operators in this field are addition and multiplication modulo P.

According to finite-field theory, if you start with a nonzero element E and repeatedly multiply by a "primitive" element α, after $P-2$ steps you will generate the rest of the field's nonzero elements in the field before getting back to E. It turns out that in a field with P elements, any integer in the range $2, \ldots, P-1$ is primitive. You can try this yourself using $P = 7$ and $\alpha = 2$, for example. The elements of the field are $0, 1, \ldots, 6$, and the operations are addition and subtraction modulo 7.

The paragraph above gives the basic idea behind maximum-length sequence generators. However, to apply them to a digital circuit, you need a field with 2^n elements, where n is the number of bits required by the application. On one hand, we're in luck, because Galois proved that there exist finite fields with P^n elements for any integer n, as long as P is prime, including $P = 2$. On the other hand, we're out of luck, because when $n > 1$, the operators in fields with P^n (including 2^n) elements are quite different from ordinary integer addition and multiplication. Also, primitive elements are harder to find.

If you enjoy math, as I do, you'd probably be fascinated by the finite-field theory that leads to the LFSR circuits for maximum-length sequence generators and other applications; see the References. Otherwise, you can confidently follow the "cookbook" approach in this section.

Figure 8-69 A 3-bit LFSR counter; modifications to include the all-0s state are shown in color.

The logic diagram for a 3-bit LFSR counter is shown in Figure 8-69. The state sequence for this counter is shown in the first three columns of Table 8-22. Starting in any nonzero state, 100 in the table, the counter visits seven states before returning to the starting state.

An LFSR counter can be modified to have 2^n states, including the all-0s state, as shown in color for the 3-bit counter in Figure 8-69. The resulting state sequence is given in the last three columns of Table 8-22. In an n-bit LFSR

Table 8-22
State sequences for the 3-bit LFSR counter in Figure 8-69.

Original Sequence			Modified Sequence		
X2	X1	X0	X2	X1	X0
1	0	0	1	0	0
0	1	0	0	1	0
1	0	1	1	0	1
1	1	0	1	1	0
1	1	1	1	1	1
0	1	1	0	1	1
0	0	1	0	0	1
1	0	0	0	0	0
.	.	.	1	0	0
.	.	.	.	.	.

counter, an extra XOR gate and an $n - 1$ input NOR gate connected to all shift-register outputs except X0 accomplishes the same thing.

The states of an LFSR counter are not visited in binary counting order. However, LFSR counters are typically used in applications where this characteristic is an advantage. A major application of LFSR counters is in generating test inputs for logic circuits. In most cases, the "pseudorandom" counting sequence of an LFSR counter is more likely than a binary counting sequence to detect errors. LFSRs are also used in the encoding and decoding circuits for certain error-detecting and error-correcting codes, including CRC codes, which we introduced in Section 2.15.4.

In data communications, LFSR counters are often used to "scramble" and "descramble" the data patterns transmitted by high-speed modems and network interfaces, including 100-Mbps Ethernet. This is done by XORing the LFSR's output with the user data stream. Even when the user data stream contains a long run of 0s or 1s, combining it with the LFSR's pseudorandom output improves the DC balance of the transmitted signal and creates a rich set of transitions that allows clocking information to be recovered more easily at the receiver.

8.5.9 Shift Registers in ABEL and PLDs

General-purpose shift registers can be specified quite easily in ABEL and fit nicely into typical sequential PLDs. For example, Figure 8-70 and Table 8-23 show how to realize a function similar to that of a 74x194 universal shift register using a 16V8. Notice that one of the I/O pins of the 16V8, pin 12, is used as an input.

The 16V8 realization of the '194 differs from the real thing in just one way—in the function of the CLR_L input. In the real '194, CLR_L is an asynchronous input, while in the 16V8 it is sampled along with other inputs at the rising edge of CLK.

Figure 8-70
PLD realizations of a 74x194-like universal shift register with synchronous clear.

Table 8-23 ABEL program for a 4-bit universal shift register.

```
module Z74x194
title '4-bit Universal Shift Register'
Z74X194 device 'P16V8R';

" Input and output pins
CLK, RIN, A, B, C, D, LIN                pin 1, 2, 3, 4, 5, 6, 7;
S1, S0, CLR_L                            pin 8, 9, 12;
QA, QB, QC, QD                           pin 19, 18, 17, 16 istype 'reg';

" Active-level translation
CLR = !CLR_L;

" Set definitions
INPUT   = [ A,  B,  C, D ];
LEFTIN  = [ QB, QC, QD, LIN];
RIGHTIN = [RIN, QA, QB, QC ];
OUT     = [ QA, QB, QC, QD ];

CTRL  = [S1,S0];
HOLD  = (CTRL == [0,0]);
RIGHT = (CTRL == [0,1]);
LEFT  = (CTRL == [1,0]);
LOAD  = (CTRL == [1,1]);

equations
OUT.CLK = CLK;

OUT := !CLR & (
          HOLD & OUT
        # RIGHT & RIGHTIN
        # LEFT & LEFTIN
        # LOAD & INPUT);

end Z74x194
```

If you really need to provide an asynchronous clear input, you can use the 22V10, which provides a single product line to control the reset inputs of all of its flip-flops. This requires only a few changes in the original program (see Exercise 8.69).

The flexibility of ABEL can be used to create shift-register circuits with more or different functionality. For example, Table 8-24 defines an 8-bit shift register that can be cleared, loaded with a single 1 in any bit position, shifted left, shifted right, or held. The operation to be performed at each clock tick is specified by a 4-bit operation code, OP[3:0]. Despite the large number of "WHEN" cases, the circuit can be synthesized with only five product terms per output.

Table 8-24 ABEL program for a multi-function shift register.

```
module shifty
title '8-bit shift register with decoded load'

" Inputs and Outputs
CLK, OP3..OP0              pin;
Q7..Q0                     pin istype 'reg';

" Definitions

Q = [Q7..Q0];
OP = [OP3..OP0];

HOLD = (OP == 0);
CLEAR = (OP == 1);
LEFT = (OP == 2);
RIGHT = (OP == 3);
NOP = (OP >= 4) & (OP < 8);
LOADQ0 = (OP == 8);
LOADQ1 = (OP == 9);
LOADQ2 = (OP == 10);
LOADQ3 = (OP == 11);
LOADQ4 = (OP == 12);
LOADQ5 = (OP == 13);
LOADQ6 = (OP == 14);
LOADQ7 = (OP == 15);

Equations

Q.CLK = CLK;

WHEN HOLD THEN Q := Q;
ELSE WHEN CLEAR THEN Q := 0;
ELSE WHEN LEFT THEN Q := [Q6..Q0, Q7];
ELSE WHEN RIGHT THEN Q := [Q0, Q7..Q1];
ELSE WHEN LOADQ0 THEN Q := 1;
ELSE WHEN LOADQ1 THEN Q := 2;
ELSE WHEN LOADQ2 THEN Q := 4;
ELSE WHEN LOADQ3 THEN Q := 8;
ELSE WHEN LOADQ4 THEN Q := 16;
ELSE WHEN LOADQ5 THEN Q := 32;
ELSE WHEN LOADQ6 THEN Q := 64;
ELSE WHEN LOADQ7 THEN Q := 128;
ELSE Q := Q;

end shifty
```

Table 8-25 Program for an 8-bit ring counter.

```
module Ring8
title '8-bit Ring Counter'

" Inputs and Outputs
MCLK, CNTEN, RESTART                    pin;
S0..S7                                  pin istype 'reg';

equations

[S0..S7].CLK = MCLK;

S0 := CNTEN & !S0 & !S1 & !S2 & !S3 & !S4 & !S5 & !S6  " Self-sync
    # !CNTEN & S0                                      " Hold
    # RESTART;                                         " Start with one 1
[S1..S7] := !RESTART & ( !CNTEN & [S1..S7]      " Shift
                      # CNTEN & [S0..S6] );  " Hold
end Ring8
```

ABEL can be used readily to specify shift register counters of the various types that we introduced in previous subsections. For example, Table 8-25 is the program for an 8-bit ring counter. We've used the extra capability of the PLD to add two functions not present in our previous MSI designs: counting occurs only if CNTEN is asserted, and the next state is forced to S0 if RESTART is asserted.

Ring counters are often used to generate multiphase clocks or enable signals in digital systems, and the requirements in different systems are many and varied. The ability to reprogram the counter's behavior easily is a distinct advantage of an HDL-based design.

Figure 8-71 shows a set of clock or enable signals that might be required in a digital system with six distinct phases of operation. Each phase lasts for two ticks of a master clock signal, MCLK, during which the corresponding active-low phase-enable signal Pi_L is asserted. We can obtain this sort of timing from a ring counter if we provide an extra flip-flop to count the two ticks of each phase, so that a shift occurs on the *second* tick of each phase.

The timing generator can be built with a few inputs and outputs of a PLD. Three control inputs are provided, with the following behavior:

RESET   When this input is asserted, no outputs are asserted. The counter always goes to the first tick of phase 1 after RESET is negated.

RUN   When asserted, this input allows the counter to advance to the second tick of the current phase, or to the first tick of the next phase; otherwise, the current tick of the current phase is extended.

RESTART   Asserting this input causes the counter to go back to the first tick of phase 1, even if RUN is not asserted.

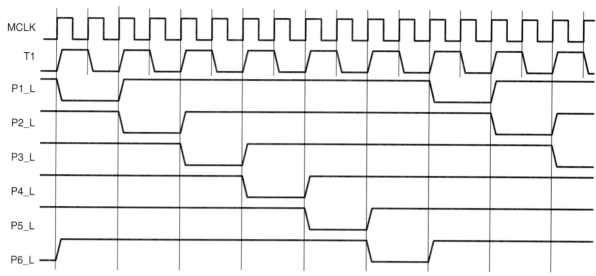

Figure 8-71 Six-phase timing waveforms required in a certain digital system.

■ **Table 8-26** Program for a six-phase waveform generator.

```
module␣TIMEGEN6␣
title␣'Six-phase␣Master␣Timing␣Generator'

"␣Input␣and␣Output␣pins␣
MCLK,␣RESET,␣RUN,␣RESTART␣␣␣␣␣␣␣␣␣␣␣␣␣␣␣␣␣␣␣␣␣pin;
T1,␣P1_L,␣P2_L,␣P3_L,␣P4_L,␣P5_L,␣P6_L␣␣␣␣␣pin␣istype␣'reg';␣

"␣State␣definitions
PHASES␣=␣[P1_L,␣P2_L,␣P3_L,␣P4_L,␣P5_L,␣P6_L];
NEXTPH␣=␣[P6_L,␣P1_L,␣P2_L,␣P3_L,␣P4_L,␣P5_L];
SRESET␣=␣[1,␣1,␣1,␣1,␣1,␣1];
P1␣=␣␣␣␣␣[0,␣1,␣1,␣1,␣1,␣1];

equations
T1.CLK␣=␣MCLK;␣PHASES.CLK␣=␣MCLK;

WHEN␣RESET␣THEN␣{T1␣:=␣1;␣PHASES␣:=␣SRESET;}
ELSE␣WHEN␣(PHASES==SRESET)␣#␣RESTART␣THEN␣{T1␣:=␣1;␣PHASES␣:=␣P1;}
ELSE␣WHEN␣RUN␣&␣T1␣THEN␣{T1␣:=␣0;␣PHASES␣:=␣PHASES;}
ELSE␣WHEN␣RUN␣&␣!T1␣THEN␣{T1␣:=␣1;␣PHASES␣:=␣NEXTPH;}
ELSE␣{T1␣:=␣T1;␣PHASES␣:=␣PHASES;}

end␣TIMEGEN6
```

Table 8-26 is a program that creates the required behavior. Notice the use of sets to specify the ring counter's behavior very concisely, with the RESET, RESTART, and RUN having the specified behavior in any counter phase.

Table 8-27 Alternate program for the waveform generator.

```
module TIMEGN6A
title 'Six-phase Master Timing Generator'

" Input and Output pins
MCLK, RESET, RUN, RESTART                    pin;
T1, P1_L, P2_L, P3_L, P4_L, P5_L, P6_L      pin istype 'reg';

" State definitions
TSTATE = [T1, P1_L, P2_L, P3_L, P4_L, P5_L, P6_L];
SRESET = [1, 1, 1, 1, 1, 1, 1];
P1F =    [1, 0, 1, 1, 1, 1, 1];
P1S =    [0, 0, 1, 1, 1, 1, 1];
P2F =    [1, 1, 0, 1, 1, 1, 1];
P2S =    [0, 1, 0, 1, 1, 1, 1];
P3F =    [1, 1, 1, 0, 1, 1, 1];
P3S =    [0, 1, 1, 0, 1, 1, 1];
P4F =    [1, 1, 1, 1, 0, 1, 1];
P4S =    [0, 1, 1, 1, 0, 1, 1];
P5F =    [1, 1, 1, 1, 1, 0, 1];
P5S =    [0, 1, 1, 1, 1, 0, 1];
P6F =    [1, 1, 1, 1, 1, 1, 0];
P6S =    [0, 1, 1, 1, 1, 1, 0];

equations
TSTATE.CLK = MCLK;
WHEN RESET THEN TSTATE := SRESET;

state_diagram TSTATE

state SRESET: IF RESET THEN SRESET ELSE P1F;

state P1F: IF RESET THEN SRESET ELSE IF RESTART THEN P1F
          ELSE IF RUN THEN P1S ELSE P1F;

state P1S: IF RESET THEN SRESET ELSE IF RESTART THEN P1F
          ELSE IF RUN THEN P2F ELSE P1S;

state P2F: IF RESET THEN SRESET ELSE IF RESTART THEN P1F
          ELSE IF RUN THEN P2S ELSE P2F;

state P2S: IF RESET THEN SRESET ELSE IF RESTART THEN P1F
          ELSE IF RUN THEN P3F ELSE P2S;

state P3F: IF RESET THEN SRESET ELSE IF RESTART THEN P1F
          ELSE IF RUN THEN P3S ELSE P3F;

state P3S: IF RESET THEN SRESET ELSE IF RESTART THEN P1F
          ELSE IF RUN THEN P4F ELSE P3S;

state P4F: IF RESET THEN SRESET ELSE IF RESTART THEN P1F
          ELSE IF RUN THEN P4S ELSE P4F;

state P4S: IF RESET THEN SRESET ELSE IF RESTART THEN P1F
          ELSE IF RUN THEN P5F ELSE P4S;
```

Table 8-27 (continued) Alternate program for the waveform generator.

```
state P5F: IF RESET THEN SRESET ELSE IF RESTART THEN P1F
           ELSE IF RUN THEN P5S ELSE P5F;

state P5S: IF RESET THEN SRESET ELSE IF RESTART THEN P1F
           ELSE IF RUN THEN P6F ELSE P5S;

state P6F: IF RESET THEN SRESET ELSE IF RESTART THEN P1F
           ELSE IF RUN THEN P6S ELSE P6F;

state P6S: IF RESET THEN SRESET ELSE IF RESTART THEN P1F
           ELSE IF RUN THEN P1F ELSE P6S;


end TIMEGN6A
```

The same timing-generator behavior in Figure 8-71 can be specified using a state-machine design approach, as shown in Table 8-27. This ABEL program, though longer, generates the same external behavior during normal operation as the previous one, and from a certain point of view it may be easier to understand. However, its realization requires 8 to 20 AND terms per output, compared to only 3 to 5 per output for the original ring-counter version. This is a good example of how you can improve circuit efficiency and performance by adapting a standard, simple structure to a "custom" design problem, rather than grinding out a brute-force state machine.

RELIABLE RESET   Notice in Table 8-27 that TSTATE is assigned a value in the equations section of the program, as well as being used in the state_diagram section. This was done for a very specific purpose, to ensure that the program goes to the SRESET state from any undefined state, as explained below.

ABEL augments the on-set for an output each time the output appears on the lefthand side of an equation, as we explained for combinational outputs on page 254. In the case of registered outputs, ABEL also augments the on-set of each state variable in the state vector for each "state" definition in a state_diagram. For each state-variable output, all of the input combinations that cause that output to be 1 in each state are added to the output's on-set.

The state machine in Table 8-27 has 2^7 or 128 states in total, of which only 13 are explicitly defined and have a transition into SRESET. Nevertheless, the WHEN equation ensures that anytime that RESET is asserted, the machine goes to the SRESET state. This is true regardless of the state definitions in the state_diagram. When RESET is asserted, the all-1s state encoding of SRESET is, in effect, ORed with the next state, if any, specified by the state_diagram. This approach to reliable reset would not be possible if SRESET were encoded as all 0s, for example.

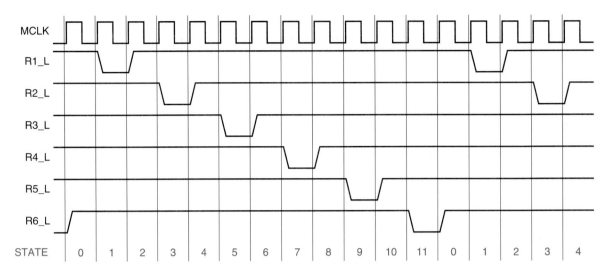

Figure 8-72 Modified timing waveforms for a digital system.

Now let's look at a variation of the previous timing waveforms that might be required in a different system. Figure 8-72 is similar to the previous waveforms, except that each phase output Ri is asserted for only one clock tick per phase. This change has a subtle but important effect on the design approach.

In the original design, we used a 6-bit ring counter and one auxiliary state bit T1 to keep track of the two states within each phase. With the new waveforms this is not possible. In the states between active-low pulses (STATE = 0, 2, 4, etc. in Figure 8-72) the phase outputs are all negated, so they can no longer be used to figure out which state should be visited next. Something else is needed to keep track of the state.

There are many different ways to solve this problem. One idea is to start with the original design in Table 8-26 but use the phase outputs P1_L, P2_L, and

Table 8-28 Additions to Table 8-26 for a modified six-phase waveform generator.

```
module TIMEG12K
...
R1_L, R2_L, R3_L, R4_L, R5_L, R6_L            pin istype 'com';
...
OUTPUTS = [R1_L, R2_L, R3_L, R4_L, R5_L, R6_L];

equations
...
!OUTPUTS = !PHASES & !T1;

end TIMEG12K
```

so on as internal states only. Then, each phase output Ri_L can be defined as a Moore-type combinational output that is asserted when the corresponding Pi_L is asserted *and* we are in the second tick of a phase. The additional ABEL code to support this first approach is shown in Table 8-28.

This first approach is easy, and it works just fine if the Pi_L signals are going to be used only as enables or other control inputs. However, it's a bad idea if these signals are going to be used as clocks, because they may have glitches, as we'll now explain. The Pi_L and T1 signals are all outputs from flip-flops clocked by the same master clock MCLK. Although these signals change at approximately the same time, their timing is never quite exact. One output may change sooner than another; this is called *output timing skew*. For example, suppose that on the transition from state 1 to 2 in Figure 8-71, P2_L goes LOW before T1 goes HIGH. In this case, a short glitch could appear on the R2_L output.

output timing skew

To get glitch-free outputs, we should design the circuit so that each phase output is the a registered output. One way to do this is to build a 12-bit ring counter and use only alternate outputs to yield the desired waveforms; an ABEL program using this approach is shown in Table 8-29.

Table 8-29 ABEL program for a modified six-phase waveform generator.

```
module TIMEG12
title 'Modified six-phase Master Timing Generator'

" Input and Output pins
MCLK, RESET, RUN, RESTART                pin;
P1_L, P2_L, P3_L, P4_L, P5_L, P6_L       pin istype 'reg';
P1A, P2A, P3A, P4A, P5A, P6A             pin istype 'reg';

" State definitions
PHASES = [P1A, P1_L, P2A, P2_L, P3A, P3_L, P4A, P4_L, P5A, P5_L, P6A, P6_L];
NEXTPH = [P6_L, P1A, P1_L, P2A, P2_L, P3A, P3_L, P4A, P4_L, P5A, P5_L, P6A];
SRESET = [1, 1, 1, 1, 1, 1, 1, 1, 1, 1, 1, 1];
P1 =    [0, 1, 1, 1, 1, 1, 1, 1, 1, 1, 1, 1];

equations

PHASES.CLK = MCLK;

WHEN RESET THEN PHASES := SRESET;
ELSE WHEN RESTART # (PHASES == SRESET) THEN PHASES := P1;
ELSE WHEN RUN THEN PHASES := NEXTPH;
ELSE PHASES := PHASES;

end TIMEG12
```

███ **Table 8-30** Counter-based program for six-phase waveform generator.

```
module TIMEG12A
title 'Counter-based six-phase master timing generator'

" Input and Output pins
MCLK, RESET, RUN, RESTART                pin;
P1_L, P2_L, P3_L, P4_L, P5_L, P6_L       pin istype 'reg';
CNT3..CNT0                               pin istype 'reg';

" Definitions
CNT = [CNT3..CNT0];
P_L = [P1_L, P2_L, P3_L, P4_L, P5_L, P6_L];

equations

CNT.CLK = MCLK;   P_L.CLK = MCLK;

WHEN RESET THEN CNT := 15
ELSE WHEN RESTART THEN CNT := 0
ELSE WHEN (RUN & (CNT < 11)) THEN CNT := CNT + 1
ELSE WHEN RUN THEN CNT := 0
ELSE CNT := CNT;

P1_L := !(CNT == 0);
P2_L := !(CNT == 2);
P3_L := !(CNT == 4);
P4_L := !(CNT == 6);
P5_L := !(CNT == 8);
P6_L := !(CNT == 10);

end TIMEG12A
```

Still another approach is to recognize that, since the waveforms cycle through 12 states, we can build a modulo-12 binary counter and decode the states of that counter. An ABEL program using this approach is shown in Table 8-30. The states of the counter correspond to the "STATE" values shown in Figure 8-72. Since the phase outputs are registered, they are glitch free. Note that they are decoded one cycle early, to account for the one-tick decoding delay. Also, during reset, the counter is forced to state 15 rather than 0, so that the P1_L output is not asserted during reset.

8.5.10 Shift Registers in VHDL

Shift registers can be specified structurally or behaviorally in VHDL; we'll look at a few behavioral descriptions and applications. Table 8-31 is the function table for an 8-bit shift register with an extended set of functions. In addition to the hold, load, and shift functions of the 74x194 and 74x299, it performs circular

Table 8-31 Function table for an extended-function 8-bit shift register.

Function	Inputs			Next state							
	S2	S1	S0	Q7*	Q6*	Q5*	Q4*	Q3*	Q2*	Q1*	Q0*
Hold	0	0	0	Q7	Q6	Q5	Q4	Q3	Q2	Q1	Q0
Load	0	0	1	D7	D6	D5	D4	D3	D2	D1	D0
Shift right	0	1	0	RIN	Q7	Q6	Q5	Q4	Q3	Q2	Q1
Shift left	0	1	1	Q6	Q5	Q4	Q3	Q2	Q1	Q0	LIN
Shift circular right	1	0	0	Q0	Q7	Q6	Q5	Q4	Q3	Q2	Q1
Shift circular left	1	0	1	Q6	Q5	Q4	Q3	Q2	Q1	Q0	Q7
Shift arithmetic right	1	1	0	Q7	Q7	Q6	Q5	Q4	Q3	Q2	Q1
Shift arithmetic left	1	1	1	Q6	Q5	Q4	Q3	Q2	Q1	Q0	0

and arithmetic shift operations. In the *circular shift* operations, the bit that "falls off" one end during a shift is fed back into the other end. In the *arithmetic shift* operations, the edge input is set up for multiplication or division by 2; for a left shift, the right input is 0, and for a right shift, the leftmost (sign) bit is replicated.

circular shift
arithmetic shift

A behavioral VHDL program for the extended-function shift register is shown in Table 8-32. As in previous examples, we define a process and use the `event` attribute on the CLK signal to obtain the desired edge-triggered behavior. Several other features of this program are worth noting:

- An internal signal, IQ, is used for what eventually becomes the Q output, so it can be both read and written by process statements. Alternatively, we could have defined the Q output as type "buffer."

- The CLR input is asynchronous; because it's in the process sensitivity list, it is tested whenever it changes. And the IF statement is structured so that CLR takes precedence over any other condition.

- A CASE statement is used to define the operation of the shift register for the eight possible values of the select inputs S(2 downto 0).

- In the CASE statement, the "when others" case is required to prevent the compiler from complaining about approximately 2^{32} uncovered cases!

- The "null" statement indicates that no action is taken in certain cases. In case 1, note that no action is required; the default is for a signal to hold its value unless otherwise stated.

- In most of the cases the concatenation operator "&" is used to construct an 8-bit array from a 7-bit subset of IQ and one other bit.

Table 8-32 VHDL program for an extended-function 8-bit shift register.

```vhdl
library IEEE;
use IEEE.std_logic_1164.all;

entity Vshftreg is
    port (
        CLK, CLR, RIN, LIN: in STD_LOGIC;
        S: in STD_LOGIC_VECTOR (2 downto 0); -- function select
        D: in STD_LOGIC_VECTOR (7 downto 0); -- data in
        Q: out STD_LOGIC_VECTOR (7 downto 0) -- data out
    );
end Vshftreg;

architecture Vshftreg_arch of Vshftreg is
signal IQ: STD_LOGIC_VECTOR (7 downto 0);
begin
process (CLK, CLR, IQ)
  begin
    if (CLR='1') then IQ <= (others=>'0'); -- Asynchronous clear
    elsif (CLK'event and CLK='1') then
      case CONV_INTEGER(S) is
        when 0 => null;                           -- Hold
        when 1 => IQ <= D;                        -- Load
        when 2 => IQ <= RIN & IQ(7 downto 1);    -- Shift right
        when 3 => IQ <= IQ(6 downto 0) & LIN;    -- Shift left
        when 4 => IQ <= IQ(0) & IQ(7 downto 1); -- Shift circular right
        when 5 => IQ <= IQ(6 downto 0) & IQ(7); -- Shift circular left
        when 6 => IQ <= IQ(7) & IQ(7 downto 1); -- Shift arithmetic right
        when 7 => IQ <= IQ(6 downto 0) & '0';    -- Shift arithmetic left
        when others => null;
      end case;
    end if;
    Q <= IQ;
  end process;
end Vshftreg_arch;
```

- Because of VHDL's strong requirements for type matching, we used the CONV_INTEGER function that we defined in Section 4.7.4 to convert the STD_LOGIC_VECTOR select input S to an integer in the CASE statement. Alternatively, we could have written each case label as a STD_LOGIC_VECTOR (e.g., ('0','1','1') instead of integer 3).

One application of shift registers is in ring counters, as in the six-phase waveform generator that we described on page 736 with the waveforms in Figure 8-71. A VHDL program that provides the corresponding behavior is

Table 8-33 VHDL program for a six-phase waveform generator.

```
library IEEE;
use IEEE.std_logic_1164.all;

entity Vtimegn6 is
    port (
        MCLK, RESET, RUN, RESTART: in STD_LOGIC; -- clock, control inputs
        P_L: out STD_LOGIC_VECTOR (1 to 6)        -- active-low phase outputs
    );
end Vtimegn6;

architecture Vtimegn6_arch of Vtimegn6 is
signal IP: STD_LOGIC_VECTOR (1 to 6); -- internal active-high phase signals
signal T1: STD_LOGIC;                         -- first tick within phase
begin
process (MCLK, IP)
  begin
    if (MCLK'event and MCLK='1') then
      if (RESET='1') then
        T1 <= '1'; IP <= ('0','0','0','0','0','0');
      elsif ((IP=('0','0','0','0','0','0')) or (RESTART='1')) then
        T1 <= '1'; IP <= ('1','0','0','0','0','0');
      elsif (RUN='1') then
        T1 <= not T1;
        if (T1='0') then IP <= IP(6) & IP(1 to 5); end if;
      end if;
    end if;
    P_L <= not IP;
  end process;
end Vtimegn6_arch;
```

shown in Table 8-33. As in the previous VHDL example, an internal active-high signal vector, IP, is used for reading and writing what eventually becomes the circuit's output; this internal signal is conveniently inverted in the last statement to obtain the required active-low output signal vector. The rest of the program is straightforward, but notice that it has three levels of nested IF statements.

A possible modification to the preceding application is to produce output waveforms that are asserted only during the second tick of each two-tick phase; such waveforms were shown in Figure 8-72 on page 740. One way to do this is to create a 12-bit ring counter and use only alternate outputs. In the VHDL realization only the six phase outputs, P_L(1 to 6), would appear in the entity definition. The additional six signals, which we name NEXTP(1 to 6), are local to the architecture definition. Figure 8-73 shows the relationship of these signals for shift-register operation, and Table 8-34 is the VHDL program.

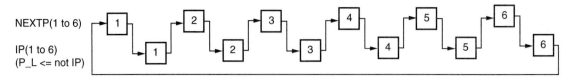

Figure 8-73 Shifting sequence for waveform generator 12-bit ring counter.

As in the previous program, a 6-bit active-high signal, IP, is declared in the architecture body and used for reading and writing what eventually becomes the circuit's active-low output, P_L. The additional 6-bit signal, NEXTP, holds the remaining 6 bits of state. Constants IDLE and FIRSTP are used to improve the program's readability.

Table 8-34 VHDL program for a modified six-phase waveform generator.

```
library IEEE;
use IEEE.std_logic_1164.all;

entity Vtimeg12 is
    port (
        MCLK, RESET, RUN, RESTART: in STD_LOGIC; -- clock, control inputs
        P_L: out STD_LOGIC_VECTOR (1 to 6)        -- active-low phase outputs
    );
end Vtimeg12;

architecture Vtimeg12_arch of Vtimeg12 is
signal IP, NEXTP: STD_LOGIC_VECTOR (1 to 6); -- internal active-high phase signals
begin
process (MCLK, IP, NEXTP)
  variable TEMP: STD_LOGIC_VECTOR (1 to 6);  -- temporary for signal shift
  constant IDLE: STD_LOGIC_VECTOR (1 to 6) := ('0','0','0','0','0','0');
  constant FIRSTP: STD_LOGIC_VECTOR (1 to 6) := ('1','0','0','0','0','0');
  begin
    if (MCLK'event and MCLK='1') then
      if (RESET='1') then IP <= IDLE;  NEXTP <= IDLE;
      elsif (RESTART='1') or (IP=IDLE and NEXTP=IDLE) then IP <= IDLE;  NEXTP <= FIRSTP;
      elsif (RUN='1') then
        if (IP=IDLE) and (NEXTP=IDLE) then NEXTP <= FIRSTP;
        else TEMP := IP;  IP <= NEXTP;  NEXTP <= TEMP(6) & TEMP(1 to 5);
        end if;
      end if;
    end if;
    P_L <= not IP;
  end process;
end Vtimeg12_arch;
```

Notice that a 6-bit variable, TEMP, is used just as a temporary place to hold the old value of IP when shifting occurs—IP is loaded with NEXTP, and NEXTP is loaded with the shifted, old value of IP. Because the assignment statements in a process are executed *sequentially*, we couldn't get away with just writing "IP <= NEXTP; NEXTP <= IP(6) & IP(1 to 5);". If we did that, then NEXTP would pick up the *new* value of IP, not the old. Notice also that since TEMP is a local variable, not a signal, its value is not visible outside the process. Also, a process invocation never uses the value of TEMP from a previous invocation. Therefore, the VHDL compiler need not synthesize any flip-flops to hold TEMP's value.

*8.6 Iterative versus Sequential Circuits

We introduced iterative circuits in Section 5.9.2. The function of an *n*-module iterative circuit can be performed by a sequential circuit that uses just one copy of the module but requires *n* steps (clock ticks) to obtain the result. This is an excellent example of a space/time trade-off in digital design.

As shown in Figure 8-74, flip-flops are used in the sequential-circuit version to store the cascading outputs at the end of each step; the flip-flop outputs are used as the cascading inputs at the beginning of the next step. The flip-flops must be initialized to the boundary-input values before the first clock tick, and they contain the boundary-output values after the *n*th tick.

Since an iterative circuit is a combinational circuit, all of its primary and boundary inputs may be applied simultaneously, and its primary and boundary outputs are all available after a combinational delay. In the sequential-circuit version, the primary inputs must be delivered sequentially, one per clock tick, and the primary outputs are produced with similar timing. Therefore, serial-out shift registers are often used to provide the inputs, and serial-in shift registers are used to collect the outputs. For this reason, the sequential-circuit version of an "iterative widget" is often called a "serial widget."

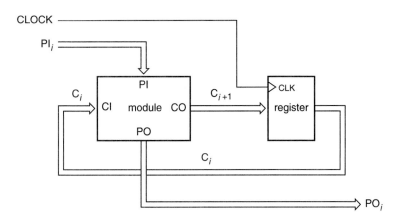

Figure 8-74
General structure of the sequential-circuit version of an iterative circuit.

Figure 8-75
Simplified serial
comparator circuit.

serial comparator   For example, Figure 8-75 shows the basic design for a *serial comparator* circuit. The shaded block is identical to the module used in the iterative comparator of Figure 5-80 on page 423. The circuit is drawn in more detail using SSI chips in Figure 8-76. In addition, we have provided a synchronous reset input that, when asserted, forces the initial value of the cascading flip-flop to 1 at the next clock tick. The initial value of the cascading flip-flop corresponds to the boundary input in the iterative comparator.

With the serial comparator, an n-bit comparison requires $n + 1$ clock ticks. RESET_L is asserted at the first clock tick. RESET_L is negated and data bits are applied at the next n ticks. The EQI output gives the comparison result during the clock period following the last tick. A timing diagram for two successive 4-bit comparisons is shown in Figure 8-77. The spikes in the EQO waveform indicate the time when the combinational outputs are settling in response to new X and Y input values.

Figure 8-76
Detailed serial
comparator circuit.

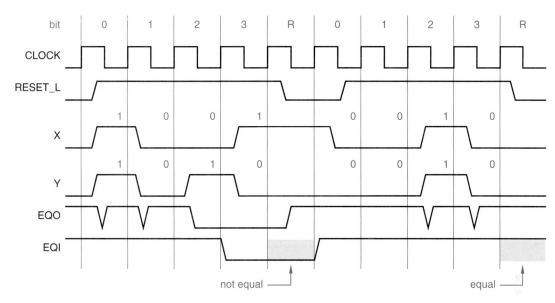

Figure 8-77 Timing diagram for serial comparator circuit.

A *serial binary adder* circuit for addends of any length can be constructed   *serial binary adder*
from a full adder and a D flip-flop, as shown in Figure 8-78. The flip-flop, which
stores the carry between successive bits of the addition, is cleared to 0 at reset.
Addend bits are presented serially on the A and B inputs, starting with the LSB,
and sum bits appear on S in the same order.

Because of the large size and high cost of digital logic circuits in the early
days, many computers and calculators used serial adders and other serial
versions of iterative circuits to perform arithmetic operations. Even though these
arithmetic circuits aren't used much today, they are an instructive reminder of
the space/time trade-offs that are possible in digital design.

Figure 8-78
Serial binary adder
circuit.

8.7 Synchronous Design Methodology

In a *synchronous system*, all flip-flops are clocked by the same common clock signal, and preset and clear inputs are not used, except for system initialization. Although all the world does not march to the tick of a common clock, within the confines of a digital system or subsystem we can make it so. When we interconnect digital systems or subsystems that use different clocks, we can usually identify a limited number of asynchronous signals that need special treatment, as we'll show in Section 8.8.3.

Races and hazards are not a problem in synchronous systems, for two reasons. First, the only fundamental-mode circuits that might be subject to races or essential hazards are predesigned elements, such as discrete flip-flops or ASIC cells, that are guaranteed by the manufacturer to work properly. Second, even though the combinational circuits that drive flip-flop control inputs may contain static or dynamic or function hazards, these hazards have no effect, since the control inputs are sampled only *after* the hazard-induced glitches have had a chance to settle out.

Aside from designing the functional behavior of each state machine, the designer of a practical synchronous system or subsystem must perform just three well-defined tasks to ensure reliable system operation:

1. Minimize and determine the amount of clock skew in the system, as discussed in Section 8.8.1.

2. Ensure that flip-flops have positive setup- and hold-time margins, including an allowance for clock skew, as described in Section 8.1.4.

3. Identify asynchronous inputs, synchronize them with the clock, and ensure that the synchronizers have an adequately low probability of failure, as described in Sections 8.8.3 and 8.9.

Before we get into these issues, in this section we'll look at a general model for synchronous system structure and an example.

8.7.1 Synchronous System Structure

The sequential-circuit design examples that we gave in Chapter 7 were mostly individual state machines with a small number of states. If a sequential circuit has more than a few flip-flops, then it's not desirable (and often not possible) to treat the circuit as a single, monolithic state machine, because the number of states would be too large to handle.

Fortunately, most digital systems or subsystems can be partitioned into two or more parts. Whether the system processes numbers, digitized voice signals, or a stream of spark-plug pulses, a certain part of the system, which we'll call the *data unit,* can be viewed as storing, routing, combining, and generally processing "data." Another part, which we'll call the *control unit,* can be viewed as starting and stopping actions in the data unit, testing conditions, and deciding what to do next according to circumstances. In general, only the control unit must be designed as a state machine. The data unit and its components are typically handled at a higher level of abstraction, such as:

- *Registers.* A collection of flip-flops is loaded in parallel with many bits of "data," which can then be used or retrieved together.
- *Specialized functions.* These include multibit counters and shift registers, which increment or shift their contents on command.
- *Read/write memory.* Individual latches or flip-flops in a collection of the same can be written or read out.

The first two topics above were discussed earlier in this chapter, and the last is discussed in Chapter 10.

Figure 8-79 is a general block diagram of a system with a control unit and a data unit. We have also included explicit blocks for input and output, but we could have just as easily absorbed these functions into the data unit itself. The control unit is a state machine whose inputs include *command inputs* that indicate how the machine is to function, and *condition inputs* provided by the data unit. The command inputs may be supplied by another subsystem or by a user to set the general operating mode of the control state machine (RUN/HALT, NORMAL/TURBO, etc.), while the condition inputs allow the control state-machine unit to change its behavior as required by circumstances in the data unit (ZERO_DETECT, MEMORY_FULL, etc.).

command input
condition input

A key characteristic of the structure in Figure 8-79 is that the control, data, input, and output units all use the same common clock. Figure 8-80 illustrates the operations of the control and data units during a typical clock cycle:

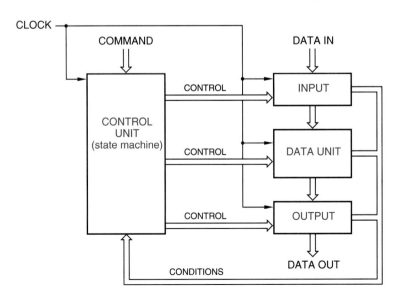

Figure 8-79
Synchronous system structure.

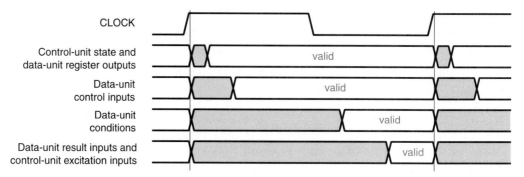

Figure 8-80 Operations during one clock cycle in a synchronous system.

1. Shortly after the beginning of the clock period, the control-unit state and the data-unit register outputs are valid.

2. Next, after a combinational logic delay, Moore-type outputs of the control-unit state machine become valid. These signals are control *inputs* to the data unit. They determine what data-unit functions are performed in the rest of the clock period—for example, selecting memory addresses, multiplexer paths, and arithmetic operations.

3. Near the end of the clock period, data-unit condition outputs such as zero-or overflow-detect are valid and are made available to the control unit.

4. At the end of the clock period, just before the setup-time window begins, the next-state logic of the control-unit state machine has determined the next state based on the current state and command and condition inputs. At about the same time, computational results in the data unit are available to be loaded into data-unit registers.

5. After the clock edge, the whole cycle may repeat.

Data-unit control inputs, which are control-unit state-machine outputs, may be of the Moore, Mealy, or pipelined Mealy type; timing for the Moore type was shown in Figure 8-80. Moore-type and pipelined-Mealy-type outputs control the data unit's actions strictly according to the current state and past inputs, which do not depend on *current* conditions in the data unit. In contrast, Mealy-type outputs may select different actions in the data unit according to *current* conditions in the data unit. This increases flexibility, but typically also

| PIPELINED MEALY OUTPUTS | Some state machines have pipelined Mealy outputs, discussed in Section 7.3.2. In Figure 8-80, pipelined Mealy outputs would typically be valid early in the cycle, at the same time as control-unit state outputs. Early validity of these outputs, compared to Moore outputs that must go through a combinational logic delay, may allow the entire system to operate at a faster clock rate. |

increases the minimum clock period for correct system operation, since the delay path may be much longer. Also, Mealy-type outputs must not create feedback loops. For example, a signal that adds 1 to an adder's input if the adder output is nonzero causes an oscillation if the adder output is −1.

8.7.2 A Synchronous System Design Example

To give you an overview of several elements of synchronous system design, this subsection presents a representative example of a synchronous system. The example is a *shift-and-add multiplier* for unsigned integers using the algorithm of Section 2.8. Its data unit uses standard combinational and sequential building blocks, and its control unit is described by a state diagram

shift-and-add multiplier

Figure 8-81 illustrates data-unit registers and functions that are used to perform an 8-bit multiplication:

MPY/LPROD   A shift register that initially stores the multiplier, and accumulates the low-order bits of the product as the algorithm is executed.

HPROD   A register that is initially cleared, and accumulates the high-order bits of the product as the algorithm is executed.

MCND   A register that stores the multiplicand throughout the algorithm.

F   A combinational function equal to the 9-bit sum of HPROD and MCND if the low-order bit of MPY/LPROD is 1, and equal to HPROD (extended to 9 bits) otherwise.

The MPY/LPROD shift register serves a dual purpose, holding both yet-to-be-tested multiplier bits (on the right) and unchanging product bits (on the left) as the algorithm is executed. At each step it shifts right one bit, discarding the multiplier bit that was just tested, moving the next multiplier bit to be tested to the rightmost position, and loading into the leftmost position one more product bit that will not change for the rest of the algorithm.

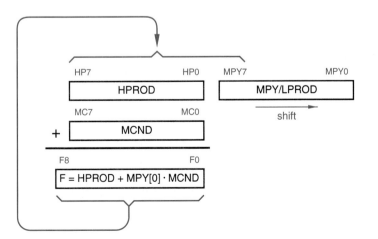

Figure 8-81
Registers and functions used by the shift-and-add multiplication algorithm.

Figure 8-82
Data unit of an 8-bit shift-and-add binary multiplier.

Figure 8-82 is an MSI design for the data unit. The multiplier, MPY[7:0], and the multiplicand, MCND[7:0], are loaded into two registers before a multiplication begins. When the multiplication is completed, the product appears on HP[7:0] and LP[7:0]. The data unit uses the following control signals:

LDMCND_L When asserted, enables the multiplicand register U1 to be loaded.

LDHP_L When asserted, enables the HPROD register U6 to be loaded.

MPYS[1:0] When 11, these signals enable the MPY/LPROD register U2 and U3 to be loaded at the next clock tick. They are set to 01 during the multiplication operation to enable the register to shift right, and they are 00 at other times to preserve the register's contents.

Figure 8-83
Control unit for an
8-bit shift-and-add
binary multiplier.

SELSUM When this is asserted, the multiplexers U7 and U8 select the output of the adders U4 and U5, which is the sum of HPROD and the multiplicand MC. Otherwise, they select HPROD directly.

CLEAR When asserted, the output of multiplexers U7 and U8 is zero.

The multiplier uses a control unit, shown along with the data-unit block in Figure 8-83, to initialize the data unit and step through a multiplication. The control unit is decomposed into a counter (U10) and a state machine with the state diagram shown in Figure 8-84.

The state machine has the following inputs and outputs:

RESET A reset input that is asserted at power-up.

START An external command input that starts a multiplication.

MPY0 A condition input from the data unit, the next multiplier bit to test.

CLEAR A control output that zeroes the multiplexer output and initializes the counter.

LDMCND A control output that enables the MCND register to be loaded.

LDHP A control output that enables the HPROD register to be loaded.

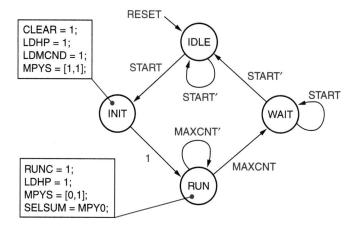

Figure 8-84
State diagram for the control state machine for a shift-and-add binary multiplier.

RUNC   A control output that enables the counter to count.

MPYS[1:0]   Control outputs for MPY/LPROD shifting and loading.

SELSUM   A control output that selects between the shifted adder output or shifted HPROD to be loaded back into HPROD.

The state diagram can be converted into a corresponding state machine using any of a variety of methods, from turn-the-crank (a.k.a. hand-crafted) design to automatic synthesis using a corresponding ABEL or VHDL description. The state machine has mostly Moore-type outputs; SELSUM is a Mealy-type output. Two boxes in the state diagram list outputs that are asserted in the INIT and RUN states; all outputs are negated at other times. The machine is designed so that asserting RESET in any state takes it to the IDLE state.

After the START signal is asserted, a multiplication begins in the INIT state. In this state, the counter is initialized to 1000_2, the multiplier and multiplicand are loaded into their respective registers, and HPROD is cleared. The RUN state is entered next, and the counter is enabled to count. The state machine stays in the RUN state for eight clock ticks, to execute the eight steps of the 8-bit shift-and-add algorithm. During the eighth tick, the counter is in state 1111_2, so MAXCNT is asserted and the state machine goes to the WAIT state. The machine waits there until START is negated, to prevent a multiplication from restarting until START is asserted once again.

The design details of the data and control units are interesting, but the most important thing to see in this example is that all of the sequential circuit elements for both data and control are edge-triggered flip-flops clocked by the same common CLOCK signal. Thus, its timing is consistent with the model in Figure 8-80, and the designer need not be concerned about races, hazards, and asynchronous operations. Unless the state machine realization is very slow, the overall circuit's maximum clock speed will be limited mainly by the propagation delays through the data unit.

8.8 Impediments to Synchronous Design

Although the synchronous approach is the most straightforward and reliable method of digital system design, a few nasty realities can get in the way. We'll discuss them in this section.

8.8.1 Clock Skew

Synchronous systems using edge-triggered flip-flops work properly only if all flip-flops see the triggering clock edge at the same time. Figure 8-85 shows what can happen otherwise. Here, two flip-flops are theoretically clocked by the same signal, but the clock signal seen by FF2 is delayed by a significant amount relative to FF1's clock. This difference between arrival times of the clock at different devices is called *clock skew*.

clock skew

We've named the delayed clock in Figure 8-85(a) "CLOCKD." If FF1's propagation delay from CLOCK to Q1 is short, and if the physical connection of Q1 to FF2 is short, then the change in Q1 caused by a CLOCK edge may actually reach FF2 *before* the corresponding CLOCKD edge. In this case, FF2 may go to an incorrect next state determined by the *next* state of FF1 instead of the current state, as shown in (b). If the change in Q1 arrives at FF2 only slightly early relative to CLOCKD, then FF2's hold-time specification may be violated, in which case FF2 may become metastable and produce an unpredictable output.

If Figure 8-85 reminds you of the essential hazard shown in Figure 7-101, you're on to something. The clock-skew problem may be viewed simply as a manifestation of the essential hazards that exist in all edge-triggered devices.

We can determine quantitatively whether clock skew is a problem in a given system by defining t_{skew} to be the amount of clock skew and using the other timing parameters defined in Figure 8-1. For proper operation, we need

$$t_{\text{ffpd(min)}} + t_{\text{comb(min)}} - t_{\text{hold}} - t_{\text{skew(max)}} > 0$$

In other words, clock skew subtracts from the hold-time margin that we defined in Section 8.1.4.

Figure 8-85 Example of clock skew.

(a)

(b)

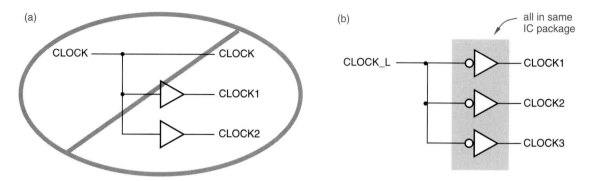

Figure 8-86 Buffering the clock: (a) excessive clock skew; (b) controllable clock skew.

Viewed in isolation, the example in Figure 8-85 may seem a bit extreme. After all, why would a designer provide a short connection path for data and a long one for the clock, when they could just run side by side? There are several ways this can happen; some are mistakes, while others are unavoidable.

In a large system, a single clock signal may not have adequate fanout to drive all of the devices with clock inputs, so it may be necessary to provide two or more copies of the clock signal. The buffering method of Figure 8-86(a) obviously produces excessive clock skew, since CLOCK1 and CLOCK2 are delayed through an extra buffer compared to CLOCK.

A recommended buffering method is shown in Figure 8-86(b). All of the clock signals go through identical buffers and thus have roughly equal delays. Ideally, all the buffers should be part of the same IC package, so that they all have similar delay characteristics and are operating at identical temperature and power-supply voltage. Some manufacturers build special buffers for just this sort of application and specify the worst-case delay variation between buffers in the same package, which can be as low as a few tenths of a nanosecond.

Even the method in Figure 8-86(b) may produce excessive clock skew if one clock signal is loaded much more heavily than the other; transitions on the more heavily loaded clock appear to occur later because of increases in output-transistor switching delay and signal rise and fall times. Therefore, a careful designer tries to balance the loads on multiple clocks, looking at both DC load (fanout) and AC load (wiring and input capacitance).

Another bad situation can occur when signals on a PCB or in an ASIC are routed automatically by CAD software. Figure 8-87 shows a PCB or ASIC with many flip-flops and larger-scale elements, all clocked with a common CLOCK signal. The CAD software has laid out CLOCK in a serpentine path that winds its way past all the clocked devices. Other signals are routed point-to-point between an output and a small number of inputs, so their paths are shorter. To make matters worse, in an ASIC some types of "wire" may be slower than others (polysilicon vs. metal in CMOS technology). As a result, a CLOCK edge may indeed arrive at FF2 quite a bit later than the data change that it produces on Q1.

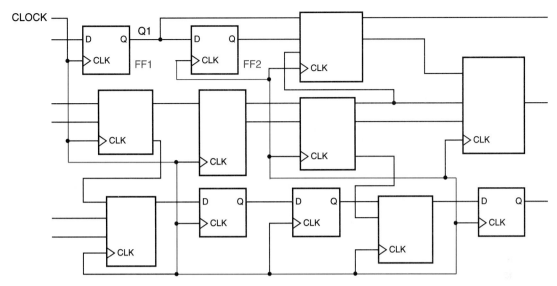

Figure 8-87 A clock-signal path leading to excessive skew in a complex PCB or ASIC.

One way to minimize this sort of problem is to arrange for **CLOCK** to be distributed in a tree-like structure using the fastest type of wire, as illustrated in Figure 8-88. Usually, such a "clock tree" must be laid out by hand or using a specialized CAD tool. Even then, in a complex design it may not be possible to guarantee that clock edges arrive everywhere before the earliest data change. A CAD timing analysis program is typically used to detect these problems, which

Figure 8-88 Clock-signal routing to minimize skew.

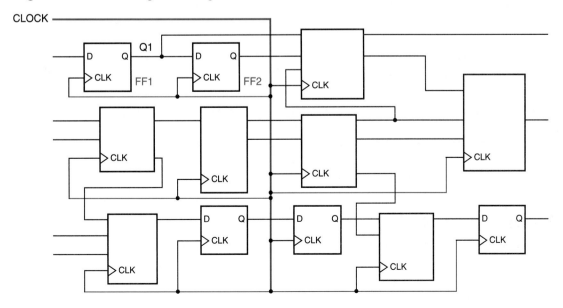

HOW NOT TO GET SKEWERED

Unbalanced wire lengths and loads are the most obvious sources of clock skew, but there are many other subtle sources. For example, *crosstalk*, the coupling of energy from one signal line into another, can cause clock skew. Crosstalk is inevitable when parallel wires are packed together tightly on a printed circuit board or in a chip, and energy is radiated during signal transitions. Depending on whether an adjacent signal is changing in the same or opposite direction as a clock, the clock's transition can be accelerated or retarded, making its transition appear to occur earlier or later.

In a large PCB or ASIC design, it's usually not feasible to track down all the possible sources of clock skew. As a result, most ASIC manufacturers require designers to provide extra setup- and hold-time margin, equivalent to many gate delays, over and above the known simulation timing results to accommodate such unknown factors.

generally can be remedied only by inserting extra delay (e.g., pairs of inverters) in the too-fast data paths.

Although synchronous design methodology simplifies the conceptual operation of large systems, we see that clock skew can be a major problem when edge-triggered flip-flops are used as the storage elements. To control this problem, many high-performance systems and VLSI chips use a *two-phase latch design,* discussed in the References. Such designs effectively split each edge-triggered D flip-flop into its two component latches, controlling them with two nonoverlapping clock phases. The nonoverlap between the phases accommodates clock skew.

two-phase latch design

8.8.2 Gating the Clock

Most of the sequential MSI parts that we introduced in this chapter have synchronous function-enable inputs. That is, their enable inputs are sampled on the clock edge, along with the data. The first example that we showed was the 74x377 register with synchronous load-enable input; other parts included the 74x163 counter and 74x194 shift register with synchronous load-enable, count-enable, and shift-enable inputs. Nevertheless, many MSI parts, FPGA macros, and ASIC cells do not have synchronous function-enable inputs; for example, the 74x374 8-bit register has three-state outputs but no load-enable input.

So, what can a designer do if an application requires an 8-bit register with both a load-enable input *and* three-state outputs? One solution is to use a 74x377 to get the load-enable, and follow it with a 74x241 three-state buffer. However, this increases both cost and delay. Another approach is to use a larger, more expensive part, the 74x823, which provides both required functions as well as an asynchronous CLR_L input. A riskier alternative sometimes employed by designers who don't know any better is to use a '374, but to suppress its clock input when it's not supposed to be loaded. This is called *gating the clock*.

gating the clock

Figure 8-89 How not to gate the clock: (a) simple-minded circuit; (b) timing diagram.

Figure 8-89 illustrates an obvious but wrong approach to gating the clock. A signal CLKEN is asserted to enable the clock and is simply ANDed with the clock to produce the gated clock GCLK. This approach has two problems:

1. If CLKEN is a state-machine output or other signal produced by a register clocked by CLOCK, then CLKEN changes some time *after* CLOCK has already gone HIGH. As shown in (b), this produces glitches on GCLK and false clocking of the registers controlled by GCLK.
2. Even if CLKEN is somehow produced well in advance of CLOCK's rising edge (e.g., using a register clocked with the *falling* edge of CLOCK, an especially nasty kludge), the AND-gate delay gives GCLK excessive clock skew, which causes more problems all around.

A method of gating the clock that generates only minimal clock skew is shown in Figure 8-90. Here, both an ungated clock and several gated clocks are generated from the same active-low master clock signal. Gates in the same IC package are used to minimize the possible differences in their delays. The CLKEN signal may change arbitrarily whenever CLOCK_L is LOW, which is when CLOCK is HIGH. That's just fine; a CLKEN signal is typically produced by a state machine whose outputs change right after CLOCK goes HIGH.

The approach of Figure 8-90 is acceptable in a particular application only if the clock skew that it creates is acceptable. Furthermore, note that CLKEN

Figure 8-90 An acceptable way to gate the clock: (a) circuit; (b) timing diagram.

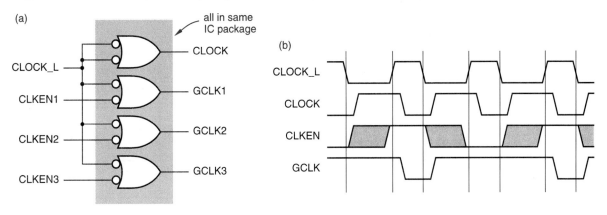

must be stable during the entire time that **CLOCK_L** is **HIGH** (**CLOCK** is **LOW**). Thus, the timing margins in this approach are sensitive to the clock's duty cycle, especially if **CLKEN** suffers significant combinational-logic delay (t_{comb}) from the triggering clock edge. A truly synchronous function-enable input, such as the 74x377's load-enable input in Figure 8-13, can be changed at almost any time during the entire clock period, up until a setup time before the triggering edge.

8.8.3 Asynchronous Inputs

Even though it is theoretically possible to build a computer system that is fully synchronous, you couldn't do much with it, unless you could synchronize your keystrokes with a 500-MHz clock. Digital systems of all types inevitably must *asynchronous input* deal with *asynchronous input signals* that are not synchronized with the system *signal* clock.

Asynchronous inputs are often requests for service (e.g., interrupts in a computer) or status flags (e.g., a resource has become available). Such inputs normally change slowly compared to the system clock frequency, and they need not be recognized at a particular clock tick. If a transition is missed at one clock tick, it can always be detected at the next one. The transition rates of asynchronous signals may range from less than one per second (the keystrokes of a slow typist) to 100 MHz or more (access requests for a 500-MHz multiprocessor system's shared memory).

synchronizer   Ignoring the problem of metastability, it is easy to build a *synchronizer,* a circuit that samples an asynchronous input and produces an output that meets the setup and hold times required in a synchronous system. As shown in Figure 8-91, a D flip-flop samples the asynchronous input at each tick of the system clock and produces a synchronous output that is valid during the next clock period.

Figure 8-91
A single, simple synchronizer:
(a) logic diagram;
(b) timing.

Figure 8-92
Two synchronizers for the same asynchronous input: (a) logic diagram; (b) possible timing.

It is essential for asynchronous inputs to be synchronized at only *one place* in a system; Figure 8-92 shows what can happen otherwise. Because of physical delays in the circuit, the two flip-flops will not see the clock and input signals at precisely the same time. Therefore, when asynchronous input transitions occur near the clock edge, there is a small window of time during which one flip-flop may sample the input as 1 and the other may sample it as 0. This inconsistent result may cause improper system operation, as one part of the system responds as if the input were 1, and another part responds as if it were 0.

Combinational logic may hide the fact that there are two synchronizers, as shown in Figure 8-93. Since different paths through the combinational logic will inevitably have different delays, the likelihood of an inconsistent result is even

Figure 8-93 An asynchronous input driving two synchronizers through combinational logic.

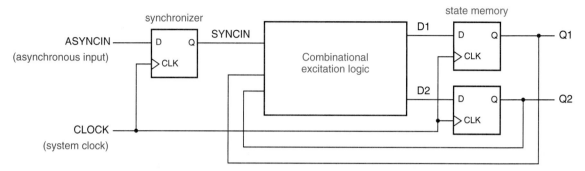

Figure 8-94 An asynchronous state-machine input coupled through a single synchronizer.

greater. This situation is especially common when asynchronous signals are used as inputs to state machines, since the excitation logic for two or more state variables may depend on the asynchronous input. The proper way to use an asynchronous signal as a state-machine input is shown in Figure 8-94. All of the excitation logic sees the same synchronized input signal, SYNCIN.

WHO CARES?   As you probably know, even the synchronizers in Figures 8-91 and 8-94 sometimes fail. The reason is that the setup and hold times of the synchronizing flip-flop are sometimes violated, because the asynchronous input can change at any time. "Well, who cares?" you may say. "If the D input changes near the clock edge, then the flip-flop will either see the change this time or miss it and pick it up next time; either way is good enough for me!" The problem is, there is a third possibility, discussed in the next section.

8.9 Synchronizer Failure and Metastability

We showed in Section 7.1 that when the setup and hold times of a flip-flop are not met, the flip-flop may go into a third, *metastable* state halfway between 0 and 1. Worse, the length of time it may stay in this state before falling back into a legitimate 0 or 1 state is theoretically unbounded. When other gates and flip-flops are presented with a metastable input signal, some may interpret it as a 0 and others as a 1, creating the sort of inconsistent behavior that we showed in Figure 8-92. Or the other gates and flip-flops may produce metastable outputs themselves (after all, they are now operating in the *linear* part of their operating range). Luckily, the probability of a flip-flop output remaining in the metastable state decreases exponentially with time, though never all the way to zero.

8.9.1 Synchronizer Failure

Synchronizer failure is said to occur if a system uses a synchronizer output while the output is still in the metastable state. The way to avoid synchronizer failure is to ensure that the system waits "long enough" before using a synchronizer's output, "long enough" so that the mean time between synchronizer failures is several orders of magnitude longer than the designer's expected length of employment.

synchronizer failure

Metastability is more than an academic problem. More than a few experienced designers of high-speed digital systems have built (and released to production) circuits that suffer from intermittent synchronizer failures. In fact, the initial versions of several commercial ICs are said to have suffered from metastability problems, for example, the AMD 9513 system timing controller, the AMD 9519 interrupt controller, the Zilog Z-80 Serial I/O interface, the Intel 8048 single-chip microcomputer, and the AMD 29000 RISC microprocessor. It makes you wonder, are the designers of these parts still employed?

There are two ways to get a flip-flop out of the metastable state:

1. Force the flip-flop into a valid logic state using input signals that meet the published specifications for minimum pulse width, setup time, and so on.

2. Wait "long enough," so the flip-flop comes out of metastability on its own.

Inexperienced designers often attempt to get around metastability in other ways, and they are usually unsuccessful. Figure 8-95 shows an attempt by a designer who thinks that since metastability is an "analog" problem, it must have an "analog" solution. After all, Schmitt-trigger inputs and capacitors can normally be used to clean up noisy signals. However, rather than eliminate metastability, this circuit enhances it—with the "right" components, the circuit will oscillate forever, once it is excited by negating S_L and R_L simultaneously. (*Confession:* It was the author who tried this over 20 years ago!) Exercises 8.97 and 8.94 give examples of valiant but also failed attempts to eliminate metastability. These examples should give you the sense that synchronizer problems can be very subtle, so you must be careful. The only way to make synchronizers reliable is to wait long enough for metastable outputs to resolve. We answer the question "How long is 'long enough'?" later in this section.

Figure 8-95
A failed attempt to build a metastable-proof $\overline{S}$-$\overline{R}$ flip-flop.

8.9.2 Metastability Resolution Time

If the setup and hold times of a D flip-flop are met, the flip-flop output settles to a new value within time t_{pd} after the clock edge. If they are violated, the flip-flop output may be metastable for an arbitrary length of time. In a particular system design, we use the parameter t_r, called the *metastability resolution time,* to denote the maximum time that the output can remain metastable without causing synchronizer (and system) failure.

t_r
metastability resolution time

For example, consider the state machine in Figure 8-94 on page 764. The available metastability resolution time is

$$t_r = t_{clk} - t_{comb} - t_{setup}$$

t_{clk}
t_{comb}
t_{setup}

where t_{clk} is the clock period, t_{comb} is the propagation delay of the combinational excitation logic, and t_{setup} is the setup time of the flip-flops used in the state memory.

8.9.3 Reliable Synchronizer Design

The most reliable synchronizer is one that allows the maximum amount of time for metastability resolution. However, in the design of a digital system, we seldom have the luxury of *slowing down* the clock to make the system work more reliably. Instead, we are usually asked to *speed up* the clock to get higher performance from the system. As a result, we often need synchronizers that work reliably with very short clock periods. We'll present several such designs, and show how to predict their reliability.

We showed previously that a state machine with an asynchronous input, built as illustrated in Figure 8-94, has $t_r = t_{clk} - t_{comb} - t_{setup}$. In order to maximize t_r for a given clock period, we should minimize t_{comb} and t_{setup}. The value of t_{setup} depends on the type of flip-flops used in the state memory; in general, faster flip-flops have shorter setup times. The minimum value of t_{comb} is zero and is achieved by the synchronizer design of Figure 8-96, whose operation we explain next.

Inputs to flip-flop **FF1** are asynchronous with the clock and may violate the flip-flop's setup and hold times. When this happens, the **META** output may become metastable and remain in that state for an arbitrary time. However, we

Figure 8-96 Recommended synchronizer design.

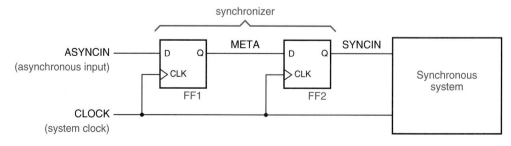

assume that the maximum duration of metastability after the clock edge is t_r. (We show how to calculate the probability that our assumption is correct in the next subsection.) As long as the clock period is greater than t_r plus the FF2's setup time, SYNCIN becomes a synchronized copy of the asynchronous input on the next clock tick without ever becoming metastable itself. The SYNCIN signal is distributed as required to the rest of the system.

8.9.4 Analysis of Metastable Timing

Figure 8-97 shows the flip-flop timing parameters that are relevant to our analysis of metastability timing. The published setup and hold times of a flip-flop with respect to its clock edge are denoted by t_s and t_h, and they bracket an interval called the *decision window*, when the flip-flop samples its input and   *decision window* decides to change its output if necessary. As long as the D input changes outside the decision window, as in (a), the manufacturer guarantees that the output will change and settle to a valid logic state before time t_{pd}. If D changes inside the decision window, as in (b), metastability may occur and persist until time t_r.

Theoretical research suggests, and experimental research has confirmed, that when asynchronous inputs change during the decision window, the duration of metastable outputs is governed by an exponential formula:

$$\text{MTBF}(t_r) = \frac{\exp(t_r/\tau)}{T_o \cdot f \cdot a}$$

Figure 8-97 Timing parameters for metastability analysis: (a) normal flip-flop operation; (b) metastable behavior.

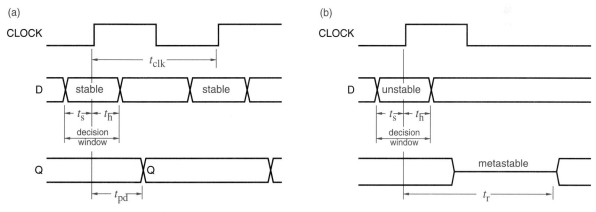

> **DETAILS,**   In our analysis of the synchronizer in Figure 8-96 we do not allow metastability,
> **DETAILS**   even briefly, on the output of FF2, because we assume that the system has been
> designed with zero timing margins. If the system can in fact tolerate some increase
> in FF2's propagation delay, the MTBF will be somewhat better than predicted.

f
a
T_o
τ

Here $\text{MTBF}(t_r)$ is the mean time between synchronizer failures, where a failure occurs if metastability persists beyond time t_r after a clock edge, where $t_r \geq t_{pd}$. This MTBF depends on f, the frequency of the flip-flop clock; a, the number of asynchronous input changes per second applied to the flip-flop; and T_o and τ, constants that depend on the electrical characteristics of the flip-flop. For a typical 74LS74, $T_o \approx 0.4$ s and $\tau \approx 1.5$ ns.

Now suppose that we build a microprocessor system with a 10-MHz clock and use the circuit of Figure 8-96 to synchronize an asynchronous input. If ASYNCIN changes during the decision window of FF1, the output META may become metastable until time t_r. If META is still metastable at the beginning of the decision window for FF2, then the synchronizer fails, because FF2 may have a metastable output; system operation is unpredictable in that case.

Let us assume that the D flip-flops in Figure 8-96 are 74LS74s. The setup time t_s of a 74LS74 is 20 ns, and the clock period in our example microprocessor system is 100 ns, so t_r for synchronizer failure is 80 ns. If the asynchronous input changes 100,000 times per second, then the synchronizer MTBF is

$$\text{MTBF(80 ns)} = \frac{\exp(80/1.5)}{0.4 \cdot 10^7 \cdot 10^5} = 3.6 \cdot 10^{11}\,\text{s}$$

That's not bad, about 100 centuries between failures! Of course, if we're lucky enough to sell 10,000 copies of our system, one of them will fail in this way every year. But, no matter, let us consider a more serious problem.

Suppose we upgrade our system to use a faster microprocessor chip with a clock speed of 16 MHz. We may have to replace some components in our system to operate at the higher speed, but 74LS74s are still perfectly good at 16 MHz. Or are they? With a clock period of 62.5 ns, the new synchronizer MTBF is

$$\text{MTBF(42.5 ns)} = \frac{\exp(42.5/1.5)}{0.4 \cdot 1.6 \cdot 10^7 \cdot 10^5} = 3.1\,\text{s}!$$

**UNDERSTANDING
A AND *F***

Although a flip-flop output can go metastable *only* if D changes during the decision window, the MTBF formula does not explicitly specify how many such input changes occur. Instead, it specifies the total number of asynchronous input changes per second, a, and assumes that asynchronous input changes are uniformly distributed over the clock period. Therefore, the fraction of input changes that actually occur during the decision window is "built in" to the clock-frequency parameter f—as f increases, the fraction goes up.

If the system design is such that input changes might be clustered in the decision window rather than being uniformly distributed (as when synchronizing a slow input with a fixed but unknown phase difference from the system clock), then a useful rule of thumb is to use a frequency equal to the reciprocal of the decision window (based on published setup and hold times), times a safety margin of 10.

The only saving grace of this synchronizer at 16 MHz is that it's so lousy, we're likely to discover the problem in the engineering lab before the product ships! Thank goodness the MTBF wasn't one year.

8.9.5 Better Synchronizers

Given the poor performance of the 74LS74 as a synchronizer at moderate clock speeds, we have a couple of alternatives for building more reliable synchronizers. The simplest solution, which works for most design requirements, is simply to use a flip-flop from a faster technology. Nowadays much faster technologies are available for flip-flops, whether discrete or embedded in PLDs, FPGAs, or ASICs.

Based on published data discussed in the References, Table 8-35 lists the metastability parameters for several common logic families and devices. These numbers are very much circuit-design and IC-process dependent. Thus, unlike guaranteed logic signal levels and timing parameters, published metastability numbers can vary dramatically among different manufacturers of the same part and must be used conservatively. For example, one manufacturer's 74F74 may give acceptable metastability performance in a design while another's may not.

Table 8-35 Metastability parameters for some common devices.

Reference	Device	τ (ns)	T_o (s)	t_r (ns)
Chaney (1983)	74LS74	1.50	$4.0 \cdot 10^{-1}$	77.71
Chaney (1983)	74S74	1.70	$1.0 \cdot 10^{-6}$	66.14
Chaney (1983)	74S174	1.20	$5.0 \cdot 10^{-6}$	48.62
Chaney (1983)	74S374	0.91	$4.0 \cdot 10^{-4}$	40.86
Chaney (1983)	74F74	0.40	$2.0 \cdot 10^{-4}$	17.68
TI (1997)	74LSxx	1.35	$4.8 \cdot 10^{-3}$	63.97
TI (1997)	74Sxx	2.80	$1.3 \cdot 10^{-9}$	90.33
TI (1997)	74ALSxx	1.00	$8.7 \cdot 10^{-6}$	41.07
TI (1997)	74ASxx	0.25	$1.4 \cdot 10^{3}$	14.99
TI (1997)	74Fxx	0.11	$1.9 \cdot 10^{8}$	7.90
TI (1997)	74HCxx	1.82	$1.5 \cdot 10^{-6}$	71.55
Cypress (1997)	PALC16R8-25	0.52	$9.5 \cdot 10^{-12}$	14.22*
Cypress (1997)	PALC22V10B-20	0.26	$5.6 \cdot 10^{-11}$	7.57*
Cypress (1997)	PALCE22V10-7	0.19	$1.3 \cdot 10^{-13}$	4.38*
Xilinx (1997)	7300-series CPLD	0.29	$1.0 \cdot 10^{-15}$	5.27*
Xilinx (1997)	9500-series CPLD	0.17	$9.6 \cdot 10^{-18}$	2.30*

Note that different authors and manufacturers may specify metastability parameters differently. For example, author Chaney and manufacturer Texas Instruments measure the metastability resolution time t_r from the triggering clock edge, as in our previous subsection. On the other hand, manufacturers Cypress and Xilinx define t_r as the *additional* delay beyond a normal clock-to-output delay time t_{pd}.

The last column in the table gives a somewhat arbitrarily chosen figure of merit for each device. It is the metastability resolution time t_r required to obtain an MTBF of 1000 years when operating a synchronizer with a clock frequency of 25 MHz and with 100,000 asynchronous input changes per second. For the Cypress and Xilinx devices, their parameter values yield a value of t_r, marked with an asterisk, consistent with their own definition as introduced above.

As you can see, the 74LS74 is one of the worst devices in the table. If we replace FF1 in the 16-MHz microprocessor system of the preceding subsection with a 74ALS74, we get

$$\text{MTBF}(42.5 \text{ ns}) = \frac{\exp(42.5/0.87)}{12.5 \cdot 10^{-3} \cdot 1.6 \cdot 10^{7} \cdot 10^{5}} = 8.2 \cdot 10^{10} \text{s}$$

If you're satisfied with a synchronizer MTBF of about 25 centuries per system shipped, you can stop here. However, if FF2 is also replaced with a 74ALS74, the MTBF gets better, since the 'ALS74 has a shorter setup time than the 'LS74, only 10 ns. With the 'ALS74, the MTBF is about 100,000 times better:

$$\text{MTBF}(52.5 \text{ ns}) = \frac{\exp(52.5/0.87)}{12.5 \cdot 10^{-3} \cdot 2 \cdot 10^{7} \cdot 10^{5}} = 8.1 \cdot 10^{15} \text{s}$$

Even if we ship a million systems containing this circuit, we (or our heirs) will see a synchronizer failure only once in 240 years. Now that's job security!

Actually, the margins above aren't as large as they might seem. (How large does 240 years seem *to you*?) Most of the numbers given in Table 8-35 are *averages* and are seldom specified, let alone guaranteed, by the device manufacturer. Furthermore, calculated MTBFs are extremely sensitive to the value of τ, which in turn may depend on temperature, voltage, and the phase of the moon. So the operation of a given flip-flop in an actual system may be much worse (or much better) than predicted by our table.

For example, consider what happens if we increase the clock in our 16-MHz system by just 25%, to 20 MHz. Your natural inclination might be to think that metastability will get 25% worse, or maybe 250% worse, just to be conservative. But, if you run the numbers, you'll find that the MTBF using 'ALS74s for both FF1 and FF2 goes down from $8.1 \cdot 10^{15}$ s to just $3.7 \cdot 10^{9}$ s, over a million times worse! The new MTBF of about 118 years is fine for one system, but if you ship a million of them, one will fail every hour. You've just gone from generations of job security to corporate goat!

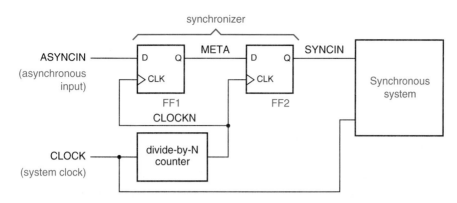

Figure 8-98
Multiple-cycle
synchronizer.

8.9.6 Other Synchronizer Designs

We promised to describe a couple of ways to build more reliable synchronizers. The first way was to use faster flip-flops, that is, to reduce the value of τ in the MTBF equation. Having said that, the second way is obvious—to *increase* the value of t_r in the MTBF equation.

For a given system clock, the best value we can obtain for t_r using the circuit of Figure 8-96 is t_{clk}, if FF2 has a setup time of 0. However, we can get values of t_r on the order of $n \cdot t_{clk}$ by using the *multiple-cycle synchronizer* circuit of Figure 8-98. Here we divide the system clock by n to obtain a slower synchronizer clock and longer $t_r = (n \cdot t_{clk}) - t_{setup}$. Usually a value of $n = 2$ or $n = 3$ gives adequate synchronizer reliability.

*multiple-cycle
synchronizer*

In the figure, note that the edges of CLOCKN will lag the edges of CLOCK because CLOCKN comes from the Q output of a counter flip-flop that is clocked by CLOCK. This means that SYNCIN, in turn, will be delayed or skewed relative to other signals in the synchronous system that come directly from flip-flops clocked by CLOCK. If SYNCIN goes through additional combinational logic in the synchronous system before reaching its flip-flop inputs, their setup time may be inadequate. If this is the case, the solution in Figure 8-99 can be used. Here, SYNCIN is reclocked by CLOCK using FF3 to produce DSYNCIN, which will

Figure 8-99 Multiple-cycle synchronizer with deskewing.

have the same timing as other flip-flop outputs in the synchronous system. Of course, the delay from CLOCK to CLOCKN must still be short enough that SYNCIN meets the setup time requirement of FF3.

In an *n*-cycle synchronizer, the larger the value of *n*, the longer it takes for an asynchronous input change to be seen by the synchronous system. This is simply a price that must be paid for reliable system operation. In typical microprocessor systems, most asynchronous inputs are for external events—interrupts, DMA requests, and so on—that need not be recognized very quickly, relative to synchronizer delays. In the time-critical area of main memory access, experienced designers use the processor clock to run the memory subsystem too, if possible. This eliminates the need for synchronizers and provides the fastest possible system operation.

At higher frequencies, the feasibility of the multiple-cycle synchronizer design shown in Figure 8-98 tends to be limited by clock skew. For this reason, *cascaded synchronizers* rather than use a divide-by-*n* synchronizer clock, some designers use *cascaded synchronizers*. This design approach simply uses a cascade (shift register) of *n* flip-flops, all clocked with the high-speed system clock. This approach is shown in Figure 8-100.

With cascaded synchronizers, the idea is that metastability will be resolved with some probability by the first flip-flop, and failing that, with an equal probability by each successive flip-flop in the cascade. So the overall probability of failure is on the order of the *n*th power of the failure probability of a single-flip-flop synchronizer at the system clock frequency. While this is partially true, the MTBF of the cascade is poorer than that of a multiple-cycle synchronizer with the same delay ($n \cdot t_{clk}$). With the cascade, the flip-flop setup time t_{setup} must be subtracted from t_r, the available metastability resolution time, *n* times, but in a multiple-cycle design, it is subtracted only once.

PLDs that contain internal flip-flops can be used in synchronizer designs, where the two flip-flops in Figure 8-96 on page 766 are simply included in the PLD. This is very convenient in most applications, because it eliminates the need for external, discrete flip-flops. However, a PLD-based synchronizer typically has a poorer MTBF than a discrete circuit built with the same or similar

Figure 8-100 Cascaded synchronizer.

technology. This happens because each flip-flop in a PLD has a combinational logic array on its D input that increases its setup time and thereby reduces the amount of time t_r available for metastability resolution during a given system clock period t_{clk}. To maximize t_r without using special components, FF2 in Figure 8-96 should be a short-setup-time discrete flip-flop.

8.9.7 Metastable-Hardened Flip-Flops

In the late 1980s, Texas Instruments and other manufacturers developed SSI and MSI flip-flops that are specifically designed for board-level synchronizer applications. For example, the *74AS4374* was similar to the 74AS374, except that each individual flip-flop was replaced with a pair of flip-flops, as shown in Figure 8-101. Each pair of flip-flops could be used as a synchronizer of the type shown in Figure 8-96, so eight asynchronous inputs could be synchronized with one 74AS4374.

74AS4374

    The internal design of the 'AS4374 was improved to reduce τ and T_o compared to other 74AS flip-flops, but the biggest improvement in the 'AS4374 was a greatly reduced t_{setup}. Because the entire synchronizer of Figure 8-96 is built on a single chip, there are no input or output buffers between FF1 and FF2, and t_{setup} for FF2 is only 0.5 ns. Compared to a conventional 74AS flip-flop with a 5 ns t_{setup}, and assuming that $\tau = 0.40$ ns, this improves the MTBF by a factor of exp(4.5/.40), or about 77,000.

    In recent years, the move toward faster CMOS technologies and higher integration has largely obsoleted specialized parts like the 'AS4374. As you can see from the last few rows in Table 8-35 on page 769, fast PLDs and CPLDs are available with values of τ that rival the fastest discrete devices while offering the convenience of integrating synchronization with many other functions. Still, the approach used by 'AS4374 is worth emulating in FPGA and ASIC designs. That is, whenever you have control over the layout of a synchronizer circuit, it pays to locate FF1 and FF2 as close as possible to each other and to connect them with the fastest available wires, in order to maximize the setup time available for FF2.

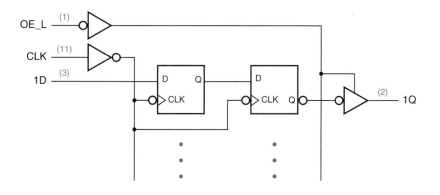

Figure 8-101
Logic diagram for the 74AS4374 octal dual-rank D flip-flop.

8.9.8 Synchronizing High-Speed Data Transfers

A very common problem in computer systems is synchronizing external data transfers with the computer system clock. A simple example is the interface between personal computer's network interface card and a 100-Mbps Ethernet link. The interface card may be connected to a PCI bus, which has a 33.33-MHz clock. Even though the Ethernet speed is an approximate multiple of the bus speed, the signal received on the Ethernet link is generated by another computer whose transmit clock is not synchronized with the receive clock in any way. Yet the interface must still deliver data reliably to the PCI bus.

Figure 8-102 shows the problem. NRZ serial data RDATA is received from the Ethernet at 100 Mbps. The digital phase-locked loop (DPLL) recovers a 100-MHz clock signal RCLK which is centered on the 100-Mbps data stream and

Figure 8-102
Ethernet synchronization problem.

ONE NIBBLE AT A TIME   The explanation of 100-Mbps Ethernet reception above is oversimplified, but sufficient for discussing the synchronization problem. In reality, the received data rate is 125 Mbps, where each 4 bits of user data is encoded as a 5-bit symbol using a so-called 4B5B code. By using only 16 out of 32 possible 5-bit codewords, the 4B5B code guarantees that regardless of the user data pattern, the bit stream on the wire will have a sufficient number of transitions to allow clock recovery. Also, the 4B5B code includes a special code that is transmitted periodically to allow nibble (4-bit) and byte synchronization to be accomplished very easily.

As a result of nibble synchronization, a typical 100-Mbps Ethernet interface does not see an unsynchronized 100-MHz stream of bits. Instead, it sees an unsynchronized 25-MHz stream of nibbles. So, the details of a real 100-Mbps Ethernet synchronizer are different, but the same principles apply.

Figure 8-103
Ethernet link and
system clock timing.

allows data to be clocked bit-by-bit into an 8-bit shift register. At the same time, a byte synchronization circuit searches for special patterns in the received data stream that indicate byte boundaries. When it detects one of these, it asserts the SYNC signal and does so on every eighth subsequent RCLK tick, so that SYNC is asserted whenever the shift register contains an aligned 8-bit byte. The rest of the system is clocked by a 33.33 MHz clock SCLK. We need to transfer each aligned byte RBYTE[7:0] into a register SREG in SCLK's domain. How can we do it?

Figure 8-103 shows some of the timing. We immediately see that the byte-aligned signal, SYNC, is asserted for only 10 ns per byte. We have no hope of consistently detecting this signal with the asynchronous SCLK, whose period is a much longer 30 ns.

The strategy almost universally followed in this kind of situation is to transfer the aligned data first into a holding register HREG in the *receive* clock (RCLK) domain. This gives us a lot more time to sort things out—80 ns in this case. Thus, the "?" box in Figure 8-102 can be replaced by Figure 8-104, which shows HREG and a box marked "SCTRL." The job of SCTRL is to assert SLOAD during exactly one 30-ns SCLK period, so that the output of HREG is valid and stable for the setup and hold times of register SREG in the SCLK domain. SLOAD also serves as a "new-data available" signal for the rest of the interface, indicating that a new data byte will appear on SBYTE[7:0] during the next SCLK

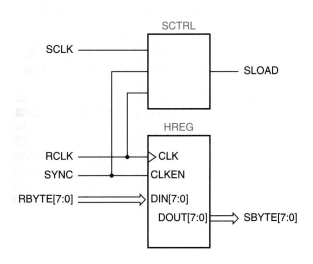

Figure 8-104
Byte holding register
and control.

Figure 8-105
Timing for SBYTE
and possible timing
for SLOAD.

period. Figure 8-105 shows possible timing for **SLOAD** based on this approach
and the previous timing diagram.

Figure 8-106 is a circuit that can generate **SLOAD** with the desired timing.
The idea is to use **SYNC** to set an **S-R** latch as a new byte becomes available. The
output of this latch, **NEWBYTE**, is sampled by **FF1** in the **SCLK** domain. Since
NEWBYTE is not synchronized with **SCLK**, **FF1**'s output **SM** may be metastable,
but it is not used by **FF2** until the next clock tick, 30 ns later. Assuming that the
AND gate is reasonably fast, this gives plenty of metastability resolution time.
FF2's output is the **SLOAD** signal. The **AND** gate ensures that **SLOAD** is only
one **SCLK** period wide; if **SLOAD** is already 1, it can't be set to 1 on the next tick.
This gives time for the **S-R** latch to be reset by **SLOAD** in preparation for the next
byte.

A timing diagram for the overall circuit with "typical" timing is shown in
Figure 8-107. Since **SCLK** is asynchronous to **RCLK**, it can have an arbitrary
relationship with **RCLK** and **SYNC**. In the figure, we've shown a case where the
next **SCLK** rising edge occurs well after **NEWBYTE** is set. Although the figure
shows a window in which **SM** and **SM1** could be metastable in the general case,
metastability doesn't actually happen when the timing is as drawn. Later, we'll
show what can happen if the **SCLK** edge occurs when **NEWBYTE** is changing.

We should make several notes about the circuit in Figure 8-106. First, the
SYNC signal must be glitch free, since it controls the S input of a latch, and it
must be wide enough to meet the minimum pulse-width requirement of the latch.

Figure 8-106
SCTRL circuit for
generating SLOAD.

Figure 8-107
Timing for the SCTRL circuit in Figure 8-106.

Since the latch is set on the leading edge of SYNC, we actually cheated a little; NEWBYTE may be asserted a little *before* a new byte is actually available in HREG. This is OK, because we know that it takes two SCLK periods from when NEWBYTE is sampled until SREG is loaded. In fact, we might have cheated even more if an earlier version of SYNC had been available (see Exercise 8.95).

Assuming that t_{su} is the setup time of a D flip-flop and t_{pd} is the propagation delay of the AND gate in Figure 8-106, the available metastability resolution time t_r is one SCLK period, 30 ns, minus $t_{su} + t_{pd}$, as shown in Figure 8-107. The timing diagram also shows why we can't use SM directly as the reset signal for the S-R latch. Since SM can be metastable, it could wreak havoc. For example, it could be semi-HIGH long enough to reset the latch but then fall back to LOW; in that case, SLOAD would not get set and we would miss a byte. By using instead the output of the synchronizer (SLOAD) both for the latch reset and for the load signal in the SCLK domain, we ensure that the new byte is detected and handled consistently in both clock domains.

The timing that we showed in Figure 8-107 is nominal, but we also have to analyze what happens if SCLK has a different phase relationship with RCLK and SYNC than what is shown. You should be able to convince yourself that if the SCLK edge occurs earlier, so that it samples NEWBYTE just as it's going HIGH, everything still works as before, and the data transfer just finishes a little sooner. The more interesting case is when SCLK occurs later, so that it just misses NEWBYTE as it's going HIGH and catches it one SCLK period later. This timing is shown in Figure 8-108.

Figure 8-108 Maximum-delay timing for SCTRL circuit.

In the timing diagram, we have shown NEWBYTE going high around the same time as the SCLK edge—less than FF1's t_{su} before the edge. Thus, FF1 may not see NEWBYTE as HIGH or its output may become metastable, and it does not solidly capture NEWBYTE until one SCLK period later. Two SCLK periods after that, we get the SCLK edge that loads SBYTE into SREG.

This timing scenario is bad news, because by the time the load occurs, SBYTE is already changing to the *next* received byte. In addition, SLOAD happens to be asserted during and a little bit after the SYNC pulse for this next received byte. Thus, the latch has both S and R asserted simultaneously. If they are removed simultaneously, the latch output may become metastable. Or, as we've shown in the timing diagram, if NEWBYTE (R) is negated last, then the latch is left in the reset state, and this next received byte is never detected and loaded into the SCLK domain.

Thus, we need to analyze the maximum-delay timing case carefully to determine if a synchronizer will work properly. Figure 8-108 shows a starting reference point t_{start} for the SCTRL circuit, namely the RCLK edge on which a byte is loaded into HREG, at end of SYNC pulse. The ending reference point t_{end} is the SCLK edge on which SBYTE is loaded into SREG. The maximum delay between these two reference points, which we'll call t_{maxd}, is the sum of the following components:

$-t_{RCLK}$ Minus one RCLK period, the delay from t_{start} back to the edge on which SYNC was asserted. This number is negative because SYNC is asserted one clock tick before the tick that actually loads HREG.

t_{CQ} One flip-flop CLK-to-Q maximum delay. Assuming that SYNC is a direct flip-flop output in the RCLK domain, this is delay from the RCLK edge until SYNC is asserted.

t_{SQ} Maximum delay from S to Q in the S-R latch in Figure 8-106. This is the delay for NEWBYTE to be asserted.

t_{su} Setup time of FF1 in Figure 8-106. NEWBYTE must be asserted at or before the setup time to guarantee detection.

t_{SCLK} One SCLK period. Since RCLK and SCLK are asynchronous, there may be a delay of up to one SCLK period before the next SCLK edge comes along to sample NEWBYTE.

t_{SCLK} After NEWBYTE is detected by FF1, SLOAD is asserted on the next SCLK tick.

t_{SCLK} After SLOAD is asserted, SBYTE is loaded into SREG on the next SCLK tick.

Thus, $t_{maxd} = 3t_{SCLK} + t_{CQ} + t_{SQ} + t_{su} - t_{RCLK}$. A few other parameters must be defined to complete the analysis:

t_h The hold time of SREG.

$t_{CQ(min)}$ The minimum CLK-to-Q delay of HREG, conservatively assumed to be 0.

t_{rec} The recovery time of the S-R latch, the minimum time allowed between negating S and negating R (see box on page 537).

To be loaded successfully into SREG, SBYTE must be remain valid until at least time $t_{end} + t_h$. The point at which SBYTE changes and becomes invalid is 8 RCLK periods after t_{start}, plus $t_{CQ(min)}$. Thus, for proper circuit operation we must have

$$t_{end} + t_h \leq t_{start} + 8 t_{RCLK}$$

For the maximum-delay case, we substitute $t_{end} = t_{start} + t_{maxd}$ into this relation and subtract t_{start} from both sides to obtain

$$t_{maxd} + t_h \leq 8 t_{RCLK}$$

Substituting the value of t_{maxd} and rearranging, we obtain

$$3t_{SCLK} + t_{CQ} + t_{SQ} + t_{su} + t_h \leq 9t_{RCLK} \qquad (8\text{-}1)$$

as the requirement for correct circuit operation. Too bad. Even if we assume very short component delays (t_{CQ}, t_{SQ}, t_{su}, t_h), we know that $3t_{SCLK}$ (90 ns) plus anything is going to be more than $9t_{RCLK}$ (also 90 ns). So this design will never work properly in the maximum-delay case.

Even if the load-delay analysis gave a good result, we would still have to consider the requirements for proper operation of the SCTRL circuit itself. In particular, we must ensure that when the SYNC pulse for the next byte occurs, it is not negated until time t_{rec} after SLOAD for the previous byte was negated. So, another condition for proper operation is

$$t_{end} + t_{CQ} + t_{rec} \leq t_{start} + 8t_{RCLK} + t_{CQ(min)}$$

Substituting and simplifying as before, we get another requirement that isn't met by our design:

$$3t_{SCLK} + 2t_{CQ} + t_{SQ} + t_{su} + t_{rec} \leq 9t_{RCLK} \qquad (8\text{-}2)$$

There are several ways that we can modify our design to satisfy the worst-case timing requirements. Early in our discussion, we noted that we "cheated" by asserting SYNC one RCLK period before the data in HREG is valid, and that we actually might get away with asserting SYNC even sooner. Doing this can help us meet the maximum-delay requirement, because it reduces the "$8t_{RCLK}$" term on the righthand side of the relations. For example, if we asserted SYNC two RCLK periods earlier, we would reduce this term to "$6t_{RCLK}$". However, there's no free lunch; we can't assert SYNC arbitrarily early. We must also consider a *minimum-delay* case, to ensure that the new byte is actually available in HREG when SBYTE is loaded into SREG. The minimum delay t_{maxd} between t_{start} and t_{end} is the sum of the following components:

$-nt_{RCLK}$ Minus n RCLK periods, the delay from t_{start} back to the edge on which SYNC was asserted. In the original design, $n = 1$.

$t_{CQ(min)}$ This is the minimum delay from the RCLK edge until SYNC is asserted, conservatively assumed to be 0.

t_{SQ} This is the delay for NEWBYTE to be asserted, again assumed to be 0.

$-t_h$ Minus the hold time of FF1 in Figure 8-106. NEWBYTE might be asserted at the end of the hold time and still be detected.

$0t_{SCLK}$ Zero times the SCLK period. We might get "lucky" and have the SCLK edge come along just as the hold time of FF1 is ending.

t_{SCLK} A one-SCLK-period delay to asserting SLOAD, as before.

t_{SCLK} A one-SCLK-period delay to loading SBYTE into SREG, as before.

In other words, $t_{mind} = 2t_{SCLK} - t_h - nt_{RCLK}$.

For this case, we must ensure that the new byte has propagated to the output of HREG when the setup time window of SREG begins, so we must have

$$t_{end} - t_{su} \geq t_{start} + t_{co},$$

where t_{co} is the maximum clock-to-output delay of HREG. Substituting $t_{end} = t_{start} + t_{mind}$ and subtracting t_{start} from both sides, we get

$$t_{mind} - t_{su} \geq t_{co}.$$

Substituting the value of t_{mind} and rearranging, we get the final requirement,

$$2t_{SCLK} - t_h - t_{su} - t_{co} \geq nt_{RCLK} \qquad (8\text{-}3)$$

If, for example, t_h, t_{su}, and t_{co} are 10 ns each, the maximum value of n is 3; we can't generate SYNC more than two clock ticks before its original position in Figure 8-108. This may or may not be enough to solve the maximum-delay problem, depending on other delay values; this is explored for a particular set of components in Exercise 8.95.

Moving the SYNC pulse earlier may not give enough delay improvement or may not be an available option in some systems. An alternative solution that can always be made to work is to increase the time between successive data transfers from one clock domain to the other. We can always do this because we can always transfer more bits per synchronization. In the Ethernet-interface example, we could collect 16 bits at a time in the RCLK domain and transfer 16 bits at a time to the SCLK domain. This changes the previously stated $8t_{RCLK}$ terms to $16t_{RCLK}$, providing a lot more margin for the maximum-delay timing requirements. Once 16 bits have been transferred into the SCLK domain, we can still break them into two 8-bit chunks if we need to process the data a byte at a time.

It may also be possible to improve performance by modifying the design of the SCTRL circuit. Figure 8-109 shows a version where SLOAD is generated directly by the flip-flop that samples NEWBYTE. In this way, SLOAD appears one SCLK period sooner than in our original SCTRL circuit. Also, the S-R latch is cleared sooner. This circuit works only if a couple of key assumptions are true:

1. A reduced metastability resolution time for FF1 is acceptable, equal to the time that SCLK is HIGH. Metastability must be resolved before SCLK goes LOW, because that's when the S-R latch gets cleared if SLOAD is HIGH.

2. The setup time of SREG's CLKEN input (Figure 8-102) is less than or equal to the time that SCLK is LOW. Under the previous assumption, the SLOAD signal applied to CLKEN might be metastable until SCLK goes LOW.

3. The time that SCLK is LOW is long enough to generate a reset pulse on RNEW that meets the minimum pulse-width requirement of the S-R latch.

Note that these behaviors makes proper circuit operation dependent on the duty cycle of SCLK. If SCLK is relatively slow and its duty cycle is close to 50%, this circuit generally works fine. But if SCLK is too fast or has a very small, very large, or unpredictable duty cycle, the original circuit approach must be used.

Figure 8-109
Half-clock-period SCTRL circuit for generating SLOAD.

All of these synchronization schemes require the clock frequencies to be within a certain range of each other for proper circuit operation. This must be considered for testing, where the clocks are usually run slower, and for upgrades, where one or both clocks may run faster. For example, in the Ethernet interface example, we wouldn't change the frequency of standard 100-Mbps Ethernet, but we might upgrade the PCI bus from 33 to 66 MHz.

The problems caused by clock-frequency changes can be subtle. To get a better handle on what can go wrong, it's useful to consider how a synchronizer works (or doesn't work!) if one clock frequency changes by a factor of 10 or more.

For example what happens to the synchronizer timing in Figure 8-107 if we change RCLK from 100 MHz to 10 MHz? At first glance, it would seem that all is well, since a byte now arrives only once every 800 ns, giving much more time for the byte to be transferred into the SCLK domain. Certainly, Eq. (8-1) on page 779 and Eq. (8-2) on page 780 are satisfied with much more margin. However, Eq. (8-3) is no longer satisfied, unless we reduce n to zero! This could be accomplished by generating SYNC one RCLK tick later than is shown in Figure 8-107.

But even with this change, there's *still* a problem. Figure 8-110 shows the new timing, including the later SYNC pulse. The problem is that the SYNC pulse is now 100 ns long. As before, NEWBYTE (the output of the S-R latch in

Figure 8-110 Synchronizer timing with slow (10-MHz) RCLK.

Figure 8-111
Synchronizer with
edge-triggered
SYNC detection.

Figure 8-106 on page 776) is set by SYNC and is cleared by SLOAD. The problem is that when SLOAD goes away, SYNC is still asserted, as shown in the new timing diagram. Thus, the new byte will be detected and transferred twice!

The solution to the problem is to detect only the leading edge of SYNC, so that the circuit is not sensitive to the length of the SYNC pulse. A common way of doing this is to replace the S-R latch with an edge-triggered D flip-flop, as shown in Figure 8-111. The leading edge of SYNC sets the flip-flop, while SLOAD is used as an asynchronous clear as before.

The circuit in Figure 8-111 solves the slow-RCLK problem, but it also changes the derivation of Eqns. 8-1 through 8-3 and may make timing more constrained in some areas (see Exercise 8.96). Another disadvantage that this circuit cannot be realized in a typical PLD, which has all flip-flops controlled by the same clock; instead, a discrete flip-flop must be used to detect SYNC.

After reading almost ten pages to analyze just one "simple" example, you should have a strong appreciation of the difficulty of correct synchronization-circuit design. Several guidelines can help you:

- Minimize the number of different clock domains in a system.

- Clearly identify all clock boundaries and provide clearly identified synchronizers at those boundaries.

- Provide sufficient metastability resolution time for each synchronizer so that synchronizer failure is rare, much more unlikely than other hardware failures.

- Analyze synchronizer behavior over a range of timing scenarios, including faster and slower clocks that might be applied as a result of system testing or upgrades.

- Simulate system behavior over a wide range of timing scenarios as well.

The last guideline above is a catch-all for modern digital designers, who usually rely on sophisticated, high-speed logic simulators to find their bugs. But it's not a substitute for following the first four guidelines. Ignoring them can lead to problems that cannot be detected by a typical, small number of simulation scenarios. Of all digital circuits, synchronizers are the ones for which it's most important to be "correct by design"!

References

Probably the first comprehensive set of examples of sequential MSI parts and applications appeared in *The TTL Applications Handbook*, edited by Peter Alfke and Ib Larsen (Fairchild Semiconductor, 1973). This highly practical and informative book was invaluable to this author and to many others who developed digital design curricula in the 1970s.

Another book that emphasizes design with larger-scale combinational and sequential logic functions is Thomas R. Blakeslee's *Digital Design with Standard MSI and LSI,* second edition, (Wiley, 1979). Blakeslee's coverage of the concept of space/time trade-offs is excellent, and he also one of the first to introduce the microprocessor as "a universal logic circuit."

Moving quickly from the almost forgotten to the yet-to-be discovered, manufacturers' Web sites are an excellent source of information on digital design practices. For example, a comprehensive discussion of bus-hold circuits can be found in Texas Instruments' "Implications of Slow or Floating CMOS Inputs" (publ. SCBA004B, December 1997), available on TI's web site at `www.ti.com`. Another discussion appears in Fairchild Semiconductor's "Designing with Bushold" (*sic*, publ. AN-5006, April 1999), at `www.fairchildsemi.com`.

Announcements and data sheets for all kinds of new, improved MSI and larger parts can also be found on the Web. Following a certain automobile manufacturer's proclamation in the 60s and then again in the late 90s that "wider is better," logic manufacturers have also introduced "wide-bus" registers, drivers, and transceivers that cram 16, 18, or even 32 bits of function into a high-pin-count surface-mount package. Descriptions of many such parts can be found at the Texas Instruments Web site (search for "widebus"). Other sites with a variety of logic data sheets, application notes, and other information include Motorola (`www.mot.com`), Fairchild Semiconductor (`www.fairchildsemi.com`), and Philips Semiconductor (`www.philipslogic.com`).

The field of logic design is fast moving, so much so that sometimes I wish that I wrote fiction, so that I wouldn't have to revise the "practices" discussions in this book every few years. Luckily for me, this book does cover some unchanging theoretical topics (a.k.a. "principles"), and this chapter is no exception. Logic hazards have been known since at least the 1950s, and function hazards were discussed by Edward J. McCluskey in *Logic Design Principles* (Prentice Hall, 1986). Galois fields were invented centuries ago, and their applications to error-correcting codes, as well as to the LFSR counters of this chapter, are described in introductory books on coding theory, including *Error-Control Techniques for Digital Communication* by A. M. Michelson and A. H. Levesque (Wiley-Interscience, 1985). A mathematical theory of state-machine decomposition has been studied for years; Zvi Kohavi devotes a chapter to the topic in his classic book *Switching and Finite Automata Theory*, second edition (McGraw-Hill, 1978). But let us now return to the less esoteric.

As we mentioned in Section 8.4.5, some PLDs and CPLDs contain XOR structures that allow large counters to be designed without a large number of product terms. This requires a somewhat deeper understanding of counter excitation equations, as described in Section 10.5 of the second edition of this book. This material can also be found on the web at `www.ddpp.com`.

The general topics of clock skew and multiphase clocking are discussed in McCluskey's *Logic Design Principles*, while an illuminating discussion of these topics as applied to VLSI circuits can be found in *Introduction to VLSI Systems* by Carver Mead and Lynn Conway (Addison-Wesley, 1980). Mead and Conway also provide an introduction to the important topic of *self-timed systems* that eliminate the system clock, allowing each logic element to proceed at its own rate. To give credit where credit is due, we note that all of the Mead and Conway material on system timing, including an outstanding discussion of metastability, appears in their Chapter 7 written by Charles L. Seitz.

self-timed systems

Thomas J. Chaney spent decades studying and reporting on the metastability problem. One of his more important works, "Measured Flip-Flop Responses to Marginal Triggering" (*IEEE Trans. Comput.*, Vol. C-32, No. 12, December 1983, pp. 1207–1209), reports some of the results that we showed in Table 8-35.

For the mathematically inclined, Lindsay Kleeman and Antonio Cantoni have written "On the Unavoidability of Metastable Behavior in Digital Systems" (*IEEE Trans. Comput.*, Vol. C-36, No. 1, January 1987, pp. 109–112); the title says it all. The same authors posed the question, "Can Redundancy and Masking Improve the Performance of Synchronizers?" (*IEEE Trans. Comput.*, Vol. C-35, No. 7, July 1986, pp. 643–646). Their answer in that paper was "No," but a response from a reviewer caused them to change their minds to "Maybe." Obviously, they've gone metastable themselves! Having two authors and a reviewer didn't improved their performance, so the obvious answer to their original question is "No"! In any case, Kleeman and Antonio's papers provide a good set of pointers to mainstream scholarly references on metastability.

The most comprehensive set of early references on metastability (not including Greek philosophers or Devo) is Martin Bolton's "A Guided Tour of 35 Years of Metastability Research" (*Proc. Wescon 1987*, Session 16, "Everything You Might Be Afraid to Know about Metastability," Wescon Session Records, `www.wescon.com`, 8110 Airport Blvd., Los Angeles, CA 90045).

In recent years, as system clock speeds have steadily increased, many IC manufacturers have become much more conscientious about measuring and publishing the metastability characteristics of their devices, often on the Web. Texas Instruments (`www.ti.com`) provides a very good discussion including test circuits and measured parameters for ten different logic families in "Metastable Response in 5-V Logic Circuits" by Eilhard Haseloff (TI pub. SDYA006, 1997). Cypress Semiconductor (`www.cypress.com`) publishes an application note, "Are Your PLDs Metastable?" (1997) that is an excellent reference including some history (going back to 1952!), an analog circuit analysis of metastability, test and measurement circuits, and metastability parameters for Cypress PLDs. Another recent note is "Metastability Considerations" from Xilinx Corporation (www.xilinx.com, publ. XAPP077, 1997), which gives measured parameters for their XC7300 and XC9500 families of CPLDs. Of particular interest is the clever circuit and methodology that allows them to count metastable events *inside* the device, even though metastable waveforms are not observable on external pins.

Most digital design textbooks now give good coverage to metastability, prompted by the existence of metastability in real circuits and perhaps also by competition—since 1990, the textbook you're reading has been promoting the topic by introducing metastability in its earliest coverage of sequential circuits. On the analog side of the house, Howard Johnson and Martin Graham provide a nice introduction and a description of how to observe metastable states in their *High-Speed Digital Design: A Handbook of Black Magic* (Prentice Hall, 1993).

Drill Problems

8.1   Suppose that in Table 8-2, the second RAM bank (RAMCS1) is decoded instead using the expression ((ABUS >= 0x010) & (ABUS < 0x020)). Does this yield the same results as the original expression, (ABUS == RAMBANK0)? Explain.

8.2   Determine the number of fuses in each of the PAL devices in Table 8-9.

8.3   How many fuses are contained in the 16V8 as described in the text? (The commercial device has additional fuses, not described in the text, for user-specific information and security.)

8.4   How many fuses are contained in the 22V10 as described in the text? (The commercial device has additional fuses, not described in the text, for user-specific information and security.)

8.5   Determine f_{max} with external feedback for all of the devices in Table 8-10.

8.6   Determine f_{max} with internal feedback for all of the GAL devices in Table 8-10.

8.7   Write an ABEL program for a 16V8 that gives it exactly the same functionality as a 74x374.

8.8   Write an ABEL program for a GAL16V8 or GAL20V8 that gives it exactly the same functionality as a 74x377.

8.9   Compare the propagation delays from AVALID to a chip-select output for the two decoding approaches shown in Figures 8-15 and 8-16. Assume that 74FCT373 latches and 10-ns 16V8C devices are used in both designs. Repeat for the delay from ABUS to a chip-select output.

8.10   What would happen if you replaced the edge-triggered D flip-flops in Figure 7-38 with D latches?

8.11   Modify the ABEL program in Table 8-12 to perform the function of a 74X162 decade counter.

8.12   Modify the ABEL program in Table 8-12 to perform the function of a 74X169 up/down counter. Does it still fit in a 16V8?

Figure X8.13

8.13 What is the counting sequence of the circuit shown in Figure X8.13?

8.14 What is the behavior of the counter circuit of Figure 8-36 if it is built using a 74x161 instead of a 74x163?

8.15 What is the behavior of the counter circuit of Figure 8-36 if the bottom input of U2 is connected to Q0 instead of Q1?

8.16 A 74x163 counter is hooked up with inputs ENP, ENT, A, and D always HIGH, inputs B and C always LOW, input LD_L = (QB · QC)', and input CLR_L = (QC · QD)'. The CLK input is hooked up to a free-running clock signal. Draw a logic diagram for this circuit. Assuming that the counter starts in state 0000, write the output sequence on QD QC QB QA for the next 15 clock ticks.

8.17 Determine the widths of the glitches shown in Figure 8-43 on the Y2_L output of a 74x138 decoder, assuming that the '138 is internally structured as shown in Figure 5-37(a) on page 358, and that each internal gate has a delay of 10 ns.

8.18 Starting with state 0001, write the sequence of states for a 4-bit LFSR counter designed according to Figure 8-68 and Table 8-21.

8.19 Calculate the minimum clock period of the data unit in Figure 8-82. Use the maximum propagation delays given in Table 5-3 for LS-TTL combinational parts. Unless you can get the real numbers from a TTL data book, assume that all registers require a 10-ns minimum setup time on inputs and have a 20-ns propagation delay from clock to outputs. Indicate any assumptions you've made about delays in the control unit.

8.20 Calculate the MTBF of a synchronizer built according to Figure 8-96 using 74F74s, assuming a clock frequency of 25 MHz and an asynchronous transition rate of 1 MHz. Assume the setup time of an 'F74 is 5 ns and the hold time is zero.

8.21 Calculate the MTBF of the synchronizer shown in Figure X8.21, assuming a clock frequency of 25 MHz and an asynchronous transition rate of 1 MHz. Assume that the setup time t_{setup} and the propagation delay t_{pd} from clock to Q or QN in a 74ALS74 are both 10 ns.

Figure X8.21

Exercises

8.22   What does the TTL Data Book have to say about momentarily shorting the outputs of a gate to ground as we do in the switch debounce circuit of Figure 8-5?

8.23   Investigate the behavior of the switch debounce circuit of Figure 8-5 if 74HCT04 inverters are used; repeat for 74AC04 inverters.

8.24   Suppose you are asked to design a circuit that produces a debounced logic input from an SPST (single-pole, *single*-throw) switch. What inherent problem are you faced with?

8.25   Explain why CMOS bus holder circuits don't work well on three-state buses with TTL devices attached. (*Hint:* Consider TTL input characteristics.)

8.26   Write a single VHDL program that combines the address latch and latching decoder of Figure 8-16 and Table 8-2. Use the signal name LA[19:0] for the latched address outputs.

8.27   Design a 4-bit ripple counter using four D flip-flops and no other components.

8.28   What is the maximum propagation delay from clock to output for the 4-bit ripple counter of Exercise 8.27 using 74HCT flip-flops? Repeat, using 74AHCT and 74LS74 flip-flops.

8.29   Design a 4-bit ripple *down* counter using four D flip-flops and no other components.

8.30   What limits the maximum counting speed of a ripple counter, if you don't insist on being able to read the counter value at all times?

8.31   Based on the design approach in Exercise 8.27 and the answer to Exercise 8.30, what is the maximum counting speed (frequency) for a 4-bit ripple counter using 74HCT flip-flops? Repeat, using 74AHCT and 74LS74 flip-flops.

8.32   Write a formula for the maximum clock frequency of the synchronous serial binary counter circuit in Figure 8-28. In your formula, let t_{TQ} denote the propagation delay from T to Q in a T flip-flop, t_{setup} the setup time of the EN input to the rising edge of T, and t_{AND} the delay of an AND gate.

8.33   Repeat Exercise 8.32 for the synchronous parallel binary counter circuit shown in Figure 8-29, and compare results.

8.34 Repeat Exercise 8.32 for an *n*-bit synchronous serial binary counter.

8.35 Repeat Exercise 8.32 for an *n*-bit synchronous parallel binary counter. Beyond what value of *n* is your formula no longer valid?

8.36 Using a 74x163 4-bit binary counter, design a modulo-11 counter circuit with the counting sequence 3, 4, 5, …, 12, 13, 3, 4, ….

8.37 Look up the internal logic diagram for a 74x162 synchronous decade counter in a data book, and write its state table in the style of Table 8-11, including its counting behavior in the normally unused states 10–15.

8.38 Devise a cascading scheme for 74x163s, analogous to the synchronous parallel counter structure of Figure 8-29, such that the maximum counting speed is the same for any counter with up to 36 bits (nine '163s). Determine the maximum counting speed using worst-case delays from a manufacturer's data book for the '163s and any SSI components used for cascading.

8.39 Design a modulo-129 counter using two 74x163s and a single inverter.

8.40 Write an ABEL program for an 8-bit module-*N* counter with load input using a PAL22v10, where the value of *N* is specified by a constant N in the program.

8.41 Design a clocked synchronous circuit with four inputs, N3, N2, N1, and N0, that represent an integer *N* in the range 0–15. The circuit has a single output Z that is asserted for exactly *N* clock ticks during any 16-tick interval (assuming that *N* is held constant during the interval of observation). (*Hints:* Use combinational logic with a 74x163 set up as a free-running divide-by-16 counter. The ticks in which Z is asserted should be spaced as evenly as possible, that is, every second tick when *N* = 8, every fourth when *N* = 4, and so on.)

8.42 Modify the circuit of Exercise 8.41 so that Z produces *N transitions* in each 16-tick interval. The resulting circuit is called a *binary rate multiplier* and was once sold as a TTL MSI part, the 7497. (*Hint:* Gate the clock with the level output of the previous circuit.)

binary rate multiplier

8.43 Repeat Exercises 8.41 and 8.42 using an 8-bit input N7..N0, and realize the circuit using an ABEL program for a single PAL22V10.

8.44 Repeat Exercises 8.41 and 8.42 using an 8-bit input N7..N0, and describe the desigbn using a behavioral VHDL program.

8.45 A digital designer (the author!) was asked at the last minute to add new functionality to a PCB that had room for just one more 16-pin MSI IC. The PCB already had a 16-MHz clock signal, MCLK, and a spare microprocessor-controlled select signal, SEL. The designer was asked to provide a new clock signal, UCLK, whose frequency would be 8 MHz or 4 MHz depending on the value of SEL. To make things worse, the PCB had no spare SSI gates, and UCLK was required to have a 50% duty cycle at both frequencies. It took the designer about five minutes to come up with a circuit. Now it's your turn to do the same. (*Hint:* The designer had long considered the 74x163 to be the fundamental building block of tricky sequential-circuit design.)

8.46 Design a modulo-16 counter, using one 74x169 and at most one SSI package, with the following counting sequence: 7, 6, 5, 4, 3, 2, 1, 0, 8, 9, 10, 11, 12, 13, 14, 15, 7, ….

8.47   Write an ABEL program for an 8-bit counter that realizes a counting sequence similar to the one in Exercise 8.46.

8.48   Design a binary up/down counter for the elevator controller in a 20-story building, using a single 16V8. The counter should have enable and up/down control inputs. It should stick at state 1 when counting down, stick at state 21 when counting up, and skip state 13 in either mode. Draw a logic diagram and write ABEL equations for your design.

8.49   Repeat the preceding exercise using VHDL.

8.50   Write a VHDL program for an n-bit counter that realizes a counting sequence similar to the one in Exercise 8.46. Write the program in such a way that the size of the counter can be changed by changing the value of a single constant N.

8.51   Modify the VHDL program in Table 8-14 so that the type of ports D and Q is STD_LOGIC_VECTOR, including conversion functions as required.

8.52   Modify the program in Table 8-16 to use structural VHDL, so it conforms exactly to the circuit in Figure 8-45, including the signal names shown in the figure. Define and use any of the following entities that don't already exist in your VHDL library: AND2, INV, NOR2, OR2, XNOR2, Vdffqqn.

8.53   Modify the program in Table 8-17 to use VHDL's generic statement, so that the counter size can be changed using the generic definition.

8.54   Design a parallel-to-serial conversion circuit with eight 2.048-Mbps, 32-channel serial links and a single 2.048-MHz, 8-bit, parallel data bus that carries 256 bytes per frame. Each serial link should have the frame format defined in Figure 8-55. Each serial data line SDATAi should have its own sync signal SYNCi; the sync pulses should be staggered so that SYNCi + 1 has a pulse one tick after SYNCi.

8.55   Show the timing of the parallel bus and the serial links, and write a table or formula that shows which parallel-bus timeslots are transmitted on which serial links and timeslots. Draw a logic diagram for the circuit using MSI parts from this chapter; you may abbreviate repeated elements (e.g., shift registers), showing only the unique connections to each one.

8.56   Repeat Exercise 8.54, assuming that all serial data lines must reference their data to a single, common SYNC signal. How many more chips does this design require?

8.57   Show how to enhance the serial-to-parallel circuit of Exercise 8-57 so that the byte received in each timeslot is stored in its own register for 125 μs, until the next byte from that timeslot is received. Draw the counter and decoding logic for 32 timeslots in detail, as well as the parallel data registers and connections for timeslots 31, 0, and 1. Also draw a timing diagram in the style of Figure 8-58 that shows the decoding and data signals associated with timeslots 31, 0, and 1.

8.58   Suppose you are asked to design a serial computer, one that moves and processes data one bit at a time. The first decision you must make is which bit to transmit and process first, the LSB or the MSB. Which would you choose, and why?

8.59   Design an 8-bit self-correcting ring counter whose states are 11111110, 11111101, ..., 01111111, using only two SSI/MSI packages.

8.60 Design two different 2-bit, 4-state counters, where each design uses just a single 74x74 package (two edge-triggered D flip-flops) and no other gates.

8.61 Design a 4-bit Johnson counter and decoding for all eight states using just four flip-flops and eight gates. Your counter need not be self-correcting.

8.62 Prove that an even number of shift-register outputs must be connected to the odd-parity circuit in an n-bit LFSR counter if it generates a maximum-length sequence. (Note that this is a necessary but not a sufficient requirement. Also, although Table 8-21 is consistent with what you're supposed to prove, simply quoting the table is not a proof!)

8.63 Prove that X0 must appear on the righthand side of any LFSR feedback equation that generates a maximum-length sequence. (*Note:* Assume the LFSR bit ordering and shift direction are as given in the text; that is, the LFSR counter shifts right, toward the X0 stage.)

8.64 Suppose that an n-bit LFSR counter is designed according to Figure 8-68 and Table 8-21. Prove that if the odd-parity circuit is changed to an even-parity circuit, the resulting circuit is a counter that visits $2^n - 1$ states, including all of the states except $11\ldots11$.

8.65 Find a feedback equation for a 3-bit LFSR counter, other than the one given in Table 8-21, that produces a maximum-length sequence.

8.66 Given an n-bit LFSR counter that generates a maximum-length sequence ($2^n - 1$ states), prove that an extra XOR gate and an $n - 1$ input NOR gate connected as suggested in Figure 8-69 produce a counter with 2^n states.

8.67 Prove that a sequence of 2^n states is still obtained if a NAND gate is substituted for a NOR above, but that the state sequence is different.

8.68 Design an iterative circuit for checking the parity of a 16-bit data word with a single even-parity bit. Does the order of bit transmission matter?

8.69 Modify the shift-register program in Table 8-23 to provide an asynchronous clear input using a 22V10.

8.70 Write an ABEL program that provides the same functionality as a 74x299 shift register. Show how to fit this function into a single 22V10, or explain why it does not fit.

8.71 Determine the number of product terms required for each output of the RING8 PLD in Table 8-25. Does it fit in a 16R8 or 16V8R?

8.72 In what situations do the ABEL programs in Tables 8-26 and 8-27 give different operational results?

8.73 Modify the ABEL program in Table 8-26 so that the phases are always at least two clock ticks long, even if RESTART is asserted at the beginning of a phase. RESET should still take effect immediately.

8.74 Repeat the preceding exercise for the program in Table 8-27.

8.75 Suppose the timing generator of Table 8-26 is used to control a dynamic memory system, such that all six phases must be completed to read or write the memory. If the timing generator is reset during a write operation without completing all six phases, the memory contents may be corrupted. Modify the equations in Table 8-26 to avoid this problem.

8.76 A student proposed to create the timing waveforms of Figure 8-72 by starting with the ABEL program in Table 8-27 and changing the encoding of each of states P1F, P2F, ..., P6F so that the corresponding phase output is 1 instead of 0, so that the phase output is 0 only during the second tick of each phase, as required. Is this a good approach? Comment on the results produced by the ABEL compiler when you try this.

8.77 The output waveforms produced by the ABEL programs in Tables 8-29 and 8-30 are not identical when the RESTART and RUN inputs are changed. Explain the reason for this, and then modify the program in Table 8-30 so that its behavior matches that of Table 8-29.

8.78 The ABEL ring-counter implementation in Table 8-26 is not self-synchronizing. For example, describe what happens if the outputs [P1_L..P6_L] are initially all 0, and the RUN input is asserted without ever asserting RESET or RESTART. What other starting states exhibit this kind of non-self-synchronizing behavior? Modify the program so that it *is* self-synchronizing.

8.79 Repeat the preceding exercise for the VHDL ring-counter implementation in Table 8-33.

8.80 Design an iterative circuit with one input B_i per stage and two boundary outputs X and Y such that X = 1 if at least two B_i inputs are 1 and Y = 1 if at least two *consecutive* B_i inputs are 1.

8.81 Design a combination-lock machine according to the state table of Table 7-14 on page 582 using a single 74x163 counter and combinational logic for the LD_L, CLR_L, and A–D inputs of the '163. Use counter values 0–7 for states A–H.

8.82 Write an ABEL program corresponding to the state diagram in Figure 8-84 for the multiplier control unit.

8.83 Write a VHDL program corresponding to the state diagram in Figure 8-84 for the multiplier control unit.

8.84 Write a VHDL program that performs with the same inputs, outputs, and functions as the multiplier data unit in Figure 8-82.

8.85 Write a VHDL program that combines the programs in the two preceding exercises to form a complete 8-bit shift-and-add multiplier.

8.86 The text stated that the designer need not worry about any timing problems in the synchronous design of Figure 8-83. Actually, there is one slight worry. Look at the timing specifications for the 74x377 and discuss.

8.87 Determine the minimum clock period for the shift-and-add multiplier circuit in Figure 8-83, assuming that the state machine is realized in a single GAL16V8-20 and that the MSI parts are all 74LS TTL. Use worst-case timing information from the tables in this book. For the '194, t_{pd} from clock to any output is 26 ns, and t_s is 20 ns for serial and parallel data inputs and 30 ns for mode-control inputs.

8.88 Design a data unit and a control-unit state machine for multiplying 8-bit two's-complement numbers using the algorithm discussed in Section 2.8.

Figure X8.91

8.89   Design a data unit and control-unit state machine for dividing 8-bit unsigned numbers using the shift-and-subtract algorithm discussed in Section 2.9.

8.90   Suppose that the SYNCIN signal in Drill 8.21 is connected to a combinational circuit in the synchronous system, which in turn drives the D inputs of 74ALS74 flip-flops that are clocked by CLOCK. What is the maximum allowable propagation delay of the combinational logic?

8.91   The circuit in Figure X8.91 includes a deskewing flip-flop so that the synchronized output from the multiple-cycle synchronizer is available as soon as possible after the edge of CLOCK. Ignoring metastability considerations, what is the maximum frequency of CLOCK? Assume that for a 74F74, $t_{setup} = 5$ ns and $t_{pd} = 7$ ns.

8.92   Using the maximum clock frequency determined in Exercise 8.91, and assuming an asynchronous transition rate of 4 MHz, determine the synchronizer's MTBF.

8.93   Determine the MTBF of the synchronizer in Figure X8.91, assuming an asynchronous transition rate of 4 MHz and a clock frequency of 40 MHz, which is less than the maximum determined in Figure X8.91. In this situation, "synchronizer failure" really occurs only if DSYNCIN is metastable. In other words, SYNCIN may be allowed to be metastable for a short time, as long as it doesn't affect DSYNCIN. This yields a better MTBF.

8.94   Look up U.S. patent number 4,999,528, "Metastable-proof flip-flop," and describe why it doesn't always work as advertised. (*Hints:* Patents can be found at www.patents.ibm.com. There's enough information in this patent's abstract to figure out how the circuit can fail.)

8.95   In the synchronization circuit of Figures 8-102, 8-104, and 8-106, you can reduce the delay of the transfer of a byte from the RCLK domain to the SCLK domain if you use an earlier version of the SYNC pulse to start the synchronizer. Assuming that you can generate SYNC during any bit of the received byte, which bit should you use to minimize the delay? Also determine whether your solution satisfies the maximum-delay requirements for the circuit. Assume that all the components have 74AHCT timing and that the S-R latch is built from a pair of cross-coupled NOR gates, and show a detailed timing analysis for your answers.

8.96 Instead of using a latch in the synchronization control circuit of Figure 8-106, some applications use an edge-triggered D flip-flop as shown in Figure 8-111. Derive the maximum-delay and minimum-delay requirements for this circuit, corresponding to Eqs. (8-1) through (8-3), and discuss whether this approach eases or worsens the delay requirements.

8.97 A famous digital designer devised the circuit shown in Figure X8.97(a), which is supposed to eliminate metastability within one period of a system clock. Circuit M is a memoryless analog voltage detector whose output is 1 if Q is in the metastable state, 0 otherwise. The circuit designer's idea was that if line Q is detected to be in the metastable state when CLOCK goes low, the NAND gate will clear the D flip-flop, which in turn eliminates the metastable output, causing a 0 output from circuit M and thus negating the CLR input of the flip-flop. The circuits are all fast enough that this all happens well before CLOCK goes high again; the expected waveforms are shown in Figure X8.97(b).

Unfortunately, the synchronizer still failed occasionally, and the famous digital designer is now designing pockets for blue jeans. Explain, in detail, how it failed, including a timing diagram.

Figure X8.97

(a)

(b)

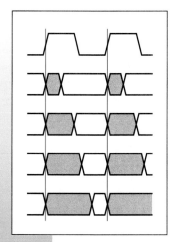

chapter 9

Sequential-Circuit Design Examples

J ust about every real digital system is a sequential circuit—all it takes is one feedback loop, latch, or flip-flop to make a circuit's present behavior depend on its past inputs. However, if we were to design or analyze a typical digital system as a single sequential circuit, we would end up with an enormous state table. For example, strictly speaking, a PC with only 16 Mbytes of main memory is a sequential circuit with well over $2^{128,000,000}$ states!

We noted in previous chapters that we typically deal with digital systems in smaller chunks, for example, partitioning them into data paths, registers, and control units (Section 8.7.1). In fact, a typical system has multiple functional units with well-defined interfaces and connections between them (as supported by VHDL, Section 4.7.2), and each functional unit may contain a hierarchy with several layers of abstraction (as supported by both VHDL and typical schematic drawing packages [Section 5.1.6]). Thus, we can deal with sequential circuits in much smaller chunks.

After all of this build-up, I have to admit that the design of complex, hierarchical digital systems is beyond the scope of this text. However, the heart of any system or any of its subsystems is typically a state machine or other sequential circuit, and that's something we *can* study here. So, this chapter will try to reinforce your understanding of sequential circuit and state-machine design by presenting several examples.

Early digital designers and many designers through the 1980s wrote out state tables by hand and built corresponding circuits using the synthesis methods that we described in Section 7.4. However, hardly anyone does that anymore. Nowadays, most state tables are specified with a hardware description language (HDL) such as ABEL, VHDL, or Verilog. The HDL compiler then performs the equivalent of our synthesis methods and realizes the specified machine in a PLD, CPLD, FPGA, ASIC, or other target technology.

This chapter gives state-machine and other sequential-circuit design examples in two different HDLs. In the first section we give examples in ABEL, and we target the resulting machines to small PLDs. Some of these examples illustrate the partitioning decisions that are needed when an entire circuit does not fit into a single component. In the second section we give examples in VHDL, which can be targeted to just about any technology.

Like Chapter 6, this chapter has as prerequisites only the chapters that precede it. Its two sections are written to be pretty much independent of each other, and the rest of the book is written so that you can read this chapter now or skip it and come back later.

9.1 Design Examples Using ABEL and PLDs

In Section 7.4 we designed several simple state machines by translating their word descriptions into a state table, choosing a state assignment, and finally synthesizing a corresponding circuit. We repeated one of these examples using ABEL and a PLD in Table 7-27 on page 634 and another in Table 7-31 on page 637. These designs were much easier to do using ABEL and a PLD, for two reasons:

- You don't have to be too concerned about the complexity of the resulting excitation equations, as long as they fit in the PLD.

- You may be able to take advantage of ABEL language features to make the design easier to express and understand.

Before looking at more examples, let's examine the timing behavior and packaging considerations for state machines that are built from PLDs.

9.1.1 Timing and Packaging of PLD-Based State Machines

Figure 9-1 shows how a generic PLD with both combinational and registered outputs might be used in a state-machine application. Timing parameters t_{CO} and t_{PD} were explained in Section 8.3.3; t_{CO} is the flip-flop clock-to-output delay and t_{PD} is the delay through the AND-OR array.

State variables are assigned to registered outputs, of course, and are stable at time t_{CO} after the rising edge of CLOCK. Mealy-type outputs are assigned to combinational outputs and are stable at time t_{PD} after any input change that

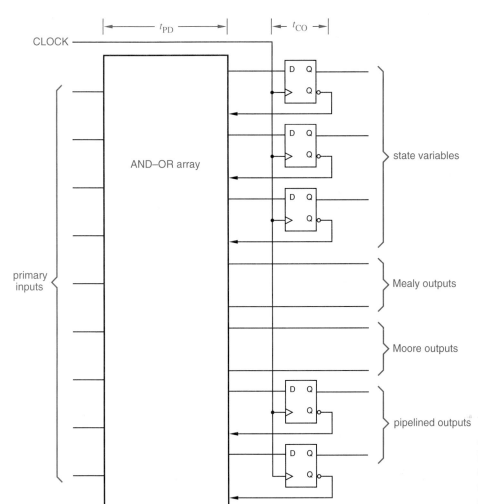

Figure 9-1
Structure and timing
of a PLD used as a
state machine.

affects them. Mealy-type outputs may also change after a state change, in which case they become stable at time $t_{CO} + t_{PD}$ after the rising edge of CLOCK.

A Moore-type output is, by definition, a combinational logic function of the current state, so it is also stable at time $t_{CO} + t_{PD}$ after the CLOCK. Thus, Moore outputs may not offer any speed advantage over Mealy outputs in a PLD realization. For faster propagation delay, we defined and used pipelined outputs in Sections 7.3.2 and 7.11.5. In a PLD realization, these outputs come directly from a flip-flop output and thus have a delay of only t_{CO} from the clock. Besides having a shorter delay, they are also guaranteed to be glitch free, important in some applications.

PLD-based state-machine designs are often limited by the number of input and output pins available in a single PLD. According to the model of Figure 9-1,

Figure 9-2
Splitting a state-machine design into three PLDs.

one PLD output is required for each state variable and for each Mealy- or Moore-type output. For example, the T-bird tail-lights machine of Section 7.5 starting on page 585 requires three registered outputs for the state variables and six combinational outputs for the lamp outputs, too many for most of the PLDs that we've described, except for the 22V10.

On the other hand, an output-coded state assignment (Section 7.3.2) usually requires a smaller total number of PLD outputs. Using an output-coded state assignment, the T-bird tail-lights machine can be built in a single 16V8, as we'll show in Section 9.1.3.

Like any state variable or Moore-type output, an output-coded state variable is stable at time t_{CO} after the clock edge. Thus, an output-coded state assignment improves speed as well as packaging efficiency. In T-bird tail lights, turning on the emergency flashers 10 ns sooner doesn't matter, but in a high-performance digital system, an extra 10 ns of delay could make the difference between a maximum system clock rate of 100 MHz and a maximum of only 50 MHz.

If a design *must* be split into two or more PLDs, the best split in terms of both design simplicity and packaging efficiency is often the one shown in Figure 9-2. A single sequential PLD is used to create the required next-state behavior, and combinational PLDs are used to create both Mealy- and Moore-type outputs from the state variables and inputs.

RELIEF FOR A SPLITTING HEADACHE	Modern software tools for PLD-based system design can eliminate some of the trial and error that might otherwise be associated with fitting a design into a set of PLDs. To use this capability, the designer enters the equations and state-machine descriptions for the design, without necessarily specifying the actual devices and pinouts that should be used to realize the design. A software tool called a *partitioner* attempts to split the design into the smallest possible number of devices from a given family, while minimizing the number of pins used for interdevice communication. Partitioning can be fully automatic, partially user controlled, or fully user controlled.

Larger devices—CPLDs and FPGAs—often have internal, architectural constraints that may create headaches in the absence of expert software assistance. It appears to the designer, based on input, output, and total combinational logic and flip-flop requirements, that a design will fit in a single CPLD or FPGA. However, the design must still be split among multiple PLDs or logic blocks inside the larger device, where each block has only limited functionality.

For example, an output that requires many product terms may have to steal some from physically adjacent outputs. This may in turn affect whether adjacent pins can be used as inputs or outputs, and how many product terms are available to them. It may also affect the ability to interconnect signals between nearby blocks and the worst-case delay of these signals. All of these constraints can be tracked by *fitter* software that uses a heuristic approach to find the best split of functions among the blocks within a single CPLD or FPGA.

In many design environments, the partitioner and fitter software work together and interactively with the designer to find an acceptable realization using a set of PLDs, CPLDs, and FPGAs.

9.1.2 A Few Simple Machines

In Section 7.4 we designed several simple state machines using traditional methods. We presented an ABEL- and PLD-based design for the first of these in Table 7-25 on page 631, and then we showed how we could make the machine's operation more understandable by making better use of ABEL features in Table 7-27 on page 634.

Our second example was a "1s-counting machine" with the following specification:

Design a clocked synchronous state machine with two inputs, X and Y, and one output, Z. The output should be 1 if the number of 1 inputs on X and Y since reset is a multiple of 4, and 0 otherwise.

We developed a state table for this machine in Table 7-12 on page 580. However, we can express the machine's function much more easily in ABEL, as shown in Table 9-1. Notice that for this machine we were better off not using ABEL's "state diagram" syntax. We could express the machine's function more naturally

Table 9-1
ABEL program for
ones-counting
state machine.

```
module onesctsm
title 'Ones-counting State Machine'
ONESCTSM device 'P16V8R';

" Inputs and outputs
CLOCK, RESET, X, Y     pin 1, 2, 3, 4;
Z                      pin 13 istype 'com';
COUNT1..COUNT0         pin 14, 15 istype 'reg';

" Sets
COUNT = [COUNT1..COUNT0];

equations
COUNT.CLK = CLOCK;
WHEN RESET THEN COUNT := 0;
ELSE WHEN X & Y THEN COUNT := COUNT + 2;
ELSE WHEN X # Y THEN COUNT := COUNT + 1;
ELSE COUNT := COUNT;

Z = (COUNT == 0);

end onesctsm
```

using a nested WHEN statement and the built-in addition operation. The first
WHEN clause forces the machine to its initial state and count of 0 upon reset, and
the succeeding clauses increase the count by 2, 1, or 0 as required when the
machine is not reset. Note that ABEL "throws away" the carry bit in addition,
which is equivalent to performing addition modulo-4. The machine easily fits
into a GAL16V8 device. It has four states, because there are two flip-flops its
realization.

Another example from Section 7.4 is a combination-lock state machine
(below we omit the HINT output in the original specification):

> Design a clocked synchronous state machine with one input, X, and one
> output, UNLK. The UNLK output should be 1 if and only if X is 0 and
> the sequence of inputs received on X at the preceding seven clock ticks
> was 0110111.

**RESETTING
BAD HABITS**

In Chapter 7 we started the bad habit of designing state tables without including an
explicit reset input. There was a reason for this—each additional input potentially
doubles the amount work we would have had to do to synthesize the state machine
using manual methods, and there was little to be gained pedagogically.

Now that we're doing language-based state-machine design using automated
tools, we should get into the habit of always providing an explicit reset input that
sends the machine back to a known state. It's a requirement in real designs!

We developed a state table for this machine in Table 7-14 on page 582, and we wrote an equivalent ABEL program in Table 7-31 on page 637. However, once again we can take a different approach that is easier to understand. For this example, we note that the output of the machine at any time is completely determined by its inputs over the preceding eight clock ticks. Thus, we can use a so-called "finite-memory" approach to design this machine, where we explicitly keep track of the past seven inputs and then form the output as a combinational function of these inputs.

The ABEL program in Table 9-2 uses the finite-memory approach. It is written using sets to make modifications easy, for example, changing the combination. However, note that the HINT output would be just as difficult to provide in this version of the machine as in the original (see Exercise 9.2).

Table 9-2 Finite-memory program for combination-lock state machine.

```
module comblckf
title 'Combination-Lock State Machine'
"COMBLCKF device 'P16V8R';

" Input and output pins
CLOCK, RESET, X            pin 1, 2, 3;
X1..X7                     pin 12..18 istype 'reg';
UNLK                       pin 19;

" Sets
XHISTORY = [X7..X1];
SHIFTX   = [X6..X1, X];

equations

XHISTORY.CLK = CLOCK;
XHISTORY := !RESET & SHIFTX;

UNLK = !RESET & (X == 0) & (XHISTORY == [0,1,1,0,1,1,1]);

END comblckf
```

FINITE-MEMORY DESIGN	The finite-memory design approach to state-machine design can be generalized. Such a machine's outputs are completely determined by its current input and outputs during the previous n clock ticks, where n is a finite, bounded integer. Any machine that can be realized as shown in Figure 9-3 is a finite-memory machine. Note that a finite-*state* machine need not be a finite-*memory* machine. For example, the ones-counting machine in Table 9-1 has only four states but is not a finite-memory machine, its output depends on every value of X and Y since reset.

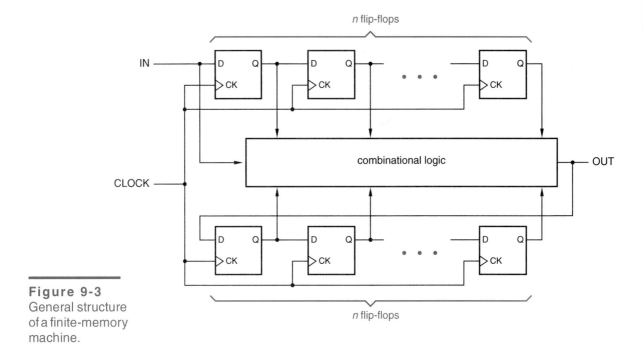

Figure 9-3
General structure
of a finite-memory
machine.

9.1.3 T-Bird Tail Lights

We described and designed a "T-bird tail-lights" state machine in Section 7.5. Table 9-3 is an equivalent ABEL "state diagram" for the T-bird tail-lights machine. There is a close correspondence between this program and the state diagram of Figure 7-64 on page 589 using the state assignment of Table 7-16 on page 590. Except for the added RESET input, the program produces exactly the same reduced equations as the explicit transition equations resulting from the transition list, which we worked out by hand in Section 7.6 on page 592.

The program in Table 9-3 handles only the state variables of the tail-lights machine. The output logic requires six combinational outputs, but only five more outputs are available in the 16V8 specified in the program. A second PLD could be used to decode the states, using the kind of partitioning that we showed in Figure 9-2. Alternatively, a larger PLD, such as the 22V10, could provide enough outputs for a single-PLD design.

An even better approach is to recognize that the output values of the tail-lights machine are different in each state, so we can also use an output-coded state assignment. This requires only six registered outputs and no combinational outputs of a 16V8, as shown in Figure 9-4. Only the device, pin, and state definitions in the previous ABEL program must be changed, as shown in Table 9-4. The six resulting excitation equations each use four product terms.

Table 9-3
ABEL program for
the T-bird tail-lights
machine.

```
module tbirdsd
title 'State Machine for T-Bird Tail Lights'
TBIRDSD device 'P16V8R';

" Input and output pins
CLOCK, LEFT, RIGHT, HAZ, RESET    pin 1, 2, 3, 4, 5;
Q0, Q1, Q2                        pin 14, 15, 16 istype 'reg';

" Definitions
QSTATE = [Q2,Q1,Q0];          " State variables
IDLE   = [ 0, 0, 0];          " States
L1     = [ 0, 0, 1];
L2     = [ 0, 1, 1];
L3     = [ 0, 1, 0];
R1     = [ 1, 0, 1];
R2     = [ 1, 1, 1];
R3     = [ 1, 1, 0];
LR3    = [ 1, 0, 0];

equations
QSTATE.CLK = CLOCK;

state_diagram QSTATE
state IDLE: IF RESET THEN IDLE
            ELSE IF (HAZ # LEFT & RIGHT) THEN LR3
            ELSE IF LEFT THEN L1 ELSE IF RIGHT THEN R1
            ELSE IDLE;
state L1:   IF RESET THEN IDLE ELSE IF HAZ THEN LR3 ELSE L2;
state L2:   IF RESET THEN IDLE ELSE IF HAZ THEN LR3 ELSE L3;
state L3:   GOTO IDLE;
state R1:   IF RESET THEN IDLE ELSE IF HAZ THEN LR3 ELSE R2;
state R2:   IF RESET THEN IDLE ELSE IF HAZ THEN LR3 ELSE R3;
state R3:   GOTO IDLE;
state LR3:  GOTO IDLE;

end tbirdsd
```

Figure 9-4
A single-PLD design for
T-bird tail lights.

Table 9-4
Output-coded state assignment for the T-bird tail-lights machine.

```
module tbirdsdo
title 'Output-Coded T-Bird Tail Lights State Machine'
TBIRDSDO device 'P16V8R';

" Input and output pins
CLOCK, LEFT, RIGHT, HAZ, RESET      pin 1, 2, 3, 4, 5;
L3Z, L2Z, L1Z, R1Z, R2Z, R3Z        pin 18..13 istype 'reg';

" Definitions
QSTATE = [L3Z,L2Z,L1Z,R1Z,R2Z,R3Z];   " State variables
IDLE  = [  0,  0,  0,  0,  0,  0];  " States
L3    = [  1,  1,  1,  0,  0,  0];
L2    = [  0,  1,  1,  0,  0,  0];
L1    = [  0,  0,  1,  0,  0,  0];
R1    = [  0,  0,  0,  1,  0,  0];
R2    = [  0,  0,  0,  1,  1,  0];
R3    = [  0,  0,  0,  1,  1,  1];
LR3   = [  1,  1,  1,  1,  1,  1];
```

9.1.4 The Guessing Game

A "guessing game" machine was defined in Section 7.7.1 starting on page 594, with the following description:

> Design a clocked synchronous state machine with four inputs, G1–G4, that are connected to pushbuttons. The machine has four outputs, L1–L4, connected to lamps or LEDs located near the like-numbered pushbuttons. There is also an ERR output connected to a red lamp. In normal operation, the L1–L4 outputs display a 1-out-of-4 pattern. At each clock tick, the pattern is rotated by one position; the clock frequency is about 4 Hz.
>
> Guesses are made by pressing a pushbutton, which asserts an input Gi. When any Gi input is asserted, the ERR output is asserted if the "wrong" pushbutton was pressed, that is, if the Gi input detected at the clock tick does not have the same number as the lamp output that was asserted before the clock tick. Once a guess has been made, play stops and the ERR output maintains the same value for one or more clock ticks until the Gi input is negated, then play resumes.

As we discussed in Section 7.7.1, the machine requires six states—four in which a corresponding lamp is on, and two for when play is stopped after either a good or a bad pushbutton push. An ABEL program for the guessing game is shown in Table 9-5. Two enhancements were made to improve the testability and robustness of the machine—a RESET input that forces the game to a known starting state, and the two unused states have explicit transitions to the starting state.

The guessing-game machine uses the same state assignments as the original version in Section 7.7.1. Using these assignments, the ABEL compiler

```
module ggame
Title 'Guessing-Game State Machine'
GGAME device 'P16V8R';

" Inputs and outputs
CLOCK, RESET, G1..G4          pin 1, 2, 3..6;
L1..L4, ERR                   pin 12..15, 19 istype 'com';
Q2..Q0                        pin 16..18 istype 'reg';

" Sets
G = [G1..G4];
L = [L1..L4];

" States
QSTATE = [Q2,Q1,Q0];
S1     = [ 0, 0, 0];
S2     = [ 0, 0, 1];
S3     = [ 0, 1, 1];
S4     = [ 0, 1, 0];
SOK    = [ 1, 0, 0];
SERR   = [ 1, 0, 1];
EXTRA1 = [ 1, 1, 0];
EXTRA2 = [ 1, 1, 1];

state_diagram QSTATE

state S1:   IF RESET THEN SOK ELSE IF G2 # G3 # G4 THEN SERR
            ELSE IF G1 THEN SOK ELSE S2;

state S2:   IF RESET THEN SOK ELSE IF G1 # G3 # G4 THEN SERR
            ELSE IF G2 THEN SOK ELSE S3;

state S3:   IF RESET THEN SOK ELSE IF G1 # G2 # G4 THEN SERR
            ELSE IF G3 THEN SOK ELSE S4;

state S4:   IF RESET THEN SOK ELSE IF G1 # G2 # G3 THEN SERR
            ELSE IF G4 THEN SOK ELSE S1;

state SOK:  IF RESET THEN SOK
            ELSE IF G1 # G2 # G3 # G4 THEN SOK ELSE S1;

state SERR:  IF RESET THEN SOK
            ELSE IF G1 # G2 # G3 # G4 THEN SERR ELSE S1;

state EXTRA1:  GOTO SOK;
state EXTRA2:  GOTO SOK;

equations

QSTATE.CLK = CLOCK;

L1  = (QSTATE == S1);
L2  = (QSTATE == S2);
L3  = (QSTATE == S3);
L4  = (QSTATE == S4);
ERR = (QSTATE == SERR);

end ggame
```

Table 9-5
ABEL program for the guessing-game machine.

Table 9-6
Product-term usage
in the guessing-game
state-machine PLD.

P-Terms	Fan-in	Fan-out	Type	Name
1/3	3	1	Pin	L1
1/3	3	1	Pin	L2
1/3	3	1	Pin	L3
1/3	3	1	Pin	L4
1/3	3	1	Pin	ERR
6/2	7	1	Pin	Q2.REG
1/7	7	1	Pin	Q1.REG
11/8	8	1	Pin	Q0.REG

```
=========
  23/32              Best P-Term Total: 16
                          Total Pins: 14
                 Average P-Term/Output: 2
```

cranks out minimized equations with the number of product terms shown in
Table 9-6. The Q0 output just barely fits in a GAL16V8 (eight product terms). If
we needed to save terms, the way in which we've written the program allows us
to try alternate state assignments (see Exercise 9.4).

A more productive alternative might be to try an output-coded state assignment. We can use one state/output bit per lamp (L1..L4), and use one more bit
(ERR) to distinguish between the SOK and SERR states when the lamps are all off.
This allows us to drop the equations for L1..L4 and ERR from Table 9-5. The
new assignment is shown in Table 9-7. With this assignment, L1 uses two
product terms and L2..L4 use only one product term each. Unfortunately, the
ERR output blows up into 16 product terms.

Table 9-7
ABEL definitions for
the guessing-game
machine with an
output-coded state
assignment.

```
module ggameoc
Title 'Guessing-Game State Machine'
"GGAMEOC device 'P16V8R';

" Inputs and outputs
CLOCK, RESET, G1..G4          pin 1, 2, 3..6;
L1..L4, ERR                   pin 12..15, 18 istype 'reg';

" States
QSTATE = [L1,L2,L3,L4,ERR];
S1     = [ 1, 0, 0, 0,  0];
S2     = [ 0, 1, 0, 0,  0];
S3     = [ 0, 0, 1, 0,  0];
S4     = [ 0, 0, 0, 1,  0];
SOK    = [ 0, 0, 0, 0,  0];
SERR   = [ 0, 0, 0, 0,  1];

...
```

Part of our problem with this particular output-coded assignment is that we're not taking full advantage of its properties. Notice that it is basically a "one-hot" encoding, but the state definitions in Table 9-7 require all five state bits to be decoded for each state. An alternate version of the coding using "don't-cares" is shown in Table 9-8.

In the new version, we are assuming that the state bits never take on any combination of values other than the ones we originally defined in Table 9-7. Thus, for example, if we see that state bit L1 is 1, the machine must be in state S1 regardless of the values of any other state bits. Therefore, we can set these bits to "don't care" in S1's definition in Table 9-8. ABEL will set each X to 0 when encoding a next state, but will treat each X as a "don't-care" when decoding the current state. Thus, we must take *extreme* care to ensure that decoded states are in fact mutually exclusive, that is, that no legitimate next state matches two or more different state definitions. Otherwise, the compiled results will not have the expected behavior.

The reduced equations that result from the output coding in Table 9-8 use three product terms for L1, one each for L2..L4, and only seven for ERR. So the

```
X = .X.;
QSTATE = [L1,L2,L3,L4,ERR];
S1    = [ 1, X, X, X,  X];
S2    = [ X, 1, X, X,  X];
S3    = [ X, X, 1, X,  X];
S4    = [ X, X, X, 1,  X];
SOK   = [ 0, 0, 0, 0,  0];
SERR  = [ X, X, X, X,  1];
```

Table 9-8
Output coding for the guessing-game machine using "don't cares."

DON'T-CARE, HOW IT WORKS

To understand how the don't-cares work in a state encoding, you must first understand how ABEL creates equations internally from state diagrams. Within a given state S, each transition statement (IF-THEN-ELSE or GOTO) causes the on-sets of certain state variables to be augmented according to the transition condition. The transition condition is an expression that must be true to "go to" that target, including being in state S. For example, all of the conditions specified in a state such as S1 in Table 9-5 are implicitly ANDed with the expression "QSTATE==S1". Because of the way S1 is defined using don't-cares, this equality check generates only a single literal (L1) instead of an AND term, leading to further simplification later.

For each target state in a transition statement, the on-sets of only the state variables that are 1 in that state are augmented according to the transition condition. Thus, when a coded state such as S1 in Table 9-8 appears as a target in any transition statement, only the on-set of L1 is augmented. This explains why the actual coding of state S1 as a target is 100000.

change was worthwhile. However, we must remember that the new machine is different from the one in Table 9-7. Consider what happens if the machine ever gets into an unspecified state. In the original machine with fully specified output coding, there are no next-states for the $2^5 - 6 = 26$ unspecified states, so the state machine will always go to the state coded 00000 (SOK) from unspecified states. In the new machine, "unspecified" states aren't really unspecified; for example, the state coded 11111 actually matches five coded states, S1–S4 and SERR. The next state will actually be the "OR" of next-states for the matching coded states. (Read the box on the previous page to understand why these outcomes occur.) Again, you need to be careful.

RESETTING EXPECTATIONS   Reading the guessing-game program in Table 9-5, you would expect that the RESET input would force the machine to the SOK state, and it does. However, the moment that you have unspecified or partially coded states as in Tables 9-7 or 9-8, don't take anything for granted.

Referring to the box on the previous page, remember that transition statements in ABEL state machines augment the on-sets of state variables. If a particular, unused state combination does not match any of the states for which transition statements were written, then no on-sets will be augmented. Thus, the only transition from that state will be to the state with the all-0s coding.

For this reason, it is useful to code the reset state or a "safe" state as all 0s. If this is not possible, but the all-0s state is still unused, you can explicitly provide a transition from the all-0s state to a desired safe state.

9.1.5 Reinventing Traffic-Light Controllers

Our final example is from the world of cars and traffic. Traffic-light controllers in California, especially in the fair city of Sunnyvale, are carefully designed to *maximize* the waiting time of cars at intersections. An infrequently used intersection (one that would have no more than a "yield" sign if it were in Chicago) has the sensors and signals shown in Figure 9-5. The state machine that controls the traffic signals uses a 1-Hz clock and a timer and has four inputs:

NSCAR   Asserted when a car on the north-south road is over either sensor on either side of the intersection.

EWCAR   Asserted when a car on the east-west road is over either sensor on either side of the intersection.

TMLONG   Asserted if more than five minutes has elapsed since the timer started; remains asserted until the timer is reset.

TMSHORT   Asserted if more than five seconds has elapsed since the timer started; remains asserted until the timer is reset.

Figure 9-5 Traffic sensors and signals at an intersection in Sunnyvale, California.

The state machine has seven outputs:

NSRED, NSYELLOW, NSGREEN Control the north-south lights.

EWRED, EWYELLOW, EWGREEN Control the east-west lights.

TMRESET When asserted, resets the timer and negates TMSHORT and TMLONG. The timer starts timing when TMRESET is negated.

A typical, municipally approved algorithm for controlling the traffic lights is embedded in the ABEL program of Table 9-9. This algorithm produces two frequently seen behaviors of "smart" traffic lights. At night, when traffic is light, it holds a car stopped at the light for up to five minutes, unless a car approaches on the cross street, in which case it stops the cross traffic and lets the waiting car go. (The "early warning" sensor is far enough back to change the lights before the approaching car reaches the intersection.) During the day, when traffic is heavy and there are always cars waiting in both directions, it cycles the lights every five seconds, thus minimizing the utilization of the intersection and maximizing everyone's waiting time, thereby creating a public outcry for more taxes to fix the problem.

The equations for the TMRESET output are worth noting. This output is asserted during the "double-red" states, NSDELAY and EWDELAY, to reset the timer in preparation for the next green cycle. The desired output signal could be generated on a combinational output pin by decoding these two states, but we have chosen instead to generate it on a registered output pin by decoding the *predecessors* of these two states.

■ **Table 9-9** Sunnyvale traffic-lights program.

```
module svaletl
title 'State Machine for Sunnyvale, CA, Traffic Lights'
SVALETL device 'P16V8R';

" Input and output pins
CLOCK, !OE                      pin 1, 11;
NSCAR, EWCAR, TMSHORT, TMLONG   pin 2, 3, 8, 9;
Q0, Q1, Q2, TMRESET_L           pin 17, 16, 15, 14 istype 'reg';

" Definitions
LSTATE  = [Q2,Q1,Q0];          " State variables
NSGO    = [ 0, 0, 0];          " States
NSWAIT  = [ 0, 0, 1];
NSWAIT2 = [ 0, 1, 1];
NSDELAY = [ 0, 1, 0];
EWGO    = [ 1, 1, 0];
EWWAIT  = [ 1, 1, 1];
EWWAIT2 = [ 1, 0, 1];
EWDELAY = [ 1, 0, 0];

state_diagram LSTATE
state NSGO:                     " North-south green
   IF (!TMSHORT) THEN NSGO      " Minimum green is 5 seconds.
   ELSE IF (TMLONG) THEN NSWAIT " Maximum green is 5 minutes.
   ELSE IF (EWCAR & !NSCAR)     " If E-W car is waiting and no one
         THEN NSGO              "  is coming N-S, make E-W wait!
   ELSE IF (EWCAR & NSCAR)      " Cars coming in both directions?
         THEN NSWAIT            " Thrash!
   ELSE IF (!NSCAR)             " Nobody coming N-S and not timed out?
         THEN NSGO              " Keep N-S green.
   ELSE NSWAIT;                 " Else let E-W have it.

state NSWAIT: GOTO NSWAIT2;     " Yellow light is on for two ticks for safety.
state NSWAIT2: GOTO NSDELAY;    " (Drivers go 70 mph to catch this turkey green!)
state NSDELAY: GOTO EWGO;       " Red in both directions for added safety.

state EWGO: " East-west green; states defined analogous to N-S
   IF (!TMSHORT) THEN EWGO
   ELSE IF (TMLONG) THEN EWWAIT
   ELSE IF (NSCAR & !EWCAR) THEN EWGO
   ELSE IF (NSCAR & EWCAR) THEN EWWAIT
   ELSE IF (!EWCAR) THEN EWGO ELSE EWWAIT;

state EWWAIT: GOTO EWWAIT2;
state EWWAIT2: GOTO EWDELAY;
state EWDELAY: GOTO NSGO;

equations

LSTATE.CLK = CLOCK;  TMRESET_L.CLK = CLOCK;
!TMRESET_L := (LSTATE == NSWAIT2)  " Reset the timer when going into
           + (LSTATE == EWWAIT2); "   state NSDELAY or state EWDELAY.
end svaletl
```

Figure 9-6 Sunnyvale traffic-light controller using two PLDs.

■ **Table 9-10** Output logic for Sunnyvale traffic lights.

```
module svaletlo
title 'Output logic for Sunnyvale, CA, Traffic Lights'
"SVALETLO device 'P16V8C';

" Input pins
FLASHCLK, OVERRIDE, Q0, Q1, Q2   pin 1, 2, 4, 5, 6;

" Output pins
NSRED, NSYELLOW, NSGREEN          pin 19, 18, 17 istype 'com';
EWRED, EWYELLOW, EWGREEN          pin 14, 13, 12 istype 'com';

" Definitions (same as in state machine SVALETL)
...

equations

NSRED = !OVERRIDE & (LSTATE != NSGO) & (LSTATE != NSWAIT) & (LSTATE != NSWAIT2)
      # OVERRIDE & FLASHCLK;
NSYELLOW = !OVERRIDE & ((LSTATE == NSWAIT) # (LSTATE == NSWAIT2));
NSGREEN  = !OVERRIDE & (LSTATE == NSGO);

EWRED = !OVERRIDE & (LSTATE != EWGO) & (LSTATE != EWWAIT) & (LSTATE != EWWAIT2)
      # OVERRIDE & FLASHCLK;
EWYELLOW = !OVERRIDE & ((LSTATE == EWWAIT) # (LSTATE == EWWAIT2));
EWGREEN  = !OVERRIDE & (LSTATE == EWGO);

end svaletlo
```

The ABEL program in Table 9-9 defines only the state variables and one registered Moore output for the traffic controller. Six more Moore outputs are needed for the lights, more than remain on the 16V8. Therefore, a separate combinational PLD is used for these outputs, yielding the complete design shown in Figure 9-6 on the preceding page. An ABEL program for the output PLD is given in Table 9-10. We've taken this opportunity to add an OVERRIDE input to the controller. This input may be asserted by the police to disable the controller and put the signals into a flashing-red mode (at a rate determined by FLASHCLK), allowing them to manually clear up the traffic snarls created by this wonderful invention.

A traffic-light state machine including output logic can be built in a single 16V8, shown in Figure 9-7, if we choose an output-coded state assignment. Only the definitions in the original program of Table 9-9 must be changed, as shown in Table 9-11. This PLD does not include the OVERRIDE input and mode, which is left as an exercise (9.7).

Table 9-11 Definitions for Sunnyvale traffic-lights machine with output-coded state assignment.

```
module svaletlb
title 'Output-Coded State Machine for Sunnyvale Traffic Lights'
"SVALETLB device 'P16V8R';

" Input and output pins
CLOCK, !OE                              pin 1, 11;
NSCAR, EWCAR, TMSHORT, TMLONG           pin 2, 3, 8, 9;
NSRED, NSYELLOW, NSGREEN                pin 19, 18, 17 istype 'reg';
EWRED, EWYELLOW, EWGREEN                pin 16, 15, 14 istype 'reg';
TMRESET_L, XTRA                         pin 13, 12 istype 'reg';

" Definitions
LSTATE   = [NSRED,NSYELLOW,NSGREEN,EWRED,EWYELLOW,EWGREEN,XTRA]; " State vars
NSGO    = [   0,      0,      1,    1,      0,      0,   0]; " States
NSWAIT  = [   0,      1,      0,    1,      0,      0,   0];
NSWAIT2 = [   0,      1,      0,    1,      0,      0,   1];
NSDELAY = [   1,      0,      0,    1,      0,      0,   0];
EWGO    = [   1,      0,      0,    0,      0,      1,   0];
EWWAIT  = [   1,      0,      0,    0,      1,      0,   0];
EWWAIT2 = [   1,      0,      0,    0,      1,      0,   1];
EWDELAY = [   1,      0,      0,    1,      0,      0,   1];
```

Figure 9-7
Traffic-light state
machine using
output-coded state
assignment in a
single PLD.

9.2 Design Examples Using VHDL

As we explained Section 7.12, the basic VHDL language features that we
introduced way back in Section 4.7, including processes, are just about all that
is needed to model sequential-circuit behavior. Unlike ABEL, VHDL does not
provide any special language elements for modeling state machines. Instead,
most programmers use a combination of existing "standard" features—most
notably enumerated types and case statements—to write state-machine descriptions. We'll use this method in the examples in this section.

9.2.1 A Few Simple Machines

In Section 7.4.1 we illustrated the state-table design process using the
simple design problem below:

Design a clocked synchronous state machine with two inputs, A and B, and
a single output Z that is 1 if:

— A had the same value at each of the two previous clock ticks, *or*
— B has been 1 since the last time that the first condition was true.

Otherwise, the output should be 0.

In an HDL-based design environment, there are many possible ways of writing
a program that meets the stated requirements. We'll look at several.

The first approach is to construct a state and output table by hand and then
manually convert it into a corresponding program. Since we already developed
such a state table in Section 7.4.1, why not use it? We've written it again in
Table 9-12, and we've written a corresponding VHDL program in Table 9-13.

Table 9-12
State and output table for the example state machine.

| | A B | | | | |
S	00	01	11	10	Z
INIT	A0	A0	A1	A1	0
A0	OK0	OK0	A1	A1	0
A1	A0	A0	OK1	OK1	0
OK0	OK0	OK0	OK1	A1	1
OK1	A0	OK0	OK1	OK1	1
			S*		

Table 9-13
VHDL program for state-machine example.

```
library IEEE;
use IEEE.std_logic_1164.all;

entity smexamp is
  port ( CLOCK, A, B: in STD_LOGIC;
         Z: out STD_LOGIC );
end;

architecture smexamp_arch of smexamp is
type Sreg_type is (INIT, A0, A1, OK0, OK1);
signal Sreg: Sreg_type;
begin

  process (CLOCK) -- state-machine states and transitions
  begin
    if CLOCK'event and CLOCK = '1' then
      case Sreg is
        when INIT => if    A='0' then Sreg <= A0;
                     elsif A='1' then Sreg <= A1;   end if;
        when A0 =>   if    A='0' then Sreg <= OK0;
                     elsif A='1' then Sreg <= A1;  end if;
        when A1 =>   if    A='0' then Sreg <= A0;
                     elsif A='1' then Sreg <= OK1;  end if;
        when OK0 =>  if    A='0' then Sreg <= OK0;
                     elsif A='1' and B='0' then Sreg <= A1;
                     elsif A='1' and B='1' then Sreg <= OK1;  end if;
        when OK1 =>  if    A='0' and B='0' then Sreg <= A0;
                     elsif A='0' and B='1' then Sreg <= OK0;
                     elsif A='1' then Sreg <= OK1;             end if;
        when others => Sreg <= INIT;
      end case;
    end if;
  end process;

  with Sreg select  -- output values based on state
    Z <= '0' when INIT | A0 | A1,
         '1' when OK0 | OK1,
         '0' when others;

end smexamp_arch;
```

As usual, the VHDL entity declaration specifies only inputs and outputs—CLOCK, A, B, and Z in this example. The architecture definition specifies the state machine's internal operation. The first thing it does is to create an enumerated type, Sreg_type, whose values are identifiers corresponding to the state names. Then it declares a signal, Sreg, which will be used to hold the machine's current state. Because of the way Sreg is used later, it will map into an edge-triggered register in synthesis.

The statement part of the architecture has two concurrent statements—a process and a selected-assignment statement. The process is sensitive only to CLOCK and establishes all of the state transitions, which occur on the rising edge of CLOCK. Within the process, an "if" statement checks for the rising edge, and a case statement enumerates the transitions for each state.

The case statement has six cases, corresponding to the five explicitly defined states and a catch-all for other states. For robustness, the "others" case sends the machine back to the INIT state. In each case we've used a nested "if" statement to explicit cover all combinations of inputs A and B. However, it's not strictly necessary to include the combinations where there is no transition from the current state; Sreg will hold its current value if no assignment is made on a particular process invocation.

The selected-assignment statement at the end of Table 9-13 handles the machine's single Moore output, Z, which is set to a value as a function of the current state. It would be easy to define Mealy outputs within this statement as well. That is, Z could be a function of the inputs as well as the current state. Since the "with" statement is a concurrent state, any input changes will affect the Z output as soon as they occur.

We really should have included a reset input in Table 9-13 (see box on page 808). A RESET signal is easily accommodated by modifying the entity declaration and adding a clause to the "if" statement in the architecture definition. If RESET is asserted, the machine should go to the INIT state, otherwise the case statement should be executed. Depending on whether we check RESET before or after the clock-edge check, we can create either an asynchronous or a synchronous reset behavior (see Exercise 9.10).

What about the state assignment problem? Table 9-13 gives no information on how state-variable combinations are to be assigned to the named states, or even how many binary state variables are needed in the first place!

A synthesis tool is free to associate any integer values or binary combinations it likes with the identifiers in an enumerated type, but a typical tool will assign integers in the order the state names are listed, starting with 0. It will then use the smallest possible number of bits to encode those integers, that is, $\lceil \log_2 s \rceil$ bits for s states. Thus, the program in Table 9-13 will typically synthesize with the same, "simplest" state assignment that we chose in the original example in Table 7-7 on page 571. However, VHDL supports a couple of ways that we can force a different assignment.

Table 9-14

Using an attribute to force an enumeration encoding.

```
library IEEE;
use IEEE.std_logic_1164.all;
library SYNOPSYS;
use SYNOPSYS.attributes.all;
...
architecture smexampe_arch of smexamp is
type Sreg_type is (INIT, A0, A1, OK0, OK1);
attribute enum_encoding of Sreg_type: type is
        "0000 0001 0010 0100 1000";
signal Sreg: Sreg_type;
...
```

One way to force an assignment is to use VHDL's "attribute" statement as shown in Table 9-14. Here, "enum_encoding" is a user-defined attribute whose value is a string that specifies the enumeration encoding to be used by the synthesis tool. The VHDL language processor ignores this value, but passes the attribute name and its value to the synthesis tool. The attribute "enum_encoding" is defined and known by most synthesis tools, including tools from Synopsys, Inc. Notice that a Synopsys "attributes" package must be "used" by the program; this is necessary for the VHDL compiler to recognize "enum_encoding" as a legitimate user-defined attribute. By the way, the state assignment that we've specified in the program is equivalent to the "almost one-hot" coding in Table 7-7 on page 571.

Another way to force an assignment, without relying on external packages or synthesis attributes, is to define the state register more explicitly using standard logic data types. This approach is shown in Table 9-15. Here, Sreg is defined as a 4-bit STD_LOGIC_VECTOR, and constants are defined to allow the states to be referenced by name elsewhere in the program. No other changes in the program are required.

Going back to our original VHDL program in Table 9-13, one more interesting change is possible. As written, the program defines a conventional Moore-type state machine with the structure shown in Figure 9-8(a). What

Table 9-15

Using standard logic and constants to specify a state encoding.

```
library IEEE;
use IEEE.std_logic_1164.all;
...
architecture smexampc_arch of smexamp is
subtype Sreg_type is STD_LOGIC_VECTOR (1 to 4);
constant INIT: Sreg_type := "0000";
constant A0  : Sreg_type := "0001";
constant A1  : Sreg_type := "0010";
constant OK0 : Sreg_type := "0100";
constant OK1 : Sreg_type := "1000";
signal Sreg: Sreg_type;
...
```

happens if we convert the output logic's selected-assignment statement into a
`case` statement and move it into the state-transition process? By doing this, we
create a machine that will most likely be synthesized with the structure shown in
(b). This is essentially a Mealy machine with pipelined outputs whose behavior
is indistinguishable from that of the original machine, except for timing. We've
reduced the propagation delay from CLOCK to Z by producing Z directly on a
register output, but we've also increased the setup-time requirements of A and B
to CLOCK, because of the extra propagation delay through the output logic to the
D input of the output register.

Figure 9-8 State-machine structures implied by VHDL programs: (a) Moore
machine with combinational output logic; (b) pipelined Mealy
machine with output register.

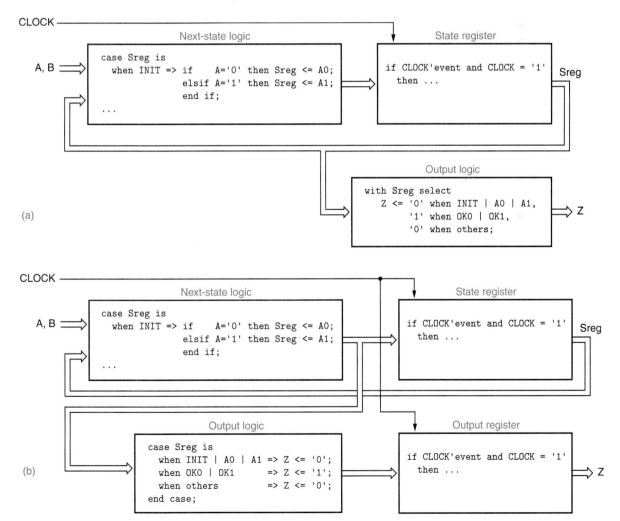

TRICKY TIMING   When we write a VHDL architecture corresponding to Figure 9-8(b), it is very important to add Sreg to the sensitivity list of the process. Notice that the output-logic case statement determines the value of Z as a function of Sreg. Throughout the first execution of the process after a rising clock edge, Sreg contains the old state value. This is true because Sreg is a signal, not a variable. As explained in Section 4.7.9, signals that are changed in a process do not acquire their new values until at least one delta delay *after* the process has executed. By putting Sreg in the sensitivity list, we ensure that the process is executed a second time, so the final value of Z is based on the new value of Sreg.

All of the solutions to the example state-machine design problem that we've shown so far rely on the state table that we originally constructed by hand in Section 7.4.1. However, it is possible to write a VHDL program directly, without writing out a state table by hand.

Based on the original problem statement on page 813, the key simplifying idea is to remove the last value of A from the state definitions, and instead to have a separate register that keeps track of it (LASTA). Then only two non-INIT states

Table 9-16
Simplified state machine for VHDL example problem.

```
architecture smexampa_arch of smexamp is
type Sreg_type is (INIT, LOOKING, OK);
signal Sreg: Sreg_type;
signal lastA: STD_LOGIC;
begin

  process (CLOCK) -- state-machine states and transitions
  begin
    if CLOCK'event and CLOCK = '1' then
      lastA <= A;
      case Sreg is
        when INIT    =>                         Sreg <= LOOKING;
        when LOOKING => if    A=lastA then Sreg <= OK;
                        else              Sreg <= LOOKING;
                        end if;
        when OK      => if    B='1'   then Sreg <= OK;
                        elsif A=lastA then Sreg <= OK;
                        else              Sreg <= LOOKING;
                        end if;
        when others  =>                         Sreg <= INIT;
      end case;
    end if;
  end process;

  with Sreg select  -- output values based on state
    Z <= '1' when OK,
         '0' when others;

end smexampa_arch;
```

must be defined: LOOKING ("still looking for a match") and OK ("got a match or B has been 1 since last match"). A VHDL architecture based on this approach is shown in Table 9-16. In the CLOCK-driven process, the first assignment statement creates the LASTA register, and the case statement creates the 3-state machine. At the end of the program, the Z output is defined as a simple combinational decode of the OK state.

Another simple state-machine example is a "ones-counting machine" with the following specification:

Design a clocked synchronous state machine with two inputs, X and Y, and one output, Z. The output should be 1 if the number of 1 inputs on X and Y since reset is a multiple of 4, and 0 otherwise.

We developed a state table for this machine in Table 7-12 on page 580. However, we can make use of the counting capabilities in the IEEE std_logic_arith package to write a VHDL program for this problem directly.

Table 9-17 shows our solution. As always, there are many different ways to solve the problem, and we have picked a way that illustrates several different language features. Within the architecture, we declare a subtype COUNTER which is a 2-bit UNSIGNED value. We then declare a signal COUNT of this type to hold the ones count, and a constant ZERO of the same type for initializing and checking the value of COUNT.

Table 9-17
VHDL program for a ones-counting machine.

```
library IEEE;
use IEEE.std_logic_1164.all;
use IEEE.std_logic_arith.all;

entity Vonescnt is
  port ( CLOCK, RESET, X, Y: in STD_LOGIC;
         Z: out STD_LOGIC );
end;

architecture Vonescnt_arch of Vonescnt is
subtype COUNTER is UNSIGNED (1 downto 0);
signal COUNT: COUNTER;
constant ZERO: COUNTER := "00";
begin

process (CLOCK)
  begin
    if CLOCK'event and CLOCK = '1' then
      if RESET = '1' then COUNT <= ZERO;
      else COUNT <= COUNT + ('0', X) + ('0', X);
      end if;
    end if;
  end process;

Z <= '1' when COUNT = ZERO else '0';

end Vonescnt_arch;
```

Table 9-18
Alternative VHDL
process for ones-
counting machine.

```
process (CLOCK)
variable ONES: STD_LOGIC_VECTOR (1 to 2);
begin
  if CLOCK'event and CLOCK = '1' then
    ONES := (X, Y);
    if RESET = '1' then COUNT <= ZERO;
    else case ONES is
            when "01" | "10" => COUNT <= COUNT + "01";
            when "11"        => COUNT <= COUNT + "10";
            when others      => null;
         end case;
    end if;
  end if;
end process;
```

Within the process, we use the usual method to check for a rising edge on CLOCK. The "if" clause performs a synchronous reset when required, and the "else" clause elegantly adds 0, 1 or 2 to COUNT depending on the values of X and Y. Recall that an expression such as "('0', X)" is an array literal; here we get an array of two STD_LOGIC elements, '0' and the current value of X. The type of this literal is compatible with UNSIGNED, since the number and type of elements are the same, so they can be combined using the "+" operation defined in the std_logic_arith package. Outside the process, the concurrent signal-assignment statement sets the Moore output Z to 1 when COUNT is zero.

For synthesis purposes, the "if" statement and assignment to COUNT in Table 9-17 don't necessarily yield a compact or speedy circuit. With a simple-minded synthesis tool, it could yield two 2-bit adders connected in series. Another approach is shown in Table 9-18. An intelligent tool may be able to synthesize a more compact incrementer for each of the two additions. In any case, formulating the choices in a case statement allows the two adders or incrementers to operate in parallel, and a multiplexer can be used to select one of their outputs according to the choices.

A final example for this subsection is the combination-lock state machine from Section 7.4 (below we omit the HINT output in the original specification):

> Design a clocked synchronous state machine with one input, X, and one output, UNLK. The UNLK output should be 1 if and only if X is 0 and the sequence of inputs received on X at the preceding seven clock ticks was 0110111.

Table 9-19
VHDL program for
finite-memory design
of combination-lock
state machine.

```
library IEEE;
use IEEE.std_logic_1164.all;

entity Vcomblck is
  port ( CLOCK, RESET, X: in STD_LOGIC;
         UNLK: out STD_LOGIC );
end;

architecture Vcomblck_arch of Vcomblck is
signal XHISTORY: STD_LOGIC_VECTOR (7 downto 1);
constant COMBINATION: STD_LOGIC_VECTOR (7 downto 1) := "0110111";
begin

  process (CLOCK)
  begin
    if CLOCK'event and CLOCK = '1' then
      if RESET = '1' then XHISTORY <= "0000000";
      else XHISTORY <= XHISTORY(6 downto 1) & X;
      end if;
    end if;
  end process;

  UNLK <= '1' when (XHISTORY=COMBINATION) and (X='0') else '0';

end Vcomblck_arch;
```

We developed a state table for this machine in Table 7-14 on page 582. But once again we can take a different approach that is easier to understand. Here, we note that the output of the machine at any time is completely determined by its inputs over the preceding eight clock ticks. Thus, we can use the so-called "finite-memory" approach to design this machine (see box on page 801). With this approach, we explicitly keep track of the past seven inputs and then form the output as a combinational function of these inputs.

The VHDL program in Table 9-19 is a finite-memory design. Within the architecture, the process merely keeps track of the last seven values of X using what's essentially a shift register, bit 7 being the oldest value of X. (Recall that the "&" operator in VHDL is array concatenation.) Outside of the process, the concurrent signal-assignment statement sets the Mealy output UNLK to 1 when X is 0 and the 7-bit history matches the combination.

9.2.2 T-Bird Tail Lights

We described and designed a "T-bird tail-lights" state machine in Section 7.5. Table 9-20 is an equivalent VHDL program for the T-bird tail-lights machine. The state transitions in this machine are defined exactly the same as in the state diagram of Figure 7-64 on page 589. The machine uses an output-coded state assignment, taking advantage of the fact that the tail-light output values are different in each state.

Table 9-20 VHDL program for the T-bird tail-lights machine.

```
entity Vtbird is
  port ( CLOCK, RESET, LEFT, RIGHT, HAZ: in STD_LOGIC;
         LIGHTS: buffer STD_LOGIC_VECTOR (1 to 6) );
end;

architecture Vtbird_arch of Vtbird is
constant IDLE: STD_LOGIC_VECTOR (1 to 6) := "000000";
constant L3  : STD_LOGIC_VECTOR (1 to 6) := "111000";
constant L2  : STD_LOGIC_VECTOR (1 to 6) := "110000";
constant L1  : STD_LOGIC_VECTOR (1 to 6) := "100000";
constant R1  : STD_LOGIC_VECTOR (1 to 6) := "000001";
constant R2  : STD_LOGIC_VECTOR (1 to 6) := "000011";
constant R3  : STD_LOGIC_VECTOR (1 to 6) := "000111";
constant LR3 : STD_LOGIC_VECTOR (1 to 6) := "111111";
begin
  process (CLOCK)
  begin
    if CLOCK'event and CLOCK = '1' then
      if RESET = '1' then LIGHTS <= IDLE; else
        case LIGHTS is
          when IDLE => if HAZ='1' or (LEFT='1' and RIGHT='1') then LIGHTS <= LR3;
                       elsif LEFT='1'                          then LIGHTS <= L1;
                       elsif RIGHT='1'                         then LIGHTS <= R1;
                       else                                         LIGHTS <= IDLE;
                       end if;
          when L1   => if HAZ='1' then LIGHTS <= LR3; else LIGHTS <= L2; end if;
          when L2   => if HAZ='1' then LIGHTS <= LR3; else LIGHTS <= L3; end if;
          when L3   => LIGHTS <= IDLE;
          when R1   => if HAZ='1' then LIGHTS <= LR3; else LIGHTS <= R2; end if;
          when R2   => if HAZ='1' then LIGHTS <= LR3; else LIGHTS <= R3; end if;
          when R3   => LIGHTS <= IDLE;
          when LR3  => LIGHTS <= IDLE;
          when others => null;
        end case;
      end if;
    end if;
  end process;
end Vtbird_arch;
```

IDLE MUSINGS   In VHDL state machines, it's not necessary to make an explicit assignment of a next state if it's the same state that the machine is already in. In the execution of a process, a VHDL signal keeps its value if no assignment is made to it. Thus, in Table 9-20, the final "else" clause could be omitted in the IDLE state, with no effect on the machine's behavior.

Separately, the robustness of the state machine in Table 9-20 could be improved by replacing the "null" statement in the "when others" case with a transition to the IDLE state.

9.2.3 The Guessing Game

A "guessing-game" machine was defined in Section 7.7.1 starting on page 594, with the following description:

Design a clocked synchronous state machine with four inputs, G1–G4, that are connected to pushbuttons. The machine has four outputs, L1–L4, connected to lamps or LEDs located near the like-numbered pushbuttons. There is also an ERR output connected to a red lamp. In normal operation, the L1–L4 outputs display a 1-out-of-4 pattern. At each clock tick, the pattern is rotated by one position; the clock frequency is about 4 Hz.

Guesses are made by pressing a pushbutton, which asserts an input Gi. When any Gi input is asserted, the ERR output is asserted if the "wrong" pushbutton was pressed, that is, if the Gi input detected at the clock tick does not have the same number as the lamp output that was asserted before the clock tick. Once a guess has been made, play stops and the ERR output maintains the same value for one or more clock ticks until the Gi input is negated, then play resumes.

As we discussed in Section 7.7.1, the machine requires six states—four in which a corresponding lamp is on, and two for when play is stopped after either a good or a bad pushbutton push. A VHDL program for the guessing game is shown in Table 9-21. This version also includes a RESET input that forces the game to a known starting state.

The program is pretty much a straightforward translation of the original state diagram in Figure 7-66 on page 597. Perhaps its only noteworthy feature is in the "SOK | SERR" case. Since the next-state transitions for these two states are identical (either go to S1 or stay in the current state), they can be handled in one case. However, this tricky style of saving typing isn't particularly desirable from the point of view of state-machine documentation or maintainability. In the author's case, the trick's primary benefit was to help fit the program on one book page!

■ Table 9-21 VHDL program for the guessing-game machine.

```
library IEEE;
use IEEE.std_logic_1164.all;

entity Vggame is
  port ( CLOCK, RESET, G1, G2, G3, G4: in  STD_LOGIC;
         L1, L2, L3, L4, ERR:          out STD_LOGIC );
end;

architecture Vggame_arch of Vggame is
type Sreg_type is (S1, S2, S3, S4, SOK, SERR);
signal Sreg: Sreg_type;
begin

  process (CLOCK)
  begin
    if CLOCK'event and CLOCK = '1' then
      if RESET = '1' then Sreg <= SOK; else
        case Sreg is
          when S1 => if    G2='1' or G3='1' or G4='1' then Sreg <= SERR;
                     elsif G1='1'                      then Sreg <= SOK;
                     else                                   Sreg <= S2;
                     end if;
          when S2 => if    G1='1' or G3='1' or G4='1' then Sreg <= SERR;
                     elsif G1='1'                      then Sreg <= SOK;
                     else                                   Sreg <= S3;
                     end if;
          when S3 => if    G1='1' or G2='1' or G4='1' then Sreg <= SERR;
                     elsif G1='1'                      then Sreg <= SOK;
                     else                                   Sreg <= S4;
                     end if;
          when S4 => if    G1='1' or G2='1' or G3='1' then Sreg <= SERR;
                     elsif G1='1'                      then Sreg <= SOK;
                     else                                   Sreg <= S1;
                     end if;
          when SOK | SERR => if G1='0' and G2='0' and G3='0' and G4='0'
                                then Sreg <= S1; end if;
          when others => Sreg <= S1;
        end case;
      end if;
    end if;
  end process;

  L1  <= '1' when Sreg = S1   else '0';
  L2  <= '1' when Sreg = S2   else '0';
  L3  <= '1' when Sreg = S3   else '0';
  L4  <= '1' when Sreg = S4   else '0';
  ERR <= '1' when Sreg = SERR else '0';

end Vggame_arch;
```

The program in Table 9-21 does not specify a state assignment; a typical synthesis engine will use three bits for Sreg and assign the six states in order to binary combinations 000–101. For this state machine, it is also possible to use an output coded state assignment, using just the lamp and error output signals that are already required. VHDL does not provide a convenient mechanism for grouping together the entity's existing output signals and using them for state, but we can still achieve the desired effect with the changes shown in Table 9-22. Here we used a comment to document the correspondence between outputs and the bits of the new, 5-bit Sreg, and we changed each of the output assignment statements to pick off the appropriate bit instead of fully decoding the state.

```
architecture Vggameoc_arch of Vggame is
signal Sreg: STD_LOGIC_VECTOR (1 to 5);
-- bit positions of output-coded assignment: L1, L2, L3, L4, ERR
constant S1:   STD_LOGIC_VECTOR (1 to 5) := "10000";
constant S2:   STD_LOGIC_VECTOR (1 to 5) := "01000";
constant S3:   STD_LOGIC_VECTOR (1 to 5) := "00100";
constant S4:   STD_LOGIC_VECTOR (1 to 5) := "00010";
constant SERR: STD_LOGIC_VECTOR (1 to 5) := "00001";
constant SOK:  STD_LOGIC_VECTOR (1 to 5) := "00000";
begin

  process (CLOCK)
  ...             (no change to process)
  end process;

  L1  <= Sreg(1);
  L2  <= Sreg(2);
  L3  <= Sreg(3);
  L4  <= Sreg(4);
  ERR <= Sreg(5);

end Vggameoc_arch;
```

Table 9-22
VHDL architecture for guessing game using output-coded state assignment.

9.2.4 Reinventing Traffic-Light Controllers

If you read the ABEL example in Section 9.1.5, then you've already heard me rant about the horrible traffic-light controllers in Sunnyvale, California. They really *do* seem to be carefully designed to *maximize* the waiting time of cars at intersections. In this section we'll design a traffic-light controller with distinctly Sunnyvale-like behavior.

An infrequently used intersection (one that would have no more than a "yield" sign if it were in Chicago) has the sensors and signals shown in

■ **Table 9-23** VHDL program for Sunnyvale traffic-light controller.

```vhdl
library IEEE;
use IEEE.std_logic_1164.all;

entity Vsvale is
  port ( CLOCK, RESET, NSCAR, EWCAR, TMSHORT, TMLONG: in  STD_LOGIC;
         OVERRIDE, FLASHCLK:                          in  STD_LOGIC;
         NSRED, NSYELLOW, NSGREEN:                    out STD_LOGIC;
         EWRED, EWYELLOW, EWGREEN, TMRESET:           out STD_LOGIC );
end;

architecture Vsvale_arch of Vsvale is
type Sreg_type is (NSGO, NSWAIT, NSWAIT2, NSDELAY,
                   EWGO, EWWAIT, EWWAIT2, EWDELAY);
signal Sreg: Sreg_type;
begin

process (CLOCK)
begin
  if CLOCK'event and CLOCK = '1' then
    if RESET = '1' then Sreg <= NSDELAY; else
      case Sreg is
        when NSGO =>                                      -- North-south green.
          if    TMSHORT='0' then Sreg <= NSGO;            -- Minimum 5 seconds.
          elsif TMLONG='1'  then Sreg <= NSWAIT;          -- Maximum 5 minutes.
          elsif EWCAR='1' and NSCAR='0' then Sreg <= NSGO;   -- Make EW car wait.
          elsif EWCAR='1' and NSCAR='1' then Sreg <= NSWAIT; -- Thrash if cars both ways.
          elsif EWCAR='0' and NSCAR='1' then Sreg <= NSWAIT; -- New NS car? Make it stop!
          else                           Sreg <= NSGO;      -- No one coming, no change.
          end if;
        when NSWAIT  => Sreg <= NSWAIT2;                  -- Yellow light,
        when NSWAIT2 => Sreg <= NSDELAY;                  --   two ticks for safety.
        when NSDELAY => Sreg <= EWGO;                     -- Red both ways for safety.
        when EWGO =>                                      -- East-west green.
          if    TMSHORT='0' then Sreg <= EWGO;            -- Same behavior as above.
          elsif TMLONG='1'  then Sreg <= EWWAIT;
          elsif NSCAR='1' and EWCAR='0' then Sreg <= EWGO;
          elsif NSCAR='1' and EWCAR='1' then Sreg <= EWWAIT;
          elsif NSCAR='0' and EWCAR='1' then Sreg <= EWWAIT;
          else                           Sreg <= EWGO;
          end if;
        when EWWAIT  => Sreg <= EWWAIT2;
        when EWWAIT2 => Sreg <= EWDELAY;
        when EWDELAY => Sreg <= NSGO;
        when others  => Sreg <= NSDELAY;                  -- "Reset" state.
      end case;
    end if;
  end if;
end process;
```

Figure 9-5 on page 809. The state machine that controls the traffic signals uses a
1-Hz clock and a timer and has four inputs:

NSCAR   Asserted when a car on the north-south road is over either sensor
on either side of the intersection.

EWCAR   Asserted when a car on the east-west road is over either sensor on
either side of the intersection.

TMLONG   Asserted if more than five minutes has elapsed since the timer
started; remains asserted until the timer is reset.

TMSHORT   Asserted if more than five seconds has elapsed since the timer
started; remains asserted until the timer is reset.

The state machine has seven outputs:

NSRED, NSYELLOW, NSGREEN   Control the north-south lights.

EWRED, EWYELLOW, EWGREEN   Control the east-west lights.

TMRESET   When asserted, resets the timer and negates TMSHORT and
TMLONG. The timer starts timing when TMRESET is negated.

A typical, municipally approved algorithm for controlling the traffic lights
is embedded in the VHDL program of Table 9-23. This algorithm produces two
frequently seen behaviors of "smart" traffic lights. At night, when traffic is light,
it holds a car stopped at the light for up to five minutes, unless a car approaches
on the cross street, in which case it stops the cross traffic and lets the waiting car
go. (The "early warning" sensor is far enough back to change the lights before
the approaching car reaches the intersection.) During the day, when traffic is
heavy and there are always cars waiting in both directions, it cycles the lights
every five seconds, thus minimizing the utilization of the intersection and
maximizing everyone's waiting time, thereby creating a public outcry for more
taxes to fix the problem.

Table 9-23   (continued) VHDL program for Sunnyvale traffic-light controller.

```
TMRESET  <= '1' when Sreg=NSWAIT2 or Sreg=EWWAIT2 else '0';
NSRED    <= FLASHCLK when OVERRIDE='1' else
            '1' when Sreg/=NSGO and Sreg/=NSWAIT and Sreg/=NSWAIT2 else '0';
NSYELLOW <= '0' when OVERRIDE='1' else
            '1' when Sreg=NSWAIT or Sreg=NSWAIT2 else '0';
NSGREEN  <= '0' when OVERRIDE='1' else '1' when Sreg=NSGO else '0';
EWRED    <= FLASHCLK when OVERRIDE='1' else
            '1' when Sreg/=EWGO and Sreg/=EWWAIT and Sreg/=EWWAIT2 else '0';
EWYELLOW <= '0' when OVERRIDE='1' else
            '1' when Sreg=EWWAIT or Sreg=EWWAIT2 else '0';
EWGREEN  <= '0' when OVERRIDE='1' else '1' when Sreg=EWGO else '0';

end Vsvale_arch;
```

While writing the program, we took the opportunity to add two inputs that weren't in the original specification. The OVERRIDE input may be asserted by the police to disable the controller and put the signals into a flashing-red mode at a rate determined by the FLASHCLK input. This allows them to manually clear up the traffic snarls created by this wonderful invention.

Table 9-24 Definitions for Sunnyvale traffic-lights machine with output-coded state assignment.

```
library IEEE;
use IEEE.std_logic_1164.all;

entity Vsvale is
  port ( CLOCK, RESET, NSCAR, EWCAR, TMSHORT, TMLONG: in  STD_LOGIC;
         OVERRIDE, FLASHCLK:                          in  STD_LOGIC;
         NSRED, NSYELLOW, NSGREEN:                    out STD_LOGIC;
         EWRED, EWYELLOW, EWGREEN, TMRESET:           out STD_LOGIC );
end;

architecture Vsvaleoc_arch of Vsvale is
signal Sreg: STD_LOGIC_VECTOR (1 to 7);
-- bit positions of output-coded assignment:  (1) NSRED, (2) NSYELLOW, (3) NSGREEN,
--                                   (4) EWRED, (5) EWYELLOW, (6) EWGREEN, (7) EXTRA
constant NSGO:    STD_LOGIC_VECTOR (1 to 7) := "0011000";
constant NSWAIT:  STD_LOGIC_VECTOR (1 to 7) := "0101000";
constant NSWAIT2: STD_LOGIC_VECTOR (1 to 7) := "0101001";
constant NSDELAY: STD_LOGIC_VECTOR (1 to 7) := "1001000";
constant EWGO:    STD_LOGIC_VECTOR (1 to 7) := "1000010";
constant EWWAIT:  STD_LOGIC_VECTOR (1 to 7) := "1000100";
constant EWWAIT2: STD_LOGIC_VECTOR (1 to 7) := "1000101";
constant EWDELAY: STD_LOGIC_VECTOR (1 to 7) := "1001001";

begin

process (CLOCK)
...               (no change to process)
end process;

TMRESET   <= '1' when Sreg=NSWAIT2 or Sreg=EWWAIT2 else '0';
NSRED     <= Sreg(1);
NSYELLOW <= Sreg(2);
NSGREEN   <= Sreg(3);
EWRED     <= Sreg(4);
EWYELLOW <= Sreg(5);
EWGREEN   <= Sreg(6);

end Vsvaleoc_arch;
```

Like most of our other examples, Table 9-23 does not give a specific state assignment. And like many of our other examples, this state machine works well with an output-coded state assignment. Many of the states can be identified by a unique combination of light-output values. But there are three pairs of states that are not distinguishable by looking at the lights alone: (NSWAIT, NSWAIT2), (EWWAIT, EWWAIT2), and (NSDELAY, EQDELAY). We can handle these by adding one more state variable, "EXTRA", that has different values for the two states in each pair. This idea is realized in the modified program in Table 9-24.

Exercises

9.1   Write an ABEL program for the state machine described in Exercise 7.30.

9.2   Modify the ABEL program of Table 9-2 to include the HINT output from the original state-machine specification in Section 7.4.

9.3   Redesign the T-bird tail-lights machine of Section 9.1.3 to include parking-light and brake-light functions. When the BRAKE input is asserted, all of the lights should go on immediately, and stay on until BRAKE is negated, independent of any other function. When the PARK input is asserted, each lamp is turned on at 50% brightness at all times when it would otherwise be off. This is achieved by driving the lamp with a 100-Hz signal DIMCLK with a 50% duty cycle. Draw a logic diagram for the circuit using one or two PLDs, write an ABEL program for each PLD, and write a short description of how your system works.

9.4   Find a 3-bit state assignment for the guessing-game machine in Table 9-5 that reduces the maximum number of product terms per output to 7. Can you do even better?

9.5   The operation of the guessing game in Section 9.1.4 is very predictable; it's easy for a player to learn the rate at which the lights change and always hit the button at the right time. The game is more fun if the rate of change is more variable.

         Modify the ABEL state machine in Table 9-5 so that in states S1–S4, the machine advances only if a new input, SEN, is asserted. (SEN is intended to be hooked up to a pseudorandom bit-stream generator.) Both correct and incorrect button pushes should be recognized whether or not SEN is asserted. Determine whether your modified design still fits in a 16V8.

9.6   In connection with the preceding exercise, write an ABEL program for an 8-bit LFSR using a single 16V8, such that one of its outputs can be used as a pseudo-random bit-stream generator. After how many clock ticks does the bit sequence repeat? What is the maximum number of 0s that occur in a row? of 1s?

9.7   Add an OVERRIDE input to the traffic-lights state machine of Figure 9-7, still using just a single 16V8. When OVERRIDE is asserted, the red lights should flash on and off, changing once per second. Write a complete ABEL program for your machine.

9.8   Modify the behavior of the ABEL traffic-light-controller machine in Table 9-9 to have more reasonable behavior, the kind you'd like to see for traffic lights in your own home town.

9.9   Write a VHDL program for the state machine described in Exercise 7.30.

9.10   Show how to modify Table 9-13 to provide an asynchronous RESET input that forces the state machine to the INIT state. Repeat for a synchronous version that forces the state machine to the INIT state if RESET is asserted on the rising clock edge.

9.11   Write and test a VHDL program corresponding to Figure 9-8(b). Using an available synthesis tool, determine whether the circuit resulting from this version differs from the one resulting from Figure 9-8(a), and how.

9.12   Write a VHDL program for the ones-counting state machine as described by the state table in Table 7-12.

9.13   Write a VHDL program for the combination-lock state machine as described by the state table in Table 7-14.

9.14   Modify the VHDL program of Table 9-19 to include the HINT output from the original state-machine specification in Section 7.4.

9.15   Repeat Exercise 9.3 using VHDL, assuming you are targeting your design to a single CPLD or FPGA.

9.16   Modify the VHDL guessing-game program of Table 9-21 according to the idea in Exercise 9.5. Add another process to the program to provide a pseudorandom bit-stream generator according to Exercise 9.6.

9.17   Modify the behavior of the VHDL traffic-light-controller machine in Table 9-23 to have more reasonable behavior, the kind you'd like to see for traffic lights in your own home town.

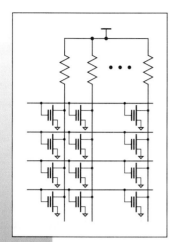

Memory, CPLDs, and FPGAs

A ny sequential circuit has memory of a sort, since each flip-flop or latch stores one bit of information. However, we usually reserve the word "memory" to refer to bits that are stored in a structured way, usually as a two-dimensional array in which one row of bits is accessed at a time.

This chapter describes several different types of memory organizations and commercially available memory chips. The same kinds of memory may be embedded into larger VLSI chips, where they are combined with other circuits to perform a useful function.

The applications of memory are many and varied. In a microprocessor's central processing unit (CPU), a "read-only memory" may be used to define the primitive steps that are performed to execute instructions in the CPU's instruction set. Alongside the CPU, a fast "static memory" may serve as a cache to hold recently used instructions and data. And a microprocessor's main-memory subsystem may contain hundreds of millions of bits in "dynamic memory" that store complete operating systems, programs, and data.

Applications of memory are not limited to microprocessors or even to purely digital systems. For example, equipment in the public telephone system uses read-only memories to perform certain transformations on digitized voice signals, and fast "static memories" as a "switching fabric" to route digitized voice between subscribers. Many portable audio compact-

831

disc players "read ahead" and store a few seconds of audio in a "dynamic memory" so that the unit can keep playing even if it is physically jarred (this requires over 1.4 million bits per second of stored audio). And there are many examples of modern audio/visual equipment that use memories to temporarily store digitized signals for enhancement through digital signal processing.

This chapter begins with a discussion of read-only memory and its applications and then describes two different types of read/write memory. Optional subsections discuss the internal structure of the different memory types.

The last two sections in this chapter discuss CPLDs and FPGAs. These devices are akin to memories in that they are large, regular structures that can be used in a variety of applications. By enabling very fast development of customized logic functions, they have become essential building blocks in modern digital design.

10.1 Read-Only Memory

read-only memory (ROM)

address input

data output

A *read-only memory (ROM)* is a combinational circuit with n inputs and b outputs, as shown in Figure 10-1. The inputs are called *address inputs* and are traditionally named A0, A1, ... , An−1. The outputs are called *data outputs* and are typically named D0, D1, ... , Db-1.

A ROM "stores" the truth table of an n-input, b-output combinational logic function. For example, Table 10-1 is the truth table of a 3-input, 4-output combinational function; it could be stored in a $2^3 \times 4$ (8×4) ROM. Neglecting propagation delays, a ROM's data outputs at all times equal the output bits in the truth-table row selected by the address inputs.

Since a ROM is a combinational circuit, you would be correct to say that it's not really a memory at all. In terms of digital circuit operation, you can treat

Figure 10-1
Basic structure of a $2^n \times b$ ROM.

MEMORY—NOT!   Most types of read-only memory are not really memory in the strictest sense of the word, because they are combinational, not sequential, circuits. They are called "memory" because of the organizational paradigm that describes their function.

Inputs			Outputs			
A2	**A1**	**A0**	**D3**	**D2**	**D1**	**D0**
0	0	0	1	1	1	0
0	0	1	1	1	0	1
0	1	0	1	0	1	1
0	1	1	0	1	1	1
1	0	0	0	0	0	1
1	0	1	0	0	1	0
1	1	0	0	1	0	0
1	1	1	1	0	0	0

Table 10-1
Truth table for a
3-input, 4-output
combinational logic
function.

a ROM like any other combinational logic element. However, you can also think
of information as being "stored" in the ROM when it is manufactured or
programmed (we'll say more about how this is done in Section 10.1.4).

Although we think of ROM as being a type of memory, it has an important
difference from most other types of integrated-circuit memory. ROM is *non-
volatile memory*; that is, its contents are preserved even if no power is applied.   *nonvolatile memory*

10.1.1 Using ROMs for "Random" Combinational Logic Functions

Table 10-1 is actually the truth table of a 2-to-4 decoder with an output-polarity
control, a function that can be built with discrete gates as shown in Figure 10-2.
Thus, we have two different ways to build the decoder—with discrete gates, or
with an 8×4 ROM that contains the truth table, as shown in Figure 10-3.

The assignment pattern of decoder inputs and outputs to ROM inputs and
outputs in Figure 10-3 is a consequence of the way that the truth table in
Table 10-1 is constructed. Thus, the physical ROM realization of the decoder is

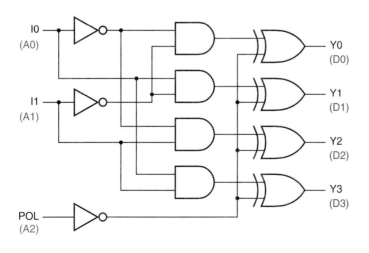

Figure 10-2
A 2-to-4 decoder
with output-polarity
control.

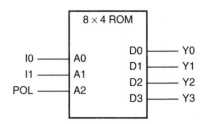

Figure 10-3
Connections to build the
2-to-4 decoder using an 8 × 4
ROM that stores Table 10-1.

not unique. That is, we could write the rows or columns of the truth table in a
different order and use a physically different ROM to perform the same logic
function, simply by assigning the decoder signals to different ROM inputs and
outputs. Another way to look at this is that we can rename the individual address
inputs and data outputs of the ROM.

For example, swapping the bits in the D0 and D3 columns of Table 10-1
would give us the truth table for a physically different ROM. However, the new
ROM could still be used to build the 2-to-4 decoder simply by swapping the Y0
and Y3 labels in Figure 10-3. Likewise, if we shuffled the data rows of the truth
table as shown in Table 10-2, we would get another different ROM, but it could
still be used as the 2-to-4 decoder with a rearrangement of the address inputs,
A0 = POL, A1 = I0, A2 = I1.

When constructing a ROM to store a given truth table, input and output
signals reading from right to left in the truth table are normally assigned to ROM
address inputs and data outputs with ascending labels. Each address or data
combination may then be read as a corresponding binary integer with the bits
numbered in the "natural" way. A data file is typically used to specify the truth
table to be stored in the ROM when it is manufactured or programmed. The data
file usually gives the address and data values as hexadecimal numbers. For
example, a data file may specify Table 10-2 by saying that ROM addresses 0–7
should store the values E, 1, D, 2, B, 4, 7, 8.

Table 10-2
Truth table with data
rows shuffled.

Inputs			Outputs			
A2	A1	A0	D3	D2	D1	D0
0	0	0	1	1	1	0
0	0	1	0	0	0	1
0	1	0	1	1	0	1
0	1	1	0	0	1	0
1	0	0	1	0	1	1
1	0	1	0	1	0	0
1	1	0	0	1	1	1
1	1	1	1	0	0	0

Figure 10-4
Connections to perform a 4×4 unsigned binary multiplication using a 256×8 ROM.

Another simple example of a function that can be built with ROM is 4×4 unsigned binary multiplication. We showed an ABEL program for this function in Section 5.11.2 and found that the number of product terms required (36) was too high to obtain with just one pass through a conventional PLD's AND-OR array. Alternatively, we can realize the function with one pass through a $2^8 \times 8$ (256×8) ROM with the connections shown in Figure 10-4.

A ROM's contents are normally specified by a file that contains one entry for every address in the ROM. For example, Table 10-3 is a hexadecimal listing of the 4×4 multiplier ROM contents. Each row gives a starting address in the ROM and specifies the 8-bit data values stored at 16 successive addresses.

The nice thing about ROM-based design is that you can usually write a simple program in a high-level language to calculate what should be stored in the ROM. For example, it took only a few minutes to write a C program, shown in Table 10-4, that generated the contents of Table 10-3.

Table 10-3 Hexadecimal text file specifying the contents of a 4×4 multiplier ROM.

```
00:  00 00 00 00 00 00 00 00 00 00 00 00 00 00 00 00
10:  00 01 02 03 04 05 06 07 08 09 0A 0B 0C 0D 0E 0F
20:  00 02 04 06 08 0A 0C 0E 10 12 14 16 18 1A 1C 1E
30:  00 03 06 09 0C 0F 12 15 18 1B 1E 21 24 27 2A 2D
40:  00 04 08 0C 10 14 18 1C 20 24 28 2C 30 34 38 3C
50:  00 05 0A 0F 14 19 1E 23 28 2D 32 37 3C 41 46 4B
60:  00 06 0C 12 18 1E 24 2A 30 36 3C 42 48 4E 54 5A
70:  00 07 0E 15 1C 23 2A 31 38 3F 46 4D 54 5B 62 69
80:  00 08 10 18 20 28 30 38 40 48 50 58 60 68 70 78
90:  00 09 12 1B 24 2D 36 3F 48 51 5A 63 6C 75 7E 87
A0:  00 0A 14 1E 28 32 3C 46 50 5A 64 6E 78 82 8C 96
B0:  00 0B 16 21 2C 37 42 4D 58 63 6E 79 84 8F 9A A5
C0:  00 0C 18 24 30 3C 48 54 60 6C 78 84 90 9C A8 B4
D0:  00 0D 1A 27 34 41 4E 5B 68 75 82 8F 9C A9 B6 C3
E0:  00 0E 1C 2A 38 46 54 62 70 7E 8C 9A A8 B6 C4 D2
F0:  00 0F 1E 2D 3C 4B 5A 69 78 87 96 A5 B4 C3 D2 E1
```

Table 10-4 Program to generate the text file specifying
the contents of a 4×4 multiplier ROM.

```
#include <stdio.h>

/* Procedure to print d as a hex digit. */
void PrintHexDigit(int d)
{
    if (d<10) printf("%c", '0'+d);
    else printf("%c", 'A'+d-10);
}

/* Procedure to print i as two hex digits.  */
void PrintHex2(int i)
{
    PrintHexDigit((i / 16) % 16);
    PrintHexDigit(i % 16);
}

void main()
{
    int x, y;

    for (x=0; x<=15; x++) {
        PrintHex2(x*16); printf(":");
        for (y=0; y<=15; y++) {
            printf(" ");
            PrintHex2(x*y);
        }
        printf("\n");
    }
}
```

*10.1.2 Internal ROM Structure

The mechanism used by ROMs to "store" information varies with different
ROM technologies. In most ROMs, the presence or absence of a diode or
transistor distinguishes between a 0 and a 1.

Figure 10-5 is the schematic of a primitive 8×4 ROM that you could build
yourself using an MSI decoder and a handful of diodes. The address inputs
select one of the decoder outputs to be asserted. Each decoder output is called a

word line

word line because it selects one row or word of the table stored in the ROM. The
figure shows the situation with A2–A0 = 101 and ROW5_L asserted.

bit line

Each vertical line in Figure 10-5 is called a *bit line* because it corresponds
to one output bit of the ROM. An asserted word line pulls a bit line LOW if a
diode is connected between the word line and the bit line. There is only one

*Throughout this book, optional sections are marked with an asterisk.

Figure 10-5
Logic diagram of a simple 8×4 diode ROM.

diode in row 5, and the corresponding bit line (D1_L) is pulled LOW. The bit lines are buffered through inverters to produce the D3–D0 ROM outputs, 0010 for the case shown.

In the ROM circuit of Figure 10-5, each intersection between a word line and a bit line corresponds to one bit of "memory." If a diode is present at the intersection, a 1 is stored; otherwise, a 0 is stored. If you were to build this circuit in the lab, you would "program" the memory by inserting and removing diodes at each intersection. Primitive though it may seem, owners of the DEC PDP-11 minicomputer (circa 1970) made use of similar technology in the M792 32×16 "bootstrap ROM module." The module was shipped with 512 diodes soldered in

DETAILS, DETAILS   Inverters with CMOS input thresholds are used in Figure 10-5 to improve the circuit's noise margins, because the 0.7-volt drop across the diodes creates a LOW level on the bit lines that is not really so low. Fortunately, with the 74HC04, a LOW level is anything less than 1.35 V. Of course, you'd never build this circuit except in an academic lab—it's easier just to buy and program a commercial ROM chip.

SNEAK PATHS   The ROM circuit of Figure 10-5 must use diodes, not direct connections, at each location where a 1 is to be stored. Figure 10-6 shows what happens if just a few diodes—such as the ones in row 3—are replaced with direct connections. Suppose the address inputs are 101; then ROW5_L is asserted, and only D1_L is supposed to be pulled LOW to create an output of 0010. However, the direct connections allow current to flow along *sneak paths*, so that bit lines D2_L and D0_L are also pulled LOW, and the (incorrect) output is 0111. When diodes are used, sneak paths are blocked by reverse-biased diodes, and correct outputs are obtained.

place, and owners programmed it by clipping out the diode at each location where a 0 was to be stored.

The diode pattern shown in Figure 10-5 corresponds to the 2-to-4-decoder truth table of Table 10-1. This doesn't seem very efficient—we used a 3-to-8 decoder and a bunch of diodes to build the ROM version of a 2-to-4 decoder. We could have used a subset of the 3-to-8 decoder directly! However, we'll show a more efficient ROM structure and some more useful design examples later.

Figure 10-6
Sneak paths
in a ROM.

*10.1.3 Two-Dimensional Decoding

Suppose you wanted to build a 128 × 1 ROM using the kind of structure described in the preceding subsection. Have you ever thought about what it would take to build a 7-to-128 decoder? Try 128 7-input NAND gates to begin with, and add 14 buffers and inverters with a fanout of 64 each! ROMs with a million bits or more are available commercially; trust me, they do not contain 20-to-1,048,576 decoders. Instead, a different structure, called *two-dimensional decoding*, is used to reduce the decoder size to something on the order of the square root of the number of addresses.

two-dimensional decoding

The basic idea in two-dimensional decoding is to arrange the ROM cells in an array that is as close as possible to square. For example, Figure 10-7 shows a possible internal structure for a 128 × 1 ROM. The three high-order address bits, A6–A4, are used to select a row. Each row stores 16 bits starting at address (A6, A5, A4, 0, 0, 0, 0). When an address is applied to the ROM, all 16 bits in the selected row are "read out" in parallel on the bit lines. A 16-input multiplexer selects the desired data bit based on the low-order address bits.

Figure 10-7 Internal structure of a 128 × 1 ROM using two-dimensional decoding.

By the way, the diode pattern in Figure 10-7 was not chosen at random. It performs a very useful 7-input combinational logic function that would require 35 4-input AND gates to build as a minimal two-level AND-OR circuit (see Exercise 10.7). The ROM version of this function would actually save a fair amount of engineering effort and space in a board-level design.

Two-dimensional decoding allows a 128×1 ROM to be built with a 3-to-8 decoder and a 16-input multiplexer (whose complexity is comparable to that of a 4-to-16 decoder). A $1M \times 1$ ROM could be built with a 10-to-1024 decoder and a 1024-input multiplexer—not easy, but a lot simpler than the one-dimensional alternative.

Besides reducing decoding complexity, two-dimensional decoding has one other benefit—it leads to a chip whose physical dimensions are close to square, important for chip fabrication and packaging. A chip with a $1M \times 1$ physical array would be *very* long and skinny and could not be built economically.

In ROMs with multiple data outputs, the storage arrays corresponding to each data output may be made narrower in order to achieve an overall chip layout that is closer to square. For example, Figure 10-9 shows the possible layout of a $32K \times 8$ ROM.

Figure 10-8
MOS transistors as storage elements in a ROM.

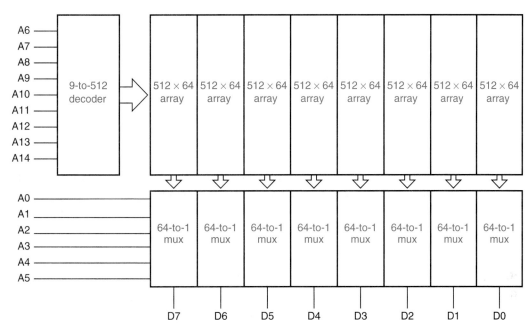

Figure 10-9 Possible layout of a 32K × 8 ROM.

10.1.4 Commercial ROM Types

Unless you visit computer museums, it's impossible nowadays to find ROM modules built with discrete diodes. A modern ROM is fabricated as a single IC chip; one that stores four megabits can be purchased for under $5. Various methods are used to "program" the information stored in a ROM, as discussed below and summarized in Table 10-5.

Most of the early integrated-circuit ROMs were *mask-programmable* *ROMs* (or, simply, *mask ROMs*). A mask ROM is programmed by the pattern of connections and no-connections in one of the *masks* used in the IC manufacturing process. To program or write information into the ROM, the customer gives

mask-programmable ROM

mask ROM

mask

Table 10-5 Commercial ROM types.

Type	Technology	Read cycle	Write cycle	Comments
Mask ROM	NMOS, CMOS	10–200 ns	4 weeks	Write once; low power
Mask ROM	Bipolar	< 100 ns	4 weeks	Write once; high power; low density
PROM	Bipolar	< 100 ns	10–50 μs/byte	Write once; high power; no mask charge
EPROM	NMOS, CMOS	25–200 ns	10–50 μs/byte	Reusable; low power; no mask charge
EEPROM	NMOS	50–200 ns	10–50 μs/byte	10,000–100,000 writes/location limit

the manufacturer a listing of the desired ROM contents, using a floppy disk or other transfer medium. The manufacturer uses this information to create one or more customized masks to manufacture ROMs with the required pattern. ROM

manufacturers impose a *mask charge* of several thousand dollars for the "customized" aspects of mask-ROM production. Because of mask charges and the four-week delay typically required to obtain programmed chips, mask ROMs are normally used today only in very high-volume applications. For low-volume applications there are more cost-effective choices, discussed next.

programmable read-only memory (PROM)

PROM programmer

A *programmable read-only memory (PROM)* is similar to a mask ROM, except that the customer may store data values (i.e., "program the PROM") in just a few minutes using a *PROM programmer*. A PROM chip is manufactured with all of its diodes or transistors "connected." This corresponds to having all bits at a particular value, typically 1. The PROM programmer can be used to set desired bits to the opposite value. In bipolar PROMs, this is done by vaporizing

fusible link

tiny *fusible links* inside the PROM corresponding to each bit. A link is vaporized by selecting it using the PROM's address and data lines, and then applying a high-voltage pulse (10–30 V) to the device through a special input pin.

Early bipolar PROMs had reliability problems. Sometimes the stored bits changed because of incompletely vaporized links that would "grow back," and sometimes intermittent failures occurred because of floating shrapnel inside the IC package. However, these problems were worked out, and reliable fusible-link technology is used nowadays not only in bipolar PROMs, but also in the bipolar PLD circuits that we described in Section 5.3.4.

erasable programmable read-only memory (EPROM)

An *erasable programmable read-only memory (EPROM)* is programmable like a PROM, but it can also be "erased" to the all-1s state by exposing it to ultraviolet light. No, the light does not cause fuses to grow back! Rather, EPROMs use a different technology, called "floating-gate MOS."

floating-gate MOS transistor

As shown in Figure 10-10, an EPROM has a *floating-gate MOS transistor* at every bit location. Each transistor has two gates. The "floating" gate is not connected and is surrounded by extremely high-impedance insulating material. To program an EPROM, the programmer applies a high voltage to the non-floating gate at each bit location where a 0 is to be stored. This causes a temporary breakdown in the insulating material and allows a negative charge to accumulate on the floating gate. When the high voltage is removed, the negative charge remains. During subsequent read operations, the negative charge prevents the MOS transistor from turning on when it is selected.

EPROM manufacturers "guarantee" that a properly programmed bit will retain 70% of its charge for at least 10 years, even if the part is stored at 125°C, so EPROMs definitely fall into the category of "nonvolatile memory." However,

EPROM erasing

they can also be erased. The insulating material surrounding the floating gate becomes slightly conductive if it is exposed to ultraviolet light with a certain wavelength. Thus, EPROMs can be erased by exposing the chips to ultraviolet

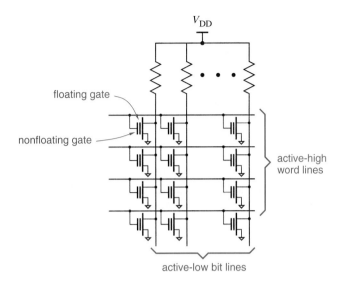

Figure 10-10
Storage matrix in an EPROM using floating-gate MOS transistors.

light, typically for 5–20 minutes. An EPROM chip is normally housed in a package with a quartz lid through which the chip may be exposed to the erasing light.

Probably the most common application of EPROMs is to store programs in microprocessor systems. EPROMs are typically used during program development, where the program or other information in the EPROM must be repeatedly changed during debugging. However, ROMs and PROMs usually cost less than EPROMs of similar capacity. Therefore, once a program is finalized, a ROM or PROM may be used in production to save cost. Most of today's PROMs are actually EPROMS housed in inexpensive packages without quartz lids; these are sometimes called *one-time programmable (OTP) ROMs*.

one-time programmable (OTP) ROM

An *electrically erasable programmable read-only memory (EEPROM)* is like an EPROM, except that individual stored bits may be erased electrically. The floating gates in an EEPROM are surrounded by a much thinner insulating layer and can be erased by applying a voltage of the opposite polarity as the charging voltage to the nonfloating gate. Large EEPROMs (1 Mbit or larger) allow erasing only in fixed-size blocks, typically 128–512 Kbits (16–64 Kbytes) at a time. These memories are usually called *flash EPROMs* or *flash memories*, because erasing occurs "in a flash."

electrically erasable programmable read-only memory (EEPROM)

flash EPROM
flash memory

As indicated in Table 10-5, programming or "writing" an EEPROM location takes much longer than reading it, so an EEPROM is no substitute for the read/write memories discussed later in this chapter. Also, because the insulating layer is so thin, it can be worn out by repeated programming operations. As a result, EEPROMs can be reprogrammed only a limited number of times, as few as 10,000 times per location. Therefore, EEPROMs are typically used for storing data that must be preserved when the equipment is not powered, but that doesn't change very often, such as the default configuration data for a computer.

Popular ROMs for microprocessors and other moderate-speed ROM applications include the *2764*, *27128*, *27256*, and *27512* EPROMs, whose logic symbols are shown in Figure 10-11. The figure also indicates pins that must be connected to a constant signal during normal read-only operation. Inputs labeled V_{CC} must be connected to +5 volts, inputs labeled V_{IH} must be connected to a valid HIGH logic signal, and "N.C." means "no connection." Different input configurations are used to program and test the device; the pin labeled VPP is used to apply the programming voltage.

Larger ROMs are available with more bits and in some cases wider data outputs of 16 or 32 bits. In 1999, one of the most impressive was the AMD 29LV640 flash memory, with 64 Mbits in 256 32-Kbyte sectors. Much smaller ROMs are also produced with 3-bit serial interfaces for specialized applications, such as downloading the programming information into FPGAs.

Multiple flash memories are often packaged into a single credit-card-size module for applications requiring large amounts of nonvolatile storage. The most common application of these modules is in digital cameras, where storing a single high-resolution image may require as much as 4 Mbytes. In 1999, the largest flash card was sold by industry leader SanDisk Corporation and contained 192 Mbytes (1536 Mbits) of memory.

Figure 10-11 Logic symbols for standard EPROMs in 28-pin dual in-line packages.

10.1.5 ROM Control Inputs and Timing

The outputs of a ROM must often be connected to a three-state bus, where different devices may drive the bus at different times. Therefore, most commercial ROM chips have three-state data outputs and an *output-enable (OE) input* that must be asserted to enable the outputs.

output-enable (OE) input

Many ROM applications, especially program storage, have multiple ROMs connected to a bus, where only one ROM drives the bus at a time. Most ROMs have a *chip-select (CS) input* to simplify the design of such systems. In addition to OE, a ROM's CS input must be asserted to enable the three-state outputs.

chip-select (CS) input

Figure 10-12 shows how the OE and CS inputs could be used when connecting four 32K × 8 ROMs to a microprocessor system that requires 128 Kbytes of ROM. The microprocessor has an 8-bit data bus and a 20-bit address bus, for a maximum address space of 1 Mbyte (2^{20} bytes). The ROM is supposed to be located in the highest 128K of the address space. To obtain this behavior, a NAND gate is used to produce the HIMEM_L signal, which is asserted when the address bus contains an address in the highest 128K (A19–A17 = 111). A 74x139 decoder then selects one of the four 32K × 8 ROMs. The selected ROM drives

Figure 10-12 Address decoding and ROM enabling in a microprocessor system.

Figure 10-13
Internal ROM
structure, showing
use of control inputs.

the data bus only when the microprocessor requests a read operation by asserting READ, which is connected to all of the OE inputs.

As we've described it so far, a CS input is no more than a second output-enable input that is ANDed with OE to enable the three-state outputs. However, in many ROMs, CS also serves as a *power-down input*. When CS is negated, power is removed from the ROM's internal decoders, drivers, and multiplexers. In this *standby mode* of operation, a typical ROM consumes less than 10% of the power it uses in *active mode* with CS asserted. In Figure 10-12, at most one ROM is selected at any time, so the total power consumption of all the ROM chips is much closer to that of one chip than four.

Figure 10-13 shows how CS and OE inputs are used inside a typical ROM. Figure 10-14 shows typical ROM timing, including the following parameters:

power-down input

standby mode
active mode

t_{AA}
access time from address

t_{AA} *Access time from address.* The access time from address of a ROM is the propagation delay from stable address inputs to valid data outputs. When designers talk about "a 100-ns ROM," they are usually referring to this parameter.

t_{ACS}
access time from chip select

t_{ACS} *Access time from chip select.* The access time from chip select of a ROM is the propagation delay from the time CS is asserted until the data outputs are valid. In some chips, this is longer than the access time from address, because the chip takes a little while to "power up." In others, this time is shorter because CS controls only output enabling.

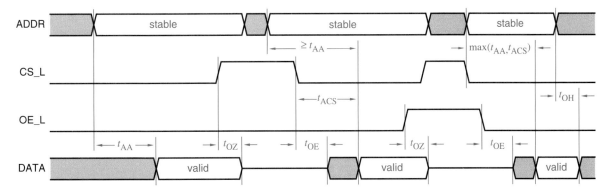

Figure 10-14 ROM timing.

t_{OE} *Output-enable time.* This parameter is usually much shorter than access time. The output-enable time of a ROM is the propagation delay from OE and CS both asserted until the three-state output drivers have left the Hi-Z state. Depending on whether the address inputs have been stable long enough, the output data may or may not be valid at that point.

t_{OZ} *Output-disable time.* The output-disable time of a ROM is the propagation delay from the time OE or CS is negated until the three-state output drivers have entered the Hi-Z state.

t_{OH} *Output-hold time.* The output-hold time of a ROM is the length of time that the outputs remain valid after a change in the address inputs, or after OE_L or CS_L is negated.

t_{OE}
output-enable time

t_{OZ}
output-disable time

t_{OH}
output-hold time

As with other components, the manufacturer specifies maximum and, sometimes, typical values for all timing parameters. Usually, minimum values are also specified for t_{OE} and t_{OH}. The minimum value of t_{OH} is usually specified to be 0; that is, the minimum combinational-logic delay through the ROM is 0.

NOT ALL INPUTS ARE CREATED EQUAL   Given the decoding and multiplexing structure in Figure 10-13, you might think that the address access time is shorter from some address inputs than it is from others. In a given application, if some address-input signals were delayed relative to the others, you could recover this delay by connecting the slower signals to the "faster" ROM inputs. After all, any input signal can be connected to any ROM address input if you're willing to rearrange the ROM contents accordingly.

However, ROM manufacturers don't specify which, if any, inputs are faster. In fact, the internal electrical characteristics of most ROMs may be such that the difference is not enough to bother with.

10.1.6 ROM Applications

As we mentioned earlier, the most common application of ROMs is for program storage in microprocessor systems. However, in many applications a ROM can provide a low-cost realization of a complex or "random" combinational logic function. In this section we'll give a couple of examples of ROM-based circuits that are used in the digital telephone network.

When an analog voice signal enters a typical digital telephone system, it is sampled 8,000 times per second and converted into a sequence of 8-bit bytes representing the analog signal at each sampling point. In Section 8.5.3 we showed how these digital voice samples could be converted between parallel and serial formats, but we did not describe the coding of the 8-bit bytes themselves. We'll do that now, and then show how ROM-based circuits can easily deal with this highly encoded information.

The simplest 8-bit encoding of the sign and amplitude of an analog signal would be an 8-bit integer in the two's-complement or signed-magnitude system; this is called a *linear encoding*. However, an 8-bit linear encoding yields a *dynamic range*—the ratio between range of representable numbers and the smallest representable difference—of only 2^8 or 256. For you audiophiles, this corresponds to a dynamic range in signal power of 20 log 256 or about 48 dB. By comparison, compact audio discs use a 16-bit linear encoding with a theoretical dynamic range of 20 log 2^{16} or about 96 dB.

Instead of a linear encoding, the North American telephone network uses an 8-bit *companded encoding* called μ-law PCM (pulse-code modulation). (The European network uses a different 8-bit companding formula called *A-law*.) Figure 10-15 shows the format of an 8-bit coded byte, a sort of floating-point representation containing sign (S), exponent (E), and mantissa (M) fields. The analog value V represented by a byte in this format is given by the formula

linear encoding
dynamic range

companded encoding
μ-law PCM
A-law PCM

$$V = (1-2S) \cdot [(2^E) \cdot (2M + 33) - 33]$$

Figure 10-15
Format of a μ-law PCM byte.

sign   exponent   mantissa

MONEY FOR NOTHING. . .   With μ-law PCM you don't get something for nothing, of course. In the companded encoding, the difference between successive coded values is greater at high amplitudes than it is at low amplitudes, so the encodings of large analog signals suffer from more quantizing distortion (caused by the difference between the sampled analog value and the nearest available coded representation). Expressed as a percentage of peak signal amplitude, though, quantizing distortion is roughly constant across the entire range of representable values.

An analog signal represented in this format can range from $-8159 \cdot k$ to $+8159 \cdot k$, where k is an arbitrary scale factor. The range of signals is $2 \cdot 8159$ and the smallest difference that can be represented is only 2 (when $E = 0$), so the dynamic range is 20 log 8159 or about 78 dB, quite an improvement over an 8-bit linear code.

Now let's look at some ROM applications involving μ-law coded voice signals. Believe it or not, in many types of phone connections your voice is purposely attenuated by a few decibels to make things work better. In an analog phone network, attenuation is performed by a simple, passive analog circuit, but not so in the digital world. Given a μ-law PCM byte, a *digital attenuator* must produce a different PCM byte that represents the original analog signal multiplied by a specified attenuation factor.

digital attenuator

One way to build a digital attenuator is shown in Figure 10-16. The input byte is applied to a μ-law decoder that expands the byte according to the formula given earlier to produce a 14-bit signed-magnitude integer. This 14-bit linear value is then multiplied by a 14-bit binary fraction corresponding to the desired attenuation amount. The fractional bits of the product are discarded, and the result is reencoded into a new 8-bit μ-law PCM byte. Each block in the figure could be built with perhaps a dozen MSI chips or a CPLD or FPGA.

Figure 10-17 is the logic diagram for a digital attenuator that uses a single, inexpensive 8K × 8 ROM instead. This ROM can apply any of 32 different attenuation factors to a μ-law input byte—it simply stores 32 different attenuation tables. The high-order address bits select a table, and the low-order address bits select an entry. Each entry is a precomputed μ-law byte corresponding to the

Figure 10-16 Block diagram of a digital attenuator using discrete components.

Figure 10-17
A digital attenuator.

given attenuation code and input byte. Table 10-6 is a C program that generates the ROM contents. The details of the `UlawToLinear` and `LinearToUlaw` functions are left as an exercise.

Another ROM application in digital telephony is a digital conference circuit. In the analog telephone network, it's fairly easy to make a conference connection between three or more parties. Just connect the analog phone wires together and you get an analog summing junction, so each person hears everyone else (it's not quite that easy to do it right, but you get the idea). In the digital network, of course, chaos would result from just shorting together digital output signals. Instead, a digital conference circuit must include a digital *adder* that produces output samples corresponding to sums of the input samples.

. . . AND THE CHECK'S FOR FREE

The digital attenuator is another good example of the many cost-effective space/time trade-offs that became possible when the telephone network "went digital." A single analog voice signal produces an 8-bit μ-law PCM sample once every 125 μs, but the digital attenuator ROM can produce a valid output in a few hundred nanoseconds or less. Thus, in a digital telephone system, a single ROM chip can attenuate hundreds of digital voice streams, where hundreds of analog attenuator networks would have been required in the old days.

Table 10-6 Program to generate the contents of an 8K × 8, 32-position
attenuator ROM for μ-law coded bytes.

```c
#include <stdio.h>
#include <math.h>

extern int UlawToLinear(int in);
extern int LinearToUlaw(int x);

void main()
{
  int i, j, position;
  int pcmIN, linearOUT, pcmOUT;
  double atten, attenDB, fpcmOUT;

  for (position=0; position<=31; position++) {  /* Make 32 256-byte tables. */
    printf("%i attenuation (dB): ");  /* Get amount in dB from designer,      */
    scanf("%f\n", attenDB);    /*  negative for attenuation, positive for gain. */
    atten = exp(log(10)*attenDB/10); /* Convert to fraction. */
    for (i=0; i<=15; i++) {  /* Construct output file in rows of 16. */
      printf("%04x:", position*256 + i*16);
      for (j=0; j<=15; j++) {
        pcmIN = i*16 + j;
        fpcmOUT = atten * UlawToLinear(pcmIN);
        if (fpcmOUT >=0) linearOUT = floor(fpcmOUT + 0.5);  /* Rounding */
        else linearOUT = ceil(fpcmOUT - 0.5);
        pcmOUT = LinearToUlaw(linearOUT);
        printf(" %2x", pcmOUT);
      }
      printf("\n");
    }
  }
}
```

You know how to build a binary adder for two 8-bit operands, but binary
adders cannot process μ-law PCM bytes directly. Instead, the 8-bit PCM bytes
must be converted into 14-bit linear format, then added, and then reencoded. As
in the digital attenuator circuit of Figure 10-16, an MSI design for this function
would require many chips. Alternatively, the function could be performed by a
single 64K × 8 ROM, as shown in Figure 10-18. The ROM has 16 address
inputs, accommodating two 8-bit μ-law PCM operands. For each pair of operand
values, the corresponding ROM address contains the precomputed μ-law PCM
sum. Table 10-7 is a C program that can be used to generate the ROM contents.

Figure 10-18
An adder circuit for
μ-law coded bytes.

These examples illustrate the many advantages of building complex combinational functions with ROMs. We usually consider a function to be "complex" if we know that a gate-level circuit for the function will be a real pain to design. However, most "complex" functions, like the examples in this subsection, have fairly straightforward word descriptions. Such a description can generally be translated into a computer program that "figures out" what the function's outputs should be for every possible input combination, so the function can be built simply by dropping its truth table into a ROM.

In addition to ease and speed of design, a ROM-based circuit has other important advantages:

- For a moderately complex function, a ROM-based circuit is usually faster than a circuit using multiple SSI/MSI devices and PLDs, and often faster than an FPGA or custom LSI chip in a comparable technology.

- The program that generates the ROM contents can easily be structured to handle unusual or undefined cases that would require additional hardware in any other design. For example, the adder program in Table 10-7 easily handles out-of-range sums. (Also see Exercise 10.22.)

- A ROM's function is easily modified just by changing the stored pattern, usually without changing any external connections. For example, the PCM attenuator and adder ROMs in this subsection can be changed to use 8-bit A-law PCM, the standard digital voice coding in Europe.

Table 10-7 Program to generate the contents of a 64K × 8 adder ROM
for μ-law coded bytes.

```
#include <stdio.h>
#include <math.h>
#define MINLINEAR -8159
#define MAXLINEAR 8159

void main()
{
  int i, j, linearSum;
  int pcmINX, pcmINY;

  for (pcmINY=0; pcmINY<=255; pcmINY++) {  /* For all Y samples... */
    for (i=0; i<=15; i++) {  /* Construct output file in rows of 16. */
      printf("%04x:", pcmINY*256 + i*16);
      for (j=0; j<=15; j++) {  /* For all X samples... */
        pcmINX = i*16 + j;
        linearSum = UlawToLinear(pcmINX) + UlawToLinear(pcmINY);
        /* The next two lines perform "clipping" on overflow. */
        if (linearSum < MINLINEAR) linearSum = MINLINEAR;
        if (linearSum > MAXLINEAR) linearSum = MAXLINEAR;
        printf(" %02x", LinearToUlaw(linearSum));
      }
      printf("\n");
    }
  }
}
```

- The prices of ROMs and other structured logic devices are always
  dropping, making them more economical, and their densities are always
  increasing, expanding the scope of problems that can be solved with a
  single chip.

  There are a few disadvantages of ROM-based circuits, too:

- For simple to moderately complex functions, a ROM-based circuit may
  cost more, consume more power, or run slower than a circuit using a few
  SSI/MSI devices and PLDs or a small FPGA.

- For functions with more than 20 inputs, a ROM-based circuit is impractical
  because of the limit on ROM sizes that are available. For example, you
  wouldn't want to build a 16-bit adder in ROM—it would require billions
  and billions of bits.

10.2 Read/Write Memory

read/write memory (RWM)

random-access memory (RAM)

The name *read/write memory (RWM)* is given to memory arrays in which we can store and retrieve information at any time. Most of the RWMs used in digital systems nowadays are *random-access memories (RAMs)*, which means that the time it takes to read or write a bit of memory is independent of the bit's location in the RAM. From this point of view, ROMs are also random-access memories, but the name "RAM" is generally used only for read/write random-access memories.

static RAM (SRAM)

dynamic RAM (DRAM)

In a *static RAM (SRAM)* ("S-ram"), once a word is written at a location, it remains stored as long as power is applied to the chip, unless the same location is written again. In a *dynamic RAM (DRAM)* ("D-ram"), the data stored at each location must be refreshed periodically by reading it and then writing it back again, or else it disappears. We'll discuss both types in this section.

volatile memory

nonvolatile memory

ferroelectric RAM

Most RAMs lose their memory when power is removed; they are a form of *volatile memory*. Some RAMs retain their memory even when power is removed; they are called *nonvolatile memory*. Examples of nonvolatile RAMs are old-style magnetic core memories and modern CMOS static memories in an extra-large package that includes a lithium battery with a 10-year lifetime. Recently, nonvolatile *ferroelectric RAMs* have been introduced; these devices combine magnetic and electronic elements on a single IC chip that retains its state even when power is not applied, just like the old-style core memories.

10.3 Static RAM

10.3.1 Static-RAM Inputs and Outputs

Like a ROM, a RAM has address and control inputs and data outputs, but it also has data inputs. The inputs and outputs of a simple $2^n \times b$-bit static RAM are shown in Figure 10-19. The control inputs are comparable to those of a ROM, with the addition of a *write-enable (WE) input*. When WE is asserted, the data inputs are written into the selected memory location.

write-enable (WE) input

The memory locations in a static RAM behave like D latches, rather than edge-triggered D flip-flops. This means that whenever the WE input is asserted, the latches for the selected memory location are "open" (or "transparent"), and input data flows into and through the latch. The actual value stored is whatever is present when the latch closes.

Static RAM normally has just two defined access operations:

Read   An address is placed on the address inputs while CS and OE are asserted. The latch outputs for the selected memory location are delivered to DOUT.

Write   An address is placed on the address inputs and a data word is placed on DIN; then CS and WE are asserted. The latches in the selected memory location open, and the input word is stored.

SERIAL-ACCESS MEMORY

Random-access memory can be contrasted with *serial-access memory*, where some location is immediately accessible at any time, but other locations take additional steps to access.

Some early computers used electromechanical serial-access memory devices, such as delay lines and rotating drums. Instructions and data were stored in a rotating medium with only one location under the "read/write head" at any time. To access a random location, the machine would have to wait until the constant rotation brought that location under the head.

In the 1970s, electronic equivalents of serial-access rotating memories were developed, including memories based on charge-coupled devices (CCDs) and others that used magnetic bubbles. Both types of devices were roughly equivalent to very large serial-in, serial-out shift registers with their serial output connected back into the serial input. This connection point was the logical equivalent of a hard disk's "read/write head." To read a particular location, you would clock the shift register until the desired bit appeared at the serial output, and to write the location, you would substitute the desired new value at the serial input.

Although they offered higher density (more bits) than DRAMs at the time they were developed, CCD and magnetic-bubble memories never gained much commercial acceptance. One reason for this was the enormous inconvenience of serial access. Another was that they were never more than a couple years ahead of DRAMs in the densities that they could achieve.

Figure 10-19
Basic structure of a $2^n \times b$ RAM.

A certain amount of care is needed when accessing SRAM, because it is possible to inadvertently "clobber" one or more locations while writing to a selected one, if the SRAM's timing requirements are not met. The following subsection gives details on SRAM internal structure to show why this is so, and the next one explains the actual timing behavior and requirements.

10.3.2 Static-RAM Internal Structure

SRAM cell

Each bit of memory (or *SRAM cell*) in a static RAM has the same functional behavior as the circuit in Figure 10-20. The storage device in each cell is a D latch. When a cell's SEL_L input is asserted, the stored data is placed on the cell's output, which is connected to a bit line. When both SEL_L and WR_L are asserted, the latch is open and a new data bit is stored.

SRAM cells are combined in an array with additional control logic to form a complete static RAM, as shown in Figure 10-21 for an 8 × 4 SRAM. As in a simple ROM, a decoder on the address lines selects a particular row of the SRAM to be accessed at any time.

Although Figure 10-21 is a somewhat simplified model of internal SRAM structure, it accurately portrays several important aspects of SRAM behavior:

- During read operations, the output data is a combinational function of the address inputs, as in a ROM. No harm is done by changing the address lines while the output data bus is enabled. The access time for read operations is specified from the time that the last address input becomes stable.

- During write operations, the input data is stored in *latches*. This means that the data must meet certain setup and hold times with respect to the *trailing* edge of the latch enable signal. That is, the input data at a latch's D input need not be stable at the moment WR_L is asserted internally; it must only be stable a certain time before WR_L is negated.

- During write operations, the address inputs *must* be stable for a certain setup time before WR_L is asserted internally and for a hold time after WR_L is negated. Otherwise, data may be "sprayed" all over the array because of the glitches that may appear on the SEL_L lines when the address inputs of the decoder are changed.

write cycle

- Internally, WR_L is asserted only when both CS_L and WE_L are asserted. Therefore, a *write cycle* begins when both CS_L and WE_L are asserted and ends when either is negated. Setup and hold times for address and data are specified with respect to these events.

Figure 10-20 Functional behavior of a static-RAM cell.

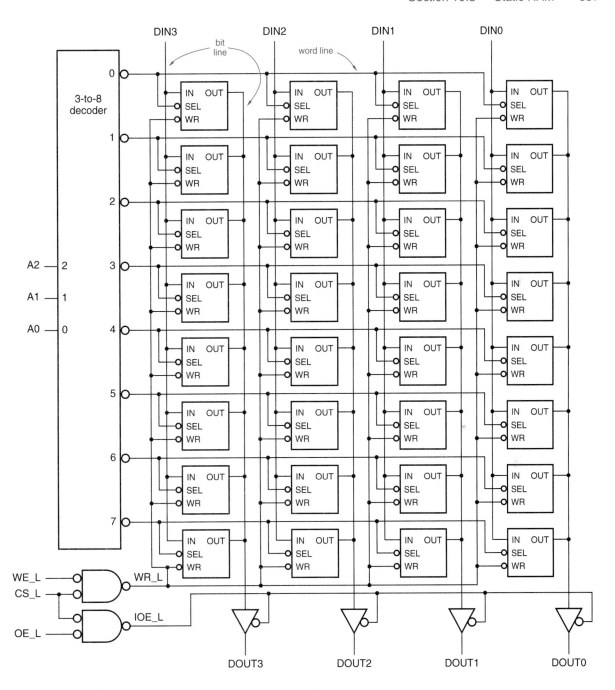

Figure 10-21 Internal structure of an 8 × 4 static RAM.

10.3.3 Static-RAM Timing

Figure 10-22 shows the timing parameters that are typically specified for read operations in a static RAM; they are described below:

t_{AA}
access time from address

t_{AA} *Access time from address.* Assuming that OE and CS are already asserted, or will be soon enough not to make a difference, this is how long it takes to get stable output data after a change in address. When designers talk about a "70-ns SRAM," they're usually referring to this number.

t_{ACS}
access time from chip select

t_{ACS} *Access time from chip select.* Assuming that the address and OE are already stable, or will be soon enough not to make a difference, this is how long it takes to get stable output data after CS is asserted. Often this parameter is identical to t_{AA}, but sometimes it's longer in SRAMs with a "power-down" mode and shorter in SRAMs without one.

t_{OE}
output-enable time

t_{OE} *Output-enable time.* This is how long it takes for the three-state output buffers to leave the high-impedance state when OE and CS are both asserted. This parameter is normally less than t_{ACS}, so it is possible for the RAM to start accessing data internally before OE is asserted; this feature is used to achieve fast access times while avoiding "bus fighting" in many applications.

t_{OZ}
output-disable time

t_{OZ} *Output-disable time.* This is how long it takes for the three-state output buffers to enter the high-impedance state after OE_L or CS_L is negated.

t_{OH}
output-hold time

t_{OH} *Output-hold time.* This parameter specifies how long the output data remains valid after a change in the address inputs.

If you've been paying attention, you may have noticed that the timing diagram and timing parameters for SRAM read operations are identical to what we discussed for ROM read operations in Section 10.1.5. That's the way it is; when they're not being written, SRAMs can be used just like ROMs. The same is not true for DRAMs, as we'll see later.

Figure 10-22 Timing parameters for read operations in a static RAM.

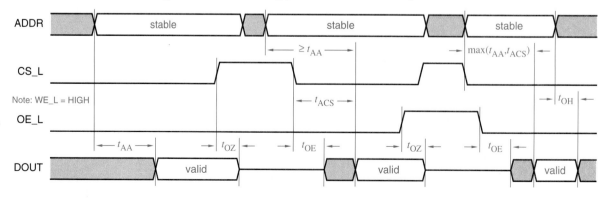

Timing parameters for write operations are shown in Figure 10-23 and are described below:

t_{AS} *Address setup time before write.* All of the address inputs must be stable at this time before both CS and WE are asserted. Otherwise, the data stored at unpredictable locations may be corrupted.

t_{ACS} *Address hold time after write.* Analogous to t_{AS}, all address inputs must be held stable until this time after CS or WE is negated.

t_{CSW} *Chip-select setup before end of write.* CS must be asserted at least this long before the end of the write cycle in order to select a cell.

t_{WP} *Write-pulse width.* WE must be asserted at least this long to reliably latch data into the selected cell.

t_{DS} *Data setup time before end of write.* All of the data inputs must be stable at this time before the write cycle ends. Otherwise, the data may not be latched.

t_{DH} Data hold time after end of write. Analogous to t_{DS}, all data inputs must be held stable until this time after the write cycle ends.

Manufacturers of SRAMs specify two write-cycle types, *WE-controlled* and *CS-controlled*, as shown in the figure. The only difference between these cycles is whether WE or CS is the last to be asserted and the first to be negated when enabling the SRAM's internal write operation.

The write-timing requirements of SRAMs could be relaxed somewhat if, instead of using latches, the cells contained edge-triggered D flip-flops with a common clock input and enable inputs tied controlled by SEL and WR. However, this just isn't done, because it would at least double the chip area of each cell, since a D flip-flop is built from two latches. Thus, the logic designer is left to reconcile the SRAM's latch-type timing with the edge-triggered register and state-machine timing used elsewhere in a system.

t_{AS}
address setup time

t_{AH}
address hold time

t_{CSW}
chip-select setup

t_{WP}
write-pulse width

t_{DS}
data setup time

t_{DH}
data hold time

WE-controlled write
CS-controlled write

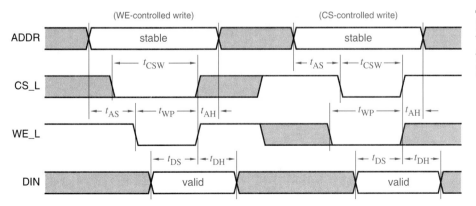

Figure 10-23
Timing parameters for write operations in a static RAM.

A large SRAM does not contain a physical array whose dimensions equal the logical dimensions of the memory. As in a ROM, the SRAM cells are laid out in an almost square array, and an entire row is read internally during read operations. For example, the layout of a 32K × 8 SRAM chip might be very similar to that of a 32K × 8 ROM shown in Figure 10-9 on page 841. During read operations, column multiplexers pass the required data bits to the output data bus, as specified by a subset of the address bits (A5–A0 in the ROM example). For write operations, the write-enable circuitry is designed so that only one column in each subarray is enabled, as determined by the same subset of the address bits.

10.3.4 Standard Static RAMs

Static RAMs are available in many sizes and speeds. In 1999, the largest conventional SRAMs used CMOS technology, were organized as 256K × 16, 512K × 8, or 1M × 4 bits (4 Mbits), and had access times as fast as 10 ns. The fastest TTL/CMOS compatible SRAMs also used CMOS technology and contained only 1K × 4 bits (4 Kbits) but had access times of 2.7 ns!

Part numbers for SRAM components are not standardized among manufacturers, although the components themselves often have identical pinouts and are interchangeable. Figure 10-24 shows part numbers and pinouts for Hitachi

Figure 10-24
Logic symbols for standard SRAMs in 28- and 32-pin DIP packages.

SRAMs, ranging in size from standard $8K \times 8$ to $512K \times 8$, that you might use in a digital design lab project; they all use a 5-V power supply and come in DIP packages. These devices are also available in more compact packages, and newer SRAM devices are typically available only in compact packages and use lower power-supply voltages, such as 3.3 V or 2.5 V.

The first two SRAMs in the figure use 28-pin DIP packages and have pinouts that are patterned after the like-size EPROMs in Figure 10-11. The *HM6264* $8K \times 8$ SRAM is similar to a 2764 EPROM, except that it has a write-enable input and two chip-select inputs instead of one; CS1_L and CS2 both must be asserted to enable the chip. The *HM62256* $32K \times 8$ SRAM is similar to a 27256 EPROM, with the addition of a write-enable input. Both chips are available from many manufacturers, often with different part numbers, and have access times in the range of 15–150 ns. The last two SRAMs in the figure use 32-pin DIP packages. The *HM628128* is a $128K \times 8$ SRAM and the *HM628512* is a $512K \times 8$ SRAM

HM6264

HM62256

HM628128

The SRAMs in Figure 10-24 have bidirectional data buses—that is, they use the same data pins for both reading and writing. This necessitates a slight change in their internal control logic, shown in Figure 10-25. The output buffer is automatically disabled whenever WE_L is asserted, even if OE_L is asserted. However, the timing parameters and requirements for read and write operations are practically identical to what we described in the preceding subsection.

A common use of SRAMs is to store data in small microprocessor systems, often in "embedded" applications (telephones, toasters, electric bumpers, etc.). General-purpose computers more often use DRAMs, discussed in Section 10-33, because their density is greater and their cost per bit lower.

Very fast SRAMs are often used in the "cache" memories of high-performance computers to store frequently used instructions and data. In fact, the "need for speed" in PC cache-memory applications led to the development and widespread deployment of a faster, clocked SRAM interface. The standard SRAMs in this subsection are now called *asynchronous SRAMs* to distinguish them from the new style discussed next.

asynchronous SRAM

Figure 10-25 Output-buffer control in an SRAM with a bidirectional data bus.

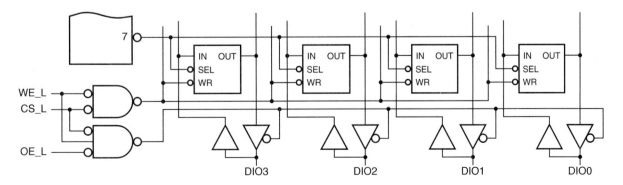

*10.3.5 Synchronous SRAM

*synchronous SRAM
(SSRAM)*

A new variety of SRAM, called a *synchronous SRAM (SSRAM)* ("S-S-ram"), still uses latches internally but has a clocked interface for control, address, and data. As shown in Figure 10-26, internal edge-triggered registers AREG and CREG are placed on the signal paths for address and control. As a result, an operation that is set up before the rising edge of the clock is performed internally during a subsequent clock period. Register INREG captures the input data for write operations and, depending on whether the device has "pipelined" or "flow-through" outputs, register OUTREG is or is not provided to hold the output data from a read operation.

*late-write SSRAM with
flow-through outputs*

The first variety of SSRAM to be introduced was the *late-write SSRAM with flow-through outputs*. For a read operation, shown in Figure 10-27(a), the control and address inputs are sampled at the rising edge of the clock, and the

Figure 10-26
Internal structure of a synchronous SRAM.

internal address register AREG is loaded only if ADS_L is asserted. During the next clock period, the internal SRAM array is accessed and read data is delivered to the device's DIO data-bus pins. The device also supports a burst mode, in which data at sequence of addresses is read. In this mode AREG behaves as a counter, eliminating the need to apply a new address at each cycle. (The control signals that support burst mode are not shown in Figures 10-26 or 10-27.)

For a write operation, shown in (b), the write data is stored temporarily in an on-chip register INREG, which is sampled one clock tick *after* the address register is loaded. Therefore, ADS_L must be inhibited for at least one tick after loading the address, so that the address in AREG is still valid when the write takes place. The write takes place during the clock period following the edge on which the "global write" control signal GW_L is asserted. As with reading, the device has a burst mode where a sequence of addresses can be written without supplying a new address.

Note that the "late-write" protocol makes it impossible to write to two different, nonsequential addresses in successive clock periods; the SRAM array is idle for one clock period between writes (except in burst mode). From the point of view of internal chip capabilities, this behavior is not necessary. However, the late-write protocol was designed this way to match the bus protocols of microprocessors that use these SSRAMs in their cache subsystems.

A *late-write SSRAM with pipelined outputs* is like the previous version, except that a register OUTREG is placed between the SRAM array output and the device output for read operations. As shown in Figure 10-28, this delays the read output data at the device pins until the beginning of the next clock period,

late-write SSRAM with pipelined outputs

Figure 10-27 Timing behavior for a late-write SSRAM with flow-through outputs: (a) read operations; (b) write operations.

Figure 10-28
Read-timing behavior
for a late-write
SSRAM with
pipelined outputs.

but it also provides the benefit that the data is now valid for almost the entire clock period. The write cycle behaves the same as with flow-through outputs. Compared to flow-through outputs, pipelined outputs provide much better setup time for the device receiving the read data, and therefore may allow operation at higher clock frequencies.

As we showed in Figure 10-26, conventional SSRAMs share the same pins for both input data and output data. During a given clock period the data I/O pins can be used for reading or writing but not both. If you study the pattern of data-bus and SRAM-array use in both styles of late-write SSRAM, you'll find cases where it's not possible to initiate a read one clock cycle after initiating a write or vice versa, due to resource conflict (see Exercise 10.32). Thus, late-write

turn-around penalty

SSRAMs suffer a *turn-around penalty*, a clock period in which the internal SRAM array must be idle when a read is followed by a write or vice versa.

zero-bus-turn-around
(ZBT) SSRAM

The turn-around penalty is eliminated in so called *zero-bus-turn-around (ZBT) SSRAMs*. The timing for a *ZBT SSRAM with flow-through outputs* is shown in Figure 10-29. The type of operation (read or write) is selected by a control signal R/$\overline{\text{W}}$ that sampled at the same clock edge as the address. Regardless of whether the operation is a read or a write, the DIO bus is used during the next clock period to transfer the read or write data. As a result, there is no data-bus-usage conflict, as long as OE is controlled properly to avoid bus-fighting between successive cycles. However, if a write is followed by a read, both operations would like to use the SRAM array during the same clock period. To avoid this resource conflict, the write operation is deferred until the next available SRAM cycle. This opportunity occurs when either another write operation or no operation is initiated on the address and control lines.

Although a ZBT SSRAM can use access the internal SRAM array on every clock cycle, this performance improvement is not without a price. While a write operation is pending, the write address and related information must be stored in another register, WAREG, since AREG is reused by other operations; this costs

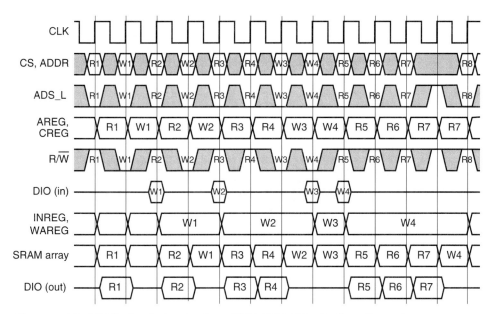

Figure 10-29 Timing behavior for a ZBT SSRAM with flow-through outputs.

chip area. More significantly for some applications, a write operation may be deferred indefinitely if it is immediately followed by a continuous series of read operations. This anomaly may require tricky controller design to detect the case where one of these read operations attempts to access the address that was just written, since the value stored in the SRAM array is "stale"!

A *ZBT SSRAM with pipelined outputs* adds OUTREG to the read data path but is otherwise similar to the previous device. In this device, both reads and writes use the DIO bus during the second clock period following the clock edge in which the operation was initiated. As in the previous device, writes to the internal SRAM array are deferred until an available cycle, so that reads can take precedence. Timing is shown in Figure 10-30. As implied by the timing, two levels of internal registers are needed for write address and data, since up to two writes may be deferred while a sequence of reads occurs.

Synchronous SRAM access protocols are very beneficial in high-speed systems. For example, address, control, and write inputs can be applied with more-or-less conventional setup and hold times with respect to the system clock, and read data on pipelined output pins is available for almost a complete clock cycle. Very importantly, the designer does not have to worry about the tricky circuits and timing paths that are otherwise needed to enable conventional SRAM latch-style operation.

SSRAMs were available in 1999 with clock frequencies as high as 166 MHz. Note that among the four styles of SSRAM that we described, no single one is the "best." The best SSRAM is the one that best fits the bus protocol and other requirements of the system in which it is used.

Figure 10-30 Timing behavior for a ZBT SSRAM with pipelined outputs.

10.4 Dynamic RAM

The basic memory cell in an SRAM, a *D latch*, requires four gates in a discrete design, and four to six transistors in a custom-designed SRAM LSI chip. In order to build RAMs with higher density (more bits per chip), chip designers invented memory cells that use as little as one transistor per bit.

10.4.1 Dynamic-RAM Structure

dynamic RAM (DRAM)

It is not possible to build a bistable element with just one transistor. Instead, the memory cells in a *dynamic RAM (DRAM)* store information on a tiny capacitor accessed through a MOS transistor. Figure 10-31 shows the storage cell for one bit of a DRAM, which is accessed by setting the word line to a HIGH voltage. To store a 1, a HIGH voltage is placed on the bit line, which charges the capacitor through the "on" transistor. To store a 0, LOW voltage on the bit line discharges the capacitor.

precharge

To read a DRAM cell, the bit line is first *precharged* to a voltage halfway between HIGH and LOW, and then the word line is set HIGH. Depending on whether the capacitor voltage is HIGH or LOW, the precharged bit line is pulled

sense amplifier

slightly higher or slightly lower. A *sense amplifier* detects this small change and recovers a 1 or 0 accordingly. Note that reading a cell destroys the original

Figure 10-31
Storage cell for one bit
in a DRAM.

voltage stored on the capacitor, so that the recovered data must be written back into the cell after reading.

The capacitor in a DRAM cell has a very small capacitance, but the MOS transistor that accesses it has a very high impedance. Therefore, it takes a relatively long time (many milliseconds) for a HIGH voltage to discharge to the point that it looks more like a LOW voltage. In the meantime, the capacitor stores one bit of information.

Naturally, using a computer would be no fun if you had to reboot every few milliseconds because its memory contents disappeared (the typical behavior of Windows notwithstanding). Therefore, DRAM-based memory systems use *refresh cycles* to update every memory cell periodically, once every four milliseconds in early DRAMs. This involves sequentially reading the somewhat degraded contents of each cell into a D latch and writing back a nice solid LOW or HIGH value from the latch. Figure 10-32 illustrates the electrical state of a cell after a write and a sequence of refresh operations.

refresh cycle

The first DRAMs, introduced in the early 1970s, contained only 1,024 bits, but modern DRAMs are available containing 256 *megabits* or more. If you had to refresh every cell, one at a time, in four milliseconds, you'd have a problem—that works out to far less than 1 ns per cell, and includes no time for useful read and write operations. Fortunately, as we'll show, DRAMs are organized using two-dimensional arrays, and a single operation refreshes an entire row of the array. Early DRAM arrays had 256 rows, requiring 256 refresh operations every four milliseconds, or one about every 15.6 μsec. Newer arrays have 4096 rows but need to be refreshed only once every 64 msec, which still works out to one row per 15.6 μsec. A refresh operation typically takes less than 100 ns, so the DRAM is available for useful read and write operations over 99% of the time.

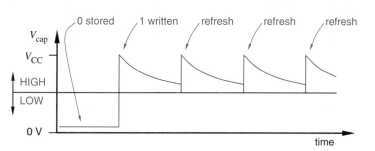

Figure 10-32
Voltage stored in a
DRAM cell after
writing and refresh
operations.

Figure 10-33
Internal structure of
a 64K × 1 DRAM.

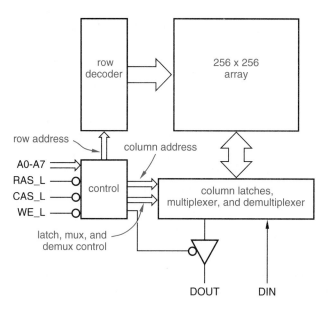

multiplexed address inputs

RAS_L (row address strobe)

CAS_L (column address strobe)

Figure 10-33 is a block diagram of the internal structure of a 64K × 1 DRAM. The logical array has 64K × 1 bits, but the physical array is square, containing 256 × 256 bits. Although the memory has 64K locations, the chip has only eight *multiplexed address inputs*. A complete 16-bit address is presented to the chip in two steps controlled by two signals—*RAS_L (row address strobe)* and *CAS_L (column address strobe)*. Multiplexing the address inputs saves pins, important for compact design of memory systems, and also fits quite naturally with the two-step DRAM access methods that we'll describe shortly.

Larger DRAMs have larger arrays and often have multiple arrays. One advantage of multiple arrays is to ease the electrical and physical design problems that would occur with an extremely large array. But even more important is the parallelism that can occur with multiple arrays. As we'll see in the next subsection, DRAM operation is much more complicated than SRAM operation. Taking advantage of the multiple arrays in larger, high-speed DRAMs, a modern DRAM controller can perform several operations in parallel—for example, completing a write operation in one array while initiating a read operation in another. This increases the effective throughput of the memory.

10.4.2 Dynamic-RAM Timing

There are many different timing scenarios for different DRAM types and operations. In this section we'll describe the most common cycles for conventional DRAM and relate them to the internal structure of the device. The most striking aspect of conventional DRAM timing is that there is no clock. Instead, DRAM operations are initiated and completed on both the rising and falling edges of RAS_L and CAS_L. This is very trick timing indeed, but the industry has somehow lived with it for over 25 years!

Figure 10-34
DRAM RAS-only
refresh-cycle timing.

The timing for a *RAS*-only refresh cycle is shown in Figure 10-34. This
cycle is used to refresh a row of memory without actually reading or writing any
data at the external pins of the DRAM chip. The cycle begins when a row
address is applied to the multiplexed address inputs (eight bits in the case of a
$64K \times 1$ DRAM) and RAS_L is asserted. The DRAM stores the row address in
an internal *row-address register* on the falling edge of RAS_L and reads the
selected row of the memory array into an on-chip *row latch*. When RAS_L is
negated, the contents of the row are written back from the row latch. To refresh
an entire $64K \times 1$ DRAM, the system designer must ensure that 256 such cycles,
using all 256 possible row addresses, are executed every four milliseconds. An
external 8-bit counter may be used to generate the row addresses, and a timer is
used to initiate a refresh cycle once every 15.6 μs.

A *read cycle*, shown in Figure 10-35, begins like a refresh cycle, where a
selected row is read into the row latch. Next, a column address is applied to the
multiplexed address inputs and is stored in an on-chip *column-address register*
on the falling edge of CAS_L. The column address is used to select one bit of the

RAS-only refresh cycle

row-address register
row latch

read cycle

column-address register

Figure 10-35
DRAM read-cycle
timing.

TRICKY TIMING   You may have noticed that the valid intervals of row and column addresses in
Figure 10-35 are skewed to the right with respect to the falling edges of RAS_L and
CAS_L, which load the addresses into on-chip edge-triggered registers. DRAMs
typically have a short setup-time requirement for the address inputs (often 0 ns) but
a relatively large, positive hold-time requirement (7–20 ns). This can cause some
problems in the design of an otherwise fully synchronous system that uses a single
clock and flip-flops with zero hold times, and is often solved by distasteful methods
involving tapped delay lines or using the opposite edge of the clock. This is one of
the reasons that many designers consider DRAM system design a tricky, black art.

Figure 10-36
DRAM write-cycle
timing.

just-read row, which is made available on the DRAM's DOUT pin. A three-state
pin, DOUT is output enabled as long as CAS_L is asserted. In the meantime, the
entire row is written back into the array as soon as RAS_L is negated.

write cycle
WE_L (write enable)

A *write cycle*, shown in Figure 10-36, also begins like a refresh or read
cycle. However, *WE_L (write enable)* must be asserted before CAS_L is asserted
to select a write cycle. One thing this accomplishes is to disable DOUT for the
rest of the cycle, even though CAS_L will be asserted subsequently. Once the
selected row is read into the row latch, WE_L also forces the input bit on DIN to
be merged into the row latch, in the bit position selected by the column address.
When the row is subsequently written back into the array on the rising edge of
RAS_L, it contains a new value in the selected column.

Several other types of cycles, not shown in the figure, are available in
typical DRAMs:

CAS-before-RAS
 refresh cycle

- A *CAS-before-RAS refresh cycle* performs a refresh operation without
  requiring a row address to be supplied by an external counter. Instead, the
  DRAM chip uses an internal row-address counter. If CAS_L is asserted
  before RAS_L is asserted, the DRAM refreshes the row selected by the
  counter and then increments the counter. This simplifies DRAM system
  design, eliminating the external refresh counter and reducing from three
  (row, column, refresh) to two the number of multiplexed sources that must
  drive the DRAM address inputs.

read-modify-write cycle

- A *read-modify-write cycle* begins like a read cycle and allows data to be
  read on DOUT when CAS_L is asserted. However, WE_L may be asserted
  subsequently to write new data back at the same location.

page-mode read cycle

- A *page-mode read cycle* allows up to an entire row ("page") of data to be
  read without repeating the entire RAS-CAS cycle. Since the entire row is
  already stored in the row latch, this cycle simply requires CAS_L to be
  pulsed LOW multiple times while RAS_L remains asserted continuously.
  A new column address is provided and a new bit becomes available on
  DOUT on each falling edge of CAS_L. This cycle provides much faster

memory access during a sequence of reads to consecutive or otherwise nearby addresses; such sequences often occur in microprocessors when they're fetching instructions or filling caches.

- Similarly, a *page-mode write cycle* allows multiple bits of a row to be written with a single RAS_L and multiple CAS_L cycles.

page-mode write cycle

Two other multiple-access modes, *static-column mode* and *nibble mode*, were provided at one time in certain DRAM chips, but these became obsolete with the marketplace dominance of so-called *extended-data-out (EDO) DRAMs*. In these devices, CAS_L no longer controls the output enable during read cycles; instead, a separate OE_L control input is provided. This is important for high-speed page-mode read cycles, as it provides valid output data for a longer period of time. Instead turning off between successive CAS_L pulses, the output is left on continuously (as controlled by OE_L), and page-mode read data is valid from the beginning of one CAS_L pulse to the beginning of the next.

static-column mode
nibble mode
extended-data-out (EDO) DRAM

10.4.3 Synchronous DRAMs

The RAS/CAS edge-based access protocol of conventional DRAMs is not only tricky, it's difficult to make it go fast and still meet timing margins when interfacing with the rest of a system. Thus, the early 1990s saw the development of *synchronous DRAMs (SDRAMs)* that use a more conventional clocked interface, and by the late 1990s these devices dominated the PC memory market.

synchronous DRAM (SDRAM)

Synchronous DRAMs retain the multiplexed addressing style of conventional DRAMs—row and column addresses are delivered in two steps. However, an SDRAM's control signals as well as its address inputs are sampled only on the rising edge of a common clock signal (CLK) with a frequency as high as 133 MHz. In addition, an SDRAM has a clock-enable signal (CKE) so that inputs are ignored if CKE is not asserted. For write operations, data is sampled on the clock edge, and for reads, output data is delivered on the clock edge.

Just like a conventional DRAM, an SDRAM requires multiple internal steps to perform an operation, and this means multiple clock ticks externally. Internally, SDRAMs have multiple DRAM banks, typically four, and they can have multiple operations pending simultaneously.

At each clock tick, an SDRAM's control signals RAS_L, CAS_L, and WE_L are interpreted as a "command word," as opposed to having individual significance. At the same time, the high-order address bits are interpreted as a "bank select" to indicate which bank a command applies to. For example, an intelligent SDRAM controller can use four ticks to initiate read operations in four different banks, and then come back and read the results one per clock tick as they complete.

Internal timing in an SDRAM is derived from the clock input CLK. Typically, a RAS signal to the internal array will be asserted immediately after the clock tick when a read or write command is given. To meet internal timing requirements, the chip generates an internal CAS signal at a later clock tick. How many ticks later is a function of the CLK frequency and the speed grade of the SDRAM chip itself. To accommodate differing requirements, the RAS-to-CAS delay, called the CAS latency, is programmable. This and several other important operating parameters must be downloaded into an SDRAM at initialization. Downloading is fairly easy; the "load parameters" command is recognized when RAS_L, CAS_L, and WE_L control signals are all asserted simultaneously, and the parameters themselves are transferred on the address lines.

10.5 Complex Programmable Logic Devices

Since their introduction years ago, programmable logic devices such as the 16V8 and 22V10 have been very flexible workhorses of digital design. As IC technology advanced, there was naturally great interest in creating larger PLD architectures to take advantage of increased chip density. The question is, why didn't manufacturers just scale the existing architectures?

For example, if DRAM densities increased by a factor of 64 over the last 10 years, why couldn't manufacturers scale the 16V8 to create a "128V64"? Such a device would have 64 input pins, 64 I/O pins, and some number of 128-variable product terms (say, 8) for each of its 128 logic macrocells. It could combine the functions of a larger collection of 16V8s and offer terrific performance and flexibility in using any input in any output function.

Or could it? Yes, a 128V64 would be very flexible, but it would not have very good performance. Unlike the 16V8, which has 32 inputs (16 signals and their complements) per AND term, this device would have 256. Due to capacitive effects, leakage currents, and so on, such a large wired-AND structure would be at least eight times slower than the 16V8's AND array.

Figure 10-37
General CPLD
architecture.

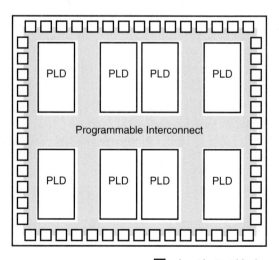

☐ = input/output block

Worse from a manufacturer's point of view, a 128V64 would not make very cost-effective use of chip area. It would use about 64 times the chip area of a 16V8, but provide the same number of inputs and outputs as only eight 16V8s. That is, for n times as much logic (in terms of inputs, outputs and AND terms), the 128V64 uses n^2 as much chip area. In terms of efficiency of silicon use, a clever designer would be better off partitioning a desired function into eight 16V8s, if such a partitioning were possible.

This is where the idea of complex programmable logic devices (CPLDs) came from. As shown in Figure 10-37, a CPLD is just a collection of individual PLDs on a single chip, accompanied by a programmable interconnection structure that allows the PLDs to be hooked up to each other on-chip in the same way that a clever designer might do with discrete PLDs off-chip. Here, the chip area for n times as much logic is only n times the area of a single PLD plus the area of the programmable interconnect structure.

Different manufacturers have taken many different approaches to the general architecture shown in the figure. Areas in which they differ include the individual PLDs (AND array and macrocells), the input/output blocks, and the programmable interconnect. We'll discuss each of these areas in the rest of this section, using the Xilinx 9500-series CPLD architecture as a representative example.

10.5.1 Xilinx XC9500 CPLD Family

The Xilinx XC9500 series is a family of CPLDs with a similar architecture but varying numbers of external input/output (I/O) pins and internal PLDs (which Xilinx calls function blocks—FBs). As we'll see later, each internal PLD has 36 inputs and 18 macrocells and outputs and might be called a "36V18." As shown in Table 10-8, devices in the family are named according to the number of macrocells they contain. The smallest has 2 FBs and 36 macrocells, and the largest has 16 FBs and 288 macrocells.

Another important feature of this and most CPLD families is that a given chip, such as the XC95108, is available in several different packages. This is important not only to accommodate different manufacturing practices but also to provide some choice and potential savings in the number of external I/O pins provided. In most applications, it is not necessary for all internal signals of a state machine or subsystem to be visible to and used by the rest of the system.

So, even though the XC95108 has 108 internal macrocells, the outputs of at most 69 of them can be connected externally in the 84-pin-PLCC version of the device. In fact, many of the 69 I/O pins would typically be used for inputs, in which case even fewer outputs would be visible externally. That's OK; the remaining macrocell outputs are still quite usable internally, since they can be hooked up internally through the CPLD's programmable interconnect. Macro-cells whose outputs are usable only internally are sometimes called *buried*   *buried macrocell*
macrocells.

Table 10-8 Function blocks and external I/O pins in Xilinx 9500-series CPLDs.

	Part Number					
	XC9536	XC9572	XC95108	XC95144	XC95216	XC95288
FBs / macrocells	2 / 36	4 / 72	6 / 108	8 / 144	12 / 216	16 / 288
Package	Device I/O Pins					
44-pin VQFP	34					
44-pin PLCC	34	34				
48-pin CSP	34					
84-pin PLCC		69	69			
100-pin TQFP		72	81	81		
100-pin PQFP		72	81	81		
160-pin PQFP			108	133	133	
208-pin HQFP					166	168
352-pin BGA					166	192

Another important dimension of Table 10-8 is the horizontal one. All but two of the packages support at least two different devices in the same package and, as it turns out, with compatible pinouts. This is a life-saver in making last-minute design changes. For example, suppose you were to target a design to an XC9572 in an 84-pin PLCC. You might find that the 69 I/O pins of the device are quite adequate. You would like to use the XC9572 for its low cost, but you might be a little nervous if your initial design used 68 out of the 72 available internal macrocells (I know that I would be!). Based on Table 10-8, you can sleep well knowing that if bug fixes or design-specification changes necessitate a more complex design internally, you can always move up to the XC95108 in the same package and pick up another 36 macrocells.

Figure 10-38 is a block diagram of the internal architecture of a typical XC9500-family CPLD. Each external I/O pin can be used as an input, an output, or a bidirectional pin according to the device's programming, as discussed later. The pins at the bottom of the figure can also be used for special purposes. Any of three pins can be used as "global clocks" (GCK); as we'll see later, each macrocell can be programmed to use a selected clock input. One pin can be used as a "global set/reset" (GSR); again, each macrocell can programmably use this signal as an asynchronous preset or clear. Finally, two or four pins (depending on device) can be used as "global three-state controls" (GTS); one of these signals can be selected in each macrocell to output-enable the corresponding output driver when the macrocell's output is hooked up to an external I/O pin.

Only four FBs are shown in the figure, but the XC9500 architecture scales to accommodate 16 FBs in the XC95288. Regardless of the specific family

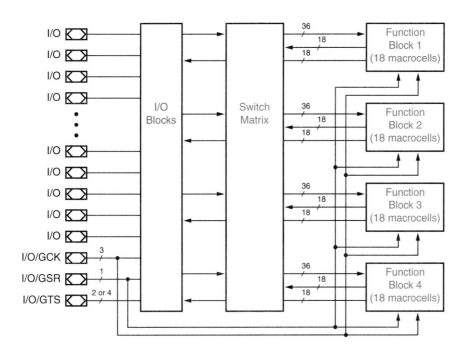

Figure 10-38
Architecture of Xilinx
9500-family CPLDs.

member, each FB programmably receives 36 signals from the switch matrix. The inputs to the switch matrix are the 18 macrocell outputs from each of the FBs and the external inputs from the I/O pins. We'll say more about how the switch matrix hooks things up in Section 10.5.4.

Each FB also has 18 outputs that run "under" the switch matrix as drawn in Figure 10-38 and connect to the I/O blocks. These are merely the output-enable signals for the I/O-block output drivers; they're used when the FB macrocell's output is hooked up to an external I/O pin.

10.5.2 Function-Block Architecture

The basic structure of an XC9500 FB is shown in Figure 10-39. The programmable AND array has just 90 product terms. Compared to 16V8- and 22V10-style PLDs, the XC9500 and most CPLD macrocells have fewer AND terms per macrocell—where the 16V8 has 8 and the 22V10 has 8–16, the XC9500 has only 5. However, this is not all that bad, because of *product-term allocation*. The

product-term allocation

Figure 10-39
Architecture of
function block (FB).

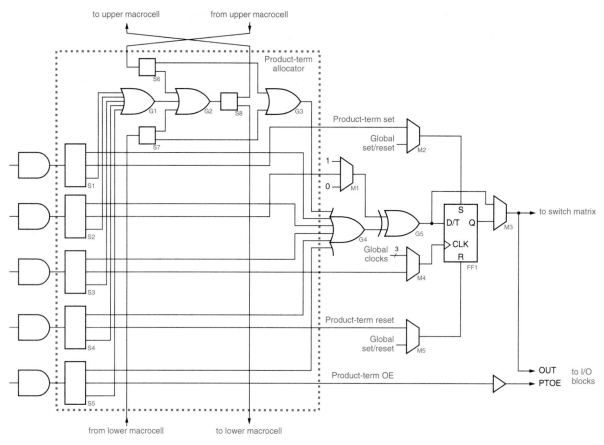

Figure 10-40 XC9500 product-term allocator and macrocell.

product-term allocator   XC9500 and other CPLDs have *product-term allocators* that allow a macrocell's unused product terms to be used by other nearby macrocells in the same FB.

Figure 10-40 is a logic diagram of the XC9500 product-term allocator and macrocell. In this figure, the rectangular boxes labeled S1–S8 are programmable signal-steering elements that connect their input to one of their two or three outputs. The trapezoidal boxes labeled M1–M5 are programmable multiplexers that connect one of their two to four inputs to their output.

The five AND gates associated with the macrocell appear on the lefthand side of the figure. Each one is connected to a signal-steering box whose top output connects the product term to the macrocell's main OR gate G4. Considering just this, only five product terms are available per macrocell. However, the top, sixth input of G4 connects to another OR gate G3 that receives product terms from the macrocells above and below the current one.

Any of the macrocell's product terms that are not otherwise used can be steered through S1–S5 to be combined in an OR gate G1 whose output can eventually be steered to the macrocell above or below by S8. Before steering,

ONE WAY   In a given XC9500 macrocell, you wouldn't normally steer daisy-chained product terms back in the direction from which they came. For example, if product terms are arriving at S6 from above, we can use them locally by having S6 steer them to G3, or S6 can steer them to G2. In the latter case, S8 should steer the output of G2 down to the lower macrocell; there's no point in steering the product terms back up to the upper macrocell. In fact, if S6 and S8 in this macrocell were steering product terms up, and S7 and S8 the macrocell above were steering product terms down, we would have a nasty loop.

these product terms may be combined with product terms from below or above through S6, S7, and G2. Thus, product terms can be "daisy-chained" through successive macrocells to create larger sums of products. In principle, all 90 product terms in the FB could be combined and steered to one macrocell, although that would leave 17 out of the FB's 18 macrocells with no product terms at all.

Besides depriving other macrocells of product terms, daisy-chaining terms has an additional price. A small additional delay is incurred for each "hop" made by steered product terms. This delay can be minimized by careful allocation of product-term-hungry macrocells so they are adjacent to macrocells with low product-term requirements. For example, a macrocell can make use of 13 product terms with only one extra hop delay if it is positioned between two macrocells that use only one product term each.

The third, middle choice for each of the steering boxes S1–S5 is to use the product term for a "special function." The special functions are flip-flop clock, set, and reset; XOR control; and output enable. Most of these special functions are not normally used.

Getting closer to the heart of the macrocell, OR gate G4 forms a sum-of-products expression using all selected product terms and feeds it into XOR gate G5. The other input of G5 can be 0, 1, or a product term, as selected by multiplexer M1. Setting this input to 1 inverts G4's sum-of-products expression, so the macrocell can be configured to use either polarity of minimized logic equations. Setting this input to a product term is useful in the design of counters. The product term is arranged to be 1 when the lower-order counter bits are 1 and counting is enabled, and the output of G4 is arranged to be the current value of the counter bit; thus, the counter bit is complemented as required.

The macrocell's flip-flop FF1 can be programmed to behave either as a D flip-flop or as a T flip-flop with enable; the latter is also useful in some styles of counter realization. Multiplexer M4 selects the flip-flop's clock input from one of four sources—the CPLD's three global clock inputs or a product term. This last choice is a no-no in synchronous design methodologies, except in well-defined synchronization applications, as in Figure 8-111 on page 783.

The flip-flop also has asynchronous set and reset inputs with input sources controlled by multiplexers M2 and M5. In most applications, set or reset would be connected to the CPLD's global set/reset input and would be used only at system initialization. However, these inputs can also be used to access an S-R latch in which the CLK input is not used, or, if you're very careful, you can use CLK and S or R in synchronization applications as in Figure 8-111.

A final multiplexer M3 selects either the flip-flop output or its data input to be used as the macrocell output, OUT. This signal is sent to the switch matrix, where it can be used by any other macrocell. It is also sent to the I/O blocks, along with a product term selected by S5 that can be used as the output-enable signal PTOE if needed.

10.5.3 Input/Output-Block Architecture

I/O block (IOB)

The structure of the XC9500 *I/O block (IOB)* is shown in Figure 10-41. There are seven, count them, seven choices of output-enable signals for the three-state driver buffer. It can be always on, always off, controlled by the product term PTOE from the corresponding macrocell, or controlled by any of up to four global output enables. The global output enables are selectable as active-high or active-low versions of the external GTS pins.

The XC9500's IOB is a good example of an important trend in CPLD and FPGA I/O architectures—providing many "analog" controls in addition to "logic" ones like output enables. Three different analog controls are provided:

- *Slew-rate control.* The rise and fall time of the output signals can be set to be fast or slow. The fast setting provides the fastest possible propagation delay, while the slow setting helps to control transmission-line ringing and system noise at the expense of a small additional delay.

- *Pull-up resistor.* When enabled, the pull-up resistor prevents output pins from floating as the CPLD is powered up. This is useful if the outputs are used to drive active-low enable inputs of other logic that is not supposed to be enabled during power up.

- *User-programmable ground.* This feature actually reallocates an I/O pin to be a ground pin, not a signal pin at all. This is useful in high-speed, high-slew-rate applications. Extra ground pins are needed to handle the high dynamic currents that flow when multiple outputs switch simultaneously.

In addition to these features, the XC9500 family provides compatibility with both 5-V and 3.3-V external devices. The input buffer and the internal logic run from a 5-V power supply (V_{CCINT}). Depending on the operating voltage of external devices, the output driver uses either a 5-V or a 3.3-V supply (V_{CCIO}). Notice that the pull-up resistor pulls to the I/O supply voltage, V_{CCIO}. Diodes $D1$ and $D2$ are used to clamp voltages above V_{CCINT} or below ground that can occur due to transmission-line ringing.

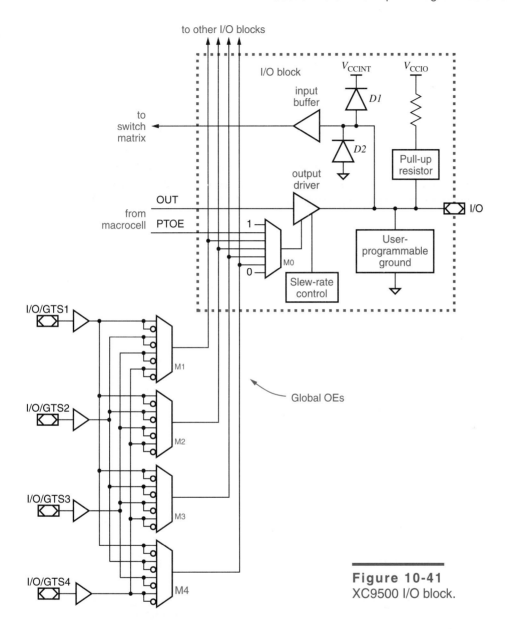

Figure 10-41
XC9500 I/O block.

10.5.4 Switch Matrix

Theoretically, a CPLD's programmable interconnect should allow any internal PLD output or external input pin to be connected to any internal PLD input. Likewise, it should allow any internal PLD output to connect to any external output pin. But if you think about this, you'll see that we're back to the same n^2 problem that we would have in building a 128V64.

The case of a typical Xilinx XC9500 family member, the XC95108, is shown in Figure 10-42. There are 108 internal macrocell outputs and 108 external-pin inputs, a total of 216 signals which should be connected as inputs to the switch matrix. Since the XC95108 has 6 FBs with 36 inputs each, the switch matrix should have (coincidentally) 216 outputs, each of which is a 216-input multiplexer driving one input of an FB's AND array.

A switch matrix such as the one shown in the figure can be built in a chip as a rectangular structure, with a column for each input, a row for each output, and a pass transistor (or transmission gate) at each crosspoint to control whether a given input is connected to a given output. But this is still a big structure—216 rows and 216 columns in the example.

With today's high-density IC technology, the problem is not so much size but speed. Having a large number of transistors connected to each row or column makes for high capacitance, which makes for slow speed. Therefore, CPLD manufacturers look for ways to reduce the size of the switch matrix.

For example, we can observe in Figure 10-42 that not every switch-matrix input must be connectable to every output. We only need every input to be connectable to *some* input of each FB, since any FB input can connect to any AND gate within the FB's AND array.

Then again, the requirement is not quite as simple as we just stated. Suppose that the switch-matrix output for FB input i ($0 \leq i \leq 35$) is created by a simple 8-input multiplexer whose inputs are switch-matrix inputs $8i$ through $8i + 7$. There are many interconnection patterns that cannot be accommodated by such a switch matrix. For example, connecting switch-matrix inputs 0–35 to the same FB would not be possible. As soon as switch-matrix input 0 is connected to FB input 0, switch-matrix inputs 1–7 are blocked.

Thus, the switch-matrix connectivity requirement must be stated more broadly: For each FB, any combination of switch-matrix inputs must be connectable to some combination of the FB's inputs.

Figure 10-42
XC95108 switch-matrix requirements.

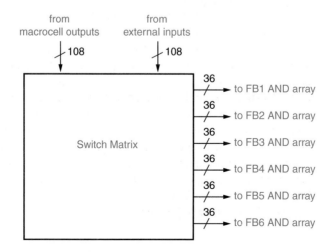

A typical CPLD switch matrix is a compromise between the minimal multiplexer scheme and a full, nonblocking crosspoint array. With anything less than a nonblocking crosspoint array, the problem of allocating input-output connections in the switch matrix is nontrivial. For each different CPLD-based design, a set of switch-matrix connections must be found by "fitter" software provided to the designer by the CPLD's manufacturer.

Finding a complete set of connections through a sparse switch matrix is one of those *NP*-complete problems that you hear about in computer science. In lay terms, that means that for some designs, the fitter software may have to run a lot longer than you care to wait to find out whether there is a solution. If the switch matrix has too few crosspoints, as in our minimal multiplexer example, even the best fitter software running "forever" will not be able to find a complete set of connections for some designs.

Thus, the design of a CPLD's switch matrix is a compromise between chip performance (speed, area, cost) and fitter-software capabilities. The fitter software usually determines not only the final connections through the switch matrix, but also the assignment of CPLD inputs and outputs to FBs, macrocells, and external pins, and of "buried logic" to FBs and macrocells. These assignments interact in turn with both the switch-matrix-connection and product-term allocation problems. The solutions to these problems are the "secret sauce" of CPLD chip and software design, and are not typically disclosed by CPLD manufacturers.

PIN LOCKING   Another important issue in CPLD chip and software design is *pin locking*. In most CPLD applications, it's OK to let the fitter software pick any pins that it likes for the device's external input and output signals. However, once the design is complete and a PCB has been fabricated, the designer would like to "lock down" the pin assignments so they remain the same even if small (or large!) changes are made to the design for bug-fixing purposes. This spares everyone the time, expense, and hassle of reworking or redesigning and refabricating the PCB.

"Locked" pin assignments are typically specified in a file that is read by the fitter software. With early CPLDs and FPGAs, locking down pins before making a small change did not guarantee success—the fitter would throw up its hands and complain that it was too constrained. If you unlocked the pins, the fitter could find a new allocation that worked, but it might be completely scrambled from the original.

These problems were not necessarily the fault of the fitter software; the CPLDs and FPGAs simply did not have rich enough internal connectivity to support frequent design changes under the constraint of pin locking. The device manufacturers have learned from this experience and improved their internal device architectures to accommodate frequent design changes. For example, some devices contain an "output switch matrix" that guarantees that any internal macrocell signal can be connected to any external I/O pin.

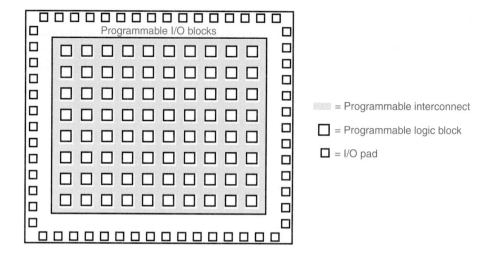

= Programmable interconnect

= Programmable logic block

= I/O pad

Figure 10-43
General FPGA chip
architecture.

10.6 Field-Programmable Gate Arrays

A field-programmable gate array (FPGA) is kind of like a CPLD turned inside-out. As shown in Figure 10-43, the logic is broken into a large number of programmable logic blocks that are individually smaller than a PLD. They are distributed across the entire chip in a sea of programmable interconnections, and the entire array is surrounded by programmable I/O blocks. An FPGA's programmable logic block is less capable than a typical PLD, but an FPGA chip contains a lot more logic blocks than a CPLD of the same die size has PLDs.

Xilinx, Inc. invented FPGAs, and in this section we'll use one of their popular families, the XC4000E, to illustrate FPGA architecture.

10.6.1 Xilinx XC4000 FPGA Family

configurable logic block
(CLB)

The programmable logic blocks in the Xilinx XC4000E family of FPGAs are called *configurable logic blocks (CLBs)*. The smallest part, the XC3003E, contains a 10×10 array of CLBs, and the largest, the XC4025E, contains a 32×32 array for a total of 1,024 CLBs. Xilinx also created extended XC4000EX and XC4000XL families, based on the XC4000E family, that have additional resources and features not discussed here. The largest member of the extended families, the XC4085XL, has 3,136 CLBs. Table 10-9 from Xilinx shows the family members that were available in 1999.

Like a CPLD family, the XC4000 family spans a range of device sizes and input/output capabilities. The table column labeled "Max. User I/O" refers to the maximum number of input/output blocks that are provided on-chip. However, XC4000 devices are available in a variety of packages, and not all of the I/Os are brought out to external pins of the smaller packages. As with CPLDs, the FPGA user has the opportunity to migrate a design from smaller to larger devices in the same package, or from a smaller package to a larger one.

Table 10-9 Resources in Xilinx XC4000-series FPGAs.

Device	CLB Matrix	Total CLBs	Max. User I/O	Flip-Flops	Max. RAM bits (no logic)	Max. Gates (no RAM)	Typical Gate Range (Logic and RAM)
XC4002XL	8 × 8	64	64	256	2,048	1,600	1,000–3,000
XC4003E	10 × 10	100	80	360	3,200	3,000	2,000–5,000
XC4005E/XL	14 × 14	196	112	616	6,272	5,000	3,000–9,000
XC4006E	16 × 16	256	128	768	8,192	6,000	4,000–12,000
XC4008E	18 × 18	324	144	936	10,368	8,000	7,000–15,000
XC4010E/XL	20 × 20	400	160	1,120	12,800	10,000	7,000–20,000
XC4013E/XL	24 × 24	576	192	1.536	18,432	13,000	10,000–30,000
XC4020E/XL	28 × 28	784	224	2,016	25,088	20,000	13,000–40,000
XC4025E	32 × 32	1,024	256	2,560	32,768	25,000	15,000–45,000
XC4028EX/XL	32 × 32	1,024	256	2,560	32,768	28,000	18,000–50,000
XC4036EX/XL	36 × 36	1,296	288	3,168	41,472	36,000	22,000–65,000
XC4044XL	40 × 40	1,600	320	3,840	51,200	44,000	27,000–80,000
XC4052XL	44 × 44	1,936	352	4,576	61,952	52,000	33,000–100,000
XC4062XL	48 × 48	2,304	384	5,376	73,728	62,000	40,000–130,000
XC4085XL	56 × 56	3,136	448	7,168	100,352	85,000	55,000–180,000

The "Flip-Flops" column counts all of the flip-flops in the device, two per CLB and two per I/O block, as we'll see later. Only a fraction of the available flip-flops are used in a typical design, but this number is a figure of merit that designers look at when roughly sizing an FPGA for an design. The "Max. RAM Bits" column is another such figure of merit. As we'll see, instead of being used for logic, each CLB can be configured as a small SRAM storing up to 32 bits.

The "Max. Gates" number is fuzzy. For the XC4002XL, the table claims that each CLB can perform the same function as about 25 gates in a discrete design. The XC4003E is even better, at 30 gates per CLB. Is this reasonable? And just what is a "gate"? Do XORs or wide NAND gates count as one gate? You can decide for yourself later, after you know more about the CLB architecture. At the same time, you can decide whether this column was created by engineering or by marketing people.

The last column of the table must certainly have been written by marketing folks, since the high end of each "typical" gate range is higher than the maximum in the column before it! Actually, there *is* a reasonable explanation for this seeming inconsistency. This column assumes that 20–30% of the CLBs are used for memory rather than logic, at 32 bits of SRAM per CLB. The minimum

Figure 10-44 XC4000 configurable logic block.

gate-level implementation of an SRAM cell is a D latch using four gates (Figure X7.27 on page 650), so we can count each CLB as 128 gates when it's used as SRAM. The number in the last column, therefore, is the number of gates used for logic *plus* the number of equivalent gates used for SRAM.

10.6.2 Configurable Logic Block

Since an FPGA can have lots and lots of CLBs, it's important that we understand them first! Figure 10-44 shows the internal structure of an XC4000-series CLB.

The CLBs most important programmable elements are the logic-function generators F, G, and H. Both F and G can perform any combinational logic function of their four inputs, and H can perform any combinational logic function of its three inputs.

How do F, G, and H work? Given the task of building a "universal" function generator for 4-input logic functions, what kind of gate-level circuit would you come up with? Think about it, and we'll come back to it later.

As in our CPLD discussion, the trapezoidal boxes in Figure 10-44 represent programmable multiplexers. Notice that the outputs of F and G as well as additional CLB inputs can be directed to inputs of H by multiplexers M1–M3, so it's possible to realize some functions of more than four inputs. A taxonomy of the functions that can be realized by F, G, and H in a single CLB is given below:

* Any function of up to four variables, plus any second function of up to four unrelated variables, plus any third function of up to three unrelated variables.
* Any single function of five variables (see Exercise 10.34).
* Any function of four variables, plus some second functions of six unrelated variables.
* Some functions of up to nine variables, including parity checking and a cascadable equality checker for two 4-bit inputs (see Exercises 10.35 and 10.36).

With appropriate programming of multiplexers M7–M8 and M12–M13, the outputs of the function generators can be directed to CLB outputs X and Y, or they can be captured in edge-triggered D flip-flops FF1 and FF2. The flip-flops can use the rising or falling edge of a common clock input, K, as selected by multiplexers M9 and M14. They can also make use of a clock-enable signal, EC, selected by M10 and M15. The sources of EC and three other internal signals are selected from a set of four miscellaneous inputs C1–C4 by multiplexers M3–M6 at the top of the CLB.

The XQ and YQ outputs of the CLB carry the flip-flop outputs out of the CLB. If a flip-flop is not used in the CLB, multiplexer M11 or M16 can select XQ or YQ to be a "bypass output" that is simply a copy of a CLB input selected by M4 or M6.

The block labeled "S/R control" from each flip-flop determines whether the flip-flop is set or reset at configuration. It also determines whether the flip-flop responds to a global set/reset signal (not shown) or to the CLB's SR signal selected by multiplexer M5.

Wow, that's a lot of programmability! Naturally, the configuration of CLBs within an XC4000 part, whether it has 3,136 CLBs or only 64, is not carried out by hand. The manufacturer provides a fitter tool that allocates, configures, and connects CLBs to match a higher-level design description written in ABEL, VHDL, Verilog, or schematic form.

Let's come back to our question of how to build a universal function generator for 4-input logic functions. It's a hard problem if you think about it at the gate level, but pretty easy if you think about it from another point of view. Any 4-input logic function can be described by its truth table, which has 16 rows. Suppose we store the truth in a 16-word by 1-bit-wide memory. When we apply the function's four input bits to the memory's address lines, its data output is the value of the function for that input combination.

This is exactly the approach that was taken by the FPGA designers at Xilinx. Function generators F and G are actually just very compact and fast 16×1 SRAMs, and H is an 8×1 SRAM. When a CLB is used to perform logic, the truth tables of logic functions F, G, and H are loaded into SRAM at configuration time from an external read-only memory. The programmable multiplexers in Figure 10-44 are also controlled by "SRAM," actually individual D latches that are also loaded at configuration time. This programming is done for all of the CLBs in the FPGA.

Besides convenience, using memory to store the truth tables has another important benefit. Any XC4000 CLB can be configured at start-up to be used as memory rather than logic. Several different modes can be configured:

- *Two 16×1 SRAMs.* F and G are used as SRAMs with independent address and write-data inputs. They share a common write-enable input, however.

- *One 32×1 SRAM.* The same four address bits are used for F and G, and a fifth address bit is applied to the H function generator and the write-enable circuitry to select between F and G, the upper and lower halves of the memory.

- *Asynchronous or synchronous.* For write operations, the SRAMS above can be configured to have "normal" asynchronous latching behavior, or they can be configured to occur on a designated edge of the K clock signal.

- *One 16×1 dual-port SRAM.* The two sets of address inputs are used to independently read and write different locations in the same SRAM. Only synchronous write operations are supported in this mode.

In these modes, function inputs F1–F4 and G1–G4 supply address, other CLB inputs H0–H2 provide data inputs and the write-enable signal, and data outputs are provided on the F and G generator outputs and can be captured in flip-flops FF1 and FF2 or exit at CLB outputs X and Y.

10.6.3 Input/Output Block

I/O block (IOB)

The structure of the XC4000 *I/O block (IOB)* is shown in Figure 10-45. An I/O pin can be used for input or output or both.

The XC4000 IOB has more "logic" controls than its cousin in the XC9500 CPLD. In particular, its input and output paths contain edge-triggered D flip-flops selectable by multiplexers M5–M7. Placing input and output flip-flops "up close" to the device I/O pins is especially useful in FPGAs. On output, relatively long delays from internal CLB flip-flop outputs to the IOBs can make it difficult to connect to external synchronous systems at very high clock rates. On input,

Figure 10-45
XC4000 I/O block.

long delays from the I/O pins to CLB flip-flop inputs can make it difficult to meet external system setup and hold times if external inputs are clocked directly into a CLB flip-flop without being captured first by a flip-flop at the IOB pin. Of course, using IOB flip-flops is possible only if the FPGA's external interface specifications allow "pipelining" of inputs and outputs.

For pipelined inputs, the XC4000 IOB actually goes one step further by providing a delay element, selectable by M8, in series with the D input of input flip-flop FF2. The effect of this element is to delay the D input relative to the FPGA's internal copies of the system clock, guaranteeing that the input will have a zero hold-time requirement with respect to the external system clock. This benefit comes at the expense of increased setup time, of course.

The IOB's other logic controls are selectable polarity, using multiplexers M1–M4, for its four inputs that come from the CLB array via the programmable interconnect. These inputs, OUT, T, OCLK, and ICLKEN, are the output bit, its three-state enable, the output clock, and the input clock enable, respectively.

Like the XC9500 IOB, the XC4000 IOB also has analog controls. The output driver's slew rate is programmable, and a pull-up or pull-down resistor may be connected to the I/O pin.

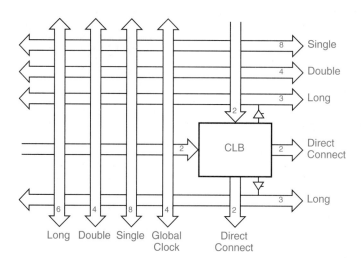

Figure 10-46
XC4000 general
interconnect
structure.

10.6.4 Programmable Interconnect

Well, we saved the best for last. The XC4000 programmable interconnect architecture is fascinating example of a structure that provides rich, symmetric connectivity in a small silicon area.

As we showed in Figure 10-43 on page 882, each CLB in an FPGA is embedded in the interconnect structure, which is really just wires with programmable connections to them. Figure 10-46 gives a little more detail of the XC4000's connection scheme. Wires are not really "owned" by any one CLB, but an XC4000 CLB array is created by tiling the chip with exactly the structure shown in the figure. For example, 100 copies of this figure make the 10×10 CLB array of an XC4003.

The number in each arrow indicates the number of wires in that signal path. Thus, you can see that a CLB has two wires (outputs) each going to the CLBs immediately below and to the right of it. It also connects to three groups of wires above, one below, and four to its left. Signals on these wires can flow in either direction.

The four wires in the group labeled "Global Clock" are optimized for use as clock inputs to the CLBs, providing short delay and minimal skew. The two "Singles" groups are optimized for flexible connectivity between adjacent blocks, without the small number and unidirectional limitation of wires in the "Direct Connect" groups.

It's possible to connect a CLB to another that's more than one hop away using "Single" wires, but they have to go through a programmable switch for each hop, which adds delay. Wires in the "Doubles" groups travel past two CLBs before hitting a switch, so they provide shorter delays for longer connections.

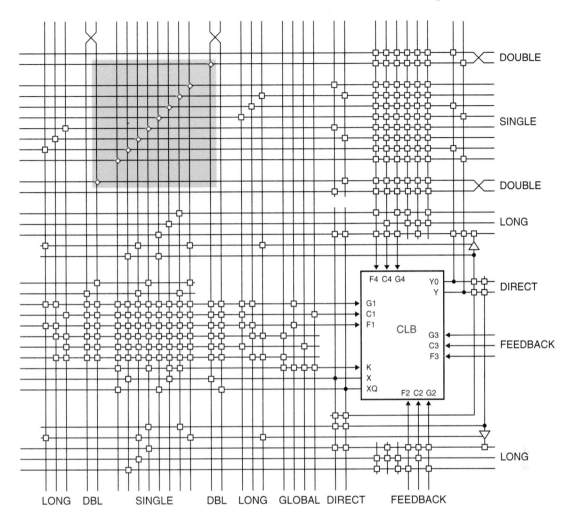

Figure 10-47 XC4000 CLB and wire connection details.

For really long connections, the "Long" groups don't go through any programmable switches at all; instead, they travel all the way across or down a row or column and are driven by three-state drivers near the CLB.

Figure 10-47 shows a CLB and wires in a lot more detail. The small squares are programmable connections—a horizontal wire is connected or not to a vertical one, depending on the state of the programming bit (again, in a latch) for that switch. Additional, specialized programmable interconnect is provided at the edges of the CLB array for connections to the IOBs.

In the figure, the shaded area in color is called a *programmable switch matrix (PSM)*. The PSM is shown in more detail in Figure 10-48. Each of the

programmable switch matrix (PSM)

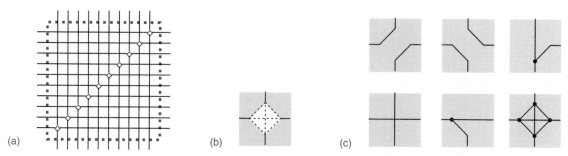

Figure 10-48 XC4000 programmable connections: (a) programmable switch matrix (PSM); (b) programmable switch element (PSE); (c) a few possible connections.

programmable switch element (PSE)

diamonds in (a) is a *programmable switch element (PSE)* that can connect any line to any other, as shown in (b). With four lines, there are 6 possible pairwise connections as shown in (b), and the PSE has a transmission gate for each one of them. Some, none, or all of the transmission gates in a PSE may be enabled—again, by configuration bits stored in latches. Thus, many different connection patterns are possible, as shown in (c).

The PSM is essential for hooking things up in the wiring structure of Figure 10-47. By enabling and disabling connections, PSEs extend or isolate wire segments in the "Single" and "Double" groups. More importantly, the PSM allows signals to "turn the corner" by connecting a horizontal wire to a vertical one. Without this, CLBs would not be able to connect to others in a different row or column of the array.

While the PSM is essential, using it has a price—signals incur a small delay each time they hop through a PSE. Therefore, high-quality FPGA fitter software searches for not just any CLB placement and wire connections that work. The "placement and routing" tool spends a lot of time trying to optimize device performance by finding a placement that allows short connections, and then routing the connections themselves.

Like CPLDs, FPGAs are judged by the flexibility of their architectures and the consistency of the results obtained from a fitter after small design changes are made. There's nothing more frustrating than making a small change to a large design and finding that it no longer meets timing requirements. Thus, FPGA manufacturers have learned to provide "extra" resources in their architectures to help ensure consistent results.

GOOD PRACTICE   Placement and routing is actually a pretty well understood problem, because it is the major part of the "back end" of any custom chip design. Thus, the same kind of tools and the same tool vendors are involved with placement and routing for both FPGAs and ASICS. So, you might like to consider any FPGA experience that you get to be good practice for ASIC design!

References

Manufacturers of ROMs and RAMs publish data books with their device specifications for their parts, and most publish individual data sheets and application notes on their web sites. Information on legacy ultraviolet-erasable EPROMs can be found at Fairchild Semiconductor's site, `www.fairchildsemi.com`. Advanced Micro Device's site, `www.amd.com`, has comprehensive information on more recent EPROMs and flash memory.

Web sites that list all kinds of SRAMs and DRAMs, old and new, include Hitachi Semiconductor (`semiconductor.hitachi.com`) and NEC Electronics (`www.necel.com`). The latest and greatest SSRAMs are offered by companies like Integrated Device Technology (`www.idt.com`), Motorola Semiconductor (`www.mot.com`), and Micron Technology (`www.micron.com`).

Industry publications such as *Electronic Systems* (`www.estd.com`) and *Electronic Design* carry tutorial and update articles from time to time that describe the capabilities and applications of the latest memory chips; just go to the Web sites and search for "SDRAM" to get started. New DRAM architectures include Rambus DRAM (RDRAM), enhanced synchronous DRAM (ESDRAM), SyncLink DRAM (SLDRAM), and double-data-rate synchronous DRAM (DDR-DRAM).

In addition to the memories discussed in this chapter, several types of "specialty" memory devices have widespread use. Probably the most common are *first-in, first-out (FIFO) memories*; these are typically used to transfer data from one processor or clock domain to another. The web sites of IDT, Motorola, and Texas Instruments (`www.ti.com`) are good sources of information on FIFOs. Although it might seem that FIFOs magically solve the metastability problems that occur when crossing clock domains, they do not. For example, see "FIFO Memories: Solution to Reduce FIFO Metastability" by Tom Jackson (Texas Instruments publ. SCAA011A, March 1996).

first-in, first-out (FIFO) memories

Another type of specialty memory is the dual-port memory, which has two independent sets of address, data, and control lines and allows independent operations to be performed on both ports simultaneously. Integrated Device Technology is a leading source for these devices; in addition to data sheets, their Web site has an excellent set of application notes on the devices.

Several manufacturers introduced different CPLD and FPGA architectures at about the same time. While Xilinx (`www.xilinx.com`) was and is the leader in FPGAs, the CPLD field is much more crowded. The first two CPLD families were introduced in the early 1990s by Altera Corporation (`www.altera.com`) and Advanced Micro Devices. AMD's CPLD product line is now owned by former competitor Lattice Semiconductor (`www.latticesemi.com`). Cypress Semiconductor (`www.cypress.com`) is another important player in the CPLD market. All of these companies have excellent resources available on their Web sites.

Drill Problems

10.1   Determine the ROM size needed to realize the combinational logic function in each of the following figures: 4-39(b), 5-39, 5-77, 6-1, and 6-6.

10.2   Determine the ROM size needed to realize the combinational logic function performed by each of the following MSI parts: 74x49, 74x139, 74x153, 74x257, 74x381, 74x682.

10.3   Draw a logic symbol for and determine the size of a ROM that realizes a combinational logic function that can perform the function of either a 74x381 or a 74x382 depending on the value of a single MODE input.

10.4   Draw a logic symbol for and determine the size of a ROM that realizes an 8×8 combinational multiplier.

10.5   Show how to design a 2M × 8 SRAM using HM628512 SRAMs and a combinational MSI part as building blocks.

Exercises

10.6   Our discussion of ROM sneak paths in connection with Figure 10-6 claimed that with A2–A0 = 101, bit lines D2_L and D0_L are pulled LOW through the direct connections. That's not really correct, unless the 74x138 is replaced with a decoder with open-collector outputs. Explain.

10.7   Describe the logic function of seven variables that is performed by the 128 × 1 ROM in Figure 10-7. Starting with the ROM pattern, one way to describe the logic function is to write the corresponding truth table and canonical sum. However, since the canonical sum has 64 7-variable product terms, you might want to look for a simple but precise word description of the function.

10.8   For the two-output logic function of Figure 4-39(b), compare the number of diodes or transistors needed in the AND-OR, AND, or storage array for PLA, PAL, and ROM realizations.

10.9   Show how to double the number of different attenuation amounts that can be selected in the digital attenuator of Figure 10-17 without increasing the size of the ROM.

10.10   Write the C functions UlawToLinear and LinearToUlaw that are required in Table 10-6. In LinearToUlaw, you should select the μ-law byte whose theoretical value is closest to the linear input. (*Hint:* The most efficient approach builds, at program initialization, a 256-entry table that is used by both functions.) Both functions should report out-of-range inputs.

10.11   Modify the C program in Table 10-6 to perform clipping in cases where the designer has specified a particular attenuation amount to be a gain, and multiplying an input value by the attenuation factor produces an out-of-range result.

10.12   Write a C program to generate the contents of a 256 × 8 ROM that converts from 8-bit binary to 8-bit Gray code. (*Hint:* Your program should embody the second Gray-code construction method in Section 2.11.)

10.13 Write a C program to generate the contents of a 256×8 ROM that converts from 8-bit Gray code to 8-bit binary. (*Hints:* You may use the results of Exercise 10.12. It doesn't matter if your program is slow.)

10.14 A certain communication system has been designed to transmit ASCII characters serially on a medium that requires an average signal level of zero, so a 5-out-of-10 code is used to code the data. Each 7-bit ASCII input character is transmitted as a 10-bit word with five 0s and five 1s. Write a C program to generate the contents of a 128×10 ROM that converts ASCII characters into coded words.

10.15 The receiving end of the system in Exercise 10.14 must convert each 10-bit coded word back into a 7-bit ASCII character. Write a C program to generate the contents of a $1K \times 8$ ROM that converts coded words onto ASCII characters. The extra output bit should be an "error" flag that indicates when a noncode word has been received.

10.16 How many ROM bits would be required to build a 16-bit adder/subtracter with mode control, carry input, carry output, and two's-complement overflow output? Be more specific than "billions and billions," and explain your answer.

10.17 Repeat Exercise 10.16 assuming that you may use *two* ROMs, so that the delay for a 16-bit addition or subtraction is twice the delay through one ROM. Assume that the two ROMs must be identical in both size and programming. Try to minimize the total number of ROM bits required, and sketch the resulting adder/subtracter circuit. Can the number of ROM bits be further reduced if the two ROMs are allowed to be different?

10.18 Show how, using additional SSI/MSI parts, a 2764 can be used as a $64K \times 1$ ROM. What is the access time of the $64K \times 1$ ROM?

10.19 Show how, using additional SSI/MSI parts, a 2764 can be used as a $2K \times 32$ ROM. You may assume the existence of a free-running clock signal whose period is slightly longer than the access time of the 2764. What is the access time of the $2K \times 32$ ROM?

10.20 Show how to build the μ-law adder circuit of Figure 10-18 with a $32K \times 8$ ROM and two XOR gates. Also, write a C program to generate the ROM contents.

10.21 Determine the ROM size needed to build the fixed-point to floating-point encoder of Figure 6-3. Draw a logic diagram using a commercially available ROM.

10.22 Write a C program to generate the ROM contents for Exercise 10.21. Unlike the original MSI solution, your C program should perform rounding; that is, for each fixed-point number it should generate the nearest possible floating-point number.

10.23 Draw a complete logic diagram for a ROM-based circuit that performs combinational multiplication of a pair of 8-bit unsigned or two's-complement integers. Signed versus unsigned operation should selected by a single input, SIGNED. You may use any of the commercial ROMs in Figure 10-11, as well as discrete gates.

10.24 Write and test a C program that generates the contents of the ROM(s) in Exercise 10.23.

10.25 Write a C program that generates a $256K \times 4$ ROM that computes the next move in a Tic-Tac-Toe game, using the input and output encodings of Section 6.2.7. Your program must be smart enough to pick a winning move whenever possible.

10.26   Repeat Exercise 10.25 using a 32K × 4 ROM. To accomplish this, the board state must be encoded in only 15 bits. Explain your coding algorithm, and write C functions to translate a cell number in either direction between your encoding and the encoding of Section 6.2.7.

10.27   For each of the timing parameters defined in Section 10.3.3, determine whether its value for the 128K × 8 SRAM you designed in Drill 10.5 is the same as the HM628512's. If it is different, indicate the new value. Use the worst-case values in the 74FCT column of Table 5-3 to determine the delays for the MSI part.

10.28   Using an HM6264 8K × 8 SRAM, a handful of MSI parts, and a PLD as building blocks, design an 8K × 8 late-write SSRAM with flow-through outputs.

10.29   Using an HM6264 8K × 8 SRAM, a handful of MSI parts, and a PLD as building blocks, design an 8K × 8 ZBT SSRAM with pipelined outputs.

10.30   Define the relevant timing parameters for the synchronous SRAM as defined in Exercise 10.28.

10.31   Calculate the values of the timing parameters you defined in Exercise 10.30 for your solution to Exercise 10.28.

10.32   In the style of Figure 10-27, draw the timing diagram for a late-write SSRAM with flow-through outputs for a series of interleaved reads and writes in the pattern R-R-W-W-R-W-R-W. Run the individual cycles as close together as possible, but be sure to account for resource conflicts that prevent back-to-back cycles. What is the average utilization of the SRAM array if the SSRAM is presented with a continuous streams of R-W-R-W-R-W requests?

10.33   Repeat the preceding exercise for a late-write SSRAM with pipelined outputs.

10.34   Using one of the theorems in Section 4.1, prove that the Xilinx XC4000-series CLB can realize any logic function of five variables.

10.35   Show how to use the function generators in an XC4000-series CLB to realize a 9-bit even parity function.

10.36   Show how to use the function generators in an XC4000-series CLB to realize an equality checker for two 4-bit operands. You should be able to cascade your design to check $4n$-bit operands with just n CLBs.

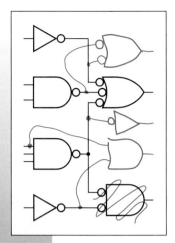

chapter

11

Additional Real-World Topics

This chapter contains a collection of "real-world" topics that are certain to interest you at some point if you are (or become) a working digital designer. We barely scratch the surface of each of these topic areas, so be sure to consult the References in areas that are of particular interest.

11.1 Computer-Aided Design Tools

"If it wasn't hard, they wouldn't call it hardware."

Many digital designers with twenty years of experience consider this statement to be indisputable. Yet more and more, digital design is being carried out using *software*, and it's getting easier as a result.

The terms *computer-aided design (CAD)* and *computer-aided engineering (CAE)* are used to refer to software tools that aid the development of circuits, systems, and many other things. "CAD" is the more general term and applies to tools both inside and outside the electronics area, including architectural and mechanical design tools, for example. Within electronics, "CAD" often refers to *physical-design* tools, such as IC- and PCB-layout programs. "CAE" is used more often to refer to *conceptual-design* tools, such as schematic editors, circuit simulators, and PLD compilers. However, a lot of people in electronics (including the author) tend to use the two terms interchangeably. In this section, we'll discuss some CAD/CAE tools used by digital designers.

IS HARDWARE NOW EASY-WARE?   Since more and more hardware design and debugging is being carried out using software tools, is it really getting easier? Not necessarily.

In the author's experience, the increasing use of CAD means that instead of spending time fighting with soldering irons and test clips, many designers can now spend their time fighting buggy programs running in a buggy software environment.

11.1.1 Hardware Description Languages

In previous decades, most logic design was performed graphically, using block diagrams and schematics. However, the rise of synthesizable HDLs coupled with programmable logic devices and very-large-scale ASIC technology in the 1990s has radically changed the way that the largest digital designs are done.

In traditional software design, high-level programming languages like C, C++, and Java have raised the level of abstraction so that programmers can design larger, more complex systems with some sacrifice in performance compared to hand-tuned assembly-language programs. The situation for hardware design is similar. The circuit produced by a VHDL or Verilog synthesis tool may not be as small or fast as one designed and tweaked by hand by an experienced designer, but in the right hands these tools can support much larger system designs. This is, of course, a requirement if we're ever to take advantage of the millions of gates offered by the most advanced CPLD, FPGA, and ASIC technologies.

register-transfer language

The use of HDLs has taken off with the availability of synthesis tools, but non-synthesizable languages have been around for a while. Most prominent are *register-transfer languages*, which have been used for decades to describe the operation of synchronous systems. Such a language combines the control-flow notation of a state-machine description language with a means for defining and operating on multibit registers. Register-transfer languages have been especially useful in computer design, where individual machine-language instructions are defined as a sequence of more primitive steps involving loading, storing, combining, and testing registers.

11.1.2 Schematic Capture

Except in homework assignments, your first step in designing a circuit usually is to convince someone that your proposed approach is the right one. That means preparing block diagrams and presentation slides, and discussing your ideas with managers and peers in a preliminary *design review*. Once your project has been approved, you can safely start on the "fun stuff," that is, drawing the schematic.

design review

schematic editor
schematic capture

Schematics used to be drawn by hand, but most are now prepared using *schematic editors,* CAD programs that run on engineering workstations. The process of creating a schematic on a computer is often called *schematic capture*.

WHAT'S A WORKSTATION?	An "engineering workstation" nowadays can be anything from a $500 PC to a $50,000 parallel processor. The "canonical" engineering workstation in terms of cost (what a typical company will spend per engineer) and performance (among the best) is a $5,000 Sun workstation, whatever model you get for that price this year.

(Huh? How did it get loose?) This term is used because the schematic editor captures more information than just a drawing. Most information in the schematic has a "type" associated with it, so that selected information can be retrieved automatically as required later in the design process.

In order to simplify both entry and retrieval of information in a schematic, a schematic editor typically provides *fields* for each information type, either hidden or drawn automatically nearby the associated schematic element. Some typical fields and their uses are described below.

fields

- *Component type.* By specifying a component type (e.g., resistor, capacitor, 74FCT374, GAL16V8), the designer can call out a predrawn symbol from a *component library*. Some components have user-selectable alternate symbols (e.g., DeMorgan equivalents for gates). The component type is used both for manufacturing documentation and for simulation.

  component type

  component library

- *Component value.* Most analog components have a value that must be specified by this field (e.g., 2.7 kΩ). Additional distinguishing information may appear in a "rating" or "tolerance" field (e.g., 1/4 W, 1%). Digital components may give a speed rating here (e.g., -10 for a 10-ns PLD).

  component value

- *Approved-part number.* Using all of the distinguishing information from the previous fields, the CAD software may automatically select a part number from the company's "approved parts list" (e.g., 126-10117-0272 for a 2.7-kΩ, 1/4-W, 1% metal-film resistor), or a warning if no such part is available. (No, in large companies you can't use just any part in the catalog; all parts must be approved through a *parts-qualification* process.)

  approved-part number

  parts qualification

- *Reference designator.* This alphanumeric label distinguishes among multiple instances of similar parts and is used just about everywhere.

  reference designator

- *Component location.* Using a coordinate system, this field may indicate the physical position of the component on an assembled PCB, simplifying the task of finding it during debugging. Of course, the final component location generally is not known during the initial circuit design. However, CAE software with *back-annotation* capability can insert this information into the schematic when the PCB layout is completed.

  component location

  back annotation

- *Pin numbers.* These specify the assignment of the signal pins on each component and are usually prespecified in the symbol that is called up from the component library. However, for SSI and other parts that have

  pin number

multiple identical sections in a single package, pin numbers (and even reference designators) might not be specified until the PCB layout is completed; this is another case where back annotation is handy.

wire type

- *Wire type.* Usually the only wire types are "ordinary" and "bus," where a bus is just a set of ordinary wires that are grouped for drawing purposes. However, other wire types, such as extra-wide PCB traces for analog signals that carry high current, may be indicated in some systems; such information is passed to the PCB-layout program.

signal name

- *Signal name.* The user may (and should) associate a name with each signal. The name is especially useful in simulation and debugging, but it does not affect the physical PCB layout.

connection flag

- *Connection flag.* This drawing element indicates the connection of one signal line to another one on the same or a different page. Software can ensure that all outgoing connections are matched by incoming ones. In a hierarchical schematic, software can use the connection flags to generate signal names in a "logic symbol" that represents the entire page.

Once you've "captured" a circuit's schematic in a CAD system, there are lots of things you can do with it besides printing it. At least two documents can be generated for manufacturing purposes:

parts list

- *Parts list.* This is a list of components and their reference designators.

net
net list

- *Net list.* A *net* is a set of pins that are all connected to the same electrical node or signal. A *net list* specifies all such connections that are required by the schematic, typically using an alphabetically sorted list of signal names. For each signal name, it lists the device pins (identified by reference designator and pin number) where that signal is connected.

  Alternatively, the net list may be sorted by reference designator and pin number, and show the signal name connected to each pin; sometimes this

pin list

  is called a *pin list.*

The parts list and the net list are the main inputs to the PCB-layout process. The designer may also use the parts list to estimate the cost and reliability of the circuit, and the manufacturing department obviously can use it to order the parts.

11.1.3 Timing Drawings and Specifications

Block diagrams and schematics are not the only drawings used in digital system design. Timing diagrams are an essential part of almost any documentation package.

Larger systems are decomposed into smaller subsystems that communicate through well-defined interfaces. The definition of an interface typically contains not only signal names and functions, but also specifications of expected timing behavior. The starting point is typically a specification of maximum and sometimes minimum clock frequencies. Inputs to a subsystem have their setup- and hold-time requirements with respect to a clock edge specified. Outputs of a subsystem have their minimum and maximum delays from a clock edge specified.

Programs such as TimingDesigner (`www.chronology.com`) can automate the tedious task of creating complicated timing diagrams and specifications, where changes in one timing parameter may propagate to many others.

11.1.4 Circuit Analysis and Simulation

The component library of a sophisticated CAD system contains more than just a symbol for each component. For ICs, the library may contain a *component* *model* describing the device's logical and electrical operation. In fact, HDLs like VHDL and Verilog were originally invented strictly for the purpose of modeling the logical behavior of components and their interconnection in systems. Using these models, logical and timing errors can be found in simulation.

component model

As a minimum, the model for an IC indicates whether each pin is an input or output. With just this information, a *design-rule checker* can detect some of the more common "stupid mistakes" in a design, such as shorted outputs and floating inputs. With the addition of input loading and output driving characteristics for each pin, the program can also determine whether any output's fanout capability is exceeded.

design-rule checker

The next step is *timing verification*. Even without a detailed model of an IC's logical behavior, the component library can provide a worst-case delay value for each input-to-output path, and setup and hold times for clocked devices. Using this information, a *timing verifier* can find the worst-case delay paths in the overall circuit, so the designer can determine whether the timing margins are adequate.

timing verification

timing verifier

Finally, the library may contain a detailed model of each component's logical behavior, in which case *simulation* may be used to predict the overall circuit's behavior for any given sequence of inputs. The designer provides an input sequence, and a *simulator* program determines how the circuit would respond to that sequence. The simulator output is usually displayed graphically, in the form of timing waveforms that the designer could see on an oscilloscope or logic analyzer if the same inputs were applied to a real circuit. In such an environment, it is possible to debug an entire circuit without "breadboarding" it, and to assemble a PCB that works on the very first try.

simulation

simulator

Simulators use different models for different types of components. Analog simulators like SPICE may use mathematical models with just one parameter for a resistor and with anywhere from a few to dozens of parameters for a bipolar transistor. Based on these models and the circuit's structure, the simulator develops and solves equations to determine the circuit's behavior.

Digital simulators normally use one of the two following models for each logic element:

behavioral model

- *Behavioral model.* In such a model, the software "understands" the basic function of the logic element. For example, an AND gate is understood to produce a 1 output with delay t_{pHL} after all inputs are 1, and a 0 output with delay t_{pHL} after any input is 0.

structural model

- *Structural model.* Larger elements may be modeled by connecting groups of smaller elements that have behavioral models.

We saw that VHDL supports both types of model. For example, Table 5-15 on page 368 gave a behavioral model for a 74x138 3-to-8 decoder. By using one of VHDL's clauses ("`after`" and other mechanisms not covered in this book), we could specify the delay from input to output. Alternatively, Table 5-14 gave a structural model of a 2-to-4 decoder as a set of gates connected according to the internal logic diagram of Figure 5-32 on page 352. Assuming that timing information already exists for the individual gates in the structure (`inv` and `and3`), the larger structural model can predict circuit timing more accurately, since it can follow the actual circuit path taken each different input transition.

software model

Both of the foregoing models are called *software models* because the chip's operation is simulated entirely by software. Some chips, including microprocessors and other LSI parts, require huge software models. In many cases, the chip manufacturers don't provide such models (they're considered to be proprietary information), and the models would take months or years for a user to develop. To get around this problem, innovative CAD companies have developed a third type of model:

physical model
hardware model

- *Physical model*, sometimes called a *hardware model*. A real, working copy of the chip is connected to the computer that's running the simulation. The simulator applies inputs to the real chip and observes its outputs, which are then used as inputs to the other models in the simulation.

A digital designer's input to a logic simulator is a schematic and a sequence of input vectors to be applied to the circuit in simulated time. The first output that the designer usually sees is "xxxxx," meaning that the simulator can't figure

uninitialized sequential circuits

out what the output will be. This usually occurs because the circuit contains sequential circuits (flip-flops and latches) that are not initialized. To obtain meaningful simulator output, the designer must ensure that all circuits are explicitly reset to a known value, even if reset is not required for correct operation in the real system. For example, the self-correcting ring counters in Figures 8-62 and 8-64 on page 726 cannot be simulated, because the simulator has no way of knowing their starting state.

Once the simulated circuit has been initialized, the designer's ability to probe its operation is limited only by the simulator's speed. Simulation of complex circuits can be very slow, compared to the speed of the real thing.

simulation inefficiency

Depending on the level of detail being simulated, the simulation time can be 10^3 to 10^8 times longer than the interval being simulated. A strictly functional simulation, in which all components are assumed to have zero delay, is fastest. A much more realistic simulation, one that calculates and considers a worst-case range of delays for all signal paths, is the slowest.

11.1.5 PCB Layout

Printed-circuit-board layout can be carried out by a program or by a person (a PCB designer) or by a combination of the two. PCB designers are usually people who like to solve puzzles. Their job is to fit all of the components in the schematic onto a board of a given size, and then to hook up all the connections as required by the schematic—and usually to do so with the smallest possible PCB size and number of interconnection layers.

PCB design can be done automatically by software, but the results aren't always so good. Both the PCB designer and the circuit designer normally must guide the initial component-placement steps, taking into account many practical concerns, including the following:

- *Mechanical constraints.* Certain connectors, indicators, and other components may have to be placed at a particular location on the PCB because of a predetermined mechanical requirement. *mechanical constraints*

- *Critical signal paths.* An automatic placement program can place components in a way that minimizes the average length of interconnections, but designers usually have a more intuitive feel for what areas of the circuit and which specific signals have the most critical requirements. In fact, the circuit designer normally provides the PCB designer with a *floorplan,* a suggested placement for groups of components that reflects the designer's ideas about which things need to be close to each other. *critical signal paths*

  floorplan

- *Thermal concerns.* Some areas of the PCB may have better cooling than others when the board is installed in a system, and certainly some components run hotter than others. The designers must ensure that the operating temperature range of the components is not exceeded. *thermal concerns*

  The temperature anywhere on a well-designed PCB in a properly cooled system typically will not exceed 10°C above ambient, but a power-hungry chip (such as a microprocessor) placed in an area with no air flow could create a temperature rise of 30°C or more. If the ambient temperature were 40°C, the actual temperature at the microprocessor would be 70°C, which is at the limit of the commercial operating range for most chips.

- *Radio-frequency emissions.* Electronic equipment radiates radio-frequency energy as a side effect. Government agencies in the U.S. and other countries specify the maximum permissible level of emissions; in the U.S. the limits are spelled out in *FCC Part 15* regulations. *radio-frequency emissions*

  FCC Part 15

  A circuit's emissions levels usually depend not only on component performance and circuit configuration, but also on the physical layout of the

PCB. During layout, special attention must therefore be paid to component placement, grounding, isolation, and decoupling, in order to minimize unwanted (and illegal!) radio-frequency emissions.

stupid mistakes

- *Stupid mistakes.* Perhaps the most important step in PCB design, especially when using new or unfamiliar components, is checking for stupid mistakes. Almost every digital designer, even successful ones, can tell you stories of laying out a PCB using the mirror image of a new part's pinout, putting mounting holes or connectors in the wrong place, using the wrong version of an IC package, using the wrong sex of a connector, leaving out the inner connection layers when the board is fabricated, and so on. Remember, "Computers don't make mistakes, humans do!"

routing

Once component placement is done, all of the signals must be hooked up; this is called *routing*. This phase of PCB layout is most easily automated; in fact, a decent "auto-router" can route a moderate-sized PCB overnight. Still, design engineers may have to give special attention to the routing of critical-path traces and to other issues such as placement of signal terminations and test points, which are discussed later in this chapter.

11.2 Design for Testability

When you buy a new piece of equipment, you expect it to work properly "right out of the box." However, even with modern automated equipment, it's impossible for an equipment manufacturer to guarantee that every unit produced will be perfect. Some units don't work because they contain individual components that are faulty, or because they were assembled incorrectly or sloppily, or because they were damaged in handling. Most manufacturers prefer to discover these problems in the factory, rather than get a bad reputation when their customers receive a faulty product. This is the purpose of *testing*.

testing
go/no-go test

The most basic test, a *go/no-go test*, yields just one bit of information—is the system 100% functional or not? If the answer is yes, the system can be shipped. If not, the next action depends on the size of the system. The appropriate response for a digital watch would be to throw it out. If many watches are turning up faulty, it's more efficient to repair what must be a consistent problem in the assembly process or an individual component than to try to salvage individual units.

diagnostic test

On the other hand, you wouldn't want to toss out a $10,000 computer that fails to boot. Instead, a more detailed *diagnostic test* may be run to locate the particular subsystem that is failing. Depending on cost, the failed subsystem may be either repaired or replaced. A typical digital subsystem is a PCB whose assembled cost with components ranges from $50 to $1,000. The repair/replace decision is an economic trade-off between the cost of the assembled PCB and the estimated time to locate and repair the failure.

This is where *design for testability (DFT)* comes in. "DFT" is a general term applied to design methods that lead to more thorough and less costly testing. Many benefits ensue from designing a system or subsystem so that failures are easy to detect and locate:

design for testability (DFT)

- The outcome of a go/no-go test is more believable. If fewer systems with hidden faults are shipped, fewer customers get upset, which yields obvious economic as well as psychological benefits.

- Diagnostic tests run faster and produce more accurate results. This reduces the cost of salvaging a subsystem that fails the go/no-go test, making it possible to manufacture more systems at lower cost.

- Both go/no-go and diagnostic tests require less test-engineering time to develop.

- Although the savings in test-engineering time may be offset by added design-engineering effort to include DFT, any increase in overall product development cost usually can be offset by decreased manufacturing cost.

11.2.1 Testing

Digital circuits are tested by applying *test vectors* which consist of inputs combinations and expected output combinations. A circuit "passes" if its outputs match what's expected. In the worst case, an n-input combinational circuit requires 2^n test vectors. However, if we know something about the circuit's physical realization and make some assumptions about the type of failures that can occur, the number of vectors required to test the circuit fully can be greatly reduced. The most common assumption is that failures are *single stuck-at faults*, that is, that they can be modeled as a single input or output signal stuck at logic 0 or logic 1. Under this assumption, an 8-input NAND gate, which might otherwise require 256 test vectors, can be fully tested with just nine—11111111, 01111111, 10111111, …, 11111110.

test vector

single stuck-at fault

Under the single-fault assumption, it's easy to come up with test vectors for individual logic elements. However, the problem in practice is *applying* the test vectors to logic elements that are buried deep in a circuit, and *seeing* the results. For example, suppose that a circuit has a dozen combinational and sequential logic elements between its primary inputs and the inputs of an 8-input NAND gate that we want to test. It's not at all obvious what primary-input vector, or sequence of primary-input vectors, must be applied to generate the test vector 11111111 at the NAND-gate inputs. Furthermore, it's not obvious what else might be required to propagate the NAND gate's output to a primary output of the circuit.

Sophisticated *test-generation programs* deal with this complexity and try to create a *complete test set* for a circuit, that is, a sequence of test patterns that fully tests each logic element in the circuit. However, the computation required can be huge, and it's quite often just not possible to generate a complete test set.

test-generation program

complete test set

controllability

observability

test points

DFT methods attempt to simplify test-pattern generation by enhancing the "controllability" and "observability" of logic elements in a circuit. In a circuit with good *controllability*, it's easy to produce any desired values on the internal signals of the circuit by applying an appropriate test-vector input combination to the primary inputs. Similarly, good *observability* means that any internal signal can be easily propagated to a primary output for comparison with an expected value by the application of an appropriate primary-input combination. The most common method of improving controllability and observability is to add *test points,* additional primary inputs and outputs that are used during testing.

11.2.2 Bed-of-Nails and In-Circuit Testing

test fixture
nail
bed of nails
automatic tester
in-circuit testing

In a digital circuit that is fabricated on a single PCB, the "ultimate" in observability is obtained by using every pin of every IC as a test point. This is achieved by building a special *test fixture* that matches the layout of the PCB and contains a spring-loaded pin (*nail*) at each IC-pin position. The PCB is placed on this *bed of nails,* and the nails are connected to an *automatic tester* that can monitor each pin as required by a test program.

overdrive

Going one step further, *in-circuit testing* also achieves the "ultimate" in controllability. This method not only monitors the signals on the bed of nails for observability, but also connects each nail to a very low impedance driving circuit in the tester. In this way, the tester can override (or *overdrive*) whatever circuit on the PCB normally drives each signal, and directly generate any desired test vector on the internal signals of the PCB. Overdriving an opposing gate output causes excessive current flow in both the tester and the opposing gate, but the tester gets away with it for short periods (milliseconds).

To test an 8-input NAND gate, an in-circuit tester needs only to provide the nine test vectors mentioned previously, and can ignore whatever values the rest of the circuit is trying to drive onto the eight input pins. And the NAND-gate output can be observed directly on the output pin, of course. With in-circuit testing, each logic element can be tested in isolation from the others.

HITTING THE NAIL ON THE PIN

The rise of high-density surface-mount packaging has made bed-of-nails testing considerably more difficult than it was with thru-hole components like DIPs. Since components may be mounted on both sides of the PCB, a special test fixture called a clam shell may be needed to connect nails to both sides of the PCB.

Furthermore, the pins of many surface-mount devices are so small and their spacing is so tight (25 mils or less), that it may be impossible to reliably land a test pin on them. In such cases, the PCB designer may have to explicitly provide test pads, extra copper patches that are large enough for a test pin to contact (e.g., 50 mils in diameter). A separate test pad must be provided for each signal that does not connect to a larger (e.g., 62-mil through-hole) component pad somewhere on the PCB.

Although in-circuit testing greatly enhances the controllability and observability of a PCB-based circuit, logic designers must still follow a few DFT guidelines for the approach to be effective. Some of these are listed below.

- *Initialization.* It must be possible to initialize all sequential circuit elements to a known state.

Since the preset and clear input pins of registers and flip-flops are available to an in-circuit tester, you would think that this is no problem. However, Figure 11-1(a) shows a classic case of a circuit (a Gray-code counter) that cannot be initialized, since the flip-flops go to an unpredictable state when PR and CLR are asserted simultaneously. The correct way to handle the preset and clear inputs is shown in (b).

- *Clock generation.* The tester must be able to provide its own clock signal *without* overdriving the on-board clock signals.

Testers usually must override an on-board clock, for several reasons: The speed at which they can apply test vectors is limited; they must allow extra time for overdriven signals to settle; and sometimes they must stop the clock. However, overdriving the clock is a no-no. An overdriven signal may "ring" and make several transitions between LOW and HIGH before finally settling to the level that the tester wants. On a clock signal, such transitions can create unwanted state changes.

(a)

(b)

Figure 11-1
Flip-flops with pull-up resistors for unused inputs: (a) untestable; (b) testable.

Figure 11-2
Clock driver circuit
that allows a tester to
cleanly override the
system clock.

Figure 11-2 shows a recommended clock driver circuit. To inject its own clock, the tester pulls CLKEN LOW and inserts its clock on TESTCLK_L. Since the tester is not overdriving any gate outputs, the resulting CLOCK signal is clean. In general, any normally glitch-free signal that is used as a clock input or other asynchronous input must not be overdriven by the tester, and would have to be treated in a way similar to Figure 11-2. This is another reason why synchronous design with a single clock is so desirable.

- *Grounded inputs.* In general, ground should not be used as a constant-0 logic source.

The in-circuit tester can overdrive *most* signals, but it can't overdrive ground. Therefore, signals that require a constant-0 input during normal operation should be tied to ground through a resistor, which allows the tester to set them to 1 as required during testing. For example, consider what would happen if we used a GAL16V8 PLD to generate a customized set of clock waveforms from a master clock, as we did in the ABEL program of Table 8-26 on page 737. If not for testing considerations, we could connect the 16V8's pin 11, its active-low global output-enable pin, directly to ground. However, it should be connected through a resistor, so that the tester can float the PLD outputs and drive the clock signals P1_L–P6_L itself. Although the tester could theoretically overdrive these signals, it shouldn't if they are used as clocks.

- *Bus drivers.* In general, it should be possible to disable the drivers for wide buses so that the tester can drive the bus without having to overdrive all the signals on the bus.

That is, it should be possible to output-disable all of the three-state drivers on a bus, so that the tester drives a "floating" bus. This reduces electrical stress both on the tester and on multibit drivers (e.g., 74x244) that might otherwise be overheated and damaged by having all of their outputs overdriven simultaneously.

11.2.3 Scan Methods

In-circuit testing works fine, up to a point. It doesn't do much good for custom VLSI chips and ASICs, because the internal signals simply aren't accessible. Even in board-level circuits, high-density packaging technologies such as surface mounting greatly increase the difficulty of providing a test point for every signal on a PCB. As a result, an increasing number of designs are using "scan methods" to provide controllability and observability.

A *scan method* attempts to control and observe the internal signals of a circuit using only a small number of test points. A *scan-path method* considers any digital circuit to be a collection of flip-flops or other storage elements interconnected by combinational logic, and is concerned with controlling and observing the state of the storage elements. It does this by providing two operating modes: a normal mode, and a *scan mode* in which all of the storage elements are reorganized into a giant shift register. In scan mode, the state of the circuit's n storage elements can be read out by n shifts (*observability*), and a new state can be loaded at the same time (*controllability*).

scan method

scan-path method

scan mode

Figure 11-3 shows a circuit designed using a scan-path method. Each storage element in this circuit is a scan flip-flop (Section 7.2.7) that can be loaded from one of two sources. The test enable (TE) input selects the source—normal data (D) or test data (T). The T inputs are daisy-chained to create the scan path shown in color. By asserting ENSCAN for 11 clock ticks, a tester can read out the current state of the flip-flops and load a new state. The test engineer is left with the job of deriving test sets for the individual combinational logic blocks, which can be fully controlled and observed using the scan path and the primary inputs and outputs.

Scan-path design is used most often in custom VLSI and ASIC design, because of the impossibility of providing a large number of conventional test points. However, the two-port flip-flops used in scan-path design do increase

Figure 11-3 Circuit containing a scan path, shown in color.

chip area. For example, in LSI Logic Corp.'s LCA500K series of CMOS gate arrays, an **FD1QP D** flip-flop macrocell uses 7 "gate cells," while an **FD1SQP D** scan flip-flop macrocell uses 9 gate cells, almost a 30% increase in silicon area. However, the overall increase in chip area is much less, since flip-flops are only a fraction of the chip, and large "regular" memory structures (e.g., RAM) may be tested by other means. In any case, the improvement in testability may actually *reduce* the cost of the packaged chip when the cost of testing is considered. For large ASIC designs with rich, complicated control structures, scan-path design should be considered a requirement.

11.3 Estimating Digital System Reliability

reliability

Qualitatively, the *reliability* of a digital system is the likelihood that it works correctly when you need it. Marketing and sales people like to say that the systems they sell have "high reliability," meaning that the systems are "pretty likely to keep working." However, savvy customers ask questions that require more concrete answers, such as, "If I buy 100 of these systems, how many will fail in a year?" To provide these answers, digital design engineers are often responsible for calculating the reliability of the systems they design, and in any case they should be aware of the factors that affect reliability.

Quantitatively, reliability is expressed as a mathematical function of time:

$$R(t) = \text{Probability that the system still works correctly at time } t$$

Reliability is a real number between 0 and 1; that is, at any time $0 \le R(t) \le 1$. We assume that $R(t)$ is a monotonically decreasing function; that is, failures are permanent and we do not consider the effects of repair. Figure 11-4 is a typical reliability function.

The foregoing definition of reliability assumes that you know the mathematical definition of probability. If you don't, reliability and the equivalent probability are easiest to define in terms of an experiment. Suppose that we were to build and operate N identical copies of the system in question. Let $W_N(t)$ denote the number of them that would still be working at time t. Then,

$$R(t) = \lim_{n \to \infty} W_N(t)/N$$

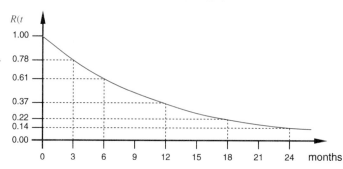

Figure 11-4
Typical reliability function for a system.

That is, if we build *lots* of systems, $R(t)$ is the fraction of them that are still working at time t. When we talk about the reliability of a single system, we are simply using our experience with a large population to estimate our chances with a single unit.

It would be very expensive to compute $R(t)$ if the only way to do it was to perform an experiment—to build and monitor N copies of the system. Worse, for any t, we wouldn't know the value of $R(t)$ until time t had elapsed in real time. Thus, to answer the customer's question posed earlier, we'd have to build a bunch of systems and wait a year; by then, our potential customer would have purchased something else.

Instead, we can estimate the reliability of a system by combining reliability information for its individual components, using a simple mathematical model. The reliability of mature components (e.g., 74FCT CMOS chips) may be known and published based on actual experimental evidence, while the reliability of new components (e.g., a Sexium microprocessor) may be estimated or extrapolated from experience with something similar. In any case, a component's reliability is typically described by a single number, the "failure rate" described next.

11.3.1 Failure Rates

The *failure rate* is the number of failures that occur per unit time in a component or system. In mathematical formulas, failure rate is usually denoted by the Greek letter λ. Since failures occur infrequently in electronic equipment, the failure rate is measured or estimated using many identical copies of a component or system. For example, if we operate 10,000 microprocessor chips for 1,000 hours, and eight of them fail, we would say that the failure rate is

failure rate

λ

$$\lambda = \frac{8 \text{ failures}}{10^4 \text{ chips} \cdot 10^3 \text{ hours}} = (8 \cdot 10^{-7} \text{ failures/hour})/\text{chip}$$

That is, the failure rate of a single chip is $8 \cdot 10^{-7}$ failures/hour.

The actual process of estimating the reliability of a batch of chips is not nearly as simple as we've just portrayed it; for more information, see the References. However, we can use individual component failure rates, derived by whatever means, in a straightforward mathematical model to predict overall system reliability, as we'll show later in this section.

Since the failure rates for typical electronic components are so small, there are several scaled units that are commonly used for expressing them: percent failures per 10^3 hours, failures per 10^6 hours, and failures per 10^9 hours. The last unit is called a *FIT*:

FIT

$$1 \text{ FIT} = 1 \text{ failure}/(10^9 \text{ hours})$$

In our earlier example, we would say that $\lambda_{\text{microprocessor}} = 800$ FITs.

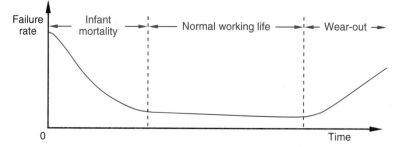

Figure 11-5
The "bathtub curve"
for electronic-
component failure
rates.

The failure rate of a typical electronic component is a function of time. As shown in Figure 11-5, a typical component has a high failure rate during its early life, during which most manufacturing defects make themselves visible; failures during this period are called *infant mortality*. Infant mortality is why manufacturers of high-quality equipment perform *burn-in*—operating the equipment for 8 to 168 hours before shipping it to customers. With burn-in, most infant mortality occurs at the factory rather than at the customer's premises. Even without a thorough burn-in period, the seemingly stingy 90-day warranty offered by most electronic equipment manufacturers does in fact cover most of the failures that occur in the first few *years* of the equipment's operation (but if it fails on the 91st day, that's tough!). This is quite different from the situation with an automobile or other piece of mechanical equipment, where "wear and tear" increases the failure rate as a function of time.

infant mortality

burn-in

Once an electronic component has successfully passed the burn-in phase, its failure rate can be expected to be pretty much constant. Depending on the component, there may be *wear-out* mechanisms that occur late in the component's life, increasing its failure rate. In the old days, vacuum tubes often wore out after a few thousand hours because their filaments deteriorated from thermal stress. Nowadays, most electronic equipment reaches obsolescence before its solid-state components start to experience wear-out failures. For example, even though it's been over 25 years since the widespread use of EPROMs began, many of which were guaranteed to store data for only 10 years, we haven't seen a rash of equipment failures caused by their bits leaking away. (Do you know anyone with a 10-year-old PC or VCR?)

wear-out

Thus, in practice, the infant-mortality and wear-out phases of electronic-component lifetime are ignored, and reliability is calculated on the assumption that failure rate is constant during the normal working life of electronic equipment. This assumption, which says that a failure is equally likely at any time in a component's working life, allows us to use a simplified mathematical model to predict system reliability, as we'll show later in this section.

There *are* some other factors that can affect component failure rates, including temperature, humidity, shock, vibration, and power cycling. The most significant of these for ICs is temperature. Many IC failure mechanisms involve chemical reactions between the chip and some kind of contaminant, and these are accelerated by higher temperatures. Likewise, electrically overstressing a transistor, which heats it up too much and eventually destroys it, is worse if the device temperature is high to begin with. Both theoretical and empirical evidence support the following widely used rule of thumb:

- An IC's failure rate roughly doubles for every 10°C rise in temperature.

This rule is true to a greater or lesser degree for most other electronic parts.

Note that the temperature of interest in the foregoing rule is the *internal* temperature of the IC, not the ambient temperature of the surrounding air. A power-hogging component in a system without forced-air cooling may have an internal temperature as much as 40–50°C higher than ambient. A well-placed fan may reduce the temperature rise to 10–20°C, reducing the component's failure rate by perhaps a factor of 10.

11.3.2 Reliability and MTBF

For components with a constant failure rate λ, it can be shown that reliability is an exponential function of time:

$$R(t) = e^{-\lambda t}$$

The reliability curve in Figure 11-4 is such a function; it happens to use the value $\lambda = 1$ failure/ year.

Another measure of the reliability of a component or system is the *mean time between failures (MTBF)*, the average time that it takes for a component to fail. For components with a constant failure rate λ, it can be shown that MTBF is simply the reciprocal of λ:

mean time between failures (MTBF)

$$\text{MTBF} = 1/\lambda$$

11.3.3 System Reliability

Suppose that we build a system with m components, each with a different failure rate, $\lambda_1, \lambda_2, ..., \lambda_m$. Let us assume that for the system to operate properly, *all* of its components must operate properly. Basic probability theory says that the system reliability is then given by the formula

$$R_{\text{sys}}(t) = R_1(t) \cdot R_2(t) \cdot \cdots \cdot R_m(t)$$
$$= e^{-\lambda_1 t} \cdot e^{-\lambda_{2(1)} t} \cdot \cdots \cdot e^{-\lambda_m t}$$
$$= e^{-(\lambda_1 + \lambda_2 + \cdots + \lambda_m) t}$$
$$= e^{-\lambda_{\text{sys}} t}$$

where

$$\lambda_{\text{sys}} = \lambda_1 + \lambda_2 + \cdots + \lambda_m$$

Thus, system reliability is also an exponential function, using a composite failure rate λ_{sys} that is the sum of the individual component failure rates.

The constant-failure-rate assumption makes it very easy to determine the reliability of a system—simply add the failure rates of the individual components to get the system failure rate. Individual component failure rates can be obtained from manufacturers, reliability handbooks, or company standards.

For example, a portion of one company's standard is listed in Table 11-1. Since failure rates are just estimates, this company simplifies the designer's job by considering only broad categories of components, rather than giving "exact" failure rates for each component. Other companies use more detailed lists, and some CAD systems maintain a component failure-rate database, so that a circuit's composite failure rate can be calculated automatically from its parts list.

Suppose that we were to build a single-board system with a VLSI microprocessor, 16 memory and other LSI ICs, 2 SSI ICs, 4 MSI ICs, 10 resistors, 24 decoupling capacitors, and a 50-pin connector. Using the numbers in Table 11-1, we can calculate the composite failure rate,

$$\lambda_{sys} = 1000 + 16 \cdot 250 + 2 \cdot 90 + 4 \cdot 160 + 10 \cdot 10 + 24 \cdot 15 + 50 \cdot 10 + 500 \text{ FITs}$$
$$= 7280 \text{ failures} / 10^9 \text{ hours}$$

The MTBF of our single-board system is $1/\lambda$, or about 15.6 years. Yes, small circuits really can be that reliable. Of course, if we include a typical power supply, with a failure rate of over 10,000 FITs, the MTBF is more than halved.

Table 11-1 Typical component failure rates at 55°C.

Component	Failure Rate (FITs)
SSI IC	90
MSI IC	160
LSI IC	250
VLSI microprocessor	500
Resistor	10
Decoupling capacitor	15
Connector (per pin)	10
Printed-circuit board	1000

11.4 Transmission Lines, Reflections, and Termination

Nothing happens instantly, especially where digital circuits are concerned. In particular, consider the fact that the "speed-of-light" propagation delay of electrical signals in *wire* is on the order of 1.5–2 ns per foot (the exact delay depends on characteristics of the wire). When wire delays are similar to the transition times of the signals that they carry, we must treat wires not as zero-delay, perfect conductors, but as the "transmission lines" that they really are.

Transmission-line behavior includes signal changes that would not be predicted by a "DC" analysis of circuit operation. The most significant changes occur during an interval of approximately $2T$ after an output changes state, where T is the delay from the output to the far end of the wire that it drives.

Transmission-line behavior usually doesn't affect the logical operation of devices whose signal transition times and propagation delays are much longer than $2T$. Thus, transmission-line behavior typically is considered only for 74LS connections over a couple feet in length, for 74AS, 74AC and 74ACT connections over a foot, and for 74FCT and ECL connections over 3–6 inches. Naturally, you can sometimes get away with longer connections depending on the application, but there are also situations where transmission-line theory must be applied to even shorter connections. Even some VLSI chips achieve high enough speeds that transmission-line behavior must be considered in the internal chip design.

11.4.1 Basic Transmission-Line Theory

Two conductors in parallel constitute the simplest *transmission line*. Consider a pair of conductors with infinite length, as shown in Figure 11-6(a). If we instantaneously put a voltage source across the pair, a certain current flows to create a voltage wave that travels along the pair. The ratio of voltage to current, V_{out}/I_{out}, depends on the physical characteristics of the conductors and is called the *characteristic impedance Z_0* of the conductor pair.

transmission line

characteristic impedance Z_0

Figure 11-6 Transmission lines: (a) with infinite length; (b) with finite length, terminated with characteristic impedance.

The magnitude of V_{out} is determined by viewing the series combination of R_{src} and Z_0 as a voltage divider, so that

$$V_{out} = V_{src} \cdot \frac{Z_0}{R_{src} + Z_0}$$

Of course, we don't have any infinitely long conductors. However, suppose that we have a 5-foot-long pair of conductors, and we place a resistance equal to Z_0 across the far end. If we instantaneously place a voltage source across the pair as shown in (b), the same current flows forever as in the infinite-length case. Thus, when a line is terminated by its characteristic impedance, we needn't consider transmission-line effects any further.

The situation is different for a transmission line of finite length that is not terminated in its characteristic impedance. An extreme case, in which the far end is short-circuited, is shown in Figure 11-7(a). For simplicity, we assume that $R_{src} = Z_0$ in this example. Initially, all the driver sees is the line's characteristic impedance, and a voltage wave with amplitude $V_{src}/2$ happily propagates down the line. However, when the wave hits the far end at time T, it sees the short-circuit. In order to satisfy Kirchhoff's laws, a voltage wave of the *opposite*

Figure 11-7 Transmission lines: (a) short-circuited; (b) open-circuited.

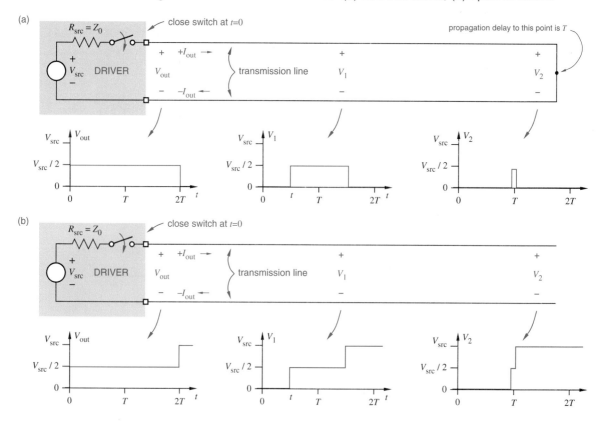

polarity propagates back down the line, canceling the original wave. The far end has *reflected* the original wave, and the driver sees the short-circuit at time 2*T*.

reflection

Another extreme case, in which the far end is open-circuited, is shown in (b). Everything starts out as before. However, when the initial voltage wave hits the far end, the current has nowhere to go, and so a voltage wave of the *same* polarity propagates back up the line, adding to the original voltage. When the reflected wave reaches the driver at time 2*T*, the voltage everywhere is V_{src}, and nothing more happens.

In the general case, the amplitude of the wave reflected at the end of a transmission line is determined by the *reflection coefficient, ρ* (rho). The value of ρ depends on Z_0 and Z_{term}, the *termination impedance* that appears at the end of the line:

reflection coefficient, ρ
termination
   impedance

$$ \rho = \frac{Z_{\text{term}} - Z_0}{Z_{\text{term}} + Z_0} $$

When a voltage wave with amplitude V_{wave} hits the end of a transmission line, a wave with amplitude $\rho \cdot V_{\text{wave}}$ is reflected. Note that three simple cases match our previous discussion:

$Z_{\text{term}} = Z_0$ When a transmission line is terminated in its characteristic impedance, the reflection coefficient is 0.

$Z_{\text{term}} = 0$ The reflection coefficient of a short-circuited line is −1, producing a reflection of equal magnitude and opposite polarity.

$Z_{\text{term}} = \infty$ The reflection coefficient of an open-circuited line is +1, producing a reflection of equal magnitude and the same polarity.

If the source impedance (R_{src}) does not equal Z_0, then reflections occur at the near end of the line as well as at the far end. Each end of the line has its own value of ρ. The *principle of superposition* applies, so that voltage at any point on the line and instant in time is the sum of that point's initial condition and all waves that have passed it so far. The end of a transmission line is said to be *matched* if it is terminated with its characteristic impedance.

principle of
   superposition
transmission-line
   matching

Figure 11-8 shows the behavior of a transmission line that is not matched at either end. (Note that the driver's ideal voltage source V_{src} is considered to have a resistance of 0 Ω for this analysis.) Reflections occur at both ends of the line, with smaller and smaller waves reflecting back and forth. The voltage everywhere on the line asymptotically approaches $0.9V_{\text{src}}$, the value that would be predicted by a "DC" analysis of the circuit using Ohm's law.

11.4.2 Logic-Signal Interconnections as Transmission Lines

So, how does transmission-line theory affect logic signals? Let's consider the case of a 74HC CMOS output driving a CMOS input at the end of a transmission line consisting of a signal line and ground, as shown in Figure 11-9. An "on"

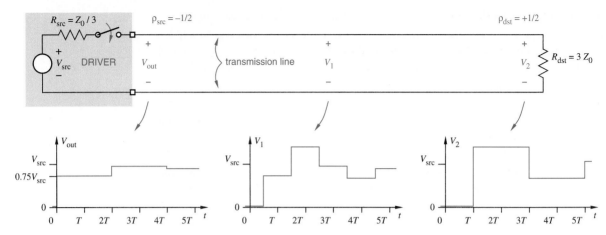

Figure 11-8 A transmission line that is not matched at either end.

p-channel or n-channel output transistor in the HCT family has resistance of roughly 100–200 Ω; for simplicity let's assume 150 Ω in either the HIGH or the LOW state. The characteristic impedance of a typical PCB trace with respect to ground is on the order of 100–150 Ω; let's assume 150 Ω for convenience, so the reflection coefficient on the driving end of the line is 0. A typical CMOS input has an impedance over 1 MΩ, so the reflection coefficient on the receiving end is effectively +1.

In the figure, we consider the case of a LOW-to-HIGH transition. When the CMOS driver switches from LOW to HIGH, the 5 V source in the driver sees the 150-Ω resistance of the driver in series with the 150-Ω Z_0 of the line, so a 2.5-V wave propagates down the line. After time T, this wave reaches the receiving gate U2 on the far end and is reflected. After time $2T$, the reflected wave reaches the sending end and is absorbed without a reflection, because $\rho = 0$ at that end.

Figure 11-9 Reflections on a logic signal line changing from LOW to HIGH.

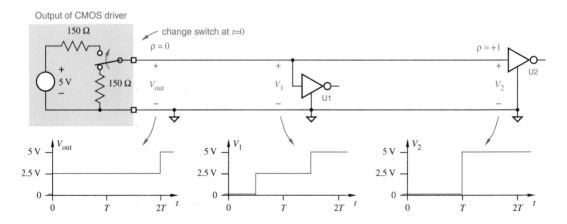

Everything works fine as far as receiving gate U2 is concerned—it sees an instantaneous transition from 0 V to 5 V at time T after the driver switches. However, consider the waveform seen by another receiving gate U1 positioned halfway between the driver and U2. As shown in the figure, U1 sees an input of only 2 V for an interval T. A receiving gate positioned closer to the driver would see this input voltage even longer. This is a problem, because 2.5 V is right at the switching threshold for 5-V CMOS inputs. If this input voltage were maintained for long, U2's output could oscillate or produce a nonlogic voltage.

Any logic-signal connection, no matter how long or how short, is a transmission line. However, the foregoing analysis is quite idealized. In practice, transmission-line behavior causes no problem if T is somewhat shorter than the transition times of logic signals or the propagation delays of logic gates. In such a case, the reflections typically settle out before the receivers on the line have a chance to notice them.

A somewhat nastier case occurs when a high-performance output switches from **HIGH** to **LOW**. For example, as shown in Figure 11-10, in the **LOW** state an FCT CMOS output looks roughly like a 10-Ω resistor to ground. Thus, the reflection coefficient on the sending end is about −0.88. When the output is first switched **LOW**, it sees a voltage divider consisting of the 10-Ω resistor in series with the line impedance of 150 Ω. Since the line was originally at 5 V, the new voltage at the output of the divider must be

$$V_{out} = \frac{10\Omega}{150\Omega + 10\Omega} \cdot 5 \text{ V} = 0.31 \text{ V}$$

Therefore, a voltage wave $(0.31 - 5.0) = -4.69$ V is propagated down the line. When this wave hits the receiving end at time T, it produces a reflection of equal sign and magnitude (since $\rho = 1$). Thus, the voltage at the receiving end is now $V_2 = 5.0 - 4.69 - 4.69 = -4.38$ V, a negative voltage! This is called *undershoot*.   *undershoot*

Figure 11-10 Reflections on a logic signal line changing from HIGH to LOW.

When the −4.69 V reflection gets back to the sending end at time $2T$, yet another reflection occurs. This time, since ρ is negative, the reflected wave has a positive amplitude, $-0.88 \cdot -4.69 = +4.10$ V. The output voltage is now modified by the sum of the incoming reflection and the new outgoing reflection: $V_{out} = 0.31 - 4.49 + 4.10 = -0.08$ V. This is no problem. However, when the +4.10 V reflection reaches the receiving end at time $3T$, the ensuing positive reflection yields $V_2 = -4.38 + 4.10 + 4.10 = 3.82$ V, back to positive again!

As suggested by the figure, the reflections continue, with the voltage at both the sending and receiving ends asymptotically approaching 0 V, the value one would predict with a "DC" analysis of the circuit. This oscillating pattern is called *ringing*.

ringing

The large magnitude of the ringing on the receiving end may be a problem, since V_2 does not settle to a range that is less than the LOW-state input voltage (0.8 V) until time $27T$. Thus, the effective propagation delay of the circuit has been increased by many times the wire delay. Worse, if the signal is a clock, extra edges will be detected on the ringing transitions, falsely clocking the flip-flops at the receiving end.

Once again, the transmission-line behavior that we've described causes no problems if T is much shorter than the transition times of logic signals or the propagation delays of logic gates. In addition, TTL and many CMOS input structures include *clamping diodes*, the normally reverse-biased diodes from each input to ground that we showed in Figure 3-75 on page 157. Since these diodes change the receiver's input impedance from very high to very low for negative voltages, they limit the negative excursion at time T to about 1 V. This reduces the reflection back to the sending end, which in turn reduces the excursion at time $3T$ to less than 1 V. The inputs of some devices also have diodes connected to V_{CC} to handle the overshoot that occurs when low-impedance drivers make a LOW-to-HIGH transition.

clamping diodes

11.4.3 Logic-Signal Terminations

Reflections can be eliminated by terminating transmission lines in their characteristic impedance. Two methods are commonly used.

In a *Thévenin termination*, a pair of resistors is placed at the end of the line as shown in Figure 11-11(a). This termination has the Thévenin equivalent circuit shown in (b). Several factors are considered when choosing the resistor values:

Thévenin termination

1. The Thévenin equivalent resistance R_{Thev} should equal or be close to the line's characteristic impedance, Z_0. The Thévenin resistance is the parallel combination of $R1$ and $R2$, that is, $(R1 \cdot R2) / (R1 + R2)$.

2. The Thévenin equivalent voltage V_{Thev} may be chosen to optimize the current sinking and sourcing requirements of the driving gate. Here, V_{Thev} equals $V_{CC} \cdot R2 / (R1 + R2)$.

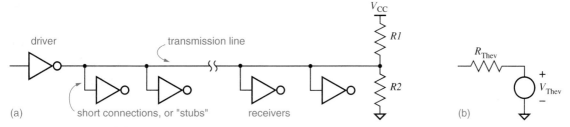

Figure 11-11 Thévenin termination: (a) circuit; (b) equivalent circuit for termination.

- For example, if the driving gate can source and sink equal amounts of current in the HIGH and LOW states (as in standard CMOS, not TTL compatible), then V_{Thev} can be halfway between V_{OL} and V_{OH}.
- On the other hand, if the driving gate can sink more current than it can source (as in TTL and TTL-compatible CMOS), then a higher value of V_{Thev} is chosen to lessen the driver's current sourcing requirements in the HIGH state, at the expense of increasing its current sinking requirements in the LOW state.

3. For three-state buses, the Thévenin equivalent voltage V_{Thev} may be chosen to yield a valid logic level when the bus is not being driven. It is especially important in this case to choose a voltage that is not close to the switching threshold of the receivers, lest they draw excessive current or oscillate.

4. The final resistor values are chosen to match standard component values (e.g., 150, 220, 270, 330, 390, 470 ohms).

For example, a "standard" termination in some TTL applications has $R1 = 220\,\Omega$ and $R2 = 330\,\Omega$, yielding $R_{Thev} = 132\,\Omega$ and $V_{Thev} = 3.0$ V. In the LOW state, a driver must sink $(3.0\text{ V})/(132\,\Omega) = 22.7$ mA, and in the HIGH state, no current is needed to maintain a valid logic level.

With a well-matched Thévenin termination at the end of a transmission line, little or no reflection will occur. The downside of this termination is that it consumes DC power at all times, and relatively high driver currents are required. A *source termination* (sometimes called a *series termination*) overcomes these problems.

source termination
series termination

As shown in Figure 11-12, a resistance equal to Z_0 minus R_d, the typical output impedance of the driver, is placed in series with the output, physically close to the output itself. Timing waveforms for this termination are illustrated in

Figure 11-12
Source termination.

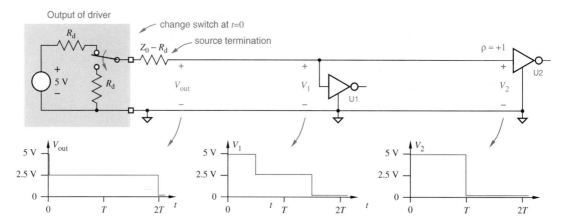

Figure 11-13 Reflections on a source-terminated line changing from LOW to HIGH.

Figure 11-13. On a transition from V_{CC} to 0 (or vice versa), the driver sees a series resistance of $2Z_0$—the series resistor plus the line itself. Half of the voltage change is developed across the series resistor and the other half across the line itself. Thus, after a one-way transmission delay T, the far-end receiver U2 sees a voltage change of $V_{CC}/2$. Assuming a reflection coefficient of 1, this full value is immediately reflected back toward the driver, bringing the voltage at U2 all the way up to V_{CC}. When this reflection reaches the driver, it is absorbed in the matched series resistance Z_0, and the line is stable.

Source termination works well for drivers whose output impedances in the LOW and HIGH states are close to equal for practical purposes (e.g., CMOS). Typical designs use series resistors of 15–40 Ω for 50–100-Ω transmission lines.

Source termination is undesirable if any receiving gates, such as U1 in the figure, are positioned only partway down the transmission line. On a transition, these "middle" gates see an initial transition of only $V_{CC}/2$; they see the full transition only after the reflection from the far end goes by. Since $V_{CC}/2$ is likely to be near the switching threshold of the receiving gates, this is not a good place to be for very long. Therefore, source termination is normally used only when a PCB is physically laid out in a way similar to Figure 11-12, with a relatively long line driving one load or a few that are clustered close together.

References

CAD tools are just tools, and you won't find any scholarly references on how to use them. However, *IEEE Design & Test of Computers* (*D&T*) is an excellent ongoing source of information about CAD tools. The best source of information for a particular CAD tool is, naturally, the supplier's documentation. When a particular tool is used *very* widely, an independent tutorial guide may appear; a good example is *SPICE: A Guide to Circuit Simulation and Analysis Using Pspice®* by Paul W. Tuinenga (Prentice Hall, 1988).

Register-transfer languages are discussed in most computer design books, including *Computer Architecture and Organization,* second edition, by John P. Hayes (McGraw-Hill, 1988). *IEEE Computer* magazine had a special issue on hardware description languages in February 1985, and *D&T* had a special issue on VHDL in April 1986.

The authoritative treatment of design for testability in the context of logic design is in Edward J. McCluskey's *Logic Design Principles* (Prentice Hall, 1986). A thorough treatment of the subject from the point of view of manufacturing and test engineers is *Digital Test Engineering* by J. Max Cortner (Wiley, 1987). Once again, *D&T* is an excellent ongoing information source. In the August 1989 special issue on built-in self-test methods, an article by Kenneth P. Parker, "The Impact of Boundary Scan on Board Test," gives a very practical perspective on the growing importance of scan methods in board-level design.

The required mathematical background and general methods for analyzing system reliability can be found in a handy little book by J. C. Cluley, *Electronic Equipment Reliability* (Halsted Press/Wiley, 1974). To see how ICs actually fail, and how their failure rates can be predicted in less than a lifetime (yours or an IC's), you should read *Failure Mechanisms in Semiconductor Devices* by E. A. Amerasekera and D. S. Campbell (Wiley, 1987).

An excellent introduction to transmission-line theory and practice from the point of view of a digital designer can be found in Thomas R. Blakeslee's *Digital Design with Standard MSI and LSI,* second edition (Wiley, 1979).

Highly recommended reading and an authoritative treatment of the analog aspects of digital design that I love to harp on is *High-Speed Digital Design* by Howard W. Johnson and Martin Graham (Prentice Hall, 1993). Although the book's subtitle is *A Handbook of Black Magic*, the authors go a long way toward demystifying the analog behavior that all high-speed digital system designers must cope with.

INDEX

Note: Page numbers for defining references are given in bold color.

XILINX STUDENT EDITION

Access the Xilinx Design Series Home Page for useful information and help

www.xilinx.com/programs/xds1.htm

This page will always contain the latest version of this information.

This CD-ROM set is designed for use with Prentice Hall's "Xilinx Design Series" educational products. This software is designed for first time installation of the student version of Xilinx Foundation Series Express F1.5 software, as well as to upgrade Xilinx Foundation Series F1.3 and F1.4 software to version F1.5 and to provide other electronic materials that may be useful. To properly execute the VHDL and Verilog synthesis software module-Synopsys FPGA Express ™ module of the Xilinx Foundation Series-software a "license.dat" file is required. If you are upgrading from 1.3 or 1.4 please access: http://www.xilinx.com/programs/xsefaq1.htm for information on making a simple modification to your existing license file. If this is a new installation then using the serial number from the coupon which is included in your Xilinx Design Series package from Prentice Hall (ISBN 0130205869), you can obtain a "license.dat" file from the Xilinx Design Series Home Page. Purchase of the Prentice Hall package is required to obtain a license file.

Basic installation instructions are printed here. Complete instructions are on the Design Environment CD-ROM disk 1 of 2 at D:\readme.wri. These instructions are complete at press time. Updated versions of these instructions may be found at

www.xilinx.com/programs/xds1.htm

Section I—Xilinx Foundation Series Express F1.5

Section II—Other tools
   a. (Optional) Xilinx CORE Generator 1.5
   b. (Optional) XESS Tools for XS40 or XS95 boards
   c. (Optional) Aldec's Active-VHDL 3.1 simulator (Student Version)
   d. (Optional) SimuCad's SILOSIII Verilog simulator (Demo Version)

Section III – License file setup

Section IV - (Optional) Adobe Acrobat and datasheets

To upgrade from F1.3 or F1.4 to 1.5, start at Section I A.

To install on a new computer, start at Section I B.

Section I—Xilinx Foundation Series Express F1.5

Section IA—Uninstall F1.3 or F1.4

1. Uninstall the Foundation Project Manager (Design Environment): Click on Start, Settings, Control Panel, Add/Remove Programs

2. Uninstall the Foundation Design Implementation Tools.

3. If uninstall does not work, then delete everything including and below C:\Xilinx directory (assuming that was the directory under which you installed your F1.3 or F1.4 Design Implementation Tools).

Section I.B—Install F1.5

1. **Insert and install Disk 1 of 2, Design Environment.**

   Begin installation: Click on Start, Run and select D:\ setup.exe When "Xilinx Foundation F1.5 Setup" window appears, click "Next" and then follow the prompts.

   Your CD KEY is: | FXAA2743390 | Do not lose this number!

2. **If you are upgrading go to http://www.xilinx.com/programs/xds1.htm for full instructions.**

   Note: You may wish to deselect XC4000E and SpartanXL devices as this will save you over 50MB of disk space. You can run setup.exe again to add these devices later.

3. **Insert and install Disk 2 of 2, Documentation.**

   Begin installation: Click on Start, Run and select D:\ setup.exe

   When "Xilinx Foundation F1.5 Documentation Setup" window appears, click "Next" and then follow the prompts. The CD KEY is not required for this step.

4. **To save disk space, you can deselect the "Multimedia Foundation Demo" in the Select Products to Install dialog box.**

5. **Under "Registry Settings Options" do not change any options.**

6. **When finished, reboot your computer to allow Registry settings to take place.**

Section II—Other tools files for F1.5

There are other tools available on disk 2.

In general, the setup is similar to what you have just done with the foundation tools. Click on the Start button, then RUN and move to the directory for the tool that you wish to install. Double click on the "setup.exe" file and follow the prompts.

The labs in the Practical Xilinx Designer lab book may be found on disk 2 in D:\XESS\ XLabs\... There is a subdirectory for each lab.

Information on system requirements, are on the Design Environment Disk CD-ROM disk 1 of 2 at D:\readme.wri as well as on the software packaging

The complete installation guide can be found on disk 2 at D:\Finstall\fndinst.pdf. Acrobat reader must be installed to read this file. Acrobat reader can be installed from the ACROREAD directory on disk 2.

Section III—License file for F1.5

For version F1.5, the "license.dat" file is required to run the Synopsys FPGA Express VHDL and Verilog synthesis tools.

If you are upgrading an existing Xilinx Foundation Student Series F1.3 or F1.4 installation up to F1.5, you do not need to make any changes to your system. Your existing "license.dat" is sufficient. If using the Express tool see the next paragraph.

If you are installing the Xilinx Foundation Series Express F1.5 software on a new computer, or you just did not obtain a "license.dat" file previously, then go to the web to obtain your license file and read the instructions at: http://www.xilinx.com/programs/xds1.htm

Section IV—(Optional) Adobe Acrobat and datasheets

There are several documents included in Adobe Acrobat format on this CD-ROM set for your reference. For more comprehensive documentation on the Xilinx Foundation Series Express F1.5 software and installation, the commercial F1.5 Quick Start Guide is available at D:\installdoc\fndinst.pdf.

These are the key documents that are included on "Disk 2 of 2, Foundation Documentation":

D:\Applinx\Databook.pdf   Xilinx 1998 Databook
D:\Applinx\coregen.pdf   Xilinx Core Solutions Databook

Other key documents which can be found using the Dynatext Document viewer in the Xilinx Book directory:
Foundations Series Quickstart Guide
Foundation Series User Guide
Verilog Reference Guide
VHDL Reference Guide

If you need to install the Acrobat reader, run D:\ACRORED\WIN32\Ar32e30.exe and follow the instructions to install the Acrobat Reader software.

In addition, there are several lab projects available at the Xilinx University Program home page under "Presentation Materials and Lab Files".

See **www.xilinx.com/programs/univ.htm**

Minimum PC Requirements:

Pentium® Processor recommended   Windows '95 or NT 4.0

48MB RAM recommended 350MB (plus 100MB swap space) free hard drive space before installation.

Depending on options you select, the F1.5 Foundation Series Express "Base" installation will consume 250 to 500 Mbytes of hard drive space. Does not include Active-HDL or Silos III.

Please note that all support for the Xilinx Student Edition software is web based. Access: **http://www.xilinx.com/programs/univ.htm** for troubleshooting and other information.

XILINIX END USER LICENSE

Attention: By opening this case, you are consenting to be bound by and are becoming a party to this agreement. If you do not agree to all the terms of this agreement, return the unopened case, along with the entire package withwithj proof of payment to authorized dealer where you took delivery within ten days; and get a full refund.

1. License. XILINX, Inc. ("XILINX") hereby grants you a nonexclusive license to use the application, demonstration, and system software included on this disk, diskette, tape or CD ROM, and related documentation (the "Software") solely by the number of simultaneous users for which you have paid XILINX a license fee, and solely for your use in developing designs for XILINX Programmable Logic devices or internal business purposes. You own the media on which the Software is recorded, but XILINX and its licensors retain title to the Software and to any patents, copyrights, trade secrets and other intellectual property rights therein. This License allows you to use the Software on a single computer. In addition, you may make up to the number of the copies of the Software, if used on separate computers, or permit up to the number of simultaneous users to use the Software, if used in a network environment, as permitted in a separate written agreement between you and XILINX, and make one copy of the Software in machine-readable form for backup purposes only. You must reproduce on each copy of the Software the copyright and any other proprietary legends that were on the original copy of the Software. You may also transfer the Software, including any backup copy of the Software you may have made, the related documentation, and a copy of this License to another party provided the other party reads and agrees to accept the terms and conditions of this License prior to your transfer of the Software to the other party, and provided that you retain no copies of the Software yourself.

2 Restrictions. The Software contains copyrighted material, trade secrets, and other proprietary information. In order to protect them you may not decompile, reverse engineer, disassemble, or otherwise reduce the Software to a human-perceivable form. You may not modify or prepare derivative works of the Software in whole or in part. You may not publish any data or information that compares the performance of the Software with software created or distributed by others.

3. Termination. This License is effective until terminated. You may terminate this License at any time by destroying the Software and all copies thereof. This License will terminate immediately without notice from XILINX if you fail to comply with any provision of this License. Upon termination you must destroy the Software and all copies thereof.

4. Governmental Use. The Software is commercial computer software developed exclusively at Xilinx's expense. Accordingly, pursuant to the Federal Acquisition Regulations (FAR) Section 12.212 and Defense FAR Supplement Section 227.2702, use, duplication and disclosure of the Software by or for the Government is subject to the restrictions set forth in this License Agreement. Manufacturer is XILINX, INC., 2100 Logic Drive, San Jose, California 95124.

5. Limited Warranty and Disclaimer. XILINX warrants that, for a period of ninety (90) days from the date of delivery to you of the Software as evidenced by a copy of your receipt, the media on which the Software is furnished will, under normal use, be free from defects in material and workmanship. XILINX's and its Licensors' and Distributor's entire liability to you and your exclusive remedy under this warranty will be for XILINX, at its option, after return of the defective Software media, to either replace such media or to refund the purchase price paid therefor and terminate this Agreement. EXCEPT FOR THE ABOVE EXPRESS LIMITED WARRANTY, THE SOFTWARE IS PROVIDED TO YOU "AS IS". XILINX AND ITS LICENSORS AND DISTRIBUTORS MAKE AND YOU RECEIVE NO OTHER WARRANTIES OR CONDITIONS, EXPRESS, IMPLIED, STATUTORY OR OTHERWISE, AND XILINX SPECIFICALLY DISCLAIMS ANY IMPLIED WARRANTIES OF MERCHANTABILITY, NONINFRINGEMENT, OR FITNESS FOR A PARTICULAR PURPOSE. XILINX does not warrant that the functions contained in the Software will meet your requirements, or that the operation of the Software will be uninterrupted or error free, or that defects in the Software will be corrected. Furthermore, XILINX does not warrant or make any representations regarding use or the results of the use of the Software in terms of correctness, accuracy, reliability or otherwise.

6. Limitation of Liability. IN NO EVENT WILL XILINX OR ITS LICENSORS OR DISTRIBUTORS BE LIABLE FOR ANY LOSS OF DATA, LOST PROFITS, COST OF PROCUREMENT OF SUBSTITUTE GOODS OR SERVICES, OR FOR ANY SPECIAL, INCIDENTAL, CONSEQUENTIAL OR INDIRECT DAMAGES ARISING FROM THE USE OR OPERATION OF THE SOFTWARE OR ACCOMPANYING DOCUMENTATION, HOWEVER CAUSED AND ON ANY THEORY OF LIABILITY. THIS LIMITATION WILL APPLY EVEN IF XILINX HAS BEEN ADVISED OF THE POSSIBILITY OF SUCH DAMAGE. THIS LIMITATION SHALL APPLY NOTWITHSTANDING THE FAILURE OF THE ESSENTIAL PURPOSE OF ANY LIMITED REMEDIES HEREIN.

7. Export Restriction. You agree that you will not export or reexport the Software, reference images or accompanying documentation in any form without the appropriate United States and foreign government licenses. Your failure to comply with this provision is a material breach of this Agreement.

8. Third Party Beneficiary. You understand that portions of the Software and related documentation have been licensed to XILINX from third parties and that such third parties are intended third party beneficiaries of the provisions of this license Agreement.

9. General. This License shall be governed by the laws of the State of California, and without reference to conflict of laws principles. If for any reason a court of competent jurisdiction finds any provision of this License, or portion thereof, to be unenforceable, that provision of the License shall be enforced to the maximum extent permissible so as to effect the intent of the parties, and the remainder of this License shall continue in full force and effect. This License constitutes the entire agreement between the parties with respect to the use of this Software and related documentation, and supersedes all prior or contemporaneous understandings or agreements, written or oral, regarding such subject matter.